Fundamentals of Comparative Vertebrate Endocrinology

Fundamentals of Comparative Vertebrate Endocrinology

Edited by

I. Chester-Jones

and

P. M. Ingleton

Sheffield University
Sheffield, United Kingdom

and

J. G. Phillips

Loughborough University of Technology
Loughborough, United Kingdom

Plenum Press • New York and London

Library of Congress Cataloging in Publication Data

Fundamentals of comparative vertebrate endocrinology.

Includes bibliographies and index.
1. Endocrinology, Comparative. 2. Vertebrates—Physiology. I. Chester-Jones, I. II. Ingleton, P. M. III. Phillips, J. G. (John Guest)
QP187.F795 1986 596'.0142 86-22703
ISBN 0-306-42314-6

A Division of Plenum Publishing Corporation
233 Spring Street, New York, N.Y. 10013

Printed in the United States of America

Contributors

R. J. Balment • Department of Zoology, University of Manchester, Manchester M13 9PL, England

T. F. C. Batten • Department of Cardiovascular Studies, Medical School, University of Leeds, Leeds LS2 9JT, England

Ian P. Callard • Department of Biology, Biological Science Center, Boston University, Boston, Massachusetts 02215

I. Chester-Jones • Department of Zoology, Sheffield University, Sheffield S10 2TN, England

D. A. Gapp • Biology Department, Hamilton College, Clinton, New York 13323

S. Harvey • Wolfson Institute, University of Hull, Hull HU6 7RK, England

I. W. Henderson • Department of Zoology, University of Sheffield, Sheffield S10 2TN, England

S-m. Ho • Department of Biology, Tufts University, Boston, Massachusetts 02155

P. M. Ingleton • Department of Pathology, Medical School, University of Sheffield, Sheffield S10 2RX, England

D. E. Kime • Department of Zoology, Sheffield University, Sheffield S10 2TN, England

S. M. Kleis • Department of Biology, Biological Science Center, Boston University, Boston, Massachusetts 02215

Frank L. Moore • Department of Zoology, Oregon State University, Corvallis, Oregon 97331

J. G. Phillips • Wolfson Institute, University of Hull, Hull HU6 7RK, England

J. A. Pudney • Biological Science Center, Boston University, Boston, Massachusetts 02215

C. G. Scanes • Department of Animal Science, Rutgers University, New Brunswick, New Jersey 08903

A. P. Scott • Ministry of Agriculture, Fisheries and Food, Directorate of Fisheries Research, Fisheries Laboratory, Lowestoft, Suffolk NR33 OHT, England

Preface

Endocrinology, as a discipline, was a late arrival in the corpus of established subjects. Its growth in recent years has been prodigious, extending from morphology to molecular levels. Most of the major endocrine glands were noted by the early anatomists, although the adrenal glands were not described until 1563 by Bartholomaeus Eustachius (1520–1574). On the other hand, elucidation of the function of these glands was extremely slow. Key work by A. A. Berthold (1849), although overlooked at the time, showed that comb atrophy in castrated fowl was prevented by testis transplantation. The idea that glands produced substances reaching the bloodstream directly and not via excretory ducts stemmed from Claude Bernard, who first used the term *internal secretion* in 1855. The clinical observations of Thomas Addison at Guy's Hospital—published as a monograph in 1855 entitled *The Constitutional and Local Effects of Disease of the Suprarenal Capsules* —were seminal. However, the stimulus of this early research did not bring immediate widespread further investigations.

Upon the discovery of secretin in 1902, Bayliss and Starling considered the term "internal secretion" to be clumsy, and the term "hormone" was coined (from ὁρμάω—I excite or arouse) and it was first used by Starling in his Croonian Lecture of 1905. The relatively simple statement that endocrine glands secrete hormones, chemical agents circulating in the bloodstream through which an event in one part of the body influences or initiates an event in another part, now involves more complicated definitions. Indeed they become blurred, as in the case of cellular "inductors" and growth factors with local actions. Nevertheless the study of glands of internal secretion demanded a general term for the subject. This became *endocrinology* (ἔνδον—within; κρίνω—separate), first used in 1909 by the Italian physician Pende and established in England by Crookshank in 1914.

Endocrinology in its origins placed predominant emphasis on medical conditions and paramedical experiments. The term *general endocrinology* was introduced to indicate that endocrine examination of any member of the animal kingdom, both vertebrate and invertebrate, was legitimate in its own right. Such an approach gave rise to textbooks which included general endocrinology in the title. The subject of the present volume is *comparative endocrinology*. Comparative is an adjective derived from comparison, and the examination of organs in the light of corresponding features in another was practiced in the earliest days of comparative anatomy, for example by Aristotle (384–322 B.C.) and Galen (130–200 B.C.). However, these early studies were concerned only with descriptive anatomy (zootomy) and the comparison of one species with another. It was the theory of evolution which gave conceptual substance to comparative anatomy. Comparative could then be applied, in the modern sense, to various disciplines such as comparative physiology. Now comparative endocrinology is an established part of the corpus of knowledge comprising endocrinology and relates to the evolutionary status of the animals considered. In practical terms, comparative endocrinology is concerned with control, at all levels from molecule to social group, and may be expected to provide some interesting insights into the evolution of intercellular "paracrine" control. For example, in the typical mammalian mosaic of pituitary cells, adjacent, but different, cells appear to influence the activity of each other; in fishes, cell types are not all intermingled and perhaps the only "paracrine" influence is via a nerve axon.

Comparative vertebrate endocrinology is an expanding area of knowledge, and the five parts of this book provide the essential building blocks of information and understanding upon which further study must be based. For this reason the editors have brought together specialist authors to present the fundamental facts, so that each chapter is designed for the clearest presentation of the particular subject area.

Steroids are very ancient molecules which have been utilized throughout the evolution of living forms. Here, in Part I, we describe the anatomy and morphology of steroid-secreting tissues in vertebrates, as well as the metabolism of steroid molecules. The physiological roles of steroids have evolved into many systems, as described particularly in Part II: Reproduction in Nonmammalian Vertebrates, Part IV: Secretion of Endocrine Glands and Their Relationship to Osmoregulation, and Part V: Endocrine and Related Factors in the Control of Metabolism in Nonmammalian Vertebrates. Modern evolutionary theories suggest that persistence of the genome into subsequent generations provides the fundamental reproductive drive. Part II describes the endocrine control of reproductive functions and mechanisms in nonmammalian vertebrates, which have evolved to ensure successful inheritance of genotype. Living organisms must consistently respond to changes in their environment to maintain their own integrity. The hypothalamus and pituitary gland together form the major integrative endocrine system which coordinates signals from the environment, principally via the brain, integument, and gastrointestinal tract, with internal stimuli from peripheral endocrine organs; the structure, functions, and secretions of this system are described in Part III. The role of hormones in maintaining internal water and ion balance in vertebrates faced with hostile hypertonic or hypotonic media is described in Part IV, while in Part V the evolution of mechanisms of action of the gut hormones is presented. The gut forms a major interface between an animal and its environment and is the site of uptake of essential amino acids, lipids, and sugars needed to maintain somal integrity; thus, hormones in this system play a truly vital role.

The level of presentation throughout is that required for undergraduate and graduate students pursuing courses in endocrinology. Thus, while further reading lists are provided, all statements and opinions are not documented as they would be in research papers and reviews. Apart from the three editors, thirteen authors of these chapters have provided the student with the necessary background from which to move forward to greater understanding of the patterns of life.

The style of this book has been prepared to accord with American usage based on *Webster's Third New International Dictionary* and *Dorland's Medical Dictionary*. Thus, for example, ''hypophys*e*al'' is the preferred spelling although ''hypophys*i*al'' is correct: The authoritative statement and explanation of the error were given by David McRioch, George B. Wislocki, and James L. O'Leary in 1940 in *Res. Publ. Assoc. Nerv. Ment. Dis.* **20**:3.

We thank Dr. David J. Groves for the original drawings in Part I, Chapters 2 and 3, which are not otherwise acknowledged.

I. Chester-Jones
P. M. Ingleton
J. G. Phillips

Contents

I. STRUCTURE OF STEROIDOGENIC TISSUES AND THEIR MODES OF SECRETION

1. The Steroids **3**
D. E. Kime

1. General Introduction 3
2. Steroid Structure and Biosynthesis 4
3. Steroids of the Testis 19
4. Steroids of the Ovary 30
5. Steroids of the Adrenal and Interrenal Glands 38
6. The Liver and Steroid Catabolism 50
7. General Conclusions and Comments 53
8. Selected Readings 56

2. Overall View of the Organization of the Vertebrate Gonad and the Structure of Steroidogenic Cells **57**
J. A. Pudney

1. Introduction 57
2. Morphology of Steroid-Producing Cells 57
3. Correlation of Structure with Function 67
4. Steroid Secretion 69
5. Organization of the Gonads 70
6. Selected Readings 94

3. Structure of the Adrenal and Interrenal Glands **95**
I. Chester-Jones

1. Introduction 95
2. Mammalia 96
3. Aves (Birds) 106
4. Reptilia 108
5. Amphibia 111
6. Pisces (Fish) 113
7. General Observations 118
8. Selected Readings 120

II. REPRODUCTION IN NONMAMMALIAN VERTEBRATES

4. Avian Reproduction **125**
S. Harvey, C. G. Scanes, and J. G. Phillips

1. Introduction 125
2. Maturation and Development 126
3. Male Reproductive System 132
4. Female Reproductive System 136
5. Environmental Interactions 153
6. Reproductive Behavior 159
7. Selected Readings 185

5. Reproduction in Reptiles **187**
Ian P. Callard and S. M. Kleis

1. Introduction 187
2. Female Reproduction 190
3. Male Reproduction 195
4. Regulation of Reptilian Gonadal Cycles 198
5. Hormone–Oviductal Interactions 202
6. Regulation of Egg Laying and Parturition 203

6. Reproductive Endocrinology of Amphibians **207**
Frank L. Moore

1. General Introduction 207
2. Amphibian Neuroendocrinology 208
3. Gonad Structure and Function 212
4. Gonadotropins and Gonadal Steroid Hormones 216
5. Prolactin and Corticosterone in Amphibian Reproduction 220

6. Hormonal Control of Amphibian Reproductive Behavior 220
7. Selected Readings 221

7. Reproductive Endocrinology of Fish **223**
A. P. Scott

1. Introduction 223
2. Class Osteichthyes: Superorder Teleostei 223
3. Class Osteichthyes: Superorder Chondrostei 252
4. Class Chondrichthyes (Cartilaginous Fishes) 252
5. Class Cyclostomes 254
6. Conclusions 255
7. Selected Readings 256

8. Vitellogenesis and Viviparity **257**
Ian P. Callard and S-m. Ho

1. General Introduction 257
2. Viviparity and Placentation in Nonmammalian Species 258
3. Evolutionary Aspects of Viviparity and Vitellogenesis: The Reptilian–Therian Transition 267
4. Comparison of Reptilian and Mammalian Viviparity: Egg Membranes and Trophoblast 267
5. The Vertebrate Egg in Relation to Vitellogenesis and Viviparity ... 268
6. Structure, Biochemistry, and Molecular Biology of Vitellogenin (Yolk) 269
7. Vitellogenin Genes and Vitellogenin Messenger mRNA Synthesis . 272
8. Posttranslational Modification of the Vitellogenin Polypeptide 274
9. Plasma Vitellogenin Levels 275
10. Hormonal Requirements of Vitellogenesis 276
11. Uptake of Vitellogenin by the Growing Oocytes 279
12. Conclusion 280
13. Selected Readings 281

III. THE HYPOTHALAMUS AND PITUITARY GLAND
T. F. C. Batten and P. M. Ingleton

9. The Structure and Function of the Hypothalamus and Pituitary Gland **285**

1. General Introduction 285
2. Anatomical Relationships of Brain and Pituitary 286
3. Protochordates 305

4. Agnatha . . . 306
5. Elasmobranchs . . . 312
6. Primitive Bony Fishes . . . 316
7. Teleosts . . . 320
8. Crossopterygian Fishes . . . 329
9. Amphibians . . . 333
10. Reptiles . . . 338
11. Birds . . . 345
12. Mammals . . . 349
13. Chemistry of Pituitary and Hypothalamic Hormones . . . 355
14. Hypothalamic Control of Pituitary Functions . . . 374
15. Actions of Pituitary Hormones . . . 391
16. Techniques for Identifying Pituitary and Hypothalamic Cells . . . 399
17. Selected Readings . . . 408

IV. SECRETION OF ENDOCRINE GLANDS AND THEIR RELATIONSHIP TO OSMOREGULATION
R. J. Balment and I. W. Henderson

10. Environmental Constraints and Adaptive Mechanisms . . . 413

1. Introduction . . . 413
2. The Osmoregulatory Process . . . 417
3. The Vertebrate Series . . . 419

11. Endocrine Glands and Their Secretions . . . 445

1. Introduction . . . 445
2. Pituitary Hormones . . . 447

12. Integration of Hormonal Actions to Regulate Extracellular Fluid Volume and Composition . . . 481

1. Introduction . . . 481
2. Calcium . . . 482
3. Water . . . 488
4. Sodium . . . 491
5. General Conclusions . . . 493

13. A Phylogenetic Perspective **497**

1. Introduction 497
2. The Renal Apparatus 498
3. The Radiation of Vertebrates 501
4. Endocrine Controls 503
5. Selected Readings 507

V. ENDOCRINE AND RELATED FACTORS IN THE CONTROL OF METABOLISM IN NONMAMMALIAN VERTEBRATES
D. A. Gapp

14. Introduction **511**

15. Gastroenteropancreatic Cells **515**

1. Characteristics and Origins 515
2. The GEP Hormones 528
3. Modes of GEP Hormone Delivery 539

16. The Gastrointestinal Endocrine System **541**

1. Introduction 541
2. Hormones of the Mammalian Gastrointestinal Tract 542
3. Possible Gut Hormones in Invertebrates and Protochordates 551
4. Gut Hormones in Nonmammalian Vertebrates 554
5. Hormonal Control of Gastrointestinal Function 561
6. Conclusions 577

17. The Endocrine Pancreas **579**

1. Introduction 579
2. Organization of the Pancreatic Islets 579
3. Comparative Aspects of Islet Morphology 584
4. Evolutionary Considerations of Morphology 599
5. Control of Synthesis and Release of Pancreatic Hormones 604
6. Comparative Aspects of Pancreatic Endocrine Secretion 610
7. Summary and Conclusions 620

18. Hormonal Control of Metabolism **623**

1. Metabolic Pathways 623
2. Regulation of Metabolic Pathways 630
3. Hormonal Control of Intermediary Metabolism in Vertebrates 634
4. Cyclostomes 636
5. Elasmobranchs 637
6. Teleosts 639
7. Amphibia 645
8. Reptiles 648
9. Birds 652
10. Appendix: Classification of Chordata 658
11. Selected Readings 660

Index 661

Part I

Structure of Steroidogenic Tissues and Their Modes of Secretion

1

The Steroids

D. E. Kime

1. General Introduction

This chapter gives a general idea of the steroid hormones found in nonmammalian vertebrates both in their similarity and their differences from mammals. An evolutionary pattern may be discerned. Section 1.1 deals with the structure of the steroids and the rules for naming different hormones to enter the steroid "language." Since endocrine studies rely on information obtained from both *in vitro* incubations and measurement of steroid levels in plasma, the advantages and limitations of these methods are given. Discussion of steroids is often divided into major classes, such as androgens, estrogens, corticosteroids, but here it is more helpful to discuss separately the steroids secreted by the individual endocrine glands. It must be emphasized, though, that it is no longer possible, especially in nonmammalian vertebrates, to link particular steroids to specific glands or to attach a particular biological activity in the sense, say, of male/female sex hormones. Even among the mammals there are major interspecific differences in the nature of steroids secreted by the endocrine glands. It is therefore to be expected that in the nonmammalian vertebrates, which include all species from the lampreys and hagfishes to the birds, that the variety will be enormous. In many classes of vertebrates only one or two species have been examined, and any conclusions deduced from these may well not be typical of the class as a whole. In a volume of this size it is not possible to cover all the information available, and we have therefore selected the data with the aim of giving an indication of the patterns of steroidogenesis within nonmammalian vertebrates and showing how they fit into a comparative or evolutionary context.

D. E. Kime • Department of Zoology, Sheffield University, Sheffield S10 2TN, England.

2. Steroid Structure and Biosynthesis

This section gives the general structure of the steroid hormones, their rules of nomenclature, and their biosynthetic interrelationships. Since structure is often correlated with biological activity, some knowledge of the shapes of the different steroids is necessary to understand properly their function explicitly. Although these facets of steroid hormones have been described in many volumes on endocrinology, these concern predominantly mammalian steroids. This section, placing special emphasis on the hormones of nonmammalian vertebrates, attempts to remedy this deficiency and acts as a reference to the structures and terminology of steroids used in other chapters of this volume. To understand how steroids act biologically we must also understand how the different classes of steroids, androgens, estrogens, progestogens, and corticosteroids are synthesized and how they interrelate. This section includes an introduction to their biosynthesis and some of the methods used in the investigations as an introduction to the more detailed treatment in the later sections on specific groups of steroids.

2.1. Structure and Nomenclature

It is not surprising, perhaps, that steroid nomenclature often confuses the newcomer to the field since the names have been assigned to hormones in a rather haphazard fashion over the last half-century. Some of the corticosteroids, for example, may have up to six different names, some names are ambiguous and could refer to several steroids, and different names may often be used in various papers in the same volume of a journal. To overcome these difficulties, a systematic nomenclature has been devised for the steroids which gives each steroid an unambiguous name and which enables a reader familiar with some simple rules to write down the full structure of the steroid (IUPAC-IUB, 1972, Definitive rules for nomenclature of steroids. *Pure Appl. Chem.* **31:** 285–322). Unfortunately, some of these names are very lengthy and are not readily usable in either conversation or in scientific writing. It is therefore becoming usual for journals to print both the systematic name and the common, or "trivial," name used in the text as a footnote to the publication. This section gives the basic rules of nomenclature and a table of the systematic and trivial names of all steroids mentioned (Table I).

2.1.1. Steroid Structure

The basic structure of a steroid consists of four fused rings, cyclopentanoperhydrophenanthrene, which are usually written in the shorthand form shown in Figure 1. Each angle represents a carbon atom and sufficient hydrogen atoms for the four valency links of carbon. All of the steroid hormones contain this basic structure to which oxygen and carbon substituents are added at different positions. Three basic steroid hydrocarbons are derived from this simple structure by addition of methyl or ethyl substituents to give estrane (oestrane is usually used in Britain), androstane, and pregnane, which are the core structures of the estrogens, androgens, and progestogens and corticosteroids, respectively. The fourth basic structure, cholestane, is derived by addition of an eight carbon chain to androstane (Figure 2).

So that we can define more clearly the position of a substituent, the rings are labeled A to D, and the carbon atoms are labeled from 1 to 27 (Figure 3). This labeling is used for all steroids. At the majority of carbon atoms on the ring there are two hydrogen atoms, and substituents such as alcohol groups may replace either of these. Since the bonds of a carbon atom are arranged tetrahedrally in space, the hydrogen atoms at each carbon will tend to lie either above the plane of the paper (referred to as β and drawn with a full line) or below the plane of the paper (referred to as α and drawn with a broken line) (Figure 4). If the orientation is not known, the bond is shown as a wavy line (~) and referred to as ξ (Greek xi). At each ring junction of the steroid molecule there is a single hydrogen atom attached to each of the carbon atoms forming the junction. These hydrogens may be oriented on opposite sides of the plane of the ring (α, β), in which case the junction is referred to as *trans,* or they may be both on the same side of the ring plane (α,α or β,β), which is referred to as *cis*. A *trans* junction results in a planar molecule, whereas a *cis* junction gives a bent molecule. This is much more clearly seen if the reader is able to obtain some simple molecular models, but is depicted diagrammatically in Figure 4.

In all naturally occurring vertebrate hormones, the ring junctions B/C and C/D are *trans* so that rings B, C, and D always form a planar structure. The A/B ring junction may, however, be either *cis* or *trans,* which results in two very differently shaped molecules. A model with A/B *trans* is planar and will lie flat on a table, whereas that with A/B *cis* has a right-angle bend at the A/B junction (Figure 5). Since biological activity must be expressed through binding of the steroid molecule to a receptor, the shape of this receptor must of necessity be very different for these two types of steroid, and it is therefore not surprising to find that the biological activities of these two forms are indeed very different. In all naturally occurring steroids the 19-methyl substituent is orientated β, and thus *cis* and *trans* equate to 5β- and 5α-hydrogen orientation at this ring junction (Figure 5). Models also show that although the two hydrogens on the carbon atoms at other parts of the ring are oriented above or below the plane of the ring, they are in fact either at 30° or 90° to this plane as a result of the tetrahedral valency angle of approximately 120°. Those substituents that are nearest to the plane of the ring are referred to as equatorial, and those at right angles to it are referred to as axial. Notice that in the five-membered ring D, the substituents at position 16 are disposed equiangularly with respect to both rings C and D and that this terminology does not apply. In general, there is a marked difference in chemical reactivity between axial and equatorial substituents, and it would not be surprising if this was also reflected in their binding affinity to receptor sites and their rates of metabolism, both of which will influence their biological activity.

Table I
Trivial and Systematic Names of Steroids

Trivial name[a]	Systematic name	See figure
Estrogens (C_{18})		
Estrone, estrone	3-Hydroxy-1,3,5,(10)-estratrien-17-one	18
Estradiol, estradiol-17β	1,3,5,(10)-estratriene-3,17β-diol	18
Estriol, estriol	1,3,5(10)-estratriene-3,16α,17β-triol	—
15α-Hydroxyestrone	3,15α-Dihydroxy-1,3,5(10)-estratrien-17-one	—
15β-Hydroxyestrone	3,15β-Dihydroxy-1,3,5(10)-estratrien-17-one	—
15α-Androxyestradiol	1,3,5(10)-estratriene-3,15α,17β-triol	—
15β-Androxyestradiol	1,3,5(10)-estratriene-3,15β,17β-triol	—
Androgens (C_{19})		
Testosterone	17β-Hydroxy-4-androsten-3-one	11
6β-Hydroxytestosterone	6β,17β-Dihydroxy-4-androsten-3-one	12
11β-Hydroxytestosterone	11β,17β-Dihydroxy-4-androsten-3-one	14
15α-Hydroxytestosterone	15α,17β-Dihydroxy-4-androsten-3-one	13
15β-Hydroxytestosterone	15β,17β-Dihydroxy-4-androsten-3-one	13
19-Hydroxytestosterone	17β,19-Dihydroxy-4-androsten-3-one	18
11-Ketotestosterone	17β-Hydroxy-4-androstene-11,17-dione	14
Epitestosterone	17α-Hydroxy-4-androsten-3-one	15
Androstenedione	4-Androstene-3,17-dione	11
11β-Hydroxyandrostenedione	11β-Hydroxy-4-androstene-3,17-dione	14
15α-Hydroxyandrostenedione	15α-Hydroxy-4-androstene-3,17-dione	13
19-Hydroxyandrostenedione	19-Hydroxy-4-androstene-3,17-dione	15
Androstenetrione, adrenosterone	4-Androstene-3,11,17-trione	14
5α-Dihydrotestosterone, 5αDHT	17β-Hydroxy-5α-androstan-3-one	6
5β-Dihydrotestosterone, 5βDHT	17β-Hydroxy-5β-androstan-3-one	16
5α-Androstanedione	5α-Androstane-3,17-dione	15
Androsterone	3α-Hydroxy-5α-androstan-17-one	—
Etiocholanolone	3α-Hydroxy-5β-androstan-17-one	14
Dehydroepiandrosterone, DHA	3β-Hydroxy-5-androsten-17-one	11
Progestogens (C_{21})		
Pregnenolone	3β-Hydroxy-5-pregnen-20-one	11
17-Hydroxypregnenolone	3β,17-Dihydroxy-5-pregnen-20-one	11
Pregnanolone	3α-Hydroxy-5β-pregnan-20-one	19
Progesterone	4-Pregnene-3,20-dione	11
15α-Hydroxyprogesterone	15α-Hydroxy-4-pregnene-3,20-dione	13
15β-Hydroxyprogesterone	15β-Hydroxy-4-pregnene-3,20-dione	13
17-Hydroxyprogesterone	17-Hydroxy-4-pregnene-3,20-dione	11
20α-Dihydroprogesterone	20α-Hydroxy-4-pregnen-3-one	7
20β-Dihydroprogesterone	20β-Hydroxy-4-pregnen-3-one	19
17-Hydroxy-20β-dihydroprogesterone, 17,20βP	17,20β-Dihydroxy-4-pregnen-3-one	7
Corticosteroids (C_{21})		
Deoxycorticosterone, DOC	21-Hydroxy-4-pregnene-3,20-dione	21
Corticosterone, B	11β,21-Dihydroxy-4-pregnene-3,20-dione	21
Cortisol, hydrocortisone, F	11β,17,21-Trihydroxy-4-pregnene-3,20-dione	21

(*continued*)

Table I (*Continued*)

Trivial name[a]	Systematic name	See figure
Cortisone, E	17,21-Dihydroxy-4-pregnene-3,11,20-trione	21
1α-Hydroxycorticosterone, 1αOH-B	1α,11β,21-Trihydroxy-4-pregnene-3,20-dione	21
1-Dehydrocorticosterone	11β,21-Dihydroxy-1,4-pregnadiene-3,20-dione	—
11-Dehydrocorticosterone, A	21-Hydroxy-4-pregnene-3,11,20-trione	21
Tetrahydrocortisone, THE	3α,17,21-Trihydroxy-5β-pregnane-11,20-dione	—
Tetrahydrocortisol, THF	3α,11β,17,21-Tetrahydroxy-5β-pregnan-20-one	—
11-Deoxycortisol, S	17,21-Dihydroxy-4-pregnene-3,20-dione	21
Aldosterone	18,11-hemiacetal of 11β,21-dihydroxy-3,20-dioxo-4-pregnen-18-al	21
18-Hydroxycorticosterone	11β,18,21-Trihydroxy-4-pregnene-3,20-dione	21
18-Hydroxydeoxycorticosterone	18,21-Dihydroxy-4-pregnene-3,20-dione	19
Cholestane derivatives (C_{27})		
Cholesterol	5-Cholesten-3β-ol	19

[a]Names in italics are approved by IUPAC-IUB.

2.1.2. Systematic Names

To obtain the systematic name of a steroid, we first take the name of the root structure, e.g., androstane, pregnane (Figure 2) and add to it the various substituents according to Table II. First, the configuration at C-5 is defined, e.g., - 5α-androstane, 5β-pregnane. Many steroid hormones contain a double bond between carbons 4 and 5, which is indicated as 4-androstene, for example. Notice that only the first carbon is indicated and that the ending of the root name is altered to "ene" to indicate a double bond. Double bonds in other positions are similarly defined, but if there is an ambiguity, both ends of the bond are indicated. For example, in the C_{18}, estrane, series a double bond could go from 5 to either 6 or 10, so to avoid ambiguity it is written as either 5(6)-estrene or 5(10)-estrene. Figure 6 shows the double bond in the 5(10) position and indicates the alternative direction of a 5(6) bond.

When several double bonds are present, the suffixes -adiene, -atriene, and so forth are used. The Δ nomenclature which is often found in older literature has now

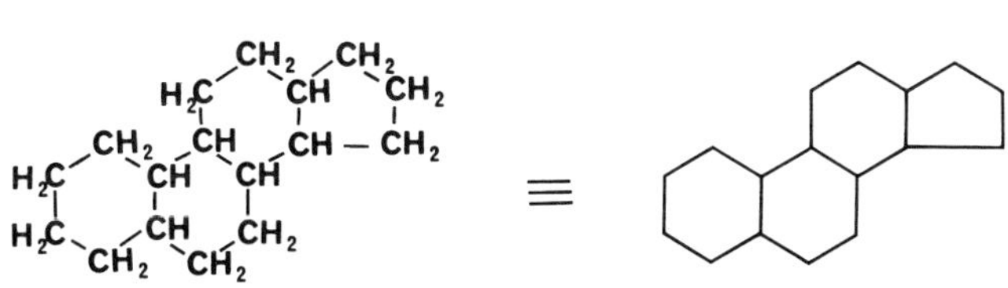

Figure 1. The steroid skeleton, cyclopentanoperhydrophenanthrene.

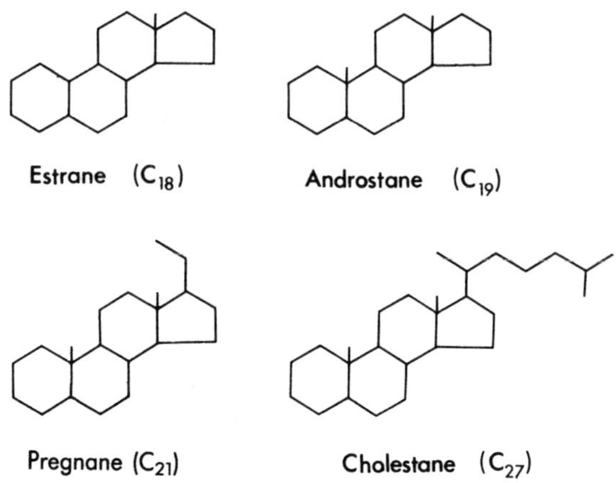

Figure 2. Basic steroid structures.

been replaced by that described here. To the name of the hydrocarbon, derived as we have described, are added the alcohol or ketone groups with their positions and orientations. The three examples given in Figure 6 illustrate this. Estrone is 3-hydroxy-1,3,5(10)-estratrien-17-one; 5α-dihydrotestosterone is 17β-hydroxy-5α-androstan-3-one; cortisol is 11β,17,21-trihydroxy-4-pregnene-3,20-dione. Only one suffix is used with the priority al > one > ol so that 5α-dihydrotestosterone should not be written as 3-oxo-5α-androstane-17β-ol, 5α-androstan-17β-ol-3-one or 17β-hydroxy-3-oxo-5α-androstane although such combinations frequently appear in the literature. The use of such nomenclature, which is unfortunately still common in the literature, serves only to confuse the purpose of systematic nomenclature, which is to give one unambiguous name to each steroid. In aromatic rings, such as in estradiol, there is only a single substituent on each carbon lying in the plane of the ring and therefore no α or β designation is required, and since the side chain of the C_{21} steroids is defined as β, it is not necessary to refer to a 17-hydroxy substituent as α in such compounds. Notice that we use en . . . one, an . . . ol, and

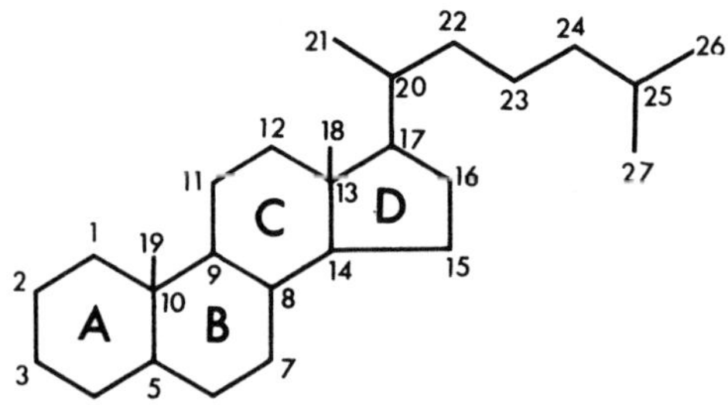

Figure 3. Numbering and lettering of the steroid rings.

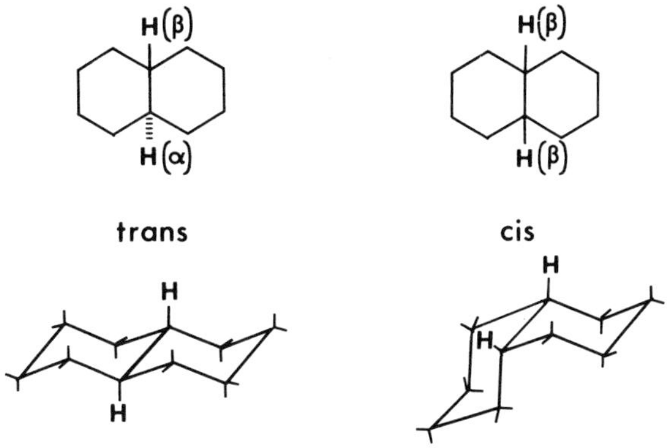

Figure 4. *trans* and *cis* isomerism.

so forth, but ene . . . dione, ane . . . triol, and so forth. Table I lists both the systematic and trivial names of all the steroids likely to be found in this volume. This should give the reader a clearer idea of the naming of the steroids and should enable him to write down the structure of any steroid by reference to the systematic name and the rules described here.

2.1.3. Trivial Names

As the trivial names were chosen in early investigations independently, some steroids have several names. Thus, cortisone, cortisol, and corticosterone are often

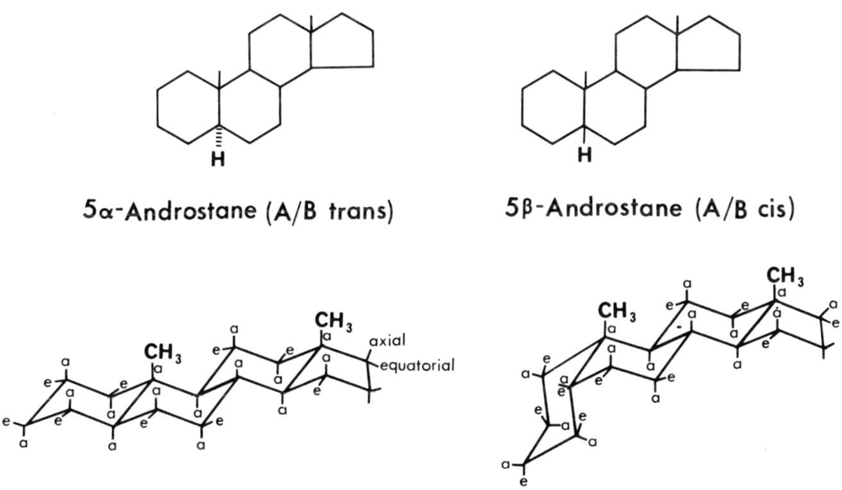

Figure 5. Steroid conformations.

Table II
Steroid Nomenclature

Substituent		Prefix	Suffix
Double bond	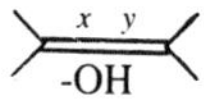	(Δ^x)	x-ene
Alcohol	-OH	Hydroxy-	-ol
Ketone	=O	Oxo-(keto-)	-one
Aldehyde	=O	—	-al

referred to as compounds E, F, and B, respectively, since these were the letters used by Kendall and his co-workers to indicate their unknown steroids during the early work of isolation and identification. Because of their brevity they are still often used. This leads to derived names such as THE for tetrahydro-Compound E or tetrahydrocortisone. Several common derivatives are: dihydro (+2H), tetrahydro (+4H), dehydro (−2H), and deoxy (−O). Although convenient, these names are ambiguous and usually give no indication as to where the oxygen or hydrogen is added to or removed from. Inversion of α to β is referred to in older literature as *epi;* for example, androsterone is 3α-hydroxy-5α-androstan-17-one, but epi-androsterone is 3β-hydroxy-5α-androstan-17-one. One danger with the use of derived trivial names is that they can refer to impossible structures. This has become especially important in recent years when new steroids have been found in nonmammalian vertebrates which do not have an established trivial name, but which have lengthy systematic ones. Two examples are 20α-hydroxy-4-pregnen-3-one and 17,20β-dihydroxy-4-pregnen-3-one (Figure 7), which are often referred to as 20α-hydroxyprogesterone and 17α,20β-dihydroxyprogesterone.

Since progesterone already possesses a 20-ketone group, addition of a hydroxyl group as implied by the name leads to an impossible structure with a pentavalent carbon at C-20, which in any case is not the required structure. These names should not be used, and to avoid the temptation of making up simpler and more pronounceable, but incorrect, names we suggest that the latter compound, which is assuming increasing importance in nonmammalian vertebrates, be referred to as 17,20βP since this and similar abbreviations are both convenient and already in general use.

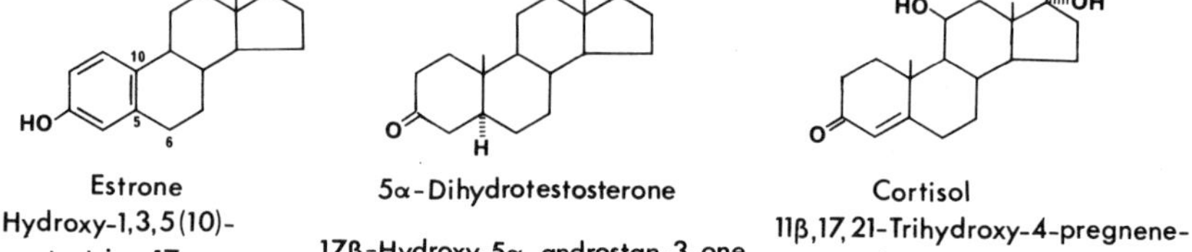

Estrone
3-Hydroxy-1,3,5(10)-
-estratrien-17-one

5α-Dihydrotestosterone
17β-Hydroxy-5α-androstan-3-one

Cortisol
11β,17,21-Trihydroxy-4-pregnene-
-3,20-dione

Figure 6. Some common steroids.

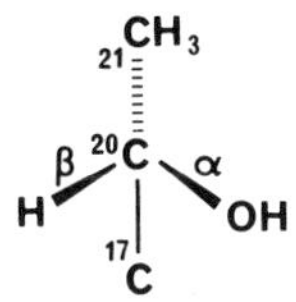

20α-Hydroxy-

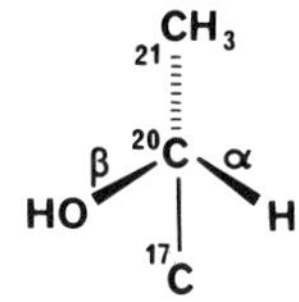

20β-Hydroxy-

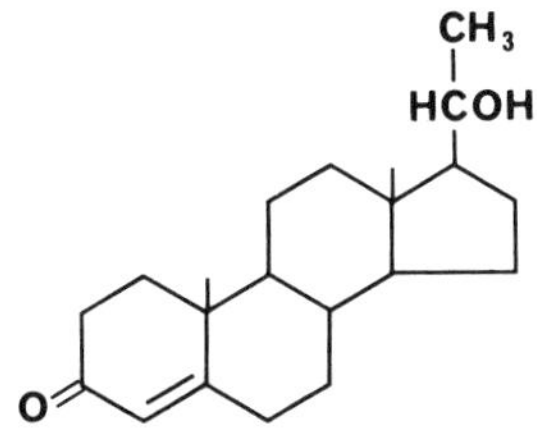

20α-Hydroxy-4-pregnen-3-one

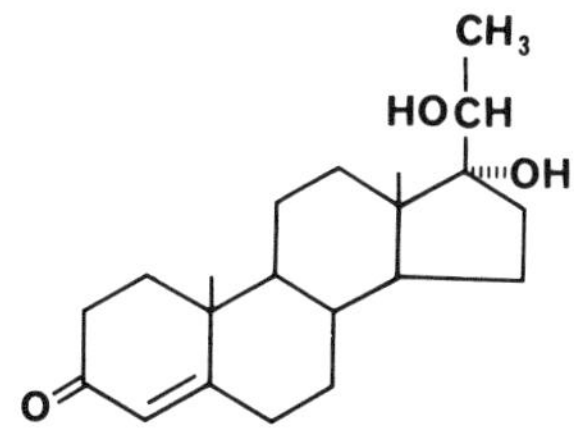

17,20β-Dihydroxy-4-pregnen-3-one (17,20βP)

Figure 7. Configuration at C-20.

The former compound is probably best referred to as 20α-dihydroprogesterone if a trivial name has to be used, although a new trivial name such as 20α-progesterol would be better to avoid ambiguity. The α and β nomenclature at C20 has a rather different meaning than that described for substituents on the main carbon nucleus since the 17-20 bond is free to rotate and thus above and below the plane of the paper becomes a meaningless concept. If the 17-20 bond is kept in the plane of the paper and the 20-21 bond bent below the plane, then the hydrogen and hydroxyl substituents at C20 will both be above the paper on either side of the main C17-C21 axis (Figure 7). Substituents to the right are α, those to the left are β, and in planar form are written as shown. Ambiguity arises by use of 20α-dihydroprogesterone for 20α-hydroxy-4-pregnen-3-one since this has a 20α-hydroxy, but a 20β-hydrogen; 20α-dihydro might reasonably be thought to refer to 20α-hydrogen and therefore a 20β-hydroxy derivative.

2.1.4. Terminology

Considerable confusion can arise from the use of the terms androgen, estrogen, progestogen, and corticosteroid in nonmammalian vertebrates. In mammals these terms have been used in two ways, chemically and biologically, and they may not always coincide. In chemical nomenclature an androgen is a C_{19} steroid of androstane structure, whereas biologically it might indicate only a steroid with androgenic properties whether or not it be a C_{19} steroid. Whereas the chemical definition is simple and unambiguous, the use of the biological definition gives rise to a number of serious problems. Even in mammals it has become increasingly recognized that biological activity can no longer be equated solely with the structure of the steroid. Estradiol, for example, was long considered the female sex steroid and equated with an estrogen by both definitions. More recent work, however, has shown that it plays an essential part in the neonatal programming of both male sexual behavior and endocrine function. Structually it is an estrogen, but its biological role in this case is masculinization, which might define it as an androgen.

In nonmammalian vertebrates, use of the biological definition becomes increasingly confusing, since female fish secrete larger amounts of testosterone than do the male and typical mammalian corticosteroids may act as inducers of ovulation in both fish and amphibians. Biological definitions of androgens, estrogens, and corticosteroids therefore become meaningless unless both the tissue and the species are defined, and it is becoming increasingly recognized that use of such definitions is no longer valid. To avoid such confusion we shall use the terms androgen and estrogen to denote derivatives of androstane and estrane irrespective of whether they have androgenic or estrogenic activity. Corticosteroids generally refer to 21-hydroxylated derivatives of progesterone and their analogs, whereas progestogens may be taken to refer to pregnane (C_{21}) derivatives not hydroxylated at position 21. Terms such as androgenic steroid, progestational hormone, and mineralocorticoid may be used to indicate specific biological activities but must be used in the context of a particular tissue and species.

2.2. Biological Activity of Steroids

This is dealt with fully in other chapters of this volume, but it is appropriate at this point to indicate the types of structure most closely associated with particular endocrine glands and types of biological activity. Much of the literature in this field relates to mammals, and in many cases, such as pregnancy hormones, this is clearly not applicable to most nonmammalian vertebrates. When investigations of steroids in nonmammalian vertebrates began in the early 1960s many workers simply looked for the mammalian hormones in fish, amphibia, reptiles, and other nonmammalian vertebrates. These were soon identified and often assumed to have a similar role to that in mammals. This approach, which is still frequently employed, gave very misleading information as to both the nature of the hormone and its biological activity. For example, testosterone was found in male teleost fish and assumed by analogy with mammals to be responsible for the male sexual characteristics of the fish. Later work, however, showed that plasma levels of testosterone were actually greater in female than male fish, and that another steroid, 11-ketotestosterone, was in fact the androgenic steroid in those teleosts examined, a factor that may easily have been missed if it had not been for a less mammalian-centered approach, with investigation of what the fish actually produced rather than being deceived by mammalian production. More examples of this will be found in later sections of this chapter, but the summary below gives a very rough guide to the different types of steroid and their activity.

2.2.1. Estrogens

These are C_{18} steroids and include estradiol and estrone and are secreted by the ovary and responsible for the secondary sexual characteristics and estrous cycle in mammals. They play an important role in sexual differentiation of the brain and are implicated in vitellogenesis in female nonmammalian vertebrates.

2.2.2. Androgens

These are C_{19} steroids and include testosterone, dihydrotestosterone, and 11-ketotestosterone secreted by the testes and sometimes ovaries and adrenals and are responsible for male sexual behavior and secondary sexual characteristics and may play an as yet undefined role in female reproduction in nonmammalian vertebrates.

2.2.3. Progestogens

These are C_{21} steroids and include progesterone and 17,20β-dihydroxy-4-pregnen-3-one(17,20βP) and are secreted mainly by the ovary but also by testis and adrenal and are responsible for the maintenance of pregnancy in mammals, final maturation and germinal vesicle breakdown of oocytes, and possibly spermiation in nonmammalian vertebrates.

2.2.4. Corticosteroids

These are 21-hydroxylated C_{21}-steroids and include cortisol, corticosterone, aldosterone, and deoxycorticosterone and are secreted by the adrenal cortex, but also formed by testes and ovaries of some species. They are responsible for the maintenance of salt and mineral balance, carbohydrate metabolism, and adaptation to salt/freshwater, and are possibly involved in spawning.

2.3. Biosynthesis of Steroids

The biosynthesis of steroids in mammals has been fully described elsewhere. Acetate is converted by a large number of steps to cholesterol, which may be considered as the precursor of all the steroid hormones. Cholesterol is further converted to pregnenolone and hence to androgens, estrogens, progestogens, and corticosteroids. Although these hormones relate generally to the distinct endocrine glands of the testis, ovary, and adrenal cortex, all of these glands have to some extent the possibility of secreting some, at least, of those generally formed by other glands, which is not perhaps surprising when the close embryological origin of the glands is considered (see Chapter 3). For example, the testicular steroid, testosterone, is the precursor of the ovarian estrogens and is secreted in large amounts by the teleost ovary, and in female mammals the adrenal is one source of androgens. The interrelationships of the different steroids are shown in Figure 8, and further details are given in the succeeding sections of this chapter.

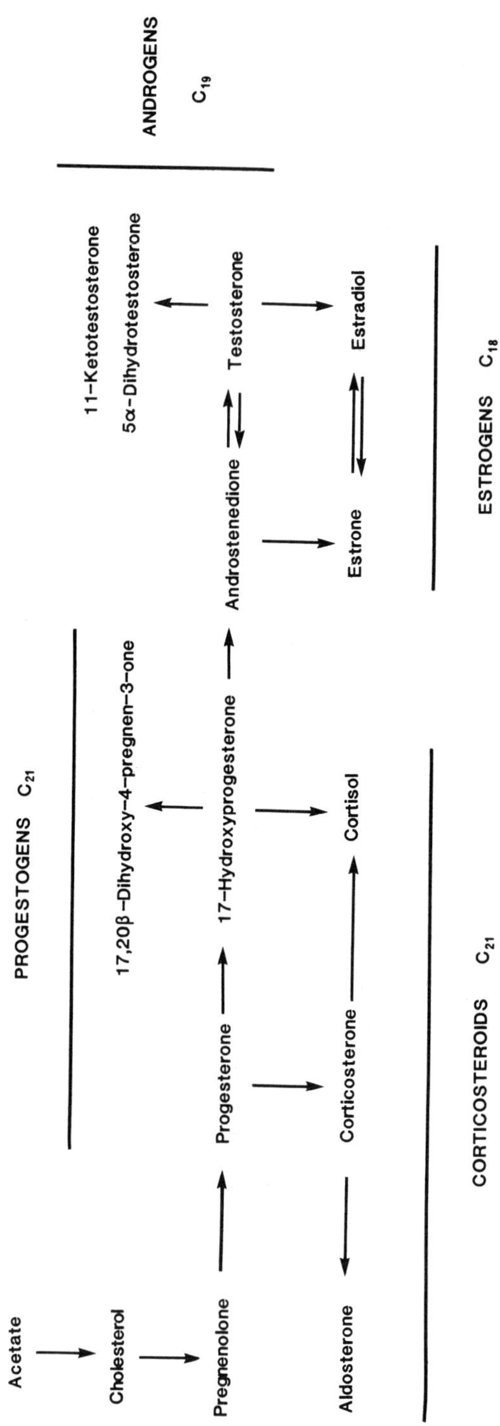

Figure 8. Interrelationships of the steroid hormones.

Although steroid biosynthesis may, at first, appear to be very complicated, it can in fact be resolved into only seven main enzyme types. Many relevant figures are given later.

2.3.1. Desmolase or Lyase

This cleaves the side chain between carbons 17 and 20 or 21 and 22, e.g., conversion of 17-hydroxyprogesterone to androstenedione, and cholesterol to pregnenolone (Figures 11, 19).

2.3.2. Hydroxylase

The most common positions for hydroxylation are 11, 16, 17, 20, 21, and 22, but in nomammalian vertebrates 1, 6, 7, and 15 hydroxylation also occurs in some species (Figures 12–14, 21).

2.3.3. Oxidoreductase or Hydroxysteroid Dehydrogenase

Several pairs of steroids exist, such as cortisol/cortisone, testosterone/androstenedione, and estradiol/estrone, in which one of the pair has an alcohol substituent and the other a ketone. Generally, the alcohol is the most biologically active, and interconversion between the pairs will affect the net biological activity of the secretion. 3β-Hydroxysteroid dehydrogenase (3βHSD) is of particular importance since it is involved in the synthesis of all hormones (in the conversion of pregnenolone to progesterone) and can be detected histochemically. It can thus be used as a measure of the steroidogenic activity of cells within a particular tissue (Figures 11, 14, 18, 21).

2.3.4. Isomerase

The oxidation of a 3β-hydroxy-5-ene to a ketone by 3βHSD is usually followed by migration of the double bond to the thermodynamically more stable 4-ene position, e.g., conversion of pregnenolone to progesterone (Figure 11).

2.3.5. Aromatase

The 4-en-3-one structure of androstenedione or testosterone may be converted into an estrogen. This involves loss of the 19-carbon atom and aromatization of the ring A to give a phenolic type of structure (Figure 18).

2.3.6. Reductase

Most of the biologically active steroids have a double bond at the 4-position. This may be converted to a saturated bond with either a 5α- or 5β-hydrogen. As we noted previously, these two possible products differ markedly in shape, and, in general, 5β products are biologically inactive, whereas 5α-steroids such as 5α-dihydrotestosterone may be more potent than their precursor. It is not perhaps surprising that 5β-reduction is common in the liver since its function is to deactivate, or catabolize, the active steroids (Figure 22).

2.3.7. Conjugation

An alcohol group may form an ester with sulfuric acid or glucuronic acid (Figure 9). Most steroids are only very slightly soluble in water, but conjugates are extremely water soluble. Conjugation both deactivates the steroids and converts them into more readily excretable forms. In mammals glucuronides are formed almost entirely by the liver, and their function is generally considered to be solely to aid excretion, but in some nonmammalian vertebrates they are also formed by endocrine tissues indicating that in such species they may perform some other function. Sulfates may, in mammals, act as intermediates in steroid biosynthesis and may be reconverted to free steroids. This may afford local protection from high concentrations of a hormone which can only be liberated in cells containing the hydrolyzing sulfatase. Such protection is especially important in the fetoplacental unit, but this type of control has not yet been found in nonmammalian vertebrates.

The cellular location of the enzymes has been extensively studied in mammals, and so far a similar distribution appears to occur in nonmammalian vertebrates (Figure 10). Some of these enzymes, particularly the lyases, are stimulated by pituitary hormones. Since these hormones are large peptides and unable to pass through the cell membrane, their effect on the steroidogenic enzymes is mediated by cyclic AMP, which acts as a messenger between the receptor on the cellular membrane and the enzyme system. Most of these enzyme systems also require the presence of a cofactor such as NAD or NADPH.

The patterns of biosynthesis will be obvious to the reader on reference to the figures in the succeeding sections. The following three sections describe in more detail the biosynthesis of the androgens, estrogens, and corticosteroids grouped under the general headings of testis, ovary, and adrenal or interrenal. Although each tissue has distinct properties, their biosynthetic capabilities show considerable overlap. Each section describes the basic steroidogenic pattern found in mammals and

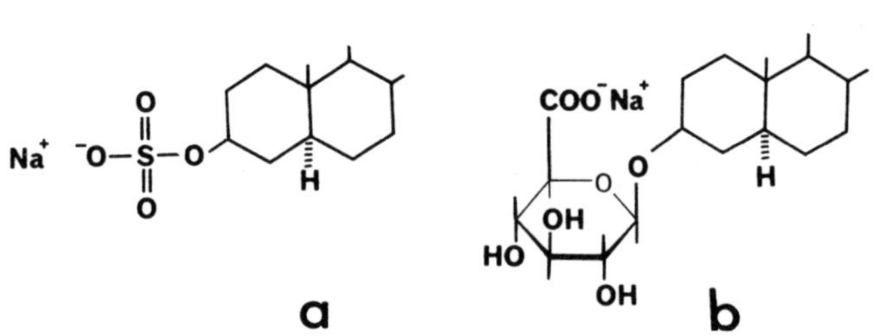

Figure 9. Steroid conjugates, (a) sulfate and (b) glucuronide.

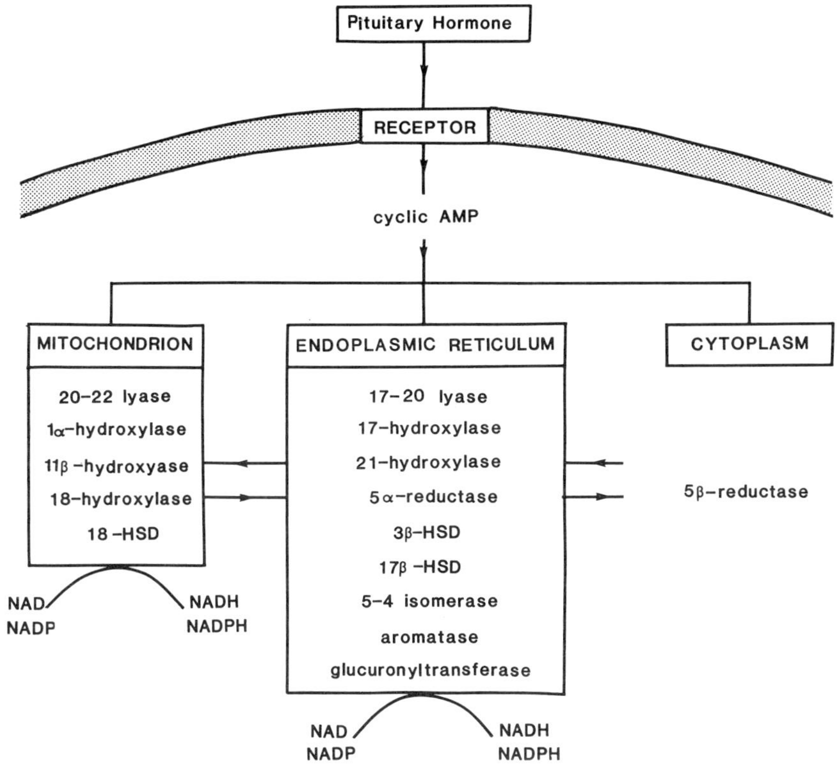

Figure 10. Cellular distribution of steroidogenic enzymes.

then discusses the variations found in different classes of nonmammalian vertebrate—the cyclostomes, teleosts, elasmobranchs, amphibians, reptiles, and birds. Although steroid biosynthesis has been well described "in mammals," this usually relates to man, the rat, and half a dozen or so domesticated species out of the several thousand extant. In nonmammalian vertebrates the problems are even more complicated, and the frequently made statement "that in teleosts," on closer examination often refers to just a salmon and a trout, which have become the teleostean equivalent of the laboratory rat. Within these limitations the following sections attempt to give some pattern to types of steroids found throughout the nonmammalian vertebrates and set them in an evolutionary context. Many of these steroids have been identified after incubation of suitable radioactive precursors with tissue *in vitro*, and it is possible that some differences may be found *in vivo*. To help understand such problems, we present briefly some of the methods employed.

2.4. Steroid Methodology

The two main questions that must be asked in any investigation are "What are the hormones?" and "How much is present under different conditions?" It is therefore essential to identify the hormones present and responsible for a particular physiological event and to quantitate the changes in production in response to internal and external stimuli.

2.4.1. Identification

This is usually done initially by incubation of an isolated tissue, such as testis or ovary, with a radioactive precursor. This precursor is a steroid which might reasonably be assumed to be formed by the tissue, such as pregnenolone, progesterone, androstenedione, or testosterone, but results must be interpreted in the light of the precursor used. Ideally, the earlier the precursor occurs in the biosynthetic pathway the better, but use of steroids from early steps such as acetate or cholesterol suffer from giving low yields, whereas later precursors do not clearly indicate that these precursors themselves could be formed by the tissue. Such experiments may, however, be valuable for identifying the unusual metabolites often found in new species. The radioactive products from such incubations are separated by chromatography and then microchemical reactions are carried out on each fraction. Comparison is made with known standards at each step, and after several such chromatograms and conversions it is usually apparent what the structure of the steroid is. Positive identification does, however, require further data, and these are usually obtained by crystallizing the radioactive metabolite with authentic material. If the specific activity (radioactivity per milligram) is constant over several crystallizations, the metabolite is considered positively identified. If authentic steroids are available in only microgram amounts, crystallization may not be possible, but other techniques such as mass spectrometry may sometimes be applied if sufficient metabolite is available. Strict criteria are applied so that a positive identification will establish a structure beyond all reasonable doubt (see Idler, D. R., 1972, *Steroids in Nonmammalian Vertebrates,* Academic Press, New York and London). We assume in all of these procedures that the radioactive and nonradioactive steroids are inseparable by all known methods. Some caution must be exercised in interpretation of results; homogenized tissue requires addition of cofactors such as NADH or NADPH and results may differ from intact cells or tissues and artifacts may be formed *in vitro* that do not occur *in vivo.*

2.4.2. Quantitation

When such steroids have been identified in *in vitro* incubations of this type, it is then possible to see whether they are formed from endogenous precursors. This can be done either by incubating isolated tissue alone, or by examining plasma taken from living animals. We can then study seasonal variations, the effect of artificial temperature and daylength regimes, or stimulation by hormones and other substances. Although a method, the double-isotope derivative assay, is available for positive identification of steroids in plasma and their quantitation, it is both tedious and time-consuming. It is, however, one of the few ways in which a steroid may be positively identified in plasma. For routine measurement, radioimmunoassay is generally used as this is an extremely rapid and easy method to use. Results must, however, be used with some caution since many antisera are not specific enough to measure only one hormone in a mixture of similar steroids. ''Androgen'' and ''testosterone'' in some cases often refer to a mixture of products, and data must be interpreted in the light of the specificity of the particular method described and in particular the cross-reactivity with other steroids and the extent of purification. All assays must be carefully validated for the particular species under investigation since interspecific differences in plasma consituents between, for example, two teleosts may be much greater than between mammals, for which most assays are designed.

The combination of methods of identification and quantitation can yield valuable information on steroids in new species, and ideally both should be used with new species. Early work on male teleosts measured only testosterone, until *in vitro* incubations showed 11-ketotestosterone to be the major hormone. Subsequent quantitative methods have shown it to be a major constituent in male fish plasma and that measurements of testosterone alone can be misleading. Assays are now available for most of the major nonmammalian hormones, but in all new species the researcher must ask himself if he is measuring the hormone he intended to measure and if this is in fact the active hormone in his species. The active hormone may not be the quantitatively most important, but little information is, as yet, available on the relative potencies of some nonmammalian hormones and their species specificity.

3. Steroids of the Testis

Testosterone may well be formed by the testis throughout the Vertebrata, but it is not the sole characteristic steroid. It is often converted to derivatives—in the testis and peripherally or by target tissues—which may be more biologically active. Furthermore testosterone may arise from the vertebrate ovary.

Testosterone is formed biosynthetically from pregnenolone by two possible pathways (Figure 11), which are referred to as the 4-ene (Δ^4) and the 5-ene (Δ^5) pathways. The 4-ene pathway proceeds via progesterone and androstenedione (4-en-3-one steroids), and the 5-ene pathway proceeds via 17-hydroxypregnenolone and dehydroepiandrosterone (5-en-3β-ol steroids). In different species, the proportion of testosterone formed via these two pathways may vary considerably.

3.1. Agnatha

Testosterone occurs in the plasma of both lampreys and hagfish of the extant Cyclostomata. Their testes can, *in vitro,* rapidly metabolize testosterone into some unusual hydroxylated compounds. There appear to be major biosynthetic differences between Petromyzonidae and Myxinidae.

Myxine glutinosa, a hagfish, is unique in the formation, *in vitro,* of both 6 and

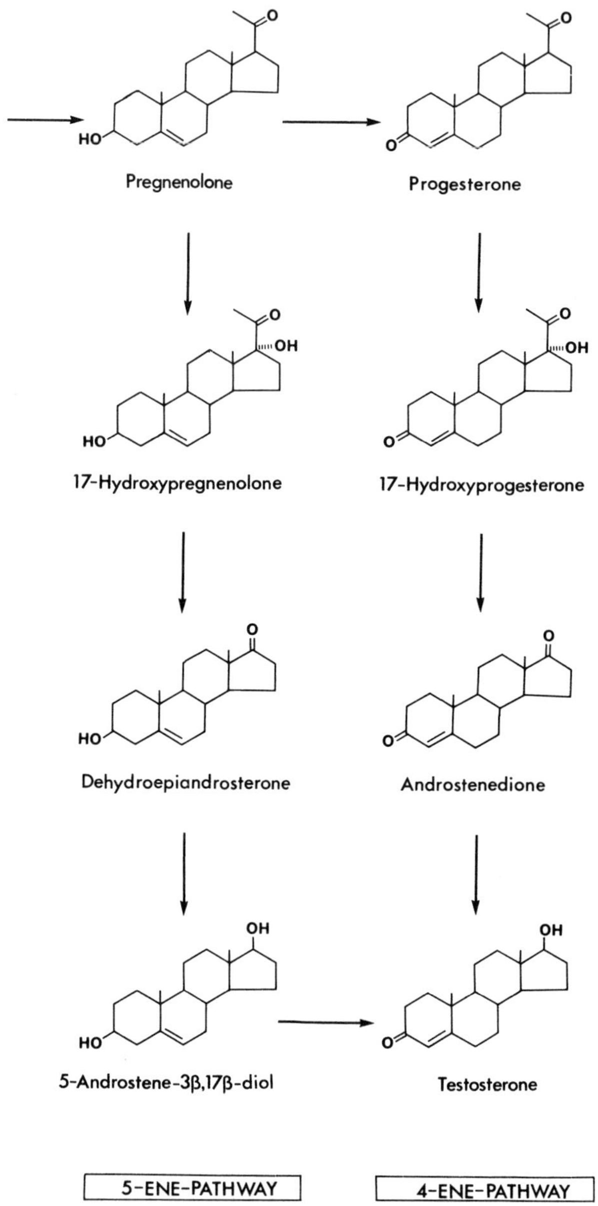

Figure 11. The biosynthesis of testosterone.

7-hydroxylated derivatives of testosterone (Figure 12); especially specific is testicular 6β-hydroxytestosterone (although it is found in other vertebrate tissues). Similarly, 7α-hydroxylation is specific except for the curious fact that it is found in the rat testis, which also has a high 5α-reductase activity to give 5α-androstanediol in the immature and 7α-hydroxytestosterone in the adult.

Other special derivatives are 15α- and 15β-hydroxylations found, *in vitro*, in the river lamprey (*Lampetra fluviatilis*) and the sea lamprey (*Petromyzon marinus*) (Figure 13). Hydroxylation at these positions is rare although it is found in the human fetus and minimally in turkey ovary (Section 4.6). Testosterone and both its 11β-hydroxy and 15-hydroxylated derivatives have been positively identified in lamprey plasma.

In their testicular metabolism of exogenous precursors *in vitro* the cyclostomes show little resemblance to any other group of vertebrates. It is, however, of particular interest that some microorganisms often lack specificity between hydroxylations at positions 6, 7, and 11 of the steroid nucleus, and it is tempting to speculate that the 11β-hydroxylase in teleost testes (see Section 3.3) may share a common origin with the 6β and 7α-hydroxylases of the hagfish testis. If models of testosterone are made with 6β-, 7α-, and 11β-substituents, it is seen that all three substituents are axially orientated and are situated in the center portion of the molecule. By rotation of the models, all three could be made to fit fairly closely to the same three receptor sites for the oxygens at 3, 17 and 6, 7, or 11 positions. This would indicate that not only are the hydroxylases very similar, but also that since biological activity is expressed via binding to a receptor, the biological activity of these three steroids could be similar. A comparison of their activities in species from the different classes would be of considerable interest.

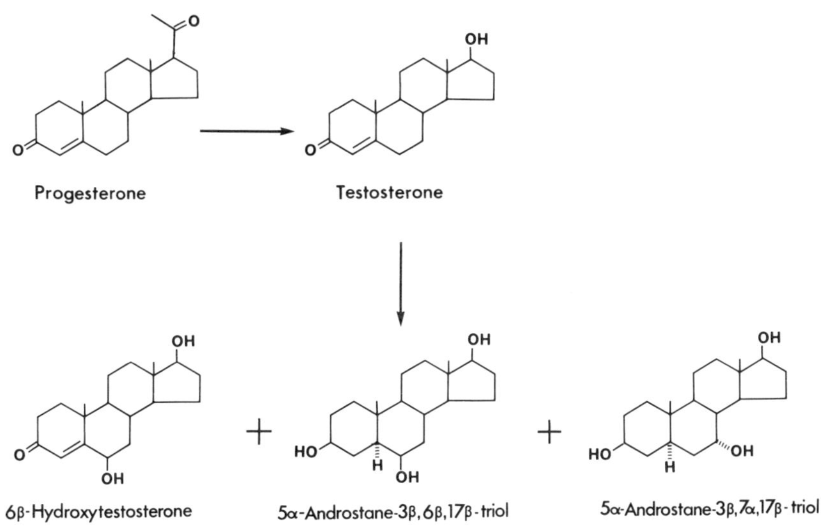

Figure 12. Testicular steroid biosynthesis in hagfish.

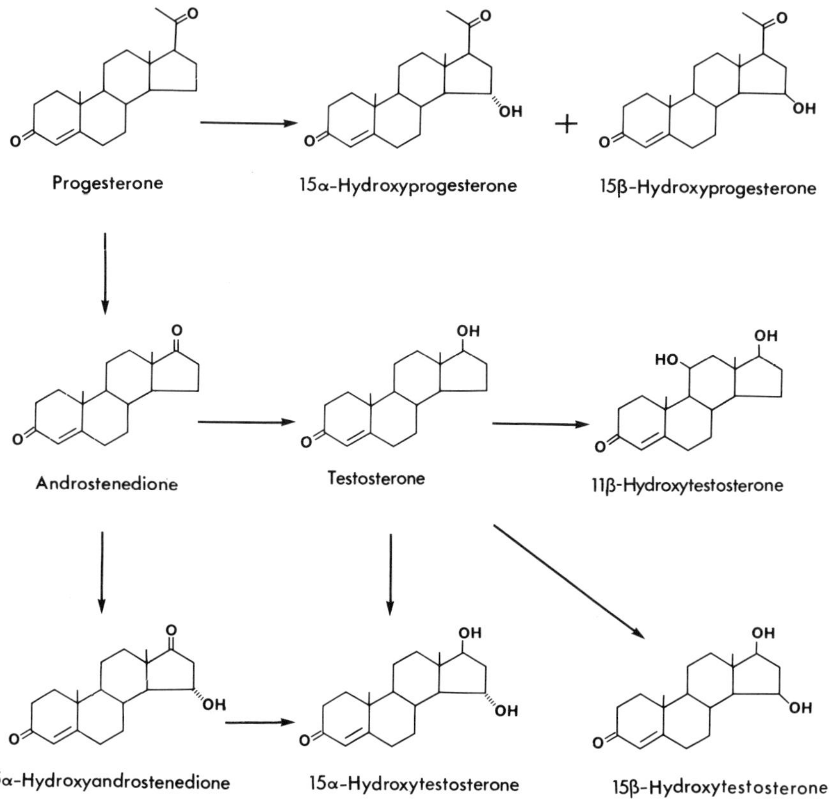

Figure 13. Testicular steroid biosynthesis in lampreys.

Do testicular steroids in cyclostomes show biological effects? Evidence is not firm. Testicular testosterone has been found in the river lamprey during the appearance of secondary sexual characters, but plasma levels are variable and not convincing. Indeed, identification is not always sure. On the one hand, testosterone has been positively identified in plasma of the Pacific hagfish (*Eptatretus stouti*) and the sea lamprey at 1.7 and 4.1 ng/ml, respectively, comparable with other vertebrates; on the other, the presence of the steroid was slight or absent. It is probable that not enough is known to make firm statements. Living cyclostomes reflect, to some degree, vertebrate emergence in evolutionary terms. The biosynthetic pathways are potentially there and, for example, indications of 11- and 21-hydroxylases and of 17,20βP fit in with their presence in teleosts.

3.2. Chondrichthyes (Subclass Elasmobranchii, Sharks, and Rays)

The pattern of testicular steroids in the few species of elasmobranch fish investigated seems to be very similar to that of mammals. Testosterone is formed from both pregnenolone and progesterone. It may be further converted to its sulfate and that of 5α-androstane-3β,17β-diol. Hydroxylated progestanes occur—deoxycorticosterone, 17-hydroxyprogesterone, 17,20β-dihydroxy-4-pregnen-3-one (17,20βP)—and these also are found in teleosts. The testis of the thorny skate, *Raja radiata,* is able to form glucuronides. The spotted dogfish, *Scyliorhinus caniculus,* has interesting but biologically unexplained features. Very high yields of an extremely polar but unidentified metabolite of both pregnenolone and testosterone were isolated in the unconjugated fraction, and it was the major constituent of glucuronide and sulfate fractions after incubation studies. Reproduction in the spotted dogfish is temperature dependent and little affected by photoperiod. The optimal temperature for division of spermatogonia (10–15°) and for formation of testosterone *in vitro* show a close correlation. In the shark testis, seminiferous tubules are responsible for the majority of the synthesis of androgens, and estrogen synthesis and aromatase activity were greatest in stage I cysts containing spermatocytes. Hydroxylase and lyase activity was greatest in the stage III cysts containing spermatids and spermatozoa (see Part II, Reproduction in Nonmammalian Vertebrates).

3.3. Teleostei

There are about 18,007 extant species, in 31 orders, 415 families, 3869 genera, which represent about 96% of known living fish. Teleost fish range from marine fishes, which release their eggs and sperm into the oceans in their millions in synchronized shoal matings, to those which produce only a few eggs that are cared for by one or both parents, from tropical fish living under almost constant daylength and warm temperature to those with extreme seasonal changes in both temperature and daylength, from hot freshwater pools to the icy ocean depths, and it includes both species with the normal sexual dimorphism (gonochoristic) and those ambisexual species in which sex change is part of the normal reproductive strategy. With such a range in habitats and reproductive strategies it would, perhaps, be surprising if the reproductive endocrinology did not also show a remarkable variety of patterns. Research has concentrated mainly on common species of North America and Europe, and especially on species of major commercial importance such as trout, salmon, and goldfish. With the more recent growth in fish farming as a protein source in the developing countries there has been an increase in work on tropical species such as tilapia and catfish, and it is becoming increasingly apparent that there may be major differences in the nature of the testicular steroids.

Testosterone is not the sole or the most potent androgen produced. Some teleost testes have active 11β-hydroxylase and 11β-hydroxy steroid dehydrogenases, which convert testosterone to 11β-hydroxy- and 11-ketotestosterone (Figure 14), the latter steroid being 10 times more androgenic than testosterone. Identification is often based on incubation studies in which 11β-hydroxyandrostenedione and androstenetrione may also be produced. However, in trout plasma, 11-ketotestosterone predominates. Probably there is a wide spectrum among fish from those which only secrete 11-ketotestosterone to those which only produce 11β-

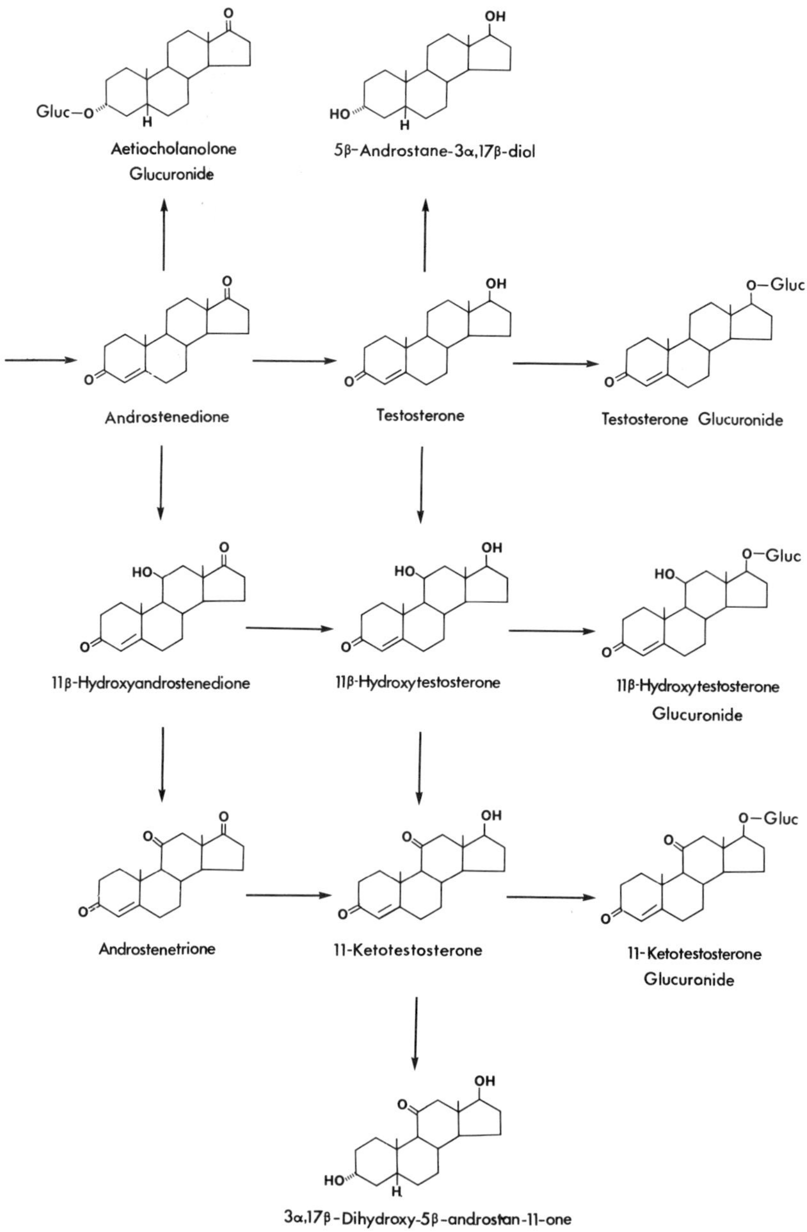

Figure 14. Androgen biosynthesis in teleost fish (Gluc, glucuronide).

hydroxytestosterone. For example, in several species in which individuals change sex, the ambisexual fish, only 11β-hydroxytestosterone is found in plasma and in tilapia both 11-oxygenated androgens are present in comparable amounts. Too few species have been examined for us to be able to relate steroidogenic patterns to reproductive strategies.

As more species are examined it is becoming more apparent that 11-oxygenated androgens may not be the major androgen in teleosts as a whole. Some species have been found in which 11β-hydroxylase appears to be completely absent and others in which it is only weak. Two other types of testicular steroid conversion, 5β-reduction and glucuronidation, are also assuming importance. In mammals these conversions are limited to the liver and are considered to be catabolic, or deactivating, steps. A pheromonal role has been demonstrated for etiocholanolone glucuronide, which is secreted by a special zone of the goby testis. In the trout, glucuronides are preferentially formed at higher temperatures and may competitively inhibit free steroid secretion, thereby delaying full sexual development until cooler environmental temperatures prevail in which gamete survival can be ensured. The role of conjugated and 5β reduced steroids in other species remains undefined.

Steroidogenesis is markedly affected by temperature, and in several species a close correlation exists between the optimum temperature for gonadotropin-stimulated steroid production *in vitro* and the normal environmental temperature prevailing during natural sexual maturation.

The gonadotropin-stimulated conversion of cholesterol to pregnenolone appears to be the most temperature-dependent step in steroidogenesis. Direct action of environmental temperature on the stimulation of gonadal steroidogenic enzymes may play an important role in the seasonal reproductive cycles of some species. Action of steroid at the target tissue is also temperature dependent.

Testosterone and 11-ketotestosterone may play different roles in reproductive development since in trout their peaks are temporally separated with 11-ketotestosterone coinciding with spermiation (Section 3.3). 11-Ketotestosterone, unlike testosterone, cannot be aromatized to an estrogen, and since brain aromatization of testosterone is essential for some of its actions at the level of the hypothalamus and central nervous system (Chapter 9), it is not surprising that testosterone can have a positive feedback on gonadotropin secretion in immature trout, whereas its 11-oxygenated derivatives are ineffective.

Plasma concentrations in teleost fish show a very wide range of levels. Salmonids, in general, have extremely high values for both testosterone and 11-ketotestosterone (20–200 ng/ml), whereas most other species appear to conform more to the mammalian range (1–10 ng/ml).

We must emphasize that out of the vast number of extant teleost species, our data are based on perhaps 30 species, which by nature of their availability may be far from representative of the class as a whole, and even of those examined, the trout, salmon, and goldfish probably account for over 80% of investigations.

3.4. Amphibia

Urodele amphibians have some features in common with teleost fish, particularly in formation of 11-ketotestosterone. The Anura differ in that species examined are unable to form 11-oxygenated androgens though they have an active 5α-reductase (Figure 15). In Part II it is noted that the urodele testicular morphology is more akin to that of teleosts, so steroidogenesis is compatible. From *in vitro* studies of incubations of urodele testes, *Necturus maculosus* (the mud puppy), plasma 11-ketotestosterone, 11β-hydroxytestosterone, and testosterone have been identified in concentrations of 1.5, 15.7, and 58.4 ng/ml, respectively. These androgens accord with sexual activity. There is, however, a peculiarity in *Necturus* in that its testis forms estrogens with aromatase activity 1000-fold greater than the ovary. Such a finding is not unusual in the vertebrate testis but in low yield, and in mammals, perhaps by the Sertoli cells under the influence of follicle-stimulating hormone

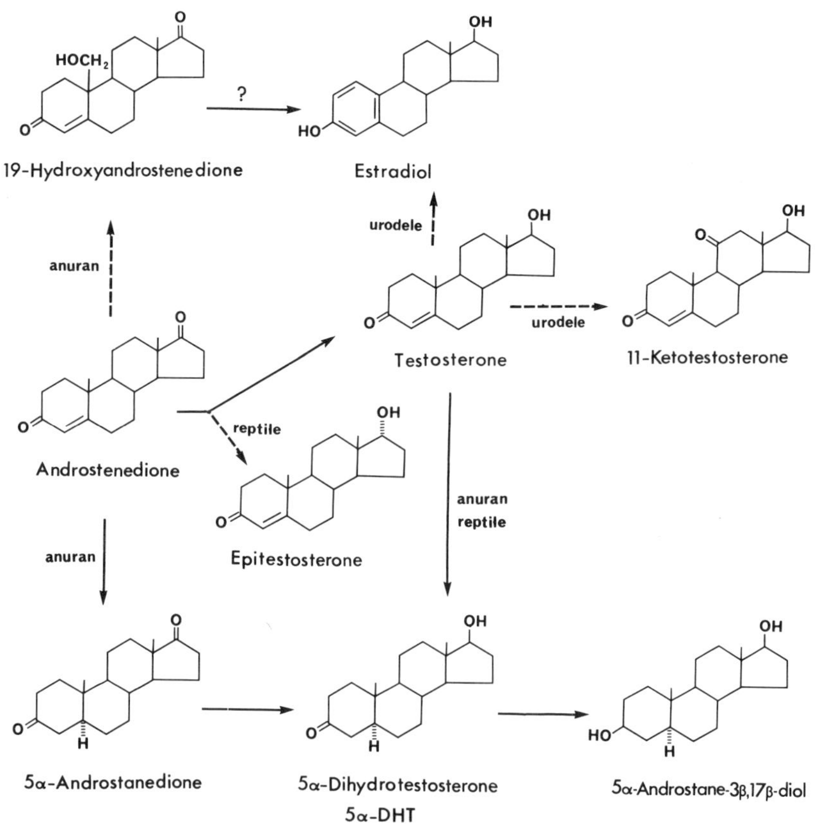

Figure 15. Steroid biosynthesis in testes of amphibians and reptiles (- - -, limited number of species).

(FSH). In *Necturus,* however, the Leydig cells produce estrogens, influenced by FSH, but for unknown reasons.

Turning to the Anura (e.g., frogs), a different steroidogenic pattern is found, *cf.* Urodela and Teleostei. The main androgen (in *in vitro* studies) is 5α-dihydrotestosterone and is probably characteristic of anuran species. It is accompanied by 5α-steroids such as 5α-androstane-3β,17β-diol and 5α-androstane-3,17-dione. Birds and mammals have only a slight 5α-reductase capacity. There are variations during a reproductive cycle. In both *Nectophrynoides occidentalis* and the bullfrog *Rana catesbeiana,* the ratio of 5α-dihydrotestosterone to testosterone is greatest during the two months leading up to spermiation. Increased pituitary gonadotropin secretion may stimulate 5α-reductase and so higher 5α-dihydrotestosterone levels at this time. Both daylength and temperature affect steroidogenesis.

Plasma androgen concentrations are generally in the mammalian range (up to 12 ng/ml), although *Bufo mauritanicus* has exceptionally high levels (600 ng/ml), the highest found in any vertebrate. 19-Hydroxylation of androgens has been observed in *Rana temporaria,* and in *Discoglossus picta,* high yields of conjugates are produced *in vitro.* The functions of these metabolites are not known.

3.5. Reptilia

In the few species examined, the major steroid in testicular incubations is testosterone. Plasma levels in most species are around 10–20 ng/ml, although *Lacerta vivipara* has 390 ng/ml. Values should be treated with caution until information is wider spread.

Like the anuran amphibia, the reptilian testis can form 5α-metabolites of androgens and progestogens, although yields are low in comparison to the amphibia. 20β-reduction of progesterone can also occur as in teleosts, and it is worth noting that this epimer is the most common in nonmammalian vertebrates, whereas in mammals the 20α-reductase is more prevalent. An unusual metabolite, epitestosterone, has been isolated from testes of *Tiliqua rugosa* and may regulate secretion of active androgens particularly during postnuptial regression when it increases at the expense of testosterone. Since epitestosterone is not readily separated from testosterone, it is possible that it may also be present in other species.

Temperature may have a marked effect on steroid production in reptiles. Unlike fish in which diurnal changes in termperature are buffered by the water, reptiles, which are commonly tropical species, may experience temperature ranges of over 30° during a single day, and moreover they may regulate their temperature to some extent by the time spent in the sun. Such changes in body temperature can be expected to be reflected in marked changes in steroidogenic activity and hence sexual behavior and development. In the turtle *Emys orbicularis,* incubation temperature may markedly affect the sex ratio of the offspring, and this may be due to changes in steroidogenic activity of the undifferentiated embryonic gonads.

3.6. Aves

Steroidogenesis in avian testes appears to be very similar to that of the mammalian and reptilian organ: testosterone is the major androgen. Other steroids have been detected *in vitro*. In the Japanese quail and the domestic fowl, reduction of the 20-oxo group of progesterone and 17-hydroxyprogesterone gives 20α- and 20β-derivatives. Both 5α- and 5β-reductases have been demonstrated in testes of several species leading to formation of 5α- and 5β-dihydrotestosterone, which may in turn be further reduced to 3α- and 3β,17β-diols (Figure 16). 5β-Reduction leads to compounds which, unlike their 5α-analogs, possess little or no biological activity. A similar reduction is found in the avian brain, and it has been suggested that 5β-reduction is a "biological inactivation shunt" acting in competition to formation of the biologically active 5α-reduced steroids and thus inhibiting the feedback effect of testosterone on the hypothalamic–pituitary system. The increased 5β-reductase activity in summer is associated with the photorefractoriness of most avian species (see Part II). This suggested action of 5β-reduced androgens is similar to that which

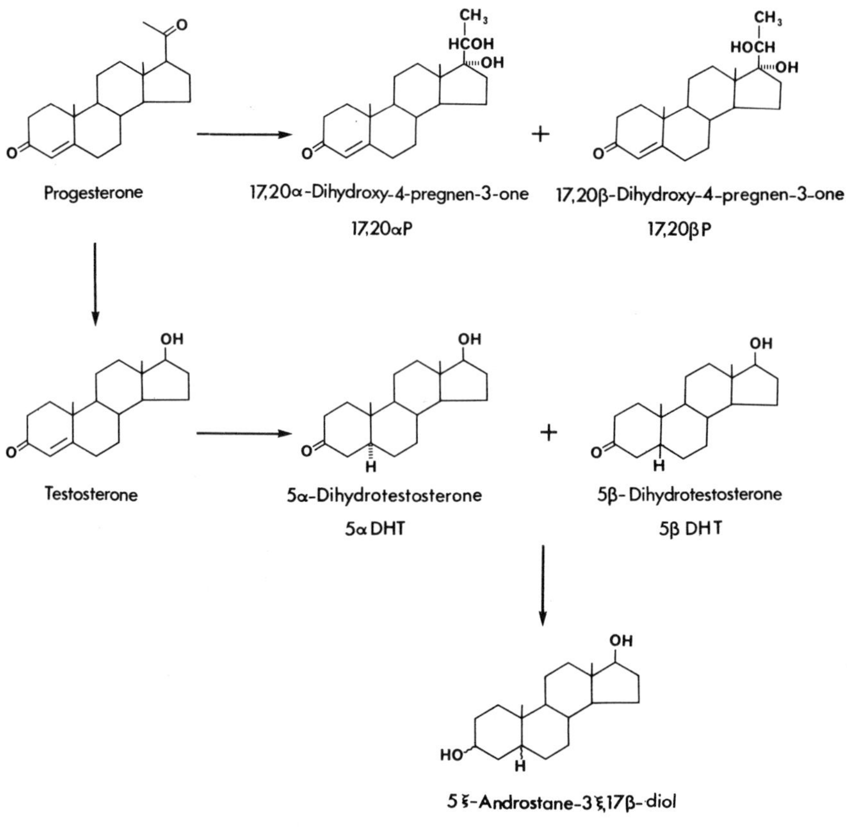

Figure 16. Steroid biosynthesis by avian testes.

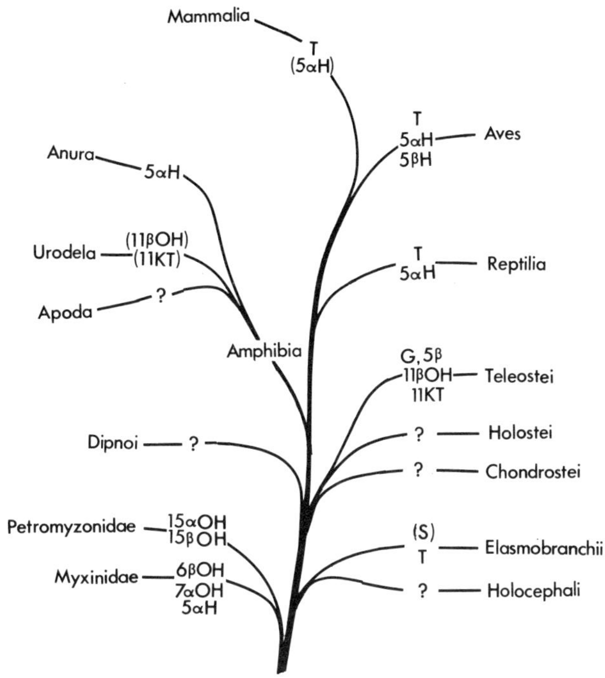

Figure 17. Phylogenetic relationship of the testicular steroids.

we have already discussed for the action of glucuronides in teleost fish and epitestosterone in reptiles.

In mammals, the major testicular site of steroidogenesis is the Leydig cell and the Sertoli cell is generally considered to be incapable of significant steroidogenic activity. As we noted earlier, this may not be true in reptiles, and in birds the Sertoli cells undergo some seasonal histological changes which indicate that they may have steroidogenic potential.

Plasma androgen levels in birds are very similar to those found in male mammals (1–10 ng/ml). Both testosterone and 5α-dihydrotestosterone concentrations vary with the season with maximum levels during the breeding season. The levels drop rapidly during the nesting phase resulting in decreased singing and aggressive behavior, which may serve to divert energy to foraging and parental activities rather than to courtship and territorial patterns.

3.7. Comparative Aspects

Since the number of species examined in many classes of vertebrates is so small, it is difficult to give a precise phylogenetic relationship of the testicular hormones. There does, however, appear to be an evolutionary trend toward simpler hormones culminating in testosterone and dihydrotestosterone in mammals (Figure

17). The most primitive species, the Agnatha, appear from current evidence to hydroxylate steroids in several positions, and we have speculated on the common origin of these hydroxylases and the 11β-hydroxylase of the teleost. Within the teleosts themselves there is considerable variety, and when more species have been examined, it will be interesting to see whether there is any trend from the more primitive to the most advanced species of this class. Urodele amphibia have closer similarities with the teleosts than with the anuran amphibians, reflecting their closer testicular morphology, and appear to be the highest class of vertebrate in which testicular hydroxylases play a significant role. Such hydroxylases, however, do not seem to be completely lost and may appear as unusual products even in mammals, where the formation of 7α-hydroxytestosterone is apparent in rats although it is atypical of the mammals as a whole.

The polar unidentified metabolite in the dogfish and epitestosterone and 19-hydroxylated derivatives, which have so far been found only in isolated species, require further confirmation before any significance can be attached to their formation. It is not immediately apparent why there should be this progressive change to simpler steroids, but since hormones must bind to receptors to express their action, it is possible that changes in shape of the receptor proteins may have resulted in more specific binding no longer necessitating so many polar groups as a means of attachment. Indeed, it would be surprising if these receptors had not themselves been greatly modified during the course of evolution, and investigations of such changes in receptor specificity may give a new perspective to changes in the steroids. The endocrine system ensures that each species reproduces at the most favorable season for the survival of its offspring. With the wide diversity of habitats and reproductive patterns found within the vertebrates it might well be expected that many different control mechanisms may be invoked in the response to seasonal cues such as daylength and temperature. Such differences may well explain the formation of many of the unusual metabolites, and it would be a mistake to dismiss completely formation of steroids that have no apparent biological activity. As we have seen in this section, several authors have suggested important regulatory roles for glucuronides, 5β-reduced androgens, and epitestosterone in the seasonal reproductive cycles of male vertebrates.

We must admit that our picture of steroidogenesis and its control is far from complete even in the major classes of vertebrates, but perhaps even more interesting in an evolutionary context will be information on those minor classes which have so far received no attention, such as the lungfish (Dipnoi), sturgeon (Chondrostei), garpike and bowfins (Holostei) and the rabbit fishes (Holocephali).

4. Steroids of the Ovary

The mammalian ovary produces two distinct types of hormone, estrogens, C_{18} steroids characterized by the aromatic ring A, and progestogens, which are C_{21} steroids. Ovarian hormones in nonmammalian vertebrates include these same two categories of hormone, namely, estrogens and progestogens, but their actual role in reproduction may be very different.

The two major estrogens, estradiol and estrone, are formed by 19-hydroxylation of testosterone or androstenedione, conversion to the 19-aldehyde, followed by loss of the 19-carbon atom (Figure 18). In the absence of other evidence this pathway is also assumed to operate in nonmammalian vertebrates.

Progesterone is formed via pregnenolone from cholesterol as in mammals, but in many nonmammalian vertebrates it is further converted to other compounds (Figure 19).

The C_{18} steroids act in most nonmammalian vertebrates in the regulation of vitellogenesis, while C_{21} steroids seem to be more implicated in germinal vesicle breakdown during the final maturation of the oocyte (see Part II).

4.1. Agnatha

Plasma concentrations of estrogens and progesterone indicated that their values are either extremely low or not detectable. *In vitro* incubations have, however, shown the ovary of the sea lamprey, *Petromyzon marinus,* to be capable of formation of estrone from tritiated androstenedione. 5α-reduced steroids were also found in these incubations, but as with the testes of both this species and of the river lamprey the major products were 15α- and 15β-hydroxylated derivatives (see Section 3.1). Testosterone, 5α-dihydrotestosterone, and 15α- and 15β-hydroxytestosterone have been positively identified in plasma of the river lamprey.

In the hagfish, *Myxine glutinosa,* ovarian steroidogenesis follows the same pattern as the testes and results in 6β- and 7α-hydroxylation, as well as 5α-reduction on a much larger scale than observed in lampreys. In another hagfish (*Ep-*

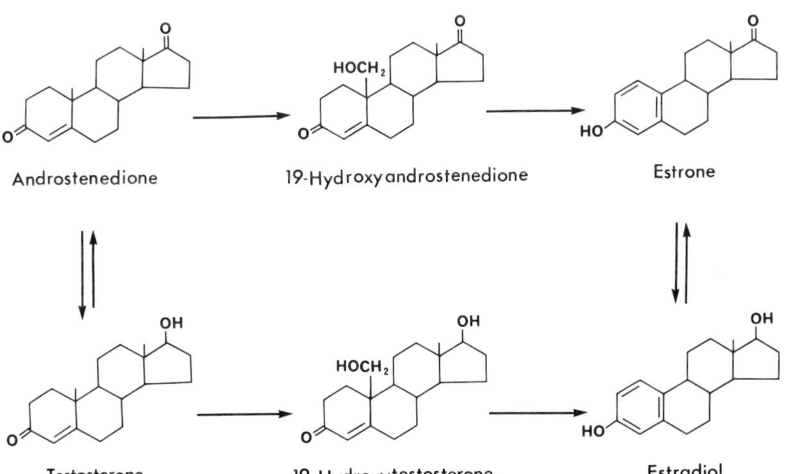

Figure 18. Biosynthesis of estrogens.

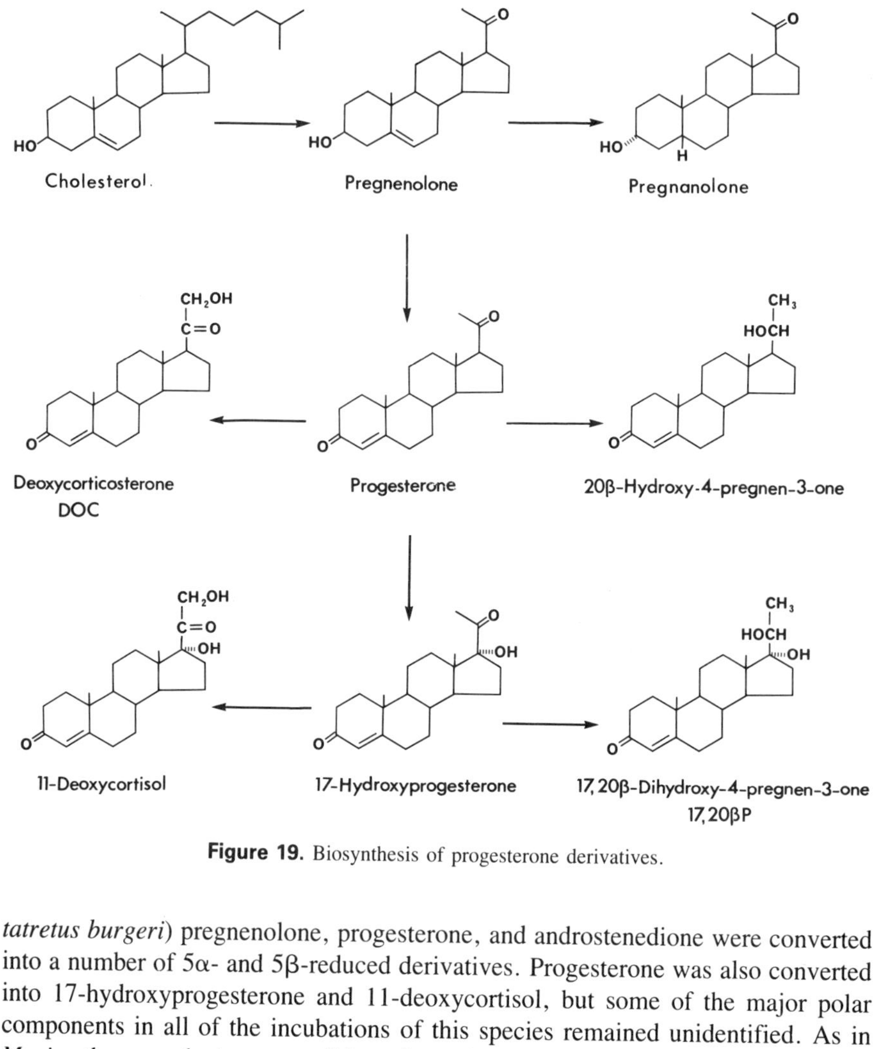

Figure 19. Biosynthesis of progesterone derivatives.

tatretus burgeri) pregnenolone, progesterone, and androstenedione were converted into a number of 5α- and 5β-reduced derivatives. Progesterone was also converted into 17-hydroxyprogesterone and 11-deoxycortisol, but some of the major polar components in all of the incubations of this species remained unidentified. As in *Myxine* these products may well have been 6β- and 7α-hydroxylated steroids. The actual active hormones in cyclostomes thus probably differ from those in other vertebrates, and it is interesting to notice that steroidogenic patterns appear to be very similar in both ovaries and testes.

4.2. Chondrichthyes

Viviparous, ovoviviparous, and oviparous species occur so that variation might be expected in the ovarian hormones during reproduction. Such information is not available except for one oviparous species, the lesser spotted dogfish, *Scyliorhinus caniculus*. Estrone and estradiol have been found in both ovaries and peripheral plasma, although estradiol predominates. Production of estradiol from androstenedione was limited to the follicular membranes of ripe follicles and ovarian stroma, and estrone production was also found in the corpora atretica and their membranes. Plasma levels of both androgen and estrogen decreased markedly after ovariectomy showing that the ovary was the major source of both hormones. Testosterone and estradiol levels in plasma peaked at 6.5 and 40 ng/ml, respectively.

4.3. Teleostei

The teleost ovary produces three major steroids, estradiol, testosterone, and a progestogen such as 17,20β-dihydroxy-4-pregnen-3-one (17,20βP). The peak plasma levels of these hormones appear at different stages of gametogenesis, which would indicate that they each affect different stages of the reproductive process (see Part II).

4.3.1. Estrogens

The major estrogens in teleosts are, like other vertebrates, estrone and estradiol with the possible formation of 16-oxygenated derivatives (estriols) in some species. Estradiol is the main inducer of vitellogenin synthesis (Part II) in the liver, but its uptake by the oocyte is stimulated by gonadotropin rather than steroids. The early stages of oogenesis therefore require gonadotropin both for initiation of estrogen synthesis, which in turn stimulates vitellogenesis, and for uptake of vitellogenin into the oocyte. The controversy of whether there are one or two gonadotropins is discussed in Part II.

Estrogens appear to be synthesized by a two-cell mechanism as in the rat. Androgens are formed in the thecal cells under the stimulus of gonadotropin and are then converted to estrogen by the granulosa cells. The aromatization does not appear to be affected by gonadotropin, although it may affect the transfer of androgen across the junction between the two follicular elements.

As in testicular steroidogenesis, temperature may have a marked effect on ovarian steroid production. In tilapia, temperature-stimulated estrogen production precedes significant increases in gonadal weight, and in female carp ovarian steroid production *in vitro* can increase as much as 20 times between 20 and 24°.

The relative proportions of the two estrogens vary with the reproductive cycle, suggesting that estrone may itself be active rather than acting simply as a precursor of estradiol. Estradiol levels in some teleosts are extremely high in comparison with mammalian values (0–0.1 ng/ml) and may reach 17 ng/ml in the trout during vitellogenesis.

4.3.2. Androgens

Unlike mammals, in which female plasma testosterone rarely exceeds one-tenth of the male values, teleosts often have higher levels in the female (200 ng/ml in trout). The role of female testosterone is not clear, but it may play an essential role in oocyte maturation. Ovarian 11-oxygenation is rare, but it has been implicated in infertility in mullets in freshwater and in sex reversal in guppies. As with teleost testes, glucuronides and 5β-reduced steroids also occur and may play a similar role to that discussed in Section 3.3.

4.3.3. Progestogens

The mammalian pregnancy hormone, progesterone, appears to be of little importance in teleost reproduction but serves only as an intermediate in the synthesis of other pregnane derivatives (Figure 19) which act as hormones during the final maturation of the oocyte and germinal vesicle breakdown prior to spawning. The most important hormone for this maturation in most of the teleost fish so far examined is 17,20βP, which is synthesized by the ovary in response to gonadotropin. In some species, for example the ayu, this progestogen is further metabolized by the ovary to 5β-reduced derivatives.

Other progestogens may also be involved in some species since ovarian production of 21-hydroxylated pregnanes such as deoxycorticosterone and 11-deoxycortisol has been demonstrated. 21-Hydroxylation is, like 11-oxygenation, limited in mammals to the adrenal gland and rarely found in gonads. These two glands may well have been steroidogenically much closer in evolutionary distant times, and it is possible that some of the common enzyme systems may be either suppressed or lost in the higher vertebrates. Adrenal–gonad interactions may also be involved in teleost ovulation, which is perhaps not surprising since spawning is frequently associated with stressful situations such as flooding or migration. Such interaction occurs in the catfish where both ovarian pregnanolone (3α-hydroxy-5β-pregnan-20-one) and adrenal cortisol are required for ovulation. The implication of a 5β-reduced steroid, pregnanolone, is of particular interest since such components are often found in teleost gonadal incubates, but in mammals, where they are hepatic catabolites, they are considered to be biologically inactive owing to their distorted conformation (Section 2.1.1). Unlike androgens and estrogens, progestogens do not appear to form conjugates in teleost ovaries.

The changes occurring in the ovary during the seasonal reproductive cycle involve a complex interplay between the ovary, the pituitary (which modulates environmental stimuli), the liver, and the interrenal. Figure 20 attempts to summarize these interrelationships, but it must be emphasized that these may not be valid for all teleosts. The interactions between the gonad and pituitary/hypothalamus are rather more complex than indicated and are discussed in more detail in Part III of this volume.

4.4. Amphibia

As far as is known, ovarian steroids are similar in urodeles and anurans. This is in contrast to testicular steroidogenesis (Section 3.4). The pattern of ovarian

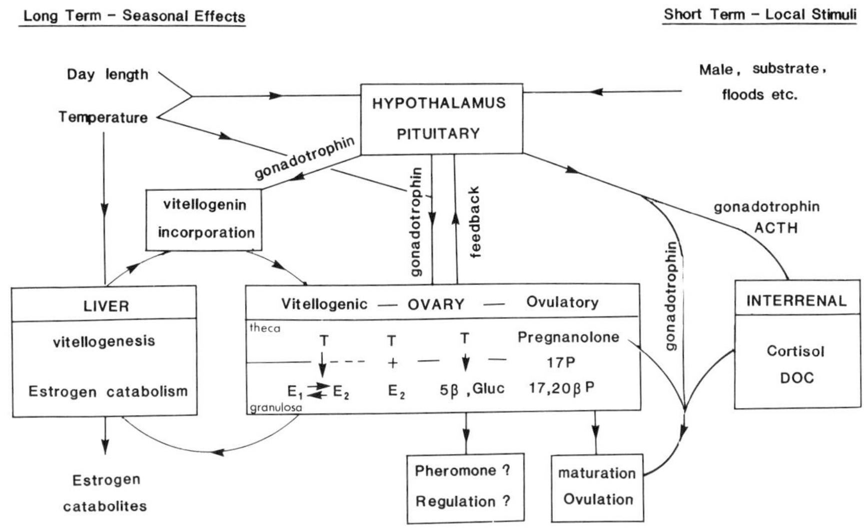

Figure 20. A summary of the factors affecting ovarian steroidogenesis during the seasonal cycle of teleosts (E_1, estrone; E_2, estradiol; T, testosterone; 5β, 5β-reduction; Gluc, glucuronide; 17P, 17 hydroxyprogesterone).

steroidogenesis in amphibia shows much greater similarity to the mammals than to fishes, with estradiol and progesterone as the main ovarian steroids. However, the high production of testosterone is more akin to the situation in teleosts. As in fish, vitellogenesis is stimulated by both gonadotropin and estradiol, the gonadotropin exerting its action via increased steroidogenesis. The steroidogenic response of the ovary varies with its stage of development, the vitellogenic follicles having 10–50 times greater estradiol production than when mature. In *Xenopus,* frog pituitary extract had a much greater effect on testosterone than progesterone synthesis, and its effect on estrogen was low. It is probable that this low response of estrogen is due to a seasonal effect since in the mud puppy, *Necturus maculosus,* gonadotropin stimulates androgen and progesterone in April but can stimulate estrogen only in the autumn.

The high plasma levels of testosterone in female amphibians, which are only slightly less than the values in males, may indicate that the 5α-dihydrotestosterone which characterizes the anuran testis (section 3.4) is the male androgenic steroid and has responsibility for the secondary sexual characteristics. Either testosterone has little such potency, or the female does not possess the necessary testosterone receptors for the expression of these characteristics. The function of testosterone in female amphibia is not clear, but since the large stage V and VI follicles are most active in testosterone synthesis, it has been suggested that progesterone may exert its effect on oocyte maturation by conversion to testosterone or that this steroid may somehow be involved in triggering maturation.

Like teleosts, the amphibian ovary may form 21-hydroxylated pregnanes such as deoxycorticosterone and deoxycortisol at least *in vitro*. Yields may exceed those of androgens and are stimulated by gonadotropin. Present evidence suggests that progesterone synthesized by the ovarian follicle is the major hormone involved in oocyte maturation. This steroid may be further converted to androstenedione, 17,20βP, and their reduced metabolites by the oocytes, themselves. This may regulate local concentrations of progesterone, which has greater maturation-inducing activity than these metabolites. Unlike other target tissues, the oocyte does not have a specific receptor for progesterone, which acts directly on the cell membrane. This may explain why androgens, corticosteroids, and even nonsteroids may in some cases be as effective as progesterone.

4.5. Reptilia

Reptilian steroids have a close resemblance to those of both amphibians and mammals, with estradiol, testosterone, and progesterone as the main ovarian hormones. A number of unusual compounds have been found in ovarian incubates from some species, and these resemble the types of steroid found in piscine and amphibian ovaries. 21-Hydroxylated pregnane derivatives such as deoxycorticosterone and deoxycortisol have been produced in some species, as have 5β-reduced pregnane derivatives, but so far these steroids have not been positively identified in plasma, nor has any biological activity been found for them although it has been suggested that pregnanolone (3α-hydroxy-5β-pregnan-20-one) may have some regulatory function in snake ovary by limiting the availability of the active progestogen.

Although estradiol has a similar function in vitellogenesis to that described for fish and amphibia, progesterone shows a clear development from a maturational to a pregnancy hormone (see Part II). Its role and secretion in viviparous reptiles shows similarities to that in mammals.

4.6. Aves

Estradiol is formed by the growing follicle in the vitellogenic phase, and as the follicle grows, it produces increasing quantities of estradiol. In the hen and turkey there is always present a hierarchy of follicles, and estradiol secretion and aromatase activity are greatest between the third and fourth of the hierarchy, with a decline as the follicle rises to become second in this hierarchy. Since follicles at all stages are always present in the ovary of these species, plasma concentrations of the estrogens show relatively small fluctuations. The major products may, in fact, be 15α- and 15β-hydroxy derivatives. These products are present in substantial amounts in the turkey with highest levels just before egg laying. In ovarian homogenates of this species as much as 40% of the estrogen was present as the 15-hydroxylated derivative, and it was suggested that estradiol may exert its physiological effects through these derivatives.

Progesterone is the major avian progestogen and is implicated in ovulation (Part II).

The majority of work on birds has been carried out on domesticated species, predominantly fowl, turkey, and ducks, and few results are available for seasonally breeding birds in which territoriality, courtship, and nesting play a more important role. The phalarope is particularly interesting in this respect since there is a major reversal in the normal reproductive behavior of the two sexes. The female initiates courtship and pair formation, is the more aggressive, and has the most colorful plumage. The ovarian content of testosterone is greater than the testicular content in this species, which is not surprising since this hormone has been implicated in such ''masculine'' behavior in other species. It appears that female testosterone production is directly related to the extent to which the female of the species partakes in the traditional male roles of courtship initiation, territorial defense, and plumage display. In some gulls, where both sexes have a similar role and there is no sexual dimorphism in plumage, testosterone levels are indistinguishable between the sexes. In the white-crowned sparrow, testosterone may also be implicated in territorial behavior since it peaks in both sexes during the prenesting period when territorial defense and courtship, activities shared by both sexes, are most important. In another race of this sparrow in which two broods are produced, testosterone is elevated during the period leading to the first brood but not in that preceding the second brood, when the same nesting site and mate are used and territorial and courtship patterns are minimal. Androgens are also implicated in the preparation for spring migration, but the peak of 5α-dihydrotestosterone that occurs in the female (but not male) during wintering has not been satisfactorily explained. In this species estrogen is again correlated with yolk deposition, but no measurements were made of progesterone, which might again be expected to peak at ovulation.

The pattern of ovarian steroid secretion in birds is more closely related to that of the turtle, with whom they have an affinity in their failure to evolve a viviparous mode of reproduction, than to the snakes and mammals, in which luteal and placental function is more highly developed. Studies on birds have concentrated heavily on domestic species for both practical and commercial reasons, and these may be far from typical of the class as a whole. The role of ovarian androgens is not yet clearly defined in the domestic species, but it is clearly correlated with a number of behavioral patterns in wild species and may well have had similar functions in the turkey and fowl before their behavior was modified by domestication.

4.7. Comparative Aspects of Ovarian Steroids

Of the three classes of ovarian steroid, the estrogens appear to be the most conservative with very little change from cyclostomes to mammals, although considerable differences are found in plasma concentrations, which are in general much higher than mammalian values. This conservatism may be a reflection of the unique aromatic structure found in this hormone. Progestogens show considerable variation from the 17,20βP and pregnanolone of teleosts to progesterone in the terrestial tetrapods. Throughout the vertebrates, these hormones show considerable evolution in their activities, matching advances in reproductive pattern, although each maintains its role at a particular stage, estradiol in the early stages comprising vitellogenesis and progestogens in the later stages of ovulation or pregnancy. The role of ovarian testosterone remains something of a mystery. Although it has androgenic activity in birds, it may well play an essential role in the normal reproductive cycle of female teleosts. Evolution of biological role and hormone receptors may well be more important than the hormones themselves.

5. Steroids of the Adrenal and Interrenal Glands

The secretions of the mammalian adrenal cortex may be divided into three main categories, glucocorticoids (cortisol and corticosterone), which regulate carbohydrate metabolism, mineralocorticoids (deoxycorticosterone and aldosterone), which maintain salt balance, and adrenal androgens. In our discussion of the reproductive hormones we found that there was considerable overlap between the structure and function of the gonadal steroids and that mammalian definitions of "male" and "female" steroids were no longer valid in nonmammalian vertebrates. A similar situation is apparent with the adrenal or interrenal steroids, and the mammalian categories of mineralo- and glucocorticoid can no longer be considered universally valid. Steroids such as cortisol which are considered as glucocorticoids in mammals may act as mineralocorticoids in fish, and it is probable that the biological activities of the interrenal steroids may be determined more by the nature of the protein receptor that by that of the steroid itself. The physiological action of the corticosteroids in osmoregulation and control of metabolism is fully discussed in Parts IV and V of this volume, and we will limit ourselves here mainly to surveying the nature of the adrenal secretions throughout the vertebrates. The function of the adrenal androgens, even in mammals, remains something of a mystery, and even less is known of their occurrence or function in nonmammalian vertebrates.

The biosynthetic pathway of corticosteroid synthesis in mammals has been well documented and in general holds for nonmammalian vertebrates, although there are some differences which we will discuss later (Figure 21). Corticosteroids are often divided into two structural categories, 17-hydroxycorticosteroids and 17-deoxycorticosteroids, and in most species one or the other usually predominates. Although vertebrates may be divided into secretors of each of these types of corticosteroid, the boundaries are not clear cut since some species, such as the mouse, may secrete different types in infancy and adulthood. This suggests that the 17-hydroxylase is suppressed rather than lost, a conclusion that is supported by the fact that 17-hydroxycorticosteroids are sometimes found *in vitro* but not in plasma. We cannot emphasize too often that although *in vitro* incubations give valuable and essential preliminary information, no steroid can be considered functional unless it can be identified in blood.

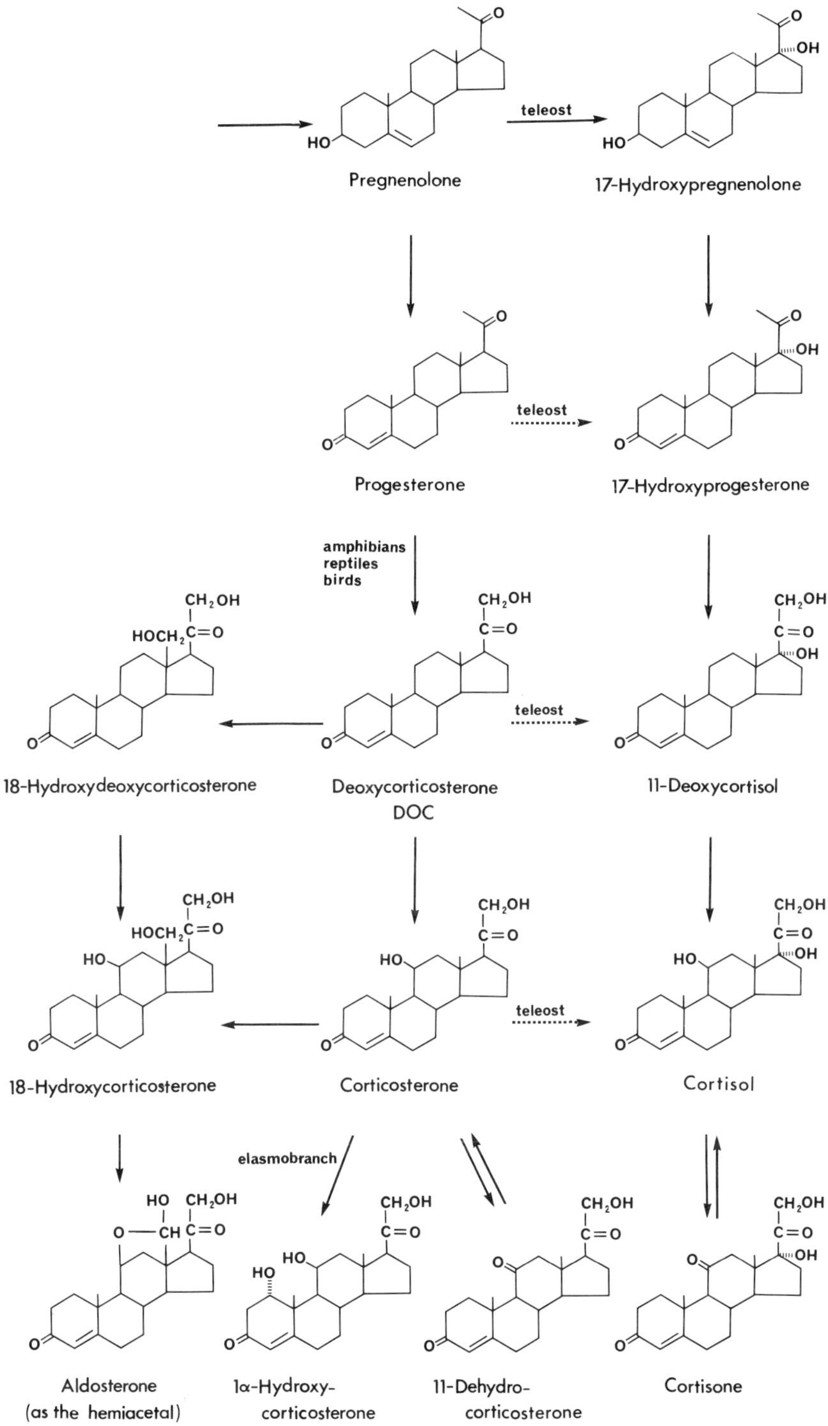

Figure 21. Biosynthesis of corticosteroids.

5.1. Agnatha

We have already seen that the nature of the gonadal steroids in the cyclostomes has been the subject of much work and that the possible formation of unusual hydroxylated steroids may explain the failure of many workers to isolate the classical steroids from plasma of these species. The nature of the corticosteroids is the subject of similar debate. Although early work indicated the presence of substantial amounts of the classical corticosteroids, cortisol and corticosterone, in plasma of the Pacific and Atlantic hagfishes, these results could not be substantiated by later work using the more specific methods developed during the 1960s. In spawning sea lampreys the formation of known corticosteroids from radioactive precursors with presumptive adrenal tissue or their presence in plasma could not be demonstrated. In the parasitic and larval sea lamprey, however, progesterone was converted into androstenedione, 17-hydroxyprogesterone, and 11-deoxycortisol, but not cortisol or corticosterone. The only identifiable steroid in blood serum taken after injection of radioactive progesterone was deoxycorticosterone, which has been identified as a metabolite of this steroid in testes. More recently it has been shown that corticosterone and cortisol are in fact present in plasmas of both the Pacific hagfish and the sea lamprey, although their concentrations (0.1 and 0.3 ng/ml in hagfish and 0.1 and 0.05 ng/ml in lamprey) were minute compared to the values of 50–200 ng/ml in mammals and teleosts. The major steroid identified in this work was 11-dehydrocorticosterone (0.6 and 3.0 ng/ml in hagfish and lampreys, respectively). ACTH stimulated all of the hagfish steroids, particularly 11-dehydrocorticosterone, which increased fourfold, but little effect was detected in the lamprey. Androstenedione was also found in lamprey plasma and, unlike progesterone and estrogens in the same samples, was stimulated by stressing the animal, suggesting that it might be a stress hormone of adrenal origin. There is evidence that lamprey plasma produces a peptide with ACTH activity in rat bioassays.

These results indicate that the classical corticosteroids, like the gonadal steroids, are present in cyclostome plasma, if at all, in concentrations much smaller than in any other vertebrate. We must therefore conclude either that the high concentrations found in other classes are unnecessary in cyclostomes or that these animals produce other, so far unidentified, corticosteroids. There is considerable evidence that the latter possibility may actually be the case. We have already seen (Sections 3.1 and 4.1) that the gonads have very active 6β-, 7α-, 15α-, and 15β-hydroxylases and in some cases also active 5α-reductases, which may explain the failure in many cases to detect known gonadal steroids. With the presumptive adrenal tissue it is probable that similar metabolites may occur since almost all of the papers in the field provide evidence of substantial amounts of unidentified products. These compounds usually have very similar chromatographic properties to authentic corticosteroids but separate from them on repeated chromatography or on attempts to crystallize them to constant specific activity. Products with similar properties were also found in low yield with incubations of progesterone with lamprey and hagfish gonads in which the hydroxylated progesterone derivatives were found. We must, therefore, consider it highly probable that the interrenal glands of the cyclostomes also produce such hydroxylated steroids and that this may explain both the high corticosteroid levels found in the early work and the failure of later workers to confirm more than trace amounts of known corticosteroids in these animals. If such compounds can be identified in incubates and their presence confirmed in plasma, we may get nearer to an understanding of the endocrine control in cyclostomes. The role of corticosteroids and corticotropin is of particular interest in the postspawning death of lampreys.

5.2. Chondrichthyes

Early investigations with elasmobranchs indicated that these fish were characterized by an extremely low production of corticosterone, 11-dehydrocorticosterone, and cortisol—a very similar situation, in fact, to the present one with the cyclostomes, which we discussed earlier. The problem in this case, however, was resolved by Idler and Truscott in 1966, who showed that the elasmobranch interrenal produced an unusual hydroxylated derivative, 1α-hydroxycorticosterone, a compound that has not been isolated from any other living animal. Idler and his co-workers have since identified this 1α-hydroxylase in 18 species of sharks, skates, and rays, and no species of elasmobranch has yet been found in which it does not occur. This enzyme may be leached out of damaged cells in tissues that have been frozen, resulting in an apparent loss of activity if preincubations are carried out to remove endogenous substrates. The enzyme activity increases from 5 to 30° but is irreversibly destroyed at 37°, which would indicate that such an enzyme could not occur in mammals. 1β-Hydroxy-5β-pregnane derivatives have been isolated from some human patients with hypertension, and although 1β-hydroxyl, in this steroid, like the 1-substituent in 1α-hydroxycorticosterone, is axial with respect to the ring A, the temperature stability of the elasmobranch enzyme and the different overall shape of the steroid would indicate that the hydroxylases themselves are very different. The plasma concentrations of 1α-hydroxycorticosterone measured in several species using fluorescence, double-isotope derivative assay, and radioimmunoassay are very variable, ranging from 1 to 140 ng/ml, comparable to corticosteroid values reported in other vertebrates. Concentrations of corticosterone and cortisol are in most cases either very low or not detectable, but surprisingly the concentrations of 1α-hydroxycorticosterone glucuronide are comparable to those of the free steroid. Neither the function of this glucuronide nor the organ in which the conjugation takes place is yet clear. Idler suggested that 1α-hydroxycorticosterone could act as a precursor for 1-dehydrocorticosterone, which might be the active steroid in elasmobranchs since the 1,4-dien-3-one is often associated with increased biological activity. This compound was not present in detectable amounts in plasma of two elasmobranchs, although it could possibly be formed from 1α-hydroxycorticosterone in target tissue. The isolation of specific glycoprotein receptors for 1α-hydroxycorticosterone in the cytosol of gills, interrenal, liver, and rectal gland of the skate, however, indicates that 1α-hydroxycorticosterone is in fact the active corticosteroid in elasmobranchs.

The biological activity of 1α-hydroxycorticosterone has not yet been clarified, partly at least because of the difficulty of performing the classical interrenalectomy and replacement experiments. 1α-Hydroxycorticosterone had 70% of the effectiveness of deoxycorticosterone in promoting sodium retention in rat and mouse bioassays and 80% of the activity of aldosterone in stimulating sodium transport in a toad bladder assay. Although it failed to promote the deposition of liver glycogen in adrenalectomized mice, it was one thid as effective as cortisol in another test for glucocorticoid activity.

In a more recent study the plasma concentration of 1α-hydroxycorticosterone in dogfish was increased by *Squalus* and mammalian ACTH and by transfer to 50% sea water. Hypophysectomy decreased, but did not totally eliminate, plasma 1α-hydroxycorticosterone, indicating that it was affected by nonpituitary factors, which was confirmed by its stimulation with dogfish renal extract and mammalian angiotensin II. The evidence thus indicates that 1α-hydroxycorticosterone acts predominantly as a mineralocorticoid in elasmobranch fish (see also Part II).

Elasmobranchs form one of the two subclasses of the Chondrichthyes, and it is particularly interesting that in the few species examined in the other subclass, the Holocephalans, there is no evidence for the formation of 1α-hydroxycorticosterone, and the major conversion product of progesterone and corticosterone is cortisol. The Holocephalans, in this respect, show a much closer relation to the teleosts than to the elasmobranchs.

5.3. Teleosts and Related Fish

Although teleosts are overwhelmingly the predominant component of the Class Osteichthyes, their steroidogenic patterns are best discussed in the context of patterns of the more primitive but related chondrostean and holostean fishes and the more advanced Sarcopterygii, which might be expected to act as a bridge between the teleosts and the phylogenetically ancient cyclostomes and terrestial tetrapods, respectively. We shall first discuss the major group, the teleosts, and later relate their steroidogenic patterns to the few available species of these other groups.

The main product of the teleost interrenal is cortisol, and there is little evidence for significant amounts of corticosterone. Although there have been some reports of 18-oxygenated corticosteroids in both incubations and plasma of isolated species, the quantities involved are generally low, and the results have not been substantiated in sufficient species for them to be classed as typical of teleosts. Although, superficially, the steroids of the teleost interrenal bear a close resemblance to those of mammals, there are a number of notable differences, and a number of biosynthetic sequences that are found in teleosts cannot occur in mammals. In the mammalian adrenal, hydroxylation at C11 or C21 precludes hydroxylation at C17 so that corticosterone cannot act as a precursor of cortisol for example. In several teleosts, however, 17-hydroxylation of corticosterone and deoxycorticosterone has been observed, although the main pathway to cortisol would still seem to go via 17-hydroxyprogesterone and 11-deoxycortisol. Several studies have also shown that these steroids are preferentially formed via 17-hydroxypregnenolone rather than via progesterone. Another peculiarity of the teleost is found in the interconversion of cortisol and cortisone. Although injected cortisol was rapidly converted into cortisone by the sockeye salmon, cortisone remained unchanged. In plasma, concentrations of cortisone often exceed those of cortisol, and it is possible that the factors affecting their interconversion may serve to regulate the availability of cortisol, which in mammals is the most active of the two hormones. Little, however, is known of their relative biological activities in fish. We noted earlier that many pairs of steroids existed of this type, and usually the hydroxylated steroid was more biologically active than the ketone. Although this holds true for estrone/estradiol, androstenedione/testosterone, and cortisone/cortisol in mammals, it may not necessarily be true in other vertebrates. A notable exception are the teleost androgens, 11-ketotestosterone/11β-hydroxytestosterone, in which the former is usually present in larger quantities and is assumed to be the most biologically active. Their relative activities are not, however, clear, but it may be significant that for this pair as for cortisone/cortisol the 11β-hydroxysteroid dehydrogenase enzyme is involved in the interconversion. Too few data are available to determine whether these two conversions are related in teleosts. Plasma concentrations of cortisol and/or cortisone have been reported in numerous teleosts, and although they show considerable variation, their general range of 100 to 400 ng/ml is comparable to that found in most mammals.

Corticosteroids are involved in many aspects of teleost life, and as might be expected, their concentrations in plasma are affected by numerous factors. Although some of these factors are similar to those in mammals and others are common to all teleosts, others may be characteristic of particular species and reflect facets of their particular habitat, or life pattern. Aspects of osmoregulation and carbohydrate metabolism are discussed in detail in Parts II and III, and we shall touch on these only briefly here and concentrate on the involvement of the interrenal in two important aspects of major importance in the life of some teleosts, that is, spawning migrations and ovulation.

Many species of teleost undertake spawning migrations as part of their normal life-cycle. In some cases, such as cod and herring, this may be an annual migration undertaken entirely in a marine environment; in other cases, the migration may be from a stream to a freshwater lake or vice versa, but the type of migration that has been of greatest interest to comparative endocrinologists is that between a freshwater and marine environment, particularly when such migrations occur, as in Pacific salmon and European eels, only once and lead almost inevitably to death. Such spawning migrations and the "programmed death" have been extensively studied and have demonstrated a close link between gonadal and interrenal steroidogenic activities and postspawning death.

Studies of the eel are severely hindered by the inability to study them during the major part of their marine migration, and there has been little success in stimulating reproduction in captive specimens.

The Pacific salmon, *Oncorhynchus nerka*, however, has been extensively studied since sexual maturation coincides with the upstream migration and spawning is rapidly followed by death. The primary cause of death is microbial infection, but this appears to result from a general tissue degeneration caused by hypercorticolism. The conditions of this upstream migration with adaptation to freshwater accompanied by starvation might well be expected to involve the interrenal steroids in osmoregulation and carbohydrate metabolism. It is, in fact, associated with increased plasma concentration of cortisol resulting from both the vigorous swimming against the current and impaired clearance of cortisol from the fish by degeneration of the liver. Studies of changes in cortisol levels are hindered by the almost inevitable stress resulting from capture, but captive fish and those caught resting in pools during the migration do not show greatly elevated levels. Vigorous exercise causes a dramatic increase in the plama cortisol concentration, and female fish had a more active adrenal and higher plasma corticosteroids than males. The cortisol secretion rate was increased by estrogens and androgens but this was reversed and life prolonged by gonadectomy. The mode of action of these gonadal hormones is not clear, but since both androgen and estrogen have an effect, it is possible that brain aromatization is involved, although other factors must be implicated since 11-ketotestosterone, a nonaromatizable androgen, is also effective. Such changes do not appear to be so extensive in the Atlantic salmon, which usually survives the spawning migration and returns afterward to the sea, and in tilapia neither ovariectomy nor estradiol treatment give a significant change in plasma cortisol. The possible involvement of hepatic catabolism of cortisol to androgen precursors is discussed in Section 6.1.

We have already seen that the teleost ovary in some species is able to synthesize some corticosteroids such as deoxycorticosterone, and we have discussed the possible interaction of ovary and interrenal in ovulation in the catfish (Section 4.3.3). In most vertebrates the interrenal synthesis of corticosteroids is stimulated by the pituitary adrenocorticotropic hormone (ACTH), and gonadotropic hormones (GTH and LH) have little effect. This applies equally well to many teleosts, but in the Indian catfish mammalian ACTH and LH and the salmon pituitary gonadotropin extract (SG.G100) were all effective in increasing plasma cortisol concentration, although the ACTH was slightly more effective than the gonadotropin preparations. That the hormones acted on the interrenal and not on the gonad was demonstrated by similar stimulation in castrate fish. This effect seems to be characteristic of the catfish since gonadotropins had no effect on salmon interrenal. Although ACTH was more effective than GTH in inducing corticosteroid synthesis in catfish, it was only nominally effective in inducing oocyte maturation *in vitro* in ovary–interrenal coculture and was ineffective even in massive doses *in vivo*. Whereas ACTH stimulates only plasma cortisol, GTH stimulates both testosterone and cortisol, which might indicate a synergism between these two steroids in the induction of ovulation in this species, and as we have already seen, ovarian pregnanolone may also be involved in this process. There is little evidence of interrenal production of 5β- or 5α-reduced pregnanes, but the body kidney of rainbow trout, although unable to synthesize corticosteroids, is able to convert progesterone into 5α- and 5β-pregnanediones and 3α and 3β-hydroxy-5α- and 5β-pregnan-20-ones.

Plasma corticosteroid concentrations in teleosts are also affected by a number of other environmental factors such as background, temperature, time of day, and season. The rainbow trout and the European eel are able to darken when placed in a black background owing to activation of melanotrops. This also results in an increase in plasma cortisol concentration, although it is not clear whether this is due to release of ACTH from the pituitary melanotrop where it acts as a precursor of αMSH. Temperature may also affect plasma corticosteroid levels in tilapia with lower values in fish kept at 28° than at 17°, but it is not clear whether this is due entirely to the higher clearance rate at 28° or to the higher interrenal activity due to lower temperature or fasting at 17°. Several teleosts show a diurnal rhythm in cortisol plasma levels, but the pattern seems nowhere near as clear cut as in mammals and may show variations with season. Such rhythms do, however, emphasize the need to take plasma samples at defined times of the day. The seasonal cycle of cortisol is also variable, but in most species there is a rise in plasma concentration toward ovulation, which might indicate that it plays some role in the reproductive cycle.

Although the Osteichthyes are comprised overwhelmingly of teleost fish, which, as we have seen, are secretors of cortisol, the more primitive subclasses, the Holosteans and Chondrosteans, and the more advanced Sarcopterygii, which are the ancestors of terrestial vertebrates, are particularly interesting from the evolutionary point of view.

Specimens from all three surviving genera of the Chondrosteans have been studied. Interrenals of the Atlantic sturgeon (*Acipenser*) incubated with cholesterol, pregnenolone, or progesterone gave cortisol as the major product with small amounts of cortisone, corticosterone, and 11-deoxycorticosterone. Plasma concentrations of cortisol, cortisone, and corticosterone were all extremely low (1.8, 0.16, and 0.07 ng/ml), but in the paddlefish (*Polyodon*) much higher values of these steroids (52, 10, and 0.2 ng/ml) were obtained, although these may have been elevated by the stress of capture. Although the steroid concentrations in the sturgeon are, as in the cyclostomes, very low, the values in the paddlefish are closer to those of teleost, and both species show, like the teleosts, a preferential secretion of 17-hydroxycorticosteroids. Two species of the third genus (*Polypterus*), which differ from the sturgeon and paddlefish in having fleshy paired fins, having lungs, and living in an environment similar to that of the lungfish, have also been studied and show a very different pattern, although plasma availability was very limited. Corticosterone was the only detectable plasma corticosteroid (4 and 10 ng/ml), and this was also the only steroid isolated from incubates of presumptive interrenal tissue. In the only holostean so far examined, the bowfin (*Amia*), cortisol (7.6 ng/ml) and corticosterone (1.1 ng/ml) were both identified in plasma.

The Sarcopterygii, linking as they do the cortisol-secreting teleosts and the corticosterone- and aldosterone-secreting amphibians, are particularly interesting from an evolutionary point of view. Representatives of three genera of Dipnoi, the Australian, African, and American lungfishes, and of the unique Crossopterygian, the coelocanth, have been studied. The American lungfish (*Lepidosiren*) showed, as expected, an intermediate pattern between teleosts and amphibians with secretion of both cortisol (6.0 ng/ml) and aldosterone (5.8 ng/ml) with a smaller amount of corticosterone (1.6 ng/ml). The African and Australian lungfishes, however, did not support these results, and the major constituent in plasma of both species was deoxycorticosterone (0.6 and 0.044 ng/ml) with a trace (0.01 ng/ml) of aldosterone only in the Australian species. No cortisol or corticosterone could be detected in either species. The coelocanth is perhaps one of the most interesting species in evolutionary terms since it is the nearest living relative to the "missing link" between fish and tetrapods, but studies of steroidogenesis in this species have, of course, been severely hampered by its lack of availability. Recently Truscott was able to obtain some frozen kidney tissue containing presumed adrenocortical cells, although the condition of the tissue was far from perfect. Evidence was obtained for the presence of deoxycorticosterone, cortisol, and corticosterone, and incubation with radioactive precursors showed the presence of 5-ene-3β-hydroxysteroid dehydrogenase, 5-ene-4-ene isomerase, 11β- and 20α-hydroxysteroid dehydrogenases, and pregnane-5α- and 5β-reductases. No evidence was found for 11β-, 17α-, or 21-hydroxylases *in vitro*, but the presence of the classical corticosteroids in tissue extracts indicates that these enzymes are present in the normal fish and may be affected by tissue degeneration during freezing. For the same reason it is not possible to confirm the presence or absence of aldosterone in this species. The difficulty in obtaining coelocanth tissue and plasma in good condition may leave the nature of its steroids unresolved for a long time.

5.4. Amphibians, Reptiles, and Birds

The transition from a completely aquatic milieu to a terrestrial environment involved the first protoamphibians in a major change in their endocrine function. A prerequisite for success in such a change was adaptation of the endocrine system to cope with the new environment, and it is therefore perhaps not surprising that it is at this stage of evolution that major changes take place in the nature of the interrenal gland and its steroids. Morphologically it is only at this stage that the interrenal tissue separates from the kidney and for the first time becomes inseparably intermingled with the chromaffin tissue, this process continuing through the reptiles until we reach the distinct adrenal gland of mammals with its discrete medullar and cortex regions (see Chapter 3). Although the gland itself continues to evolve from amphibians to mammals, the steroids secreted are very similar in amphibians, reptiles, birds, and mammals. Two major changes occur in the transition from the more primitive aquatic vertebrates, loss of 17-hydroxylase activity and acquisition of an 18-hydroxylase and an 18-hydroxysteroid dehydrogenase. The major steroids of the interrenal thus become corticosterone and aldosterone instead of cortisol. The loss of the 17-hydroxylase in the interrenal shows a remarkable parallel with the transition from 17,20βP to progesterone as the ovarian steroid and loss of the testicular 11β-hydroxylase. We have already mentioned that there have been isolated reports of the formation of aldosterone and 18-hydroxylated corticosteroids in low yield in some species of teleost, so it would seem that the ability to form these steroids may have already been present prior to the transition to land and that an increased synthesis of these steroids gave a selective advantage in the new environment. It would, in fact, be rather surprising if our protoamphibian was able to "invent" a new enzyme to produce suitable steroids to cope with its new environment, and it is perhaps more likely that the 18-hydroxylase resulted from adaptation of a preexisting enzyme.

The substitution of 18-hydroxylation for 17-hydroxylation suggests that this enzyme may in fact have been the 17-hydroxylase itself and that a change in structure of this enzyme could favor hydroxylation at C18 instead of C17. If we look at the models for cortisol and 18-hydroxycorticosterone, or 11-deoxycortisol and 18-hydroxydeoxycorticosterone, we can see that the 17α- and 18-hydroxy substituents are in very similar positions with respect to the axis through the oxygens at positions 3, 20, and 21. An enzyme with receptors for functional groups at these positions might thus be easily adapted through a relatively minor mutation to hydroxylate at C18 instead of C17. Furthermore, once hydroxylated at one of these positions, the other position would not be able to present itself to the active site of the enzyme, and in fact we find that 18 hydroxylation does preclude 17α-hydroxylation and vice versa. We know very little of the structures of steroidogenic enzymes in mammals, let alone in primitive vertebrates, so these suggestions can remain no more than speculation, but it is quite clear that the enzymes do undergo evolutionary changes since, as we have seen, the conversion of deoxycorticosterone to 11-deoxycortisol that occurs in teleosts is not possible with mammalian enzymes.

17-Hydroxylase activity is not completely lost in amphibians, reptiles, and birds, and there are isolated reports of its occurrence in plasma or incubations in all three classes, and in mammals cortisol again emerges as the predominant corticosteroid. Although some orders of mammals, particularly Rodentia, have a very similar pattern of steroids to amphibians and reptiles with high plasma levels of corticosterone and aldosterone, most mammals secrete predominantly cortisol, with low levels of aldosterone which nevertheless play an important part in maintenance of the sodium–potassium balance. The mammalian adrenal has a clear zonation, and the steroidogenic potential of the outer zone, the zona glomerulosa, shows a remarkable similarity to the amphibian and reptilian interrenal, in its synthesis of 18-hydroxylated rather than 17-hydroxylated corticosteroids. This similarity is, however, only superficial since the cells of these more primitive vertebrates have histological characters more similar to the cortisol-secreting zona fasciculata of the mammals.

Much of our information on steroid production by the interrenal glands of amphibians, reptiles, and birds has been obtained from *in vitro* incubations with radioactive precursors, and the picture at present available for the concentrations of adrenal steroids in plasma is very patchy. Corticosterone levels exhibit a considerable range, to some extent probably reflecting experimental conditions, but the majority of the values reported fall in the range of 10–100 ng/ml, which is broadly comparable to that of mammals. Far fewer results are available for aldosterone, and much of the evidence for its secretion has been obtained *in vitro* from both exogenous and endogenous precursors. The highest values, 2–160 ng/ml, have been obtained in amphibians, almost entirely anurans. In birds the aldosterone values are much lower (0.1 ng/ml) and are similar to those of mammals. Aldosterone concentrations in reptiles have rarely been measured, but in 10 species of marine, terrestrial, and freshwater snake aldosterone ranged from 0.8 to 54 ng/ml. Corticosterone in this sample varied from 1 to 400 ng/ml, but there was no correlation between the plasma levels of either hormone and the habitat. Corticosterone concentrations are stimulated by ACTH, but in some reptiles at least, this stimulation is temperature dependent, with highest stimulation at 40° and lowest at 10°. As with the action of temperature on gonadal steroidogenesis, this is also believed to act directly on the steroidogenic enzymes of the interrenal. In reptiles and amphibians steroid levels fall after hypophysectomy to low levels, but in birds hypophysectomy decreases corticosterone only to about half of the intact levels, indicating another source of stimulation. It has been suggested that the chicken hypothalamus may secrete a substance with ACTH-like activity, possibly α-MSH, which can directly stimulate the adrenal as well as corticotropin-releasing factor, which acts by stimulating release of ACTH from the pituitary.

Corticosteroids in amphibians, reptiles, and birds, like most other vertebrates, show both diurnal and seasonal variations. There are too few studies for generalizations of either the nature of the cycle or the factors that regulate it. The life-cycles of vertebrates are too diverse, and corticosteroids, which may be involved in changes in metabolism, hibernation, migration, return to aquatic spawning grounds, and ovulation to mention but a few, are affected by too many factors to rationalize these seasonal and diurnal rhythms. We shall limit ourselves to two examples from the recent literature. In the European green frog, *Rana esculenta,* both seasonal and diurnal rhythms in plasma corticosteroids occur. Corticosterone shows a clear seasonal variation with maximal levels of 26 ng/ml in the spawning season of May–June and lowest levels in February (4 ng/ml). This annual cycle is not correlated with plasma estrogen cycles and is best correlated with locomotor activity. Unlike many mammals, there is no significant difference between the sexes in plasma corticosteroid concentrations. *In vitro* studies have shown that the seasonal variation in these concentrations is related to the increased sensitivity of the interrenal to ACTH stimulation. Stimulation of corticosterone and aldosterone were highest in May–June and lowest in winter. There is also a clear diurnal rhythm in these two hormones during the reproductive season. During daylight aldosterone (10 ng/ml) is approximately twice the concentration of corticosterone, but after the onset of darkness there is a rapid rise in both hormones, with a greater increase in corticosterone (25 ng/ml) than in aldosterone (15 ng/ml). Daylight is passed in feeding activities, but during darkness sexual activity such as mating calls, territorial defense, and searching for a partner predominates. Increased adrenal activity thus correlates both during the seasonal and daily cycles with sexual activity, but unfortunately plasma sex steroids were not measured in these animals, and it is not yet possible to further clarify adrenal–gonad relations either in synthetic ability or in sensitivity to ACTH.

As we have seen in fish, corticosteroids have often been implicated in ovulation, and in some amphibians corticosteroids are formed by the ovary *in vitro*. Corticosteroids frequently show an increase during the reproductive season, but it is not clear whether this is involved in ovulation itself since this rise often occurs in both sexes. Recent work with the domestic hen has, however, shown that there is a surge in corticosterone coinciding with oviposition and a diurnal peak approximately 2 hr after the onset of light. There is no corticosteroid surge 4–8 hr preceding ovulation as found for the gonadal steroids.

5.5. Comparative Aspects

The essential characteristics of the corticosteroids, the 11β- and 21-hydroxylation of progesterone, appear to have developed at a very early stage of evolution. These characteristics are present in teleosts, and even in cyclostomes cortisol has been isolated, albeit in very low concentrations. It thus appears that the necessary enzymes for corticosteroid synthesis are phylogenetically ancient and had developed even in the most primitive vertebrates. The isolation of cortisol from gonads of invertebrates also suggests that they are indeed very ancient enzymes. The isolation of any compound is, to some extent, determined by the experimental design, and the failure to isolate a particular steroid may not necessarily denote that it is not present, rather that it could not be isolated under the conditions employed. Despite several experiments specifically designed to find aldosterone and 18-oxygenated corticosteroids in teleosts, its presence is still not confirmed in more than a few species in relatively low yield. It is only in the tetrapods that this steroid regularly occurs, and we have speculated that this may have evolved from the 17-hydroxylase, the disappearance of which also coincides with the emergence of the vertebrates from their aquatic environment. The change in oxygenation pattern may thus have been an important factor in adaptation to a terrestial environment, but we should be wary of overinterpreting the very high aldosterone levels found in present-day amphibians as indicative of similar values in the amphibian ancestor of reptiles since these results have been obtained almost entirely on the anurans, which are a highly developed form of amphibian and may bear little resemblance to those in the direct evolutionary line to the reptiles. The 18-oxygenated corticosteroids retain their importance as mineralocorticoids in all vertebrates from amphibians to mammals, but the 17α-hydroxylase potential is not entirely lost and reemerges in many mammals that secrete cortisol, which, however, now has predominantly glucocorticoid activity. Some mammalian species, particularly rodents, are notable for secreting predominantly 17-deoxycorticosteroids such as corticosterone, and the wide variety of corticosteroid patterns in mammals suggests that the major evolutionary changes, apart from the innovation of 18-oxygenation, have been in the corticosteroid receptors and the uses to which the hormones have been put rather than in the steroids themselves.

In our discussions of the gonadal hormones we showed that there were frequently very large amounts of "inactive" steroids, such as glucuronides or 5β-reduced steroids, formed alongside the normal hormones. There have been few such reports in interrenal studies, although the chicken adrenal has been reported to form glucuronides. The general absence of these steroids from the interrenal may, to some extent, be a reflection of the fact that work on the interrenal started much earlier than on the gonads and such steroids were ignored, but it is also possible that such steroids may only be of importance in the gonads where seasonal regulation and timing is of major importance.

6. The Liver and Steroid Catabolism

It is, perhaps, not surprising, that the mammalian liver has very different enzymatic activities from the interrenal and gonads since its function is to inactivate the steroids and prepare them in suitable form for excretion via the bile or urine. Whereas the endocrine glands possess mainly oxidative enzymes, the liver is chacterized by a preponderance of reductases. Hepatic catabolism is characterized by a destruction of the biologically active centers of the steroid hormones, and it is therefore not surprising that the major changes are reduction of the ketones at positions 3 and 20, reduction of the double bond, loss of the corticosteroid side chain and conjugation with sulfuric and glucuronic acid (Figure 22). Reduction of the 4-ene to a 5β-steroid, as we have seen, leads to a major change in molecular conformation, and conjugation gives increased solubility and ease of excretion. The liver also hydroxylates steroids at a number of positions, notably 1, 2, 6, 7, 15, and 16. The catabolic pathways of the steroids in mammals have been well described, but this field has been very much neglected in nonmammalian vertebrates. The liver can no longer be considered in isolation since it has been clearly shown in mammals that its rate of catabolism is affected by gonadal steroids and that this in turn may regulate endocrine steroid secretions. There is, however, little information on such interactions in nonmammalian vertebrates. We shall briefly outline what little is known of hepatic catabolism in nonmammalian vertebrates, dividing them rather arbitrarily into two categories, (1) fish and (2) amphibians, reptiles, and birds.

6.1. Fish

The hepatic catabolism of androgens in cyclostomes shows a very close resemblance to gonadal metabolism. In hagfish (Myxinidae) the main transformations are 5α-reduction and 6β- and 7α-hydroxylation. As in the gonads of these species, the 7α-hydroxylated derivatives are predominantly reduced to 5α, whereas 6β-hydroxylation appears to block this reductase activity. In lampreys (Petromyzonidae)

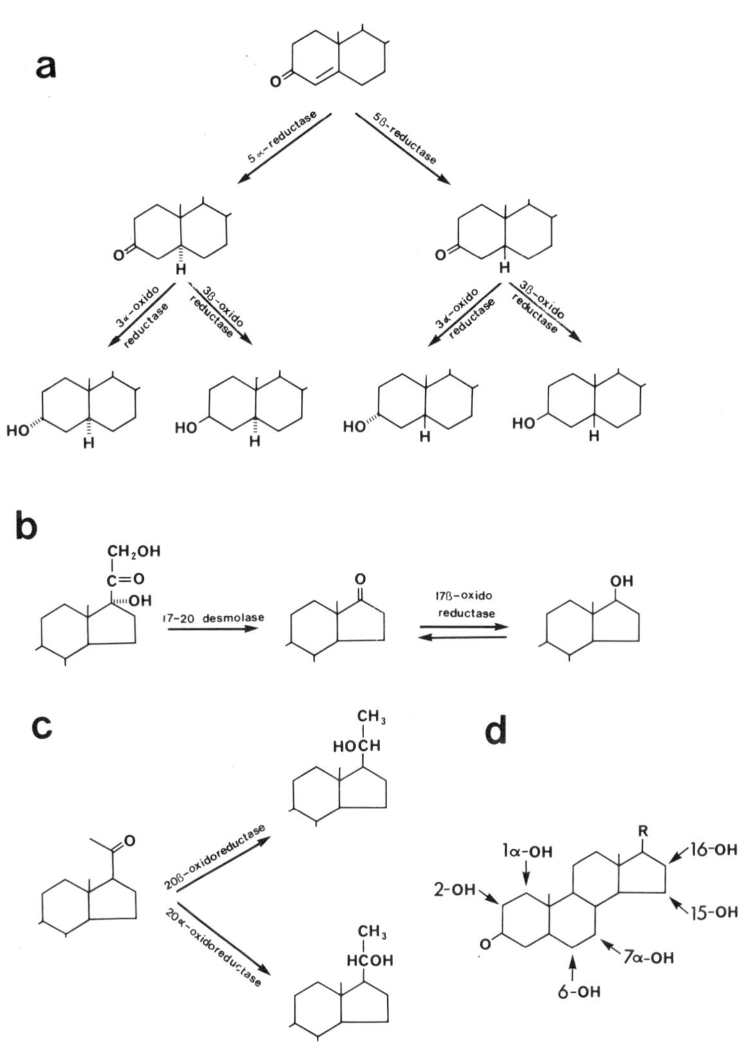

Figure 22. Catabolism of steroids.

15α-hydroxylase and 5α-reductase are the predominant liver enzymes *in vitro*. It is in fact quite remarkable how little difference there is in the enzymatic activities of the gonads and liver in these phylogenetically ancient vertebrates.

6α-, 6β-, and 16α-hydroxylases, 5α-reductase, and 3α-oxido-reductase have been found in elasmobranch liver. The characteristic elasmobranch corticosteroid, 1α-hydroxycorticosterone, is excreted via the bile as its glucuronide, but a small amount is also converted to 1-dehydrocorticosterone.

Too few species have been examined to give a general pattern for the very diverse class of teleost fish, but most of the mammalian enzymatic patterns appear to be present within the class: 1α-, 6α-, 6β-, 7α-, and 16-hydroxylases; 3α-, 3β-, 20α-, and 20β-oxidoreductases; 5α- and 5β-reductases; and glucuronyl transferase. Glucuronyl transferase activity increases with temperature and with sexual maturation in trout. The catabolic difference between teleost species appears to be comparable to the differences within mammals.

Even in mammals the liver is often neglected in endocrine studies, but it inevitably plays an essential role in steroid dynamics, which are greatly influenced by the rate of catabolism, and this may itself influence the secretion rates and morphology of the endocrine glands themselves. In nonmammalian vertebrates, and in fish in particular, where the steroid transformations of the gonads and liver are even closer it might be expected that these links may be of even greater importance.

In the trout androgens can influence the activity of the hepatic 17-oxidoreductase enzyme, which in turn regulates the interconversions of testosterone/androstenedione, estradiol/estrone, and ketotestosterone/androstendione, and since the 17-hydroxy rather than 17-keto form is usually considered to have greatest biological activity, the liver may in fact play an important role in regulating the overall biological activity of the steroids. This may be of particular importance with estrogens since in fish the liver is both the target tissue in its synthesis of vitellogenin and the site of steroid catabolism. It is thus in a unique position to regulate the amount of active estrogen at the target site.

A further possible interaction between the liver, interrenal, and gonads is suggested by the conversion of cortisol to androstenetrione by several species of teleost since this steroid may act as the precursor of the main teleost androgen, 11-ketotestosterone. Since androgens in turn affect both gonadal and adrenal biosynthesis and hepatic catabolism, it is possible that the liver may play a more direct role in the regulation of steroid synthesis than is often assumed.

6.2. Amphibians, Reptiles, and Birds

Too few species of these classes have been examined to make positive generalizations, but examples from all of them have been shown to possess 5β-reductase, 3α- and 3β-, 20α- and 20β-oxidoreductases. Hydroxylase activity at 6β, 11α, 15α, 15β, and 16α has been demonstrated in some amphibians and reptiles. Glucuronidation also occurs but may be less active than in mammals.

The liver catabolism of steroids has received very little attention in nonmammalian vertebrates, and it is not possible to make any general pattern from the metabolites isolated as differences may be as likely to arise from the different incubation techniques as from species differences. The results to date do, however, show that most of the common mammalian patterns of catabolism are present in nonmammalian vertebrates with perhaps the exception of the most primitive, the cyclostomes, in which conjugation and 5β-reduction do not appear to be of major importance. In the cyclostomes particularly, and to some extent in teleosts, the liver and gonads seem to have more enzymes in common than in the more advanced vertebrates.

7. General Conclusions and Comments

When the literature on steroids in nonmammalian vertebrates was surveyed in 1972, in what were still the early days of such investigations, it required a full volume of 500 pages to cover the then current state of knowledge (Idler, D. R., 1972, *Steroids in Nonmammalian Vertebrates,* Academic Press, New York and London). During the last 10 years there has been a rapid expansion in the literature on this subject, spurred on no doubt by Idler's book. In a volume of this size it has not been possible for us to make a comprehensive survey of the present state of the literature, and we have had to be rather selective and to make generalizations which may in time be found unjustified. The nonmammalian vertebrates cover an enormous range of species from an almost infinite number of habitats and life-styles and covering several million years of evolution. The species investigated tend to be those readily available in Europe and North America and may well not be typical of others from more remote habitats. In some classes there do appear to be very distinct characteristics which are found in most members of the class investigated and which we have indicated are typical. Such characteristics might be 11-oxygenated gonadal androgens in teleosts, 1α-hydroxycorticosterone in elasmobranch interrenals, dihydrotestosterone in anuran testes, and 17,20βP in teleost ovaries. When we try to fit these characteristics into an evolutionary pattern, we do have a more difficult problem since it is not clear whether these are primitive vertebrate attributes or whether they are adaptations for a particular habitat. Anurans are very advanced amphibians, and 5αDHT may be an adaptation for their environmental niche rather than being intermediate between fish and reptiles, although the 5α-reductase activity in reptilian testes indicates that their common ancestor may well have had an active 5α-reductase. Testicular 11-oxygenation seems to be characteristic of teleost fish and to some extent urodele amphibia, which may indicate that it is primitive, but it is equally possible that it may have increased in importance in present-day teleosts but lost importance in the ancestors of the tetrapod line.

It is possible that further study of hormones in teleosts may show some evolutionary pattern within this, the largest of all vertebrate classes. The 1α-hydroxylase of the elasmobranch interrenal has no other parallel in vertebrate endocrine glands. Even among the cyclostomes there appears to be a clear divergence in hydroxylases between Myxine and Petromyzontes with 6, 7, and 15, respectively. It is therefore perhaps in the even more primitive creatures that we must look for the origin of steroid pathways. Other chordates such as Amphioxus, sea squirts, and acorn worms may well hold some steroidogenic surprises. Studies on invertebrates have been almost entirely limited to isolation of the classical mammalian steroids, but no biological function has yet been assigned to such steroids, much greater importance being given to the insect hormones of the ecdysone family. The steroid skeleton is a very ancient structure, and Sandor and Mehdi suggest that it may be of a comparable age to other basic biomolecules such as amino acids, nucleic acids, peptides, and DNA. Steroids in some form are certainly present in all forms of plant and animal life, but we do not yet know when they acquired the necessary structure and receptors for endocrine control. We have, in this chapter, concentrated almost entirely on the characteristics of the steroids and neglected those of the receptors. This is, of course, because little is known of the structure of these receptors, but since they are essential for biological activity, the two must be considered together and we must visualize a parallel evolution of steroids, their receptors, and their synthesizing enzymes, changes in one affecting interactions with the other and some changes having a profound effect on the biological uses to which the steroid–receptor complex is put. It is therefore perhaps not surprising that it is at that stage where vertebrate evolution makes its greatest step from an aquatic to a terrestrial environment that the greatest change in steroidogenic patterns also occurs. The loss of gonadal 11-oxygenation and interrenal 17-hydroxylation and the rising importance of 18-oxygenated corticosteroids point to a traumatic change in endocrine patterns which may even have been essential for success in otherwise suitably adapted "amphibians." Much attention has focused on the physical changes of this adaptation such as changes from fins to legs and gills to lungs, but the less visible endocrine changes are perhaps even more remarkable.

The study of nonmammalian vertebrates started by demonstrating the presence of mammalian steroids in frogs, fish, and others, but it was soon clear that the gonads in particular not only produced different hormones, but also were often characterized by the production of large amounts of other apparently biologically inactive steroids. We have discussed the possible uses of such compounds as glucuronides, 5β-reduced steroids, and epitestosterone as pheromones or in the competitive regulation of "active" steroid formation, and the quantities of these steroids formed *in vitro* and the circulating levels of at least the glucuronides *in vivo* suggest that they cannot easily be ignored. Studies of pituitary stimulation, metabolic clearance, seasonal changes, and other factors must take into account the fact that other unmeasured hormones may also be stimulated and give a misleading result. A stimulation of the gonad might, for example, dramatically increase androstenedione, testosterone glucuronide, or etiocholanolone but have little effect on testosterone itself, giving the erroneous conclusion that steroidogenesis was not stimulated.

Concentrations of the steroids in plasma show enormous variation within any particular vertebrate class, and it would be very difficult to quote, for example, a typical value for testosterone in teleosts. We have attempted in Table III to summarize the steroids formed in the gonads and interrenals throughout the vertebrates and to give an approximate range for the values that might be expected. If any trend can be found in the evolution of vertebrates, then it might be toward lower concentrations of simpler steroids, which suggests that this is a reflection of changes in receptor conformation and specificity.

We have seen that although there may be typical steroids for particular vertebrate classes, unusual steroids are often formed which may be typical of a phylogenetically distant class, as, for example, the 7α-hydroxylase in both hagfish and rats. In some cases, as with gonadal aldosterone, this may be an artifact of the incubation resulting from desuppression of an enzyme and suggests that the poten-

Table III
Plasma Steroid Concentration of Ovarian, Testicular, and Interrenal Steroids throughout the Vertebrates (ng/ml)[a]

	Ovary	Testis	Interrenal		
Cyclostomes	E_2 (?) (6β 7α, 15 T)	T (0–4) (6β, 7α, 15 T)	F (0.1)	B (0.05–0.3)	(+ ?)
Elasmo-branchs	E_2 (40), T (6)	T (2–10)	—	—	1αB (1–140)
Chondrostei/Holeostei	?	T (4) ?	B/F/E (1–50)		
Teleost	E_2 (2–40), T (2–200), 17,20βP (200–400)	T (1–50), KT/βT (5–100)	F/E (100–400)		
Sarcopterygii	?	?	DOC/F/Aldo/B	(0.04–6)	
Urodele	E_2 (2), T (5–15), Prog (4–30)	T (10–50), βT (16), DHT (2–50)		?	
Anuran	E_2 (1–10), T (5–15), Prog (4)	T (2–50) DHT (1–10)		B(10–100)	Aldo (2–160)
Reptiles	E_2 (1), T (5), Prog (1–5)	T (10–40)		B(10–100)	Aldo (1–50)
Birds	E_2 (0.5), T (2), Prog (6)	T (1–40)		B (10–100)	Aldo (0.1–1)
Mammals	E_2 (0.1) Prog (1–20)	T (1–10)	F/B (50–200)		Aldo (0.2)

[a]E_2, estradiol; T, testosterone; Prog, progesterone; KT, 11-ketotestosterone; βT, 11β-hydroxytestosterone; DHT, dihydrotestosterone; F, cortisol; E, cortisone; B, corticosterone; Aldo, aldosterone.

tial of many enzymes remains in all of the steroidogenic endocrine glands throughout the vertebrates. We have dealt with steroids in isolation, but we must remember that many of these enzymes, particularly conjugating enzymes and the hydroxylases of the P-450 type, also play an important role both in the synthesis of nonhormonal compounds and in the metabolism of nonsteroidal xenobiotics. A 7α-hydroxylase and 5β-reductase, for example, are essential components in the synthesis of the bile acids.

In this chapter we have tried to illustrate the wide variety of hormonal patterns throughout the vertebrates and show how they fit into particular life-styles and habitats. It will be apparent to the reader that there remains much to be learned of the nature of the steroids themselves in individual species, their mode of control by pituitary and environmental factors, and their mode and mechanisms of action in target tissues. When we consider the few species so far examined and the range and diversity of the remaining species, we may well feel that it will be a long time before we have a full picture of the role of steroid secretions of the vertebrate endocrine glands.

8. Selected Readings

Callard, I. P., Callard, G. V., Lance, V., Bolaffi, J. L., and Rosset, J. S., 1978, Testicular regulation in nonmammalian vertebrates. *Biol. Reprod.* **18:**16–43.

Chapman, B. I., 1979, Radioimmunoassay. *Chem. Brit.* **15:**439–447.

Delrio, G., and Brachet, J., 1980, *Steroids and Their Mechanism of Action in Nonmammalian Vertebrates,* Raven Press, New York.

Epple, A., and Stetson, M. H., 1980, *Avian Endocrinology,* Academic Press, New York and London.

Hoar, W. S., Randall, D. J., and Donaldson, E. M., 1983, *Fish Physiology,* Vol. 9, Part A. *Reproduction: Endocrine Tissues and Hormones,* Academic Press, Orlando, FL.

Idler, D. R., 1972, *Steroids in Nonmammalian Vertebrates,* Academic Press, New York and London.

IUPAC-IUB, 1972, Definitive rules for nomenclature of steroids. *Pure Appl. Chem.* **31:**285–322.

Lance, V., and Callard, I. P., 1978, Hormonal control of ovarian steroidogenesis in nonmammalian vertebrates, in: *The Vertebrate Ovary* (R. E. Jones, ed.), Plenum Press, New York and London, pp. 361–407.

Richter, C. J. J., and Goos, H. J. Th., 1982, *Reproductive Physiology of Fish,* Pudoc, Wageningen, The Netherlands.

Sandor, T., Fazekas, A. G., and Robinson, B. H., 1976, The biosynthesis of corticosteroids throughout the vertebrates, in: *General, Comparative and Clinical Endocrinology of the Adrenal Cortex,* Vol. 1 (I. Chester Jones and I. W. Henderson, eds.), Academic Press, London and New York, pp. 25–142.

2

Overall View of the Organization of the Vertebrate Gonad and the Structure of Steroidogenic Cells

J. A. Pudney

1. Introduction

In the natural world success is measured in terms of both the individual and the species. Therefore, it is to be expected that a particular reproductive strategy employed by a species would be the most effective so far evolved for carrying out the propagation of its members. Concomitantly, the organization of the gonads and development of the reproductive tract of each species would have evolved to produce the most efficient morphological arrangement of structures adapted not only to the reproductive strategy employed, but to the particular environment inhabited.

It is the purpose of this chapter to describe and discuss the basic organization of the gonads throughout the vertebrates. The main emphasis will be concentrated on those cells responsible for the synthesis and secretion of steroid hormones, and thus, a large section will be devoted to the morphological characteristics of steroid-producing cells. It is intended that this chapter should serve not only as an introduction to the more specialized ones, dealing with reproduction in specific vertebrate classes, but also provide the necessary morphological background on steroid-producing cells for the chapter on steroidogenesis (see Chapter 1 and Figure 1a).

2. Morphology of Steroid-Producing Cells

It was known that the gonads of both sexes synthesized and secreted steroid hormones long before the morphological characteristics of steroid-producing cells were clearly described. For instance, the major steroids of the gonads, testosterone

J. A. Pudney • Biological Science Center, Boston University, Boston, Massachusetts 02215.

of cytoplasm and an irregular-shaped nucleus with increased amounts of heterochromatin.

2.2. Endoplasmic Reticulum

The agranular or smooth endoplasmic reticulum (SER) is a discrete compartment of the cytoplasm delineated by single membranes. It is so called because it lacks the ribosomal granules attached to the other membranous system in the cell, the granular or rough endoplasmic reticulum (RER).

During the explosion of histological investigations that followed the development of the EM, it became apparent that cells of tissue known to produce steroid hormones always contained a well-developed system of SER (Figures 1b and 2). This organelle is now recognized as the structural trademark of any cell involved in the synthesis of steroids.

Early studies on a wide variety of steroid-producing cells described the SER as a system of vacuoles or vesicles, but the SER is a highly labile system of membranes that is notoriously difficult to preserve. With improvements in preparation of tissue for EM, and the use of vascular perfusion to preserve tissue, it became apparent that this vesicular type of organization represented a fixation artifact. The agranular reticulum commonly exists as a system of tubules, 400–700 Å in diameter, often branching and anastamosing to form a complex tridimensional compartment within the cell (Figure 1). However, this organelle is morphologically versatile and can exist in a wide variety of configurations; cisternal whorls of membranes, fenestrated cisternae, sheets of tubules, and parallel arrays of tubules have all been described in many different types of steroid-producing cells (Figure 2). These various configurations often coexist with the typical tubular form, to which they are usually connected (Figure 2). It is not known whether the differing organizations of the SER represent transitions from one functional state to another within the same cell, although some membrane systems may be interconvertible.

Although the unusual arrangements of the SER are often contiguous with the tubular system, they actually appear to represent an isolated compartment of membranes (Figure 2). There have been no attempts, as yet, to correlate these different configurations with the elaboration of specific steroids, but the varying organizations of SER are unlikely to be fortuitous and may represent discrete functional specializations, either transient or permanent. An indication of how important the organization of the SER is, in relation to steroid synthesis, can be obtained by analyzing structural changes in steroid-producing cells associated with functional differences. For instance, the agranular reticulum in the Leydig cell of a ground squirrel testis occurs as a tight network of branching tubules during the breeding season when high levels of testosterone are synthesized (Figure 1). In contrast, during sexual regression, when testosterone levels decrease, the SER of the Leydig cells is reduced to a system of unbranched, isolated tubules. Thus, the morphological appearance of this organelle can often be used empirically to assess the steroidogenic capabilities of a particular cell.

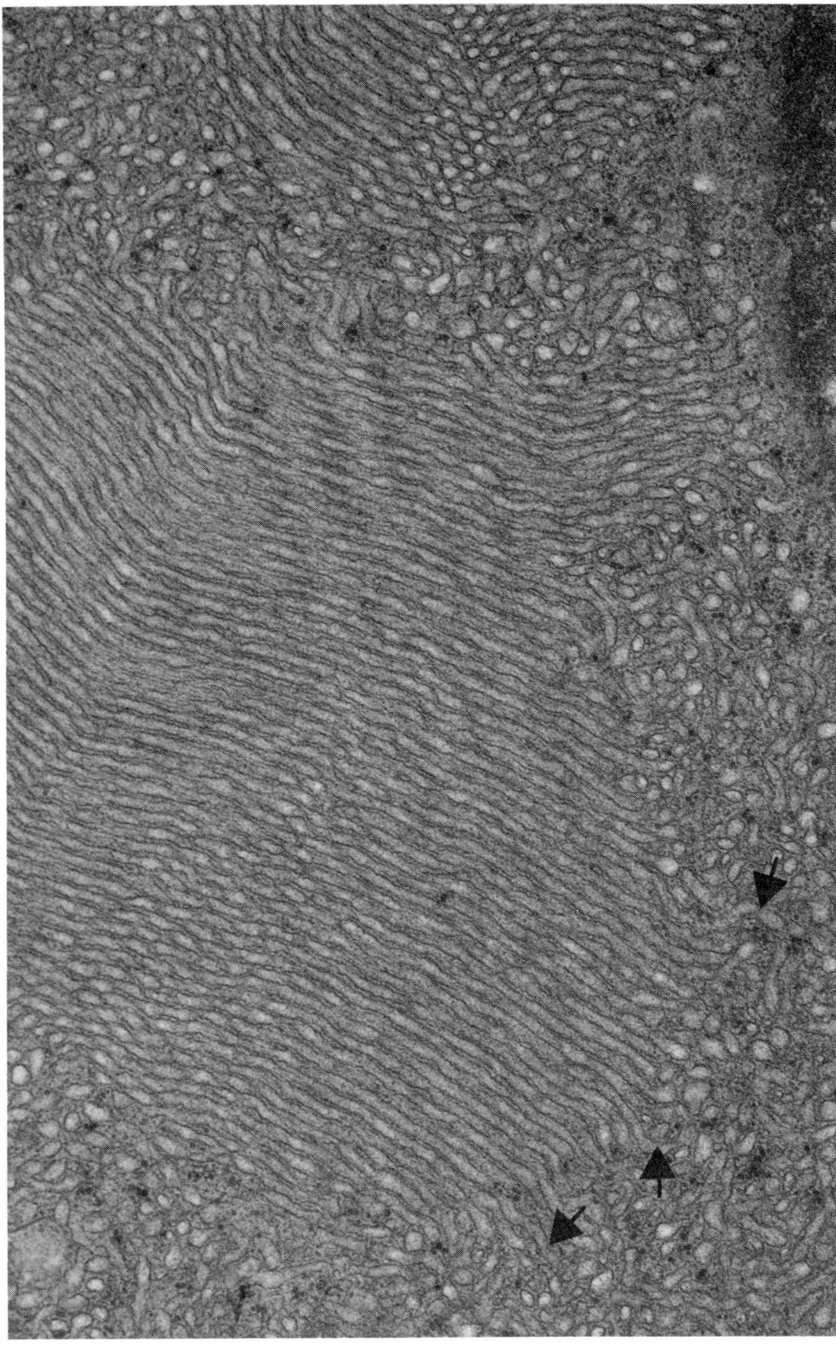

Figure 2. The agranular reticulum commonly exists as a tight network of anastomosing tubules as in this Leydig cell from the amphibian *Necturus maculosus*. It also occurs in other configurations seen here as large areas of parallel tubules seemingly isolated from the surrounding SER but actually connected to these membranes (which can be seen at the arrows). (×29,000.)

However, more attention has been given to correlating the abundance of SER with the production of steroid hormones. One of the effects of tropic stimulation is an often dramatic increase in the volume of SER, which can be related to a rise in the formation of steroids. Furthermore, it has been suggested that the amount of SER reflects the ability of steroid-producing cells to synthesize cholesterol. Cells with a well-developed SER are assumed to be capable of producing most of the cholesterol required for steroidogenesis, whereas cells with a relatively undeveloped SER are thought to rely on an exogenous source of cholesterol obtained via the circulation. Recently the correlation of the amount of SER with steroid synthesis has been substantiated using the technique of morphometry, which demonstrated a direct linear positive correlation between testosterone secreted and both the volume and surface density of the SER in Leydig cells of all species studies (Table I). Species differences in the steroidogenic capacity of a given mass of Leydig cells reside in the structural components of these cells, particularly the abundance of SER, and in no other cell organelle. Caution should be used, however, in depending on morphological observations alone as an index of steroidogenic ability without supporting biochemical data. The presence of a well-developed SER in a particular cell should be taken only as a possible indication of steroidogenic activity. Such reservations are necessary since an abundant reticulum can occur in cells without an associated increase in steroid production. Furthermore, involuted Leydig cells containing a poorly developed SER have been shown to retain tenaciously the ability to metabolize certain steroids. In fact, the abundance of SER may reflect the number of steroidogenic enzymes present in a cell needed for the production of steroids. Cells forming steroids *de novo* from acetate or cholesterol, e.g., Leydig cells, presumably require more membranes to accomplish this synthesis (see Figure 1) compared to cells where only a few steroidogenic steps occur, e.g., ovarian granulosa cells, which simply carry out the conversion of testosterone to estrogen (see Figure 11 and Chapter 1, Figure 18).

In steroid-producing cells the granular or rough endoplasmic reticulum is poorly represented, usually consisting of a few isolated cisternae. Some cells, however, such as adrenal cortical or lutein cells, contain this organelle as distinct stacks of membranes arranged in parallel arrays. Although the biochemical compositions of the membranes composing the systems differ, they are often connected to each other. This suggests that both membrane proteins and steroidogenic enzymes present in the SER are produced by ribosomes of the contiguous RER. Studies on hepatic tissue have provided much of the data concerning formation of the SER and its enzymes. During the differentiation of hepatocytes, both biogenesis of SER and its enzymatic differentiation depend on proteins synthesized by the RER. As yet, the molecular mechanisms controlling the development of the SER or how the membrane proteins and enzymes are transported to form this organelle have not been investigated.

An interesting observation is that although one effect of tropic hormones on steroid-producing cells is hypertrophy of the SER plus an increase in steroid production (often by the induction of specific enzymes), there is little evidence to suggest a

Table I
Relationship between Testosterone Production in Various Species and Volume Densities of Leydig Cell Organelles

Animal	Testosterone secreted[a]	Volume density[b]			
		SER	Mitochondria	RER	Lipid
Guinea pig	262.5 ± 38.6	60.9 ± 0.9	18.2 ± 0.7	8.9 ± 0.4	7.9 ± 0.7
Rabbit	163.6 ± 35.0	42.2 ± 1.1	17.0 ± 0.9	23.8 ± 1.2	3.8 ± 0.4
Dog	113.5 ± 20.5	36.7 ± 1.2	13.0 ± 0.7	25.5 ± 1.1	20.5 ± 1.3
Rat	96.2 ± 21.7	32.8 ± 1.2	19.7 ± 0.7	30.1 ± 1.0	0.3 ± 0.1
Hamster	16.7 ± 3.5	12.4 ± 0.7	27.9 ± 0.5	40.8 ± 0.7	0.6 ± 0.1

[a] μg/g Leydig cell per hr.
[b] Expressed as percentage of Leydig cell cytoplasmic volume. (Taken from Zirkin, B. R., Ewing, L. L., Kromann, N., and Cochran, R. C., 1980, Testosterone secretion by rat, rabbit, guinea pig, dog and hamster testis perfused *in vitro*: Correlation with Leydig cell ultrastructure. *Endocrinology* **107:**1867–1874. By copyright permission of the Endocrine Society.)

concomitant increase in RER. Presumably this increase in protein production during stimulation is accomplished by a higher rate of protein synthesis by the ribosomes.

2.3. Mitochondria

In eukaryotic cells mitochondria are sites of oxidative phosphorylation and are basically similar in structural composition, consisting of an outer membrane and an inner membrane that produces a number of internal projections called cristae. These mitochondrial cristae serve to increase the surface area of the inner membrane and are the site of many important enzymes. In addition to providing energy requirements, the mitochondria of steroid-producing cells are also involved in the process of steroidogenesis. Furthermore, the organization of cristae differs in these cells being tubular rather than straight or lamellar as in other cells. Ultrastructurally these tubules in cross-section have a vesicular profile and are referred to as tubulovesicular. Their presence in mitochondria is usually taken as an indication that the cell is involved in steroid synthesis. This correlation is not always true, however, since in some steroid-producing cells (notably the adrenal zona glomerulosa of many species), the mitochondria predominantly contain lamellar cristae. Also, mitochondria tend to be pleomorphic in cells that produce steroids, rather than regularly oval or spherical as in other cells (Figure 1).

As yet the functional relevance of the tubulovesicular type of organization is not known, but it appears to be specifically related to steroid production. For example, tropic hormone stimulation of the adrenal cortex results not only in mitochondrial division but also in a reorganization of cristae from a lamellar to tubular form; furthermore, there is a direct relationship between the internal structure of the mitochondria and the production of steroids. In cultured adrenal cells, the development of tubulovesicular cristae is closely followed by the acquisition or activation of mitochondrial steroidogenic enzymes; and the mitochondria of steroid-producing cells in the fetus or neonate tend to have lamellar cristae, but as they mature and synthesize steroids these change to the tubulovesicular type. Although this evidence indicates a possible functional relationship between the tubular configuration of mitochondrial cristae and the ability to synthesize steroids, the fact remains that some steroid-producing cells do possess mitochondria with lamellar cristae.

Like the SER the configuration of mitochondrial cristae of steroid-producing cells can also be highly diverse. Many bizarre arrangements have been described, such as extremely long lamellar cristae or tightly packed bundles of tubules giving a paracrystalline profile in cross-section. All these configurations have been associated with the steroidogenic activity of the mitochondria and have variously been interpreted to indicate either an increase or decrease in steroid synthesis, but any precise function of these different arrangements of cristae has yet to be determined.

Apart from the cristae, various inclusions have been described in the mitochondria of steroid-producing cells. Lipidlike deposits are commonly observed, which may indicate the accumulation of a sterol or steroid precursor.

2.4. Golgi Apparatus

The Golgi apparatus in steroid-producing cells resembles that present in other cells. It consists of a heterogeneous collection of smooth saccules, arranged as parallel arrays, and associated smooth and coated vesicles (Figure 3). The membranes forming this organelle exhibit a polarity with an outer concave, or *cis,* surface and an inner convex, or *trans,* surface. Although the Golgi apparatus appears, morphologically, to be well developed in steroid-producing cells, there is little evidence to suggest an involvement in the formation or secretion of steroids. Only a few observations have demonstrated an increase in the Golgi apparatus following stimulation of steroid-producing cells, but there is the intriguing possibility that this organelle may be the site of steroid conjugation to form sulfates or glucuronides. Although this hypothesis awaits verification, it is interesting to note that a number of sulfotransferases are associated with membranes of the Golgi apparatus, which is the exclusive site of protein sulfation in other cell types, and one isolated observation has shown that in the steroid-producing interstitial cell of the ovary, sulfation also occurs in the Golgi apparatus. More studies are clearly needed before the rôle of the Golgi apparatus can be adequately defined.

2.5. Lipid Droplets

Lipid droplets, varying in both number and dimension, are a common inclusion of steroid-producing cells (Figures 1b and 3). Histochemical tests for the presence of lipids are often used as a morphological indication of possible steroid synthetic activity. In seasonally breeding animals cyclical changes in lipid content of Leydig cells are taken as a functional index of androgen production. It is now

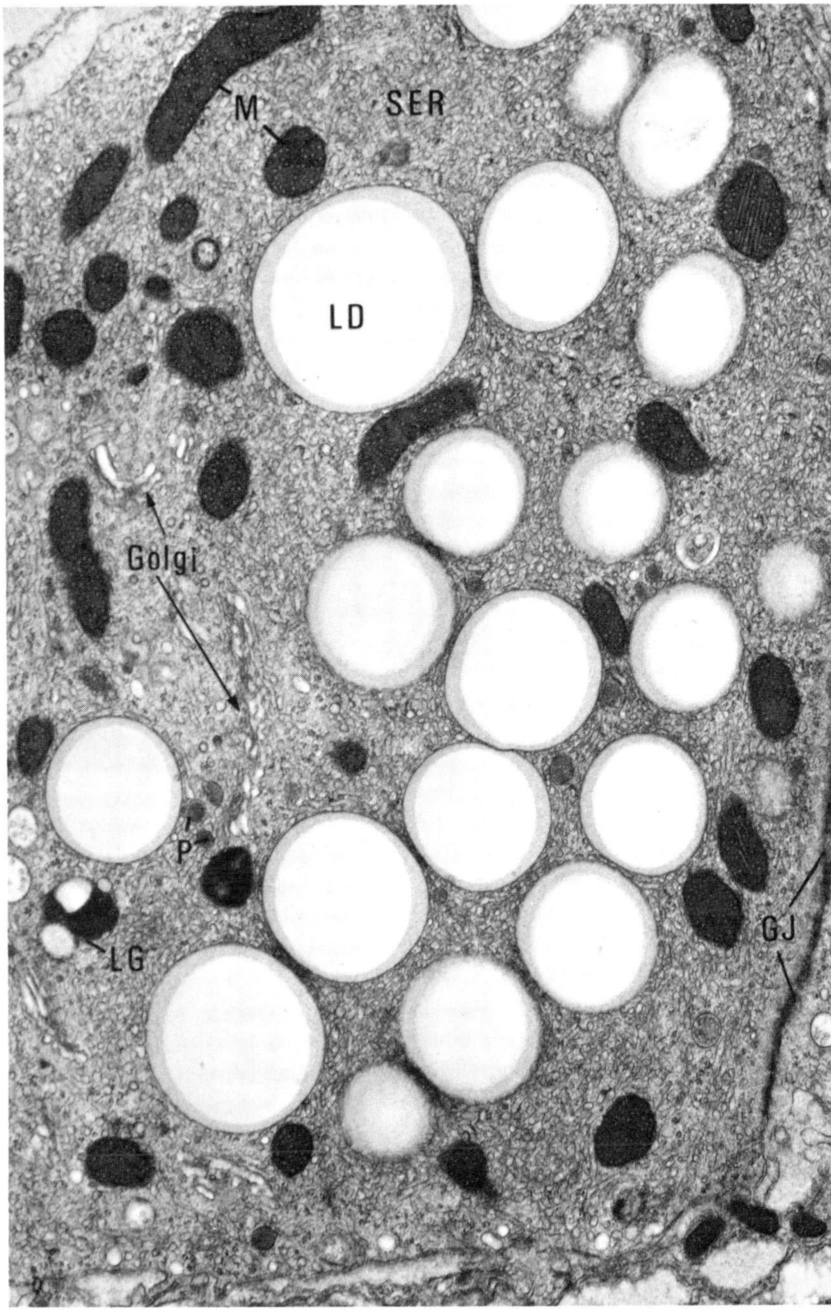

Figure 3. Leydig cell of the ground squirrel in greater detail. The SER is organized as a network of anastomosing tubules. Embedded in these membranes are lipid droplets (LD) and mitochondria (M). Also present are peroxisomes (P), lipofuscin granules (LG), several profiles of the Golgi apparatus, and an extensive gap junction (GJ) formed by adjacent Leydig cells. Note the paucity of granular reticulum. (×19,000.)

generally accepted that lipid present in steroid-producing cells represents a store of cholesterol or cholesterol esters intended for steroidogenesis, which usually are depleted during active steroid production.

Lipid droplets occur free in the cytoplasm of steroid-producing cells, and controversy exists as to whether they are bounded by a limiting membrane. Since lipid represents a store of cholesterol, there is often a close association between these inclusions and the SER (which synthesizes cholesterol) and mitochondria (which metabolize cholesterol).

2.6. Peroxisomes

These are single membrane-bound organelles, also called microbodies, that contain catalase and various other enzymes (Figure 3). They are commonly present in steroid-producing cells but, in contrast to those in other cells, usually lack the crystalline core or nucleoid, suggesting a deficiency in urate oxidase. A relationship between peroxisomes and steroidogenesis has been inferred since there is often a continuity or close association between these organelles and SER, and when cholesterol synthesis is inhibited, there is an increase in the number of peroxisomes and catalase activity. This has led to the suggestion that peroxisomes may regulate the intracellular pool of cholesterol and its utilization for steroidogenesis.

2.7. Lipofuscin

Lipofuscin granules of varying composition and morphological appearance are often found to accumulate in steroid-producing cells (Figure 3). The physiological significance of these granules is not known. It is generally assumed they represent metabolically inert cellular material not completely degraded by lysosomal hydrolases. The presence of acid phosphatase has been confirmed in lipofuscin granules of steroid-producing cells, indicating these inclusions are residual bodies resulting from lysosomal digestion. In steroid-producing cells, however, these granules could also be derived from the autooxidation of lipid into insoluble polymers during periods of low steroidogenic activity. This could account for the often exceptional number of lipofuscin granules found in these cells.

2.8. Filaments

Filaments are a common cytoplasmic constituent of cells, including those specialized for steroid production. These filaments, e.g., microtubules, 10-nm intermediate filaments, and 7-nm microfilaments (actin) have been variously associated with either a cytoskeletal rôle or the movement of cells and their organelles. In steroid-producing cells, however, evidence is accumulating for the involvement of filaments in the synthesis or secretion of steroids. Cytochalasin B (which disrupts actin filaments) inhibits the steroidogenic response of cultured Leydig cells to either luteinizing hormone or cyclic AMP, an effect due mainly to decreased transport of cholesterol to the mitochondria. This disruption of filaments, preventing the translocation of cholesterol to mitochondria, may be related to other studies where direct attachment of filaments to organelles, including lipid droplets and mitochondria, has been described. Filaments may therefore control the topographical distribution of organelles within the cytoplasm and be instrumental in controlling steroidogenesis by regulating the morphological intimacy and hence functional activity of organelles involved in steroid synthesis.

2.9. Surface Activity

As with all cells, those which produce steroids are bounded by a plasmalemma. They do, however, produce various surface specializations, albeit not unique to these cells, which are possibly related to their function of synthesizing and secreting steroid hormones. Steroid-producing cells are connected by gap junctions, which are sometimes quite extensive (Figure 3). These junctions serve as areas of communication, presumably allowing groups of cells to act as functional units. The ability to develop filopodia (microvilli) appears to be a natural function of the plasmalemma (Figure 1). In steroid-producing cells, it has been shown that stimulation causes an increase in filopodia formation. Increased cell surface/interstitial fluid contact enhances the uptake of nutrients and oxygen and increases the delivery of stimulating factors, such as lipoproteins and trophic hormones, while simultaneously removing steroid products more efficiently.

Steroid-producing cells have commonly been observed to contain coated pits on their membrane surface. Recent investigations have shown that these coated pits (so called because of their coating of the protein clathrin) are the sites of uptake of ligands bound to receptors. The movement of these complexes into coated pits results in their internalization as coated vesicles by a process called receptor-mediated endocytosis, which has been observed in steroid-producing cells. The fate of these coated vesicles containing the bound hormone or protein is variable. In many instances they fuse with lysosomes, resulting in the degradation of their contents. This may also cause the release of bound protein, as has been demonstrated for low-density lipoprotein in adrenal cells. Often the coated pits are processed by the Golgi apparatus for recycling of the plasmalemma, and the occurrence of numerous coated vesicles in the vicinity of the Golgi apparatus probably reflects this process.

3. Correlation of Structure with Function

The early investigators of the fine structure of steroid-producing cells readily appreciated that the presence of an extensive SER was possibly related to steroid synthesis. This relationship, however, was not substantiated until cell fractionation techniques became available. Using cell homogenates and differential centrifugation it was possible to (individually) isolate organelles and investigate their involvement in steroidogenesis by *in vitro* incubations with radioactive precursors. Analy-

Table II
Subcellular Localization of Some Important Steroidogenic Enzymes Found in the Testis, Ovary, and Adrenal

Enzyme	Cellular organelle
Cholesterol-20α and 22R-hydroxylases	Mitochondria
C_{20}–C_{22} Lyase	
Δ^5-3β-Hydroxysteroid dehydrogenase	Smooth endoplasmic reticulum
+Δ^5–Δ^4 Isomerase	
17α-Hydroxylase	Smooth endoplasmic reticulum
C_{17}–C_{20} Lyase	Smooth endoplasmic reticulum
17β-Hydroxysteroid dehydrogenase	Smooth endoplasmic reticulum
21-Hydroxylase	Smooth endoplasmic reticulum
Aromatase enzyme system	Smooth endoplasmic reticulum
11β-Hydroxylase	Mitochondria
18-Hydroxylase	Mitchondria
20α-Hydroxysteroid dehydrogenase	Cytosol (i.e., not associated with an organelle)

ses of these incubations provided information on the subcellular localization of enzymes concerned with steroid production (Table II).

As can be seen from Table II, the majority of steroidogenic enzymes are found on the SER. These enzymes are called microsomal enzymes because during homogenization of the cell, the SER breaks down into vesicles called microsomes. It should be remembered that microsomes *per se* are an artifact of tissue preparation.

Looking at Table II, it can be immediately appreciated that steroids have to be translocated from one organelle to another before the final hormonal product is produced (Chapter 1). Thus, cholesterol, either stored as lipid droplets, synthesized on the SER, or obtained from the circulation, must be transported to the mitochondria. Mitochondrial fractionation experiments have shown that the side-chain cleavage enzyme responsible for the formation of pregnenolene resides on the inner membrane of the cristae. Once formed, the pregnenolene must gain access to the membranes of the SER where a variety of enzymes can metabolize it to other steroid hormones, e.g., androgens and estrogens. In the case of the adrenal, however, further enzymatic steps are accomplished in the mitochondria; hence steroid intermediates must be transferred back to the mitochondria.

The translocation of steroids from one organelle to another is reflected in their topographical distribution. Lipid droplets and mitochondria usually intimately surrounded by membranes of the SER are often apposed to each other (Figure 3). Thus, the functional activity of these organelles results in close morphological relationships allowing movement of sterol and steroid intermediates to occur over short distances. Once steroids have gained access to the lumen of the SER, they can, presumably, be transported (and stored?) anywhere within this huge cytoplasmic compartment for further enzymatic steps. Shuttling of steroids from one organelle to another must require precise regulatory mechanisms, but how this movement of steroids is accomplished and what controls their destination within the cell are as yet a mystery. Some insight into this process has been gained by the detection and isolation of various carrier proteins. Much of these data have been derived from hepatocytes and show that a specific binding protein precisely controls the movement of cholesterol within the cell. Binding to carrier protein also enhances the rate of exchange of cholesterol between liver microsomes and mitochondria *in vitro*. Sterol carrier proteins necessary for the synthesis of cholesterol have also been identified. These proteins are present in the cytosol fraction (i.e., they are not bound to membranes) and are noncatalytic. The binding of sterols/steroids to carrier proteins may be a prerequisite for movement of these hydrophobic compounds through the cytoplasm. They may also play a role in the recognition of enzyme sites on the SER or mitochondrial cristae. Carrier proteins may be synthesized by the RER and subsequently transported to the membranes of the SER. Continuity between the two systems is well documented and may represent a pathway where accepter proteins could reach the SER to combine with steroids in a similar manner to the synthesis of lipoprotein by liver cells.

Although the biosynthetic pathways for steroid production have been established in many cells, it is as yet not possible to relate the distribution of steroidogenic enzymes to specific domains within the membranes of the SER. For instance, are they arranged as functional units with all the enzymatic steps at a single locus? Or are they deployed sequentially along the membrane, with steroids being passed from one enzyme to another as on an assembly line? Unfortunately, we know very little about the organization of steroidogenic enzymes in the SER or how their spatial distribution within this organelle is controlled.

Certain aspects of the morphology of steroid-producing cells raise other interesting functional questions. The SER is an often extensive organelle whose membranes provide a large surface area for the attachment of steroidogenic enzymes and cofactors. There is usually a disproportionate amount of SER compared to the number of mitochondria, and although the SER responds immediately to tropic stimulation by proliferation of membranes, in contrast the mitochondria do not react as dramatically. This has led to the suggestion that the SER may act as a site of cholesterol storage for use during periods of active steroidogenesis. In fact, the membranes of the SER do contain nearly twice as much cholesterol compared to membranes of RER.

4. Steroid Secretion

The concept of secretion as a distinct process separate from synthesis is, as yet, unknown for steroid-producing cells. The idea that steroids are secreted as they are synthesized was demonstrated by the fact that the levels of steroid present in the rat adrenal gland are slight compared to those secreted into the adrenal vein. Subsequently, it has been accepted that steroid hormones are simply removed by a process of diffusion.

To date, the numerous studies on the fine structure of steroid-producing cells have failed to reveal any morphological evidence for the method of secretion of their products. Steroids are not apparently stored in quantity and are released in small amounts, which makes detection of the release process very difficult. Various studies have claimed, on morphological grounds, that specific mechanisms for the release of hormones from steroid-producing cells do exist. Unfortunately, these studies are not particularly convincing, and our understanding of the morphological and biochemical events leading to steroid secretion is rudimentary. Recent evidence suggests that in the adrenal cortex the release of steroid hormones is controlled by novel steroid–protein complexes, and, hopefully, future efforts will be concentrated on these interesting cells to elucidate their mode of secretion.

5. Organization of the Gonads

5.1. The Testes

Originally, the testes of vertebrates were derived from metametrically arranged coelomic pouches, and this type of organization can still be seen in the protochordate *Amphioxus*. During evolution of the vertebrates these rows of testicular tissue became restricted and reduced to form localized compact organs. In all vertebrates the testes are paired, except in the cyclostomes, where both gonads are fused, and in a few teleost species where only one testis develops. The testes tend to be uniformly oval or round, but in some vertebrates, such as various urodele amphibians and teleosts, they are composed of distinct lobes.

In most vertebrates the testes lie free within the body cavity attached by a membrane, the mesorchium, to the body wall or else in a specialized region of the peritoneal cavity, the scrotal sacs, as in many mammalian species. Usually the testes are isolated structures but can be closely associated with other tissue elements such as the layer of hemopoietic tissue, the epigonal organ, which invests the testes in selachians, or the fat body attached to the testes of amphibians. This fat body is apparently essential for spermatogenesis since its removal causes testicular regression.

Basically, the testis in all vertebrates consists of two disparate compartments. One is composed of germ cells and nongerminal elements, the somatic cells, which in mammals are called Sertoli cells. The second compartment is the interstitial tissue which, apart from connective tissue elements, blood vessels, nerves, and lymphatics, contains the Leydig cells, which synthesize the male hormones or androgens. These tissue compartments are enclosed by a connective tissue testicular capsule, the tunica albuginea. As we shall see, the organization of these compartments within the testis differs enormously throughout the vertebrate series. Since the actual process of spermatogenesis appears to be similar in all vertebrates and will be dealt with in greater detail in chapters on specific classes, this account will confine itself to the general anatomical arrangement of the testes. A survey of the testes in vertebrates clearly demonstrates two distinct types of organization of the germinal tissue, one found in the anamniotes (fish and amphibia) and the other in the amniotes (reptiles, birds, and mammals).

5.1.1. Anamniotes

In these species the unit of spermatogenesis is a spermatocyst (usually referred to as a cyst), which consists of a clone of isogenetic germ cells plus a single supporting somatic cell. Development of these cysts begins with the association of primary spermatogonia with somatic cells (Figure 4). Each primary spermatogonium then divides mitotically to produce secondary spermatogonia enclosed by somatic cells (Figures 4 and 5). In some species the somatic cells also divide during this period. The secondary spermatogonia become associated with only one of the somatic cells, resulting in the formation of a cyst. Beyond this stage the somatic cell number remains constant, but the size of the cyst increases as the germ cells mature

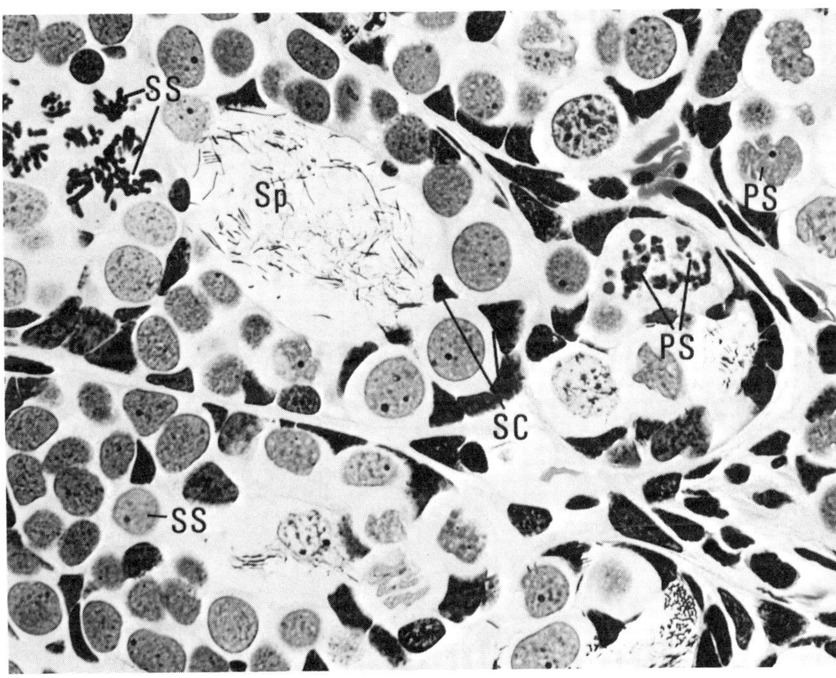

Figure 4. Testis of *Necturus maculosus* showing the formation of cysts in seminiferous lobules. Initially primary spermatogonia (PS) become associated with somatic cells (SC). The primary spermatogonia divide to form a clone of isogenetic germ cells. These germ cells along with a somatic cell make up a cyst. The primary spermatogonia then form the secondary spermatogonia (SS) that also undergo mitotic division. Note the presence of spermatozoa (Sp) in the lumen of the seminiferous lobule that were released from the distal portion of the lobule. (×480.)

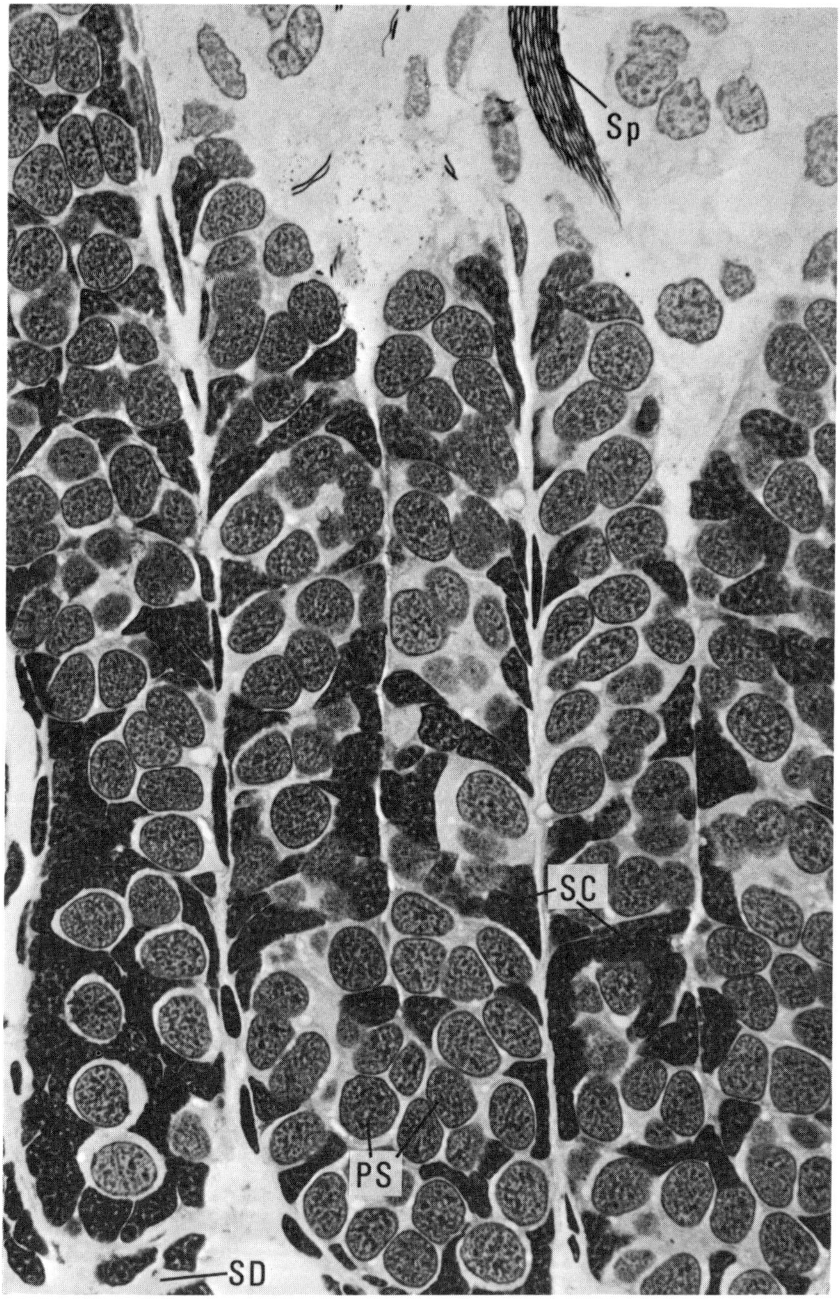

Figure 5. In *N. maculosus* as cysts mature they migrate distally away from the sperm-collecting ducts toward the blind end of the seminiferous lobule. Only these distal regions contain germ cells undergoing spermatogenesis. Spermatozoa (Sp) released into the lumen of the distal lobular segment must pass through the immature segment to reach the sperm duct (SD). The immature region contains a permanent population of spermatogonia (PS) acting as stem cells for succeeding bouts of spermatogenesis. Sertoli cell (SC). (×300.)

and often begins to develop a lumen. Spermiogenesis then ensues with the formation of spermatids, which become rearranged within the cyst. This results finally in bundles of these cells with their heads (or in some teleost species the tails) orientated toward the nucleus of the somatic cell (Figure 6). Finally a clone of spermatozoa is formed embedded in the cytoplasm of a single somatic cell (Figure 6). Quantitative data have been obtained during germ cell development in various species. For instance, in selachians four spermatogonial divisions take place, resulting in the production of 64 spermatids attached to a single somatic cell. Following spermiation the somatic cells usually degenerate and are either resorbed or evacuated through the sperm duct (Figure 6). Owing to the degeneration of cysts it has been stated that anamniotes do not possess a permanent germinal epithelium. This is not strictly correct, however, since in urodeles and teleosts there is always a portion of the germinal tissue that contains a resident population of stem cells (spermatogonia) serving as a reservoir for succeeding bouts of spermatogenesis (Figure 5). Thus, these species do have a permanent, although restricted, germinal epithelium.

The germ cell cysts are contained in larger structures formed from the dilated ends of sperm-collecting ducts. These blind-ending sacs or pouches have been variously called ampullae, follicles, lobules, or tubules, depending on the class of vertebrate investigated. They may be single and straight or branched, and are delineated by a basal lamina. Often these seminiferous compartments are bounded by a layer of myoid cells, forming a primitive boundary wall. The diverse nomenclature used to describe the germinal compartment can obviously be confusing when comparing the organization of the testis from one group of vertebrates to another. In the amniotes the term seminiferous tubules is used exclusively to describe the organization of the germinal tissue. Since these tubules are entirely different in their structural composition to those described in the anamniotes, it would be appropriate to restrict this term to the amniotes. The other descriptions of the seminiferous compartments of the anamniotes as ampullae, follicles, or lobules are anatomically correct. To bring some kind of consistency, however, to the nomenclature used to describe the organization of the testis in the anamniotes, the term lobule will be used throughout the present account.

Although the basic arrangement of germinal tissue in the testes of the anamniotes consists of cysts contained in lobules, there is variation in their development. In one type found in cyclostomes, most species of teleosts, and anura, the cysts remain stationary within the seminiferous lobule, and maintain their integrity until maturation of the spermatids. At this time the cyst wall ruptures resulting in a layer of somatic cells containing bundles of spermatozoa lining the lobule; the spermatozoa are then released into the lumen of the lobule and enter the sperm ducts. The other pattern is found in elasmobranchs, various species of teleosts, and urodeles, in which the cysts are not stationary but migrate as they develop during spermatogenesis. The direction of this migration, however, differs between the groups. In urodeles the cyst begins its development proximal to the central sperm-collecting duct and as it matures moves in a distal direction toward the blind end of

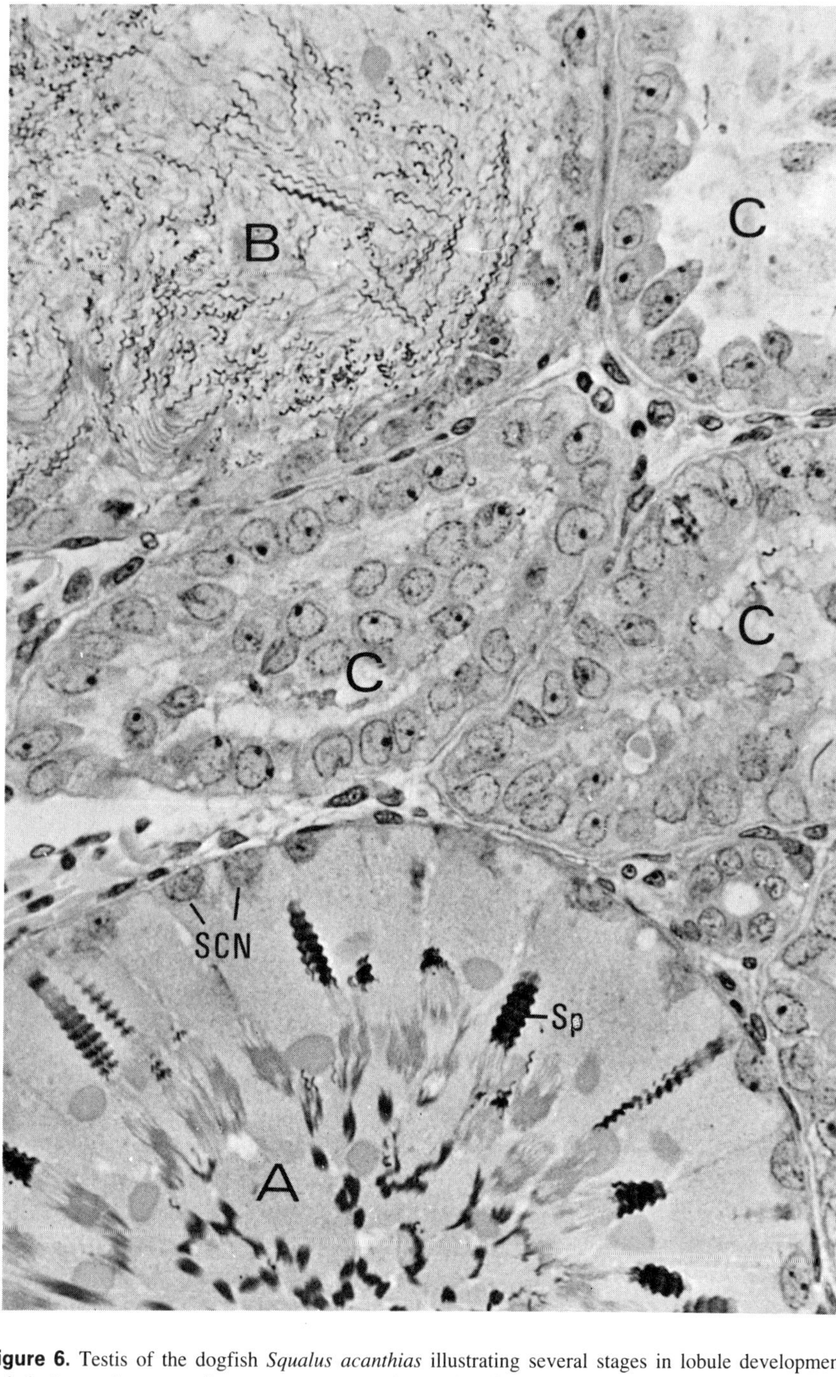

Figure 6. Testis of the dogfish *Squalus acanthias* illustrating several stages in lobule development. Lobule A contains cysts of mature spermatozoa (sperm bundles, Sp) embedded in a single somatic cell with their heads orientated toward the somatic cell nucleus (SCN). Lobule B is filled with spermatozoa that have just been released from the somatic cells. Lobules C contain somatic cells left following spermiation and which will undergo degeneration. (×250.)

the lobule (Figure 5). This region contains the spermatogenically active cysts, whereas the proximal regions of the lobule contain immature germ cells. Spermiation occurs when the cyst wall ruptures and the liberated spermatozoa pass down the lobule (through the immature regions) to reach the sperm duct (Figure 4). In other species, such as some teleosts, the cysts migrate toward the sperm duct from their origin at the blind end of the lobule. On reaching the sperm duct the cyst breaks down to release the sperm either free or as naked bundles called spermatozeugmata. The somatic cells of these cysts then transform into epithelial cells which line the sperm ducts. The elasmobranchs are an extreme case since the entire lobule and its contained cysts migrate through the testis. The cysts begin development in a germinal zone at the rostral end of the testis and subsequently, as the lobule forms, migrate toward the caudal region. By the time the lobules reach this position, the cysts are filled with mature spermatozoa, where spermiation occurs and the lobules degenerate.

Two points should be emphasized about the formation of cysts in the anamniotes. One is that in many species, e.g., urodeles, although the spermatogenically active cysts contain synchronously developing germ cells, the lobule itself may contain cysts at different stages of development (Figure 6). Furthermore, seminiferous lobules may not all contain cysts at the same stage of maturation; this may vary from lobule to lobule within the same testis, e.g., elasmobranchs. The use of the term migration to describe the movement of cysts (or lobules, as in the elasmobranchs) may be misleading; these structures do not actually move by means of any locomotor ability of their own. They are passive structures, and presumably it is the continual formation of cysts (or lobules in the elasmobranchs) that causes their migration along the length of the seminiferous lobule (or testis, as in elasmobranchs).

In many anamniote species there is a distinct wave of spermatogenesis, but the level of testicular organization manifesting this wave and the speed of maturation differs. For urodeles it is at the level of the cyst, which begins development at the caudal end of the testis and proceeds toward the cephalic region. In elasmobranchs, this wave of germ cell maturation involves the whole lobule, and since these migrate from the rostral to the caudal end of the testis, there occurs a distinct zonation of spermatogenesis. Thus, the rostral end is composed of lobules with spermatogonial cysts, the caudal end with cysts filled with mature sperm, and the intervening region with zones of cysts containing germ cells at successive stages of development.

An interesting phenomenon is the occurrence in teleosts of ambisexual species, which possess both an ovary and a testis. These gonads may be present simultaneously or develop sequentially. Those species where the testis develops first are termed protandrous, while those where the ovary develops first are protogynous.

5.1.2. Amniotes

The organization of germinal tissue in the amniotes is entirely different from that found in the anamniote testis. There is no cystic form of spermatogenesis;

instead, a permanent germinal epithelium is present containing successive stages of germ cell development. The germinal tissue occurs in seminiferous tubules. These are single structures (except in birds where extensive branching occurs) that open at both ends into a series of collecting channels, the rete testis. The number and length of these tubules vary greatly among the amniotes.

The presence of a permanent germinal epithelium results in all stages of spermatogenesis developing within the seminiferous tubule at the same time, in association with a permanent population of somatic or Sertoli cells. The immature germ cells, the spermatogonia, are located at the base of the epithelium, while the spermatocytes and spermatids occur at successively higher levels. The spermatozoa border the lumen of the tubules into which they are eventually released. As in the anamniotic cyst, during spermiogenesis the heads of the spermatozoa become oriented toward the nucleus of the Sertoli cell. This pattern of spermatogenesis results in a distinct stratification of germ cells as they develop within the seminiferous tubules. The proliferation of spermatogonia at the base of the epithelium and their displacement toward the luminal surface indicate that this is a highly dynamic tissue. As yet, however, the mechanisms controlling and effecting this movement are unknown.

The continual occurrence of different germ cell stages in the amniote testis results in two phenomena which will be briefly discussed: First is the cycle of the seminiferous epithelium. The presence of germ cells at different phases of maturation results in well-defined cellular associations or stages, which succeed each other indefinitely within a certain portion of the germinal tissue. Thus, each cellular association represents a stage in a cyclical process. The series of successive stages occurring between two appearances of the same cellular association in a given portion of the seminiferous tubule is defined as the cycle of the seminiferous epithelium. The number of stages constituting the cycle is constant for a given species; e.g., the rat has 14 different stages, whereas man has only six. It should be remembered that the stages are not static associations, but they have been defined merely for the convenience of comparing the spermatogenic process between species. The development of a generation of germ cell occurs continuously but concomitantly with the formation of earlier and later generations. Second, associated with this cycle is the wave of the seminiferous epithelium. A given stage of the cycle of the seminiferous epithelium occupies a specific segment within the seminiferous tubule. These segments occur consecutively along the length of the tubule forming the wave of seminiferous epithelium. Each wave consists of the exact number of stages forming the cycle of the epithelium, i.e., 14 segments for the rat.

5.1.3. Leydig Cells

Leydig cells were initially described within the interstitial tissue of the mammalian testis. Subsequently, these cells were recognized as the site of production of male sex hormones—the androgens. Since androgens are steroids, the identification of Leydig cells in the testes of vertebrates has relied on morphological criteria, i.e., the presence of organelles involved in steroid production or by the histochemical detection of enzymes involved in steroid synthesis. Cells occurring in the interstitial compartment of the testis that fulfill one or both of these conditions have been termed Leydig cells.

Leydig cells have been reported in the testes of representatives of all classes of vertebrates studied. Although the presence of Leydig cells in the amniote testis is undisputed, there has been some controversy over the years as to the identity of these cells in the testes of the anamniotes. In certain species of teleosts, definitive Leydig cells could not be identified, and lobule boundary cells were considered homologous in structure and function to Leydig cells of the amniotes. These cells were thought to arise as fibroblasts present in the boundary wall, which became seasonally modified to secrete steroid hormones. However, recent critical examination using the EM indicates that lobule boundary cells correspond to somatic cells of the germinal tissue and not to Leydig cells.

The identity of Leydig cells in the anamniote testis has been further complicated by the description of lobule boundary cells in urodele amphibians. These lobule boundary cells are not, however, homologous to those previously described in the teleost testis. In the urodele testis these lobule boundary cells do actually occur outside the basal lamina of the seminiferous lobule and hence represent an interstitial component (see Figure 7), and, therefore, can be considered homologous to Leydig cells of the amniote testis. The rather casual use of the term lobule boundary cells can obviously lead to misconceptions when comparing the presence of Leydig cells within the anamniote testis. Since recent studies on the urodelean testis have demonstrated that the so-called lobule boundary cells appear structurally homologous with Leydig cells of other vertebrate species, it would be more precise and less confusing simply to refer to these cells as Leydig cells. Any cell in the interstitial compartment of the testis that displays the morphological characteristics of a steroid-producing cell and can be shown to possess an endocrine function should be designated a Leydig cell. Morphologically, the Leydig cells of all vertebrate species are very similar in structure. Functionally, however, this analogy is not so close since in the amniote testis testosterone is the major androgen produced, whereas in teleosts it is 11-ketotestosterone, and dihydrotestosterone in amphibians.

Normally, Leydig cells are present in the interstitial tissue surrounding the seminiferous compartments. This topographical association reflects the functional relationship between the two compartments since androgens are essential for the maintenance of spermatogenesis. In various vertebrate species, however, Leydig cells have been found to occur in other locations. A startling arrangement occurs in Teiid lizards, which have a distinct layer of Leydig cells beneath the testicular capsule with few present round the seminiferous tubules. Leydig cells have also been detected associated with the sperm ducts of many teleost species. This extragonadal location of Leydig cells reaches a peak in one species, the black goby (*Gobius jozo*), where there is a discrete glandular region composed of Leydig cells distinctly separated from the seminiferous lobules. Another unusual arrangement of Leydig cells is found in the urodele testis, where they occur in a specialized region

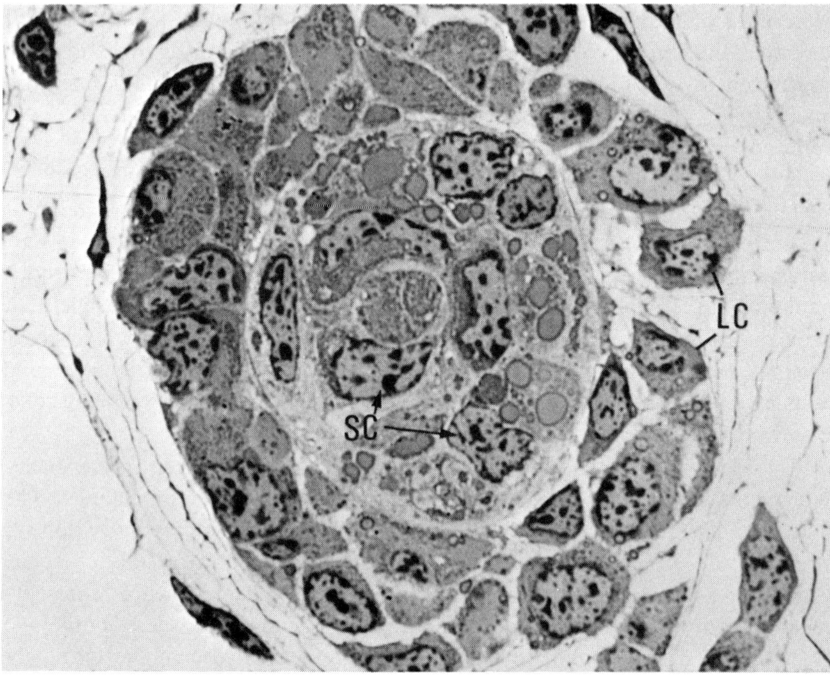

Figure 7. Glandular tissue of *N. maculosus* testis with hypertrophy of Leydig cells (LC), which occurs when somatic cells (SC) atrophy following spermiation. The seminiferous lobule is filled with a mass of lipid-laden degenerating somatic cells. (×520.)

of the testis termed the glandular tissue. The development of this glandular tissue is associated with the completion of spermatogenesis in these species. During germ cell maturation the Leydig cells remain undifferentiated, and it is not until spermiation, with degeneration of the somatic cells, that development of the Leydig cells occurs (Figure 7). This structural hypertrophy is correlated with functional activity since Leydig cell differentiation is accompanied by the synthesis of increasing amounts of steroid hormones. Among the anamniotes other examples have been reported showing that spermatogenesis and Leydig cell development are separated in time during the annual reproductive cycle.

The physiological significance of this temporal or spatial separation of Leydig cells and germinal tissue is unknown. It may be an adaptation for secreting high titers of androgens into the peripheral circulation; certainly the development of Leydig cells, in species where this is asynchronous with spermatogenesis, has been associated with a rise in plasma androgens. Presumably, the increased levels of circulating androgens stimulate and maintain accessory sex structures and initiate sex behavior.

Reproductive activity is cyclical in most vertebrates. This is reflected in the morphological appearance and functional activity of Leydig cells. It is not the purpose of this account to describe in detail these seasonal changes, and the reader is referred to a recent excellent review by G. A. Lincoln [1981, Seasonal aspects of testicular function, in: *The Testis* (H. Burger and D. de Kretser, eds.), Raven Press, New York, pp. 255–302]. In brief, during the breeding season Leydig cells are hypertrophied containing abundant organelles associated with steroid production (see Figure 1b). With the onset of sexual regression these cells atrophy, becoming reduced both in size and in amount of cytoplasmic organelles. One point to be emphasized is that Leydig cells of many vertebrate species undergo a seasonal lipid cycle. During testicular quiescence the Leydig cells contain few lipid droplets. As testicular recrudescence ensues, these lipid droplets increase in number and size. In birds, reptiles, and some teleosts this lipid becomes depleted during the breeding season as a result of increased steroidogenic activity. For other species, however, such as amphibia, the lipid content of the Leydig cells reaches a maximum after the breeding season. This postnuptial steatogenesis is due to the accumulation of sterol precursors such as cholesterol, caused by the cessation of steroidogenic activity. Many mammalian species also display a cyclical accumulation and depletion of lipid in Leydig cells during the breeding season. This lipid cycle, however, is not so intense as that occurring in other vertebrates.

5.1.4. Sertoli Cells

In all vertebrates spermatogenesis invariably proceeds in close association with a nongerminal element, the somatic cell. This interdependence appears to be an ancient evolutionary relationship occurring in the protochordate *Amphioxus*. The Sertoli cell was originally described in the human testis and subsequently has been reported to occur in all vertebrate species investigated. The Sertoli cell has been intensely studied in mammals, where it plays a pivotal rôle in maintaining and regulating spermatogenesis. Since the mammalian Sertoli cell has been the subject of many excellent reviews, it will only be discussed in comparison with the somatic cell present in other vertebrate classes. Unfortunately, the somatic cell in the germinal tissue of the nonmammalian species has not received as much attention. We are, therefore, still fairly ignorant of many of the functional and morphological aspects of this cell in many vertebrates. As we have seen throughout this discussion on the organization of the testis, the terminology used to describe the amniote testis is consistent, whereas that used for the anamniote testis is inconsistent; the somatic cell is no exception. It has been variously called a cyst, follicle, companion, or nurse cell as well as a Sertoli cell. In all cases it has been considered analogous, if not homologous, with the Sertoli cell of the amniote testis. The relationship of somatic cells with the germinal elements in anamniotes, however, is radically different from that occurring in the amniote testis. This is a reflection of the different types of spermatogenesis occurring in the two groups.

As discussed previously, the germ cells of the amniote testis mature in association with a permanent population of Sertoli cells. Following puberty these Sertoli cells do not divide, and their number remains constant until death. During this period they are always surrounded by several layers of germ cells at different stages of maturation. Furthermore, the general morphological characteristics of these cells appear reasonably constant throughout the spermatogenetic cycle. They do, however, undergo dramatic changes in structure in those species which have seasonal breeding cycles.

The cystic form of spermatogenesis results in a totally different relationship between germ cells and somatic cells. The somatic cell plays a major rôle in the formation of the cyst and undergoes numerous divisions to provide the correct ratio of somatic to germ cells. During spermatogenesis each cell is associated with a single population of isogenetic germ cells which mature in synchrony. As these germ cells develop, the somatic cells undergo great changes in structure. Initially, as the cysts form, the somatic cells are relatively undifferentiated, but by the time spermatids have been produced the somatic cell has become highly developed and at this stage is often referred to as a "glandular nurse cell" (see Figure 6). In urodeles and elasmobranchs there is a distinct migration of the somatic cell nucleus as the cyst forms; in urodeles it moves from a basal location to a more central position, and in elasmobranchs it moves in the reverse direction. In the majority of the anamniotes, once the sperm have been released, the somatic cell degenerates (see Figures 6 and 7), and so there is continual replenishment of somatic cells during each reproductive cycle.

From this rather cursory comparison it is clear that the somatic cell of the anamniotes differs tremendously in development and morphology compared to the Sertoli cells of amniotes. Despite these many differences, the somatic cell of the anamniote is often referred to as a Sertoli cell, but this interpretation is neither valid nor precise. Furthermore, this description implies they are morphologically and functionally homologous to the Sertoli cells of the mammals. This is certainly not true as far as the structure of these cells is concerned, and as yet data on their function are too sparse to permit conclusions as to their homology with Sertoli cells. The somatic cells of the anamniotes do carry out some functions similar to these of mammalian Sertoli cells; i.e., they support and nurture germ cells and phagocytose germinal elements. This superficial resemblance is not adequate to justify their acceptance as Sertoli cells. The term Sertoli cell should be restricted to the somatic cell of the amniotes, and the term somatic cell restricted to anamniotes. This is an anatomically correct but innocuous descriptive term that would bring consistent terminology for any nongerminal element present in the seminiferous lobules of the anamniote testis.

Investigations on the fine structure of somatic and Sertoli cells in many different species of vertebrates have demonstrated the presence of organelles associated with steroid synthesis (Figure 8). Such studies suggest these cells may produce steroid hormones necessary for the regulation of spermatogenesis. Furthermore, the histochemical localization of steroidogenic enzymes in these cells in various species strengthens this hypothesis.

In mammals it is generally accepted that Sertoli cells are incapable of *de novo* steroidogenesis from cholesterol. However, numerous studies have shown that they do possess the ability to metabolize steroid precursors to hormones such as an-

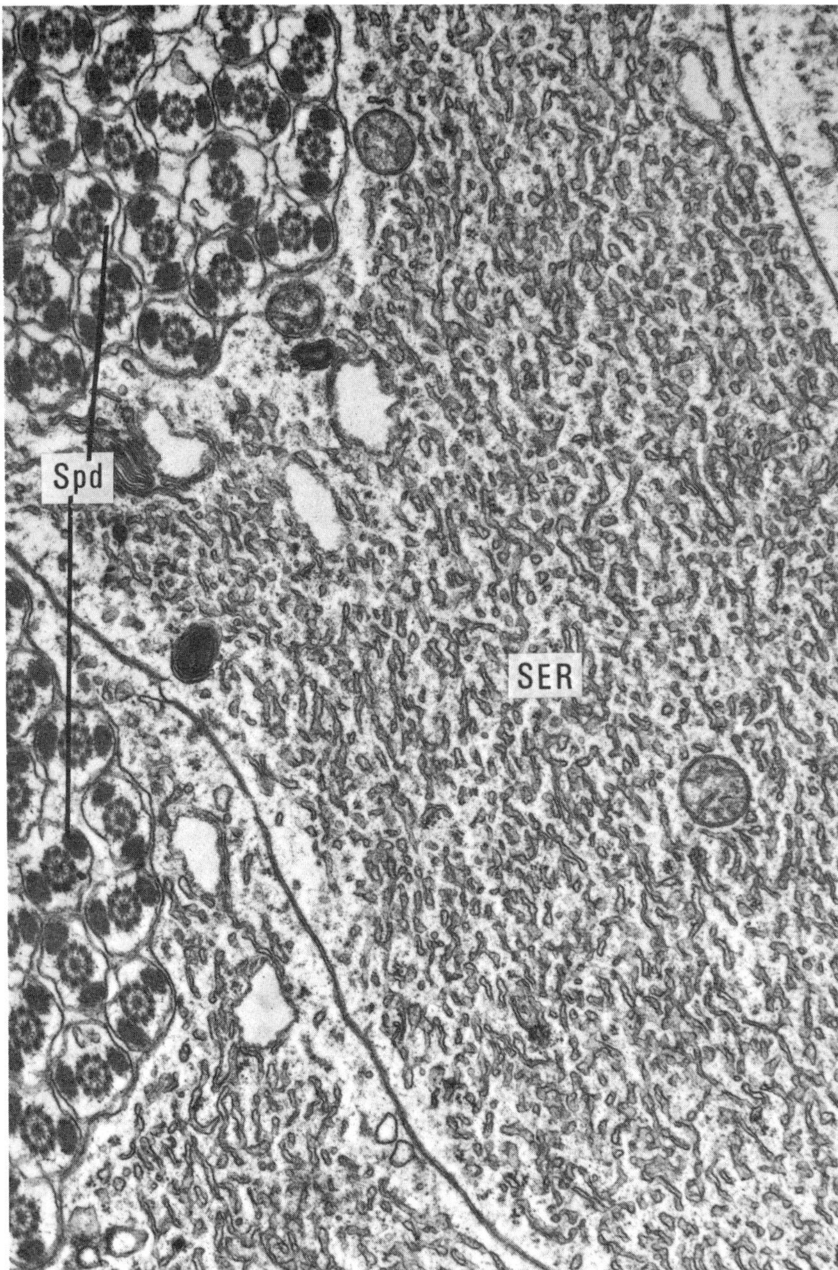

Figure 8. A somatic cell of *S. acanthias* which is associated with mature spermatids (Spd). At this stage of development the somatic cell is filled with a mass of tubular agranular reticulum. (×17,000.)

drogens. The nonmammalian somatic cell has not been as intensely investigated as far as its potential for steroidogenesis is concerned. Much of the evidence supporting claims for steroid production by these cells has come from histological observations with little biochemical data to confirm this activity. The somatic cells of nonmammals (as with the Leydig cells) undergo cyclical changes in lipid content. During the nonbreeding season the somatic cells possess few lipid droplets, but with the onset of spermatogenesis the amount of lipid increases and accumulates. At the time of spermiation the somatic cells are densely lipoidal, and this lipid has been shown to be rich in cholesterol (see Figure 7). The cells remain in this condition for varying lengths of time, depending on the species. Prior to the resurgence of spermatogenesis this lipid is reabsorbed and disappears. These observations have been interpreted as an indication that somatic cells have a steroidogenic function.

As yet there have been no definitive studies on the morphology of the somatic cell in either the anamniotes or nonmammalian amniotes, throughout their seasonal cycle of activity, and there is even less information concerning the function of these cells during this cycle. We need to know the fine structural changes that these cells undergo during spermatogenesis and what functional changes can be associated with this. An indication that both the morphology and activity of somatic cells change as germ cells mature has been seen in the elasmobranch *Squalus acanthias* (spiny dogfish). During the initial stages of cyst formation the somatic cells do not contain an abundance of organelles involved with steroidogenesis. As the cysts develop, the amount of SER in the somatic cell increases progressively. By the time spermatids have formed, the somatic cell is completely filled with tubules of SER (Figure 8). At all stages of development of the seminiferous lobule in the dogfish, the Leydig cells are sparse and poorly differentiated. Biochemical data from dissected zones of the testes demonstrate that the steroidogenic activity of the seminiferous lobules increases as the cysts mature. This functional change correlates with the somatic cell differentiation since they appear to be the only cells structurally competent for steroid production.

It is possible that in species where Leydig cells are poorly developed, e.g., dogfish, or where a temporal or spatial separation of Leydig cell development and germinal tissue occurs, e.g., urodeles, Teiid lizards, the somatic cell plays a greater rôle in steroidogenesis than Sertoli cells of mammals. Since the functional activity of the Leydig cells is out of phase with spermatogenesis in these species, the somatic cell must represent the major site of steroid hormone production, needed for the regulation of germ cell maturation.

A tremendous amount of work, both morphological and biochemical, needs to be carried out before the exact function of the somatic cell in the nonmammalian vertebrate testis can be established and before valid comparisons can be made with the Sertoli cell of mammals.

5.2. The Ovary

The ovaries of vertebrates, like the testes, are usually paired structures. Exceptions are the cyclostomes, where the ovaries are fused, many elasmobranchs, where development is asymmetrical with one ovary persisting and the other atrophies, and birds, where only the left ovary develops (although in some avian species the right ovary is occasionally present). The ovaries lie within the abdominal cavity attached to its wall by a mesentery, the mesovarium. As observed for the testes in male elasmobranchs, the ovaries are associated with the epigonal organ, and the ovaries of amphibians with a fat body.

The ovaries may be solid structures, as in cyclostomes, many elasmobranchs, a few teleosts, birds, and mammals, or "hollow," as in other elasmobranchs and teleosts and all amphibia and reptiles. This hollow ovary in elasmobranchs, amphibia, and reptiles is due to the formation of large, lymph-filled, central cavities. In teleosts, however, it results from the embryonic development of longitudinal ovarian folds that eventually fuse to enclose a coelomic cavity (cystovarian condition). This is lined by a peritoneal epithelium, unlike the ovarian cavity in the other species, which is lined by a mesothelium. Cystovarian ovaries often produce extensions of the coelomic cavity that function as oviducts and open into the cloaca.

In females, the formation and development of ova are remarkably similar in all groups of vertebrates. Modifications in the structure of the ovary are mainly dependent on the method of fertilization, whether this is internal or external, and on the degree of maternal involvement in the maintenance of the fertilized ova. Thus, whether the reproductive strategy of the female is oviparous, ovoviviparous, or viviparous will have an effect on the organization of the ovary.

5.2.1. General Organization

The ovaries contain the germinal components and endocrine elements the follicle cells responsible for the production of the female sex hormones. Whereas in the testes the endocrine compartment (Leydig cells) is distinctly separated from the germinal tissue, in the ovary the follicle cells actually surround the germ cells. These germ cells are located in the ovarian stroma, which consists of connective tissue, nerves, and blood vessels (Figure 9). This stroma provides structural support for the germ cells as they develop and is also the source of the endocrine follicle cells.

5.2.2. Oogenesis

The production of ova, or oogenesis, is initiated by an oogonium that undergoes mitotic division to produce primary oocytes; these undergo meiosis to yield the ova. For most teleosts and reptiles and for all amphibia, oogenesis continues throughout adult life. The oogonia act as stem cells dividing mitotically to produce daughter oogonia. One of these remains as a stem cell while the other enters the maturational cycle. In cyclostomes, elasmobranchs, a few teleosts, and reptiles, birds, and mammals, the early stages of meiotic division occur in the developing ovary, and the number of potential ova the female can produce is predetermined at birth.

The number and size of ova produced by the ovary vary enormously through-

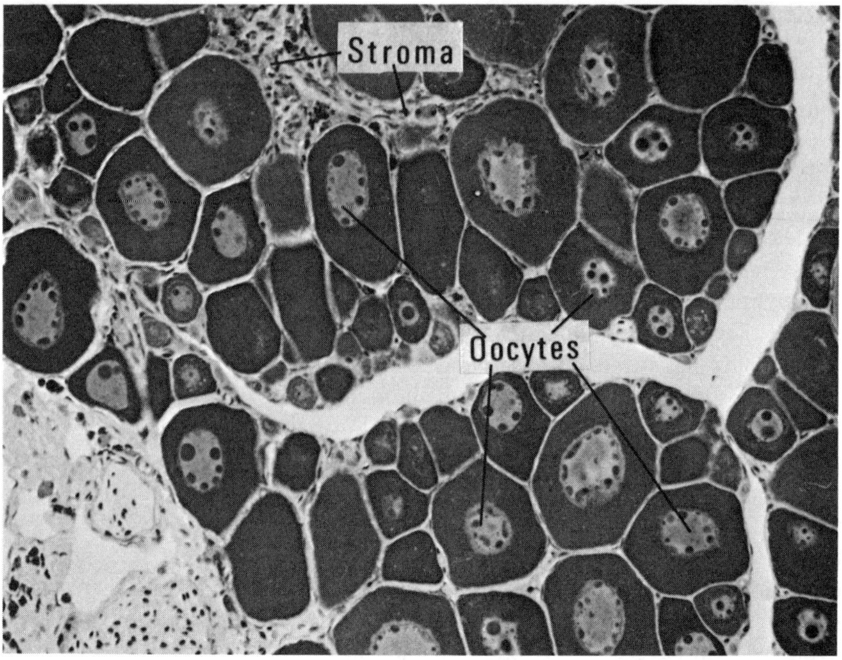

Figure 9. Ovary of the teleost *Pagellus acarne* showing numerous oocytes at different stages of maturation embedded in the ovarian stroma, which provides the endocrine follicle cells. (×300.)

out the vertebrates and reflect the reproductive strategy employed by the female. Generally, species that practice external fertilization produce large numbers of ova, e.g., teleosts, amphibia (see Figure 9). For instance, it has been calculated that in a teleost species (the Ling, Molva) during one breeding season the ovaries contained nearly 29 million ova! Females that carry out internal fertilization with an ovoviviparous or viviparous mode of reproduction produce much smaller numbers of ova. The size of the ova depends on the degree of deposition of yolk granules. The presence of yolk inclusions dominates the structure of the mature ova in nonmammalian species, and the massive accumulation of this yolk often results in the formation of huge ova. The largest ova produced (and therefore the largest cells known) are those of some species of elasmobranchs, which measure 100 mm or more, and birds, e.g., the ostrich!

5.2.3. Follicle Development

On entering meiotic prophase the oocyte acquires an investment of epithelial cells derived from the ovarian stroma (Figure 10). This forms the ovarian follicle. The primordial follicle is enclosed by a single layer of flattened undifferentiated cells, the follicular epithelium (Figure 10). The plasma membranes of oocyte and

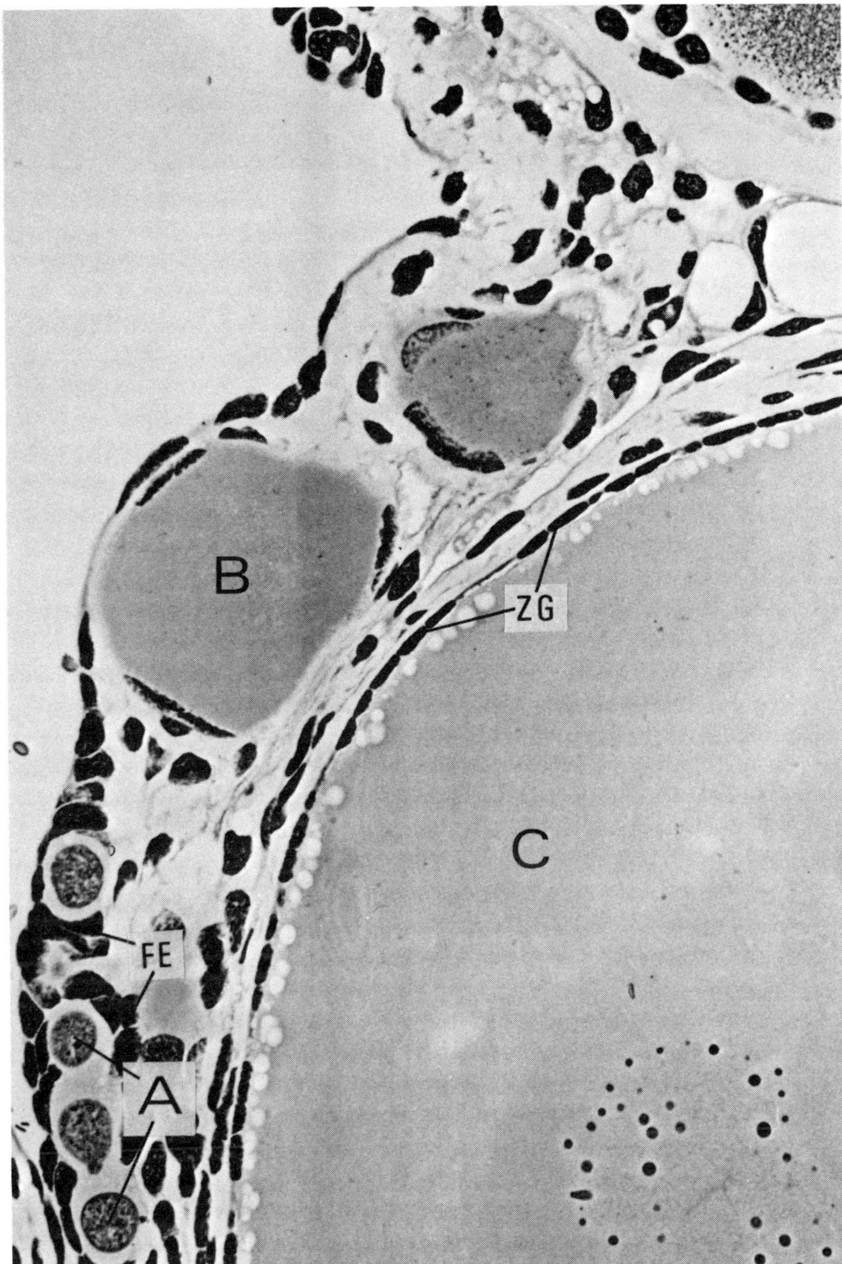

Figure 10. Ovary of *N. maculosus* depicting various stages in follicle development. Young oocytes (A) have acquired an investment of epithelial cells, the follicular epithelium (FE), derived from the ovarian stroma. As the oocytes (B) mature and enlarge, this follicular epithelium becomes stretched, appearing as a thin layer surrounding the oocyte (C). The zona granulosa (ZG) is one cell thick; in this light micrograph, only the nuclei of the granulosa cells can be resolved, and it is very difficult to detect the theca interna and externa. (×450.)

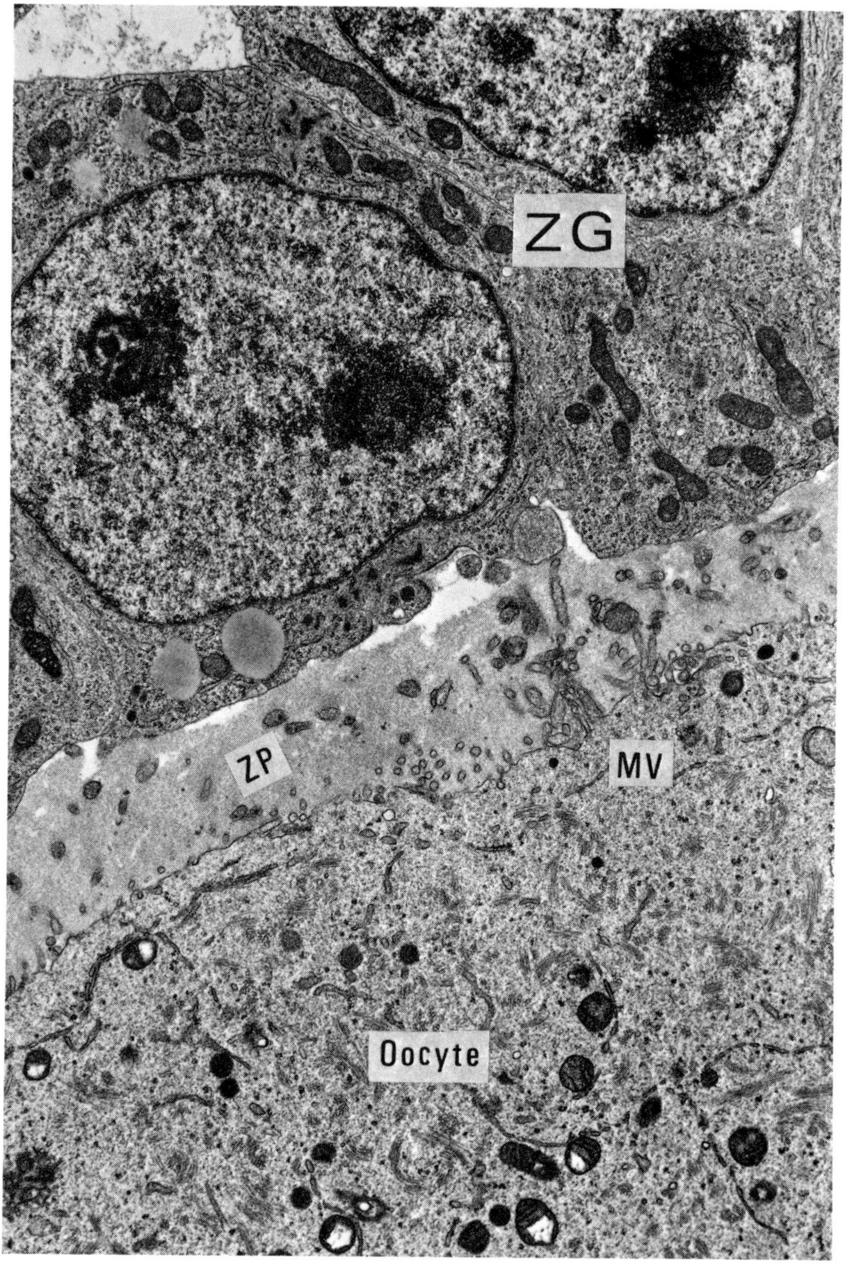

Figure 11. Ovary of the mouse. The zona granulosa (ZG) forms numerous cytoplasmic processes that penetrate the zona pellucida (ZP) to become closely associated with microvilli (MV) produced by the oocyte. Although the zona granulosa produces steroids, note the paucity of organelles associated with steroid synthesis in the granulosa cells since these cells are involved in the conversion of only one steroid—testosterone to estrogen. (This micrograph was kindly supplied by Dr. Everett Anderson, Laboratory of Human Reproduction and Reproductive Biology, Harvard Medical School, Boston, Massachusetts.) (×81,000.)

follicle cells at this stage are relatively smooth and closely apposed to each other. As the oocyte grows, the follicular epithelium increases in size and in many vertebrates undergoes mitotic division to produce a layer of cells called the membrana granulosa. Enlargement of the granulosa is accompanied by the development of an outer layer of cells (also derived from the ovarian stroma) called the theca. In most vertebrates the theca is composed of two layers, an inner, well-vascularized interna and an outer externa, formed of connective tissue elements. The membrana granulosa and theca are often separated by a distinct basal lamina.

As the follicle matures and the oocyte increases in size, a clear, highly polymerized membrane develops, called the zona pellucida in mammals (Figure 11) or zona radiata in nonmammals (Figures 12 and 13). This membrane separates the oocytes from the adjacent follicular cells. Following development of the zona pellucida/radiata the plasma membrane of the oocyte produces numerous microvilli which penetrate this membrane. At the same time irregular cytoplasmic projections are formed by adjacent granulosa cells, which also enter the membrane. The cytoplasmic extensions of both oocyte and granulosa cells then establish close contact with each other (Figure 11). An extreme condition is seen in some reptiles in which the cytoplasmic processes fuse. Often the granulosa cell processes may penetrate deep into the ooplasm of the oocyte. Thus, despite the presence of a thick, membranous zona pellucida/radiata, the oocyte and granulosa cells maintain an intimate association, which persists during oocyte maturation until ovulation. The cytoplasmic processes of the granulosa cells increase their surface area while close contact with the oocyte facilitates the transport of nutrients (and, in nonmammals, yolk, or its precursor) into the germ cell.

The organization of the developing ovarian follicle differs between the mammalian and nonmammalian vertebrates (Figure 12). In mammals both granulosa and thecal cell layers are well developed, and the oocyte only grows for a certain period and then stops, while the follicle continues to increase in size. Eventually a cavity called the follicular antrum is formed within the granulosa cells. This is filled with a fluid, liquor folliculi, secreted by the granulosa. The oocyte is suspended in the antrum, attached to one side of the follicle in a layer of granulosa cells, the cumulus oophorus. At ovulation the follicle wall ruptures and the oocyte is flushed out with the follicular fluid. In nonmammals the production of large, yolk-filled ova results in a different pattern of follicular formation. As in mammals, the follicle develops from the ovarian stromal cells forming the follicular epithelium and an outer layer of cells, which may or may not be separated into two thecal regions. These cell layers have, as in mammals, been termed granulosa and theca. Whether, however, this homology is functionally correct is not known for all species of nonmammalian vertebrates. In most cases, the development of the follicle cells is not as extensive as that in mammals. For many species the granulosa consists of only a single layer of cells, but in a few elasmobranchs, reptiles, and possibly birds it may be stratified. An extreme example occurs in some petromyzontid cyclostomes, where a single layer of granulosa cells incompletely invests the oocyte. Owing to the large size of the ova resulting from the accumulation of yolk reserves, the granulosa and thecal

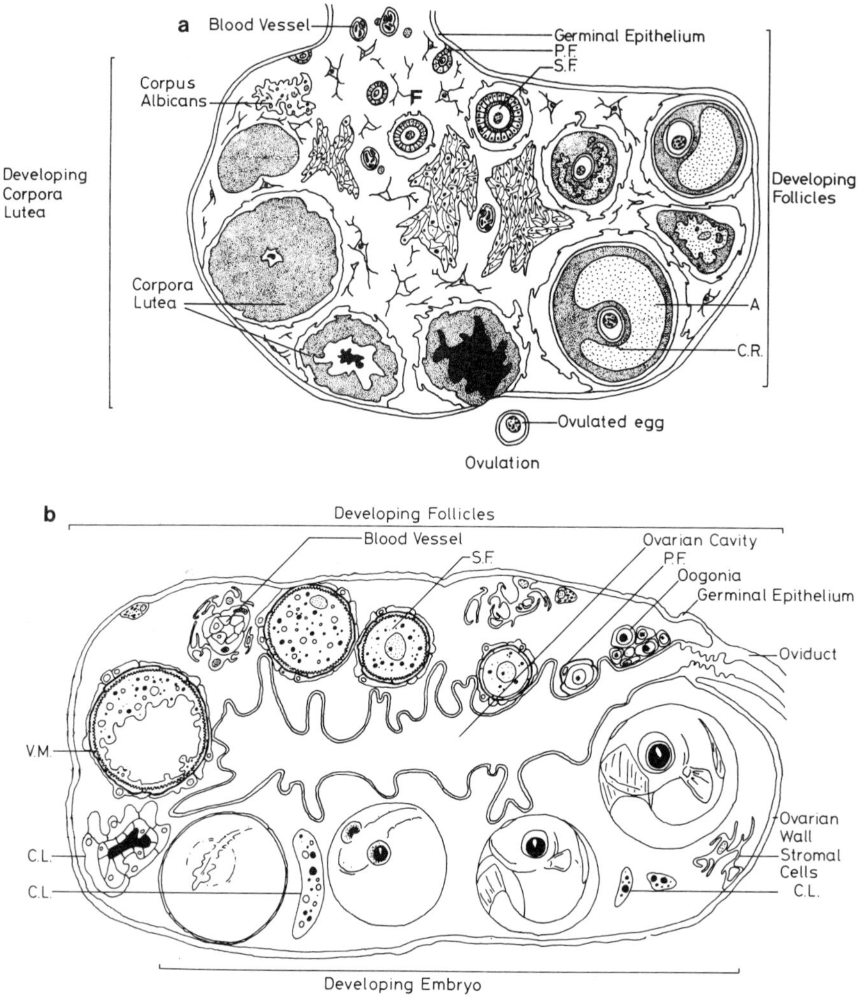

Figure 12. Diagrams illustrating the organization of mammalian follicle (a) and teleostean follicle (b). PF, primary follicle; SF, secondary follicle; A, antrum; CR, corona radiata; CL, corpus luteum; VM, vitelline membrane. (See also Figure 2, Chapter 7; Figures 5 and 6, Chapter 8.)

cells become tightly stretched and flattened as the follicle matures, making the cell layers, particularly the theca, difficult to identify in sections prepared for light microscopy (Figure 10). The large, yolk-filled ova completely fill the follicle in the nonmammalian vertebrates, and there is no formation of a follicular antrum (see Figure 13).

The follicle cells represent the endocrine tissue of the ovary, producing the female sex hormones. During the preovulatory phase the follicle cells are responsible for the secretion of estrogens. Following ovulation the follicle becomes reorganized to form the corpus luteum, which, in many vertebrates, plays a vital role

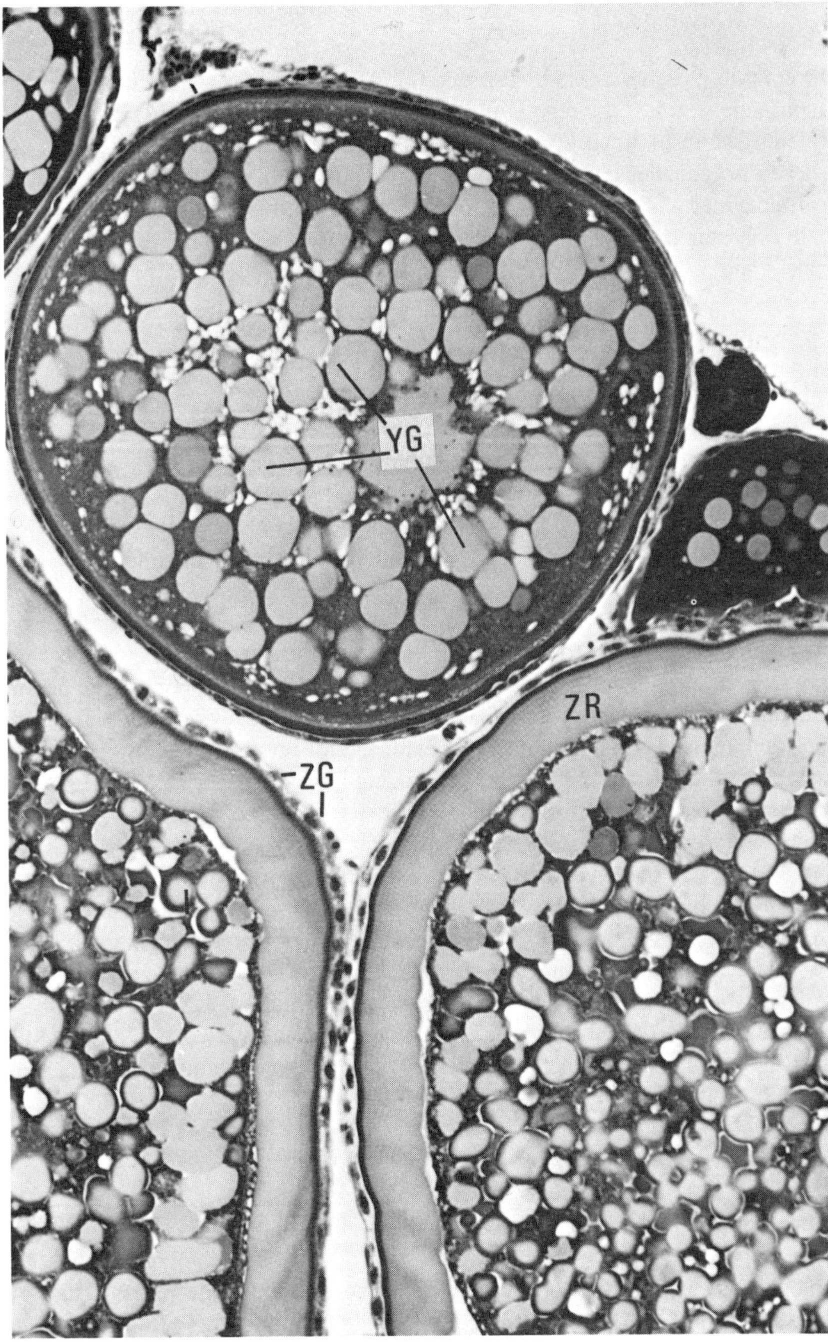

Figure 13. Ovary of the teleost *Spicara maena* containing large vitellogenic oocytes filled with yolk granules (YG). The oocytes are surrounded by a well-developed zona radiata (ZR), and the follicular epithelium consists essentially of attenuated zona granulosa (ZG) cells. (×320.)

in the control of reproduction by the secretion of progesterone. Since the female sex hormones are steroids, the recognition of ovarian endocrine tissue has relied (as did the identification of Leydig cells in the male) on histological evidence, i.e., the presence of organelles characteristic of steroid-producing cells and/or the localization of enzymes associated with steroid synthesis. The involvement, however, of follicle cells, the membrana granulosa and theca, and the importance of the corpus luteum in the production of ovarian steroids vary throughout the vertebrates.

5.2.4. Theca

In mammals the theca is represented by well-developed theca interna and theca externa. As in all vertebrates, the theca externa is composed of connective tissue elements. The theca interna, however, during the maturation of the oocyte, transforms into a highly active steroidogenic tissue. EM examination of these cells in many mammalian species has demonstrated they contain abundant organelles normally associated with steroid synthesis. Although nonmammalian vertebrates develop thecal cells, they are not as highly developed as in mammals, and some, e.g., amphibia, produce only a single cell layer. As mentioned previously, the large ova of nonmammals cause the follicular cells to become stretched and flattened. Thus, in some vertebrates the theca is not prominent at the light microscope level, and the identification of these cells can be determined only by means of EM, e.g., cyclostomes, amphibia. Unlike mammals the evidence for production of ovarian steroid hormones by the theca interna is not convincing. In cyclostomes EM observations have demonstrated a well-developed SER in the theca interna cells, and steroidogenic enzymes have also been detected in this layer. This evidence indicates that in these vertebrates the theca interna is steroidogenically active. For the rest of the nonmammalian vertebrates, although steroidogenic enzymes have been localized to the theca interna, e.g., some teleosts, elasmobranchs, and amphibia, it is at present agreed that the production of steroids by this layer is negligible. In birds, there is an unusual development of the theca into thecal glands. These consist of lipid-laden cells forming islets in the theca interna. They have been described as possessing the fine structural characteristics of steroid-producing cells, suggesting an endocrine function.

5.2.5. Membrana Granulosa

In mammals this is multilayered and during the preovulatory phase of follicle development secretes the liquor folliculi, which fills the follicular antrum. The granulosa cells are the site of estrogen production during this phase. This is a cooperative effort between the theca interna, which synthesizes androgens, and the granulosa cells, which aromatize these steroids to estrogens. Although the membrana granulosa is not highly developed in nonmammals (Figure 13), histological evidence indicates it is the primary site of estrogen production. In these species the granulosa cells also fulfill another very important function, which is the deposition of yolk in the oocyte during the process of vitellogenesis. The myxinid cyclostomes are unique in that the granulosa cells also secrete a shell around the developing oocytes, a function normally carried out by the oviducts but which are absent from myxinids.

5.2.6. Corpus Luteum

The corpus luteum is a structure that develops from the cellular components of the ovarian follicle following ovulation. Since the endocrine role of the corpus luteum is primarily concerned with postovulatory events of reproduction, its significance will vary with the mode of reproduction of the female.

In mammals this structure secretes progesterone. After ovulation the granulosa cells undergo hypertrophy and hyperplasia; there is an increase in lipid droplets and organelles associated with steroid production, particularly the SER. During the preovulatory phase the membrana granulosa is avascular, but with the formation of the corpus luteum this layer becomes invaded by blood vessels. The transformation of granulosa cells that accompanies corpus luteum formation is called luteinization, at which time these cells become known as lutein cells. The fate of the theca interna cells during luteinization of granulosa cells is uncertain. There appear to be species differences as to whether or not they are incorporated as secretory elements of the corpus luteum.

Corpora lutea have been described in all nonmammalian vertebrates except the petromyzonid cyclostomes, which die after spawning. In species of oviparous teleosts and amphibians in which fertilization is external, the corpora lutea are poorly developed transient structures, and where studied, lack histochemically detectable steroidogenic enzymes. In ovoviviparous and viviparous species, however, e.g., elasmobranchs and viviparous amphibia, corpora lutea are often highly developed, persistent, and assume an endocrine steroidogenic function. Reptiles possess highly developed corpora lutea that are fully functional. The granulosa cells undergo luteinization, accumulating cholesterol positive lipid, become highly vascular, exhibit steroidogenic enzyme activity, and have been shown to secrete progesterone. The corpora lutea persist throughout egg laying in oviparous species or gestation in the few viviparous species. Although distinct corpora lutea occur in birds, with good histological evidence for an endocrine function, they are transitory, and the physiological significance of these structures is uncertain.

5.2.7. Corpora Atretica

A predominant feature of all vertebrate ovaries is that follicles can degenerate at any stage of development leading to the production of corpora atretica (Figure 14). In nonmammalian species it appears that vitellogenic follicles, i.e., those undergoing yolk deposition in the oocyte, are more prone to atresia than previtellogenic follicles. They arise by invasion of the oocyte by granulosa cells and sometimes by thecal cells. These follicular cells phagocytose the oocyte and its

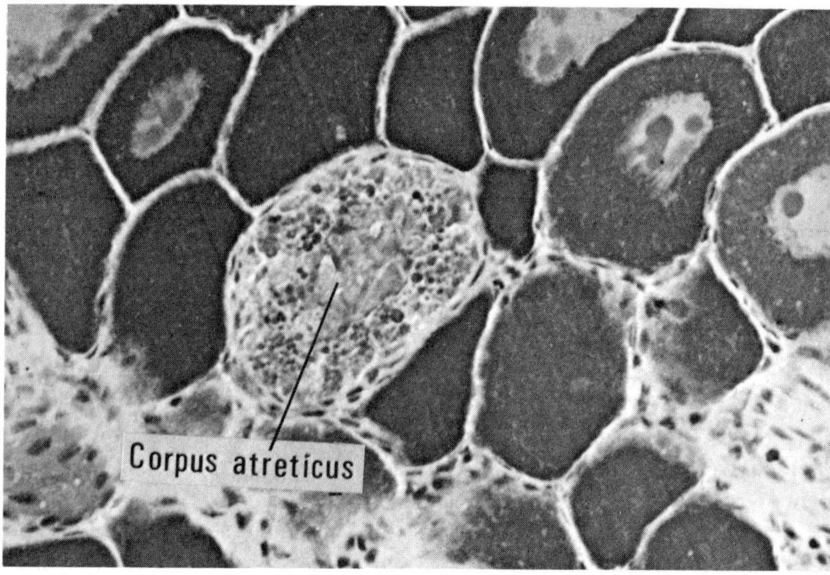

Figure 14. Ovary of the teleost *Serranus cabrilla* showing a corpus atreticus among maturing oocytes. (×250.)

yolky inclusions. When the oocyte and the yolk have been absorbed, there remains a solid, well-vascularized mass of granulosa cells. These cells then regress and are ultimately removed by macrophages.

The physiological significance of corpora atretica is unknown. In some vertebrates they are often called "preovulatory corpora lutea" and have been assigned an endocrine rôle by the secretion of steroids. Thus, in some species of amphibians the thecal cells of the corpus atreticus hypertrophy, and histochemical evidence suggests these cells may produce steroids. Despite such fragmentary reports, it appears that the corpora atretica in the majority of vertebrates are not involved in the formation of steroid hormones.

5.2.8. Ovarian Cycles

The ovaries of all vertebrates undergo distinct cycles of activity culminating in the production of ova. There is, however, a distinct difference in the pattern of this cycle between nonmammalian and mammalian species. In nonmammals the ovarian cycle essentially corresponds to the follicular phase of the cycle occurring in mammals. This is so because the majority of nonmammalian vertebrates are oviparous or ovoviviparous and, as discussed previously, do not require a functional corpus luteum. Hence the luteal phase, so essential in the mammals, is absent.

Most vertebrates also undergo seasonal cycles of ovarian activity. For cyclostomes and some teleosts there is only one ovarian cycle. The full complement of follicles is produced simultaneously. This is related to their life history since following ovulation the animals soon die. In many teleosts there is synchronous development of oocytes in the ovary, with a generation of mature follicles plus a second population of previtellogenic oocytes that will mature and ovulate either later in the same breeding season or early in the next. Ovarian cycles have been described in only a few elasmobranchs. Usually there is only one cycle, and in both ovoviviparous and viviparous species follicles destined to mature for the next breeding season begin to develop late in gestation. Amphibia tend to produce only a single population of ova during the breeding season. The reptiles possess reproductive cycles that involve either one ovarian cycle or many, which may be repeated continuously as long as environmental conditions permit. It has been suggested that the large yolk-filled ova of reptiles produce an inhibitory effect on the development of follicles and thus regulate this phase of the ovarian cycle. The ovarian cycle in birds is variable, depending on the species and the length of the breeding season. In most mammals the ovarian cycle is continuous until a successful mating occurs. The length of this cycle depends mainly on the persistence of the corpus luteum. Those species which do not ovulate spontaneously have no luteal phase, and the degenerating follicles are quickly replaced by another population of maturing oocytes. In spontaneously ovulating animals the luteal phase is longer and varies from species to species. Some develop a corpus luteum with no secretory activity, whereas others possess a functional corpus luteum that inhibits subsequent follicle development for different lengths of time. Conception in mammals, irrespective of the ovarian cycle in the unmated state, always leads to an extension of the luteal phase needed for the proper maintenance of gestation.

5.2.9. Viviparous Fish

The organization of the ovary in viviparous teleosts is unique. The majority of teleosts are oviparous, but in species that practice viviparity the ovary has become markedly modified since it is involved in a gestational role (see Figure 12b). Fertilization actually occurs within the ovary, which is capable of storing spermatozoa for varying periods of time. The ova, unlike those of oviparous species, are usually yolk-free, and the amount present depends on the length of gestation of the embryo. In some species the embryos are maintained in the ovarian cavities. The ovary becomes highly vascular and glandular, and its epithelium discharges nutrients into these cavities. Other species have a more remarkable arrangement; they do not ovulate; instead fertilization occurs in the follicle, which then transforms into a nutritive structure often called a follicular pseudoplacenta.

Another interesting characteristic of the ovaries of these viviparous teleosts is the occurrence of superfetation. This arises when two or more generation of embryos, at different stages of development, are present in the ovary at the same time. Superfetation is a consequence of intrafollicular fertilization, rendering ovulation unnecessary, and short successive breeding cycles where the maturation of oocytes overlaps gestation periods.

6. Selected Readings

Christensen, A. K., and Gillim, S. W., 1969, The correlation of fine structure and function in steroid-secreting cells with emphasis on those of the gonads, in: *The Gonads* (K. W. McKerns, ed.), Biochemical Endocrinology Series, Appleton Century Crofts, New York, pp. 415–488.

Fawcett, D. W., 1975, Ultrastructure and function of the Sertoli cell, in *Handbook of Physiology*, Section 7 vol. 5. *Endocrinology: The Male Reproductive System* (D. W. Hamilton and R. O. Greep, eds.), American Physiology Society, Washington, D.C., Chapter 2, pp. 21–55.

Grier, H. J., 1981, Cellular organization of the testis and spermatogenesis in fishes. *Am. Zoologist* **21:** 345–357.

Jones, R. E., ed., 1978, *The Vertebrate Ovary. Comparative Biology and Function,* Plenum Press, London.

Lincoln, G. A., 1981, Seasonal aspects of testicular function, in: *The Testis* (H. Burger and D. de Kretser, eds.), Raven Press, New York, pp. 255–302.

Roosen-Runge, E. C., 1977, The process of spermatogenesis in animals, in: *Developmental and Cell Biology Series* (M. Abercrombie, D. R. Newth, and J. G. Torrey, eds.), Cambridge University Press, New York.

Zuckerman, S., and Weir, B. J., eds., 1977, *The Ovary.* Vol. I. *General Aspects.* Vol. II. *Physiology.* Vol. III. *Regulation of Oogenesis and Steroidogenesis,* Academic Press, London.

3

Structure of the Adrenal and Interrenal Glands

I. Chester-Jones

1. Introduction

In Chapter 1 the types of steroids in vertebrates were described and pathways for their formation discussed. Certain steroids made by the adrenal/interrenal glands predominate; these are called corticosteroids. For eutherian mammals, they are aldosterone and cortisol, though rodents are generally corticosterone secretors; for Amphibia, reptiles and birds, aldosterone and corticosterone; for bony fish, teleosts, cortisol; and for cartilaginous fish, elasmobranchs, a special compound, discussed in Chapter 1, the unusual 1α-hydroxycorticosterone. In Pisces, that is, all groups of fish, the occurrence of aldosterone is dubious. All these hormones are formed in steroidogenic cells very similar to those described in Chapter 2 for the gonads. Morphologically these cells are gathered together and make up the adrenal and interrenal. In all classes, with the exception of the elasmobranchs, there are loose or close relationships with chromaffin tissue.

The term adrenal gland is used when the organ is near, but separate from, the kidney on each side of the body. This is characteristic of mammals, reptiles, and birds. These are jawed vertebrates (Gnathostomata) with fetal membranes—Amniota—and their significance is discussed in Section 7. Cells of the interrenal gland intermingle, to varying degrees, with kidney tissue, represented by amphibians and fish. The Agnatha (without jaws) possess living members of the order Cyclostomata (roundmouths) and may have groups of cells representing the interrenal gland. The designation suprarenal gland is confined to medical literature.

I. Chester-Jones • Department of Zoology, Sheffield University, Sheffield S10 2TN, England.

2. Mammalia

2.1. Eutheria

Much of our knowledge comes from man and laboratory and agricultural animals and spreads over comparative studies; so it is appropriate to start with this group. It is dominant with some 4500 genera. The adrenal glands are paired structures lying cephalad to the anterior pole of each kidney. They are vascularized by branches from the main abdominal vessels, chiefly the aorta and phrenic and renal arteries. Figure 1 gives the general plan. The numerous arterioles run to the surface of the gland, coming to lie in the loose external connective tissue from which their twigs, in an ill-defined plexus, vascularize the gland. For venous drainage, one vein emerges from the hilus of each gland to open into the inferior cava or the left renal vein (some species, e.g., the dog, are unusual in having up to four emergent veins). [Idelman, S., 1978, The structure of the mammalian adrenal cortex, in: *General Comparative and Clinical Endocrinology of the Adrenal Cortex* (I. Chester-Jones and I. W. Henderson, eds.), Academic Press, London, New York, San Francisco, Vol. 2, Chapter 1, pp. 1–199.]

The adrenal is, in fact, made up of two glands of diverse embryological origin though, of course, the pituitary is another example (Part III). The center of the

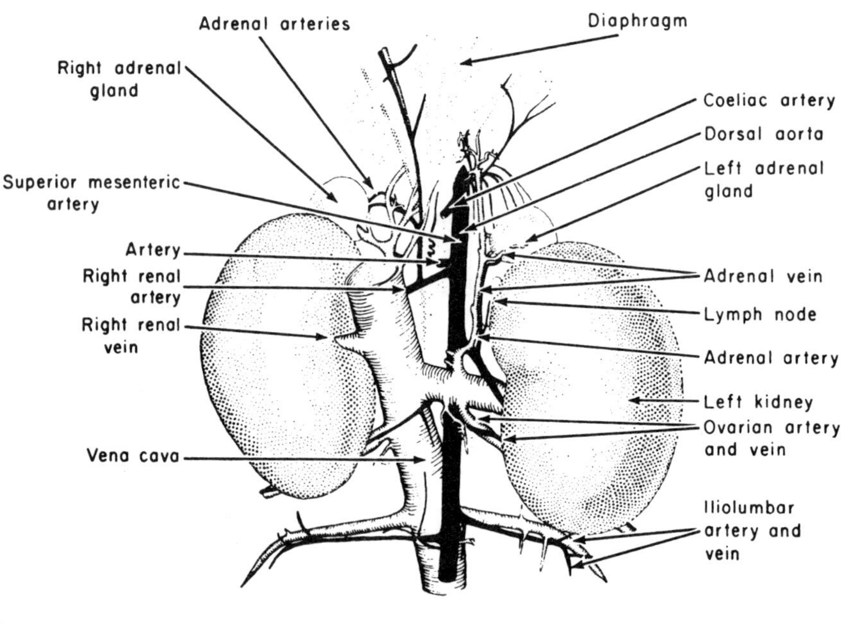

Figure 1. Ventral view of a dissection of a female rat of the Wistar albino strain showing the vascular system associated with the adrenal glands after latex injection. (×2; reproduced at 45%) (Prepared by I. Carthy and J. G. Phillips.)

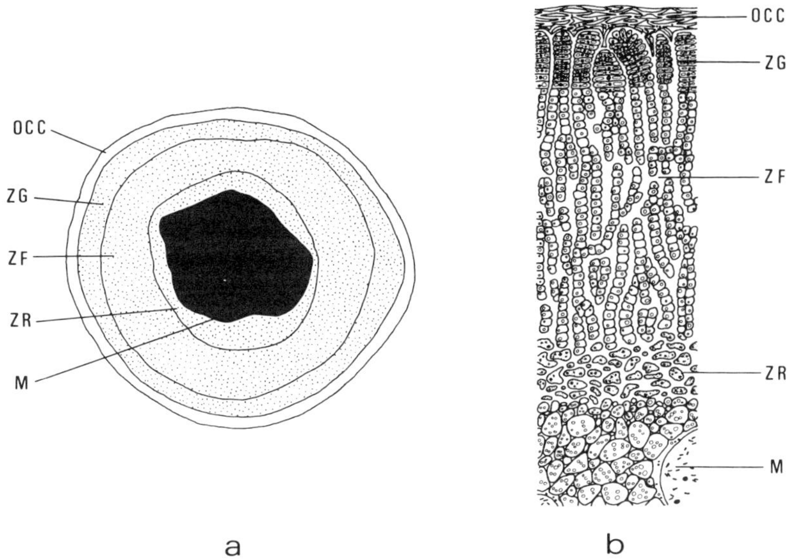

Figure 2. (a) Midsection of the adrenal gland in the Eutheria. OCC, outer connective tissue capsule; ZG, zona glomerulosa; ZF, zona fasciculata; ZR, zona reticularis; M, centrally coalesced chromaffin tissue, medulla. These are the main divisions. Subdivisions are often named, e.g., zona intermedia between the ZG and ZF. The medulla is often surrounded by a connective tissue capsule of varying thickness. (b) A portion of a, showing OCC, the outer connective tissue layer which encapsulates the adrenal gland. The ZG is shown as the looped ends of the cords of the ZF, obviously not so clearly seen in all species and dependent on the angle of sectioning. There is usually more transition from the inner ZF to the ZR, which may be in some phases a large active (on histological bases) zone.

gland, the medulla, comprises coalesced chromaffin cells—so called because of their coloration by potassium bichromate, the "chromaffin reaction" of Henle (1865) (Figure 2a). They secrete epinephrine and norepinephrine, whereas the surrounding adrenal cortex produces steroids, discussed in Chapter 1. Initially chromaffin cells arise from neighboring paraganglion cells of the neural crest. This is formed by a longitudinal band of nervous tissue lying dorsally between the neural tube and the ectoderm. In addition to the chromaffin cells, migration from the neural crest gives rise to ganglia, melanophores, and so on—a fascinating part of embryology well worth referring to a suitable textbook to see the ramifications of neural crest cellular migrations. The adrenal cortex arises from mesoderm, and the chromaffin cells pass through its anlage.

In the fully formed gland, the cortex is divided, on histological criteria, into three zones, named by Arnold (1866): zona glomerulosa, zona fasciculata, and zona reticularis (Figure 2 a,b).

The cells of the zona glomerulosa consist of ovoid groups or "balls" immediately inside the outer connective tissue capsule. This region may be particularly obvious, as in sheep, and difficult to discern in small rodents (e.g., mouse, gerbil).

The nuclei vary in shape from a sausage to an oval appearance, with one or two nucleoli. The cytoplasm contains a varying number of lipid droplets (liposomes), depending on the species and its physiological status, and these lie between the nucleus and that side of the cell which abuts on its local capillary.

The zona fasciculata appears histologically in radial lines when the sections are cut in appropriate orientations. The cells of the zone are polyhedral with round nuclei and one or two nucleoli. They are packed with cytoplasmic liposomes in some species (e.g., the laboratory rat, which provides so great a basis for endocrinology) and not in others (e.g., the hamster).

The zona reticularis presents a variety of aspects. Generally it comprises anastomosing networks of reticular cell cords surrounding the large blood sinuses. The cells are smaller than those of the rest of the cortex. At different ages and in a range of species, the reticular cells appear degenerative. Sometimes a zona reticularis is not apparent and at others a histologically "healthy" zone is to be seen.

The vasculature of the adrenal cortex is all-important in the determination of structure and activity (Figure 3). In general, in the Eutheria, arterioles (1) form a network to embrace the balls of cells of the zona glomerulosa; (2) proceed radially concomitantly with the columns of the zona fasciculata and enter the zona reticularis to splay out into sinuses; (3) pass through the adrenal cortex to vascularize the medulla in addition to the blood seeping in from reticularis sinuses; (4) occasionally loop back to the outer connective tissue capsule (e.g., in humans). Blood drains to the major medullary vein, which carries away formed steroid hormones and their metabolites and the catecholamines (epinephrine, norepinephrine) from the chromaffin tissue. The possibility of interplay between the products of the different areas can readily be envisaged but not precisely delineated.

The ultrastructure of only a few species has been studied, and much work has been concentrated on the adrenal of the laboratory rat (Figures 4–6, from Lentz, T. L., 1971, *Cell Fine Structure,* W. B. Saunders Company, Philadelphia, London, Toronto). Chapter 2 is a guide to the terms used.

The zona glomerulosa cells (Figure 4) are columnar with a spherical nucleus and a well-developed nucleolus. A basement lamina (BL) coats the perivascular surface. There is extensive, smooth-surfaced endoplasmic reticulum (SER) as a network of branching and anastomosing tubules. There are a few short segments of rough-surfaced endoplasmic reticulum and some free cytoplasmic ribosomes occurring in clusters, spirals, or linear chains. Mitochondria (M) are round, oval, or elongated with cristae in the form of lamellar infoldings of the inner mitochondrial membrane. Lipid droplets (LD) are spherical or irregular in shape. The Golgi apparatus is well developed, adjacent to the nucleus. Other structures include lysomes; multivesicular bodies and glycogen granules are also seen.

The zona fasciculata (Figure 5) cells are large, polyhedral with a central nucleus, which contains peripherally condensed chromatin, a nucleolus, and an internal fibrous lamina (FL). Branching and anastomosing tubules of smooth-surfaced endoplasmic reticulum (SER) are extensive. The rough-surfaced endoplasmic reticulum (ER) is more abundant in this zone than in the zona glomerulosa, consist-

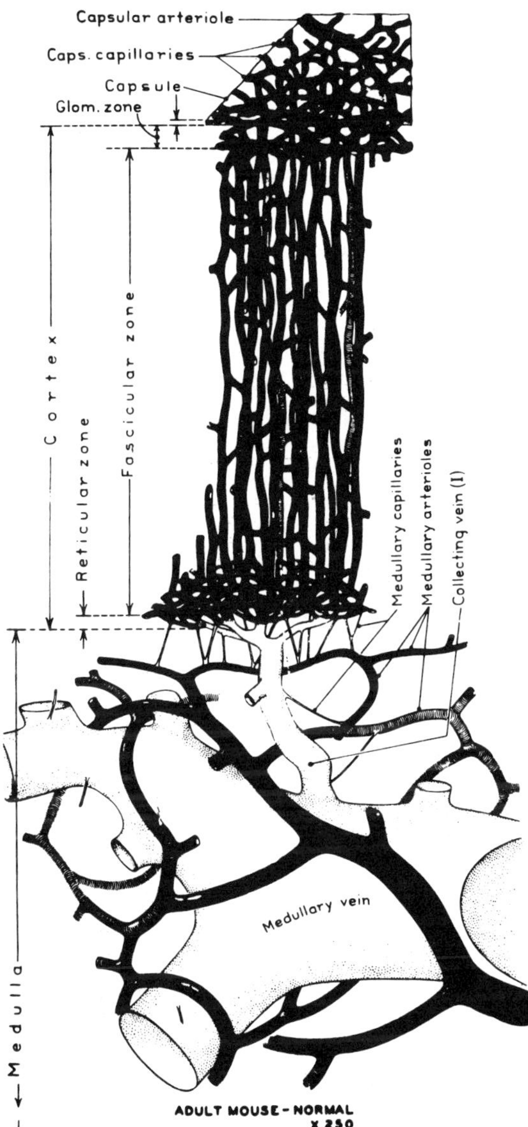

Figure 3. Vasculature displayed in part of the cortex and adjacent medulla of a normal adult mouse. (×250.) (From Gersh, I., and Grollman, A., 1941, The vascular pattern of the adrenal gland of the mouse and rat and its physiological response to changes in glandular activity, Carnegie Institute Washington Publication 525, *Contributions to Embryology*, Vol. 29, no. 183, pp. 112–125.)

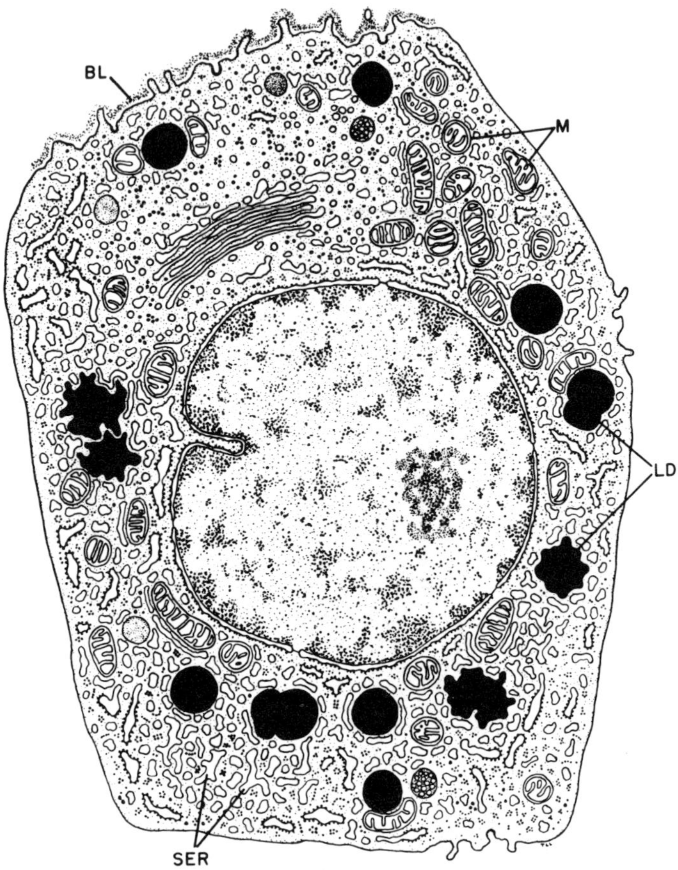

Figure 4. Diagram of the ultrastructure of a cell of the zona glomerulosa. BL, basement lamina; SER, smooth-surfaced endoplasmic reticulum; M, mitochondria; LD, lipid droplets. Further description in the text. (From Lentz, T. L., 1971, *Cell Fine Structure,* W. B. Saunders Company, Philadelphia, London, Toronto.)

ing of lamellar stacks of ribosome-studded cisternae. The mitochondria (M) are large and generally spherical. The cristae occur as vesicular inpocketings of the inner mitochondrial membranes and as vesicles apparently free in the mitochondrial matrix. The vesicular cristae of the zona fasciculata cells are in contrast to the lamellar cristae of the zona glomerulosa. Lipid droplets (LD) are frequent and large. The Golgi apparatus is well developed and there are lysomes (Ly) and lipofuscin pigment granules (LPG).

The cells of the zona reticularis (Figures 6 and 7) have much in common with those of the zona fasciculata, including tubules of smooth-surfaced endoplasmic reticulum, stacks or whorls of rough-surfaced lamellae, cytoplasmic ribosomes, Golgi apparatus, lipid droplets, and lysomes. The differences lie in the structure of the mitochondria and in the number and size of lipofuscin pigment bodies. The mitochondria are more elongated than in those of other zones. Two types of cristae are seen: vesicular cristae and lamellar invaginations of the inner mitochondrial membrane. The lipofuscin pigment granules (LPG) are present in large number and often quite big. Their shape is irregular, comprising a granular matrix within which are embedded droplets of varying size, probably lipid with a dense or translucent rim. Smaller vesicles and vacuoles containing material similar to the granular component of the lipofuscin body occur in the cytoplasm. The pigment in the granules gives the golden-brown color of these cells of the zona reticularis, as seen with the light microscope. It is interesting to see catecholamine granules in cells of the zona reticularis (Figure 7). They may have an important, though unknown, function.

The medulla (chromaffin cells) consists of pheochromocytes arranged in nests, alveoli or cords presenting a meshlike anastomosing network, between which are

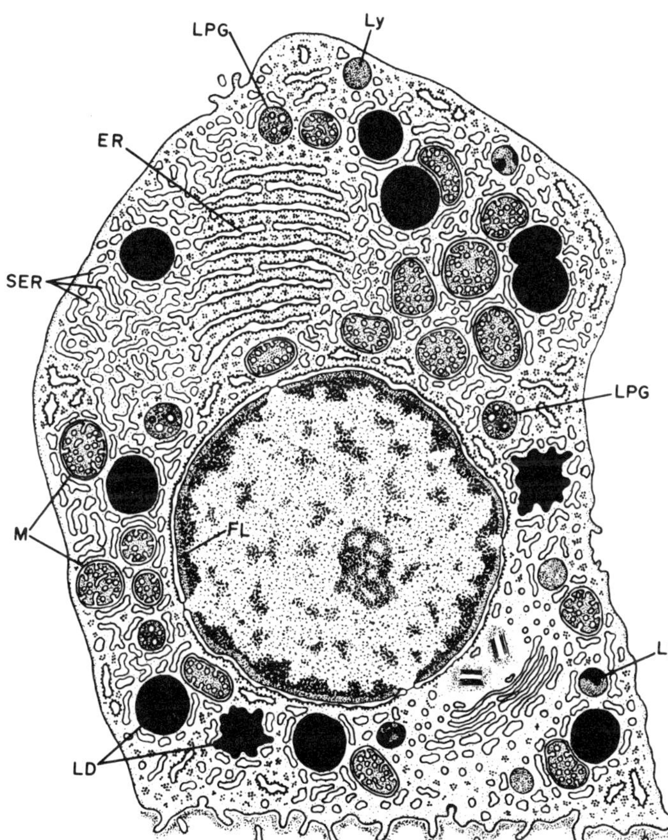

Figure 5. Diagram of the ultrastructure of a cell of the zona fasciculata. FL, nuclear internal fibrous lamina; SER, smooth-surfaced endoplasmic reticulum; ER, rough-surfaced endoplasmic reticulum; M, mitrochondria; LD, lipid droplets; Ly, lysosomes; LPG, lipofuscsin pigment granules. Further description in the text. (From Lentz, T. L., 1971, *Cell Fine Structure,* W. B. Saunders Company, Philadelphia. London, Toronto.)

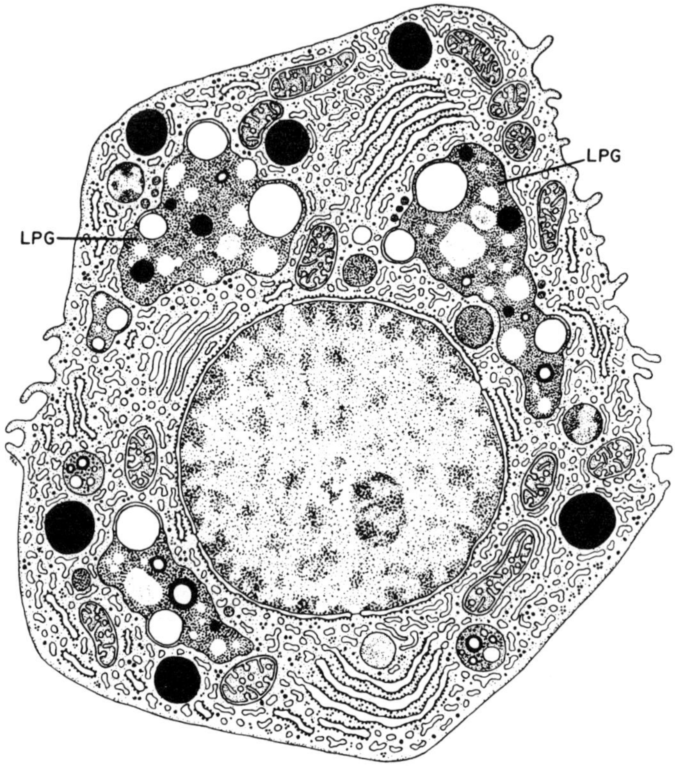

Figure 6. Diagram of the ultrastructure of one type of cell of the zona reticularis. LPG, lipofuscin pigment granules vary in number and frequently are very numerous. In some conditions a reticularis cell has close resemblance to one of the zona fasciculata. Further description in the text. (From Lentz, T. L., 1971, *Cell Fine Structure,* W. B. Saunders Company, Philadelphia, London, Toronto.).

rich vascular plexuses of capillaries, and sinusoids with walls containing elastic tissue. The cells are ovoid or polygonal; the nucleus is eccentric and may be spherical, ovoid, or elongated. There are one or more prominent nucleoli (Figures 7 and 8, from Lentz, T. L., 1971, *Cell Fine Structure,* W. B. Saunders Company, Philadelphia, London, Toronto). Sympathetic ganglion cells lie singly or in groups, and the cells are innervated by preganglionic fibers which aid the control of secretion. There are two types of chromaffin cell, though they are very similar. Cells that contain norepinephrine have opaque granules after glutaraldehyde fixation and osmium, whereas those with epinephrine are only moderately dense (Figure 8). They are membrane-bound at 1000–3000 Å in diameter. The characteristics of chromaffin cells are similar throughout the Vertebrata.

Transient Zones

A zone, the X zone, lies, in some species, between the inner zona fasciculata and the medulla and is exemplified by the mouse adrenal cortex, particularly the laboratory strain Swiss albino. The zone consists of cells with acidophilic cytoplasm and prominent basophilic nuclei. In the male, the X zone collapses by the direct action of androgens produced at maturity, giving pyknotic nuclei and shrunken cell cytoplasm. A similar picture is presented during first pregnancy, induced, perhaps, by ovarian androgens. In the virgin female, the X zone gradually declines over

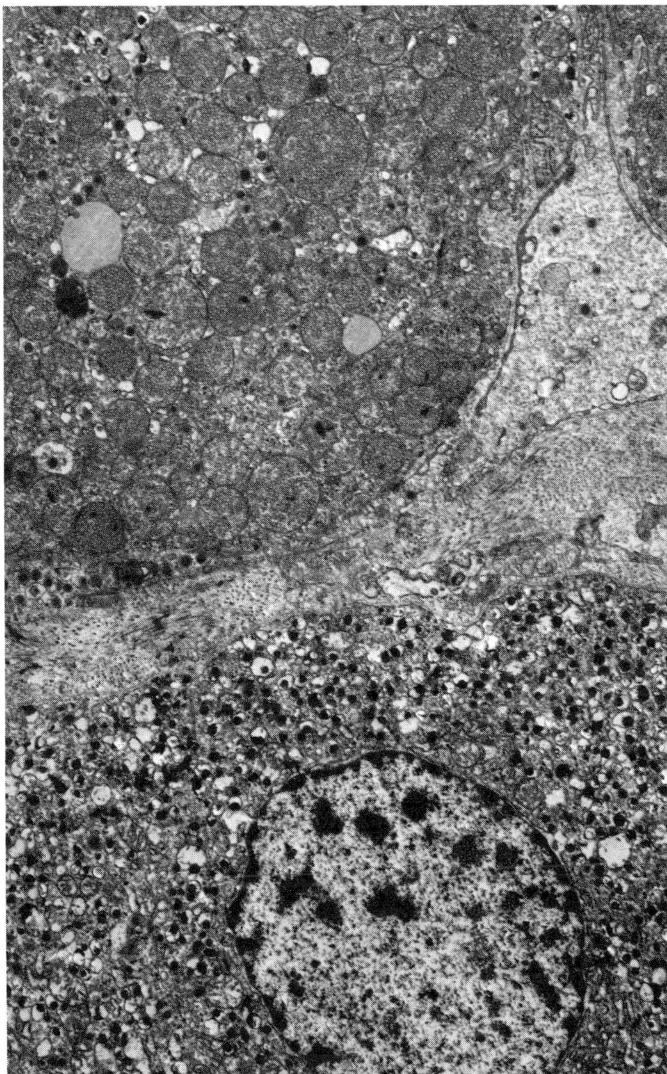

Figure 7. Photomicrograph of ultrastructure: Bottom: Part of a chromaffin cell (see Figure 8) which demonstrates the rounded nucleus and the numerous opaque, dense-cored vesicular granules containing norepinephrine. Top: Part of zona reticularis cells (see Figure 6). Note that chromaffin granules are scattered within these cells. (Courtesy of Patricia M. Ingleton.)

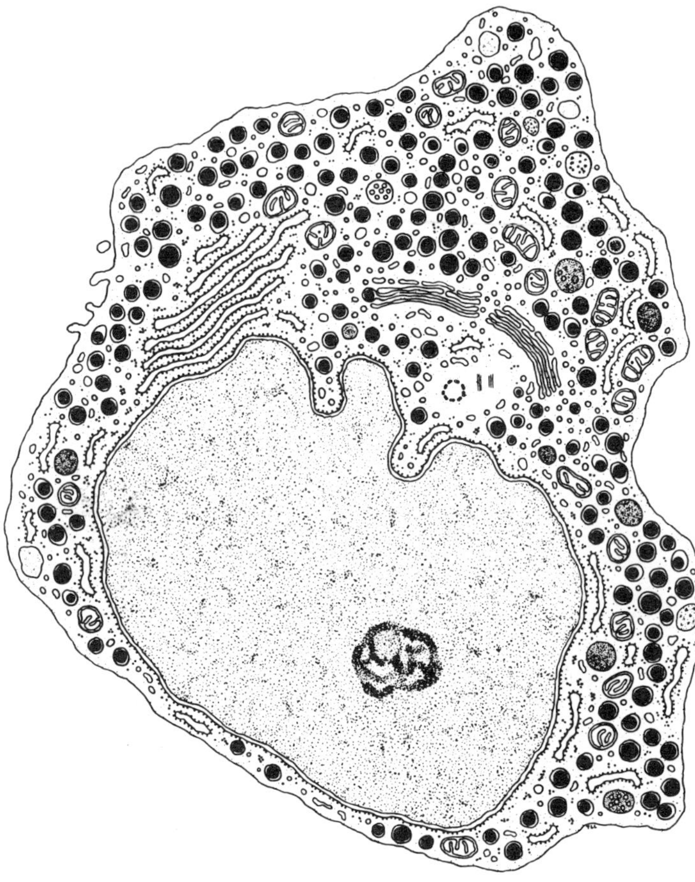

Figure 8. Diagram of the ultrastructure of a chromaffin cell—norepinephrine-storing. The granules, after glutaraldehyde and osmium fixation, are opaque. Further description in the text. Cells storing epinephrine have granules which are only moderately dense. (From Lentz, T. L., 1971, *Cell Fine Structure,* W. B. Saunders Company, Philadelphia, London, Toronto.)

many months by "fatty degeneration," that is, accumulation of lipid droplets. The gonadotropin, luteinizing hormone (LH), is tropic for the X zone and not ACTH, as is the zona fasciculata. Castration allows a persistent X zone, and when done postpubertally, a secondary X zone appears under the influence of LH not then opposed by androgens. No function has been assigned to the X zone. The pattern shown by this strain of mice is not a universal one. Changes in genetic makeup produce variations, so that, for example, female mice homozygous for the hypothyroid mutation have poor development of the X zone. Furthermore, certain wild rodents, especially voles and shrews, which have an X zone histologically similar, display varied reactions. In some species, the X zone does not involute at male puberty, and in the red squirrel (*Sciurus vulgaris*) it is a wide zone, changing markedly concomitant with breeding cycles. The X zone might be expected to produce steroids but there is no evidence for this. The mitochondria are not similar to those of known steroid-producing cells and, indeed, form peculiar complexes.

Another transient zone is seen in the adrenal gland of the human embryo. It is referred to as the fetal or transient cortex and occupies some 90% of the developing gland, the outermost part being the "permanent cortex," that is, a zone that will give rise to the adrenal of the postnatal and adult humans. The fetal cortex disappears around the time of birth. It has some relationship to the pituitary (perhaps ACTH?), as it does not fully develop in the anencephalic. No function has been proven to be associated with the zone, and it certainly does not appear to have steroidogenic properties!

2.2. Metatheria (Marsupials)

In general, the adrenal gland of marsupials has coalesced chromaffin tissue, as a medulla with a rim of adrenocortex proportionally smaller than that seen in Eutheria. Nevertheless, zonae glomerulosa, fasciculata, and reticularis can be seen, though not clearly distinctive. In the American opossum (*Didelphys virginiana*), cells of the zona glomerulosa have numerous short and rod-shaped mitochondria. Their cristae are in the shape of lamellae. The smooth endoplasmic reticulum (SER) makes up a richly developed network of long tubules with a cross diameter of about 500 Å. The rough endoplasmic reticulum occurs as tubules scattered in the cytoplasm. Liposomes are rare. In the zona fasciculata, the mitochondria have tubules rather than cristae. SER fills up the cell with an anastomosing system of tubules with a diameter between 500 and 1000 Å. There are some liposomes.

In one Australian marsupial (*Trichosurus vulpecula,* the brush-tailed possum) there is, in the female, a large additional adrenocortical zone. It lies between the definitive cortex and the medulla and may come to occupy about 45% of the whole gland. The volume of the zone may change in relation to reproductive status and produce steroids, though neither statement is conclusive. It is, therefore, another transient zone, seen in some Mammalia, without ascribed function.

2.3. Prototheria (Monotremata)

The females lay large, yolky shelled eggs. The family Tachyglossidae (Echidna or spiny anteater) has two genera and five species occurring in Australia and New Guinea. Little is known about the adrenal gland. It is an organ separate from the kidney on each side. The cortex appears homogeneous, made up of closely packed, irregularly placed cords, often separated by blood sinuses. The cells in the center of the gland are more closely packed. The chromaffin tissue is confined to the lower bulbous pole of the gland and is separated from the cortex by a marked development of connective tissue. The family Ornithorhynchidae, duck-billed platypus, occurs in Eastern Australia with one genus and one species. The separate adrenal glands are encapsulated with connective tissue. The chromaffin tissue occupies the venous pole of the gland. The cortical tissue seems to be composed of three groups of cells in a rather complicated way. As nothing of functional significance is known of these cell types, speculation is not of value. The evolutionary relationships of the monotremes have been the subject of much discussion, without a final consensus. It is, however, of interest that the structure of the adrenal of *Echidna* is very similar to that of the reptile *Sphenodon* (Figure 12).

3. Aves (Birds)

The paired adrenal glands are found, each by a kidney, as separate encapsulated organs, composed of adrenocortical and chromaffin tissues. These are intermingled, to varying degrees, depending on the species. There is no coalescence of chromaffin tissue to form a medulla (as noted earlier to be characteristic of the Eutheria and Metatheria). The glands lie, often wholly or partly covered by the gonads, at the anterior end of the kidneys (metanephric in embryological origin) and just behind the lungs. The location and vascularization of the gull (*Larus argentatus*) is shown in Figure 9, as characteristic of those genera of birds so far examined.

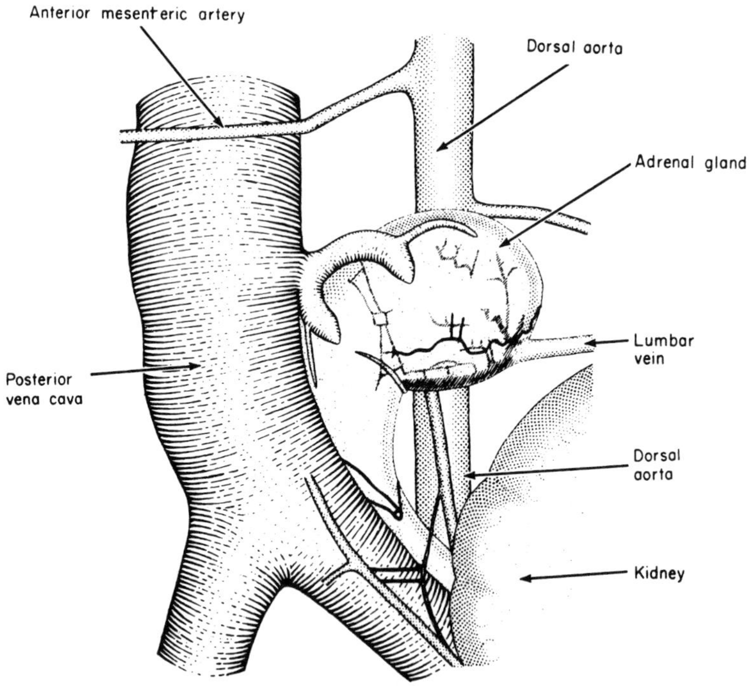

Figure 9. Ventral view of the left adrenal gland (*in situ*) of the gull (*Larus argentatus*). (Preparation by I. Carthy and J. G. Phillips.)

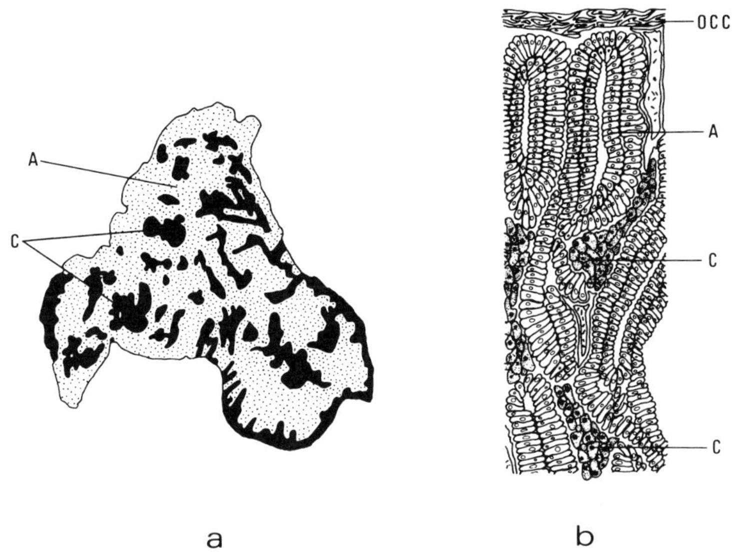

Figure 10. (a) A simple diagram to emphasize the intermingling of chromaffin tissue in the typical avian adrenal gland. C, chromaffin islets; A, adrenocortical cells. In some bird species the areas of chromaffin tissue are large and numerous, in others slight. (b) Part of a. OCC, envelopes the gland; A, the adrenocortical cords with intermingled chromaffin islets (C).

[Holmes, W. N., and Phillips, J. G., 1976, The adrenal cortex of birds, in: *General, Comparative and Clinical Endocrinology of the Adrenal Cortex* (I. Chester-Jones and I. W. Henderson, eds.), Academic Press, London, New York, San Francisco, Vol. 1, Chapter 5, pp. 293–420.]

The basic unit of avian adrenocortical tissue is a cord composed of a double row of parenchymal cells, with their long axes in the transverse plane of the strand. The cords radiate from the center of the gland, branch, and anastomose frequently. At the periphery, they loop against the inner surface of the connective tissue capsule (Figure 10a,b). The path of any one individual cord is tortuous and therefore difficult to follow throughout its length without three-dimensional pictorial display. The nucleus of each parenchymal cell is frequently situated toward the basal lamina on the outer margin of the cord. The cells at the periphery of the gland may be large with a considerable amount of cytoplasm. The nuclei are round or slightly oval, the chromatin network is thin and diffuse, and there are usually two nucleoli. Toward the middle of the gland, the cells are more elongated and smaller. Adrenal or interrenal lipid is characteristic of many species of all vertebrate groups, so that in birds, also, we find sudanophilic and osmophilic droplets in the cells, though clearly the amounts in each vary according to genus and to physiological status. In ultrastructural studies, adrenal cells are characterized by the presence of numerous liposomes, many mitochondria, and the moderate amount of SER. Frequent inter-digitations occur between the plasma membranes of adjacent cells. All cells contain many mitochondria with cristae, mainly tubular. In the matrix of a mitochondrion, paracrystalline aggregates of parallel tubules, about 120 Å in diameter, are often seen. Cisternae of SER and lesser amounts of rough endoplasmic reticulum are distributed throughout the cells. Close associations between cisternae and endoplasmic reticulum, mitochondrial surfaces, and lipid droplets are to be observed. An extensive Golgi complex, with three to six cisternae in each body, is situated close to, and on the apical side of, the nucleus. Some smooth and some coated vesicles, about 1000 Å in diameter, are found in the apical region of the cell. Microtubules, 250 Å in diameter, are found in the peripheral regions of the cell. Close to the nucleus are microfilaments, 100 Å in diameter.

Groups of chromaffin cells lie between the cords of parenchymal cells and also ensheathed within the connective tissue reticulum. In some species, the chromaffin tissue lies toward the outer portion of the gland, and in others, it is scattered as islets. A chromaffin cell has numerous neurosecretory granules and sparse lipid droplets, and mitochondria are not abundant (as for Figures 7 and 8).

The possibility of zonation of the avian adrenal has intrigued workers, as aldosterone and corticosterone are the main steroid hormones secreted. Following the theory of the basic cord (see Section 7), it is not surprising to find that the outer subcapsular cells, looping in a manner typical of the zona glomerulosa, are designated as predominantly aldosterone secretors. Those cells reaching toward the central part of the gland form corticosterone.

4. Reptilia

Reptiles are the third tetrapod group with an amnion—Amniota. It characteristically has the "new" kidney, the metanephros in the evolutionary sequence (Figures 11 and 12a,b). As a consequence, the mesodermal cells that form the adrenal gland come to lie, as in birds and mammals, near the kidney. [Lofts, B., 1978, The adrenal gland in Reptilia, in: *General, Comparative and Clinical Endocrinology of the Adrenal Cortex* (I. Chester-Jones and I. W. Henderson, eds.), Academic Press, London, New York, San Francisco, Vol. 2, Chapter 4, pp. 292–369.] It is a discrete encapsulated organ, and basically, it does not differ substantially from the varying degrees of intermingling of adrenocortical and chromaffin tissues observed in birds. In the Squamata, the adrenal glands are generally elongate yellowish-colored bodies, enclosed within the folds of the mesorchium or mesovarium, in close contiguity with the gonads. In lizards and snakes, the right adrenal is cephalad of the left (Figure 11). Crocodilia are similar, but in the Chelonia the glands are located on the ventral surface of the kidneys, though still separate organs. Vascularization of the reptilian adrenal gland falls into the general pattern seen in the Amniota. The arterial supply is provided by a series of small arteries branching off the dorsal aorta (Figure 11). Anteriorly an artery branches off from the dorsal aorta and joins the adrenal gland dorsolaterally approximately a third of the way along its length. It then divides into two main branches, one anterior, the other

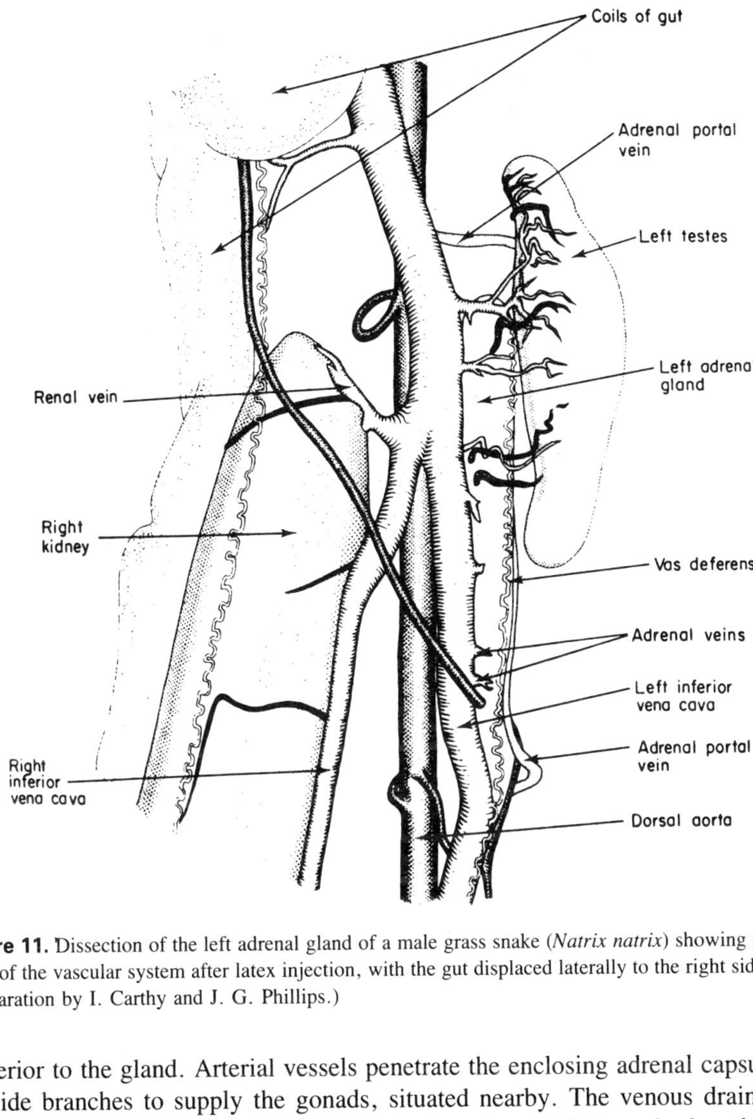

Figure 11. Dissection of the left adrenal gland of a male grass snake (*Natrix natrix*) showing a ventral view of the vascular system after latex injection, with the gut displaced laterally to the right side. (×5.) (Preparation by I. Carthy and J. G. Phillips.)

posterior to the gland. Arterial vessels penetrate the enclosing adrenal capsule and provide branches to supply the gonads, situated nearby. The venous drainage of each gland is by numerous short efferent veins, given off along the length of the gland and entering directly into the vena cava on each side. In addition, there are the portal vessels, that is, two veins, one anterior and one posterior, running from the body wall to each adrenal. The morphological schema of arterial supply and venous drainage that applies to mammals, birds, and reptiles has an exception in a few cases (such as *Alligator mississippiensis* and *Lacerta viridis*) where the adrenal venous return is by a number of small ports that interconnect the gland intimately to the venae cavae.

Structural examinations show that the reptilian adrenal is completely enclosed

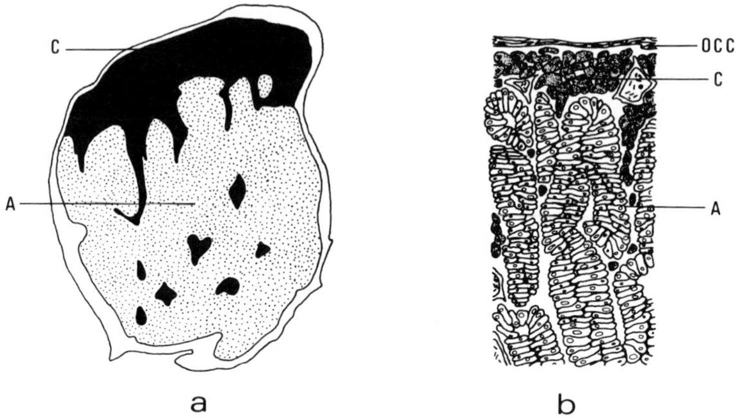

Figure 12. (a) The reptilian gland has, in most species, the similar intermingling of cortical and chromaffin tissue as demonstrated in birds. This diagram of *Sphenodon punctatus* illustrates the tendency for chromaffin tissue (C) to have peripheral aggregations in addition to the islets scattered throughout the adrenocortical cells (A). The adrenal of *Echidna* (Prototherian) is similar. (b) A portion of a with the encapsulating OCC; the adrenocortical tissue (A) cast into cords with chromaffin tissue (C).

in a thin connective tissue capsule, consisting of collagenous, reticular, and elastic fibers, surrounding the inner mass of cells (Figure 12a,b). These cells are organized into a loosely connected network of cortical cords, which are generally orientated longitudinally along the length of the gland. The avian pattern is seen once more with pronounced islets of chromaffin tissue intermixed with cortical cords. As would be expected, the degree of intermingling varies from species to species. No functional significance has been assigned to avian and reptilian variations in this regard. Once it is accepted that chromaffin cells migrate between the adrenocortical cells as they form the adrenal gland, and this for unknown reasons, then varying degrees of association must be anticipated solely on morphological considerations. Nevertheless, an expression of coalescence of chromaffin tissue into clumps is seen from a variety of species assigned to Rhynchocephalia and Ophidia. By and large, the pattern of intermix in Crocodilia and Chelonia is similar to that in birds. The whole glandular mass is permeated with numerous small lenticular blood sinuses which ramify between the cortical cords, some passing to the periphery and others coalescing to form one or more of the larger central sinuses, which eventually lead into the tributaries of the short adrenal veins entering the posterior vena cava.

Adrenocortical tissue displays a homogeneity of microscopic structure throughout the Reptilia. As we have seen in Aves, the basic unit is a cord of a double row of cortical cells to form a continuous network. These irregularly anastomosing cords run in all three planes, oriented, in general, parallel to the antero-posterior axis of the gland. The cords are separated by vascular channels, outlined by thin collagen fibers that separate the cortical cells from the endothelial lining.

Reticular fibers run irregularly to form a lattice encompassing groups of cortical cells.

The usual cortical cell is columnar and has a round or oval basophilic nucleus, but shows seasonal variations. It has a scattering of fine chromatin particles and one or two conspicuous nucleoli. Cells have numerous lipid droplets. The properties of the tetrapod cell are displayed in its ascorbic acid content and enzymatic sequences concerned with steroid hormone secretion. As to the possibility of zonation, there has been no rigorous investigation of the Reptilia. It is, however, implicit in the general theory of the basic cortical cord that peripheral cells contain the enzyme systems leading to mineralocorticoid formation and the central ones to glucocorticoids. Evidence is sparse, though the cobra (*Naja naja*) presents an easily distinguishable subcapsular zone, though it would be an extrapolation to assign mineralocorticoid secretion to this. In ultrastructural terms, the reptilian adrenocortical cell has all the characteristics of typical steroid-secreting cells. It is relevant, therefore, to mention that the peripheral zone of the cobra differs from the appearance of the inner cortical cords of the gland.

5. Amphibia

Amphibians belong to the Tetrapoda, and the major interrenal secretions are aldosterone and corticosterone (Figures 13 and 14a,b) [Hanke, W., 1978, The adrenal cortex (interrenal) of Amphibia, in: *General, Comparative and Clinical Endocrinology,* Vol. 2 (I. Chester-Jones and I. W. Henderson, eds.), Academic Press, London, New York, San Francisco, Chapter 5, pp. 419–495]. However, the metanephros has not appeared, in any evolutionary sense. Therefore, the mesoderm gives the kidneys and interrenal tissue, which, therefore, are intermingled in their embryology. As a representative of the Anamniota, the interrenal lies ventrally on the kidney (mesonephros) and intermingles with the renal tubules and interstitium. It is thus vascularized by common interrenal and renal blood vessels. Though not a discrete encapsulated gland, so characteristic of mammals, birds, and reptiles, its basic microstructure displays the general vertebrate pattern. It can be seen that the interrenal, the adrenocortical homolog, comprises discrete or scattered tissues set about the ventral surface of the functional opisthonephros (that is a term to include the mesonephros and its posterior parts) (Figures 13 and 14a,b). Bilateral branches of the dorsal aorta supply both the gland and the kidneys. The blood exits through venules into renal veins to enter the posterior vena cava. Venous blood from the renal portal vein enters numerous small vessels to reach the outer edges of the kidneys and is also carried to the vena cava by the renal veins. In Anura, islets of interrenal tissue are found on the ventromedial surface of each kidney, whereas in Urodela they are set in the midline. Anomalies occur in that, for example, the genera of the Pipidae (which includes *Xenopus laevis*) of the Anura have a urodele type of arrangement of the interrenal islets.

Histologically, the interrenal is made up of cords of polygonal cells in-

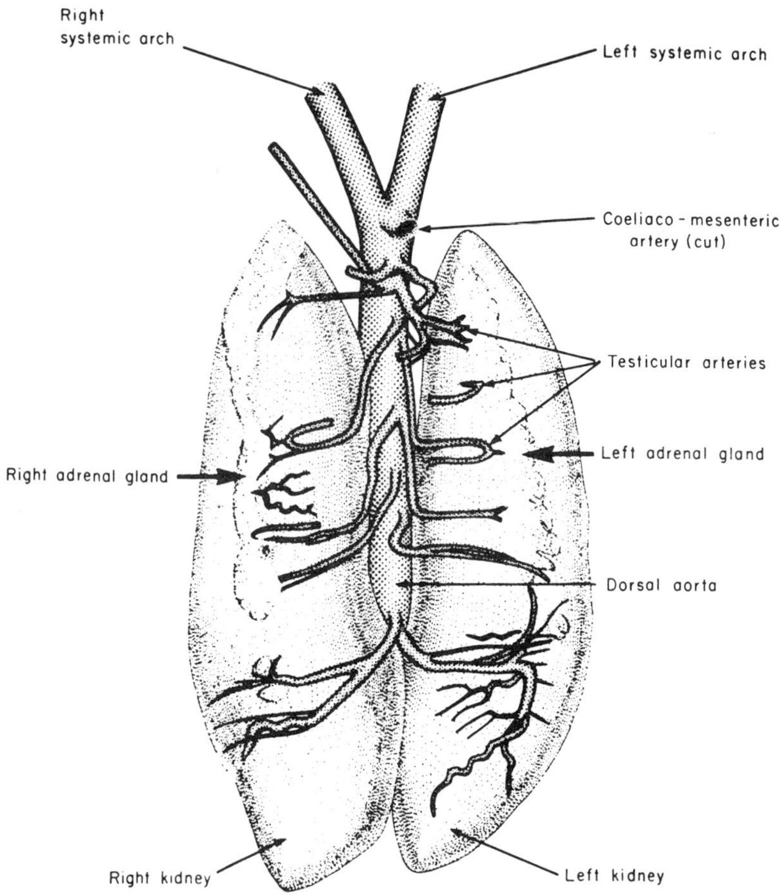

Figure 13. Dissection of the adrenal gland of a male frog (*Rana* sp.?) showing arterial and venous systems following latex injection. Ventral view with testes removed. (×5.) (Preparation by I. Carthy and J. G. Phillips.)

terspersed with blood sinuses; chromaffin cells, mostly in separate small groups, intermingle. (Stilling cells occur in Ranidae; they are clearly characterized by prominent cytoplasmic eosinophilic granulation, but their function and possible homologies remain unresolved problems.) In common with interrenal and adrenocortical cells, there are numerous lipid droplets, varying concomitantly with activity, and particularly with seasonal changes. Other familiar characteristics are to be seen: 5-ene-3β-ol steroid-dehydrogenase, cholesterol, glucose-6-phosphate dehydrogenase. The ultrastructure shows tubular mitochondria, SER, and liposomes.

The possibility that the amphibian interrenal presents zonation has been examined. In *Rana* species (e.g., *R. temporaria, esculenta*), the subcapsular zone and the zone nearest the kidney have active cells with large nuclei and nucleoli, considerable amounts of RNA, and fine lipid granules. Between these zones, the central part

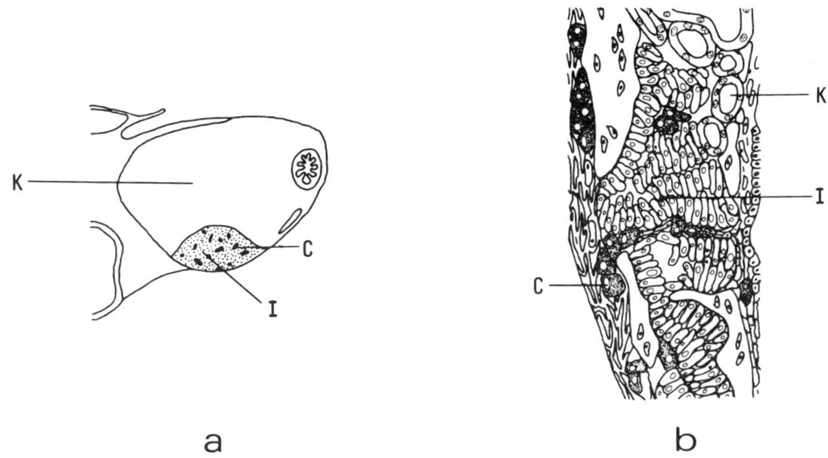

Figure 14. (a) Amphibia. There is no discrete adrenal gland, separated from the kidney. The interrenal lies on the ventral surface of the mesonephros with chromaffin tissue (C) intermingling with the interrenal tissue (I). K, kidney. (b) Amphibia. The interrenal cords of cells (I) are adiposed to the kidney tubules (K); chromaffin tissue (C) intermingles with the adrenocortical homologous cells (I).

of the gland has cells with smaller nuclei and larger lipid droplets. The whole interrenal can be given an active appearance, histologically, by seasonal activity and, experimentally, by exogenous ACTH. In times of hibernation and seasonal inactivity, zonation is not very obvious as cells throughout are small, and lipid droplets are large and coarse, indicative, in general, as tokens of low secretory capacity.

The life histories of Amphibia underline the importance of this group in functional studies. They are all dependent on water at some stage, some genera to a slight degree and others totally aquatic. Experimental studies have contributed much to osmoregulatory mechanisms, the control of water and electrolyte movements (see Part IV).

6. Pisces (Fish)

Pisces or fish is used as a convenient term for nontetrapod vertebrates, but it has no real standing in established systematics (Figures 15a,b, 16a,b, and 17a,b) [Chester-Jones, I., and Mosley, W., 1980, The interrenal gland in Pisces, in: *General, Comparative and Clinical Endocrinology of the Adrenal Cortex* (I. Chester-Jones and I. W. Henderson, eds.), Academic Press, London, New York, San Francisco, Vol. 3, Chapter 7, pp. 395–472]. The whole group comprises about half the total number of vertebrates, perhaps about 20,000 living species. Comprised within the whole lie the jawed fish, predominantly the Teleostei, with bones, and the Elasmobranchii, cartilaginous fish, and jawless fish, Agnatha. The interrenal glands cannot be considered apart from the excretory systems, the basic

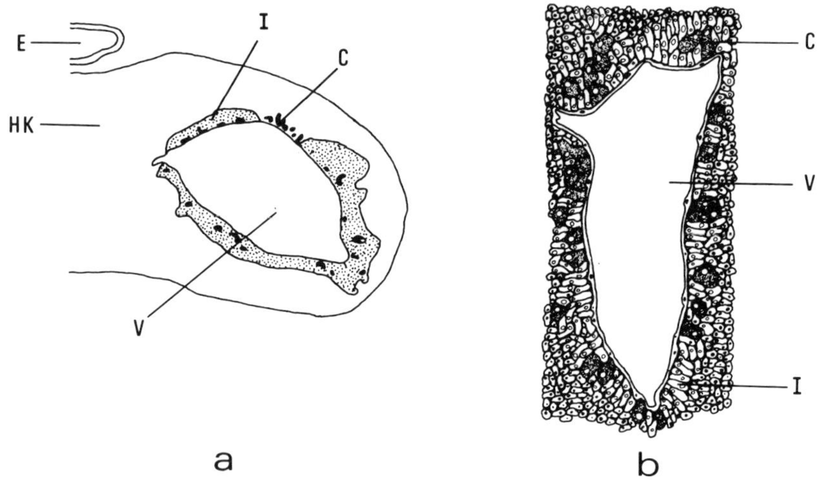

Figure 15. (a) Teleostei (bony fish). The interrenal tissue (I) lies anteriorly in the head kidney (HK), which is the nonfunctional part of the mesonephros, rarely displaying tubules, and mostly lymphocytes. The interrenal–adrenocortical homolog is disposed about branches of the posterior cardinal veins (V), and the diagram shows the fish's left side. The esophagus (E) is central. Chromaffin tissue (C), as black dots, is scattered throughout the interrenal tissue in many species, though the majority of the 32,000 species have not been examined. (b) Teleostei. A typical set of interrenal cells (I) set about a branch of the posterior cardinal vein (V) with intermingling of chromaffin cells (C).

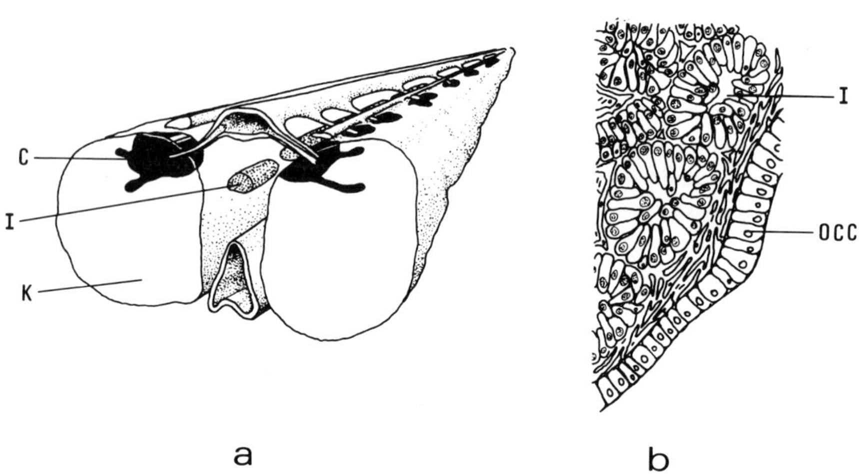

Figure 16. (a) Elasmobranchs. The main point is that the interrenal–adrenocortical homolog (I) is completely separated from chromaffin tissue (C), which is segmentally arranged. In the dogfish type the interrenal is rod-shaped and is found midway between the mesonephric groups. (b) Elasmobranchs. The interrenal–adrenocortical homolog is cast into lobules and cords of cells. It is a discrete gland bounded by a connective tissue capsule (OCC).

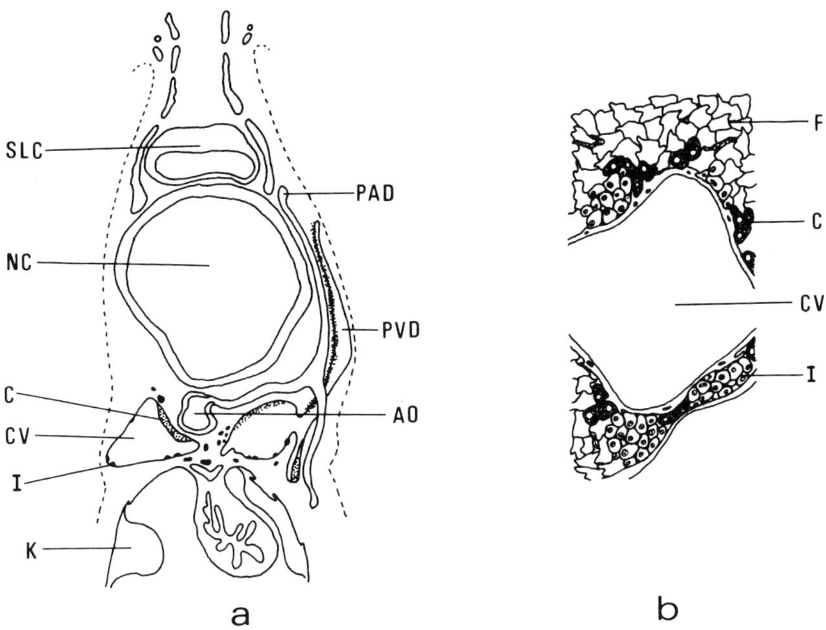

Figure 17. (a) Cyclostomes. Petromyzonidae. A transverse section of *Petromyzon* which shows the presumptive interrenal–adrenocortical homolog cells (I) arranged about the cardinal veins. Chromaffin cells (C), stippled, are nearby. SLC, spinal cord; PAD, PVD, posterior parietal arteries and veins; AO, aorta; K, kidney = mesonephros ? pronephros; NC, notochord. (b) Cyclostomes. *Lampetra.* Presumptive interrenal cells (I) occurring about a cardinal vein (CV) with adjacent chromaffin cells (C) with fatty tissue (F).

involvement of which we have seen in the wide-ranging group of Tetrapoda. Basically in vertebrate embryology, the renal, gonadal, and interrenal tissues derive, in varying proportions, from mesodermal blocks, referred to as nephrogenic blastema. The fundamental similar plan is, of course, differentially expressed. Potentially it may extend from the level of the second postotic somite to some distance behind the anus. Confusion has arisen by the use of pronephros and mesonephros (and, in Amniota, metanephros) supposedly to imply a succession along the length. However, it is only possible to distinguish a pronephros when the young occur as a larval stage, thus requiring an early functional excretory system.

6.1. Teleostei

These bony fish, an infraclass of the class Osteichthyes, are the most abundant and diversified group of all vertebrates. They are placed in, about, 31 orders, 415 families, 3869 genera, and 18,007 species to represent about 96% of known living fish.

In general, the teleost interrenal consists of groups, cords, or strands of cells set around the posterior cardinal veins and their branches, vascularized by twigs of the dorsal aorta (Figure 15a,b). The interrenal is usually confined to the anterior part of the kidney, named the "head kidney," commencing caudad to the last gill slit, usually anterior mesonephros, though with residual pronephric elements. In the majority of fish, the head kidney, representing the cephalad part of the mesonephros, is mostly lymphoid with pigment and hematopoietic cells. We can simplify the main patterns into: type I—interrenal tissue surrounds the postcardinal veins or their largest branches; type II—interrenal tissue surrounds small or medium-sized branches of the veins and is widely dispersed throughout the head kidney; type III—associated with the venous sinuses, scattered throughout the lymphoid and hemopoietic tissue; type IV—a solid mass of cells in a localized area. The histology of the interrenal cell conforms to the tetrapod pattern, though with less display of lipid droplets. The presence of cholesterol, ascorbic acid, glucose-6-phosphate dehydrogenase, and 5-ene-3β-ol-steroid dehydrogenase has been demonstrated in the few species examined. The ultrastructure is so similar to that normally occurring in a typical cell of the eutherian zona fasciculata as not to demand separate description. Hypophysectomy results in various degrees of atrophy, depending on species, the length of time after operation, and the environment. There is clear evidence for pituitary control by some type of fish adrenocorticotropin (ACTH), and mammalian ACTH is most effective—an interesting emphasis on evolutionary uniformity in the midst of morphological diversity. The major hormonal steroid is cortisol; in addition, aldosterone debately occurs. Indeed, there is a view that aldosterone, at least as a dominant mineralocorticoid, arose with the emergence of the Tetrapoda. The formidable osmoregulatory tasks facing teleosts in an enormous range of salinities may well depend, inter alia, on a type of fish prolactin plus cortisol (see Part IV).

Apparent stressful reactions may be involved in such natural processes as migration and spawning. The interrenal in teleosts undergoes seasonal variations, with enlargement at the reproductive period. The Salmonids are prime examples. The universal postspawning mortality of the Pacific salmon has been associated with hyperplasia of the interrenal. Many tissues show degenerative changes similar to those found in man, in which adrenal hyperplasia and hypersecretion are well documented. In the Atlantic salmon, immature parr show increasing interrenal activity, which increases at the smolt stage. Once in the sea, the salmon interrenal indicates maximum activity. In another genus of Salmonidae, most Ayu (*Plecoglossus altivelis*) die after spawning. Some overwinter if the degree of fatigue and debility is not too excessive. Interrenal tissue increases then in the prespawning period, to become marked in the breeding season. There is, in general, interrenal hypertrophy associated with gonadal maturation.

6.2. Dipneusti (Dipnoi) (Lungfish)

There are three extant genera of lungfishes—the Australian *Neoceratodus*, the South American *Lepidosiren*, and the African *Protopterus*. The presumptive interrenal consists of small groups of cells closely associated with the postcardinal veins and their tributaries where the vessels pass through the kidneys. The cells contain lipid, cholesterol, and 5-ene-3β-hydroxysteroid dehydrogenase. The morphological site is more amphibian than teleostean, akin to the interrenal tissue of the urodele, *Pleurodeles waltii,* and tadpoles of *Rana.*

Other groups—*Holostei (Amia calva), Chondrostei* (sturgeons), *Crossopterygii* (coelacanth)—are not well enough known to be included in this chapter.

6.3. Chondrichthyes—Elasmobranchii (Sharks and Rays), Holocephali (e.g., Ratfish); Cartilaginous Fish (Figure 16a,b)

These comprise dominant, successful fish of evolutionary antiquity. It is important to note that the interrenal gland is completely separated from the chromaffin cells, which lie in paired groups segmentally arranged. The interrenal itself lies in the region of the posterior kidney as blocks of cells, horseshoe-shaped (rays), or rod-shaped (sharks), or concentrated in one oval form (*Torpedo*). It is generally found that interrenal cells are homogeneous, grouped in cords or lobules, usually bounded by a thin connective sheath. The cells are round or polygonal, with a large prominent nucleus containing basophilic chromatin, with one or more nucleoli. Cells have lipid droplets, cholesterol, and 5-ene-3β-hydroxysteroid dehydrogenase. Ultrastructurally, two types of cell, dark and light, are indicated in immature *Raia clavata.* Dark cells are packed with liposomes and mitochondria, and in light cells the organelles are more widely spaced. SER and Golgi material are evident. Liposomes are of variable size, with or without a single limiting membrane. The numerous mitochondria are round or oval, with cristae mostly of the tubulovesicular type. Polyribosomes are dispersed throughout the cytoplasm. There are suggestions of interrenal cell changes with sexual activity and viviparity, though further investigation is needed. Zonation within the interrenal mass of cells may well exist, though it is not usually obvious. It is, however, emphasized in nurse sharks (*Ginglymostoma cirratum*). The interrenal of elasmobranchs is regarded as sluggish in its response to the usual experimental procedures. In the common species of *Scyliorhinus,* total removal of the pituitary gland may produce some histological changes about a year after operation.

6.4. Agnatha (Cyclostomata) (Jawless Fish) (Figure 17a,b)

Living jawless vertebrates divide into Petromyzonidae (lampreys) and Myxinidae (hagfishes). These two groups may represent in their living forms those types which heralded the transition from protochordate to vertebrate somatic organization, perhaps as far back as the Cambrian era. However this may be, they present difficulties in identification of interrenal tissue, which, if it exists, may or may not secrete known interrenal/adrenocortical steroids.

For *Petromyzonidae,* presumptive interrenal tissue may be concentrated within the pronephric region on the dorsal side of the pericardium and along the walls of the cardinal veins. Sterba and his colleagues [see Chester-Jones, I., and Mosley,

W., 1980, The interrenal gland in Pisces, in: *General, Comparative and Clinical Endocrinology of the Adrenal Cortex* (I. Chester-Jones and I. W. Henderson, eds.), Academic Press, London, New York, San Francisco, Vol. 3, Chapter 7, pp. 395–472] found, in *Lampetra planeri,* cells with organelles characteristic of steroid production. The definitive enzyme, 5-ene-3β-hydroxysteroid dehydrogenase, is not readily demonstrated, but it may occur at certain stages of the lamprey life-cycle, for example during migration. There may be agreed pathways of vertebrate steroidogenesis in certain lamprey cells, but the theory has yet to be proven. Lampreys clearly have to osmoregulate and may well rely on nonsteroidal mechanisms as, for example, in Amphioxus—still to be identified.

Myxinidae may be related to the *Petromyzonidae,* but differ markedly in that the serum has a similar ionic composition to that of the environmental seawater. Again, the presumptive interrenal tissue may lie in the region of the pronephros, but we are here on even weaker ground than in the discussion about the lampreys. There are no certain results.

7. General Observations

The morphology of the adrenal and interrenal glands is inextricably bound up with the embryological formation of the kidneys and gonads and with the fate of neural crest cells. Amphibia have a mesonephros as the kidney of the adult. As this comes from the same block of mesoderm as the gonads and interrenal, then they are intimately associated. Thus the interrenal intermingles with the kidney. In mammals, birds, and reptiles, another kidney, the metanephros, appears. This permanent kidney of the amniotes arises far caudad in the body. The ureter, renal pelvis calyces, and collecting tubules are all derived from a bud growing off the mesonephric duct. On the other hand, the secretory tubules and Bowman's capsules differentiate from the caudal end of the nephrogenic cord and thus have an origin similar to that of the entire mesonephric tubules, but at a lower level. Yet the common mesodermal tissue still gives rise also to the adrenal cortex.

Thus, a separate adrenal gland is seen, with the chromaffin cells intermingling with the cortical cells until in the Eutheria they coalesce to form the medulla. The reasons for this association are not known—either the varying degrees of association or the ultimate appearance as a medulla. In the dominant group of jawed fish, the teleosts, the interrenal is certainly associated with the mesonephric kidney but appears anteriorly as the "head kidney" near the cardinal veins. A scattering of chromaffin cells is seen. On the other hand, in the cartilaginous fish, the elasmobranchs, the interrenal lies posteriorly and chromaffin tissue is completely separate. These are unsolved puzzles.

However disparate the gross anatomy of the adrenal/interrenal gland presents itself, a unifying concept is the basic morphology. The cells are arranged in cords. This can be seen, for example, in the teleost interrenal. This is very evident in the Tetrapoda, and it seems that a cord consists of a double strand (Figure 18). The

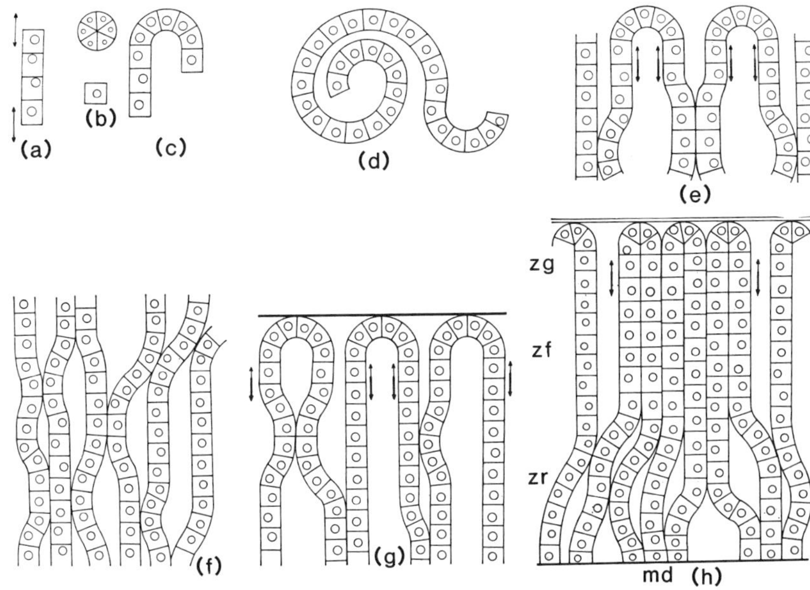

Figure 18. The organization of the basic cord of adrenocortical cells throughout the vertebrates. (a) The basic unit: a cord of cells capable of growth in each direction. (b) Transverse sections of a showing that the cord can be either a single line of cells or grouped to give a radial pattern. (c) Simple looping of the basic cord. (d) More complex looping (as in teleosts). (e) Illustrating the capacity to form loops, all parts of which are not equally capable of growth: it is easier for growth to take place in the direction away from the loop. (f) Organization of cords of cells in longitudinal runs (as in amphibians and reptiles). (g) Looping of cords peripherally against a well-formed outer connective tissue capsule (c) (as in birds) when the consideration given in e applies. (h) Looping of cords peripherally against a well-formed capsule (c) (when e again applies) and also the termination of the cords against a central medulla (md) (with necessary alteration of growth pattern) as in Metatheria and Eutheria. zg, zona glomerulosa; zf, zona fasciculata; zr, zona reticularis.

cords are haphazardly arranged in Amphibia, reptiles, and birds, but set in a positive pattern in eutherian mammals, surrounding the medulla. Here we run into difficulties of interpretation. It is clear that the adrenal cortex of the eutherian mammal comprises double strands. Yet we have the following facts to fit into any general theory:

1. In the absence of the anterior lobe of the pituitary, zonae fasciculata, and reticularis degenerate, the zona glomerulosa remains and persists actively to produce aldosterone, the major mineralocorticoid; the animal has sodium and potassium plasma levels within the normal range. Water loads are not readily excreted.
2. Appropriate administration of ACTH to the hypophysectomized animal (especially documented, the rat, mouse) leads to the reappearance of the zonae fasciculata and reticularis, and cortisol (or corticosterone) is again secreted. Water loads are readily excreted.

3. Stripping the adrenals *in situ* (called enucleation), leaving only the outer connective tissue capsule and few adherent adrenocortical cells, is followed in due course by the regrowth of the adrenal cortex, cast into zones, though without the medulla.

The zona glomerulosa responds to electrolyte changes, the angiotensins, serotonin, but can secrete aldosterone without the influence of ACTH. The zona fasciculata and zona reticularis depend on ACTH. If we accept that the basic structure is a cord, then the outer part, forming the zona glomerulosa in Eutheria, has an 18-hydroxylase, excluding 17-hydroxylation, to give the mineralocorticoid aldosterone, and the zona fasciculata in cortisol secretors, a 17-hydroxylase. There are two theories about the eutherian adrenal cortex, which can in some ways be used in examination of the lower vertebrates. The ''cell migration'' theory offers the view that there is cell division around the periphery of the gland. The new supply of cells migrates slowly centripetally, eventually to reach the zona reticularis, where degeneration occurs. There is evidence for this idea, in that most cell divisions occur peripherally and that the zona reticularis, in some species, is an area of collapse, of necrotic cells with lipofuchsin granules, a known marker of aging tissues. On the other hand, the zona glomerulosa is the primary source of the mineralocorticoid, the 18-hydroxylated aldosterone, together with corticosterone (see Chapter 1). The cells of the zona fasciculata must change to suppress aldosterone and chiefly manufacture corticosterone (e.g., rodents) or, in other Eutheria, predominantly cortisol, and become dependent on ACTH (Part III). This switch is not explained, and it would be helpful for the theory of the basic cord and for ''cell migration'' if it were so. Hence the ''zonal'' theory gained favor. Each zone is regarded as a separate entity, sustained by cell divisions within each zone. This means that two types of cell exist, aldosterone secretors and corticosterone/cortisol producers. The zona reticularis presents difficulties in encompassing it within this theory. Histologically it is not obvious in some species; in others it is degenerative; in some it may secrete sex hormones. The question is still open. This also applies to birds and reptiles, where there is some evidence of zonation and in which there may be two types of cell, one for aldosterone, the other for corticosterone.

8. Selected Readings

Chester-Jones, I., and Mosley, W., 1980, The interrenal gland in Pisces, in: *General, Comparative and Clinical Endocrinology of the Adrenal Cortex,* Vol. 3 (I. Chester-Jones and I. W. Henderson, eds.), Academic Press, London, New York, San Francisco, Chapter 7, pp. 395–472.

Handbook of Physiology. Section 7. Vol. VI. *Adrenal Gland,* American Physiological Society, Washington, D.C., 1975.

Hanke, W., 1978, The adrenal cortex (interrenal) of Amphibia, in: *General, Comparative and Clinical Endocrinology,* Vol. 2 (I. Chester-Jones and I. W. Henderson, eds.), Academic Press, London, New York, San Francisco, Chapter 5, pp. 419–495.

Holmes, W. N., and Phillips, J. G., 1976, The adrenal cortex of birds, in: *General, Comparative and*

Clinical Endocrinology of the Adrenal Cortex, Vol. 1 (I. Chester-Jones and I. W. Henderson, eds.), Academic Press, London, New York, San Francisco, Chapter 5, pp. 293–420.

Idelman, S., 1978, The structure of the mammalian adrenal cortex, in: *General, Comparative and Clinical Endocrinology of the Adrenal Cortex,* Vol. 2 (I. Chester-Jones and I. W. Henderson, eds.), Academic Press, London, New York, San Francisco, Chapter 1, pp. 1–199.

Lentz, T. L., 1971, *Cell Fine Structure,* W. B. Saunders Company, Philadelphia, London, Toronto.

Lofts, B., 1978, The adrenal gland in Reptilia, in: *General, Comparative and Clinical Endocrinology of the Adrenal Cortex,* Vol. 2 (I. Chester-Jones and I. W. Henderson, eds.), Academic Press, London, New York, San Francisco, Chapter 5, pp. 419–495.

Part II

Reproduction in Nonmammalian Vertebrates

4

Avian Reproduction

S. Harvey, C. G. Scanes, and J. G. Phillips

1. Introduction

Birds normally reproduce sexually by the fusion of male and female gametes—spermatogonia and oogonia—produced by differentiated testes and ovaries. Fertilization is internal and all birds are oviparous. Eggs are laid at the gastrula stage, and almost all embryonic development occurs outside the female's body. The egg is composed of layers that nourish or protect the embryo and are usually laid so that the young hatch when seasonally variable conditions are most favorable. The assumption of parental duties by the adults enables embryonic development to proceed optimally and contributes to the survival of the neonates. Parental investment is a primary breeding strategy of birds, to maximize the success of limited matings. These are usually only between socially bonded partners, restricted seasonally.

Avian reproduction is characteristically cyclical, usually seasonal. Gonadal activity is never continuous. Gonadal development occurs in response to stimulatory environmental cues and in turn promotes courtship and mating behavior. After breeding, gonadal regression and quiescence occur, facilitating the expression of parental behavior. The synchronization of these physiological and behavioral activities is achieved by complex neuroendocrine pathways and is largely controlled by gonadotropic and gonadal hormones. The physiological and behavioral aspects of reproduction are considered in this chapter, but ecological aspects should be considered by further reading (see Selected Readings).

S. Harvey and J. G. Phillips • Wolfson Institute, University of Hull, Hull HU6 7RK, England. *Present address for S. H.:* Department of Physiology, University of Alberta, Edmonton, Alberta, Canada T6G 2H7. **C. G. Scanes** • Department of Animal Science, Rutgers University, New Brunswick, New Jersey 08903.

2. Maturation and Development

2.1. Endocrine Control of Reproduction

Reproductive activity is controlled by the brain in response to neural information about the external stimuli: physical—daylength, temperature, food; psychological—social status, interactions. Such information is integrated into endogenous biological clocks to give circadian and circannual rhythms. Reflection in gonadal steroid activity modifies the output of gonadotropin-releasing factor (GnRH) from hypothalamic neurons. GnRH enters the hypophyseal portal circulation and regulates the synthesis and release of pituitary gonadotropins, follicle-stimulating hormone (FSH), and luteinizing hormone (LH), which in turn stimulate gonadal growth and steroidogenesis (Figure 1).

Activity of GnRH neurons is controlled by the synaptic transmission of various neurotransmitters from the central nervous system. In particular, GnRH release appears to be stimulated by adrenergic (A) or noradrenergic (NA) pathways, gonadotropin release being stimulated by increased NA or A turnover rates, but reduced by drugs being stimulated by increased NA or A turnover rates, but reduced by drugs that block NA transmission. Dopaminergic (DA) neurons may inhibit GnRH release, since DA administration suppresses gonadotropin secretion and hypothalamic DA turnover is decreased when NA or A turnover is enhanced. These neuronal pathways may, or may not, be sensitive to the feedback effects of gonadal steroids on GnRH release. While gonadal steroids accumulate in hypothalamic

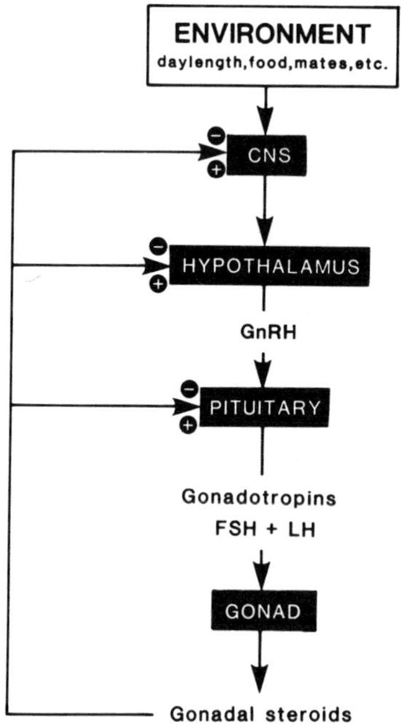

Figure 1. Schematic representation of the hypothalamo–pituitary–gonadal axis in birds. GnRH, gonadotropin-releasing hormone.

tissue and have both negative and positive effects on GnRH release, some GnRH stimulation, "hypothalamic drive," occurs via steroid-independent pathways and is thought to regulate the initiation and termination of breeding.

The dynamic equilibrium between gonadotropin secretion and gonadal steroid production is demonstrated by the stimulatory effect of castration on LH and FSH release and by the inhibitory effects of exogenous steroids. Basal gonadotropin release is tonically suppressed in females by estrogen and, to a lesser extent, by progesterone and in males by 5α-dihydrotestosterone (5αDHT) and testosterone, which can be aromatized to estrogen within the brain or converted to androstenedione, but both are also inhibitory. Steroids exert their inhibitory effects on gonadotropin release by directly inhibiting the release of GnRH, by decreasing pituitary responsiveness to GnRH stimulation, and by suppressing basal pituitary LH release. However, despite these inhibitory mechanisms gonadal steroids may conversely regulate gonadotropin release by a stimulatory feedback system. Progesterone, 5αDHT, structurally related steroids (e.g., deoxycorticosterone), and testosterone, to a lesser extent, can trigger LH release. Progesterone acts at hypothalamic sites or at extrahypothalamic control centers to stimulate GnRH release and at the pituitary gland to potentiate GnRH-stimulated LH release. Estradiol, androstendione, and 5β-dihydroprogesterone have no stimulatory effect on gonadotropin secretion, although at low concentrations estradiol sensitizes the pituitary to GnRH stimulation and may contribute to the formation of progesterone receptors in both pituitary and hypothalamic tissue. The expression of this pathway only occurs in sexually mature birds, in which it facilitates gonadal recrudescence and is responsible for the preovulatory release of LH, in laying hens.

In addition to hormonal ("long-loop") feedback from the gonads, the control of hypothalamopituitary function may also be regulated by "short-loop" feedback, from gonadotropins, and by an "ultra-short-loop" from GnRH. However, these pathways have yet to be established in birds (see Part III).

The control of the hypothalamo–pituitary–gonadal axis may also be modified by the actions of other hormones, particularly by prolactin, thyroid hormones, and adrenal corticoids. The adrenal hormones may be involved in the entrainment of rhythmical hormone secretion and have been shown to have both stimulatory and inhibitory effects on gonadotropin secretion. Reproductive function is strongly influenced by thyroid hormones, the circulating concentrations of which reflect changes in environmental conditions such as food availability, temperature, and photoperiod. The thyroid hormones can promote gonadal development, although at high doses they suppress reproductive activity. In some species gonadal regression at the end of the breeding season may be directly related to seasonal changes in thyroid function, since it is prevented by surgical removal of the thyroid gland. Prolactin also affects gonadotropin secretion, having pro- and antigonadal effects; the increase in prolactin secretion coincident with the onset of incubation is specifically implicated in gonadal regression and the maintenance of ovarian quiescence. Thus, although gonadotropin secretion declines in many species at the end of the breeding season because the hypothalamo–pituitary–gonadal axis becomes refractory to stimulatory stimuli (long daylengths), other interacting changes in the concentrations of corticosterone, thyroxine (T_4), triiodothyronine (T_3), and prolactin may contribute to the termination of breeding.

2.2. Sexual Development

Genitalia, genital function, and sexual behavior are largely determined prenatally or perinatally. Somatic growth, interacting with environmental factors, results in puberty, sexual maturation, and reproductive activity, the participation in which declines with aging.

Sex determination is dependent on chromosome constitution and differs from that in mammals in that the male sex is homogametic (ZZ), while females are heterogametic (ZW) and bearers of the H-Y antigen, which in birds is responsible for ovarian differentiation. Since the expression of the H-Y antigen is enhanced by estrogens, male genotypes may be feminized embryologically following excess exposure to estrogen secreted from embryonic ovaries or deposited maternally in the yolk. Thus, when the embryos of disparate sex develop in double-yolked eggs, the gonad of the male "twin" is feminized via estrogens secreted into the anastomosed allantoic blood vessels from the female "twin," which (unlike bovine freemartins) is only slightly masculinized.

During embryological development gonadal differentiation occurs autonomously, after the colonization of the gonadal sites by the primordial germ cells (PGCs), which subsequently induce the production of oogonia in females and spermatogonia in the males (see Part I). The localization of the PGCs within the gonadal site is regulated by the expression of the H-Y antigen or is caused by properties (e.g., chemical attractants) of the somatic tissue. The development of the presumptive gonad into ovarian or testicular tissue also appears to result from information contained within the somatic tissue or carried by the PGC chromosomes. The development and future size of the gonads depends on the number and distribution of the PGCs within the undifferentiated gonad, which has a common appearance in genetic males and genetic females. The expression of the male phenotype occurs when the PGCs invade the gonadal medulla, whereas the female phenotype occurs when the PGCs settle within the cortical tissue. In both sexes the PGCs tend to localize in the left gonad, resulting in the asymmetrical gonadal development of birds. The right testis is usually smaller than the left testis, and the right ovary is normally vestigial in adult hens, except in some raptors and rare instances in ducks and chickens. If the left ovary becomes diseased or is surgically removed, the right ovary can develop into an ovary, an ovotestis, or a testislike structure, producing androgens and spermatozoa. Egg laying may persist if sufficient ovarian function remains, but the development of testicular elements in the female can induce male plumage and masculine behavior. Sex reversal may conversely result from the development of a left ovotestis in the male, although this condition is much rarer and does not result in plumage changes in the adults.

Gonadal differentiation (and oogenesis) is complete before hatching, as is the differentiation of the gonoduct system. Prior to gonadal differentiation the embryonic gonads possess both Wolffian and Müllerian duct systems, through which the male and female gametes are subsequently passed. While the Wolffian ducts persist in both sexes, independently of gonadal hormones, both Müllerian ducts in the males and the right Müllerian duct in the female regress in response to gonadal hormones and inhibitory factors. Male and female birds are thus hatched with their gonads and gonoducts fully differentiated, but not fully developed.

The reproductive systems of the adult male and female birds (described in Sections 3 and 4) are acquired after a period of somatic growth, since the bird must be sufficiently well grown to assume parental responsibilities and to lay eggs large enough to produce hardy nestlings. Once somatic maturity has been achieved, many photoreceptive species (e.g., chickens, ducks) then attain sexual maturity, even when exposed to nonstimulatory environmental conditions (e.g., short days or constant darkness). The onset of sexual maturity is, however, usually triggered by environmental conditions that stimulate the hypothalamo–pituitary–gonadal axis, provided the axis is fully developed and responsive. The juveniles of some species (e.g., white-crowned sparrows) fail to become sexually mature in the year in which they hatch, despite the attainment of somatic maturity and exposure to long-day conditions, because their neuroendocrine pathways are "refractory" to environmental stimulation (see Section 5) until "juvenile refractoriness" is broken by exposure to short-day conditions.

The development of the hypothalamo–pituitary–gonadal axis depends on the maturation of the neuroendocrine pathway's control of GnRH release in response to daylength or steroid feedback. Brain norepinephrine (NE) or serotonin (5-HT), the neurotransmitters involved, are detectable at hatch but their concentration gradually increases toward puberty. In many species the axis is sensitive to light and to the inhibitory effects of gonadal steroids long before somatic maturity, although the positive steroidal feedback pathway only becomes functional when body growth nears completion and gonadal development has been initiated. Unlike mammalian species the onset of puberty appears to be age- rather than size-related, in that some gonadal development can be prematurely induced, but complete development, such as the production of the largest yellow, yolky ovarian follicles, cannot. The onset of puberty is thought to be triggered by a suppression of the negative feedback pathway controlling gonadotropin release, facilitating increased FSH and LH release (Figures 2 and 3) in response to environmental stimuli (e.g., light). As the gonads enlarge, sex steroid levels increase in peripheral circulation and begin to exert an inhibitory effect on gonadotropin secretion. The concentrations of LH and FSH subsequently decline before the completion of gonadal growth, although remain higher than in juveniles. The FSH concentration in males declines further and at a faster rate than LH, possibly because of the release of "inhibin" from testicular Sertoli cells. The decline in the gonadotropin level in female birds is often greater than that in males, since estrogens have greater inhibitory potency than androgens. The rise in the gonadal steroid levels sitmulates the development of accessory and secondary sexual characters, gamete formation, and sexual behavior. Reproductive

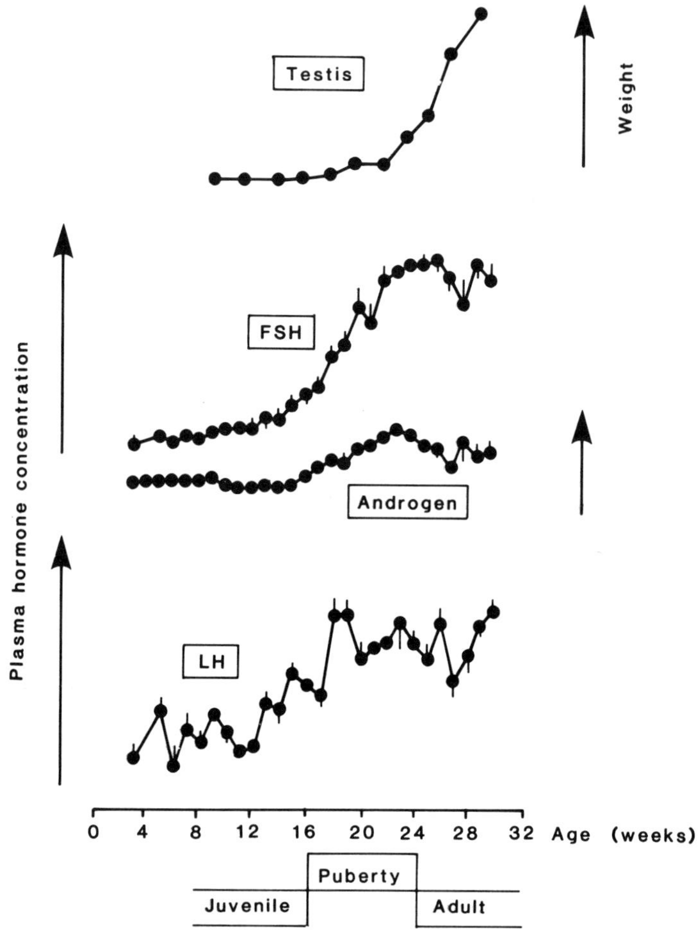

Figure 2. Maturational changes in testicular weight and the plasma concentrations of androgens, luteinizing hormone (LH), and follicle-stimulating hormone (FSH) in domestic fowl. (Based on Sharp, P. J., *et al.*, 1977, Variations in stored and plasma concentrations of androgens and luteinizing hormone during sexual development in the cockerel. *J. Endocrinol.* **74:**467–476, and Sterling, R. J., *et al.*, 1984, Changes in concentrations of plasma luteinizing hormone, follicle-stimulating hormone, androgen, growth hormone, prolactin, thyroxine and triiodothyronine during growth and sexual development in the cockerel. *Br. Poult. Sci.* **25:**353–360.)

behaviors, such as courtship displays, vocalization, and mating, are hormone-dependent and can be induced in juveniles in response to exogenous gonadal steroids.

The onset of sexual behavior normally occurs before the attainment of the circulating gonadal steroid levels of adults. The component sexual activities have different thresholds for elicitation and are sequentially induced by rising steroidal levels; crowing and courtship, for instance, are elicited before mating in response to

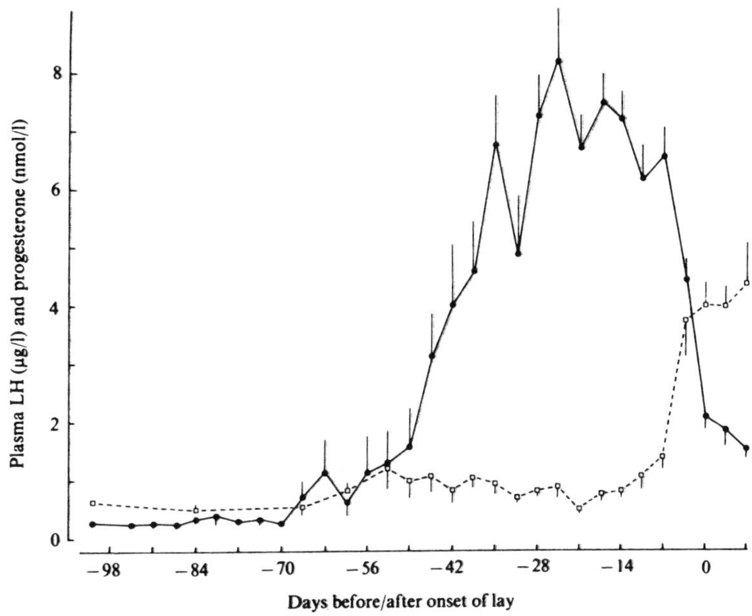

Figure 3. Plasma concentrations of luteinizing hormone (LH, ●) and progesterone (□) in relation to the onset of lay in ducks maintained on a photoperiod of 16 hr light:8 hr darkness. (From Wilson, S. C., and Morris, T. R., 1982, Concentrations of luteinizing hormone and progesterone in plasma during sexual development of the khaki Campbell duck. *J. Endocrinol.* **93**:47–53.)

lower testosterone concentrations. These behavioral effects of testosterone are potentiated at puberty by the neonatal decline in brain 5β-reductase activity, responsible for the production of inactive testosterone metabolites. Although testosterone is produced in both males and females, in which the circulating testosterone level may be similar, male behavior is not observed in females because of dimorphic differences in the intracellular metabolism of testosterone. Males and females do not have morphological differences in the areas of the brain controlling reproductive behavior, but the neural pathways involved and their sensitivity to exogenous steroids may be sexually related.

The behavioral differences between males and females are accentuated at puberty, although much of the sex-related behavior occurs prenatally or perinatally. Embryonic exposure to estrogens not only feminizes male embryos morphologically, it also feminizes the brain, resulting in the subsequent expression of female behavior. The pharmacological blockade of estrogen action prevents the feminization of the embryonic brain. Whereas normal females rarely show male (copulatory) behavior, birds treated embryonically with antiestrogens will do so in response to exogenous testosterone. Males, which are the "neutral" sex in birds, are conversely feminized by embryonic exposure to estrogens or, as a result of aromatization, to testosterone. Testosterone by itself does not induce masculine behavior in female birds. Sexual differentiation in some species (e.g., galliforms) thus appears to proceed by the loss of behavioral patterns, the embryonic exposure to endogenous estrogens resulting in the demasculinization of female birds. Neonatal exposure to gonadal steroids appears to promote the acquisition or reinforcement of appropriate behavioral patterns.

Reproductive behavior and the corresponding responses normally appear in individuals on the first occasion that the relevant circumstances arise. The onset of breeding occurs in most species when they are just under 12 months old; few species mature in the year in which they hatch. Some birds (e.g., quail) can begin breeding when only six weeks old, whereas larger birds (eagles and albatrosses) may not breed until they are 9–12 years of age, the age of first breeding partly related to growth rate. In some species breeding does not coincide with the attainment of somatic maturity because ecological conditions (such as lack of nest sites) defer their participation in reproductive activity. Prolonged immaturity is an adaptation arising out of the difficulties inherent in breeding, which make young, inexperienced birds unsuccessful and may even endanger their own survival. Deferred maturity tends to occur in species (e.g., swans) that have low clutch size, extended postfledgling parental care, high adult survival, and prolonged reproductive spans.

As birds approach the completion of their reproductive span, aging results in the laying of fewer, less fertile eggs by the females, and in suppression of mating frequency and sperm production in the males. This physiological and behavioral decline in reproductive activity does not necessarily result from reduced gonadal steroid secretion. Aging does not reduce pituitary responsiveness to GnRH, although it may suppress gonadotropin secretion by reducing the responsiveness of the hypothalamopituitary axis to positive gonadal steroid feedback. This decline in reproductive function probably results from a loss of gonadal steroid receptors at target sites and because of an increase in the threshold required for the induction of hormone action. A decrease in the 5α-reductase pathway is probably also causally involved in the suppression of reproductive activity in aging males.

3. Male Reproductive System

3.1. Structure and Function of Testes

The embryology of the bird testis is well established. During ontogeny primordial germ cells migrate from the splanchnopleure to form germinal epithelium and become associated with Sertoli cells, which are embedded in the stromal connective tissue of intermediate mesodermal origin. The proliferation of the coelomic epithelium surrounding these elements form convoluted tubules and these seminiferous tubules become closely associated with interstitial or Leydig cells, derived from fibroblast cells of the coelomic epithelium. The tubules are not divided into distinct lobules. The Sertoli and Leydig cells together form the endocrine portion of the

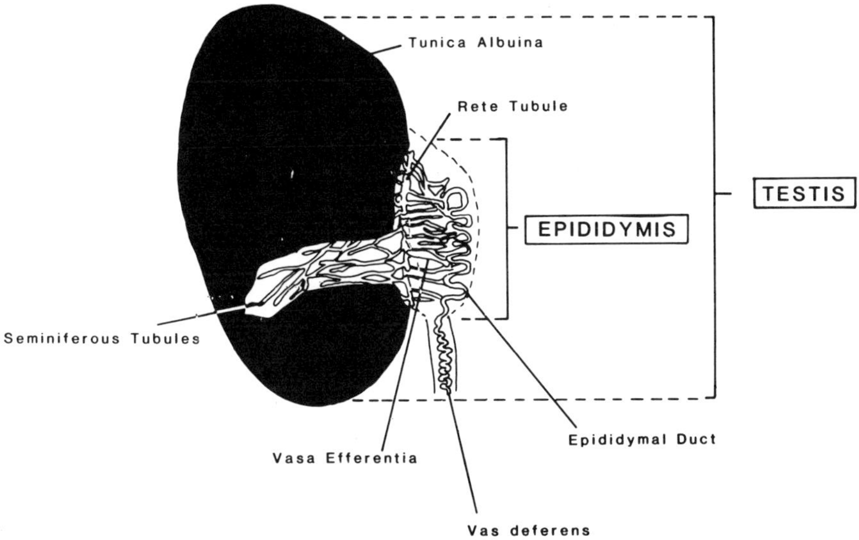

Figure 4. Diagrammatic representation of the chicken testis, showing internal structure. (Based on Sturkie, P. D., and Opel, H., 1976, *Avian Physiology,* 3rd ed., Springer-Verlag, New York, p. 331.)

testis and are responsive to gonadotropins from the pituitary gland. In most birds the testicular tissue, including accessory sex structures (e.g., vasa deferentia, vasa efferentia epididymes, and rete tubules), undergoes marked fluctuations in size and histological appearance from an inactive condition in the nonbreeding season to an expanded and fully functional testis in the breeding season (Figure 4). These changes reflect the cyclical release of gonadotropins, particularly FSH.

The paired testes are round to ovoid, situated in the abdominal cavity of the body just anterior to the kidney but ventral to it; they receive spermatic arteries from the dorsal aorta, and the venous drainage back to the vena cava is via the testicular veins. The surface of the testis is a thick, fibrous sheath, the tunica albuginea, which is bounded by a fragile coat of tissue called the tunica vaginalis. Because of the dramatic fluctuations in size of the testis, depending on season, the capsule of the testis is likely to be stressed by the changing dimensions, and lesions occur on the surface; these are repaired by the invasion of fibroblasts forming a new coat.

3.2. Spermatogenesis

The endocrine control of spermatogenesis may be similar to that in mammals, in that FSH may cause Sertoli cell hypertrophy and together with androgens cause germ cell development. The germ cells produce spermatogonia, which during breeding divide mitotically several times forming another generation of spermatogonia or primary spermatocytes (Figure 5). The meiotic divisions follow pro-

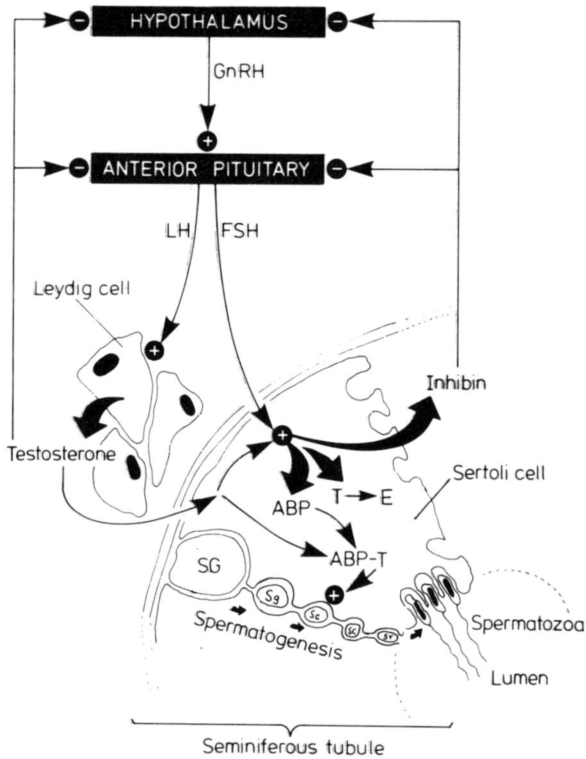

Figure 5. Diagram showing the functional dichotomy of the avian testis. Leydig cells, the target for LH, are sites of testosterone (T) biosynthesis within the seminiferous tubules. Sertoli cells support the germ cells. The stages of spermatogenesis are indicated, starting peripherally with spermatogonium (SG), spermatogonia (Sg), primary spermatocytes (SC), secondary spermatocytes (Sc), and spermatids (ST). Testosterone enters the Sertoli cells, combines with androgen-binding protein (ABP, which is FSH dependent), and secures a local concentration of androgen close to the spermatocytes to effect subsequent meiotic divisions. Testosterone may also be converted under FSH influence to estrogens within the Sertoli cells. Inhibin may be produced in the Sertoli cells and effect a negative feedback pathway to modulate pituitary FSH release. (Based on Goldsworthy, G. J., *et al.*, 1981, *Endocrinology,* Blackie, Glasgow.)

ducing, in succession, secondary spermatocytes and spermatids; differentiation of spermatids to mature spermatozoa then follows. The completion of spermatogenesis is achieved over a 2- to 3-week period, and the mature testis contains seminiferous tubules lined by a multilayered germinal epithelium containing all stages of spermatozoan development.

3.3. Transport of Semen and Copulation

Sperm pass from the seminiferous tubules to rete tubules (Fig. 4) and then to the vasa efferentia. The spermatozoa leave the testis via a long coiled duct, the epididymis, and are carried in seminal fluid. Avian semen, a product of the seminiferous tubules, and an aqueous fluid (containing salts and amino acids) added by the epididymis form a viscous ejaculate that is relatively small in volume (ca. 0.1–1.0 ml per ejaculation) but contains 0.17–3.5 billion spermatozoa. The volume and spermatozoan contents of the ejaculate decrease with increased mating frequency, which may be more than 25 times an hour in some domestic fowl. During the breeding season sperm and the epididymal fluid pass down the vas deferens, which is often expanded into a seminal sac near the cloaca. In some passerines the expanded vas deferens projects from the surface of the body and resemble scrotal sacs. Sperm are not stored in the epididymis but in the vas deferens or seminal sacs. The posterior parts of the seminal sacs protrude into the cloaca as erectile papillae. During copulation the ejaculate is deposited either via a protrusible phallus (as in Anatidae and Rhatites) directly into the vagina, which is everted into the cloaca, or by the adiposition of the grooved cloacal openings, the so-called "cloacal kiss." Spermatozoa must be ripened or matured in the epididymis before they are capable of fertilization, a process that may take ca. 24 hr. Spermatozoa suspended in avian semen remain viable for long periods of time (over 1 month) in both the male and the female tract. Maximum sperm fertility usually occurs 2–3 days after copulation, after storage and possibly nourishment in the uterus, in glands that may provide a barrier to the passage of defective sperm. Following copulation the spermatozoa may reach the infundibulum, at the top of the uterus where fertilization occurs, within 30 min, although the spermatozoa are not as fertile if they do so.

3.4. Steroidogenesis

The reproductive hormones produced by the testis are mainly androgens, the most important of which is testosterone. The testes can produce female sex steroids, estrogens and progestagens, but they may not be released into the circulation. Testicular steroids originate from two sources, the Leydig cells and the Sertoli cells, both of which share ultrastructural features typical of other steroid-producing cells. The Sertoli cells probably have a greater steroidogenic capacity than the Leydig cells. Steroidogenesis is stimulated by LH, which has a potent effect on the Leydig cells via the activation of the enzyme adenylate cyclase, which in turn catalyzes the conversion of ATP to cyclic AMP. Cyclic AMP then serves as a second messenger, stimulating further protein synthesis, principally the enzymes responsible for the stepwise conversion to the final secretory products (testosterone, androstenedione, and 5αHT) of the precursors cholesterol and pregnenolone. Testicular stores of cholesterol-rich lipid accumulate before the onset of breeding but are depleted at the height of breeding and following breeding and are low in juveniles. The evidence for a similar role of LH on the Sertoli cells is less certain. Indeed, it is most likely that the somatic components of the seminiferous tubules, the Sertoli cells, are under the control of FSH and androgens of Leydig cell origin. These two hormones act on spermatogenesis via the Sertoli cells, but the precise sequence of events is unknown, even for mammals. It is likely, however, that the Sertoli cells themselves contribute to the intratubular buildup of androgens and that, in turn, androgens are converted to estrogens by a process of aromatization, stimulated by FSH. Details of the biosynthetic pathways involved in testicular steroidogenesis are described by Kime (Chapter 1).

In photoperiodic species the increase in daylength during spring increases testicular size and androgen secretion in parallel with increased gonadotropin release. Although testosterone or its 5α-reduced metabolites provide inhibitory feedback for gonadotropin release, gonadal growth may be maintained by the preferential metabolism of testosterone through a β-reductive system, which acts as an inactivating shunt, producing inactive metabolites. The circulating testosterone level usually peaks at or before the height of the breeding season, but the concentration is not maintained and declines before the onset of gonadal regression. In some species this decline occurs from an increase in the hormone's clearance rate and by a reduction in its secretion rate. It may, however, also result from the release of a glycoprotein, inhibin, from the Sertoli cells which inhibits hypothalamic GnRH activity and GnRH-induced secretion.

3.5. Secondary Sexual Characteristics

The function of the testis is to produce spermatozoa and to secrete hormones. These hormones are essential for the growth and maintenance of accessory reproductive organs and for the induction of secondary sexual characteristics. Sexual attraction and mating are facilitated by behavioral and morphological features which are usually specific for either sex, although the degree of sexual dimorphism and the importance of these characters vary widely between species.

The most obvious secondary sexual characteristics are usually for display purposes and involve alterations in the structure or color of the plumage or plumes, changes in bill color, and the acquisition of elaborate spurs, combs, or keratinized facial adornments. Vocalization and the quality of song are usually testosterone dependent, as are other sexual behaviors. Male reproductive behavior is induced by the rising levels of circulating androgens at the onset of sexual maturation. Increased testosterone secretion is particularly associated with aggressive behavior, expressed during territorial encounters with rivals or toward females at the onset of courtship behavior, and is responsible for mating behavior, especially libido. These androgen-dependent behaviors are further discussed in Section 6.

4. Female Reproductive System

4.1. Ovarian Structure and Function

In the adult bird, the ovary is a large organ containing the developing yolk-filled follicles, comprising a mass of very small follicles, a hierarchy of large developing ones, and those regressing after ovulation (Figure 6).

Figure 6. The ovary of the domestic hen. The mature ovary of a laying hen showing two populations of developing follicles. One consists of small follicles (less than 5 mm in diameter) each containing small amounts of "white yolk." The more spectacular group is composed of larger, rapidly growing follicles containing the characteristic yellow yolk on which the vascular system is particularly prominent. The pale, avascular band seen across the larger follicles is known as the stigma. This splits at ovulation allowing the oocyte containing the germinal disc and the yolk to be liberated. A germinal disc can be seen to the left of the stigma in the middle follicle on the right of the photograph. After ovulation the "postovulatory" follicles remain in the ovary as pale, flattened structures on the left-hand side of the ovary. (From Gilbert, A. B., 1980, ARC Poultry Research Centre Report, Clark, Constable Ltd., Edinburgh.)

The wall of the mature follicle (Figure 7) is composed of several cell layers: the theca externa (connective tissue), theca interna (steroid producing), and granulosa layer, which is involved in both hormone production and yolk deposition (see Part I). Within the granulosa layer is the vitelline membrane surrounding the ovum. Both the granulosa and thecal cells play a role in steroidogenesis (see Section 4.4). The very large preovulatory follicles contribute most to ovarian weight, although consisting of only relatively few cells. The small follicles can be divided into three groups: (1) microscopic, (2) white yolk-filled, and (3) yellow yolk-filled. During growth and maturation of the follicles the somatic elements and the ovum grow and differentiate synchronously and the total weight of the ovary can increase 100-fold or more. White yolk-filled follicles develop from microscopic follicles by deposition of simple lipids (e.g., triglycerides and cholesterol). Subsequently, true,

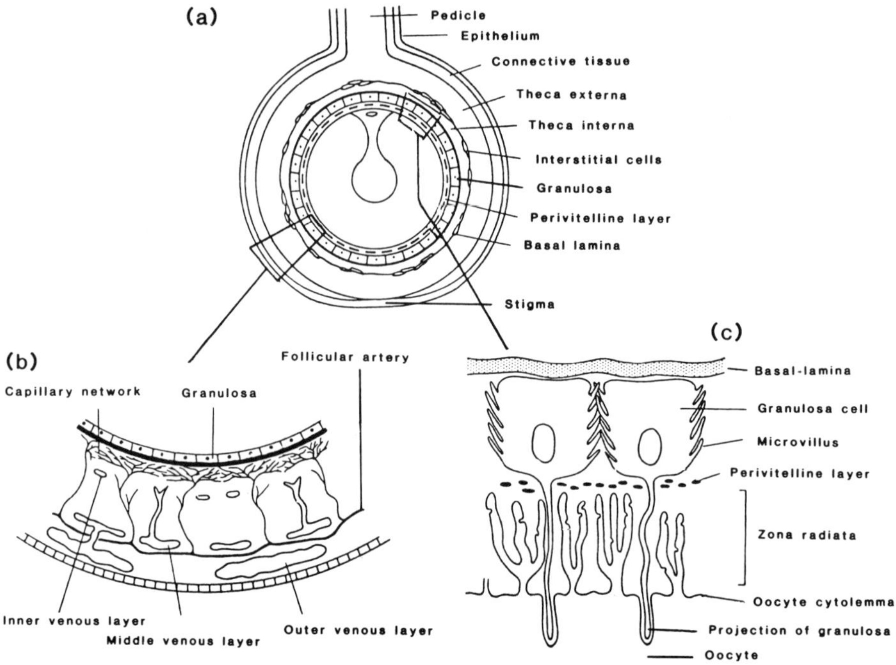

Figure 7. Structure of the follicle and follicle wall. (a) Basic structure of follicle and oocyte (not to scale), (b) vascular system of the follicle, (c) structure of the granulosa–oocyte region. [Adapted from Gilbert, A. B., 1979, in: *Form and Function in Birds,* Vol. I (A. S. King and J. McClelland, eds.), Academic Press, London, p. 237.]

yellow yolk accumulates by active transport of precursors from the blood; the granulosa cells assist in this purpose via their lipoprotein lipase enzyme systems and ion pumps. Yolk deposition is influenced by hormones, of which follicle-stimulating hormone plays a dominant role. The development of the hierarchy of follicles is rigidly controlled, so that there is an orderly progression with, in the chicken, one ready to ovulate each day. How this hierarchy of development is maintained is not fully understood; probably one of the factors is the supply of oxygen, nutrients, and hormones via the blood supply which is regulated by both adrenergic and cholinergic innervation. Hormones also play a role, because injections of pregnant mare serum gonadotropin cause a plethora of follicles to mature simultaneously. The final stage of follicular growth is linear (Figure 8).

4.2. Structure and Function of the Oviduct

In the embryo, there are two oviducts, but the right atrophies to leave the predominant left, which is held in place in the body cavity by ventral and dorsal ligaments. The oviduct enlarges with the onset of gonadal activity and can be divided on the basis of both anatomy and function into five distinct regions: infundibulum, magnum, isthmus, uterus (shell gland), and vagina (Figure 9, Table I).

During copulation, spermatozoa are deposited into the vagina, where they are generally stored, for up to 2 weeks, in specialized areas known as sperm "glands" or sperm host glands. Following oviposition, spermatozoa are released and transported up the oviduct to the infundibulum where the ovum from the next ovulation will be fertilized. The physiology of these processes is little understood. Sperm transport through the oviduct may involve their endogenous activity (due to the tails of spermatozoa) and oviductal activities which include both a ciliary component and muscular contractions. In some species sperm host glands are also located in the infundibulum.

The ovum is rapidly engulfed in ciliated fingerlike processes (fimbriae) and transported into the infundibulum (the internal opening of the oviduct) in which fertilization may occur during its short time there. Following fertilization the albumin (principally proteinaceous solids) begins to be deposited round the ovum, a process that continues for 2–3 hr in the magnum, the longest part of the oviduct (see Section 4.7). In the isthmus the membranes that surround the albumin are produced

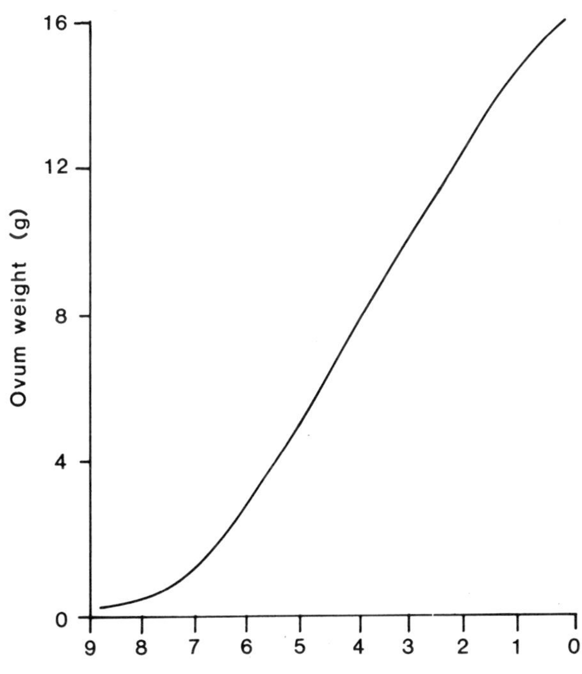

Figure 8. Rate of follicular development showing yolk deposition. (From Etches, R. J., 1984, in: *Reproductive Biology of Poultry* (F. J. Cunningham *et al.*, eds.), Brit. Poultry Sci. Ltd. Longman Group, Harlow.)

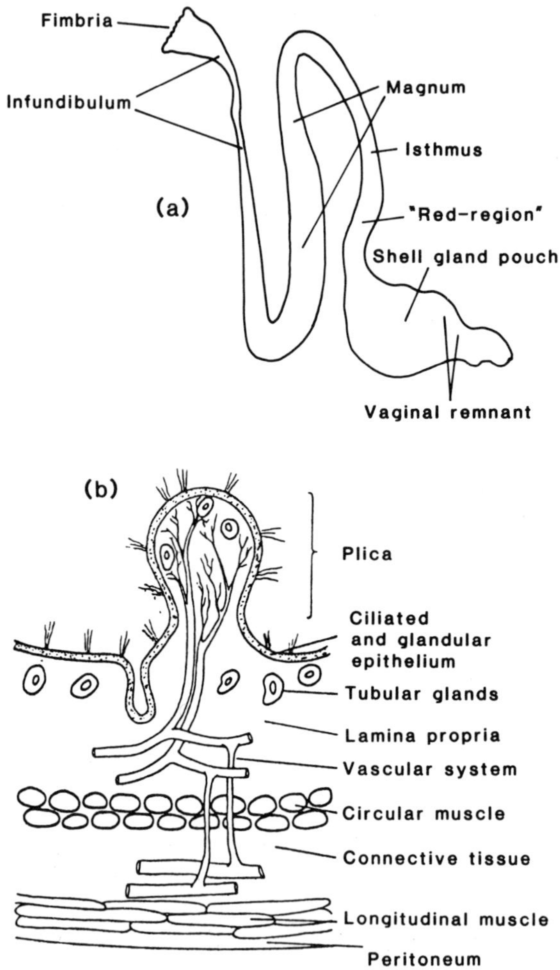

Figure 9. The structure of the oviduct in adult female birds. (a) Diagram of the oviduct in the domestic fowl, (b) structure of the oviduct wall. [Adapted from Gilbert, A. B., 1979, in: *Form and Function in Birds,* Vol. I (A. S. King and J. McClelland, eds.), Academic Press, London.]

over a 1- to 5-hr period, and deposition of the calcified shell is initiated. Finally, the fluids of the albumin layer (plumping fluid) are added in the uterus, in which the egg remains for 20 hr and in which shell formation is completed (Section 4.8). Here also the cuticle and pigments of the egg are added, and at this stage the egg is completely formed.

Movement of the ovum down the oviduct is carefully controlled by the muscles of the duct wall, and the smooth muscle of the uterus and vagina induce the release, i.e., laying, of the egg at oviposition (Section 4.9).

Table I
Transit Time of the Ovum in the Various Parts of the Oviduct of the Domestic Hen

Part of oviduct	Time (hr min)	Length (cm)	Mean velocity (cm/hr)
Infundibulum	18 min	9	30.0
Magnum	2 hr and 54 min	33	11.4
Isthmus	1 hr and 14 min	10	8.1
Uterus	20 hr and 46 min	10	0.48

4.3. Ovarian Hormones

The bird ovary produces three groups of hormones: (1) progestagens, (2) androgens, and (3) estrogens. Possibly a polypeptide, relaxin, is also produced, as it is in eutherian mammals, but there is no evidence of its involvement in avian physiology. The steroids are fully dealt with in Part I. Here we use the "trivial" names so that the chief ovarian progestagen is progesterone, the androgen is testosterone, and estrogen is estradiol-17β; others are given in Part I. The basic four-ringed structure of steroids is given very different activities by differences in side chains. All these ovarian sex steroids are required for reproduction in eutherian mammals; estrogens dominate one phase of the sexual cycle and progesterone another, luteal, phase. Pregnancy is absent from birds though present in representatives of all other Gnathostomes (see subsequent chapters). Nevertheless, the familiar ovarian hormones are needed for the production of a yolky egg surrounded by a layer of egg white, membranes, and shell (Figure 10). Synthesis of the yolk precursors, by the liver, is initiated and maintained by ovarian steroids, as is the production of egg white proteins and the massive growth and differentiation of the oviduct, an essential prerequisite for egg albumin synthesis. Alone, neither progesterone nor the androgens have a major effect on reproductive physiology of birds, but they act synergistically with estrogens to affect oviduct development and egg white production. Furthermore, some hormone actions must occur in sequence; for example, progesterone can only stimulate avidin production (an egg white constituent) after oviduct development has been stimulated by estrogens. Estrogens and androgens together play a major role in calcium homeostasis in laying birds. Production of the egg shell causes an enormous drain on calcium reserves, and it is the function of estrogens and androgens to facilitate calcium retention in bones, by provoking the formation of medullary bone and also enhancing the absorption of calcium via the small intestine (Section 4.8).

The sex steroids also affect reproductive behavior of birds directly by changing their appearance, e.g., comb development in female chickens, and by their activities via hypothalamic and pituitary peptide hormone secretion (see Section 6).

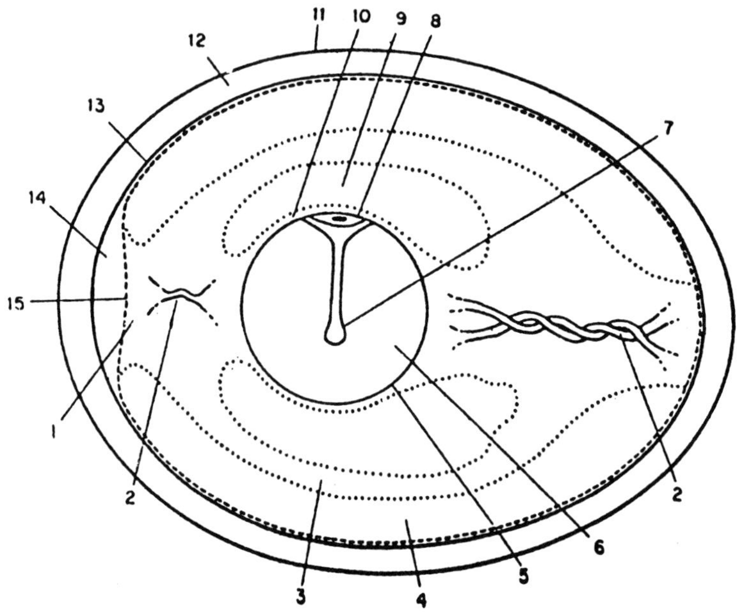

Figure 10. Structure of the egg. 1, "Ligament"; 2, chalazae; 3, middle thick white (albumen); 4, outer thin white (albumen); 5, yolk membranes; 6, yolk; 7, latebra; 8, germinal disc; 9, inner thin white (albumin); 10, chalaziferous layer; 11, cuticle; 12, calcified portion of shell; 13, outer shell membranes; 14, air space; 15, inner shell membranes. It should be noted that the same scale is not used for all structures. [Adapted from Gilbert, A. B., 1979, in: *Form and Function in Birds,* Vol. I (A. S. King and J. McClelland, eds.), Academic Press, London.]

4.4. Ovarian Steroidogenesis

Steroid hormones are produced in specific locations and cell types within the ovary (Figure 11). Steroids are synthesized from cholesterol precursors by hormones that are activated by gonadotropin (particularly LH) stimulation. The developing (preovulatory) ovarian follicles appear to be the major sites of steroidogenesis. As the yellow, yolky follicle matures, it secretes increasing quantities of estrogen, although its estrogen content falls when it becomes the second largest follicle, at a time when progesterone and testosterone synthesis increase. The mature preovulatory follicle (the F_1 follicle) has the greatest steroidogenic activity and almost exclusively produces progesterone. Progesterone is produced by the granulosa cells, the steroidogenic capacity of which increases with follicular development. Progesterone is converted to androgens and subsequently to estrogens by the theca cells, particularly those in less well-developed follicles (e.g., F_3 follicles), in which some *de novo* progesterone synthesis occurs. In the presence of androgens the granulosa layer also has some, albeit low, capacity to synthesize estrogens. The biosynthetic pathways involved during ovarian steroidogenesis are detailed by Kime (Chapter 1).

4.5. Ovulation

Ovulation is the release of the ovum following rupture of the follicular wall at the stigma, though the mechanism is not fully understood. The stigma is an area of the follicle wall where the stratum granulosum, theca, and tunica albuginea become progressively thinner from about 30 min before ovulation. There is no evidence that ''bursting'' is due to an increase in follicular hydrostatic pressure, but may be due to several factors including hemostasis, proteolysis, cell death, and reduced tissue cohesion due to collagenase action. Although epinephrine and norepinephrine can provoke ovulation *in vitro,* a physiological role for catecholamines (from either adrenal chromaffin tissue or nerve terminals in the follicle wall) in the control of ovulation has yet to be established and may merely be a pharmacological observation.

The ovulation of a mature follicle from the ovary occurs in response to LH stimulation. Prior to ovulation (4–6 hr in most species studied) a preovulatory surge of LH, but not FSH, occurs (Figure 12). The magnitude of this increase is only two- to threefold, whereas the increase in the LH concentration is frequently 100-fold in female eutherian mammals. The stimulus for the LH surge is probably progesterone, stimulating GnRH and LH release from the hypothalamopituitary axis. This stimulation occurs via a positive feedback system, since it follows the release of estrogen from the developing follicle, which may occur in response to a small, nocturnal increase in the circulating LH concentration. Estrogen primes the neuroendocrine tissues to provocative stimulation and interacts synergistically with progesterone in the induction of LH release.

The timing of ovulation is strictly controlled and normally occurs in the domestic fowl approximately 30 min following oviposition of the previous egg, although the two events are not causally linked. The timing of the preovulatory release of LH

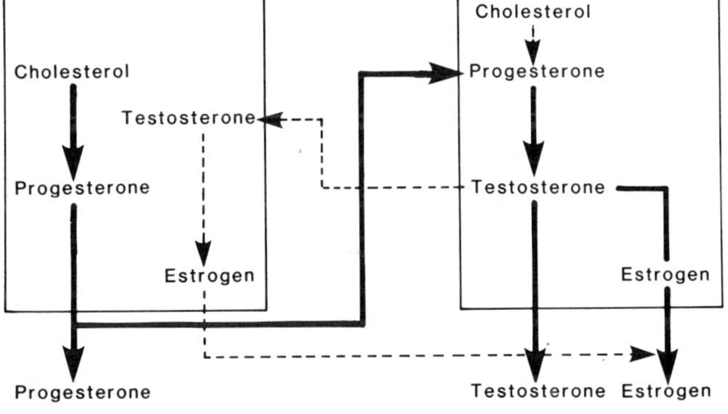

Figure 11. Two-cell theory of ovarian hormone production. The solid line indicates the principal route; the dotted line indicates minor routes.

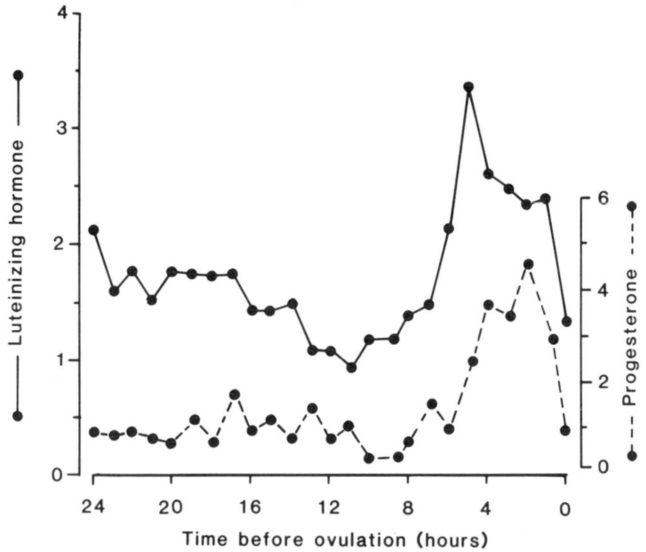

Figure 12. Plasma concentrations of luteinizing hormone and progesterone during the ovulation cycle of the domestic hen. [From Scanes, C. G., *et al.*, 1983, in: *Physiology and Biochemistry of the Domestic Fowl*, Vol. 5 (B. M. Freeman, ed.), Academic Press, London, p. 39.]

involves the circadian system, but the precise mechanisms involved are poorly understood.

In domestic chickens and other birds that produce sequences of eggs, ovulation occurs every day at progressively later times, until a clutch of eggs is laid. Then there is a missed day (equivalent to the interclutch interval) followed by another sequence of ovulations, starting at a similar time to the previous initial ovulation. This ''ratchet'' timing system (Table II) could occur if the sensitivity of the hypothalamus/pituitary to positive progesterone feedback follows a daily (24-hr) rhythm and if the ovarian follicle produces progesterone (to trigger preovulatory LH surges)

Table II
Timing of Oviposition of the Domestic Fowl (Lights on 0600–2000)

Sequence length (days)	Hour of oviposition on day									
	1	2	3	4	5	6	7	8	9	10
2	—	10	14	—	10	15	—	10	14	—
3	—	9	11	15	—	9	12	16	—	9
4	—	9	10	13	16	—	8	10	11	15
5	—	8	10	12	13	16	—	9	10	11
6	—	8	10	11	12	14	16	—	8	9

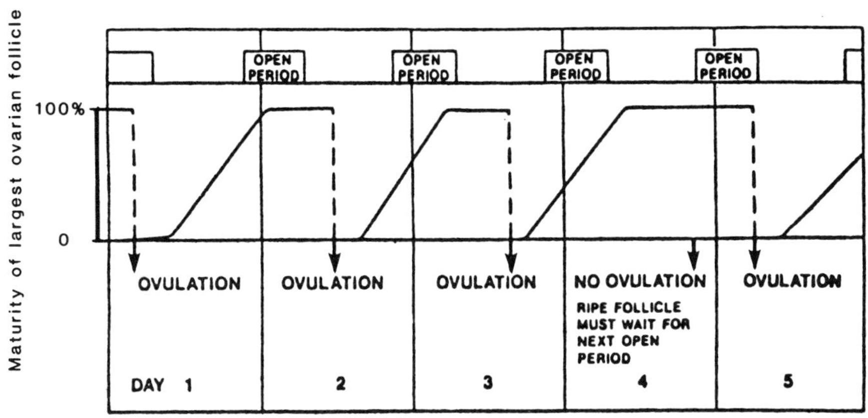

Figure 13. Sequential pattern of egg laying in domestic fowl entrained to a 24-hr lighting cycle. [From Sharp, P. J., 1983, in: *Progress in Non-mammalian Brain Research,* Vol. 3 (G. Nistico and L. Bolis, eds.), CRC Press, Boca Raton, Florida, p. 120.]

when the hypothalamus is sensitive, during a so-called "open" period. Since the final stage of follicular development requires more than 24 hr to complete, follicular maturation and increased progesterone release may not coincide with the open period, resulting in a missed day of ovulation (Figure 13). Pause days could alternatively occur if a small daily increase in LH release, capable of stimulating progesterone synthesis, occurs when the follicle is not sufficiently mature to respond, preventing the subsequent preovulatory LH surge.

4.6. Yolk

The knowledge of yolk—its chemistry structure, synthesis, and accumulation—is very largely limited to studies on a single avian species, the domestic fowl.

The yolk of the chicken egg, which weighs ca. 19 g, consists of water (48.7%) and solids (51.3%), mainly composed of lipids (32.6% of yolk weight), protein (16.6%), minerals (1.1%), and carbohydrate (1.0%). The principal minerals in the yolk are phosphorus (102 mg per yolk), potassium (64.5 mg), sodium (61.0 mg), calcium (27 mg), and magnesium (5.5 mg). Other minerals, together with vitamins, are present in trace amounts.

If yolk is diluted with water and subjected to ultracentrifugation, the solids become distributed among three phases. These are the precipitate fraction (the granules), water-soluble fractions (the livetins), and a "floating" (less dense than water) fraction (the low-density fraction). These represent respectively 13.0%, 5.3%, and 33.7% of the yolk weight. The granular fraction contains two major constituents—lipovitellin (8.3% of the yolk weight) and phosvitin (2.1%). Lipovitellin is a phospholipoprotein (27% lipid, 73% protein). On the other hand, phosvitin, which is also a phosphoprotein, contains 11% carbohydrate and this represents most of the carbohydrates of the yolk. Phosvitin, which is often complexed with calcium, is a unique protein which has a very high serine content (56% of phosvitin weight) with the serine residues phosphorylated (phosphorus making up 10% of phosvitin weight). The livetins, the water-soluble fraction of the yolk, consist of three major proteins. These are α-livetin (32%), β-livetin (34%), and γ-livetin (30%). These livetins are identical to three plasma proteins, with α-livetin being identical to plasma albumin, β-livetin to α2 glycoprotein, and γ-livetin to α-globulins. The low-density fraction is largely lipid (approximately 90%), including triglyceride, phospholipid, and cholesterol, and 10% protein.

Yolk has a clear physical structure. In the center is a small spherical mass of white yolk and this is surrounded by yellow or true yolk. The yellow yolk is made up of tightly packed yolk spheres together with an aqueous continuous phase. The yolk spheres contain both the granular and the low-density (lipid-rich) chemical fractions of the yolk, and the continuous phase contains dissolved proteins, probably the livetins.

Yolk precursors are synthesized by the liver under the stimulation of ovarian estrogens. For instance, phosvitin and lipovitellin are synthesized as part of a large precursor molecule, vitellogenin, which has a molecular weight of approximately $2.3–2.4 \times 10^5$. Following synthesis of previtellogenin from a specific mRNA there are posttranslational changes, including addition of carbohydrate (glycosylation), lipid (lipidation), and phosphate groups (phosphorylation). The modified molecule is transported to the ovary via the blood stream. During this time vitellogenin acts as carrier for calcium, the plasma concentration of which in a laying hen is far greater than that in a nonlaying or male chicken. At the ovary vitellogenin is transported into the developing ovum. This active process involves a specific ovarian vitellogenin receptor and probably micropinocytosis. Within the ovary, vitellogenin undergoes proteolytic cleavage to yield two distinct types of phosvitin (of respective molecular weights 34,000 and 28,000) and one unit of lipovitellin (molecular weight 17×10^5). The phosvitin and lipovitellin are incorporated into the yolk structures. The principal source of yolk lipid is that synthesized by the liver. Estrogens act on the liver also increasing the production of fatty acids about twofold. These are transported by the ovary largely as triglycerides bound to the carrier protein—very low-density lipoprotein (VLDL). Estrogens also increase the synthesis of VLDL by the liver. At the ovary the uptake of triglyceride involves the enzyme breakdown of the triglyceride–VLDL complex by lipoprotein lipase. Other yolk precursor proteins are also synthesized by liver under estrogen stimulation. These include riboflavin binding protein and apolipoprotein II. The synthesis and deposition of yolk are summarized in Figure 14.

4.7. Egg Albumin Formation

Between the membrane under the shell and yolk lies the white or the albumin layers of the egg, which are composed of about 12.3% solids and 87.7% water. The vast majority of the egg white solids are proteins (representing over 85% of the

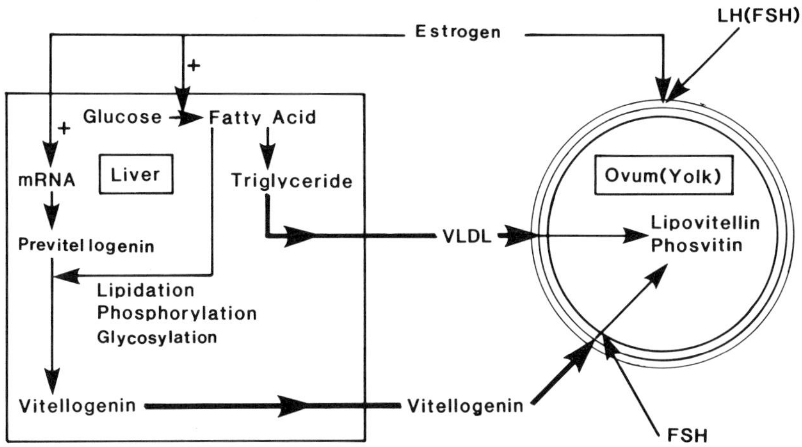

Figure 14. Schematic representation of yolk formation and deposition. Yolk precursors are synthesized by the liver under the stimulation of ovarian hormones. Proteins (vitellogenin) and triglycerides (e.g., very-low-density lipoprotein, VLDL) are transported in the blood to the ovary.

solids), together with some carbohydrates and minerals. The egg white has a distinct physical structure, being made up of four compartments: (1) the inner (thin) albumin, (2) the middle (thick) albumin layer, (3) the outer (thin) albumin layer, and (4) the chaliziferous layer, which immediately surrounds the yolk. These vary in their constituent glycoproteins and in their water content. Some of the glycoprotein (ovomucin) strands of the chaliziferous layer become twisted during passage through the oviduct, condense along the axis of the egg, and form the chalazae, the coiled albumin ropes that suspend the yolk (Figure 10). The proteins of the albumin layers include: (1) ovalbumin (54% of total solids), (2) ovotransferrin (13%), (3) ovomucoid (11%), (4) lysozyme (3.5%), (5) ovomucin (2%), (6) flavoprotein (0.8%), (7) ovoglycoprotein (0.5%), (8) ovomacroglobulin (0.5%), (9) ovoinhibitors (0.1%), and (10) avidin (0.05%). With the exception of lysozyme, these are all glycoproteins. Some of the protein constituents have well-established properties; some glycoproteins bind vitamins or minerals; ovotransferrin binds iron, flavoprotein binds riboflavin, and avidin binds biotin; others inhibit proteases (ovomucoids and ovoinhibitors) or are enzymes, in the case of lysozymes, which act to hydrolyze polysaccharides, particularly those found in bacterial cell walls.

The albumin layers are laid down by essentially a two-phase process. The first phase involves the secretion of all the albumin proteins, but only some of the albumin water content, mainly by secretory cells in the epithelium of the magnum and to a much smaller extent also in the infundibulum and isthmus. Synthesis of the proteins of the egg white is controlled by estrogen and progesterone. In the isthmus, the egg membranes are formed and calcification of the shell appears to be initiated, but formation of the albumin layers is not complete until the egg reaches the uterus or shell gland. The second phase involves addition of much of the water of the albumin and occurs in the uterus. This is referred to as "plumping," and the watery secretion is known as plumping fluid. This aqueous fluid is passed across the egg membranes so they lose their puckered appearance and become smooth. The albumin as a whole provides an aqueous medium for embryonic development, protecting the embryo from desiccation. This environment is maintained because water loss is depressed by the high-protein (solute) concentration in the albumin layers, which also provide nutrient proteins, vitamins, minerals, and water for the developing embryo. The presence of lysozyme in egg white protects the embryo from bacterial infection, and ovotransferrin binds to heavy metals.

4.8. Shell Formation

4.8.1. Structure of the Shell

The shell consists of three portions: the membranes, the calcified portion of the testa, and the cuticle (Figure 15). The membranous portion contains two membranes, which vary in thickness in different species of birds. The outer membrane is in intimate contact with the thick, calcified portion of the shell. Indeed, the testa is fused with the outer membranes in the regions of the core/cone from which the calcite crystals radiate (Figure 15). Approximately 95% of the testa is calcium carbonate, largely in the form of calcite crystals. The crystal columns are sur-

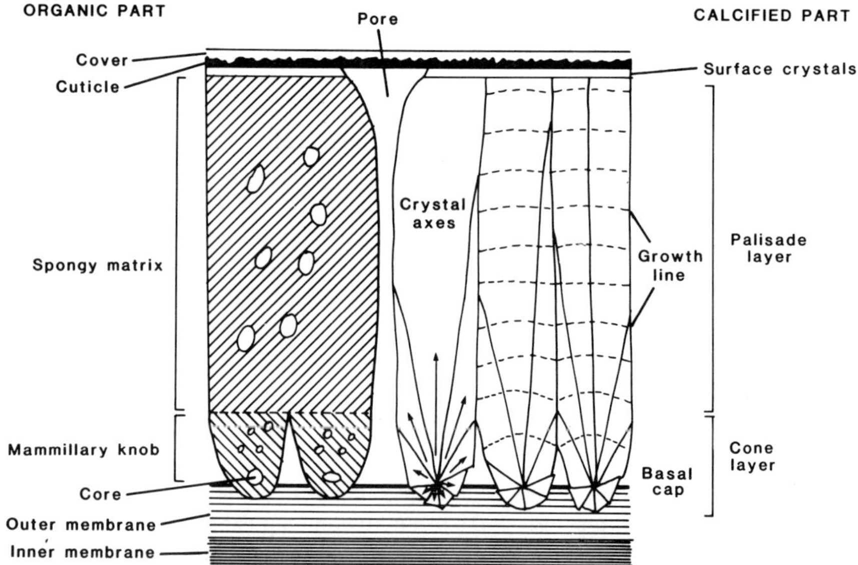

Figure 15. The structure of the shell. [From Gilbert, A. B., 1979, in: *Form and Function in Birds,* Vol. I (A. S. King and J. McClelland, eds.), Academic Press, London.]

rounded by a spongy matrix, and there are pores over the entire surface of the egg for gaseous exchange. The outside of the egg shell is covered by the cuticle. This is largely organic, containing 90% peptide and proteins and also some carbohydrates.

The function of the shell is to provide protection for the developing embryo from physical damage, predation, water loss, and bacterial invasion. Moreover, the shell is the major source of calcium for embryonic development, providing about 80% of that required. Finally, the shell also has a role in gaseous exchange.

4.8.2. Formation of Shell

The shell is formed in the uterus (shell gland), and to a limited extent also the isthmus, by processes not fully understood. During the laying down of the shell, there is a large increase in blood flow to the uterus, and serum calcium ions are taken up by the cells of the uterine epithelium. Approximately 75% of the circulating concentration of calcium in the reproductively active female birds is bound to the yolk's precursors (Ca free 1.25 mM, Ca protein bound 3.75 mM). However, there is a dynamic equilibrium between free, ionized calcium and the protein-bound component. Hence, as more calcium ions are taken up by the uterine cells, calcium dissociates from the yolk precursor proteins. Within the uterine cells calcium is bound to a protein similar to that formed in the cells of the small intestine where calcium is absorbed. Movements of calcium into the uterine cell and from the cells into the uterine lumen probably involve specific calcium pumps. Once in the area adjacent to the egg membrane, calcium is deposited, with carbonate ions, into calcite crystals.

Similarly, carbonate ions are passed into the uterine lumen. The major source of the carbonate ions appears to be metabolically produced carbon dioxide. This is converted to carbonic acid, under the influence of uterine carbonic anhydrase, which is then ionized producing carbonic ions which are shunted into the uterine lumen:

Metabolism

$$\downarrow$$

$$CO_2 + H_2O \underset{}{\overset{\text{Carbonic anhydrase}}{\rightleftarrows}} H_2CO_3 \longrightarrow \begin{array}{l} 2H^+ \rightarrow \text{Blood} \\ + \\ CO_3^{2-} \rightarrow \text{Shell} \end{array}$$

Some bicarbonate ions derived from the blood stream also contribute to shell carbonate.

The shell is formed by growth of the mammillary and palisade layers (Figure 15). A proteinaceous organic phase (less than 3% of the testa) is produced as an integral part of the shell proper. The shiny cuticle is then applied, and finally in some birds (including the chicken) some additional calcium carbonate is laid down on top of the cuticle. Pigment (porphyrins) is secreted from uterial epithelial cells shortly before laying.

4.8.3. Calcium Metabolism and Shell Formation

The formation of the egg and, in particular, that of the shell puts a tremendous load on the calcium resources of the female. For instance, in the domestic chicken, over 2 g of calcium is deposited per day into the egg shell, and this represents over 10% of the whole body stores of calcium. Thus, the hen must consume and absorb equivalent large quantities of calcium. Naturally, there are problems of short-term storage of calcium and mechanisms for its rapid mobilization. These problems are met by specific adaptations in the physiology of the female bird.

As the female comes into reproductive condition, there are increases in both calcium absorption and storage. Absorption of calcium in the small intestine is stimulated by the active metabolite of vitamin D—1,25-dihydroxy cholecalciferol (1,25-DHCC). In turn, the synthesis of 1,25-DHCC in the kidney is elevated under the influence of reproductive hormones including estrogen and prolactin. The additional calcium intake is initially stored in the medullary bone, which is laid down in the lumen of the long bones (Figure 16). The formation of medullary bone is again under hormonal control, in this case requiring both estrogens and androgens. It

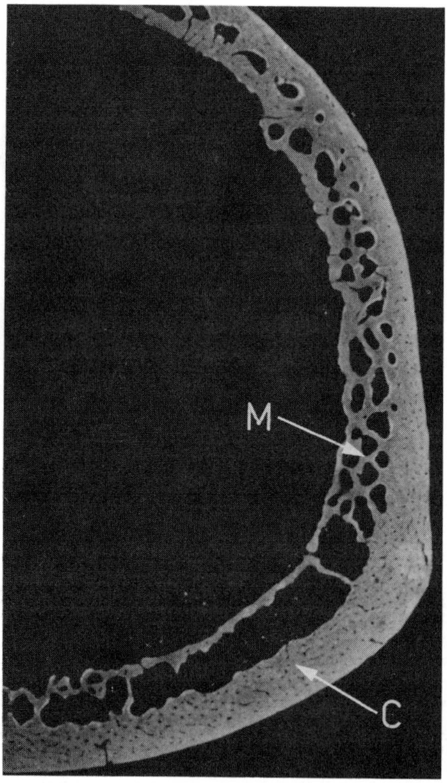

Figure 16. Microradiograph of a transverse section of the tibia of a laying hen, showing the cortical (C) and medullary (M) bone. [From Luck, M. R., and Scanes, C. G., 1982, in: *Aspects of Avian Endocrinology* (C. G. Scanes *et al.*, eds.), Texas Tech Press, Lubbock, Texas, p. 263.]

might be noted that during shell formation, calcium is derived from both medullary and cortical bone.

During shell formation there are short-term changes in calcium metabolism. As shell formation proceeds, there is a decrease in the plasma concentration of calcium ions. This directly leads to the release of parathyroid hormone (PTH) from the parathyroid glands, which in turn acts to increase osteolysis (bone resorption) and the availability of calcium for shell formation. In addition, the PTH released elevates calcium absorption from the small intestine and increases phosphate loss via the urine. It might also be noted that there is a high specific appetite for calcium at the time of shell formation (and hence also the time of decreased plasma concentration of calcium ions and of high plasma concentrations of PTH).

An interesting feature of calcium metabolism is that if a chicken has inadequate calcium levels in the diet, it will completely stop producing eggs after an initial period of "shell-less" eggs. This is due to a deficiency in hypothalamic function with a concomitant reduction in GnRH release and hence an absence of the preovulatory LH surge and ovulation itself. This is obviously an important adaption, as continuity of egg production would grossly deplete the calcium stores.

4.9. Oviposition

Oviposition is the process by which the egg is moved from the uterus through the vagina and hence to the exterior. Like so many aspects of the reproductive physiology of the female, this has been studied almost solely in the chicken. The discussion of oviposition will therefore be limited to the situation in the domestic fowl.

During oviposition, the muscular wall of the uterus contracts while both the uterovaginal sphincter and the vagina relax. This acts to push the egg into the vagina. At this point, the hen squats, and the egg is then expelled owing to contraction of oviduct muscle. Expulsion is aided by elevated abdominal pressure caused by contractions of the abdominal muscles. It might be noted that there are increases in both respiratory rate and body temperature at the time of oviposition. There is no agreement as to extent to which eggs are laid "pointed" or "blunt" end first. Estimates of between 54% and 82% pointed end first have been reported in the literature.

The timing of oviposition is precisely controlled in the chicken, approximately 26 hr following ovulation and 30 min prior to the ovulation of the next ovum in the sequence. This control is so precise that if a hen had an ovulation-to-oviposition time of, say, 25 hr 43 min for one egg, this exact time would remain remarkedly constant over successive cycles. This implies there should be some intricate mechanism controlling oviposition. Any probable scenario for the endocrine induction of oviposition involves hormones of the posterior pituitary gland (arginine vasotocin and/or mesotocin), together with prostaglandins E_1, E_2, and $F_{2\alpha}$ stimulating uterine contractions. In addition, the ovarian hormones, estrogen and progesterone, affect both the vigor of the muscular contractions of the oviduct and the ability of the

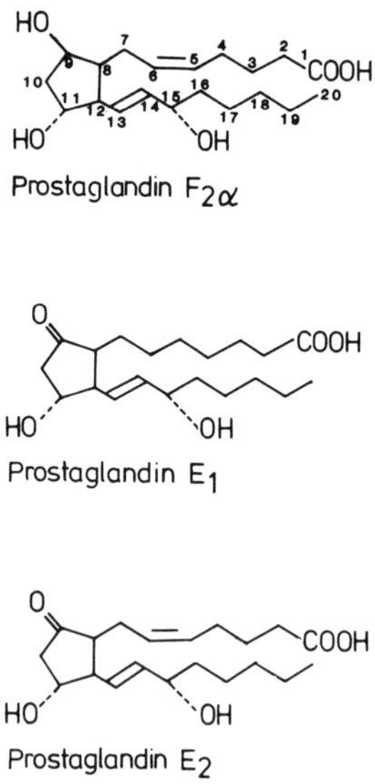

Figure 17. The structure of the prostaglandins involved in oviposition.

uterus to synthesize prostaglandins. Oviposition occurs coincident with a peak in corticosterone secretion, but this may merely reflect the stress and fatigue involved in the laying of the eggs. There is no evidence that nervous innervation of the uterus or vagina normally has a role in causing oviposition. A reasonable control mechanism might involve the release of oxytocic posterior pituitary hormones, as a neuroendocrine reflex in response to the completion of calcification and also perhaps to the preovulation development of the next ovum in the sequence. The posterior pituitary hormones appear to act by increasing uterine production of prostaglandins E_1, E_2, and $F_{2\alpha}$ (for structures see Figure 17). These, in turn, all stimulate the contraction of the smooth muscles of the uterine wall. In addition, prostaglandin E_1 (and also perhaps PGE_2) causes relaxation of the uterovaginal sphincter and the vagina. Thus, the egg moves into the vagina and thence to the exterior. The largest preovulatory follicles also produce prostaglandins of the E and F series. The relative

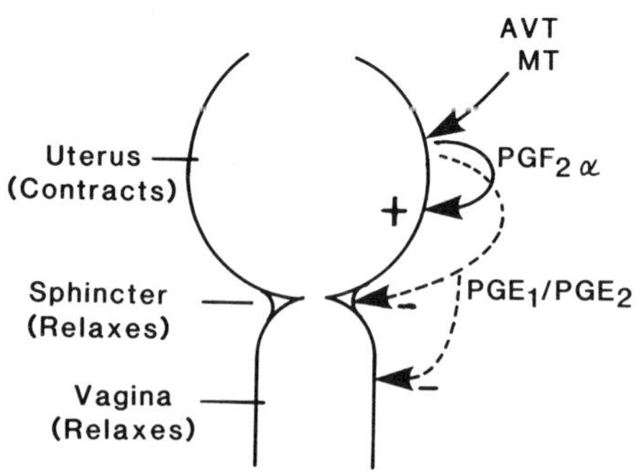

Figure 18. The control of oviposition by the posterior pituitary hormones arginine vasotocin (AVT) and mesotocin (MT) and by prostaglandins (PGE_1, PGE_2, and $PGF_{2\alpha}$).

contribution of uterine- and ovarian-derived prostaglandins in initiating and actually causing oviposition is not established. Figure 18 summarizes this model for oviposition.

5. Environmental Interactions

5.1. Seasonality

Birds produce their young at a suitable time of year, thus maximizing the chances of their survival, to reach sexual maturity, and to minimize dangers to the parents. It follows that many birds in the temperate zone breed in the spring, a period when food is available for the metabolic demands of reproduction and for the

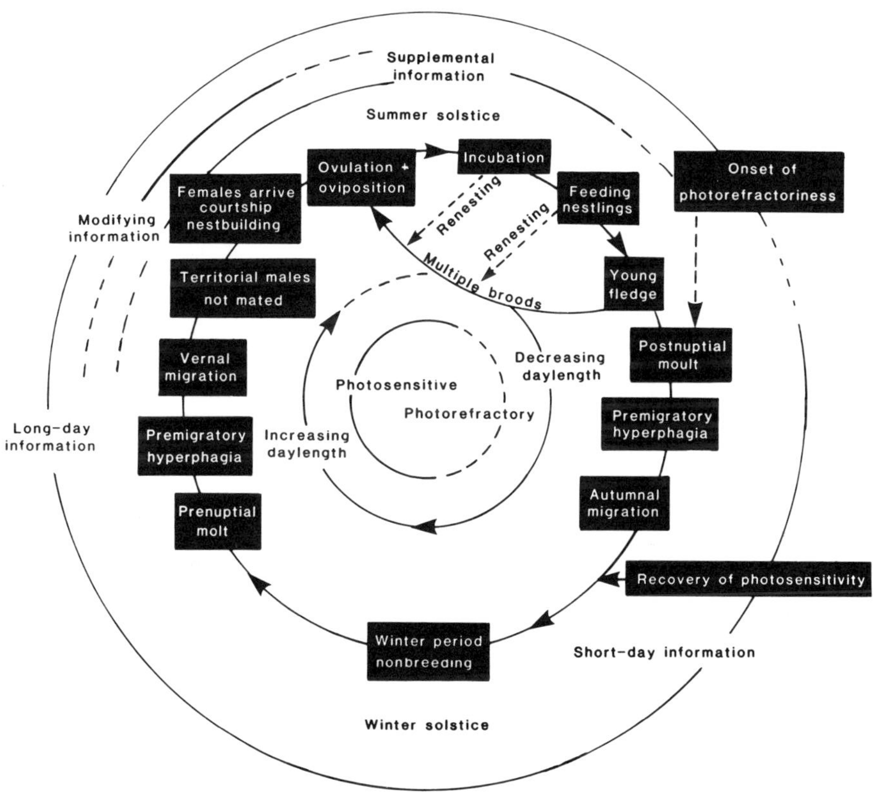

Figure 19. Schematic representation of the annual cycle of migratory species that breed at mid and high altitudes and winter north of the equator (based on *Zonotrichia leucophrys*). [From Wingfield, J. C., and Farner, D. S., 1980, in: *Seasonal Breeding in Higher Vertebrates* (B. K. Follett and R. J. Reeter, eds.), Karger, Basel.]

growth and development of the chicks. In many cases reproduction follows migration to breeding grounds, whereas desert species may breed following rainfall. There is great selective value in breeding at the optimum season, ensured by the evolutionary development of sophisticated mechanisms. Different origins (polyphyletic) may give similar physiological mechanisms of seasonal breeding in different species of bird. (See Figures 19 and 20.)

Two distinct types of environmental influences determine seasonal breeding in birds and other animals, ultimate and proximate factors. The ultimate factors (food, weather, reproductive readiness) determine the optimal breeding season. Proximate factors trigger the development of gonads and secondary reproductive organs, to anticipate the breeding season. Daylength (outside the equatorial zone) is here the prime proximate factor, and its increase in winter/spring promotes reproductive development in many avian species. Once the photoperiod has exceeded a critical number of hours, light rapidly (within 24 hr) exerts a stimulatory effect on both FSH and LH secretion, as a result of increased hypothalamic drive and GnRH release. The rate of gonadotropin secretion is proportional to both the intensity and length of

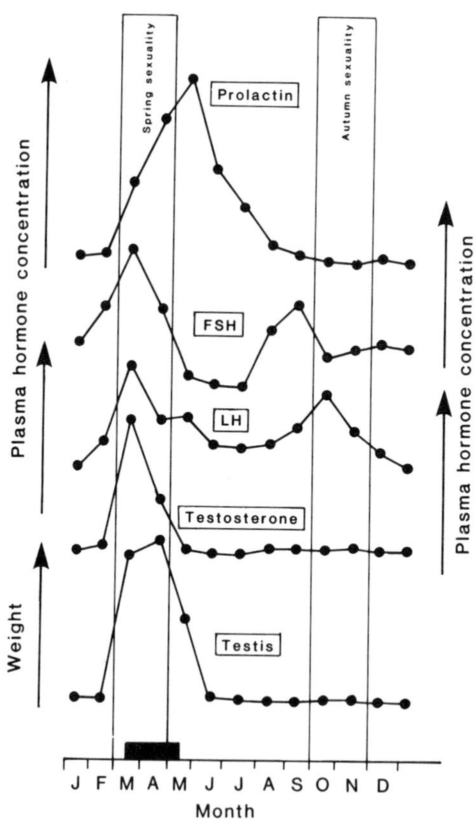

Figure 20. Endocrine changes associated with spring and autumn sexuality of the rook, *Corvus frugilegus*. Black bar indicates the period of mating. (From Lincoln, G. A., *et al.*, 1980, Endocrine changes associated with spring and autumn sexuality of the rook *Corvus frugilegus*. *J. Zool. London* **190:** 137–153.)

the photoperiod, up to a species-specific maximum. Gonadal growth and steroidogenesis are induced and facilitated by expression of positive gonadal steroid feedback and by the suppression of the negative feedback mechanism controlling gonadotropin release.

In most birds the gonads and other reproductive organs are almost completely atrophied outside the breeding season. For instance, in the Japanese quail the weight of the testes during reproduction is between 2 and 3 g, otherwise only around 10 mg. In migrating species gonadal regression reduces body weight and allows diversion of energy to muscle and fat deposits, needed for migratory flights. The atrophied gonads do not need maintenance, so that in temperate zone species the energy can be used to maintain essential body temperature and condition during cold winter months.

In some desert birds, the principal proximate factor influencing reproductive development is rainfall. In most species there are minor moderating factors influencing the exact timing of reproduction, including the possible delay of egg laying if weather conditions are bad or if the male is not present. Furthermore, nutritional status (energy, protein, micronutrients) may affect the timing and number of eggs produced.

5.1.1. Annual Cycles in the Temperate Zone

Birds of the temperate zone develop full reproductive condition at that time in the spring or early summer which is ideal for the survival and development of the hatchling. This varies with species and their geographical location. Thus, there is a tendency for breeding in any one species to occur later progressing northward, and this is applicable to both migratory and nonmigratory species.

An attractive explanation of seasonal breeding in temperate-zone birds is that the gonads develop once a critical or threshold daylength has been reached and regress in the fall (autumn) as daylength decreases below that threshold. Although symmetrical patterns of gonadal growth and daylength occur throughout the year in some species (e.g., doves), such patterns do not occur in most temperate-zone birds, since they would prohibit premigratory moult. Moreover, if the young were hatched late in the season, their chances of survival would be minimal and the drain on the parents might decrease the probabilities of successful reproduction the following year. Gonadal regression can thus occur toward the end of the breeding season despite exposure to long photoperiods. The inability to maintain gonadal development or to initiate gonadal recrudescence following breeding occurs because of the onset of photorefractoriness.

''Absolute'' photorefractoriness occurs when long photoperiods of any length, even continuous light, no longer stimulate gonadotropin secretion. ''Relative'' photorefractoriness occurs when the critical daylength for stimulating gonadal activity is increased. The onset of photorefractoriness occurs at a rate inversely related to the length of the stimulatory photoperiod and is hastened in birds with quiescent gonads, following breeding. The development of refractoriness probably occurs because of a decrease in the secretion of GnRH, although in some species the gonads may become unresponsive to gonadotropin stimulation. Photorefractoriness may be caused by increased hypothalamic sensitivity to inhibitory steroidal feedback or to increases in prolactin or thyroxine secretion or to a change in the organization of the circadian system involved in photo-induced gonadotropin release.

Photosensitivity is regained after the gonadal steroid levels have declined and after a period of exposure to short daylengths. Photoperiods short enough to break photorefractoriness may, however, be long enough to promote gonadal development via GnRH and gonadotropin secretion and result in a temporary resumption of reproductive activity in the fall (autumn sexuality), but this is usually limited to courtship behavior and pair formation (Figure 20).

5.1.2. Seasonal Breeding in the Tropical Zones

Many populations of birds in equatorial rain forests or tropical islands breed throughout the year. However, individual birds do not breed continuously. Examples of reproductive cycles of 7–12 months have been reported for individuals in these geographical zones. In areas where rainfall is seasonal or intermittent, as in hot deserts, reproduction follows rainfall; alternatively, humidity may be the proximate factor involved.

5.1.3. Seasonal Variations in Domesticated Birds

Domestic chickens produce eggs and semen throughout the year, when subjected to a natural photoperiod. This is due, in part, to intense selection of year-round egg production, although in winter reproductive efficiency decreases. However, if chickens are held in a strictly controlled environment of even humidity and temperature with a long daily photoperiod (14L:10D), then maximal egg production can be achieved and maintained irrespective to season. There is no evidence of photorefractoriness in domestic chickens, but egg production does fall after 9 months of continuous daily egg laying. On the other hand, domestic turkeys and ducks appear similar to their wild counterparts, showing the typical breeding pattern of the temperate zone including photorefractoriness following a prolonged period of long daylength.

5.2. Photoperiodism

The onset of reproductive activity in response to increasing daylength, the development of photorefractoriness and subsequent photosensitivity following short-day exposure, suggests that some birds can distinguish between long and short days and hence measure the passage of time. Daylength must therefore be measured, and the possible mechanisms involved in the measurement of daylength have been described by two hypothetical models. The first is based on the ''hourglass'' principle and involves some chemical that is synthesized during the day and broken down during the night. If the photoperiod is long enough, the chemical would accumulate and trigger the hormonal induction of reproduction. The alternative hypothesis is based on the interactions of the light/dark cycle with an internal daily rhythm or clock, either an external or internal coincidence model. The external coincidence model (Figure 21) envisages the bird having an endogenous free-running rhythm of about 24 hr (circadian) in the absence of external stimuli. This circadian rhythm may be regulated by a biological clock, possibly one involving the pineal gland, the suprachiasmatic nucleus, and the retina in an interlocking system. The daily cycle of light and dark entrains the circadian rhythm to exactly 24 hr. The daily rhythm is of sensitivity to light (photosensitivity) during a critical period of a few hours each day. When light is coincident with the sensitive phase, reproductive development occurs. However, on short daylengths light is only coincident with the insensitive phase. The internal coincidence model, at its simplest (Figure 22), involves two circadian rhythms—one is entrained by dawn and the second by dusk; thus, as the daylength changes, the two rhythms change in position relative to each other. When the correct alignment is achieved, the system is activated and causes the cascade release of GnRH, LH, and FSH, and the gonadal steroids.

Experimental evidence for and against the involvement of these hypothetical models comes particularly from Japanese quail and white-crowned sparrows. In so-called "night-interruption" studies gonadal development in birds maintained under short daylengths can be induced by short (0.5–2 hr) flashes of light during critical periods of darkness, so demonstrating the existence of the "external coincidence model" and invalidating the hourglass model. Gonadal activity fails to occur if the flashes of light fall outside this critical period, while the amount of light required for gonadal development is not itself a critical factor. The "internal coincidence model" has received little direct support, but its existence cannot be dismissed since its presence has been indicated in so-called "resonance" studies, in which flashes

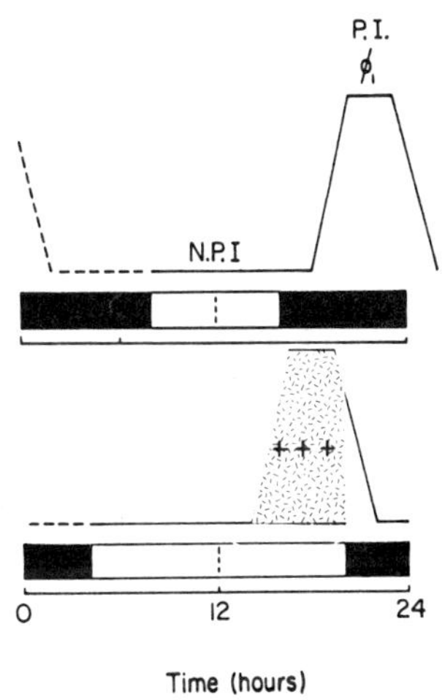

Figure 21. A diagrammatic representation of the simple external coincidence model of photoperiodism. The upper panel shows the cycle of photosensitivity with the photosensitive or photoinducible (P.I.) phase coinciding with period of night and the nonphotoinducible (N.P.I.) phase of the rhythm coinciding with the daylight hours. The midpoint of the photoinducible phase is defined as the phase angle (ϕ_1). As photoperiod increases from short to long daylength (lower panel), there is increasing coincidence between the photosensitive phase and the hours of daylight; it should be noted that the phase angle shifts in position relative to midday. Coincidence of the photosensitive phase and daylight leads to gonadotropin secretion and hence gonadal growth.

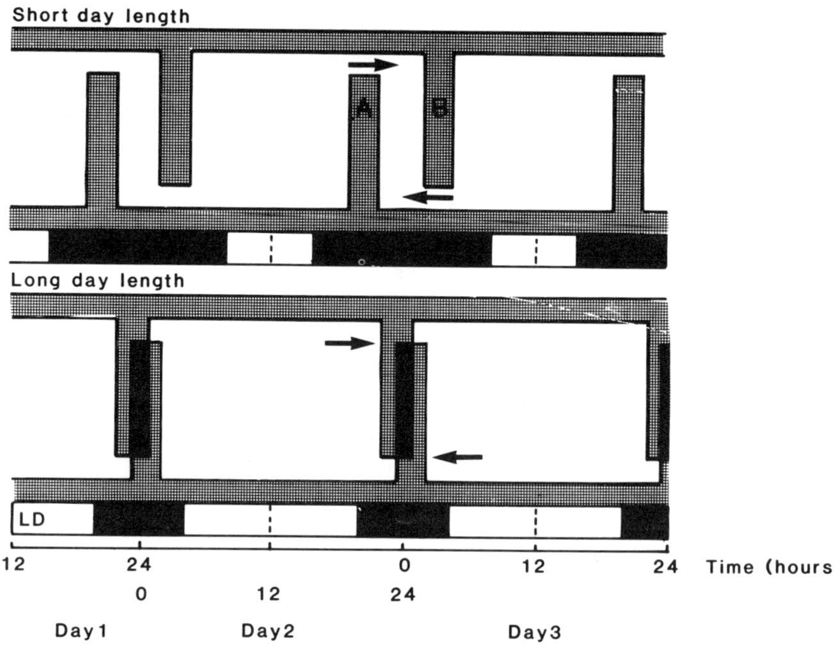

Figure 22. The internal coincidence model of photoperiodism. The upper panel represents two circadian rhythms which are linked respectively to dusk (or "lights off") (A) and to dawn (or "lights on") (B). As the daylength increases these cycles shift in their relative positions until they overlap. This coincidence is shown on the lower panel, which represents long daylength conditions.

of light stimulated LH secretion in some birds when given at 24-hr intervals at critical times during the first 3 days that they were exposed to constant darkness.

The photoperiod mechanisms that stimulate reproductive activity have probably evolved independently on a number of occasions in different avian groups. It must therefore be emphasized that different species may differ in the photoperiodic mechanisms involved.

5.2.1. Pineal Involvement

The pineal body has a major role in the photoperiodic control of reproduction in some mammalian species (e.g., hamsters) and may be of importance in birds. On short daylengths the pineal acts to suppress the secretion of reproductive hormones, probably by inhibiting the release of GnRH and hence of the gonadotropins and sex steroids. The mechanism of pineal involvement is unclear, but concerns both the quantity and the pattern of melatonin secretion. Melatonin is synthesized and released during the nighttime, and hence as the daylength becomes shorter, more melatonin will be produced over a longer period. This may directly inhibit GnRH release. Conversely, under longer daylength the reduced melatonin secretion will allow reproductive development by increasing GnRH release. Similarly, in sheep the photoperiodic induction of prolactin secretion involves the pineal and probably also melatonin. The involvement of the pineal in avian photoperiodism is possible, although lacks supportive evidence. The pineal is not required for either photoinduction or reproductive development or for the regression of the gonads which occurs on short daylengths. Furthermore, the pineal does not appear to be involved in photorefractoriness.

5.2.2. Light Detection

The stimulation of reproduction by increased daylengths necessitates mechanisms for light and/or dark detection. Surprisingly, the eyes do not seem to be required for the photostimulation of gonadal activity, although they are involved in the termination of reproduction when exposed to shortening daylengths and may play an important role in the entrainment of the circadian oscillators responsible for the development of photoperiodic sensitivity and the "open" period of the ovulation cycle. In those species examined it appears that light can be detected by extraretinal photoreceptors in the brain, possibly within the hypothalamus, and these are responsible for photoinduction of gonadal activity. The spectrum of light involved in the photostimulation of reproduction is in the red region and not in the green region of visible light.

5.2.3. Annual and Circannual Rhythms

It has been suggested that the cycle of seasonal breeding in at least some birds is due to endogenous rhythms which have a periodicity of approximately 1 year and are timed by environmental factors, particularly photoperiod. The existence of circannual rhythms in reproductive activity has been demonstrated in a number of species maintained under constant photoperiod and temperature, although under such a free-running situation the rhythms usually have a periodicity of less than 12 months. Circannual cycles of reproductive activity may thus be causally involved in the timing of poorly understood annual phenomena such as photorefractoriness, moult, and migration.

6. Reproductive Behavior

6.1. Mating Behavior

Successful breeding can be achieved by increasing parental investment during rearing or by increasing the number of mating opportunities. Mating behavior may thus be monogamous or polygamous, although aspects of both breeding strategies are usually adopted by most species.

Monogamy is the primary strategy employed, since incubation and brooding are lengthy processes during which eggs and young have to be protected against predation and temperature extremes. Shared rearing is of survival advantage for both the parents and the young, but necessitates pair bonding. When most of the parental activities (e.g., nest building, incubation, brooding) are carried out by one sex, the pair bond may only last a few days, although some species may reunite after a temporary separation. Favorable environmental conditions at the beginning of the season may enable parental care to be achieved solely by one partner, while the other prepares a second nest. Since environmental conditions for late-hatching young are usually less favorable, the pair bond has to be sufficiently strong so that both parents share in the care of the second brood despite a period of separation. Bond formation can occur within minutes of a pair approaching each other, but contact must be reinforced over several days, weeks, or breeding cycles before it becomes resistant to separation. In highly monogamous species, such as swans, gulls, or owls, the pair bond may last a lifetime, and since reproductive success improves with age and experience, the maintenance of the bond with a proven mate is of selective advantage and promotes male fidelity.

Despite monogamous bonding most species also engage in extrapair copulation as a secondary mating strategy, should the opportunity arise. Extrapair copulation is of advantage for a male if he can fertilize a female and she can rear his brood without his assistance and is of advantage to the female if it allows her to pair with a genetically superior or more fertile male. A temporary separation during the breeding season may thus promote polyandrous or polygynous pairings, although polygamous matings occur most frequently in those species in which the male takes no part in egg incubation (most passerines) and in those with brief pair bonds and minimal territorial defense (most waterfowl).

Polygynous mating as a primary strategy usually occurs when environmental or behavioral conditions cause an ecological "clumping" or excess of females enabling simultaneous or successive pairings. Polygynous associations are most successful when the females nest asynchronously and the males are free from parental duties and also occur in environments in which territorial resources (e.g., food, nest sites) are abundant and can be defended. Polygyny also occurs in environments which prohibit resource control but in which the males compete for females on the basis of social status, as demonstrated through sexual display (as in Mandarins). Polyandrous mating has also evolved from the emancipation of the female from rearing and usually arises when males outnumber females, the quality and availability of food are poor, or the female has low energy reserves after egg laying. Polyandrous females are usually "determinate layers" and as such can only increase their fecundity during good conditions by producing additional clutches, provided the breeding season is sufficiently long to allow renesting. The occurrence of polyandry as a primary mating strategy is, however, extremely rare and, apart from parasitic species (e.g., cuckoos), is mainly characteristic of species with nidifugous young that require little neonatal care.

Table III summarizes mating systems in birds. It is axiomatic that all mating

Table III
An Ecological Classification of Mating Systems

Monogamy: Neither sex has the opportunity of monopolizing additional members of the opposite sex. Fitness often maximized through shared parental care.
Polygyny: Individual males frequently control or gain access to multiple females.
Resource defense polygyny: Males control access to females *indirectly*, by monopolizing critical resources.
Female (or harem) defense polygyny: Males control access to females *directly* usually by virtue of female gregariousness.
Male dominance polygyny: Mates or critical resources are *not economically monopolizable*. Males compete for females by sorting out positions of dominance and/or directly demonstrating quality through display. Territory size and degree of male aggregation are inversely related to the size of female home range. Extreme aggregation (leks) occurs. When female home ranges are mobile, a female pursuit strategy may develop.
Rapid multiple clutch polygamy: Both sexes have substantial but relatively *equal* opportunity for increasing fitness through multiple breedings in rapid succession. Males and females each incubate separate clutches of eggs.
Polyandry: Individual females frequently control or gain access to multiple males.
Resource defense polyandry: Females control access to males *indirectly*, by monopolizing critical resources.
Female access polyandry: Females do not defend resources essential to males but through interactions among themselves, may limit access to males.
Cooperative polyandry: A single female and a group of males form a communal breeding unit in which all males have an opportunity for mating. Females are able to maintain a male group because of advantages to mates in cooperating as a breeding unit.

strategies have evolved to maximize the probability that genes from extant individuals will be transferred into subsequent generations.

6.2. Territoriality

The establishment of a territory is a prerequisite for reproductive activity in most species and is an area defended against conspecific individuals. The selection of a territory is initially induced by complex ecological stimuli (which may be species specific) but once established, breeding birds may return year after year, even if the ecological conditions gradually change. The territories offering the best conditions tend to be occupied by early migrants or resident species, by sexually experienced birds, and by birds with a high hierarchical position. The size of the territory selected is dependent on its function and varies with species and with the resources available and the degree of interspecies toleration. The territory may include areas used for feeding or be restricted to the nest site or to areas used for pairing or mating. It can vary too with sex, the female often defending the immediate nest site and the male asserting dominance over mating areas. The territorial boundaries may coincide with topographical features or be established where the birds conflict between fear and aggression is in balance. This behavioral conflict facilitates the repulsion of trespassers and the degree of aggression shown by territory holders. The territory is defended and trespass deterred by threats of varying intensity. Intimidation is achieved through signals, and actual attack is rarely necessary. These signals may be vocal or visual, in which color, posture, or balletic ritual may be involved. Vocal signals are particularly of selective advantage for nocturnal species and birds living in dense vegetation, but are only adjuncts to defensive behavior since songbirds can maintain their boundaries even when muted. Territorial vocalization is principally by male birds and is the main means of proclaiming territory by migratory temperate-zone species, many of which begin gonadal activity and establish a territory before the arrival of females and before pair formation or nest building. Territorial behavior during the breeding season is strongly correlated with gonadal activity, since gonadotropins and gonadal steroids are causal in inducing aggressive behavior, song, and song motif. Territorial vocalization is not for sexual attraction and usually differs from that during courtship. To assist in territorial defense some females also engage in singing at the beginning of the breeding season, with malelike calls, although they may be muted at other stages of the reproductive cycle. Neighboring birds learn to recognize each other by their songs and learn not to respond aggressively to them. Songbirds continue singing, to proclaim ownership or to attract mates, but do not approach rivals unless they have unfamiliar songs or sing from unfamiliar locations. In this way, time and effort are not wasted in the hazards of territorial squabbles and each bird is able to devote more time to reproductive activity. Territorial behavior is thus of adaptive significance in that it strengthens pair bonds and results in breeding dispersal, which reduces the likelihood of predation, lessens the spread of disease, and prevents the excessive exploitation of nest sites or food resources through overcrowding.

6.3. Nesting

Most birds lay and incubate their eggs in nests, which afford protection against predation and adverse weather conditions, although some lay in shallow scrapes or holes in the ground. The only exceptions are parasitic species (e.g., cuckoos), which lay in the nests of others, and some penguins that incubate their eggs on their feet.

Nest site selection and nest building normally occur before or during courtship, although in some polygamous species (e.g., turkeys) it occurs after copulation. The selection of the nest site may occur long before the onset of building, which only happens when ready for breeding. The interval between site selection and nest building may be as long as several months or as short as several hours, depending on the urge to breed. The nest site may be selected by either or both sexes, and the degree of ritual involved reflects the specificity of the ecological requirements, being more elaborate for hole nesters than for open-nesting birds and less for ground-dwelling species.

The form of the nest and materials used in its construction vary with species and nesting site. The activities involved in nest building are innate, since indi-

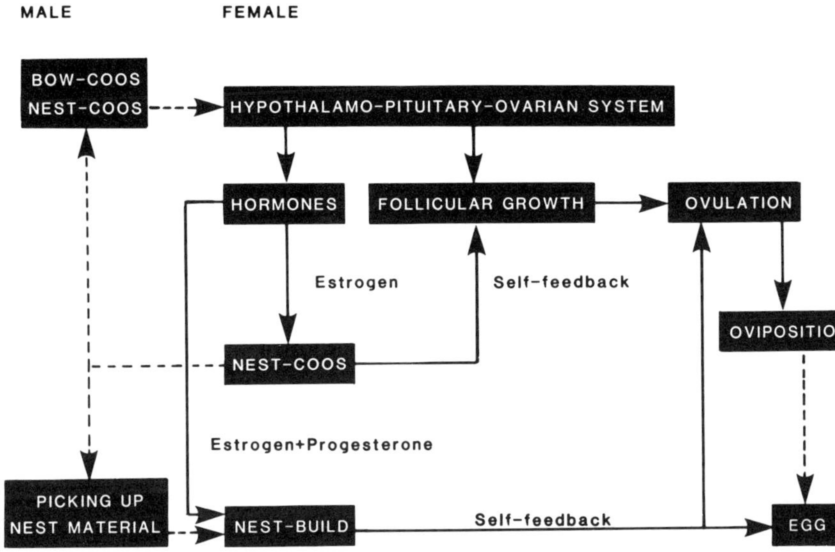

Figure 23. Environmental stimulation of gonadal activity in female ring doves. Nesting material. The presence and courtship behavior of the male stimulates gonadal development and sexual behavior of the female, the participation in which modifies further reproductive activity. (From Cheng, M.- F., 1983, in: *Hormones and Behaviour in Higher Vertebrates* (J. Balthazart *et al.*, eds.), Springer-Verlag, Berlin, p. 408.)

viduals reared away from their parents can still construct species-typical nests. The construction of the nest may nevertheless involve some degree of learning, in that actions which result in structures resembling completed nests may be reinforced and thus initially repeated. The constituent activities involved in nest building appear to be integrated into a functional sequence in response to specific internal (endocrine) and external stimuli (see Figure 23). The presence of the mate, nest-site nesting material, nest, and the act of nest building are all important stimuli, but the nature of the precise stimuli involved probably varies for each species. In ring doves, in which the male does most of the carrying of nesting material and the female does most of the nest building, nesting material appears to stimulate nesting behavior in the male, which in turn appears to stimulate the nesting behavior of the female. The initiation of a constituent nest-building activity provides negative feedback for that activity, but provides a stimulus that evokes further sequential activity. In this way gathering leads to carrying, carrying leads to the positioning of material in the nest, and so on. The proprioceptive stimuli related to the building of the nest and tactile and visual stimuli associated with its formation may also mediate nestbuilding activity. For instance, the size and texture of the nest cup constructed by a female canary regulates the frequency of her nest building movements, determines the material selected (lining or structural material), and ultimately terminates nest construction. These temporal activities may be triggered by the increase in the sensitivity of the canaries' brood patch during nest building. Since the interval between defeathering and egg laying is longer in paired canaries prevented from building a nest than in paired birds allowed to nest and as oviposition is delayed by the daily removal of nesting material from the nest, the nest provides stimuli that promote egg laying. The act of building is also important, in that the onset of laying in ring doves is related to the intensity of nest building. Unlike the canary, the daily emptying of the nest bowl can stimulate doves to build more and apparently to lay their eggs sooner.

Nest building is thus closely related to courtship activity and ovarian development and is synchronized so that egg laying begins when the nest is nearly completed or fully constructed. If the nest is built too far in advance of egg laying, it may be too tall and unstable, whereas if it built too late, it may be unsuitable for egg incubation. The synchronization of this behavior is achieved by complex endocrine interactions (Figure 23). During a normal reproductive cycle stimuli from the male may provoke gonadotropin secretion and ovarian development and the release of estrogen and progesterone, the combination of which mediates nest-building activity. While exogenous estrogen has been shown to induce all aspects of nest-building behavior, participation in nest building results in an increase in the plasma progesterone concentration and a decline in the estrogen level in doves. The first significant surge of progesterone in doves occurs on the first day of nest building, and the elevation of the progesterone level eventually results in the induction of incubation behavior. The act and intensity of nest building by the female dove, with little or no tactile stimulation from the constructed nest, may feed back on the hypothalamo-pituitary axis to cause a preovulatory surge of LH and a depression of FSH release. A depression in the FSH level is necessary for ovulation in the dove, and the acceleration of nest-building activity (by the removal of nesting material from the nest bowl) results in an earlier decline in the FSH concentration and an earlier onset of lay.

6.4. Courtship

Copulation rarely results from random encounters with the opposite sex, and courtship behavior is responsible for establishing and maintaining pair bonds, for reducing mutual aggression, and for synchronizing reproductive rhythms.

The attraction of a mate is dependent on social communication and is achieved by visual and auditory signals, the specificity of which results in reproductive isolation. Social communication in most birds (gallinaceous species excepted) is also essential for gonadal development. The presence of a female stimulates gonadotropin secretion in male birds, stimulating testicular development, testosterone production, and spermatogenesis in unpaired ring doves and starlings. Paired male pigeons have more spermatogenic activity and higher plasma LH levels than unpaired males. Once elevated, however, the plasma testosterone level may feedback on the pituitary to reduce gonadotropin secretion, since the androgen level is only elevated at the beginning of the courtship cycle.

Changes in the male androgen level during the reproductive cycle may also be related to the behavioral–physiological condition of the female mate. Paired females display less courtship behavior as the incubation period approaches, and the behavior of the female may fail to maintain elevated androgen levels or actively suppress androgen secretion. The decline in androgen level in paired male ring doves cannot be prevented if exposed to novel, courting females and is accelerated if experimentally paired with incubating females. This decline in the androgen level may be important in reducing male aggression and facilitating copulation. The early components of male courtship behavior (e.g., "chasing," "driving," "pecking") are often aggressive (to demonstrate prowess) and are induced by androgens acting synergistically with gonadotropins, particularly FSH. Since the progression of the courtship cycle (bowing to nest, cooing, Figure 23) therefore occurs when the androgen level (and aggression) is diminishing, the serial elicitation of each behavioral component may result from a change in neural sensitivity to testosterone or a temporal reduction in the hormonal threshold required or from separate androgen-dependent mechanisms.

Courtship behavior by the males stimulates, via gonadotropins, ovarian development and gonadal steroid secretion in females. The degree of ovarian development in ring doves and number of ovulations are related to the amount of courtship activity experience, two males inducing more gonadal development than one. The sight and sound of courting males (through glass partitions) is sufficient to induce oviduct development and egg laying (in canaries, budgerigars, pigeons, and doves) although this is accelerated if directed toward the females and is augmented if combined with tactile stimulation. Courtship singing alone or the sounds of other breeding pairs can promote female and male reproductive development and advance the onset of nest building in isolated females. The onset of nest building and egg laying are also related to the vocal repertoire of the male and occur earliest in females exposed to a large repertoire. Female vocalization occurs less frequently than male singing and is of limited repertoire, but during courtship does contribute to the gonadal synchrony of the pair, by stimulating male gonadal activity.

Female vocalization, "wing flipping," "nest solicitation," sexual "crouching," and other sexual behavior are initiated in most birds by estrogens, in combination with gonadotropins and possibly gonadotropin-releasing factors. The initiation of courtship behavior in polyandrous females may, however, be mediated by androgens, since their ovarian testosterone level is enhanced during the courtship period. The display of female courtship behavior usually occurs when in a preovulatory condition, and as female courtship can stimulate male spermatogenesis, reciprocal gonadal stimulation during the courtship period therefore ensures that the behavior of one mate is appropriate to the physiological condition of the other. Reciprocal gonadal stimulation also ensures that gonadal activity occurs when fertilization is most likely and thereby minimizes the energy required for successful reproduction.

Courtship behavior results in gonadal synchrony since it is composed of a stereotyped sequence of events (e.g., Figures 24 and 25); the behavior of one

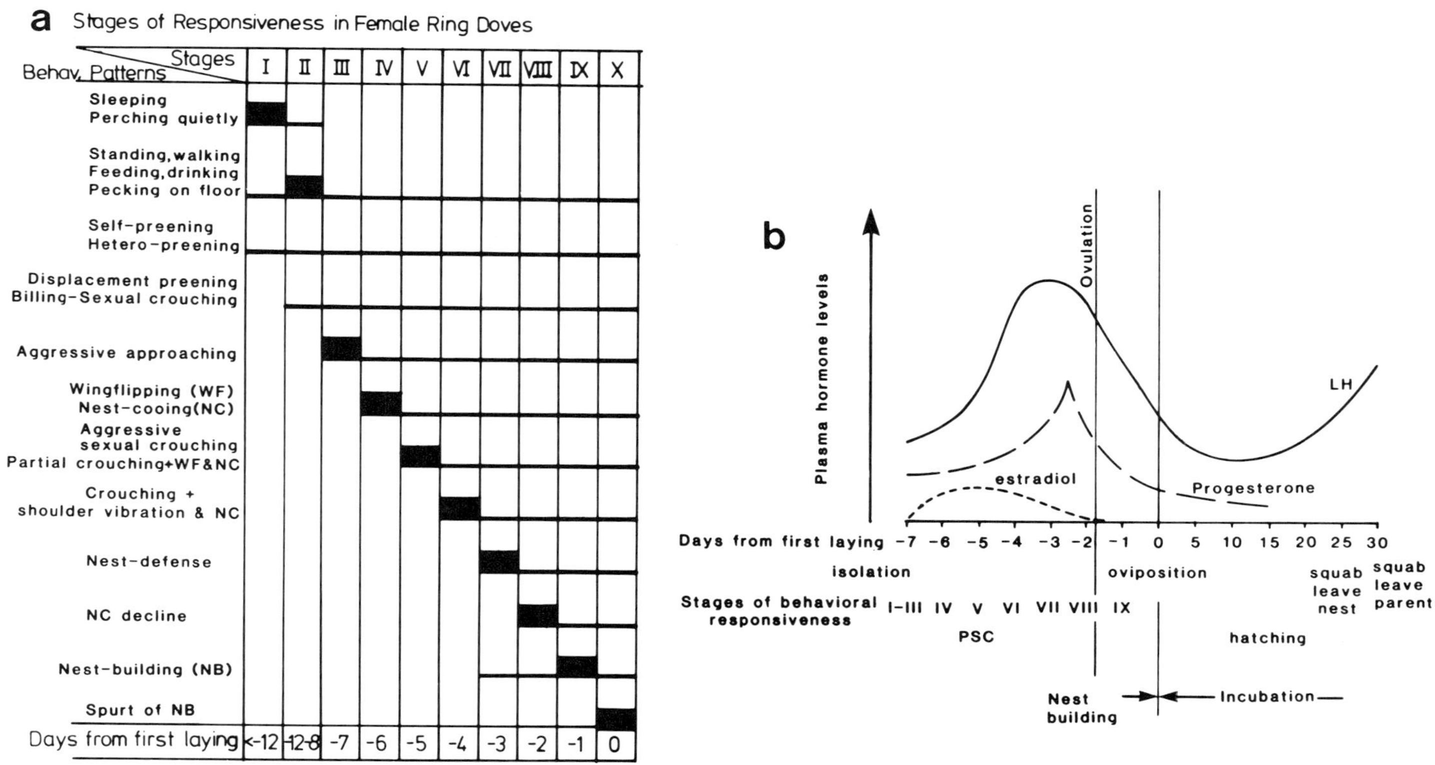

Figure 24. Hormones and behavior in female ring doves (*Streptopelia risoria*). (a) Sexual behavior during the breeding cycle. The horizontal lines indicate the days on which the respective behavior patterns occur, relative to the laying of the first egg. The solid bars indicate the day in which each behavior occurs in > 50% of the birds. *Behavioral stages:* I, nonresponsive; II, low responsive; III, approaching; IV, nest coos; V, proceptive sexual crouching; VI, crouched nest coos; VII, nest-site defense; VIII, nest-coo decline; IX, active nest building; X, egg laying. (b) Changes in female responsiveness during the courtship cycle in relation to circulating hormone levels. PSC, proceptive sexual crouching. (From Cheng, M.- F., 1977, *J. Endocrinol.* **74:** 37.)

partner elicits a specific response from the other, which in turn elicits a further response from the first partner. The duration and orderly progression of these events does depend on reciprocal reactions between the pair, but can be modified by past experience. For a compatible pair to move from one behavioral stage to the next, the behavior of the male and female must be in harmony, although sexually experienced birds asynchronously paired may be able to realign their behavior. Sexually experienced cockerels may thus approach familiar hens from the rear and mount them, without having to engage in "waltzing" behavior (Figure 25) as do strangers. Male ring doves paired with females of advanced ovarian stage can similarly move quickly from "boo-cooing" and "chasing" (characteristic of early courtship) to nest-oriented behavior, and at the same time the females may delay their onset of ovulation and egg laying. Similarly, although the progression of a courtship cycle in inexperienced birds may be dependent on their endocrine status, an experienced pair (e.g., ring doves) may progress from courtship to incubation even if the male is

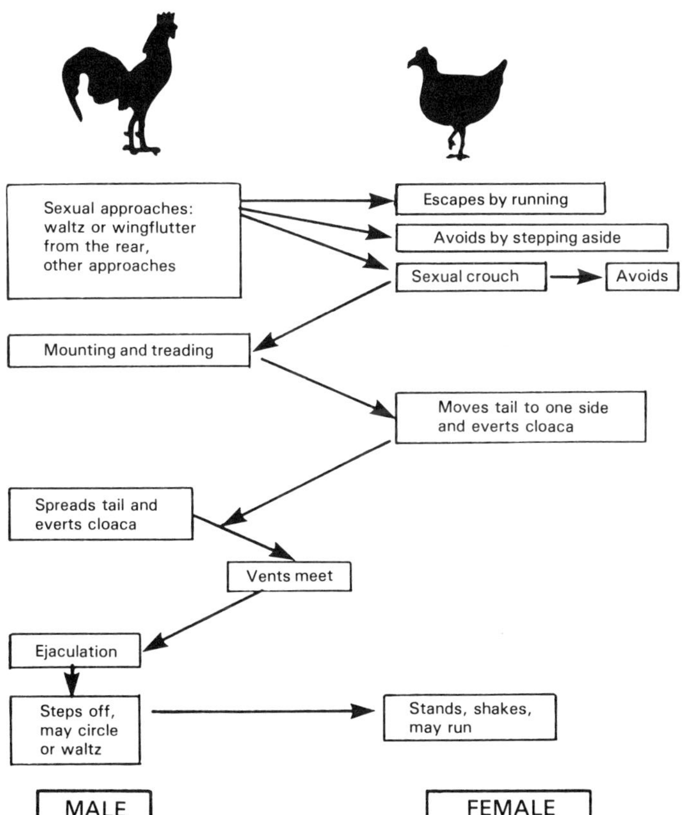

Figure 25. A simplified signal–response sequence in the mating behavior of chickens. [Adapted from Guhl, A. M., 1962, in: *The Behaviour of Domestic Animals* (E. S. S. Hafez, ed.), Ballière, London, p. 451.]

castrated prior to pairing or the normal decline in male androgen level is experimentally prevented by exogenous hormone therapy.

Mating normally only occurs after the successful completion of courtship behavior, and unless the female is aroused and receptive, the behavioral sequence of events is not initiated. These reciprocal interactions may also fail to be triggered if the male is not responsive to the female, even if she is receptive.

Copulation and egg fertilization normally follow courtship behavior, although in some species (e.g., ducks) autumnal courtship may also occur but not result in the production of offspring. In such species both sexes may engage in social displays and frequently copulate, but since the females are not fertile until early March, fertile eggs are not produced. The male may show some gonadal activity at this time, possibly as a result of interstitial cell secretory activity, but the testes remain sterile and spermatogenesis does not occur until the following February. Such autumnal courtship behavior is initiated by the congregation of male and female groups and is similar to that during spring, although it is responsible merely for the establishment of an overwintering pair bond.

6.5. Copulation

Copulation is the sexual act resulting in the fertilization of ova by sperm. Coitus occurs when the male's libido is adequate and the female is sexually receptive (Figure 26). The female crouches before the male, usually with the head low, the wings outstretched, the tail raised, and the vent exposed. The male then lowers his tail and brings his copulatory papillae (or the penis in Anatidae species) into contact with the engorged lips of the vent and ejaculates or everts his cloaca, resulting in the transfer of sperm. Successful copulation necessitates that the male be correctly oriented, and this is achieved by reference to the female's body, particularly the head. The termination of female receptivity occurs once the oviduct or vent has been fully everted and occurs in an orgasmiclike response. This may occur before male ejaculation (depending on the female's threshold of sensitivity and degree of tactile stimulation) and may occur even if the male fails to inseminate her, despite repeated copulatory attempts. The male may dismount prematurely (as a result of loss of libido, fear, preference for another female, incorrect orientation or intraspecies interference), and the female may remain in the sexual crouch. The termination of female receptivity prevents sexual crouching for some time afterward, and repeated copulation similarly suppresses male libido.

The sexual crouch is elicited in response to male courtship, although sexually experienced birds may self-initiate a "proreceptive" crouch. The proreceptive sexual crouch closely precedes ovulation and occurs just when blood estrogen levels begin to fall and before the preovulatory rise in plasma LH. The proreceptive crouch is therefore similar to the period of "heat" during the estrus cycle of mammalian species and occurs when the production of fertile eggs is maximal. The frequency of mating is thus maximal in most birds at the initiation of ovarian development and

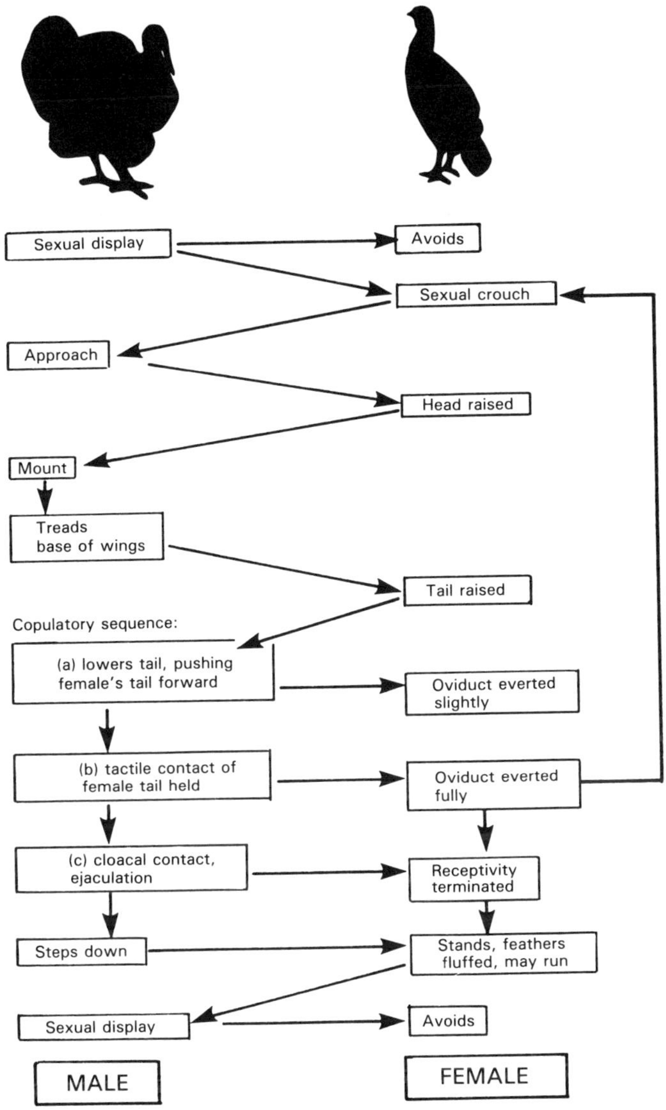

Figure 26. A simplified signal–response sequence in the mating behavior of turkeys. [Adapted from Hale, E. B., and Schein, M. W. 1962, in: *Behaviour of Domestic Animals* (E. S. E. Hafez, ed.), Ballière, London, p. 531.]

declines with the onset of lay. This decline is not due to sexual satiation but to physiological changes related to ovulation itself.

Copulation may take place throughout the courtship period, but with varying frequency according to mating strategy, clutch size, and stage of reproductive cycle. Copulation may occur at any time of the day, but in some species (e.g., chickens) occurs most frequently during late afternoon, reflecting a similar diurnal pattern of testosterone secretion and semen production. Testosterone secretion also correlates with mating frequency in polygynous species, in that dominant males with high testosterone levels have a greater libido and increased mating frequency. Increased libido does not, however, correlate with reproductive success, since the more sexually active males commonly have aspermic ejaculations and tend to be avoided by soliciting females in preference for less active males. The occurrence of a high mating frequency in female birds is not related to differences in gonadal activity, since steroid implants fail to affect female mating frequency and because hens of low social rank and small size mate frequently but lay few eggs.

6.6. Laying

Laying, the oviposition of eggs, occurs contemporaneously, or closely follows courtship and nest-building behavior in most species. Laying can, however, occur spontaneously in domesticated species or be temporally separated from courtship and nesting behavior in others (e.g., Shearwaters), which intercalate a ''honey-moon'' (feeding) period prior to laying, to build up bodily reserves. Egg production is energetically demanding and in most species necessitates a prebreeding increase in body weight, which might be met by mate feeding, by geographical migration, or (as in warblers) by local movements to insect (or seed) outbreaks.

Since the production of viable offspring necessitates a minimum amount of egg material, egg size cannot be greatly reduced when food is scarce. The ability of the female to produce eggs is therefore a major factor controlling clutch size, particularly in precocial species. Clutch size, the number of eggs laid sequentially and brooded simultaneously, is dependent in most species on geographical, seasonal, and annual variations in resources. The size of the clutch laid is also dependent on whether the species involved is a ''determinate'' or ''indeterminate'' layer. The ovary of a determinate female produces a finite number of follicles, and when all the eggs have been laid, incubation behavior is initiated. In contrast, the ovaries of indeterminate layers produce more follicles than are required, the termination of lay occurring when tactile or visual stimuli from the eggs induce the onset of incubation and gonadal collapse.

During the formation of a clutch many birds lay one egg each day until the clutch is complete, although some lay on alternate days (e.g., European cuckoos) or after intervals of 3 days (African swifts) or after 5- to 7-day intervals (e.g., hornbills). Species that produce large clutches of eggs (e.g., ducks) often lay a number of eggs on successive days and then have one or more ''pause'' days before resuming lay. The sequence of lay, as well as the intervals between the sequences, may be irregular or regular. The interval between the laying of successive eggs in a sequence varies, but in most hens ranges between 24 hr and 28 hr, depending on the time required for the formation of the egg and the length of the laying sequence. The lag period (the interval between oviposition minus 24 hr) is greater in shorter sequences than in long ones, and the lag between the first and second eggs and between the last two eggs of a sequence is greater than that between the intervening eggs, reflecting differences in the time of ovulation. The time of day the egg is laid therefore depends on daylength and the time of ovulation, the length of the sequence, and the position of the egg in the sequence.

The laying of the egg usually requires several minutes or hours and is often preceded by distinct alterations in behavior. In domesticated chickens and turkeys prelaying behavior may commence 3–5 hr before oviposition, is characterized by rapid pacing and escape movements, and may be accompanied by a distinctive, estrogen-dependent call. Such behavior may be homologous to that of wild birds during nest-site selection, which precedes nesting in both cases. The time spent on the choice of nest and in nesting is related to the lag in oviposition time. This relationship suggests it is hormonally controlled by the postovulatory follicle, the removal or ligation of which prohibits prelaying behavior at the predicted time the following day. This behavior is independent of the oviduct or the presence of an egg in the oviduct and is regulated by ovarian hormones, since it is still observed when the ovary is denervated and transplanted. Progesterone is causally involved, as it induces nesting 24 hr after its administration in estrogen-treated ovariectomized hens. A prolonged period of estrogen sensitization, as normally occurs between the onset of oviductal growth and ovulation, is, however, essential for the induction of prelaying behavior. The combination of estrogen and progesterone controls nest examination and nest entry by an interaction with neurons in the ventral hyperstriatum of the brain involved in attentional mechanisms. These neurons do not receive a direct visual input, but they do receive visual information from efferents emanating from other parts of the brain, and they cause the hen to respond and sustain its response to stimuli relevant to laying.

The oviposition of the egg occurs after its peristaltic passage into the vagina, the distention of which causes a "bearing-down" reflex that leads to strong contractions of the abdominal muscles, increased respiratory movements, and the expulsion of the egg. After laying, most birds show some interest in the egg, which heralds the onset of incubation behavior.

6.7. Incubation

Broodiness is a phenomenon unique to birds and is a term to describe two distinct aspects of parental behavior, the incubation of eggs and the brooding of young.

The onset of incubation normally begins after the laying of the penultimate or last egg of a clutch, but some species incubate after the laying of the first or antepenultimate egg. During the laying period mating and nest-building behavior gradually decline, while nesting gradually increases. This may trigger incubation behavior in indeterminate layers and, by causing follicular regression, limit clutch size.

During incubation the optimal air temperature above the egg is usually close to body temperature (34–39°C) and normally must remain fairly constant for correct embryonic development. Successful incubation is not passively achieved merely by sitting on eggs but results from various physiological and anatomical adaptations of the adult bird and by stereotyped behavior (see Figure 27). Transfer of heat from the parent is often achieved by the formation of brood patches, which result from the loss of feathers from the ventral abdomen. Blood supply to the brood patch is increased during incubation, and its efficiency in heat transfer is enhanced by the hyperplasia and folding of the skin and by the development of edema. The formation of the brood patch may occur in both sexes and rarely occurs in nonincubating birds. Its formation in females precedes incubation and results from a synergism between prolactin and estrogen, whereas in the male it may not develop until after

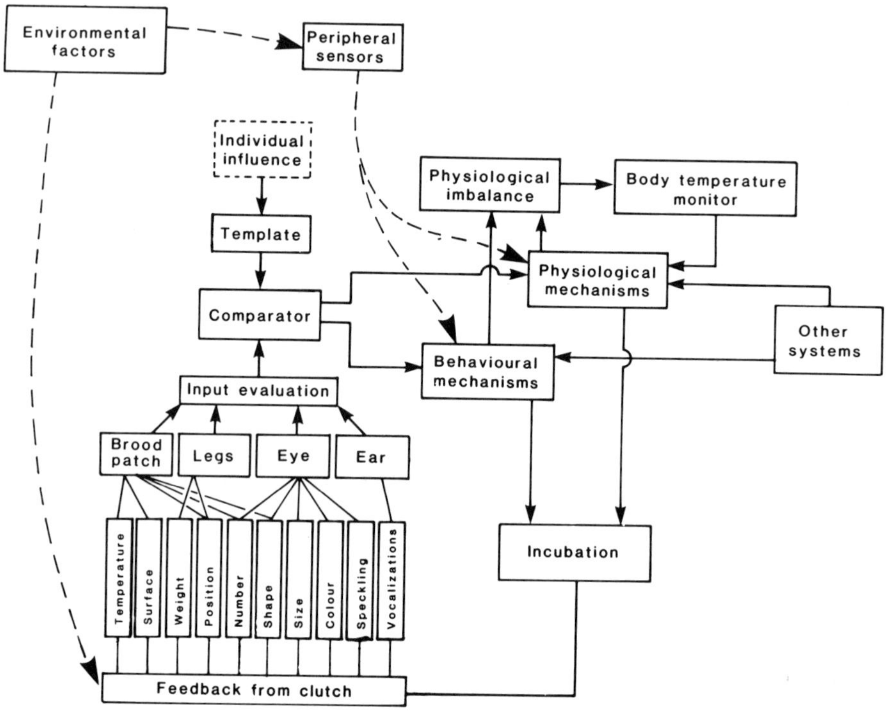

Figure 27. Functional control of incubation in the herring gull. Sensory input during incubation constitutes feedback on at least 10 characteristics. The measured state of these variables is compared to the template (expectation or "optimum"), and mismatch will result in commands activating either physiological mechanisms (primarily those concerned in body temperature control) or behavioral mechanisms or both. Events are not completely predictable; hence, the template is here placed under the influence of motivational factors. Other needs of the animal, such as foraging or predator avoidance, may modify or interrupt incubation (other systems). Environmental factors influence incubation directly, eliciting anticipatory action (feed forward) or via a change in feedback state of the clutch. [From Drent, R., 1975, in: *Avian Biology*, Vol. 5 (D. S. Farner and J. R. King, eds.), Academic Press, New York, p. 333.]

the female lays or incubates and results from a synergism between prolactin and androgens. The formation of a brood patch is not, however, essential for incubation, since some Pelecaniformes, Anseriformes, and Collumbiformes, and Passerine males incubate without them. In the absence of a brood patch the transfer of heat to eggs has been achieved in gannets and boobies by incubating the eggs in the webbing between the toes (which becomes highly vascularized during incubation) and in penguins by supporting the egg between the feet and a pouchlike fold of loose skin and feathers. Some species (e.g., Megapods) bury their eggs in hot sand, volcanic ash, earth mounds, or rotting vegetation and utilize solar or volcanic energy or the heat of fermentation for incubation, a technique also used by plovers and galliforms during periods of unattentiveness. Parental birds also prevent their eggs from overheating by providing shade or ventilation, by wetting the eggs, and by using surface blood vessels on the legs or brood patch to function as heat exchanges in the dissipation of heat.

The termination of the incubation period and onset of parental behavior is signaled by the piping of the embryos within the egg. If the eggs addle or are infertile, the adult may continue to incubate them far beyond the normal incubation period, whereas if they become damaged, the adult may desert the clutch. The maintenance of incubation behavior is therefore closely related to environmental stimuli and is synchronized by various internal (hormonal) cues.

Prolactin has historically been associated with broodiness, and exogenous preparations have been shown to induce broody behavior, especially if administered to laying hens or to estrogen-sensitized birds. In most species studied the plasma prolactin concentration gradually increases during the period of lay, and prior to its cessation the concentration rises abruptly, more so in birds that become broody rather than anovulatory. In view of its antigonadal effects, the increase in prolactin level may trigger incubation behavior. The timing of this increase varies from hen to hen and between species and may reflect differences in clutch size and sensitivity to environmental stimuli. Environmental stimuli such as heat, dim lighting, and particularly the presence of nests or eggs promote nesting and contribute to the onset of incubation in indeterminate layers, although they are of less importance in determinate layers. Birds may have a cerebral template of a completed clutch, and the accumulation of eggs may stimulate prolactin secretion. The increase in prolactin secretion is, however, dependent on gonadal steroid feedback (Figure 28) to increase hypothalamic prolactin-releasing activity and pituitary sensitivity to provocative stimuli, released in response to environmental cues. The persistence of high gonadal steroid levels during the prolonged laying periods of poultry may thus trigger prolactin secretion and induce broody behavior in hens deprived of stimulatory environmental stimuli. The endocrine mechanism initiating incubation is therefore likely to be similar in most determinate and indeterminate layers; incubation being induced by prolactin following a period of elevated gonadal steroid levels. Collumbiformes are, however, exceptional, in that progesterone appears to terminate lay and induce incubation, as a result of which the prolactin level increases to stimulate crop sac development. In species that lay a small number of eggs the sudden increase in progesterone level (which is only released from mature follicles) is an effective mechanism for the induction of incubation, but not for species with long laying cycles and chronically elevated progesterone concentrations.

In all hens that become broody the circulating LH level falls just before or after the onset of incubation, and estrogen and progesterone concentrations are concomitantly reduced. The fall in the LH level in most species commences before the cessation of lay, indicating that the gonadotropin concentration may remain sufficiently high for a few days to maintain the functional integrity of the largest yellow yolk ovulatory follicles but not the growth of smaller ovarian follicles. Prior to the onset of incubation the decline in the LH concentration may be initiated by high gonadal steroid levels feeding back negatively via increased brain dopamine turnover. Dopamine has antigonadal effects in birds, and dopaminergic transmission is increased in turkeys in a transitional state between laying and incubating. The high gonadal steroid levels may also be responsible for the increase in prolactin, and once elevated, prolactin may exert an antigonadal effect and contribute to ovarian collapse. Prolactin reduces LH secretion (possibly by inhibiting the release of hypothalamic GnRH), and since it blocks LH-induced steroidogenesis and premature ovulation, it may also act at the ovarian level to block LH binding to its receptor site or at a stage between gonadatropin binding and steroid biosynthesis. Preovulatory surges of LH normally occur during the night, and it is therefore of interest that the onset of nesting in bantam hens is preceded by nocturnal surges in prolactin secretion, possibly to inhibit the pituitary–gonadal axis.

The gradual increase in the prolactin concentration may be responsible for the parallel development of the brood patch, which is maintained throughout incubation if the prolactin level remains elevated but regresses when the level declines. The sensitivity of the brood patch increases toward incubation and is thought to promote nesting because the hen finds relief from peripheral irritation by sitting on the nest or eggs. Once incubation is initiated, the nest and clutch of eggs may provide visual and/or tactile cues that reinforce prolactin secretion by a reflex mechanism akin to that induced in mammals by the suckling of estrogen-primed nipples.

The integration of the environmental stimuli that increase prolactin secretion in incubating birds is probably mediated by a neural serotonergic mechanism in the brain. Serotonin stimulates prolactin secretion, and hypothalamic serotonin turnover is increased in incubating birds. Serotonin has some inhibitory effects on gonadal growth and gonadotropin secretion, and thus increased serotonin turnover may directly or indirectly contribute to the decline in LH secretion.

Once incubation has been initiated, the persistence of a high prolactin level may maintain brood patch development, prevent gonadal activity, and encourage nesting. The act of incubation and the presence of the nest or eggs provide positive feedback to maintain prolactin secretion, since the prolactin level declines (while the LH level rises) when incubating birds are deprived of eggs or prevented from nesting. These physiological–behavioral relationships and their integration by environmental stimuli are illustrated in Figure 28.

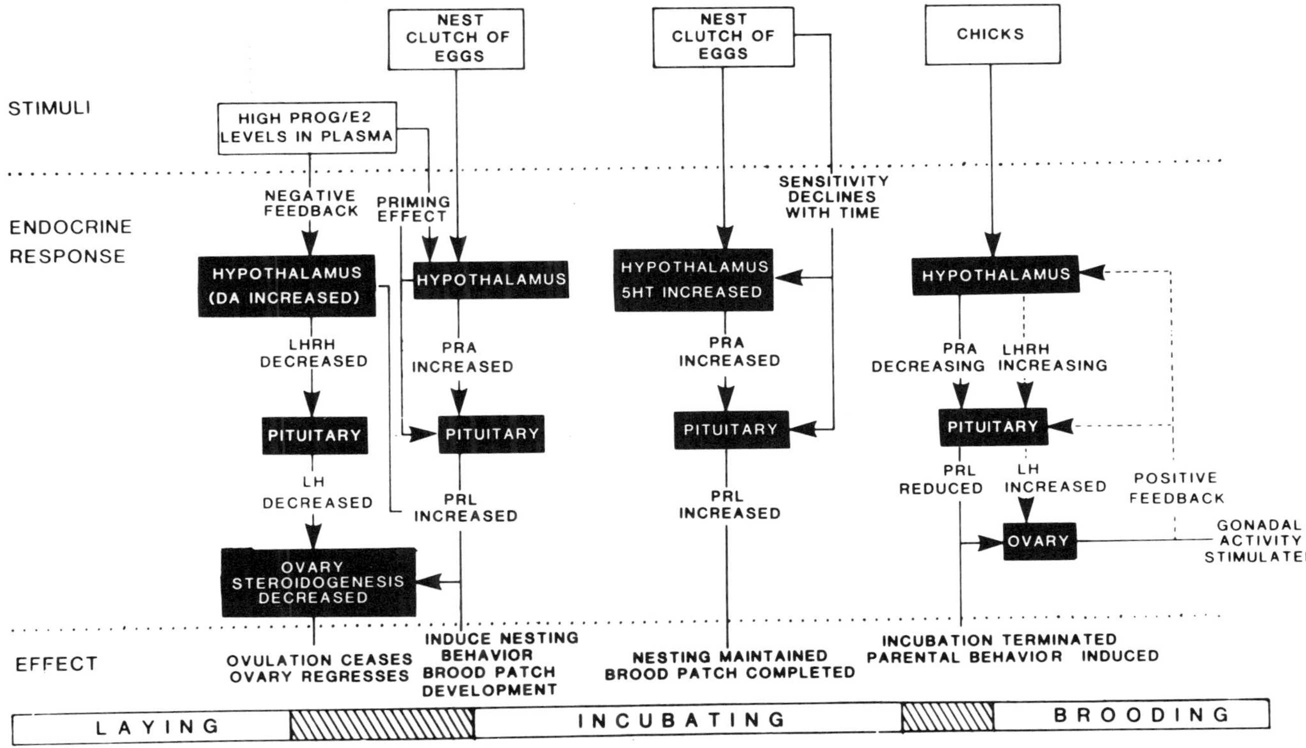

Figure 28. A model for some of the endocrine relationships during broodiness. The various stages of the reproductive cycle are indicated. Shading between adjacent stages indicates a transitional stage. Prog, progesterone; E_2, estrogen; DA, dopamine turnover; LH-RH, luteinizing-hormone-releasing hormone; LH, luteinizing hormone; PRA, prolactin-releasing activity; PRL, prolactin; 5HT, serotonin turnover. [From Harvey, S., and Bedrak, E., 1983, in: *Reproductive Biology of Poultry* (F. J. Cunningham *et al.*, eds.), Brit. Poult. Sci. Ltd., Longman Group, Harlow, p. 111.]

6.8. Parentalism

Following hatch the chicks are usually cared for by their parents. The degree of care shown depends on the physical development of the chicks at hatch and may include the feeding and sheltering of the young and their protection against predators and other dangers. The degree of care is least in species with precocial young, which are typically open-eyed at hatch, well developed, covered in down, and, after drying off, capable of some degree of locomotion. The duration of parentalism in precocial species varies with different species and climatic factors, but usually terminates long before the chick reaches adult body weight. Migratory arctic-breeding swans and geese are, however, exceptional in that they care for their offspring until the next breeding season. The prolongation of family life in these species may be of importance in the establishment of traditional flight paths and feeding grounds. Most parental behavior in altricial species (e.g., Passerines) is similar to that in precocial species, although since altricial young hatch prematurely, poorly developed with eyes closed and little or no down, and are uanble to leave the nest (nidicolous), it is of necessity more intense and limits the size of the brood.

The expression of parental behavior is initiated in response to the young and can be elicited in birds not involved in incubation. Thus, although the males of some species do not incubate eggs, they will accept and feed young placed in the nests of laying females. The females are stereotyped to a temporal sequence of laying and incubating eggs before the brooding of young and will not adopt nestlings placed in their nest until incubation has at least begun. The expression of parental behavior intensifies during the brooding period and, as with incubation, is initially not strong. If newly hatched chicks are replaced by eggs, certain birds do not abandon the nest but begin incubating again, indicating the lack of parental behavior at this time and the stereotyped nature of the response.

The induction of broody behavior is therefore dependent on environmental stimuli, principally the chicks. Thus, whereas the presence of eggs rarely induces domestic fowl to become broody, they can easily be made broody by keeping them with baby chicks. Parentalism can similarly be induced in nonlaying hens, immature pullets, and even cockerels, especially if maintained in warmth and darkness.

The initiation of parentalism is thought to reflect endocrine activity and can be induced in Galliformes by prolactin treatment. Following the hatch of chicks and turkeys, the parents emit a characteristic clucking call. This parental call can be induced by prolactin treatment, which also encourages the adoption of chicks or poults. However, the significance of these findings is obscure, since these behaviors normally occur when the endogenous prolactin level is declining.

In these precocial species the nest and nesting do not appear to stimulate prolactin secretion after the termination of incubation, since the prolactin level declines despite the continuation of nesting. The possibility that the presence of chicks may be causal in reducing prolactin secretion in turkeys is suggested by the prompt decline in the prolactin concentration when the eggs are removed and substituted with poults. The induction of parental behavior by prolactin treatment

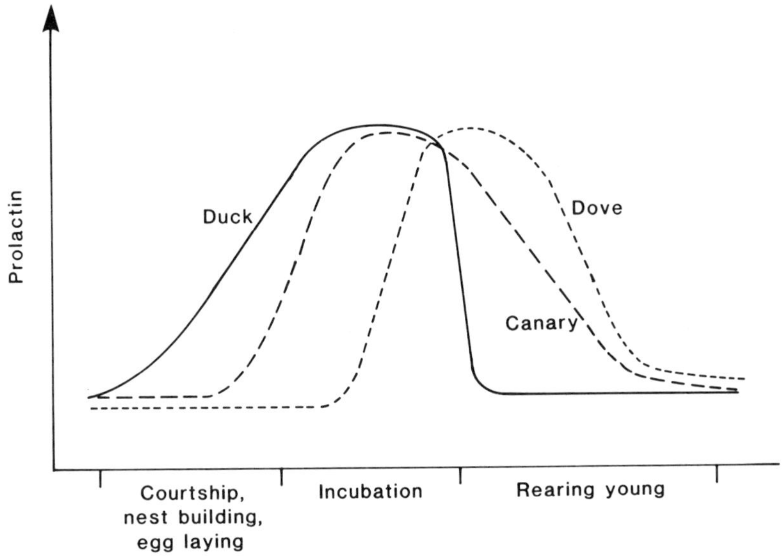

Figure 29. Schematic representation of changes in prolactin concentration in female ducks (—), canaries (— —), and doves (- - -) during the breeding cycle. (Based on Goldsmith, A. R., 1982, Plasma concentrations of prolactin during incubation and parental feeding throughout repeated breeding cycles in canaries (*Serinus canarius*). *J. Endocrinol.* **94:**51.)

may thus indicate that a prerequisite for its expression is a high prolactin level during the incubation period to sensitize brain tissues.

The decline in the prolactin level in birds brooding their young may be effective in facilitating the refeathering of the brood patch and in permitting ovarian recrudescence (Figure 29). Once the prolactin level declines during the brooding period, the resurgence of gonadal steroid secretion may terminate the broody condition (Figure 29), since broodiness can be interrupted by treatment with estrogens and androgens.

The timing of the decline in prolactin secretion appears to be related to the degree of care required by the young and is delayed in altricial species like starlings and canaries, which feed nestlings for 2 or 3 weeks after hatch. Such a delay may prevent the birds from entering a second reproductive cycle before the young are mature enough to fend for themselves. However, although parental behavior and sexual activity appear to be mutually exclusive in most species, they are not necessarily incompatible. Some double-brooded species begin a new clutch while the fledged young of the first breed are still being fed. Similarly, the males in polygamous species may simultaneously be mating with a female on one nest site while helping in the feeding of the young on another.

Prolonging the period of gonadal suppression may be effective in reducing the aggression of the parents toward their young, which they might otherwise attack.

Adult cockerels presented with neonatal chicks often attack them, and similarly, nonbroody ring doves will kill squabs placed with them. Such behavior is suppressed by prolactin treatment. This prolongation of prolactin secretion does not, however, occur in all altricial species, since there is a fairly rapid decline after hatching in pied flycatchers. In such cases, photorefractoriness may, however, prevent the onset of reproductive activity before the nestlings are brooded.

The dclay in the decline of the prolactin concentration in Collumbiformes is related to the feeding of cropmilk to the squabs. In ring doves (*Streptopelia risoria*) prolactin is not high at the start of incubation but increases around the midpoint, causing the crop sacs to hypertropy in readiness for milk production. The prolactin level and the feeding of crop contents are maintained by tactile, visual, and auditory stimuli from the chicks, although replacing growing squabs with newly hatched ones cannot maintain the prolactin level. This may indicate that the care of the young does not stimulate prolactin secretion in the way that incubation appears to do in other species or that the stimuli responsible are specific to their own young. When the young are weaned on to grain, the prolactin level declines and the crop sac regresses.

In summary, the expression of parental behavior in birds is stereotyped and normally only occurs in an orderly temporal sequence, and as with incubation, its integration in a successful reproductive cycle appears to be related to a complex interaction of environmental stimuli, changing hormone secretion, and behavioral transitions.

6.9. Molt

Molt is very closely synchronized with reproductive cycles and the seasons of the year and is the periodic process of feather loss and replacement. Molting has adaptive significance and, in addition to varying the plumage for breeding or migration purposes, occurs to replace feathers as they become worn or obsolete and to maintain a functional plumage for protection, insulation, and flight.

The acquisition of the fully feathered plumage of adult birds by downy nestlings is achieved through a series of postnatal, postjuvenile, and prenuptial molts. Once the adult plumage is attained, the feathers are replaced in most healthy birds at least once a year, and in some species much of the plumage is replaced two or even three times during a year. A single annual (postnuptial) molt is characteristic of most species; its main function is the renewal of worn or faded plumage, and it is usually complete involving all the feathers. Other keratinized epidermal secondary sexual characteristics, such as the highly colored beak plates or horns of male puffins and pelicans, are also shed at the end of the breeding season, before or at the same time as the feathers.

In addition to the postnuptial molt a second molt occurs in many species before the onset of reproductive activity (Figure 30). The prenuptial molt, when present, may involve some or all of the feathers, and its main purpose is to accentuate secondary sexual characteristics, by replacing winter plumage with brighter feathers

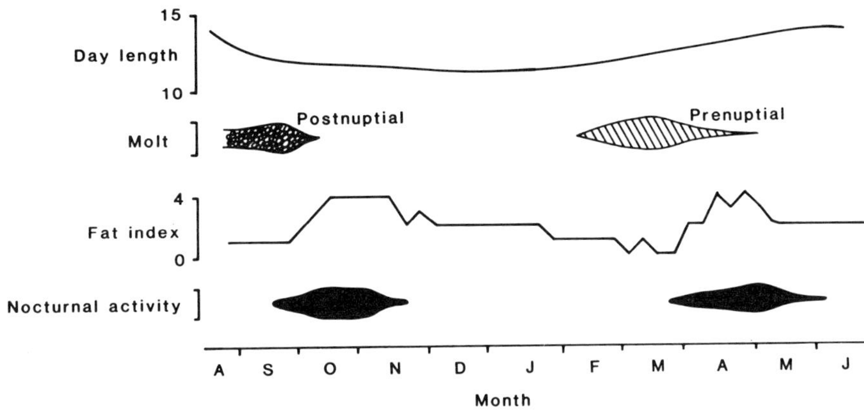

Figure 30. The occurrence of molt, fat deposition, and nocturnal activity in *Passerina cyanea*. (Based on Emlen, S. T., 1969, Bird migration: Influence of physiological state upon celestial orientation. *Science* **165:**716–718.)

or with additional plumes or other adornments. The display of nuptial plumage provides a means for attracting members of the opposite sex and for advertising territorial defense.

Molting is achieved at least cost to the bird in maintaining itself and may occur over a long or short period of time. Prolonging the period of molt reduces the daily energy demands for feather replacement, but it may subject the bird to a constant energy drain or impair some functions of the plumage. Molting over a short interval (e.g., between the end of breeding and onset of migration) may be an ecological necessity for some species, especially if the winter food supply is scarce or the wings are incapable of a prolonged migratory flight. A heavy molt may, however, result in flightlessness and impair the ability to forage for food at a time when energetic demands are greatly increased. Consequently, the advantages of both molting schedules are adopted by some species which prolong part of the molt (especially the wings to retain the ability to fly), while rapidly molting other parts of the plumage. Prolonging the molting period may reduce the daily energy requirements for feather regeneration to such an extent that breeding and molting can occur together. Although molting occurs outside the breeding season in the majority of tropical and temperate birds, and species from arid environments, reproductive activity and molt overlap in more than 100 species. An overlap of molting and breeding may occur when food availability is abundant or when the demands of incubation necessitate a period of immobility.

The timing of the molt in most temperate zone species is largely controlled by seasonally changing photoperiod, but in seasonally uniform areas of the world the rainfall or food availability may be determining factors. Unseasonal changes in temperature, rainfall, and food availability may modify molt in temperate species, but they are not causal factors since they vary irregularly over short periods and are poor indicators of future conditions. The completion of molt usually requires several weeks, and daylength is a more reliable predictor of environmental conditions, as demonstrated experimentally by the phase shifting of breeding and molt by changes in the photoperiod. Although breeding and molt may be mutually exclusive in terms of metabolic demands, they are separated during the annual cycle by similar stimuli and are probably regulated by similar mechanisms.

The stimulation of gonadal activity during spring occurs as a result of increased gonadotropin secretion, and this may also stimulate the prenuptial molt. The administration of LH to some nonbreeding finches results in previously plucked areas of the breast being refeathered with black nuptial plumage rather than with brown winter plumage. Since these effects occur during quiescent gonadal activity and are observed in castrated birds, seasonal variations in gonadotropin level may directly influence the period of molt. The development of secondary sexual plumage in dimorphic species also occurs after the onset of gonadal activity since it is androgen or estrogen dependent. The onset of reproductive activity in some species is preceded (and may be caused) by an increase in thyroid activity, and as increased thyroid function often occurs during the postnuptial molt, it may also be involved in the prenuptial molt.

The subsequent surge in gonadal steroid secretion during the breeding cycle appears to inhibit molt. During this time the replacement of any lost feathers may be delayed, as in the postnuptial molt if favorable environmental conditions prolong the breeding season. A delay in the timing of molt may accelerate the rate of its progress, to ensure its completion during favorable conditions or before migration. Molt is similarly arrested or delayed in birds treated with androgens or estrogens, whereas it is advanced (in pigeons) following castration. The removal of gonadal steroids by castration also causes an almost continuous molt in chickens and ducks. Hypophysectomy (removing gonadotropic stimulation) has a similar effect as castration, although the shape and degenerative barbules of the replacement feathers are indicative of a hypothyroid state; hypophysectomy also resulting in a thyrotropin (TSH) deficiency.

The withdrawal of hormones concerned with reproduction at the end of the breeding cycle may thus provide a stimulus for the heavy postnuptial molt observed in many temperate zone birds. The difference in the time taken for molt to be completed in tropical and temperate zone birds also suggests that this may result from a strong endocrine stimulus to ensure that molt is completed quickly in higher latitudes. A synergism between several endocrines may be involved in such phasing, one of which may be thyroxine.

The thyroid gland is involved in the timing of the annual molt since pituitary TSH content and thyroid activity increase prior to or during the molting period, as a result of the seasonal decline in gonadotropin or gonadal steroid secretion. The persistence of normal molting schedules in some castrated songbirds may thus indicate seasonal changes in thyroid function induced by changes in gonadotropin activity or a direct response to environmental stimuli. Molt occurs only when thyroid function is increased and does not occur in some species (e.g., quail) maintained on long photoperiods if gonadal regression fails to occur. This results from the inhibitory effects of sex hormones because increased thyroid activity occurs in photostimulated castrated quail and in intact species (spotted Munia) in which the increase in daylength is insufficient to stimulate gonadal growth. The increased circulating TSH and thyroxine (T_4) levels in castrated birds are reduced by androgen or estrogen treatment, which also depresses thyroid function in intact birds. The reciprocal relationship between the thyroid–gonadal axis is also demonstrated by the decline in blood estrogen levels in fowl during forced (starvation-induced) molting. Force molting increases T_4 concentrations and also elevates circulating triiodothyronine (T_3) levels during refeathering.

The administration of T_4 stimulates an unseasonal molt in many birds (after a latency of 5–7 days), but affects body feathers rather than those of the head, neck, wings, or tail. The rate and intensity of molt induced by T_4 is dose-related and at high dosage may cause nakedness (in canaries). The dose of thyroid hormone required to stimulate molt also varies seasonally and is greater during gonadal activity than before or afterward.

Thyroid hormones not only promote the metabolic changes during molt but stimulate the developing feather papillae, causing cell division and thereby accelerating molt. The thyroid hormones also increase oxidative processes in the follicle and regulate the cellular material entering the generative zone of the feather. Although feather germs can be formed in the absence of the thyroid, thyroidectomy or pathological thyroid disorders prevent further development in intact or castrated birds, effects reversed by T_4 administration. Extirpation of the thyroid after the formation of feather germs has no effect on molt. Consequently, while thyroidectomy 2 or 3 months before the onset of the postnuptial molt in starlings delayed or arrested the molt, thyroidectomy 1 month before molt was without effect. The reported insensitivity of some wild species to T_4 may thus be related with their physiological condition rather than to species differences.

The effects of thyroxine on molt may (paradoxically) be augmented by progesterone, which increases in parallel with T_4 during periods of forced molting. Progesterone may stimulate thyroid activity and may locally accelerate feather replacement in plucked regions of the plumage. At high doses (less toward the end of the breeding cycle) it may also cause a cessation of lay and induce molting, probably by inhibiting gonadotropin secretion and thereby reducing blood sex hormone levels. However, phases of increasing progesterone and T_4 levels rarely overlap in the annual cycles of temperate species and are probably of little significance in the natural environment except in species that molt during sexual maturation and in species with overlapping breeding and molting cycles or in species (e.g., rooks) that molt during gonadal recrudescence following spring breeding.

In species that have a prolonged molt its interruption during phases of the reproductive cycle may be a consequence of increased prolactin secretion. The wing molt in pigeons is interrupted during the period of parental care, at a time when endogenous prolactin levels are high. The administration of prolactin-enriched pituitary extracts to pigeons similarly delays molt. Prolactin may thus serve to separate molt and parental care, both of which impose considerable nutritional demands on the parents. Prolactin similarly delays the postnuptial molt in some migratory songbirds and may be important in phasing fat deposition and migratory restlessness in temperate species.

The mechanisms involved in the periodic molts of arid region birds have yet to be determined, but molt in these birds is probably independent of gonadotropin or gonadal steroid secretion. In such birds there may be an endogenous or internal rhythm of feather loss and renewal, which may proceed without any seasonal changes in the secretion of molt-inhibiting or stimulating hormones. In many tropical species the molting rhythm is more precise than the gonadal cycle, which may be more dependent on food availability. The molt cycle in tropical birds may be timed by some factor (endogenous or exogenous) and that, in turn, may impose a periodicity on the reproductive cycle.

In addition to hormonal influences, the shedding of feathers may also occur as a result of neural stimuli. When some birds are handled, kept in captivity, or exposed to stressful conditions, they may shed all their feathers at once. This "fright" molt may be accompanied by increased adrenocortical function, but the rapidity of its onset suggests a neural mechanism. A similar mechanism may also be responsible for the molt of one bird soon after the onset of molt by its mate, possibly in an endeavor to promote a synchronization of molt. The shedding of feathers by male swans and geese 3–4 days after the female may result from visual and/or auditory stimuli resulting from the female, and under the influence of nerve impulse, these stimuli may loosen the old feather within its follicle, stimulating the development of the feather germ.

The first feather in a series to drop is usually from a follicle more highly vascularized than the others, since it can receive more oxygen and nutrients and "molting hormone" than its neighbors. Each feather in a sequence is usually only replaced after its neighbor has partly regrown or completed a time-dependent change, and as it is shed, it may improve the vascularization of adjacent follicles. The regular shedding of feathers may also result from differing sensitivities to hormone action, the last molting feathers having a higher threshold than earlier ones requiring prior sensitization. Following feather shedding the follicle requires a specific time for papillae development and during this time is refractory to the action of molting hormones, successive applications of thyroid hormones being less able to induce molt. As a result of refractoriness the cyclicity of molting schedules may become independent from reproductive cycles if their frequency is photoperiodically increased. The timing and control of molts is thus complex, dependent on phases of reproductive activity, in most cases, and temporal patterns of feather sensitivity to molting stimuli.

6.10. Migration

The annual cycle of many avian species is characterized by cyclical periods of migration which occur with temporal precision before and after the breeding season. Migration in the postbreeding season usually occurs because of adverse changes in the environment at a time when caloric intake must be maintained for thermoregulation. Spring migration from the winter range occurs to avoid its undesirable summer conditions or to exploit the summer conditions of its breeding range. Most migrants only breed in the summer and must leave their winter quarters to do so.

In some species, "weather migrants," migration occurs in response to improvements or deteriorations in environmental conditions (ultimate factors), even interrupting breeding to do so. In "true migrants" the urge to migrate is instinctive, migration occurring every year at the same time of the month, irrespective of local weather conditions or past experience, and usually in advance of adverse seasonal changes in the environment. The inception of instinctive migration occurs in response to "proximate" changes in the internal and external environment which (by temporal association with later "ultimate" changes) predict future alterations in the external environment. Spring and autumn migration therefore normally occur in response to different proximate factors, and the proximate factors involved may be species specific and age or sex related. The onset of migration is thus dependent on the birds' physiological status, and the proximate factors act as a "fine-tuning" system to integrate different behavioral phases (reproduction, molt, migration) during the annual cycle, since these are temporally related.

The physiological state necessary for migration (*Zugdiposition*) is associated with the deposition of fat, to meet the metabolic demands of flight, and premigratory increases in fat deposition occur with circannual rhythmicity (Figure 30). The deposition of fat occurs over a few days and results from hyperphagia and hyperphagia-induced lipogenesis. Premigratory fattening is photoinducible in some species, probably by daylength effects on hypothalamic feeding centers or alterations in endocrine secretion.

Increasing daylength stimulates thyroidal activity, and migration occurs during periods of enhanced T_4 secretion, when thyroidal colloid is being resorbed and the plasma T_4 and T_3 level is elevated. Although seasonal cycles in thyroid function also occur in nonmigratory species in relation to molt and the reproductive cycle, they are generally of lower amplitude than those in migratory birds. Nonmigratory species do not accumulate fat in the spring and have a lower incidence of prenuptial molting, possibly reflecting evolutionary relationships between the thyroid hormones, molt, and fattening. Increased thyroid function results in an increase in food intake and results in the accumulation of fat. The increase in food intake also stimulates the extrathyroidal conversion of T_4 to T_3 (which is metabolically more active), and premigratory fattening occurs in redheaded buntings when the T_3/T_4 ratio is elevated. Thyroidectomy prevents premigratory fattening in this species and reduced hepatic free fatty acid content, effects reversed by the administration of exogenous T_4 or T_3. In the bunting the autumn migration occurs after the postnuptial molt at a time when T_4 rather than T_3 is elevated. The seasonal increase in T_4 during the postnuptial molt may also be responsible for the autumnal deposition of some fat in nonmigratory species. Autumnal fattening may partly occur in response to surplus energy previously used during breeding, but may occur as a result of "physiological castration," castration increasing thyroid function.

Increasing daylength has some stimulatory effects on prolactin secretion, and the deposition of fat, prior to migration, may occur as a result of increased prolactin secretion. Pituitary prolactin levels are increased during periods of spring and autumn migration, and exogenous prolactin augments photoinducible weight increases by stimulating hyperphagia and accelerating the rate of liver lipogenesis. This effect appears to be temporarily synergized with a diurnal rhythm of plasma corticosterone and with photoperiodically induced increases in LH and FSH, neither of which has any effect by itself. Prolactin, especially in combination with corticosterone, also stimulates migratory restlessness, possibly by causing an increase in nocturnal fat catabolism.

Once migratory reserves have been laid down, many migratory birds become very active, particularly during the night (Figure 30). This circannual restlessness (*Zugunruhe*) is partly stimulated by large fat reserves, but its intensity may be regulated by thryoidal function, being reduced in hypothyroid birds and stimulated by thyroid hormone therapy. The influence of the thyroid hormones on locomotor activity is probably due to their activation of oxidizing pathways and their stimulation of energy metabolism, resulting in the mobilization of fat and body weight loss. The metabolism of migratory species is higher than in nonmigrants of the same size necessitating greater activity to maintain thermoregulation. Cold weather stimulates migratory restlessness during autumn and, since this also increases thyroidal activity, may be a stimulus for the inception of movement in receptive birds.

Premigratory restlessness is also associated with flocking. Postnuptial flocking occurs because of gonadal regression and the suppression of territoriality, whereas prenuptial flocking occurs during early phases of gonadal recrudescence. Similar seasonal cycles of gonadal development occur in nonmigrants, although gonadal recrudescence in spring occurs earlier, develops at a faster rate, and is not associated with an increase in body weight.

The sexual and migratory cycles are both photoperiodic and may be regulated by a common endocrine mechanism. Gonadal recrudescence induced by spring photostimulation results in the production of sex hormones that may trigger vernal migration. Premature migration can be induced in some species by midwinter photostimulation, although vernal fattening and migratory activity do not occur in birds castrated prior to photostimulation. Testosterone may affect spring fattening at hypothalamic sites, but since it stimulates prolactin secretion, the effects of castration may be mediated by a prolactin deficiency and an inability to develop *Zugdisposition*. Castration after the onset of photostimulation consequently fails to disrupt spring fattening, especially since the persistence of *Zugdisposition* may itself stimulate migratory activity. Testosterone is, however, directly involved in the induction of nocturnal migratory restlessness, probably by changing the phase relationship of two oscillators that control the morning and evening maxima of normal daytime activity so that they extend into the dark phase.

The importance of gonadal differentiation in the induction of spring migration may also be responsible for the failure of birds that take 2 or more years to attain sexual maturity to leave their winter quarters. The incomplete gonadal development in some adults during unfavorable environmental conditions may similarly be responsible for their inability to migrate or complete their migration. Since spring migration may also be arrested after completion of gonadal development, migration may be associated with increasing titers of gonadal hormones in the peripheral circulation. Increasing the blood testosterone level at the end of the breeding season by high doses of exogenous hormone resembles the physiological condition during spring and promotes migration, even though the testes are regressing. Conversely, if the testosterone level is prevented from declining at the end of the breeding season by low doses of exogenous hormone, autumn migration is prevented (possibly by suppressing thyroid function). Autumn migration necessitates a reduction in gonadal activity, but since it occurs during or after gonadal regression and in immature birds or castrates, unlike the vernal migration, it is not dependent on the phase of gonadal activity *per se*.

Following gonadal regression the development of photorefractoriness in photoreceptive migrants may prevent them from immediately reentering a second reproductive cycle when exposed to the longer daylengths of their winter quarters. Few migratory cycles correspond symmetrically with the annual pattern of decreasing and increasing daylength, and in some cases the timing of the gonadal cycle may be controlled by an endogenous rhythm that programs the temporal sequence of reproductive activity. These circannual cycles are entrained by environmental stimuli (*Zeitgebers*) at one point in time (e.g., exposure to long photoperiods during summer), and the cycle is gradually lost when exposed to constant conditions. The overlap of gonadal enlargement, molt, and fattening in contrast (favorable) conditions suggest that the normal temporal association between them is not necessarily causal. These functions must be controlled by a distinctive endocrine *milieu*, and there is some evidence that the synthesis and release of the different pituitary hormones responsible for creating these conditions are controlled by separate oscillators phased differently to each other. Circannual cycles are thus especially adaptive for migrants that visit or move within the tropics, in which the quasistable environmental conditions are deficient in reliable proximate timing devices. In addition to determining the onset of migration, circannual rhythmicity may also be involved in determining the temporal course, distance, and direction of migration, especially in younger birds. These circannual mechanisms may be replaced in older birds by other mechanisms, including mechanisms of true navigation.

The control of migratory behavior is thus complex and, in addition to environmental factors, is dependent on changing patterns of hormone secretion, the receptivity and physiological readiness of the bird, and instinctive and acquired behavioral patterns.

7. Selected Readings

Balthazart, J., Prove, E., and Gilles, R., eds., 1983, *Hormones and Behaviour in Higher Vertebrates*, Springer-Verlag, Berlin.

Cunningham, F. J., Lake, P. E., and Hewitt, D., eds., 1983, *Reproductive Biology of Poultry*, British Poultry Science Ltd., Longman's Group, Harlow.

Farner, D. S., and King, J. R., eds., 1971–1983, *Avian Biology*, Vol. 1–7, Academic Press, New York.

Mikami, S. I., Homma, K., and Wada, M., eds., 1983, *Avian Endocrinology: Environmental and Ecological Perspectives*, Springer-Verlag, Berlin.

Perrins, C. M., and Birkhead, T. R., 1983, *Avian Ecology*, Blackie, Glasgow.

5

Reproduction in Reptiles

Ian P. Callard and S. M. Kleis

1. Introduction

Despite considerable diversity of reptilian cycles, it is possible to classify those of both males and females into a few broad categories. Most of the available information is derived from the *Squamata* and *Chelonia,* with little information available from the *Crocodilia* and *Rhyncocephalia.* The following patterns can be observed:

1. Seasonal breeding (discontinuous) with periods of reproductive activity separated by inactivity. Males show most variation in the time of onset and duration of spermatogenesis. Females may be seasonally monestrous (single clutch) or seasonally polyestrous (multiple clutch). This pattern is found in all temperate species and some from the tropics.
2. Acyclic (continuous) with a similar intensity of reproductive activity all year. Species with this pattern of reproduction tend to be tropical in distribution. There is some question as to whether truly continuous breeding in reptiles has been convincingly demonstrated in many species for which it is claimed. However, seasonal changes are much less dramatic than in temperate species, and at least one oviparous sea snake appears to have a continuous pattern of reproduction.

External environmental cues (temperature, photoperiod, rainfall) exert profound effects on the physiology and behavior of reptiles, and the appropriate synchrony and timing of the reproductive cycles and behaviors in relationship to such

Ian P. Callard and S. M. Kleis • Department of Biology, Biological Science Center, Boston University, Boston, Massachusetts 02215.

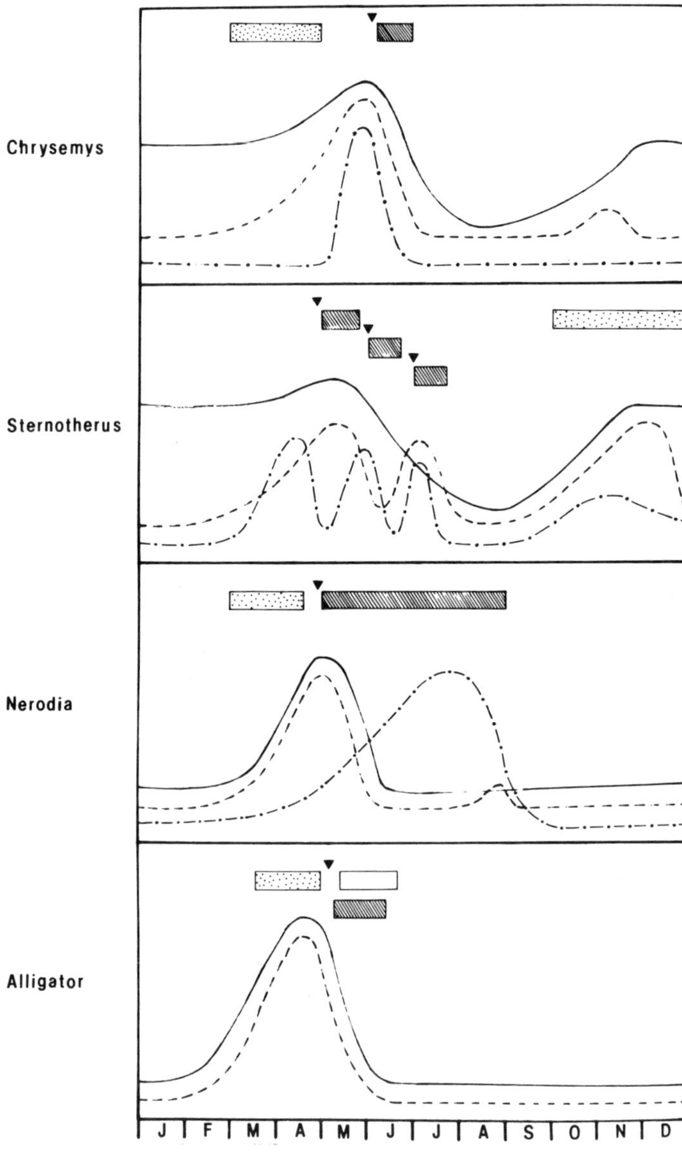

Figure 1. Endocrine correlates of reproductive cycles in female reptiles. See text for details. Stippled bar, period of preovulatory follicular growth; hatched bar, period of luteal activity; solid line, ovarian growth; dashed line, plasma estradiol level; dashed–dotted line, plasma progesterone level. Months of the year shown at the bottom of the diagram.

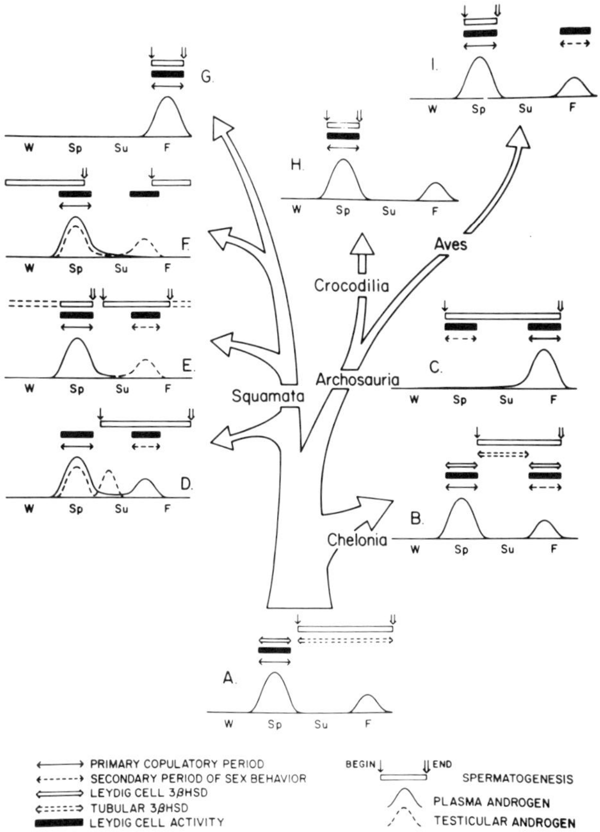

Figure 2. Endocrine correlates of reproductive cycles in male reptiles. See text for details. A, Basic temperate zone "postnuptial" pattern, similar to amphibia. B and C, Turtle pattern as in *Chrysemys picta, Trionyx sinensis* (B) and *Sternotherus oderatus* (C). D–G, Squamate patterns as in *Agkistrodon piscivorus, Nerodia* sp., *Lacerta vivipara* (D); *Vipera berus, Uromastix hardwickii* (E); *Naja naja; Tiliqua scincoides* (F); *Acrochordus granulatus* (G). H, Crocodilian pattern—*Alligator mississippiensis.* I, Temperate avian pattern.

environmental cues are crucial to ensure birth of young at an appropriate time of the year for their continued survival. In temperate regions, the combined male/female cycle in monestrous species lasts 8–10 months, beginning with spermatogenesis and followed by spring vitellogenesis, mating, ovulation, and summer embryonic development. In regions with longer, more severe winters, the complete cycle can take longer than 12 months owing to spermatogenic arrest. Sperm storage may be in the male ducts, or in the female ducts if fall matings occur prior to vitellogenesis. In some lizards, vitellogenesis and ovulation may occur in the fall with embryonic development either being delayed (oviparous species) or young ready for birth in spring after winter development (e.g., *Sceloporus cyanogenys*).

Table I
Ranges of Plasma Gonadal Steroid Levels during the Reproductive Cycle of Reptiles

Order/Species[a]	Sex	Steroid, mean range (ng/ml)		
		Progesterone	Testosterone	Estradiol
Order *Chelonia*				
Chrysemys picta[21,6,7]	M	—	0.10–50.0	—
	M	—	0.05–10.0	—
	F	0.1–2.5	0.05–5.7	ND–1.2
Sternotherus odoratus[22]	M	—	2.0–150	—
	F	0.75–4.5	0.3–1.8	0.5–6.0
Testudo hermanni[15]	M	—	0.05–25.0	—
Chelonia mydas[19,20]	M	—	12.0–33.0	—
	F	0.1–18.5	0.1–1.30	0.03– 0.1
Chelydra serpentina[18]	F	0.1–3.6	—	0.01–0.3
Order *Squamata*				
Suborder *Lactertilia*				
Uromastix hardwickii[1,2]	M	—	3.1–22.4	—
	F	1.6–13.4	0.4–1.5	0.2–0.45
Lacerta vivipara[8,25]	M	—	1.84–445	—
	F	10–400	—	—
Tiliqua rugosa[4]	M	—	2.0–47.0	—
	F	3.0–13.0	—	—
Sceloporus cyanogenys[5]	F	9.0–33.0	—	—
Sceloporus jarrovi[10]	F	0.5–3.8	—	—
Anolis carolinensis[13]	F	2.3–2.6	—	0.5–2.1
Chameleo pumilis[23]	M	0.6	—	—
	F	0.86–5.0	—	—
Suborder *Ophidia*				
Nerodia species[24,14,17]	M	—	0.05–21.0	—
	F	0.6–9.5	0.3–0.8	0.3–1.0
Thamnophis s. sirtalis[11]	F	1.0–6.2	—	—
Naja naja[3]	M	—	0.1–15.0	—
	F	1.4–25.0	0.1–0.7	0.1–0.3
Agkistrodon piscivorus[12]	M	—	0.5–4.0	—
Acrochordus granulatus[9]	M	—	0.1–80	—
	F	10–120	0.1–0.9	0.05–1.0

Types of reproductive cycle and correlated plasma steroid levels are diagrammed in Figures 1 and 2. Reported values for the three major plasma gonadal steroids in male and female reptiles are summarized in Table I.

2. Female Reproduction

Most female reptiles show a clear seasonal pattern of ovarian activity and egg production, alternating with ovarian inactivity. Generally, in temperate, monestrous

Table I *(Continued)*

Order/Species[a]	Sex	Steroid, mean range (ng/ml) Progesterone	Testosterone	Estradiol
Cerberus rhynchops[9]	M	—	0.2–4.0	—
	F	0.2–0.6	0.01–0.07	0.2–0.8
Laticauda colubrina[9]	M	—	0.5–2.5	—
	F	0.1–0.5	0.01–0.1	.05–0.6
Order *Crocodilia*				
Alligator mississippiensis[16,*]	M	—	2.0–53	—
	F	—	ND–8.0	ND–1.6

[a]Superscript numbers refer to the following references:

1. Arslan, M., Lobo, J., Zaidi, A. A., Jalali, S., Qazi, M. H., 1978a, *Gen. Comp. Endocrinol.* **36:**16–22.
2. Arslan, M., Zaidi, P., Lobo, J., Zaidi, A. A., Qazi, M. H., 1978b, *Gen. Comp. Endocrinol.* **34:**300–303.
3. Bona-Gallo, A., Licht, P., MacKenzie, D. S., Lofts, B., 1980, *Gen. Comp. Endocrinol.* **42:**477–493.
4. Bourne, A. R., Seamark, R. F., 1975, *Comp. Biochem. Physiol.* **50B:**535–536.
5. Callard, I. P., Bayne, C. G., McConnell, W. F., 1972, *Gen. Comp. Endocrinol.* **18:**175–194.
6. Callard, I. P., Callard, G. V., Lance, V., Eccles, S., 1976, *Gen. Comp. Endocrinol.* **30:**347–356.
7. Callard, I. P., Lance, V., Salhanick, A. R., Barad, D., 1978, *Gen. Comp. Endocrinol.* **35:**245–257.
8. Courty, Y., Dufaure, J. P., 1979, *Gen. Comp. Endocrinol.* **39:** 336–342.
9. Gorman, G. C., Licht, P., McCollum, F., 1981, *J. Herpetol.* **15:**335–354.
10. Guillette, L. J. Jr., Speilvogel, S., Moore, F. L., 1981, *Gen. Comp. Endocrinol.* **42:**20–29.
11. Highfill, D. R., Mead, R. A., 1975, *Gen. Comp. Endocrinol.* **27:**369–400.
12. Johnson, L. F., Jacob, J. S., Torrance, P., 1982, *Herpetologica* **38:**16–25.
13. Jones, R. E., Guillette, L. J. Jr., Summers, C. H., Tokarz, R. R., Crews, D., 1983, *J. Exp. Zool.* **227:**145–154.
14. Kleis San Francisco, S., Callard, I. P., 1986, *Gen. Comp. Endcrinol.* **61:**490–498.
15. Kuchling, G., Skolek-Winnisch, R., Bamberg, E., 1981, *Gen. Comp. Endocrinol.* **44:**194–201.
16. Lance, V., 1984, *Symp. Zool. Soc. Lond.* **52:**357–383.
   * Lance, V. (unpublished).
17. Lance, V., Callard, I. P., 1978, *Gen. Comp. Endocrinol.* **35:**295–301.
18. Lewis, J., Mahmoud, I. Y., Klicka, J., 1979, *J. Endocrinol* **80:**127–131.
19. Licht, P., Wood, J., Owens, D. W., Wood, F., 1979a, *Gen. Comp. Endocrinol.* **39:**274–289.
20. Licht, P., Rainey, W., Cliffton, K., 1979b, *Gen. Comp. Endocrinol.* **40:**116–122.
21. Licht, P., Breitenbach, G. L., Congdon, J., 1985, *Gen. Comp. Endocrinol.* **59:**130–139.
22. McPherson, R. J., Boots, L. R., MacGregor, R. III, Marion, K. R., 1982, *Gen. Comp. Endocrinol.* **48:**440–451.
23. Veith, W. J., 1974, *Zool. Africana* **9:**161–183.
24. Weil, M. R., Aldridge, R. D., 1981, *Gen. Comp. Endocrinol.* **44:**44–53.
25. Xavier, F., 1962, *Herpetologica* **38:**62–70.

squamates, seasonal ovarian growth is explosive, and the ovary remains small between growth phases. Successive generations of ovarian follicles destined to become successive clutches are much less readily discerned in squamates than in monestrous and polyestrous chelonians, in which follicular growth extends over the greater part of the year. Temperate zone species typically ovulate in the spring, but some viviparous species ovulate in the fall or winter (e.g., *Sceloporus cyanogenys, S. jarrovi*). In these, embryos develop over the winter and parturition occurs in spring. Most small lizards appear to reproduce annually, but many snakes and turtles breed every second, third, or even fourth year, as do some marine turtles.

Female squamates have been divided into two basic types regarding reproductive strategies: (1) early-maturing, multiple-brooded, and (2) late-maturing, single-brooded. Most snakes fall into the latter category. Of the early-maturing species, oviparity is almost the universal type of reproduction, and nearly all species have multiple broods (seasonally polyestrous), and mean clutch size is significantly smaller than in the late-maturing type. These also are smaller bodied at maturity and tend to be tropical and temperate in distribution, as opposed to primarily temperate for the delayed-maturity type. Viviparity is a strategy of the delayed maturity type. It has been suggested that the increased incidence of viviparity associated with increasing altitude and latitude may represent an adaptation to more uncertain climates. Maintaining the embryos in the more controllable environment of the maternal organism provides for water, thermoregulation, and protection from bacteria and predators, thereby helping to ensure reproductive success. Thus, viviparity contributes to an increase in survivorship of eggs and young, which, with increased body size at maturity and larger clutches, compensates for a lower fecundity per season as compared with multiple-brooded types.

In most viviparous reptiles, the amount of yolk deposited during vitellogenesis equals that of related oviparous reptiles of similar size and is the main source of embryonic nutrients during the usually annual gestation period of approximately 3 months. In such species yolk sac placentation is established early in development prior to that of the allantoplacenta. Three main types of allantoplacenta have been described based on histological observations; however, in all cases the embryonic–maternal tissue attachment remains superficial. In the most specialized type, an elliptical area of intimately apposed, interdigitating, glandular maternal and embryonic tissues develops, and in some cases (as in the skink, *Chalcides chalcides*), this is accompanied by a dramatic decrease in yolk stores. This suggests that the increase in placental complexity represents a transition to direct maternal nutrition of the developing embryos. In addition to water and gas exchange, the placenta of some species has been shown to be permeable to inorganic ions and steroids. More information on the development of viviparity may be found in Chapter 8, "Vitellogenesis and Viviparity."

Similar reproductive strategies appear to exist within the *Chelonia,* although viviparity is absent, as with the *Crocodilia.* Most northern varieties of *Pseudemys* and *Chrysemys* produce a single clutch in late spring (seasonally monestrous) although two clutches have been reported. In contrast, some tropical turtles, such as *Kinosternum leucostomum,* believed to be long adapted to the subtropics, produce small clutches throughout the year (seasonally polyestrous). A category intermediate between these two exists in some Chelonians. Southern varieties of *Pseudemys, Chrysemys,* and *Sternotherus odoratus* produce from one to six clutches without reduction in clutch size, and sea turtles are well known for their ability to produce multiple large broods. Such species appear to be capable of optimizing their reproductive potential in more favorable environments by increasing the number of clutches, but maintaining a large brood size. A basic difference between these chelonians and lizards may be the longevity and continued increase in body size of turtles.

2.1. Female Reproductive Cycles

Once a reproductive cycle has been initiated, the differences between the reproductive patterns may be attributable primarily to (1) the length of the intervals between successive clutches, and (2) the number of follicles that reach maturity during each clutch interval. The control of these two variables is by a complex interplay between the ovary, pituitary, and hypothalamus, which is reflected in temporal changes in plasma gonadotropins and gonadal steroids. The gonadal steroids have important intraovarian and peripheral and central actions concerned with follicular maturation, ovulation, and initiation of appropriate responses from the reproductive tract. (See Figure 1.)

2.1.1. *Chelonia*

2.1.1a. Monestrous; Single-Clutch (*Chrysemys picta*)

In this turtle, virtually synchronous peaks of the three major gonadal steroids occur associated with the final stages of follicular maturation, levels of all three steroids dropping at ovulation and remaining low during the summer phase of ovarian regression. During the fall, in this and other chelonian species, ovarian recrudescence occurs and is associated with a second but smaller peak of testosterone and estradiol, but not progesterone. The falling phase of the progesterone peak in *Chrysemys* coincides with the brief organization of the corpus luteum, which persists during the approximately 2-week period during which ovulated eggs remain in the oviduct. Plasma levels of LH also rise during the spring period of gonadal development and fall after ovulation. (See Figure 1, top panel.)

Although this species is classified as a monestrous species on the basis of a number of years of observation and published reports, the lability of this condition is indicated by observations of two clutches following an early and mild spring. This suggests that the number of clutches is imposed by environmental conditions and may exceed one if favorable conditions exist.

2.1.1b. Polyestrous; Multiclutch (*Sternotherus odoratus, Chelonia mydas*)

In *Sternotherus odoratus,* which produces two or three clutches per year, all three major steroids rise in association with the first preovulatory phase, but only progesterone and estrogen show dramatic increases with each succeeding preovulatory phase of the cycle. In *Chelonia mydas,* the green sea turtle, plasma levels of LH and progesterone show pronounced peaks just before nesting. These peaks are correlated with the timing of ovulation. FSH levels remain low except for an increase at the time of nesting. Unlike *Chrysemys* and *Sternotherus,* plasma estrogen levels were not correlated with periods of vitellogenesis which are punctuated by repeated ovulations. In turtles, unlike the squamates (next section), periods of vitellogenesis are extensive and may occur in both spring and fall. (See Figure 1, second panel.)

2.1.2. *Squamata* (Lizards, Snakes)

Squamates differ from single-clutch turtles in the duration of ovarian recrudescence and vitellogenesis. In single-clutched (monestrous), viviparous (e.g., *Nerodia sipedon*) squamates, ovarian recrudescence is explosive and mostly occurs during a 4- to 6-week period in spring. During the rest of the year, no vitellogenesis occurs. In this species and others, plasma estradiol and testosterone peak at ovulation and then fall and remain low during pregnancy. Progesterone levels, however, begin to rise during the period of follicular growth, reach maximum during the luteal phase (midpregnancy), and decline to parturition. In oviparous squamates, such as the cobra (*Naja naja*), plasma estradiol and testosterone also rise during vitellogenesis and follicular growth, and plasma progesterone rises after ovulation and remains high until oviposition. (See Figure 1, third panel.)

2.1.3. *Crocodilia*

Progress is being made in understanding the reproductive biology of this important group. For the American alligator, *Alligator mississippiensis,* courtship and copulation occur in late April–early May. Ovarian growth is rapid in April and is well correlated with high plasma estradiol levels. Ovulation takes place soon after copulation, and oviposition of a single clutch of 30–40 eggs occurs 2–3 weeks later. Corpora lutea are formed after ovulation. (See Figure 1, bottom panel.)

2.2. Sites of Steroid Biosynthesis

The granulosa cells of all reptiles are the major site of steroid synthesis during vitellogenesis. Enzyme histochemical studies have shown that the intensity of the reaction for 3β-hydroxysteroid dehydrogenase (3β-HSD), a marker enzyme for steroidogenesis, increases in the granulosa cells as the follicles enlarge, suggesting that this may be the site of steroid synthesis and secretion. However, an "interstitial gland" derived from atretic follicles has also been suggested as a possible source of steroids.

After ovulation, a corpus luteum is formed. In oviparous species, the corpus luteum is short-lived and involutes just prior to oviposition, whereas in viviparous squamates the corpus luteum remains functional at least into midgestation and is probably the source of the elevated progesterone levels that occur during pregnancy. *In vitro* studies using radioactive precursors have established that the pathways of steroid biosynthesis are basically similar to those in mammalian ovaries. However, in some squamates it was found that deoxycorticosterone (DOC) was a major metabolite of pregnenolone in ovarian tissue incubates. In a snake, the activity of

21-hydroxylase enzyme, necessary for the formation of DOC, appeared to be confined exclusively to luteal tissue.

It has been suggested that extraovarian sources of progesterone during pregnancy may exist. The adrenal glands of snakes are also capable of producing this steroid, and it may be an important contributor of the residual levels of progesterone found in nonpregnant snakes; however, the contribution of the adrenal gland to circulating steroid levels during pregnancy is unknown. In a lizard, *Sceloporus jarrovi*, the peak of plasma progesterone appears to be reached after the corpora lutea have compacted and regressed, implicating a source other than the corpus luteum. However, this is not so in the snake *Nerodia*. Atretic follicles present throughout the reproductive cycle in some species may be responsible for the release or production of those steroids, or perhaps the chorioallantoic placenta, a known source of steroids in mammals, may also serve this function in some reptiles. Although histochemical and microscopic studies of garter snake placentae indicate the presence of steroidal or phospholipid materials, removal of all embryonic tissues (yolk, embryos, embryonic membranes) during late trimester I or early trimester II in *Thamnophis elegans* did not alter the level of circulating progesterone that was normally found in pregnant animals. No enzyme histochemical evidence for steroidogenesis has been found in water snake embryonic membranes. The possibility of locally acting progesterone or other hormones from these embryonic tissues has not been addressed.

3. Male Reproduction

Much has been made of the variations in the timing of the various phases of the male reproductive cycle in temperate reptiles, but the differences are more apparent than real. Thus, the terms "vernal" (spring), "aestival" (summer), "prenuptial," and "postnuptial" have been used to describe the timing of spermatogenesis with reference to the seasons and other reproductive events. The real differences are found in (1) the time of onset of spermatogenesis, (2) the speed of completion of spermatogenesis, and (3) the timing of the major period of sexual activity. However, a basic pattern can be discerned; the relationship between cycles in different species and the basic pattern is shown in Figure 2.

3.1. Male Reproductive Cycles

The basic pattern is like that of temperate amphibians (Figure 2A) with a spring period of reproductive behavior and mating, coincident with high plasma androgens and Leydig cell activity, as indicated by 3β-HSD. This is followed after a brief period by the onset of spermatogenesis, which is completed prior to the fall period of hibernation. Testicular growth and spermatogenesis coincide, and 3β-HSD activity is present in the tubules. Sperm are retained in the testes or accessory sex ducts over winter prior to the spring mating period. Throughout the reptiles, most variation appears to occur in the time of onset and duration of spermatogenesis.

3.1.1. *Chelonia*

The basic pattern is retained in chelonian species, as seen in *Chrysemys* and *Trionyx* (Figure 2B). However, in addition, these species have a second period of Leydig cell activity in the fall associated with the appearance of a new generation of Leydig cells. This is reflected in an active interstitial 3β-HSD, and a secondary period of mating behavior has been reported at this time of year. This trend is carried further in a more southerly species of turtle, *Sternotherus* (Figure 2C). Here, Leydig cells are active in the spring and the fall, but the major secretion of androgen is in the fall when mating and sperm transfer occur. Sperm are retained in the female tract until spring, when ovulation occurs. This switchover of the primary mating period from spring to fall in this species may be important in allowing rapid onset of the female cycle in the spring without the necessity of a preovulatory period of sexual behavior. This may be important in allowing the female of this southerly species to optimize reproductive potential, producing as many as three clutches (see Section 2.1).

3.1.2. *Squamata*

Among temperate squamates, the basic reptilian pattern (as in *Chrysemys* and *Trionyx*) can be seen in at least two snakes (Agkistrodon, the cotton mouth, *Nerodia,* the water snake (Figure 2D) and one lizard (*Lacerta*). Some differences may exist, although these may be more apparent than real and might disappear as the data become more complete. In *Lacerta,* in which testicular androgen has been measured, it is notable that one testicular androgen peak correlates with the period of Leydig cell activity and peak plasma androgen levels. A second testicular androgen peak also occurs, but this is not associated with Leydig cell activity or a peak in plasma androgens. It may be important that this increase in testicular androgen correlates roughly with the onset of spermatogenesis.

A slight modification of this squamate cycle is seen, as exemplified by the European viper (*Vipera berus*) and a lizard from Pakistan (*Uromastix*) (Figure 2E). In these species, two periods of Leydig cell activity occur, and in the viper, at least, fall mating behavior occurs, and in the more southerly part of the range, fertile matings have been reported. However, plasma androgen levels have not been measured. The spermatogenic cycle correlates with the latitudinal range of individuals under study. Thus, in the northerly individuals, spermatogenesis does not complete prior to the onset of cold weather and hibernation; spermatogenesis is arrested and is completed in the following spring when mating and sperm transfer occur. In the southerly part of the range, spermatogenesis may be completed, and a successful fall mating is thus possible. As with the chelonian, *Sternotherus,* this may allow earlier production of a brood by the female in the spring and thus maximize reproductive success. It has been suggested that gestation may continue during the winter hibernation in this species. If this were so, this would be of clear adaptive advantage. However, this observation needs to be confirmed. In *Uromastix,* where plasma and testicular androgens have been measured, only an increase in intratesticular androgen has been noted in the fall.

In two squamates, the Asian cobra, *Naja naja,* and the Australian scincid lizard, *Tiliqua scincoides* (Figure 2F) (in this case, for convenience, the cycle of this Southern Hemisphere species is reversed for comparison to the other Northern Hemisphere species), the onset of spermatogenesis is significantly delayed and does not begin until the fall. However, climatic conditions are such that spermatogenesis is not arrested in the winter months and is completed in time for the spring mating period despite the late onset. The adaptive value of this pattern is not immediately apparent, but it may relate to optimal temperatures for the onset of spermatogenesis. Although both species appear to have a period of Leydig cell activity in the fall, around the time of spermatogenic onset, this is not reflected in a rise in plasma androgens in either species; however, in *Tiliqua,* testicular androgens increase at this time.

Two other reptilian testicular patterns deserve note. In *Acrochordus* (Figure 2G), a tropical marine snake from the central Philippines (approximately 10° north), a clear testicular cycle can be seen, but, as with the southern turtle *Sternotherus,* the primary reproductive season is in the fall, when plasma androgen is high. In this species, spermatogenesis is completed in a short period of time, between August and November. Thus, the cycle in this species is more like that of the Southern Hemisphere lizard, *Tiliqua.* This might reflect on opportunistic adjustment to seasonally available food supply or another environmental variable.

3.1.3. *Crocodilia*

Recently, the testicular cycle of the American alligator has been determined (Figure 2H). The alligator retains the temperate pattern of a spring mating period, associated with Leydig cell activity and high plasma androgen levels. However, unlike most other reptiles, spermatogenesis is actually begun and completed in the spring within approximately 3 months. Although it is not clear whether a fall period of Leydig cell activity exists, there is a second, smaller plasma androgen peak at this time. The similarity of the alligator testicular cycle and that of temperate avian species (Figure 2I) probably reflects a common ancestral origin among the archosaurian stock.

Thus, the several trends in testicular cycles in reptiles appears to be related to the time of onset of spermatogenesis; although this is most frequently a spring–summer occurrence, it may begin in the fall, as in *Naja.* When begun in spring or summer, completion of spermatogenesis may be in the fall or the subsequent spring, and the primary determinant of completion at one time or the other is probably external environmental temperature. When spermatogenesis begins in the fall, it may continue through the winter and still complete in time for a spring mating.

Once the onset of the cycle has been determined by temperature, endogenous (possibly testicular) hormones (steroids and peptides) determine the course of the cycle and the interaction between tubular and interstitial elements and the timing of reproductive behavior. Of primary interest is the lack of confluence (asynchrony) between the onset of spermatogenesis and a peak in plasma androgen level, because the early stages of spermatogenesis are believed to be androgen-dependent, with the androgen presumably originating in Leydig cells. In fact, most temperate reptilian species studied are *asynchronous* in regard to Leydig cell activity/plasma androgens and spermatogenic onset. This being so we must consider spermatogenesis to be independent of a concurrent high level of plasma androgen or that adequate quantities of androgens are supplied from another intratesticular source (e.g., Sertoli cells).

3.2. Sites of Steroid Synthesis

Evidence that Sertoli cells may be an androgen source, as well as Leydig cells, comes from several observations. (1) In the freshwater turtle, *Chrysemys picta,* the enzyme 3β-HSD appears in the tubules after the spring period of Leydig cell activity and concurrent high plasma androgens. The enzyme remains active through the summer period of spermatogenesis, only disappearing at spermiogenesis in the fall, when Leydig cells once again become active. (2) In the same species, gonadotropin (FSH) can induce a massive secretion of androgen within 15 min of injection even when Leydig cells are atrophic and 3β-HSD cannot be demonstrated in the interstitium. (3) In two species of lizard (*Lacerta, Uromastix*) and possibly a third (*Tiliqua*), a peak of *testicular* androgen is coincident with the onset of spermatogenesis, without a corresponding plasma androgen peak. Possibly androgen is produced by active Sertoli cells for local consumption only. In the water snake, *Nerodia,* 3β-HSD has been demonstrated simultaneously in the interstitium and the tubules, suggesting that even when Leydig cells may be actively supplying androgen for peripheral use, tubular elements may provide direct steroid support for spermatogenesis. This removes the need for synchrony of spermatogenic onset with the spring mating activity and the associated high plasma androgen of Leydig cell origin. In addition, the circulating Leydig-cell-derived androgen serves to maintain the epididymis in a condition supportive of mature sperm. In this way, in most temperate male reptiles, germ cell production goes on during the summer when temperatures are optimum.

4. Regulation of Reptilian Gonadal Cycles

4.1. Environmental Cues

Temperature, photoperiod, food availability, and rainfall have been implicated as important variables in the control of onset of squamate reproductive cycles. In *Anolis carolinensis,* it appears that temperature and photoperiod act to cue different phases of the annual testicular cycle. Thus, spermatogenesis appears to be temperature-dependent from late fall to early spring, but photoperiod is important in testicular maintenance and eventual regression. At least one important locus through which temperature may influence gonadal function is the gonad itself, since at low (20°) temperatures the testis of *Anolis* will not respond to FSH by growth, but at higher temperatures (32°) it will. In addition, the primary influences of both temperature and photoperiod are probably exerted via hypothalamic control of gonadotropin output. In *Podarcis sinscula,* a lacertid lizard, a "refractory period" occurs in late summer, when the gonad becomes refractory to the stimulatory effects of high temperature; gonadal endocrine activity, vitellogenesis, and spermatogenesis cease. This refractory period disappears in late autumn, is useful in preventing reproduction at an inopportune time, and may be important in synchronizing the reproductive cycle, as maintenance at high temperatures from the middle of the refractory period may inhibit ovarian development the next spring.

In *Cnemidophorus uniparens,* the refractory period is under the control of an endogenous circannual oscillator at the hypothalamohypophyseal level. The pineal gland may play a role in reproduction in temperate lizards through an indirect thermoregulatory mechanism, but in *Cnemidophorus* it may not be involved in this refractoriness or the control of clutch interval.

The parietal eye of lizards has been implicated in control of metabolic rate, body temperature, gonadal and locomotor activity, increased energy usage and thermal sensitivity, increased exposure to light, and dispersal rate. It has been suggested that the parietal eye is primarily a photosensory organ, and retinal responses have been demonstrated. Transmission of information related to lighting is believed to prevent overexposure to the sun and thus overheating, thus functioning primarily in thermoregulation, rather than photoperiodism.

Social environment may also play a role in the synchronization of male/female interaction culminating in gonadal development and mating behavior. In *Anolis carolinensis,* androgen-dependent courtship behavior facilitates environmental cues stimulating ovarian recrudescence, and aggressive behavior between males inhibits initiation of ovarian activity. Mating further inhibits sexual receptivity during a particular follicular cycle.

4.2. Hypothalamic–Pituitary–Gonadal Interactions

Gonadotropic hormone releasing hormones (GnRH) appear to be present in the reptilian brain, and hypothalamic content of GnRH may vary seasonally. Furthermore, injections of pure mammalian LH-RH into female *Chrysemys picta* resulted in increased plasma levels of both immunoreactive (IR) LH and progesterone. Homologous hypothalamic extracts stimulated pituitary production of IR-LH *in vitro* in *Chrysemys.* In contrast, injections of mammalian or avian GnRH failed to consistently elevate either plasma IR-LH or steroids in *Sternotherus.*

4.2.1. Steroid Feedback

The onset of seasonal ovarian growth in the lizard *Dipsosaurus dorsalis* can be prevented by estrogen implants in the hypothalamus. In *Sceloporus cyanogenys* this same treatment did not interfere with ovarian development or vitellogenesis, but had a marked inhibitory effect on ovulation and caused oviduct regression, suggesting a reduction in ovarian steroid output. Steroid implants in other regions of the brain were ineffective, demonstrating a steroid feedback mechanism at the level of the hypothalamus. Progesterone implants made early in the ovarian recrudescent phase inhibited further ovarian development and induced a follicular atresia identical to that seen in hypophysectomized lizards. These data suggest that progesterone acts to inhibit the production of hormone or hormones concerned with follicular development and growth, whereas 17β-estradiol acts to prevent production of a hormone or hormones concerned with ovulation.

Further evidence for brain areas as target tissues for gonadal steroids has been provided by autoradiography where hormone-concentrating cells occur in areas likely to be concerned with the synchrony of gonadal and associated behavioral phenomena (preoptic, septal/hypothalamic, and thalamic areas). Also, an estrogen receptor has recently been described in the brain of *Chrysemys picta*. Reptilian brain and pituitary tissues are active in converting androgens to estrogens (aromatization), and this may be an important step in the expression of various androgenic actions on sex behavior, gonadotropin secretion, and brain sex differentiation.

4.2.2. Pituitary Gonadotropins

The existence of different cell types in the reptilian hypophysis was originally demonstrated in *Anolis carolinensis,* and subsequently, cycles of activity have been observed in other squamates. Definitive identification of two distinct gonadotrope cell types has not been achieved in reptiles.

Biochemical evidence indicates that there are two chemically distinct gonadotropic molecules in the anterior pituitary gland of crocodilian and chelonian species. The LH molecule from the snapping turtle, *Chelydra serpentina,* can be dissociated into subunits devoid of biological activity, and the subunits recombined or combined with α or β subunits of ovine LH with a restoration of biological activity. However, attempts to isolate two separate gonadotropic entities from the pituitary of squamates have been unsuccessful. The single moiety that appears to be present does not show clear homology to either mammalian FSH or LH. The inability to isolate two different biological entities from snakes is compounded by the lack of homologous bioassays and extremely narrow species specificity. Distinct biological roles of FSH and LH in the Reptilia are still far from clear for similar reasons.

4.2.3. Hormone Actions and Effects

4.2.3a. Gonadotropins

Separate biological roles of FSH and LH, identified in crocodiles and turtles, are still unclear. Both turtle FSH and LH stimulate ovarian hormone production by turtle ovarian tissue. The turtle testis responds well to both mammalian and chelonian FSH, by androgen production *in vivo* and *in vitro,* and the response is quite specific. Studies have also demonstrated binding of FSH to turtle testicular preparations, suggesting the presence of receptors. The binding of FSH increases as spermatogenesis proceeds in *Chrysemys.* Autoradiographic studies suggest that the interstitium is a site of gonadotropin action. Studies using turtle ovarian cells have shown that cyclic AMP stimulates steroid synthesis, suggesting that this nucleotide may be an intermediate in gonadotropin action. The effects of gonadotropin on the ovary are dual, causing first steroid synthesis and then uptake of yolk manufactured in the liver by the oocyte. The actions of the gonadotropins on the follicle–oocyte may involve steroid receptor interaction in the regulation of gonadotropin receptors and indirect local effects via estrogens stimulating follicular vascular supply and permeability. Such mechanisms may be involved in selective growth of certain follicles destined to ovulate next.

Thus, although these studies support the concept of gonadotropin regulation of the gonads by receptor mechanisms at this level of evolution, the relative importance of the two gonadotropins in the regulation of the reproductive cycle and steroidogenesis still remains to be established. This is particularly true of the largest and most diverse reptilian group, the squamates (lizards and snakes), in which only one gonadotropin has been identified thus far.

4.2.3b. Steroid Hormone Transport and Action

There is no doubt that the steroid hormones exert important physiological actions on the brain, liver, and reproductive tract of both sexes associated with reproduction. Thus, yolk protein production by the liver stops, the reproductive tract atrophies (or fails to grow), and sex behavior is not exhibited in the absence of the gonads. The effects of the steroid hormones appear to be mediated by specific binding proteins (receptors) assisted by less specific binding proteins involved in plasma treatment of hormones. The plasma binding proteins are synthesized in the liver and are also important in determining the actual amount of steroid available to the tissue. It is generally considered that the "free" (non-protein-bound) steroid is available for immediate tissue use; however, there is evidence that steroids can also enter cells bound to proteins. The binding of steroids to these plasma proteins may also protect them from rapid degradation by various tissues of the body. Two classes of plasma steroid binding proteins have been described in most mammals: a transcortin, which binds C_{21} steroids, and a sex steroid binding protein (SBP), which specifically binds C_{18} and C_{19} steroids. Similar SBPs have been described in several nonmammalian vertebrates. In *Chrysemys picta,* an SBP with an apparent K_a (affinity constant) of 10^8 M^{-1} and a binding capacity of 10^{-12} moles/mg has been described. Both testosterone and progesterone were able to compete with estradiol for binding sites, implying that perhaps this binding protein exhibits a different steroid specificity from those found in mammals. In the lizard, *Lacerta vivipara,* a SBP similar to that of mammals has been described. A transcortin-type (adrenal steroid binding) protein was also described; however, this protein was also able to bind testosterone, and the binding of corticosterone was weak.

SBPs exist in the tissues of reptiles as well. A SBP similar to that found in the plasma has been demonstrated in *Chrysemys* oviduct. However, it is unknown whether it is of interstitial or intracellular origin. Such proteins may function in the transport and concentration of steroids within the cell cytoplasm and could function as a labile reservoir of steroid inside the cell, allowing binding and release of free steroid according to local concentration gradients.

Other SBPs have been described in the oviduct, brain, and liver of some reptiles, which fit the description of a steroid receptor, the obligatory intermediate of steroid action at the level of the genome. Estradiol receptor has been detected and characterized in the brain, liver, oviduct, testis, and epididymis of the turtle, *Chrysemys picta,* and a progesterone receptor has been reported in the oviduct of a snake, *Nerodia,* and *Chrysemys picta.* These receptor molecules have a high affinity ($K_a = 10^9 - 10^{10}\ M^{-1}$) for the steroid and a very narrow steroid specificity. In the viviparous snake, *Nerodia,* changes in oviduct progesterone receptor occur during the cycle, and correlations with plasma steroids suggest regulation of the progesterone receptor by estrogen.

Inhibitory effects of progesterone on follicular growth may be mediated by direct effects of this steroid on yolk protein uptake or via a reduction in hepatic estrogen receptor (see Chapter 8, "Vitellogenesis and Viviparity"). Progesterone may also be important in mediating the effects of gonadotropins on ovulation.

5. Hormone–Oviductal Interactions

There appears to be a reciprocal interaction between the ovary and the oviduct which is concerned with the regulation of ovulation and oviposition. Thus, ovarian factors, possibly luteal in origin, appear to delay oviposition in a turtle, *Chrysemys,* and lizard, *Sceloporus.* In addition, the removal of eggs from the oviduct appears to allow the development and ovulation of supernumary oocytes in the lizard, *Cnemidophorus uniparens.* Although the exact nature of the humoral factors or possible neural pathways involved is unknown, *in vivo* administration of progesterone was found to prevent oviposition in the lizard, *Dipsosaurus dorsalis.* However, in several viviparous reptiles, removal of the source of progesterone by luteectomy or ovariectomy resulted in delayed parturition. Other factors important to normal pregnancy maintenance and parturition may also have been eliminated by these surgical treatments. In ovariectomized *Chameleo pumilus pumilus,* a decrease in the vascularity of the uterus and abnormal spacing of embryos was noted; however, this condition was not apparent in animals supplemented with progesterone. Ovariectomy, however, did not alter maternal–fetal exchange of amino acids in *Xantusia vigilis* or tritiated water in *Mabuya capensis,* implying that the functional integrity of the placenta in these animals, unlike mammals, is not dependent on progesterone. The demonstration of progesterone and estradiol receptors in the oviduct of various reptiles suggests that it is most likely a target organ for these hormones. In *Thamnophis elegans*, it has recently been shown that the high estrogen level and not the rising progesterone level of the preovulatory phase is important for maintaining the height and secretory condition of the luminal and glandular epithelium of the uterus. In fact, this rising estradiol titer may be responsible for the increased synthesis of progesterone receptor in the oviduct of *Nerodia*, a water snake.

The presence of progesterone receptor in both snake and turtle reproductive tract indicates that progesterone, most likely in conjunction with other hormones, is important in maintaining the connective tissue framework, musculature, and vascular structure of the oviduct. Other studies suggest that progesterone may be important in preventing a new wave of follicular growth while embryonic materials are within the oviduct. In fact, the life of the corpus luteum seems to roughly correlate with the amount of time the embryo remains in the oviduct, lasting only 2 weeks in the oviparous turtle, *Chrysemys*, and its life being prolonged in viviparous squamates. This action of progesterone may be important in maintaining a quiescent uterus to prevent premature expulsion of embryos or giving the oviduct time to add other products and the eggshell in oviparous species.

6. Regulation of Egg Laying and Parturition

The process of egg laying, birth, or parturition is controlled by a complex hormonal "cascade" phenomenon that probably involves neural signals, prostaglandins, ovarian (follicular/luteal) steroids, posterior lobe and ovarian peptides (oxytocin, vasotocin), and relaxin as key elements. It is likely that the signal to expel a fully developed fetus originates in the uterus, as does the signal to retain the early zygote. From mammalian studies, we would expect that the volume of the conceptus, sensed by stretch receptors in the uterine wall, sets in motion the chain of events that results in birth. In most viviparous squamates, motility of term embryos may be most critical, as the oviducal contents are already large from very early pregnancy when the large-yolked eggs enter the oviduct.

Arginine vasotocin (AVT), a peptide of hypothalamic or ovarian origin, can induce parturition during the last several weeks of pregnancy in the lizard (*Sceloporus jarrovi*, and the high progesterone titers existing until the latter part of pregnancy may exhibit AVT secretion or decrease the sensitivity of the uterus to AVT. Steroids may induce α- and β-adrenergic receptor formation in the uterus, with the relative abundance of each type of receptor determining the appropriate contractility of the uterine musculature during pregnancy. However, experimental evidence has yet to lend support to this hypothesis. Several egg-laying species respond to posterior lobe hormones (AVT, oxytocin) by oviposition.

In *Anolis*, the uterus on the same side as an ovary with a large corpus luteum failed to contract *in vitro* in response to AVT, whereas the contralateral uterus exhibited rhythmical contractions when exposed to AVT. Unilateral deluteinization removed the unilateral inhibition of uterine contraction, implying the existence of some inhibitory substance(s) produced by the corpus luteum which could be delivered to the ipsilateral uterus via uteroovarian blood vessels. *In vivo,* administration of progesterone was found to reduce oviduct contractility *in vitro* in the turtle, *Chrysemys.*

More recently a role for a relaxinlike molecule has been postulated. Of all the hormones involved in the process of parturition and reproductive tract adaptation to large eggs, and/or increasing conceptus size, this has received the least attention. In mammalian species, this hormone has several important effects: (1) to increase the connective tissue matrix of the uterus during pregnancy, allowing uterine adaptation to the increased fetal load, (2) to cause relaxation of the uterine musculature so as to maintain a level of tonus appropriate for fetal development, (3) to allow a change in the nature of cervical collagen, an effect necessary to allow exit of the fetus from the tract at birth, and (4) to influence the ischial–interpubic ligaments allowing for flexibility of the pelvis during the process of parturition. A primary source of this hormone in females is the corpus luteum. Recent advances in the comparative biochemistry of this hormone have demonstrated the widespread occurrence of this hormone in ovarian tissues of nonmammalian species. Questions regarding its physiological action in these species are of obvious interest in any discussion of the evolution of viviparity. In turtles, estrogen and relaxin treatments can increase the functional cross-sectional area of certain regions of the reproductive tract. This may be an important physiological adjustment allowing the passage of large eggs. Although no data on the exact site of origin of this material within the reptilian ovary are available, it is possible that both this peptide and progesterone are coordinately or sequentially synthesized and secreted by the corpus luteum of reptilian as in mammalian species.

In summary, follicular/luteal steroid hormones (estrogen, progesterone) have important functions in reproductive tract preparation for egg processing, retention, and oviposition/parturition. Although great variability in the functions of these processes exists in reptiles, the endocrine controls are probably very similar. Follicular, and then luteal, estrogens induce appropriate receptors for progesterone, relaxin, and vasotocin/oxytocin. The timing of oviposition/parturition is determined by the balance of these hormones and their receptors, which in turn determine the quiescence or motility of the reproductive tract. Thus, follicular–luteal development can be viewed as a differentiation sequence involving steroid production (first estrogen, then progesterone) followed by peptide production (relaxin, vasotocin). As in mammals, the programmed sequence of follicular/luteal development is dependent on the acquisition of appropriate gonadotropin receptors resulting in appropriate ovarian activity. In parallel, the ovarian hormones induce continued differentiation of the reproductive tract and responses appropriate to the stage of the cycle, gestation or parturition.

In the evolution of viviparity within the vertebrates (Figure 4, Chapter 8), the maternal–fetal connection, reproductive tract development, and parturition have come to depend on progesterone to varying degrees. In reptiles, once the period of egg retention has been extended to cover *in utero* embryogenesis, continuance of gestation does not appear to critically rely on the ovary. This is possibly due to the relative superficiality of the maternal–fetal connection (placenta), which is correlated with the adequacy of yolk deposits for development. The mechanisms of parturition in both oviparous and viviparous reptiles appear to be similarly hormonally dependent. The key element in prolongation of the intrauterine period of development is the prolongation of the life of the corpus luteum. In turn, the common denominator in this event is the presence of eggs, embryos, or fetuses *in utero*. Through either a neurohormonal or hormonal mechanism the embryo signals the ovarian control mechanism in such a way that the corpus luteum does not degenerate early. This does not explain what causes a given animal to retain eggs beyond the normal species-specific time initially, but this may be due to adverse environmental conditions inhibiting the signaling mechanism for egg laying. In time, the signal for luteal maintenance of progesterone secretion is prolonged, and eggs are retained for longer than normal periods of intrauterine development followed by loss of eggshell, close apposition of fetal membranes with maternal structures, and placentation. The final step in the evolution from reptilian viviparity is the reduction and elimination of yolk from the egg associated with enhancement of mechanisms for maternal–fetal exchange. (See discussion in Chapter 8, "Vitellogenesis and Viviparity.")

ACKNOWLEDGMENT. The author acknowledges the support of the National Science Foundation (NSF 81-04144) during the writing of these chapters and for research on which some of the material is based.

6

Reproductive Endocrinology of Amphibians

Frank L. Moore

1. General Introduction

Amphibians have the same basic repertoire of reproductive hormones as other vertebrates. For example, their hypothalamus synthesizes a decapeptide that is structurally and functionally similar to the gonadotropin-releasing hormone (LH-RH) of mammals. LH-RH stimulates the pituitary gland of amphibians to release two distinct gonadotropins, referred to as follicle-stimulating hormone (FSH) and luteinizing hormone (LH) after the homologous mammalian gonadotropins. The gonadotropins travel through the blood to the gonads and stimulate gametogenesis and the synthesis of steroid hormones (progesterone, 17β-estradiol, testosterone, and 5α-dihydrotestosterone). Despite similarities between the reproductive hormones of mammals and amphibians, amphibians are not simply hairless laboratory rats. In many respects, the reproductive endocrinology of amphibians is interesting and unique.

Amphibians exhibit remarkable diversity in reproductive patterns. Temperate-zone species generally have discrete seasonal cycles of reproduction with pronounced changes in gonad size, gamete production, and sex accessory structures. In contrast, tropical amphibians typically have continuous gametogenesis and can reproduce during any month of the year. In some species, particularly the anurans (frogs and toads), fertilization is external; courtship and mating behaviors climax when the spermatozoa and eggs are shed simultaneously into the environment. In other species, primarily urodeles (salamanders), internal fertilization occurs when the spermatozoa are transported into the female reproductive tract via spermatophores. The spermatozoa are stored until the female ovulates, and fertilization

Frank L. Moore • Department of Zoology, Oregon State University, Corvallis, Oregon 97331.

occurs as the eggs pass down the oviducts and through the cloaca. Although most amphibians are oviparous, a number of species are ovoviviparous or viviparous, in that the fertilized eggs are retained throughout embryonic development. Developing embryos are usually maintained in an enlarged oviduct, but there are species that use specialized pouches on their backs (*Gastrotheca* spp.) or incubate the eggs in their stomachs or vocal sacs (*Rheobatrachus* spp.).

2. Amphibian Neuroendocrinology

In order for these diverse patterns to result in successful reproduction, all aspects of the reproductive process—timing of courtship, production of gametes, development of sex accessory structures, parental care—must be coordinated within an individual or between the sexual partners. Successful reproduction, in other words, requires synchronization of physiological and behavioral responses to a variety of sensory inputs. Such synchronization is the result of interactions between neural activity and hormonal milieu. (See Figure 1.)

The amphibian brain probably responds to sensory cues (for example, lengthening photoperiod or warming temperatures) by releasing neurohormones that regulate the secretion of gonadotropins. The gonadotropins, in turn, stimulate the gonads to secrete steroid hormones and to produce mature gametes. When the animal is physiologically ready to reproduce, the hormonal inputs to the brain result in the expression of mating behaviors in those animals that perceive the appropriate social stimuli. Thus, reproduction depends on the functional interdependence of the nervous and endocrine systems. Indeed, their functional and structural interdependence is so great that modern neurobiologists no longer view these systems as separate entities.

As in other vertebrates, the amphibian brain contains identifiable regions, fiber

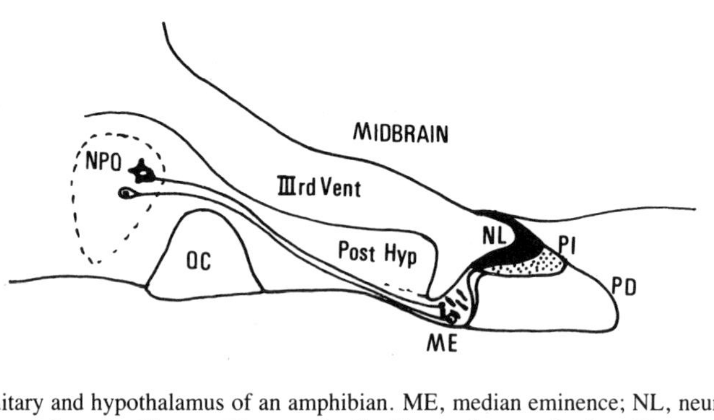

Figure 1. Pituitary and hypothalamus of an amphibian. ME, median eminence; NL, neural lobe; NPO, preoptic nucleus; OC, optic chiasma; PD, pars distalis; PI, pars intermedia; Post Hyp, posterior hypothalamus; IIIrd Vent, cavity of the third ventricle; AF+ (▲) and AF− (●), peptidergic neurons.

tracts, and nuclei. The brain region that most profoundly alters the secretion of reproductive hormones is the hypothalamus, located in the ventral diencephalon.

2.1. Hypothalamus

Neurons in the hypothalamus regulate reproduction by synthesizing and releasing different quantities of neuropeptides (LH-RH or vasotocin), catecholamines (norepinephrine or dopamine), indolamines (serotonin), and other chemical messengers. These substances are released either locally in the hypothalamus and cerebrospinal fluid or from neurohemal organs into the blood. The net result is that the hypothalamus translates neural activity into hormonal responses and thereby orchestrates changes in reproductive functions.

The anterior portion of the hypothalamus includes the preoptic area. It consists of several clusters of cells, including the preoptic recess organ and preoptic nucleus, and myriad neural fibers. Some of these fibers are dendrites that project to the third ventricle and contact the cerebrospinal fluid (CSF). In this way, it is anatomically possible for the preoptic cells to respond to chemical messengers in the CSF and to add their own secretions to the CSF. Other fibers from the preoptic area project caudad and terminate in either the median eminence or the neural lobe, which are neurohemal organs (see Section 2.2 and Part III).

LH-RH is found in cell bodies of the ventral (medial) preoptic area as well as neural fibers from the medial septal nucleus of the ventral telencephalon. The dorsal aspect of the preoptic area, the magnocellular preoptic nucleus, does not contain immunoreactive LH-RH. Instead, cell bodies in these nuclei are of at least three types and synthesize arginine vasotocin, mesotocin, and somatostatin. More will be said about these peptides in Section 2.2.

The posterior portion of the hypothalamus contains a number of areas with aggregations of secretory, aminergic cells. The dorsal infundibular nucleus, which is thought to be homologous to the arcuate nucleus of mammals, and the ventral infundibular nucleus are probably both involved in regulating the release of gonadotropins.

2.2. Amphibian Neurohypophysis and Adenohypophysis

The neurohypophysis includes the median eminence (ME) and neural lobe. The ME, usually, is located on the ventral fold of the infundibulum, whereas the neural lobe is on the dorsal fold. The degree of thickening varies with the species, but generally these structures are more conspicuous in anurans than urodeles.

The ME and neural lobe are highly vascularized neurohemal organs. Hypothalamic arteries give rise to a network of capillaries which enter the ME and then drain to the portal system of the pars distalis of the adenohypophysis. The plexus intermedius, lying between the neural lobe and pars intermedia, receives blood from capillaries in the ME as well as from a separate branch of the hypophyseal artery and from the mesencephalon of the brain; the outflow is into the systematic circulation.

2.2.1. Neural Lobe

In amphibians, some nerve terminals in the neural lobe originate from neurons of the magnocellular preoptic nucleus. These neurosecretory cells secrete structurally related nonapeptides, vasotocin and mesotocin; some amphibians also may synthesize oxytocin. These peptides are synthesized as part of a larger precursor molecule; then biochemical processing separates the biologically active peptide from the neurophysin. Neurophysins have specific binding sites for the appropriate nonapeptide and may function as carrier molecules after being secreted into the plasma.

Vasotocin and mesotocin are best known for regulating water-balance functions. However, vasotocin is associated also with amphibian reproduction in that it induces reproductive behavior and stimulates contractions in the oviduct and sperm duct. The role of vasotocin in amphibian behavior will be discussed in Section 6. It is clear that oviducts in urodeles and anurans contract strongly when exposed to vasotocin, mesotocin, or oxytocin. Oviducts usually are more sensitive to vasotocin than to the other neuropeptides, responding to picogram concentrations of vasotocin. Likewise, the vas deferens of male newts exhibit contractions when treated with vasotocin. These studies provide hints that vasotocin has a physiological role in parturition, oviposition, and sperm transport—but additional studies are needed.

2.2.2. Median Eminence

The median eminence is the primary site of release for the peptides and biogenic amines that control reproductive hormones in the pars distalis. The most important reproductive hormones in amphibians are the gonadotropins (FSH and LH) and prolactin. In amphibians, the control of the release of FSH, LH, and prolactin appears to be similar to that of mammals.

As in other vertebrates, LH-RH stimulates the release of FSH and LH in amphibians. The LH-RH of amphibians has been isolated and compared to mammalian LH-RH. Apparently, the brain of amphibians contains a decapeptide that is chemically indistinguishable from mammalian LH-RH. However, the unsolved mystery concerns the four different forms of LH-RH found in amphibian brains and sympathetic ganglia. These LH-RHs apparently are decapeptides with primary structures that differ by only one or two amino acids. The predominant type of amphibian LH-RH, in terms of total concentration in the brain, is chemically indistinguishable from mammalian LH-RH. In addition to this mammalianlike LH-RH, another LH-RH in the amphibian brain is indistinguishable from the LH-RH in salmon. It remains a mystery whether these naturally occurring LH-RH analogs have distinct physiological functions.

Injections of mammalian LH-RH have been found to stimulate spermiation and ovulation and to increase plasma testosterone levels, indicating that LH-RH affects the release of gonadotropins in amphibians. Furthermore, direct radioimmunoassay shows both FSH and LH are released following LH-RH treatment. Thus, the model for mammals, in which one releasing factor (LH-RH) controls the release of two gonadotropins (FSH and LH), seems to describe amphibians as well.

The control of prolactin release has not been thoroughly studied in amphibians, but the available data indicate that the regulators are similar to those of mammals. Because prolactin secretion increases markedly when the pars distalis is isolated from hypothalamic control (usually by transections or pituitary transplants), prolactin in amphibians is thought to be predominantly under inhibitory control. The substance from the ME that inhibits the release of prolactin is probably dopamine; the administration of dopamine suppresses the secretion of prolactin in bullfrogs, as it does in mammals. A recent study showed that injection of thyrotropin-releasing hormone (TRH) into bullfrogs stimulates the release of prolactin. Thus, there may be dual control of prolactin in amphibians, with dopamine inhibiting and TRH stimulating prolactin release.

2.2.3. Pars Distalis

The pars distalis is the largest part of the adenohypophysis. The range of hormones secreted by the pars distalis of amphibians is similar to that of other vertebrates. Available evidence indicates that the most likely products of the amphibian pars distalis are corticotropin, β-endorphin, growth hormone, prolactin, thyrotropin, FSH, and LH.

Gonadotropins are produced in the pars distalis by identifiable cells, usually referred to as type 2 basophils. These basophils are distributed throughout the pars distalis in the species that have been studied. In *Rana nigromaculata,* ultrastructural changes in type 2 basophils, indicating high levels of activity, occur during May and August when spermatogenesis and steroidogenesis, respectively, are increasing. This study and others support the conclusion that this cell type is the site of gonadotropin synthesis.

Two gonadotropins that are structurally and functionally homologous to mammalian FSH and LH have been isolated from the pituitary glands of *Ambystoma tigrinum, Rana pipiens,* and *Rana catesbeiana.* These amphibian gonadotropins have distinct biological activities in a number of bioassays, and they have chromatographic behavior similar to that of ovine FSH and LH. As with mammalian gonadotropins, the amphibian gonadotropins are glycoproteins and are made up of two subunits. The α subunits are virtually identical but the β subunit confers the specificity, with both subunits necessary for full activity.

The exact structures of *Rana* and *Ambystoma* FSH and LH have not been determined. However, it is clear that even among amphibians the structures of LH and FSH are species specific. The effectiveness of a purified gonadotropin from one species in eliciting a response in another varies from species to species. This probably reflects variation in gonadotropin structure, as well as in the characteristics of gonadotropin receptors. The pattern of variation in response to heterologous gonadotropins appears to be independent of phylogenetic relationships. For example, the stimulation of androgen secretion in rough-skinned newts (*Taricha granulosa*) is greater when ovine LH is used than when *Rana* LH is used.

There is no doubt that some amphibians possess FSH- and LH-like gonadotropins, but it is possible that, when more species are investigated, other amphibians will be found to use only one gonadotropin as in teleosts. However, these species will probably reflect a derived situation in which one of the gonadotropins has been lost.

3. Gonad Structure and Function

3.1. Spermatogenesis in the Amphibian Testis

Amphibians, like other anamniotic vertebrates, exhibit the cystic type of spermatogenesis. This means that germ cells are held in small clusters within a membranous sheath, called germinal cysts. Development of germ cells within each germinal cyst is synchronized. (See Figure 2.)

A germinal cyst forms from a single primary spermatogonium by mitotic proliferation. Primary spermatogonia are initially surrounded by follicle cells, which are fibroblastlike cells with darkly staining, crescent-shaped nuclei; as the primary spermatogonium divides to form secondary spermatogonia, the follicle cells flatten and encapsulate the germ cells. The germinal cyst enlarges by subsequent mitotic and meiotic divisions.

Germinal cysts generally remain intact until the spermatids begin to elongate and transform into spermatozoa. At this time, the cyst ruptures and the follicle cells become Sertoli cells. Sertoli cells are characteristic steroid secretory cells, with a well-developed, elaborate system of smooth endoplasmic reticulum and numerous mitochondria (Chapter 2). In frogs and toads Sertoli cells are attached, usually, to the basement membrane of the seminiferous tubule. Each cell supports a large number of spermatozoa, with the heads of late spermatids and spermatozoa embedded in deep recesses, invaginations, on the apical end of the Sertoli cell. Thus, the sperm bundles in anurans are arranged in a regular pattern with the sperm heads pointing toward the periphery and the sperm tails projecting into the lumen. In some urodeles, the Sertoli cells are not attached to the basement membrane, and the sperm bundles are scattered throughout the lumen.

Mature spermatozoa remain attached to the Sertoli cells until they are evacuated from the testis, a process referred to as spermiation. Ultrastructural studies of Sertoli cells during spermiation show that the endoplasmic reticulum enlarges and the apical portion of the Sertoli cell swells. The swelling appears to unfold the invaginations that hold the sperm heads, thus releasing the spermatozoa. The detached spermatozoa move into the sperm-transporting ducts. In most anurans, spermiation is a rapid process which occurs during or shortly before mating so that the

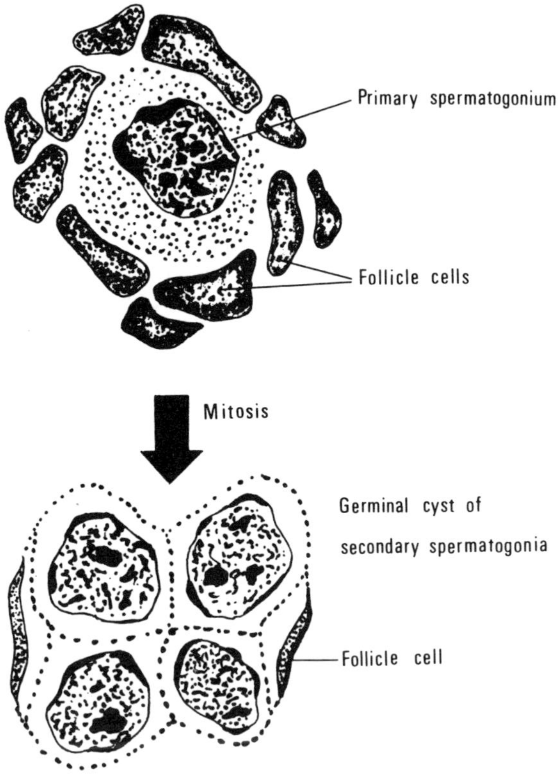

Figure 2. Development of a germinal cyst in the amphibian testis. A large primary spermatogonium is surrounded by darkly staining follicle cells, and a cluster of secondary spermatogonia is encapsulated by flattened follicle cells. All germ cells in a germinal cyst divide and mature synchronously.

male can release large quantities of sperm over a freshly laid clutch of eggs. In urodeles, spermiation is a more prolonged process and may occur some months before the mating season.

3.2. Testicular Structures of Anurans and Urodeles

3.2.1. Anatomical Differences between Anurans and Urodeles

Anurans and urodeles have distinctly different anatomical arrangements in their testes; the anuran testis more closely resembles that of reptiles, birds, and mammals, whereas the urodele testis contains structures similar to those of some teleost fishes.

The anuran testis contains a mass of convoluted seminiferous tubules and conspicuous interstitial Leydig cells. Seminiferous tubules of anurans are fairly permanent structures which serve as the site for development of germ cells. As the germinal cysts lining the tubules enlarge, the diameter of the tubule expands. At spermiation the mature spermatozoa are transported through the lumen of the tubules to the sperm ducts. A few primary spermatogonia remain in the seminiferous tubules after spermiation and give rise to the next generation of mature sperm.

In urodeles, seminiferous lobules serve the same function as seminiferous tubules of anurans. These lobules may be short tubes, crypts, or ampullae, depending on the species of urodele. Growth and maturation of spermatozoa are similar to the processes in anurans, but following spermiation the seminiferous lobules of urodeles are resorbed. However, before this happens, the lobule walls hypertrophy and differentiate into the lobule boundary cells. These lobule boundary cells are in close association with the Sertoli cells in the lumen, and together these two cell types form a very active endocrine tissue, synthesizing and secreting large quantities of steroid hormone.

3.2.2. Steroidogenic Cells in the Amphibian Testis

Steroid-synthesizing cells are identified by the following characteristic features: (1) large quantities of lipoproteins, (2) elaborate smooth endoplasmic reticulum, (3) numerous mitochondria with tubular cristae, (4) histochemical evidence of steroidogenic enzyme activity, and (5) capacity to form steroid hormones when supplied with precursors. Based on these criteria, testes from amphibians have three different cell types that are thought to be steroidogenic: the Sertoli cell, the interstitial Leydig cells of anurans, and the lobule boundary cells of urodeles.

3.3. Sertoli Cells

Sertoli cells of anurans and urodeles have lipid droplets and 3β-hydroxysteroid dehydrogenase (3β-HSDH) activity in the cytoplasm. Incubation of these cells with steroid precursors results in the production of identifiable steroid hormones. Ultrastructural and histochemical changes in Sertoli cells are related to reproductive cycles and seasonal variations in steroid levels. It is not known whether the Sertoli cell synthesizes steroid hormones only after it is fully differentiated or whether the progenitor follicle cells also synthesize steroids. There is some evidence that follicle cells from germinal cysts have 3β-HSDH activity. Considering the important role of steroids in maintaining spermatogenesis in mammals, it is conceivable that follicle cells might produce steroids that act locally on the encapsulated germ cells.

In anurans, as in other vertebrates, interstitial Leydig cells that produce steroid hormones are located in the interstices between the seminiferous tubules. Urodeles are thought to lack these cells, but this issue has not been thoroughly resolved.

Seasonal changes in ultrastructure and histochemistry of interstitial cells of anurans are similar to those of other seasonally breeding vertebrates and are correlated with seasonal changes in the concentration of testosterone in the blood. Interstitial cells exhibit seasonal accumulations and depletions of cholesterol-positive lipids, seasonal changes in 3β-HSDH activity, and seasonal variations in nuclear size. It is thought that the interstitial cells of anurans are fairly permanent, being rehabilitated during the periods of postnuptial atrophy, instead of resorbed. The types of steroid hormones secreted by the interstitial cells will be discussed in Section 4.1.1. Lobule boundary cells of the urodele testis, which develop after spermiation, contain numerous cholesterol-positive lipids as well as 3β-HSDH activity. However, seasonal changes in plasma androgen levels are somewhat independent of the development of lobule boundary cells. For example, in rough-skinned newts (*T. granulosa*), the sharp rise in plasma androgen concentration during August and September occurs in males which completely lack lobule boundary cells. On the other hand, *in vitro* studies with *T. granulosa* testes have shown that later in the season the glandular tissue, containing Sertoli and lobule boundary cells, produces almost all of the testicular androgens.

3.4. Amphibian Ovary

3.4.1. Anatomy of the Amphibian Ovary

Unlike testes, the ovaries are anatomically similar in anurans and urodeles. Amphibian ovaries are paired, hollow, saclike organs that are subdivided into lobes. The outer surface of the ovary is composed of a squamous epithelium (germinal epithelium). Inside are large numbers of ovarian follicles which, in the breeding adult, are at different stages of maturational development. A standard classification of the stages of development of amphibian ovarian follicles is as follows: *stage 1*, very small, primordial follicles; *stage 2*, previtellogenic follicles; *stage 3*, vitellogenic follicles, in which yolk is accumulating in the oocytes; *stage 4*, postvitellogenic follicles, in which yolk deposition is complete.

A postvitellogenic follicle is composed of a yolk-filled oocyte surrounded by three distinct cell layers. The outermost layer appears to be a continuation of the squamous epithelium that surrounds the entire ovary; beneath this lies the layer of thecal cells, with the granulosa cells adjacent to the oocyte.

In addition to the mature and developing follicles, there are clusters of naked germ cells (oocytes and oogonia) in the germinal epithelium, as well as corpora atretica and corpora lutea. Corpora atretica develop from mature follicles when cells from the granulosa layer involute and the yolk is resorbed. Corpora atretica are, in effect, the preovulatory follicles which are being resorbed; they are particularly prevalent in malnourished females and in females after the spawning season. There is no convincing evidence that corpora atretica serve a secretory role and produce steroid hormones. Corpora lutea develop after ovulation when the granulosa layer contracts and the cells enlarge to fill the lumen in the evacuated follicle; luteal tissue may produce steroids in the few species of ovoviviparous urodeles and anurans with brood pouches.

3.4.2. Steroidogenic Cells in the Ovary

Ultrastructural and histochemical studies of the amphibian ovary reveal several possible sites for the synthesis of steroid hormones including the ovarian follicles and, at least in a few species, the corpora lutea and interstitial gland cells.

In the vitellogenic and postvitellogenic follicles, steroid hormones are produced by cells in the thecal and granulosa layers. Although histochemical studies report stronger 3β-HSDH activity in the granulosa layer than in the thecal layer, cells in both layers exhibit characteristics of steroid synthesis. In mammals, ovarian follicles produce testosterone from cholesterol in the thecal cells; the testosterone is then aromatized to estradiol in the granulosa cells. It is not yet known whether amphibian follicles have a similar pattern of steroid synthesis.

Postovulatory follicles in most oviparous amphibians are rapidly resorbed and do not form functional corpora lutea. On the other hand, corpora lutea are more developed and last longer in ovoviviparous urodeles (e.g., *Salamandra salamandra*) and in anuran species with cutaneous brood pouches (e.g., *Nectophrynoides occidentalis*). In these species, there is evidence that the corpora lutea serve an endocrine function by secreting steroid hormones. It has been suggested that steroid hormones from the corpora lutea act on oviducts and other support tissues and also act, via negative feedback, to inhibit ovarian development during gestation. For example, during the 4-year gestation period of *Salamandra atra,* numerous corpora lutea with 3β-HSDH activity are present in the ovaries.

Interstitial gland cells, which appear as nodules in the germinal epithelium, exhibit seasonal changes in lipid content and 3β-HSDH activity, but their physiological significance has yet to be determined.

4. Gonadotropins and Gonadal Steroid Hormones

4.1. Male Reproductive Functions

4.1.1. Hormonal Control of Testicular Steroids

In amphibians, as in other vertebrates, gonadotropins from the pars distalis regulate the secretion of steroid hormones from the testis. The principal steroids are testosterone and 5α-dihydrotestosterone, but the amphibian testis also produces notable quantities of other steroids: estrone, 17β-estradiol, progesterone, 11-ketotestosterone, 11β-hydroxytestosterone, 4α-androstanedione, and 5α-androstane-3β-, 17β-diol. Administration of LH to *Necturus maculosa* causes a drop of the 11-oxygenated testosterones (11-ketotestosterone and 11β-hydroxytestosterone), and a reciprocal rise in testosterone and dihydrotestosterone.

The predominance of testosterone and dihydrotestosterone in the plasma of amphibians is comparable to the situation in birds and mammals, and the production of 11-oxygenated testosterones by some urodeles is reminiscent of steroid metabolism in teleost fishes.

The issue of which gonadotropin, LH or FSH, specifically controls steroidogenesis in amphibians has not been completely resolved. The best experimental evidence comes from studies that use homologous gonadotropins, i.e., studies in which LH or FSH from *A. tigrinum* or *R. catesbeiana* is injected into males of the same species. In these studies LH has a high specificity for stimulating testicular androgens, and FSH is relatively ineffective.

In most amphibians, when heterologous gonadotropins are administered, LH is more effective than FSH in stimulating androgen secretion. But in some amphibians, notably *Taricha* and *Xenopus,* FSH is as effective as, or better than, LH in stimulating androgen secretion. The results of studies with heterologous gonadotropins, however, are difficult to interpret, because gonadotropin receptors may not recognize heterologous gonadotropins with the same degree of specificity with which they recognize the endogenous gonadotropins.

One study with a salamander, *Necturus,* is of particular interest. The administration of LH caused a marked rise in plasma androgen levels, whereas FSH caused a rise in estrogen. Perhaps LH acts specifically on the interstitial tissue (or lobule boundary cells) to produce testosterone, and FSH acts on the Sertoli cells to aromatize the testosterone to estradiol—an explanation that also has been suggested for mammals.

4.1.2. Hormonal Control of Spermatogenesis

Spermatogenesis is controlled by gonadotropins from the pars distalis. Almost all studies have relied on injection of mammalian FSH or LH. Both hormones stimulate proliferation of spermatogonia and spermatocytes, some species reponding more to one than the other. However, several observations support the concept that both LH and FSH influence amphibian spermatogenesis. In the only homologous test, in the bullfrog, both FSH and LH stimulate spermatogenesis. For example, seasonal study of serum levels of FSH and LH in bullfrogs demonstrates parallelism between the concentrations of these two gonadotropins and the degree of testicular development. Steroid hormones may also influence spermatogenesis; for example, in *A. tigrinum* larvae testosterone and FSH synergize to stimulate spermatogenesis. The diverse observations of amphibian spermatogenesis generally support the hypothesis that LH regulates the synthesis of androgens (as suggested earlier) and that FSH influences the effects of androgens on spermatogenesis.

4.1.3. Hormonal Control of Spermiation

The hormonal control of spermiation in urodeles is virtually unstudied and is only poorly understood in anurans. In these amphibians, release of spermatozoa occurs shortly after treatment with either FSH or LH, and in the bullfrog homologous FSH and LH are equipotent. Other hormones including noradrenaline and androgens may also influence spermiation.

4.2. Female Reproductive Functions

4.2.1. Hormonal Control of Ovarian Steroidogenesis

Amphibian ovaries secrete measurable quantities of all three types of sex steroids: androgen, estrogen, and progestogen. The serum concentrations of these hormones show seasonal variation, with 17β-estradiol highest in the fall when oviducts and ovaries are enlarging; estrone and progesterone are highest in the spring just prior to spawning and ovulation. Serum testosterone concentrations also show seasonal variation, but the physiological significance of this hormone in female frogs is unknown.

Although FSH may be more important for estrogen production during the early stages of ovarian growth, final oocyte maturation and ovulation are specifically LH-dependent in amphibia. These actions of LH are mediated by progesterone produced in the ovaries. One of the effects of LH is to enhance uptake of precursor steroids, e.g., pregnenolone, which are metabolized by follicular cells to progestogens.

4.2.2. Hormonal Control of Ovarian Growth

Follicle cells of the amphibian ovary secrete 17β-estradiol under the influence of pituitary gonadotropins. A prime function of estradiol is to stimulate synthesis of vitellogenin in the liver. This lipoprotein is taken up into oocytes by active transport and there converted to yolk platelet proteins, lipovitellin and phosvitin. Gonadotropins control the uptake of vitellogenin into the oocyte, but the specific roles played by FSH and/or LH have not been defined.

4.2.3. Hormonal Control of Ovulation and Meiosis

Before ovulation, the nuclear membrane in the oocyte disintegrates, so-called "germinal vesicle breakdown," which marks the onset of the first meiotic division. At ovulation an increase in collagenolytic activity causes the thecal and granulosa layers to rupture and the mature oocyte to be discharged from the follicular capsule.

Both ovulation and germinal vesicle breakdown are stimulated by treating mature ovarian follicles with LH (not FSH) or progesterone. Most experimental evidence indicates that LH controls the synthesis of progesterone in the follicle cells of a mature ovarian follicle. Progesterone then acts directly on the oocyte to initiate meiosis and on the follicular layers to trigger ovulation.

4.3. Environmental Control of Gonadal Development

In those amphibians that exhibit seasonal reproductive cycles, hormone levels are influenced by environmental conditions, particularly temperature and pho-

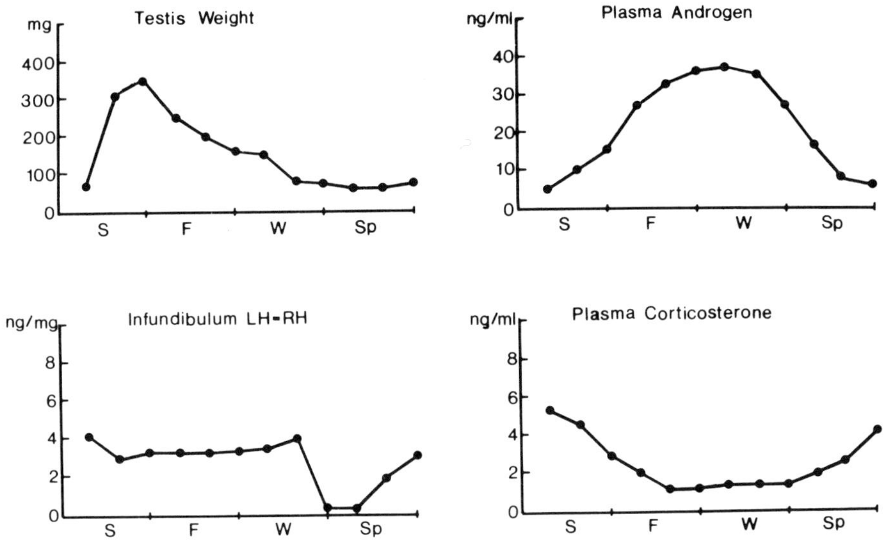

Figure 3. Seasonal changes in reproductive hormones and testis weight in male rough-skinned newts (*T. granulosa*).

toperiod. A long photoperiod of 12–16 hr of light, when associated with warm temperature (20°C), accelerates testicular development in frogs and salamanders. When temperature is cold or photoperiod is short, testicular recrudescence is delayed, although it still occurs. Thus, a combination of long photoperiod and warm temperature triggers the onset of gonadal development, even though the annual cycle of reproduction probably is driven by endogenous factors.

Seasonal changes in gonadal development are associated with changes in the concentrations of reproductive hormones. The timing of the changes in reproductive hormones is somewhat species specific, but for the few species that have been studied, a general pattern is apparent. For example, during the summer months in *T. granulosa* males (Figure 3) spermatogenesis and spermiogenesis occur and result in the development of large, mature testes. In the fall, the testes begin to secrete large quantities of testosterone and dihydrotestosterone; these androgens facilitate the development of secondary sexual characteristics. There is a decrease in testis weight, beginning in late fall and continuing into spring, because of the evacuation of spermatozoa from the testes into the Wolffian ducts for storage. Male newts are in full breeding condition in late fall, even though most matings occur in March and April. By late spring, testes are devoid of mature spermatozoa and stop secreting androgens, resulting in a precipitous decline. This decline in androgens is preceded by an equally precipitous decline in hypothalamic LH-RH. Testicular recrudescence begins in later spring, after the concentration of LH-RH in the hypothalamus has returned to high levels. These temporal relationships among the reproductive hormones indicate that the central nervous system activates the reproductive cycle by increasing the production of LH-RH, which in turn stimulates the pituitary–gonad cycle.

5. Prolactin and Corticosterone in Amphibian Reproduction

Amphibian reproductive endocrinology involves considerably more than the hormones of the hypothalamus–pituitary–gonad axis. Prolactin and corticosterone are particularly important in amphibian reproduction; prolactin having several purported functions that affect amphibian reproduction indirectly. For example, prolactin influences reproduction by controlling body growth, metabolism, osmoregulation, and metamorphosis. A more direct effect on reproduction involves the stimulation by prolactin of water drive; high levels of prolactin alter the behavior of some salamanders so that they enter the water at the beginning of the breeding season. Prolactin also synergizes with testosterone and other gonadal hormones to stimulate the development of sexual accessory structures such as the caudal fin of male newts, the secretory activity of oviducts in toads, and the formation of mucus in the cloacal glands of male urodeles.

Corticosterone is secreted by the interrenal tissue, which is homologous to the adrenal glands of mammals, in response to harsh or life-threatening environments. The neuroendocrine control of corticosterone secretion is similar to that of other vertebrates. Injections of mammalian corticotropin-releasing factor (CRF) into male newts result in a significant rise in circulating corticosterone level, presumably because CRF stimulates ACTH release and ACTH activates steroid synthesis in the interrenal tissue. These hormone changes are typical of the stress response of many vertebrates.

The effects of corticosterone on amphibian reproduction are inhibitory. Elevated levels of corticosterone tend to shut off reproduction by causing sex steroid levels and other gonadal activities to decrease. Injections of corticosterone into reproductively active male newts result in a rapid cessation of courtship behavior and a marked drop in testosterone titers in the blood. These inhibitory effects of corticosterone help to explain why reproductive activities wane and sexual accessory structures atrophy when amphibians have been stressed.

6. Hormonal Control of Amphibian Reproductive Behavior

Sexual behavior patterns of male amphibians range from amplectic clasping behaviors to vocalizations that attract receptive females to the pond. These activities are controlled by gonadal steroid hormones as well as several peptide hormones. In most species studied, the concentrations of testosterone and dihydrotestosterone in the blood are very high during the breeding season. There is good evidence that these androgens play a role in stimulating amphibian sexual activity, because castration of sexually active males causes disappearance of reproductive behaviors and replacement therapy with testosterone maintains the reproductive behaviors.

Testosterone or dihydrotestosterone, however, will not initiate behavior in sexually inactive males nor are plasma androgen levels correlated with the behavioral state of the animal. Thus, the role of androgens seems to be to maintain sexual activity in males, but not initiate it. This is where the peptide hormones come into the picture.

Three peptide hormones have been found to stimulate sexual behavior in male amphibians: arginine vasotocin, LH-RH, and prolactin. Experiments with vasotocin and LH-RH show that these two neuropeptides activate sexual behavior by stimulating cells in the brain. Injections of vasotocin are 10,000 times more effective in stimulating sexual behavior in male newts when injected directly into the third ventricle of the brain instead of into the abdominal cavity. Other experiments demonstrate that amphibian sexual behaviors can be suppressed by injections of antagonists or antibodies to vasotocin or LH-RH. It is known that systemic injections of prolactin into crested newts stimulate the expression of male courtship, but it is not known whether prolactin is acting directly on brain cells or on peripheral target tissue, like cloacal glands.

The functional relationship between testicular androgens and these peptides is presently under investigation. Perhaps testosterone and dihydrotestosterone maintain the integrity of nerve cells and their responsiveness to peptides by preventing loss of receptors. Several experiments show that behavioral responses to an injection of vasotocin are maintained in castrated males, by replacement therapy with testosterone and dihydrotestosterone. One hypothesized role for the neuropeptides is to activate specific neural pathways so that the propensity to exhibit sexual behavior is elevated; thus, neuropeptides may represent one aspect of the physiological mechanism underlying sexual arousal.

Information about the behavioral endocrinology of female, as well as male, amphibians is limited to only a few types of behaviors in a few species. In female *T. granulosa,* sexual attractiveness is enhanced by progesterone. Attractiveness and receptivity of *X. laevis* females are stimulated by estradiol, progesterone, and LH-RH. Receptivity of *R. pipiens* females is induced by arginine vasotocin and prostaglandins. Thus, the hormonal control of sexual behavior in female amphibians involves both gonadal steroids (progesterone and estradiol) and neuropeptides (vasotocin and LH-RH). However, as in the entire field of amphibian reproductive endocrinology, many questions remain unanswered.

7. Selected Readings

Ball, J. N., 1981, Hypothalamic control of the pars distalis in fishes, amphibia and reptiles. *Gen. Comp. Endocrinol.* **44:**135–170.

Licht, P., 1979, Reproductive endocrinology of reptiles and amphibians: Gonadotropins. *Annu. Rev. Physiol.* **41:**337–351.

Licht, P., Papkoff, H., Farmer, S. W., Muller, C. H., Tsui, H. W., and Crews, D., 1977, Evolution of gonadotropin structure and function. *Rec. Prog. Horm. Res.* **33:**169–248.

Lofts, B., 1974, Reproduction, in *Physiology of the Amphibia,* Vol. II (B. Lofts, ed.), Academic Press, New York, pp. 107–218.

Moore, F. L., 1983, Behavioral endocrinology of amphibian reproduction. *Bioscience* **33:**557–561.

Specker, J. L., and Moore, F. L., 1980, Annual cycle of plasma androgens and testicular composition in the rough-skinned newt, *Taricha granulosa. Gen. Comp. Endocrinol.* **42:**297–303.

7

Reproductive Endocrinology of Fish

A. P. Scott

1. Introduction

This chapter is concerned with the nature of the hormones involved in fish reproduction, their site of origin, and their effects on gonadal development during the reproductive cycle.

Since the field is so large and the literature correspondingly extensive, I have decided to leave out consideration of subjects which, though important, are peripheral to the main theme. These include viviparity and ovoviviparity, environmental control of reproduction [see Sundararaj, B. I., 1981, *Reproductive Physiology of Teleost Fishes,* Aquaculture Development and Coordination Programme, ADCP/Rep. 81/16, The Food and Agriculture Organization of the United Nations, Rome, 82 pp., and Dodd, J. M., and Sumpter, J. P., 1984, Reproductive cycles of cyclostomes, elasmobranchs and bony fishes, in *Marshall's Physiology of Reproduction* (G. E. Lamming, ed.), 4th ed., Churchill Livingstone, London, pp. 1–126], and the effects of hormones on secondary sexual characteristics (see review by Pickford, G. E., and Atz, J. W., 1957, *The Physiology of the Pituitary Gland of Fishes,* The New York Zoological Society, New York, 613 pp.).

2. Class Osteichthyes: Superorder Teleostei

2.1. The Gonads

2.1.1. Early Gonadal Development

Teleost gonads develop in the dorsolateral lining of the peritoneal cavity—one on each side of the dorsal mesentery. They are invaded by primordial germ cells at a

A. P. Scott • Ministry of Agriculture, Fisheries and Food, Directorate of Fisheries Research, Fisheries Laboratory, Lowestoft, Suffolk NR33 OHT, England.

very early stage in embryonic development, but do not differentiate till several weeks, or even months, after hatching [e.g., rainbow trout (*Salmo gairdneri*)—3 months, golden-gray mullet (*Mugil auratus*)—8 months]. The ovaries generally differentiate before the testes.

2.1.2. Ovarian Development

The ovaries develop as hollow organs and consist of many connective tissue septa (ovigerous lamellae) which project into the lumina, rather like the pages of a book. It is on these that the eggs develop.

2.1.2a. Oogonial Proliferation and Oogonial–Primary Oocyte Transformation

Premeiotic sex cells, termed *oogonia,* develop from the primordial germ cells. They appear in section as small and rounded, with large nuclei each containing a single prominent nucleolus. They are generally found in small nests.

The oogonia proliferate by mitotic division soon after the ovaries have differentiated, and following this, some of the oogonia are transformed into *primary oocytes.* In homeothermic vertebrates, cyclostomes and elasmobranchs oogonial proliferation and transformation occur once only (at an equivalent period in development), and no oogonia persist into adult life. In the majority of teleosts, reptiles, and amphibians, however, oogonial proliferation and subsequent formation of new oocytes occurs throughout reproductive life. In species with circumscribed annual breeding cycles, oogonial proliferation has been observed just before, during, or immediately after the spawning period, whereas in species with prolonged breeding cycles, oogonial proliferation may be continuous.

The transformation of oogonia into primary oocytes constitutes the first phase of meiosis; the chromosomes replicate and proceed through the leptotene, zygotene, and pachytene stages of prophase (Figure 1). This is when chromosome pairing and crossover take place. From the completion of pachytene till a few days before the eggs are to be shed, the meiotic process is arrested at diplotene. This may last for as little as 2 days in the medaka (*Oryzia latipes*) or as much as 12–16 years in the European eel (*Anguilla anguilla*). During this period oocytes usually increase in volume several thousandfold.

2.1.2b. Primary Growth Phase

At the beginning of the primary growth phase oocytes are very similar to the oogonia (they contain scant cytoplasm and large, centrally located nuclei), but each oocyte becomes invested with a layer of follicle cells.

In the course of oocyte growth, the nuclei increase in size and multiple nucleoli appear. The cytoplasm also increases in volume and becomes strongly chromophilic. This is due to the presence of aggregates of ribonucleoprotein particles.

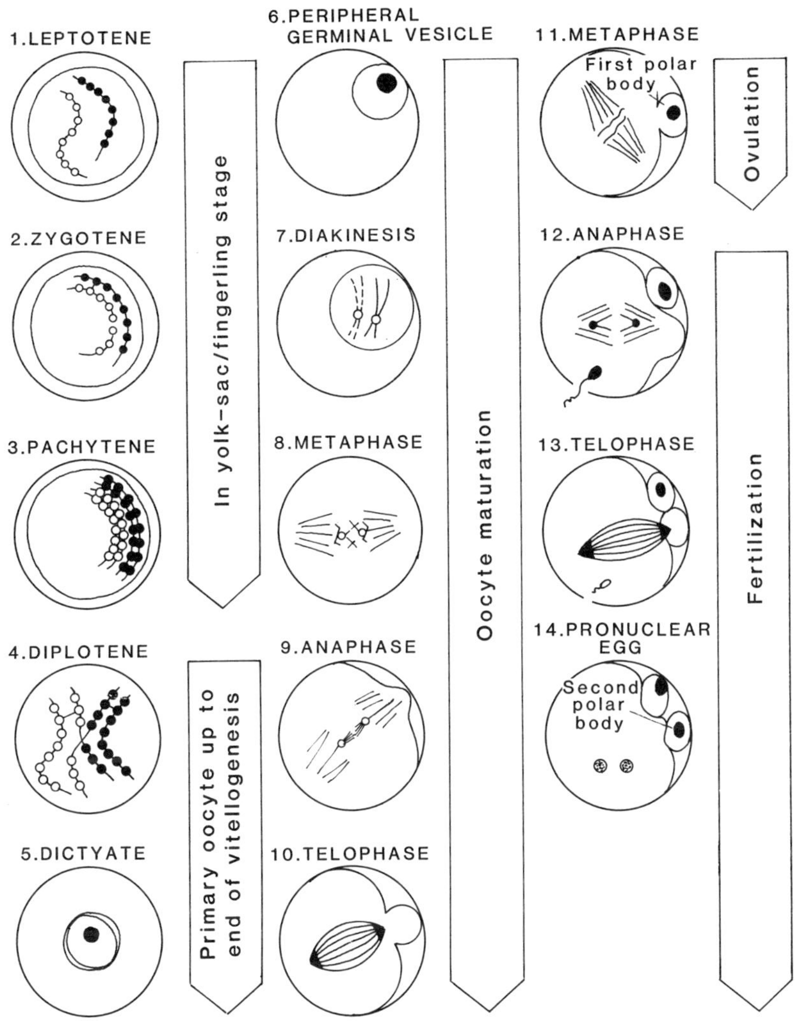

Figure 1. Meiotic stages in the oocyte.

The aggregates become surrounded by less densely stained cytoplasmic material to form the complexes known as "Balbiani bodies." The function of these is unclear. Toward the end of the primary growth phase, the Balbiani bodies dissolve and the cytoplasm becomes clear again.

2.1.2c. Secondary Growth Phase

The secondary growth phase involves the accumulation of yolk within the oocytes. It comprises three separate processes, each of which leads to the formation of a different type of "yolk." The processes are endogenous vitellogenesis, uptake of lipids (lipid yolk), and exogenous vitellogenesis.

i. Endogenous Vitellogenesis. In the teleosts, endogenous vitellogenesis is most apparent during the early stages of the secondary growth phase. It is characterized by the appearance of vesicles (vacuoles) in the cytoplasm, which contain a glycoproteinaceous material known as "intravesicular yolk." These vesicles generally increase in size and number until they fill the cytoplasm. It has been established [see Ginzburg, A. S., 1972, *Fertilisation in Fishes and the Problem of Polyspermy,* Moscow Academiya Nauk SSSR, Institut Biologii Razvitiya (translated by Israel Program for Scientific Translations Ltd., Jerusalem), 366 pp.] that these organelles subsequently become the cortical alveoli, which at fertilization release their contents into the perivitelline space.

In most species, endogenous and exogenous vitellogenesis appear to take place concomitantly. It is possible, however, to recognize a discrete stage (at the beginning of the secondary growth phase) during which yolk vesicle formation is predominant and the ovary grows very slowly. In at least one fish, the eel, there is a stage in the life-cycle that lasts for 2–3 years (the "silver-eel" stage) during which there is only yolk vesicle formation. Exogenous vitellogenesis does not occur until the eels start their marine migration.

ii. Lipid Yolk. The cytoplasm in many teleost eggs, albeit not all (see Ginzburg, 1972), contains separate lipid inclusions (usually oil droplets) in addition to vesicles and protein yolk granules.

Although oil droplet accumulation is classified under the heading "endogenous vitellogenesis," it occurs in some species well before the yolk vesicle stage and in other species well after. It therefore seems unlikely that the processes of oil droplet formation and yolk vesicle formation are functionally linked.

iii. Exogenous Vitellogenesis. In nonmammalian vertebrates, the principal events responsible for the enormous growth of oocytes, which takes place during the secondary growth phase, involve the sequestration and packaging of a hepatically derived plasma precursor, vitellogenin, into yolk protein. Vitellogenins and egg-yolk proteins have been characterized in some detail in a number of teleost species. Vitellogenins are lipoglycophosphoproteins with a high molecular weight (300,000–600,000). When taken up by the oocytes they are enzymatically cleaved to form lipovitellin (a lipid-rich protein) and phosvitin (a phosphate-rich protein).

The sequestration process has been investigated in some detail in the sheepshead minnow, *Cyprinodon variegatus,* and killifish, *Fundulus heteroclitus.* Secondary-stage oocytes in these fish are immediately invested by an acellular vitelline envelope (zona pellucida, zona radiata, chorion) which appears during the yolk vesicle stage. This is penetrated by many channels linking the ooplasm with the large intercellular spaces of the granulosa cell layer. The granulosa layer is generally a single cell thick and is separated by a distinct membrane (basal lamina) from the theca cell layer. The theca layer contains an extensive capillary network which bounds the entire follicle.

Vitellogenin molecules appear to pass through the capillary walls, between the cells of the theca layer, through the basal lamina, between the cells of the granulosa layer, and finally through the zona pellucida before being incorporated into the oocyte via micropinocytosis and then translocated to yolk granules or spheres. In the majority of teleosts the yolk accumulates in fluid-filled spheres which migrate centripetally and fuse to form a continuous yolk mass. As this occurs, the yolk vesicles are displaced toward the periphery. The yolk spheres usually fuse, a process that gives teleost eggs their characteristic transparency. This does not appear to be linked to any particular event in development.

2.1.2d. Oocyte Maturation

In a fully grown vitellogenic oocyte, the nucleus (usually referred to as the germinal vesicle)—with its chromosomes still at the diplotene or dictyate stage (Figure 1)—lies in the center of the oocyte. Maturation starts with the germinal vesicle migrating to the periphery at the animal pole. A large space filled with yolk-free cytoplasm forms at this site, and the nuclear membrane then disintegrates (germinal vesicle breakdown). The spindle of the first meiotic division forms with the chromosomes concentrated at the center (metaphasic plate). The two sets of chromosomes draw apart (anaphase and telophase) and one set is pinched off (the first polar body). The second spindle and metaphasic plate is formed and meiosis halted once again.

During oocyte maturation, many teleost eggs increase in size by taking up water. This process is especially pronounced among marine teleosts that spawn pelagic eggs: its effect being to reduce the specific gravity of the eggs so that they can float in seawater.

2.1.2e. Ovulation

Ovulation generally follows oocyte maturation so rapidly that the two events are often classified as a single event. The processes are entirely distinct, however.

Ovulation is the process whereby mature eggs are released from their follicles into either the body cavity (e.g., rainbow trout) or the lumen of the ovary (e.g., turbot). In some teleosts hydration of the eggs (see Section 2.1.2d) may be the mechanism by which the eggs are released from their follicles. In the majority, however, the process involves detachment of the follicles from the oocytes, the appearance of a hole in the follicle, and active expulsion of the egg through the hole by the contraction of smooth muscle fibers which are present in the thecal layers of the follicles. Ovulation is usually accompanied by the production of an ovarian (or coelomic) fluid.

2.1.2f. Postfertilization

Once the egg has been fertilized, meiosis resumes and one of the two haploid sets of female chromosomes is eliminated (as the second polar body). The remaining set combines with the set contributed by the sperm to reestablish the normal diploid genome. It is a matter of interest that if the eggs are subjected to high pressure or mild temperature shocks at this stage, the two female sets can be prevented from separating (due to damage to the meiotic spindle). This results in the formation of embryos with three sets of chromosomes. These are commonly referred to as triploids, and both embryos and adults are viable. Gonad development in the adults is normally impaired, however, due to the fact that the three sets of chromosomes cannot pair up properly at meiotic prophase.

2.1.3. Patterns of Oocyte Growth

Reproduction in all teleosts is a cyclical phenomenon, and most of the teleosts can be divided into three groups depending on whether they

1. Spawn once only and then die.
2. Spawn several times during their life but generally only produce one batch of eggs a year.
3. Spawn several times during their life but produce more than one batch of eggs during each spawning season.

2.1.3a. Group A

The Pacific salmon (*Oncorhynchus* spp.) is an example of a fish that spawns once only and then dies. It exhibits *synchronous* ovarian growth, in which all oocytes, once formed, grow and are released from the ovary in unison: replenishment of one stage by another does not occur.

2.1.3b. Group B

Rainbow trout, Atlantic salmon (*Salmo salar*), and Indian catfish (*Heteropneustes fossilis*) are examples of fish that produce a single batch of eggs a year. They have *group-synchronous* ovaries in which at least two distinct populations of oocytes can be distinguished at any one time—in the examples cited, a population of primary oocytes and a population of synchronously developing secondary oocytes.

Several species in group B exhibit "arrest" of oocyte development. For example, in many North European freshwater species (e.g., cyprinids) oocytes may enter the yolk vesicle and exogenous vitellogenesis stages during the autumn and become arrested halfway through their development during the winter, only to resume development when water temperatures rise again in the spring. A large number of species also hold on to fully grown secondary oocytes until such time as conditions are right for spawning.

2.1.3c. Group C

Among those fish which produce more than one batch of eggs during the course of a single spawning season there are some interesting adaptations to the pattern of oocyte development.

i. The killifish and seahorse (*Sygnathus fuscus*) have *asynchronous* ovaries, in which oocytes can be found at all stages of development from previtellogenesis to ovulation.

ii. The anchovy (*Engraulis mordax*) and stickleback (*Gasterosteus aculeatus*) have a similar pattern of development to group B fish but are able to achieve several complete cycles during the course of a single spawning season because the time required for the development of the eggs from the primary oocyte to ovulation stage is only 4–7 days (compared with 6–9 months in the rainbow trout); the number of batches produced in these species is markedly influenced by temperature and food availability.

iii. Medaka and blenny (*Blennius pholis*) ovaries contain more than one distinct batch of secondary oocytes. The medaka exhibits a daily rhythm of oviposition, with eggs being laid just before dawn almost every day during the breeding season (which may last 4 months). Recruitment of a batch of oocytes into vitellogenesis occurs at the same time as the largest batch of oocytes is recruited into maturation. Since this happens daily, and as it takes 2 days for oocytes to complete vitellogenesis and 1 day to mature, it is possible to observe three batches of oocytes in the ovary every evening. In the blenny there are as many as eight separate batches of vitellogenic oocytes in the ovary at any one time.

iv. Several marine fish that produce large numbers of pelagic eggs (e.g., haddock, *Melanogrammus aeglefinus;* turbot, *Scophthalmus maximus*) exhibit more or less synchronous growth of the oocytes up to the preoocyte maturation stage, but then the oocytes are recruited into maturation and ovulation in batches. Haddock may produce 16 batches of eggs at intervals of 2–5 days, turbot 10 or more at similar intervals that recur extremely regularly.

2.1.4. Testicular Development

The testes generally occupy the same position as the ovaries. They are connected to the exterior by a duct, the vas deferens, which opens at the genital papilla. There are two basic types of testicular structure in teleosts—termed "tubular" and "lobular." The tubular type is not common, being found only in atheriniform fish.

The common lobular type is actually composed of tubules, but is so defined because the diameters of the tubules are variable and look like lobes in histological section. All the tubules are bounded by a continuous basement membrane which effectively divides the testis into two compartments—intertubular and intratubular. The intertubular space, also known as the interstitium, contains boundary (perilobular) cells which are found in association with the membrane—they do not generally form a continuous layer, and the basement membranes of adjacent tubules may thus abut directly. The interstitium also contains connective tissue, fibroblasts, blood vessels, and Leydig (or interstitial) cells. The latter are involved in steroidogenesis (see Section 2.2.1).

The intratubular space contains only germ cells and Sertoli cells, the latter being of somatic origin.

The proliferation of germ cells takes place within cysts formed by the Sertoli cells. *Primary spermatogonia* are found lying singly or in small groups at scattered sites along the tubule walls. These are stem cells which in most teleosts are present at all seasons; *secondary spermatogonia* proliferate from these stem cells by mitotic division. The secondary spermatogonia transform into *primary spermatocytes*. Meiotic division of the primary spermatocytes produces the *secondary spermatocytes* and these the *spermatids*—all still within the cyst membrane. The spermatids metamorphose into *spermatozoa* (the process known as spermiogenesis), and then the cyst ruptures, liberating the spermatozoa into the tubule lumen. At the time of spawning the testes become hydrated, and the sperm are ejected into the sperm duct (the process known as spermiation).

After releasing the spermatozoa into the central lumen, the Sertoli cells form a tight barrier along the walls of the tubules, so that there is no direct contact (i.e., across the membrane) between the spermatozoa and the blood stream. This barrier maintains a favorable microenvironment for the survival of the spermatozoa. For example, in most freshwater fish that have been investigated the seminal fluids have been found to have a potassium content 5–10 times higher than that of blood plasma. Part of the seminal fluid probably originates in the Sertoli cells, and the rest in the epithelium of the vas efferens.

As well as having supporting, barrier-forming, and ionoregulatory functions, Sertoli cells are also involved in the phagocytosis of residual bodies cast off by the spermatids, the secretion of nutritive substances into the tubule lumen, and steroidogenesis (see Section 2.2.1.).

2.1.5. Patterns of Testicular Development

The development of germ cells within a single cyst is synchronous. Cysts at all stages of development can generally be found, however, during the period when the testes are growing rapidly.

In trout, there is a pronounced cycle with a regular succession of germ cell stages. In carp, there is no distinct cycle, and cysts containing spermatozoa and secondary spermatogonia can be found throughout the year.

2.2. Endocrine Organs and Hormones

2.2.1. The Gonads

2.2.1a. Ovarian Steroids

In teleosts, a wide variety of steroids has been identified in ovarian extracts, in *in vitro* incubates of ovaries, and in the plasma of mature or maturing females. Many of these are the same as the steroids secreted by mammalian ovaries; indeed, the biosynthetic pathways are similar (see Chapter 1).

To understand ovarian steroidogenesis in teleosts the notion that androgens are male hormones and the estrogens are female hormones must be dispelled. Testosterone and androstenedione are major secretory products of teleost ovaries. They are the obligatory precursors of 17β-estradiol and estrone, respectively. The conversion of the androgens to estrogens is effected by an enzyme complex termed aromatase. Ovarian aromatase is at its most active in females undergoing vitellogenesis.

At the time of oocyte maturation, in all fishes so far studied, androgen and more especially estrogen production decline, and the major products of steroidogenesis appear to be C21 steroids: the progestagens, 17α-hydroxy,20β-dihydroprogesterone and 17α-hydroxyprogesterone, are produced by salmonid and cyprinid ovaries, whereas the deoxycorticosteroids, 11-deoxycortisol and 11-deoxycorticosterone, are produced by the ovaries of a number of marine teleosts. The formation of steroid glucuronides (see Section 2.2.1b) may also increase at this stage.

2.2.1b. Testicular Steroids

The teleost testis differs from that of most other vertebrates in that it can synthesize derivatives of androstenedione and testosterone with either a hydroxyl or keto group at the 11 position in the steroid nucleus. These androgens are believed to be responsible for the induction of male secondary sexual characteristics (e.g., skin thickening, color changes, and kype formation) and stimulation of spermatogenesis in teleosts: i.e., they perform the role of testosterone in mammalian males. 11-Ketotestosterone is the major androgen in teleosts. However, 11β-OH testosterone predominates in ambisexual teleosts (i.e., those that change sex naturally during their life-cycle).

Another unusual feature of the teleost testis is its extremely high capacity for steroid glucuronide formation. In mammals, steroid glucuronidation is limited to the liver and is generally considered to be a means of facilitating the deactivation and excretion of steroid hormones. The activity of the testicular glucuronyl transferase increases with temperature, and it has been suggested that one of its functions in teleosts is to limit the secretion of the free and active androgens to the preferred temperature for reproduction. Another suggestion is that in teleosts specific steroid glucuronides have evolved as pheromones. Unlike the free steroids, glucuronides are freely soluble in water and would appear to be suitable for this role. One such glucuronide has in fact been identified in a teleost *Gobius jozo* where the male bears a prominent mesorchial gland, composed almost entirely of Leydig cells (see Section 2.2.1d) which synthesizes a number of 5β-reduced androgen conjugates, the major one being etiocholanolone glucuronide: this compound, when dissolved in water at low concentrations, attracts gravid females of the same species.

Testes of salmonids incubated *in vitro* also synthesize large amounts of 17α-hydroxy,20β-dihydroprogesterone. Indeed, this and/or its 20α-isomer have been shown to be major products of the testes of a wide variety of vertebrates (see Chapter 1).

2.2.1c. Cellular Origin of Ovarian Steroids

All vertebrate oocytes are surrounded by a complex multilayered follicle made up of a number of cell layers. Two of these are always present—an inner *granulosa* layer, which is separated by a basement membrane from an outer *theca* layer (Figure 2). The latter may be subdivided into a theca interna and theca externa. One particular cell type in the theca interna and the predominant cell type that makes up the granulosa have both been implicated in steroidogenesis. Histochemical studies have revealed the presence of 3β-hydroxysteroid dehydrogenase in both cell types. Ultrastructurally, only the special theca cells show the characteristics of steroidogenic tissue. These problems have been partly resolved by the recent development of reliable techniques for isolating relatively pure theca and granulosa tissue. It has

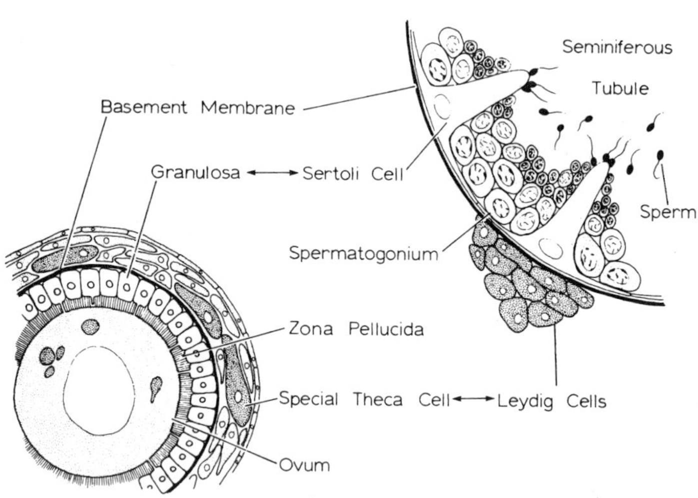

Figure 2. Diagram showing steroid-synthesizing cells in teleost gonads. (From Hoar, W. S., and Nagahama, Y., 1978, The cellular sources of sex steroids in teleost gonads. *Ann. Biol. Anim. Bioch. Biophys.* **18:**893–898, with permission.)

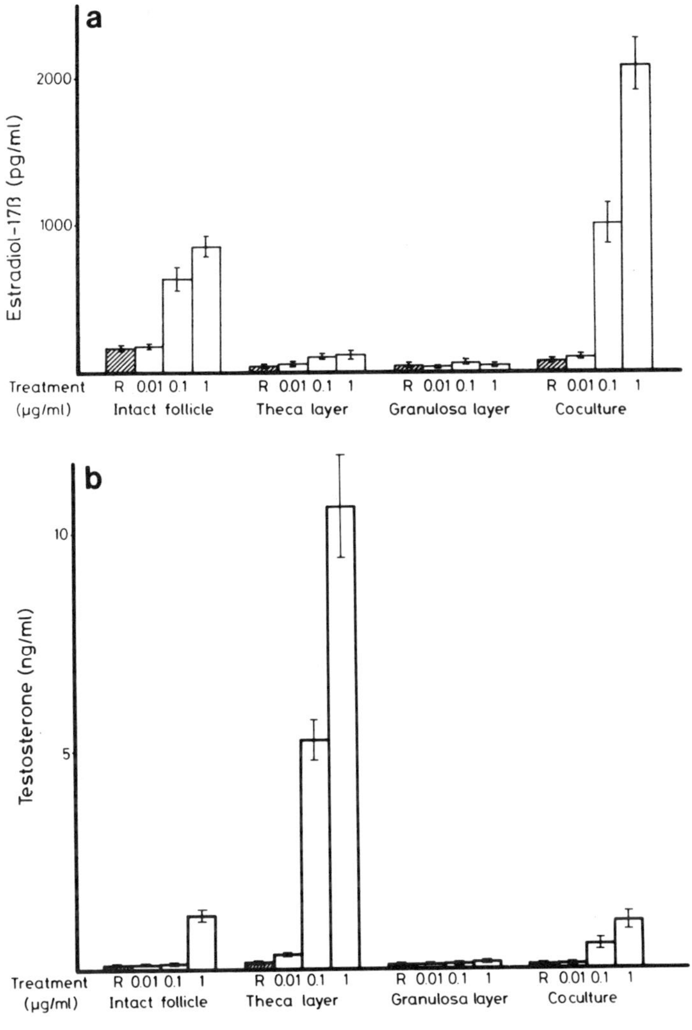

Figure 3. Effects of chinook salmon GtH on (a) 17β-estradiol and (b) testosterone secretion by amago salmon. Follicles were incubated in Ringer alone (R, shaded bars) or Ringer with various doses of GtH (0.01–1μg ml^{-1}) (open bars) for 18 hr. Vertical bars represent the mean ± SEM of three replicates. (From Kagawa, H., *et al.*, 1982, Estradiol-17β production in amago salmon (*Oncorhynchus rhodurus*) ovarian follicles: Role of the thecal and granulosa cells. *Gen. Comp. Endocrinol.* **47:**440–448, with permission.)

been possible to prepare four different follicular layers from oocytes of the amago salmon (*Oncorhynchus rhoduras*)—intact follicles, theca layers, pure granulosa layers, and theca/granulosa cocultures. Stimulation with salmon gonadotropin (0.1, 1 μg/ml) induces 17β-estradiol production by intact follicles and the coculture preparations but not by the isolated theca or granulosa layers (Figure 3a). In contrast, GtH stimulates testosterone production by the theca layers by up to 80 times but hardly at all by the granulosa layers (Figure 3b). Incubation of granulosa tissue with testosterone results in enhanced 17β-estradiol production. These results suggest a two-cell-type model of ovarian steroidogenesis (as has been demonstrated in mammals) whereby the theca layer synthesizes the androgens, which are then transferred to the granulosa layer to be aromatized to estrogens. The synthesis of 17α-hydroxy,20β-dihydroprogesterone by the amago salmon ovary appears to be almost identical; 20β-hydroxysteroid dehydrogenase is localized in the granulosa layer; in response to GtH stimulation the theca layers of the mature oocytes produce 17α-hydroxyprogesterone, and the granulosa layers convert it into 17α-hydroxy,20β-dihydroprogesterone.

2.2.1d. Cellular Origin of Testicular Steroids

Two cell types are also thought to be involved in steroidogenesis in the testis—the Leydig cells and the Sertoli cells. Many comparisons can be drawn between special theca and Leydig cells, on the one hand, and granulosa and Sertoli cells, on the other. For example, all cell types contain 3β-hydroxysteroid dehydrogenase activity; special theca and Leydig cells are outside the basement membrane, and granulosa and Sertoli cells are within the basement membrane (see Figure 2); like granulosa cells, the Sertoli cells lack ultrastructural features of steroid-secreting cells; in birds and mammals, the Sertoli cells contain 20β-hydroxysteroid dehydrogenase activity (cf. follicular cells in amago salmon). It thus seems probable that the Leydig cells are mainly concerned with androgen production and the Sertoli cells with transforming steroids derived from this source, although this remains to be proven in teleosts.

2.2.2. The Pituitary Gland

Most of the important phases in gonadal development are controlled by a hormone or hormones secreted by the pituitary gland, these being referred to as gonadotropins (see also Part III).

2.2.2a. Gonadotropins (GtHs)

GtHs have been purified from several species and shown to be similar in structure to the mammalian luteinizing and follicle-stimulating hormones (LH and FSH): i.e., they are glycoproteins, have a molecular weight of ca. 30,000 daltons, and are composed of two subunits, α and β, the biological activity being conferred by the β subunit.

Many studies have indicated the presence of only one type of GtH in teleosts, as opposed to the two in mammals, this GtH appearing to be more like LH than FSH.

However, recently, a separate type of "gonadotropin" has been identified in the pituitaries of at least four teleost species. This GtH has an apparently lower carbohydrate content than the conventional GtH, and its sole action in females appears to be to stimulate vitellogenin uptake. The conventional carbohydrate-rich GtH, on the other hand, can stimulate oocyte maturation and ovulation, spermiation, gonadal steroidogenesis, and ovarian cyclic adenosine monophosphate activity as well as the uptake of vitellogenin by the oocytes. It is emphasized that the carbohydrate-poor GtH is not considered to be homologous to FSH. To date, all physiological studies have employed only radioimmunoassays for carbohydrate-rich GtH, which in the remainder of this chapter will be referred to simply as GtH.

It is possible that there may be yet another GtH in certain teleosts (e.g., one that is more FSH-like) which has been overlooked, owing to the use of inappropriate starting material or inadequate bioassays. However, in at least one fish, the carp (*Cyprinus carpio*), there is no need to postulate its existence, as the carbohydrate-rich GtH that has been isolated and extensively purified has been shown to be capable of inducing and maintaining complete gonadal development in hypophysectomized individuals.

2.2.2b. Cellular Origin of Gonadotropin(s)

The pituitary gland in teleosts consists of a central neurohypophysis interdigitating with a shell-like adenohypophysis. The adenohypophysis is subdivided into rostral pars distalis, proximal pars distalis, and pars intermedia. The gonadotropin-secreting cells are generally found in the ventral region of the proximal pars distalis, though they can sometimes be found in the rostral pars distalis (e.g., in salmonids and eels). They can be recognized at the light microscopical level by their staining reactions and location: they are basophils (i.e., the cell contents react with basic dyes, e.g., alcian blue; the carbohydrate reacts with periodic acid/Schiff's reagent.) At the electron microscopic level, they can be recognized by the presence of relatively large granules (200–500 μm diameter), often dilated rough endoplasmic reticulum (RER), when active, and irregular small vacuoles. Variations in the number, shape, and size of these organelles have led some workers to conclude that there is a single GtH cell type in teleosts, and others, that there are two GtH cell types. These discordant results may reflect a real situation, based on species differences (i.e., some species have only one gonadotropin whereas others have two). It may equally be due to temporal differences in the appearance of cell types, however. For example, two cell types have been reported in rainbow trout—one full of globules and granules and the other full of vacuoles. Castration of adults results in the transformation of the former cell type into the latter.

2.2.3. The Hypothalamus

2.2.3a. Gonadotropin-Releasing Hormone (GnRH)

It has been established that the teleost brain contains a peptide similar in its structure and biological actions to mammalian luteinizing hormone releasing hormone (LH-RH). Teleost hypothalamic extracts induce GtH release by teleost pituitaries both *in vivo* and *in vitro;* material cross-reacting with anti-LH-RH sera can be identified in brain extracts of many teleosts (among them cod, winter flounder, *Tilapia*); the highest concentrations of LH-RH-like material are found in the hypothalamus and telencephalon (as they are in mammals).

The primary structure of GnRH from chum salmon (*Oncorhynchus keta*) is Pyro-Glu-His-Trp-Ser-Tyr-Gly-Trp-Leu-Pro-Gly-NH_2; it differs by only two amino acids Trp^7 and Leu^8, from mammalian LH-RH, in which they are Leu^7 and Arg^8. Other teleosts also possess immunoreactive GnRHs of similar size though slightly different structure. In addition, a ''big'' LH-RH immunoreactive protein, probably a GnRH precursor, has been identified in some fishes.

2.2.3b. Brain Centers Involved in the Control of GtH Secretion

Although there is no obvious pituitary portal system in teleosts, the blood supply to the adenohypophysis passes through the neurohypophysis, which contains fibers originating in two areas in the hypothalamus, the nucleus lateralis tuberis (NLT) and the preoptic region. Fibers from these regions also penetrate the adenohypophysis and abut GtH cells directly.

Cytological indications of neurosecretory activity in the NLT and preoptic region have long been known to be associated with reproductive activity. Experimental treatments designed to alter GtH secretion have been shown to modify the amount of stainable material in these regions.

The preoptic region is comprised of two nuclei—the nucleus preopticus (NPO) and nucleus preopticus paraventricularis (NPP)—both of which send fibers to the pituitary gland.

Several regions in the brain have been found to concentrate the sex steroids testosterone and 17β-estradiol. These include the NLT, ventral NPO, ventral NPP, area ventralis telencephali pars ventralis (Vv), nucleus recessus pars lateralis (NRL), and nucleus recessus posterioris (NRP). The approximate positions of all these regions in the brain are shown in Figure 4. The NRL and NRP, which are part of the same structure, the paraventricular organ, send aminergic fibers to the neurohypophysis and to the NPO/NPP and NLT regions, suggesting functional interactions (perhaps in steroid feedback control of GtH synthesis and secretion). The Vv area appears to be involved in the control of spawning behavior in the goldfish. The NPO is the source of octapeptides such as arginine vasotocin. Lesions and antiestrogen implants in this nucleus do not appear to affect gonad development.

Immunoreactive LH-RH has been localized in sections of brains and pituitaries from several species, both in perikarya (the cell bodies that are located in brain nuclei) and in nerve fibers (which generally run in well-defined tracts). Unfortunately, virtually no two studies have been in agreement concerning the anatomical position (or sometimes even the presence) of immunoreactive LH-RH. This is due in part to methodological difficulties such as differences in antiserum specificity (which can easily arise because fish GnRHs are not identical to LH-RH), to dif-

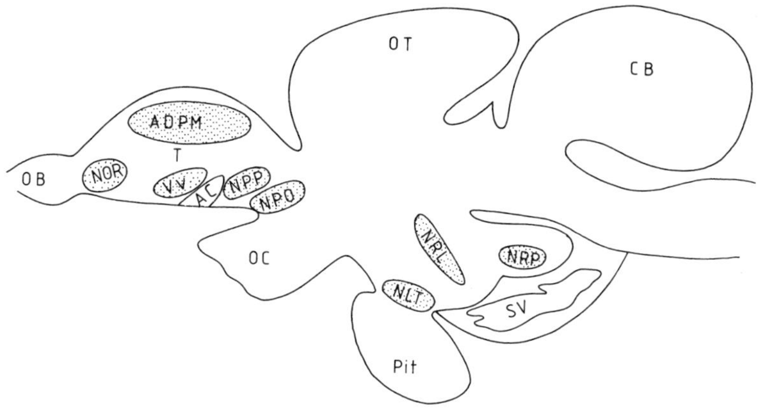

Figure 4. Parasagittal representation of the distribution of regions in the teleost brain that bind 17β-estradiol and testosterone and/or contain LH-RH-immunoreactive perikarya. AC, anterior commissure; CB, cerebellum; OB, olfactory bulb; OC, optic chiasma; OT, optic tract; Pit, pituitary; SV, saccus vasculosus; T, telencephalon; ADPM, area dorsalis pars medialis; NLT, nucleus lateralis tuberis; NOR, nucleus olfactoretinalis; NPO, nucleus preopticus; NPP, nucleus preopticus periventricularis; NRL, nucleus recessus lateralis; NRP, nucleus recessus posterioris.

ferences between species, and to incorrect descriptions of the sites of staining (this applies especially to the nuclei within the preoptic region).

Immunoreactive perikarya have so far been identified in the area dorsalis pars medialis (ADPM) in the rainbow trout brain, in the ventral–lateral border of the NPP and posterior portion of the NLT in the goldfish, and in the NPP, NLT, nucleus olfactoretinalis (NOR), and dorsal midbrain (not shown in Figure 4) of the platyfish, *Xiphophorus maculatus*.

In addition, fibers immunostained with anti-LH-RH sera have been found at sites all over the brain in several teleosts—e.g., in tracts linking the NPP, NLT, and NOR regions with each other, in the pituitary, ADPM, optic and olfactory nerves, and pineal complex.

Destruction of the posterior part of the NLT in goldfish, Atlantic salmon, and killifish has been shown to cause a decrease in the gonadosomatic index. Implantations of antiestrogens in the NLT have also been shown to cause a significant rise in plasma GtH levels. These results clearly implicate the NLT in the control of GtH secretion via GnRH.

The specific destruction of the LH-RH-containing nuclei in the ventral–lateral border of the NPP has not yet been achieved as this nucleus also contains a GtH-release inhibitory factor (see Section 2.2.3c).

Lesions in the ADPM have been found to reduce GtH secretion but at the same time accelerate the onset of ovulation in rainbow trout and appear to have no effect on the goldfish.

During the development of the platyfish, the NOR is the first region in the brain in which immunoreactive LH-RH can be identified. However, immunoreactive LH-RH only appears in the NPP and NLT regions when gonadal development is initiated. It has been suggested that the NOR may be involved in the control of the onset of gonadal development in the platyfish.

2.2.3c. GtH-Release Inhibitory Factor (GnRIF)

It has long been suspected that the brains of some teleosts contain a factor that inhibits the release of gonadotropin. Overwhelming evidence for the existence of such a factor in the goldfish brain has recently been published. It has been found, for example, that whereas small lesions in the NLT cause gonadal regression in sexually mature goldfish, large lesions generally lead to an increase in serum GtH levels and ovulation. An important feature of these large lesions is that they damage the pituitary stalk, thereby isolating the pituitary from the hypothalamus. Lesions in the anteroventral NPP have also been shown to cause a massive and prolonged release of GtH in female goldfish. These levels correspond to those found during natural ovulation and suggest that the natural GtH surge (at ovulation) may result from the combination of an abatement of a GnRIF (originating in the NPP) and stimulation by GnRF.

Further work has shown that GnRIF is probably dopamine. Injections of dopamine into mature female goldfish suppress GtH secretion. Injections of pimozide, a dopamine antagonist, followed by LH-RH injections 12 hr later induce virtually 100% ovulation and very high plasma GtH levels. Injections of LH-RH alone are far less effective.

2.3. Control of the Different Stages of Reproductive Development

2.3.1. Sex Differentiation

In most of the teleosts that have been examined sex is genetically determined. A number of fish (e.g., gilthead sea bream, *Sparus aurata*) are hermaphrodites and change sex naturally at some time during their lives. In virtually all species examined it has been found possible to override the genetic sex-determining mechanisms by treating the alevins or fry with sex steroids (in the water or the diet). Androgens generally cause genetic females to become functional males, and estrogens cause genetic males to become functional females. To effect functional sex inversion, the androgenic or estrogenic compounds must be administered during the indifferent stage of gonadal development and maintained up to the time that the gonads are fully differentiated. Dosage is also important, with undertreatment likely to yield hermaphrodite fish and overtreatment sterile fish.

The efficacy of exogenously applied steroids in altering the sex of teleosts strongly suggests that steroids are involved in the natural differentiation process. This is supported by the fact that the gonads develop a capacity for sex steroid synthesis very early in development. In rainbow trout, for example, the histologically undifferentiated gonads of first-feeding fry (at D_{50}—i.e., 50 days after hatching at 11°C) contain 3β-hydroxysteroid dehydrogenase, $\Delta_5 \rightarrow \Delta_4$ isomerase, and 17β-hydroxylase activity, indicating a capacity to synthesize progesterone, 17α-hydroxypregnenolone, and 17α-hydroxyprogesterone from pregnenolone. By D_{100}, when ovaries and testes can be distinguished, the ovaries still have only the same synthesizing capacity as D_{50} gonads. The testes, however, contain additional enzymes—17,20-desmolase and 11β-hydroxylase—showing a capacity to synthesize androstenedione and 11β-hydroxyandrostenedione. By D_{200} the testes contain 17β-hydroxysteroid dehydrogenase (enabling testosterone and 11-ketotestosterone production), and the ovaries contain aromatase (enabling estrogen production). The situation by D_{200} is thus similar to that in mature gonads.

2.3.2. The Female

2.3.2a. Oogonial Proliferation

Evidence, largely from hypophysectomy experiments, suggests that oogonial proliferation is under pituitary control. Since oogonial proliferation occurs close to the spawning period when GtH levels are usually high, it is possible (though not proven) that this hormone is involved.

2.3.2b. Oogonial–Oocyte Transformation

The transformation of oogonia into oocytes is not inhibited by hypophysectomy.

2.3.2c. Primary Oocyte Growth Phase

The growth of primary oocytes, up to a certain size, does not appear to be prevented by hypophysectomy. Their growth is probably affected, however, by factors that control body growth such as the rate of metabolism, food availability, and metabolic hormones.

2.3.2d. Endogenous Vitellogenesis

The early stages of the secondary growth phase, involving yolk vesicle formation, are considered to be gonadotropin-dependent. The evidence is not very clear, however. Generally, after hypophysectomy or transfer of ovaries into culture dishes, all oocytes revert to a pre-yolk vesicle stage. This indicates that yolk vesicle formation is under pituitary (though not necessarily GtH) control.

In rainbow trout, raised GtH levels have been found in the pituitary and plasma during the endogenous vitellogenic stage by several (though not all—see Figure 5) workers. This may be due to the pulsatile release of GtH, recently demonstrated in female rainbow trout, which, though of high amplitude, is very infrequent (Figure 6).

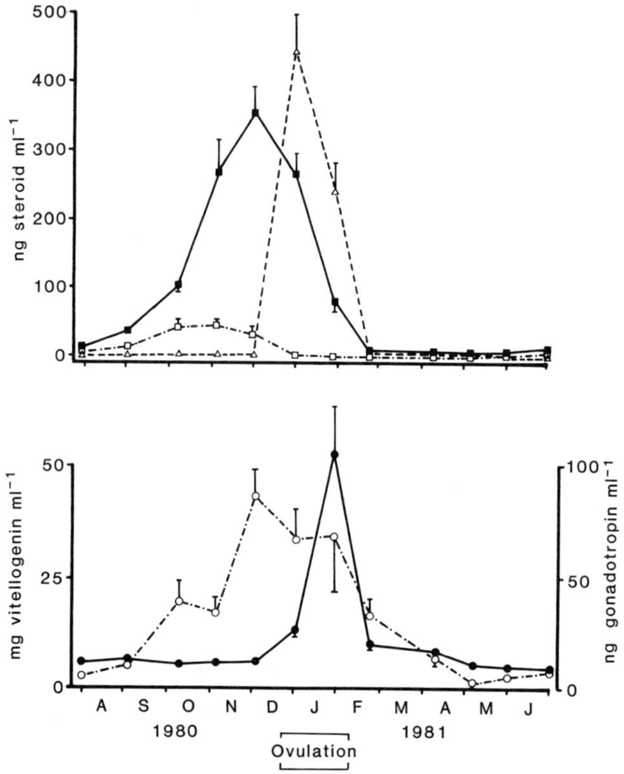

Figure 5. Annual changes in plasma concentrations of testosterone (■), 17β-estradiol (□), 17_α-hydroxy, 20β-dihydroprogesterone (△), vitellogenin (○), and gonadotropin (●) in female rainbow trout. Vertical bars indicate SEM. (From Scott, A. P., and Sumpter, J. P., 1983, Seasonal variations in sex steroids and gonadotropin in females of autumn and winter-spawning strains of rainbow trout (*Salmo gairdneri*). *Gen. Comp. Endocrinol.* **52:**79–85, with permission.)

In female silver eels (in which yolk vesicle formation is a distinct process) the content of GtH in the pituitary and plasma is extremely low. Also, the ovaries of silver eels show no capacity to synthesize 17β-estradiol; in fact, they show a capacity to synthesize mainly 11-ketoandrogens (like testes). If GtH is injected into silver eels, in the form of human chorionic gonadotropin (HCG) or salmon GtH, its effect is to induce 17β-estradiol synthesis and exogenous vitellogenesis. These experiments suggest that in the eel, at least, endogenous vitellogenesis is not controlled by the secretion of GtH. There is a strong possibility that either thyroxine or the pituitary-derived thyroid-stimulating hormone, which is not dissimilar in structure to GtH, may be involved.

2.3.2e. Exogenous Vitellogenesis

In most fish, there is definite evidence that the stage of exogenous vitellogenesis is under the control of GtH. For example, hypophysectomy during this stage leads to ovarian atrophy, and injections of GtH into immature fish are effective in the stimulation of ovarian steroid production, vitellogenesis, and oocyte growth. Full ovarian growth can also be obtained in immature fish by frequent injections of salmon GtH and/or HCG. Despite this, most attempts to show changes in the levels of GtH in the plasma of maturing females (especially rainbow trout—see Figure 5) have been unsuccessful. As with the early stages of vitellogenesis, the explanation probably lies in the pattern of release of GtH. In female rainbow trout, for example, GtH is released episodically (five to six low-amplitude pulses per day) (Figure 6). In goldfish, on a 16-hr light day, it has been found that serum GtH levels

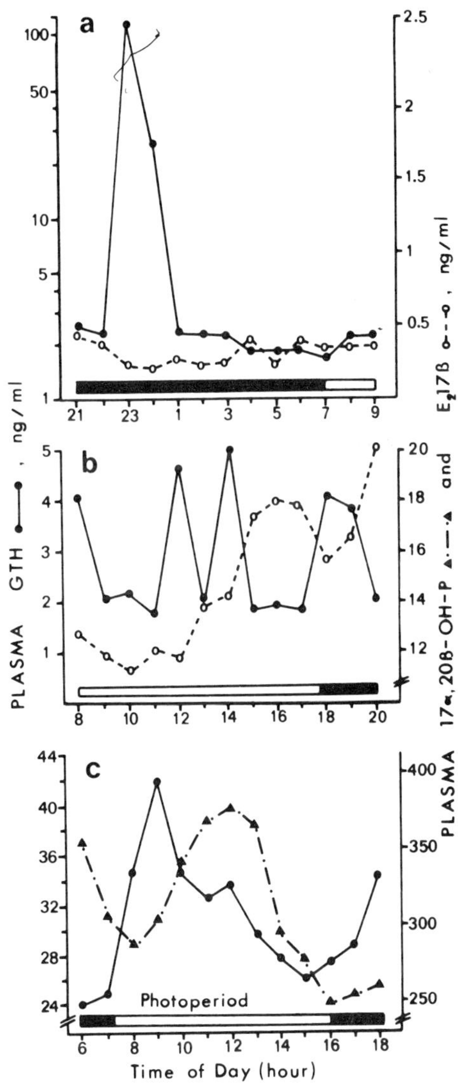

Figure 6. Isolated examples of individual profiles of plasma 17β-estradiol, 17_α-hydroxy, 20β-dihydroprogesterone, and GtH levels in female rainbow trout at different characteristic stages of sexual development: (a) beginning of secondary growth phase (March), (b) advanced exogenous vitellogenesis (October), and (c) maturation/ovulation (January). [From Zohar *et al.*, 1982, Gonadotrophic function during the reproductive cycle of the female rainbow trout, *Salmo gairdneri*, in relation to ovarian steroid secretion: *in vivo* and *in vitro* studies, in: *Reproductive Physiology of Fish* (C. J. J. Richter and H. J. Th. Goos, eds.), Pudoc, Wageningen, The Netherlands, pp. 14–18.]

are basal in fish at different stages of gonadal activity for most of the 24-hr period. However, in those fish undergoing vitellogenesis there is a significant surge in GtH levels 8–12 hr after the onset of the light period. Also, lesions in the NLT and NPO of the goldfish hypothalamus, which block ovarian growth, appear to work by removing the surge in GtH and not by completely preventing its secretion.

These observations suggest that the daily variations in GtH release, rather than overall levels of GtH, are important in the control of vitellogenesis.

There is incontrovertible evidence, from studies on a wide range of teleosts, that the production of vitellogenin by the liver is induced by ovarian estrogens—mainly 17β-estradiol and estrone. 17β-Estradiol is the more potent steroid. Livers of male and immature fish can be stimulated to produce vitellogenin by injections of 17β-estradiol. In the rainbow trout, two injections of 17β-estradiol are sufficient to induce vitellogenin synthesis lasting 1–2 months. While the secondary oocytes are growing in the rainbow trout, plasma levels of vitellogenin, 17β-estradiol, and estrone are usually considerably elevated (e.g., see Figure 5). The growth of the ovary is also considerable. In the eel, for example, it may, when fully developed, weigh up to half of the weight of the whole fish. This quite phenomenal growth requires the incorporation of very large quantities of exogenous yolk into the oocytes, which is made possible by the liver secreting vitellogenin at such a rate that blood levels reach tens of milligrams per milliliter. Vitellogenin is generally found in the plasma of females for a long time after steroids have returned to basal levels following ovulation.

There appears to be considerable species variability in the magnitude of the peak plasma levels of 17β-estradiol found in the many studies that have been published: for example, 140 ng ml^{-1} in the plaice (*Pleuronectes platessa*), 50 ng ml^{-1} in the rainbow trout, and 0.70 ng ml^{-1} in the bluefish (*Pomatomus saltator*). Although some of these differences probably reflect species differences in levels of plasma binding proteins, differences in secretion and clearance rates and in receptor sensitivity, any speculations on such matters at the moment would be fruitless as, within a single species, separate studies appear to have established peak 17β-estradiol levels that are as much as 10-fold different (e.g., in rainbow trout, reported peak levels have ranged from 5 ng ml^{-1} to 50 ng ml^{-1}). These differences have yet to be satisfactorily explained.

Perhaps the most interesting discovery that has been made (in over 15 species) concerning the secondary growth phase is the presence of high levels of testosterone in plasma (Figure 5). These range (with the previously mentioned reservations) from 250 ng ml^{-1} in the rainbow trout to 1.7 ng ml^{-1} in the bluefish. In some species, the levels have been shown to correlate with the gonadosomatic index (GSI), and in most species levels reach a peak just prior to oocyte maturation and are higher than levels found in males. Androstenedione has also been demonstrated in the plasma of female fish. The most plausible explanation for the presence of androgens in the plasma is that they are synthesized by the ovary, released into the plasma, and converted to estrogens when they reach the brain and pituitary. Both these organs have been shown to contain large amounts of aromatase.

As confirmation of a central role for testosterone and androstenedione, it has been shown that injections of these and other aromatizable androgens into immature salmonids greatly increase the synthesis and accumulation of GtH in the pituitary gland, whereas nonaromatizable androgens like 5α-dihydrotestosterone are ineffective.

Although estrogens and aromatizable androgens are capable of inducing GtH *synthesis* by the pituitary gland, they do not appear to promote its *release*. Indeed, there is evidence that ovarian steroids inhibit GtH release during the period of exogenous vitellogenesis: for example, castration of female rainbow trout at this stage results in a dramatic increase in plasma GtH levels.

2.3.2f. Oocyte Maturation

The endocrine control of oocyte maturation in teleosts is basically similar to that in higher vertebrates: GtH is released from the pituitary in response to hypothalamic signals; it is carried to the ovary, where it stimulates production of C21 steroids by follicular cells; these steroids bind to receptors in (or on) the oocyte, which trigger oocyte maturation; ovulation then follows.

2.3.2g. Artificial Induction of Oocyte Maturation and Ovulation

In many commercial species oocytes fail to mature and/or ovulate in captive broodstock and must therefore be made to do so by hormonal injections. The first studies on this problem in teleosts were carried out over 50 years ago in South America and involved injecting mature females with homologous or heterologous pituitary extracts from other mature fish. Similar methods are still widely used, although with refinements such as the use of preserved glands, partially purified gonadotropin preparations, and human chorionic gonadotropin. Advances in the understanding of the hypothalamo–pituitary–gonadal axis have resulted in the development of techniques involving either the stimulation of the production and release of endogenous GtH (e.g., by use of LH-RH or antiestrogen) or the direct stimulation of oocyte maturation and ovulation by the appropriate steroid hormones and/or prostaglandins. When using these latter compounds it has been found that a small priming injection of GtH or crude pituitary extract (given 1–3 days before) is often required in order to obtain efficient maturation and ovulation (e.g., when using LH-RH in coho salmon (*Oncorhynchus kisutch*), and when using steroids in gray mullet (*Mugil cephalus*). These priming injections appear to be necessary to stimulate the migration of the germinal vesicles from the central to subperipheral position in the oocytes, thereby increasing their sensitivity to further hormonal stimulation. Fish in which this process has already occurred do not require a priming injection.

2.3.2h. Induction of Oocyte Maturation and Ovulation *In Vitro*

Much information on the processes of oocyte maturation and ovulation has come from *in vitro* experiments. These experiments have established that GtH acts on the follicular cells to stimulate *in situ* production of the steroidal mediators of maturation, whereas steroids act directly on the oocyte. This has been shown in several species by removing the follicle layers from the oocytes either enzymically or mechanically. It has also been demonstrated in the amago salmon by treating intact follicles with cyanoketone, an analog of pregnenolone that binds irreversibly to 3β-hydroxysteroid dehydrogenase and inhibits the conversion of 5- ene to 4- ene steroids. In this situation, GtH has no effect on oocyte maturation, but exogenously applied steroids are still active.

In the rainbow trout 17α-hydroxy,20β-dihydroprogesterone and 17α-hydroxyprogesterone are effective mediators of oocyte maturation *in vitro* (i.e., at physiological levels); corticosteroids have little or no effect. This contrasts with the pike (*Esox lucius*), brook trout (*Salvelinus fontinalis*), and goldfish, in which not only the progestagens but also 11-deoxycorticosteroids (11-deoxycorticosterone and 11-deoxycortisol) are effective at physiological levels, and with the medaka and catfish, in which progestagens, 11-deoxycorticosteroids, and 11-oxygenated corticosteroids (e.g., cortisol) are all equally effective.

A problem of considerable interest is how and at what stage in development oocytes acquire the competence to mature in response to GtH and steroids. In all species the age and size of the oocytes appear to be crucial. Small, underdeveloped oocytes cannot be induced to mature. Intermediate-sized oocytes can sometimes be induced by GtH injection, but they either do not ovulate or else become nonviable eggs.

In several species it has been shown that the degree of germinal vesicle migration is an important factor. In the rainbow trout, for example, oocytes removed from females at the end of the vitellogenic phase, in which the germinal vesicles are still central, require 10–20 times more 17α-hydroxy,20β-dihydroprogesterone to induce oocyte maturation *in vitro* than oocytes in which the germinal vesicles are peripheral. The sensitivity of oocytes to steroids and GtH may also be influenced by other hormones. In the rainbow trout, for example, testosterone and cortisol enhance and 17β-estradiol decreases the efficiency of GtH and 17α-hydroxy,20β-dihydroprogesterone on oocyte maturation *in vitro*.

2.3.2i. Changes in Hormone Levels during Oocyte Maturation and Ovulation

i. Rainbow Trout. It can be seen in Figure 7 that there are marked changes in the levels of GtH and steroids in the plasmas of rainbow trout undergoing oocyte maturation and ovulation. GtH, 17α-hydroxy,20β-dihydroprogesterone, and 17α-hydroxyprogesterone levels are considerably elevated from 8 days before to approximately 28 days after ovulation. 17β-Estradiol levels reach basal levels 4 days before ovulation, and testosterone levels peak 8 days before ovulation and then fall thereafter.

Evidence that has already been discussed suggests that the rise in GtH levels is responsible for the stimulation of progestagen secretion by the ovaries and that the

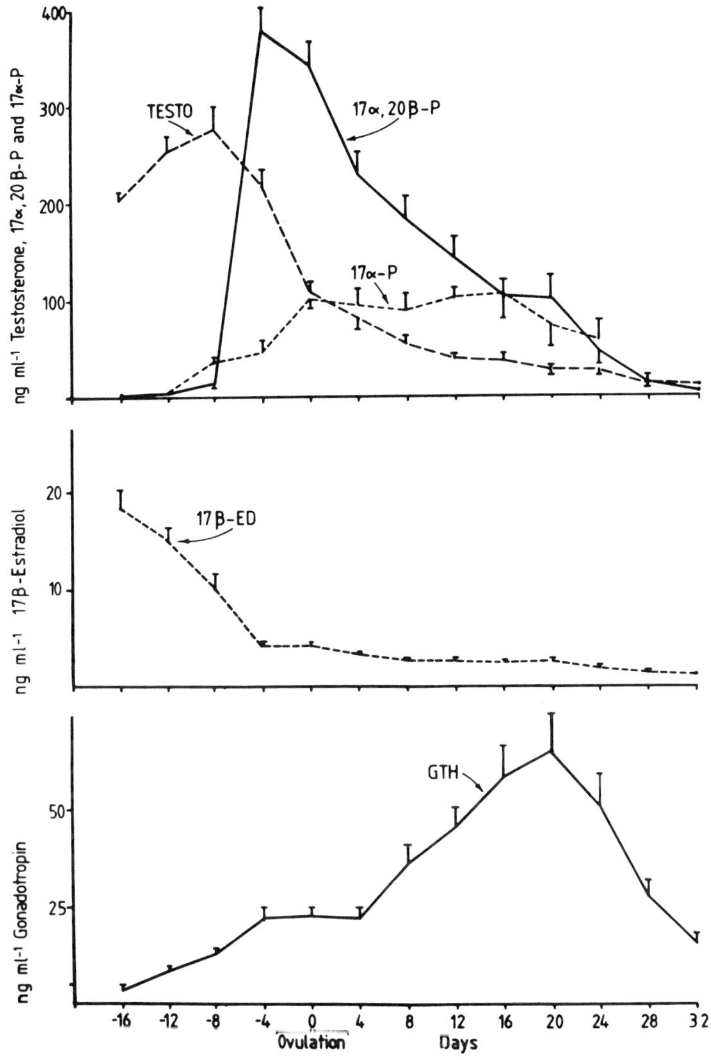

Figure 7. Changes in hormone levels about the time of ovulation in rainbow trout: testosterone (TESTO), 17β-estradiol (17β-ED), 17α-hydroxyprogesterone (17α-P), 17α-hydroxy, 20β-dihydroprogesterone (17α,20β-P), and gonadotropin (GTH). (From Scott *et al.*, 1983, Hormone changes during ovulation in the rainbow trout, *Salmo gairdneri*. *Gen. Comp. Endocrinol.* **49:**128–134; with permission.)

progestagens are, in turn, responsible for the induction of oocyte maturation. The period over which 17α-hydroxy,20β-dihydroprogesterone levels are at their highest encompasses the period during which oocyte maturation has been shown to occur: it takes about 50 hr at 10°C in the rainbow trout.

The testosterone peak prior to ovulation appears to be a common feature in all teleosts that have so far been investigated although its significance has not been established.

As with all the other stages of the reproductive cycle in female rainbow trout, there is difficulty in explaining how GtH can, on the one hand, control testosterone and 17β-estradiol secretion and, on the other, control progestagen secretion. It has been suggested that this might be achieved by changes in the frequency and amplitude of GtH pulses: low-frequency, low-amplitude pulses of GtH probably favoring stimulation of testosterone/17β-estradiol secretion, and continuous, high-amplitude release (Figure 6) favoring progestagen production.

The continued rise in GtH levels after ovulation is a well-established phenomenon in salmonids. It probably results from a reduction in negative feedback by testosterone and/or 17α-hydroxy,20β-dihydroprogesterone. Whether GtH has any physiological actions (e.g., stimulation of oogonial proliferation) at this stage is not known. Its presence may be related to the fact that salmonids can retain their eggs for several weeks following ovulation.

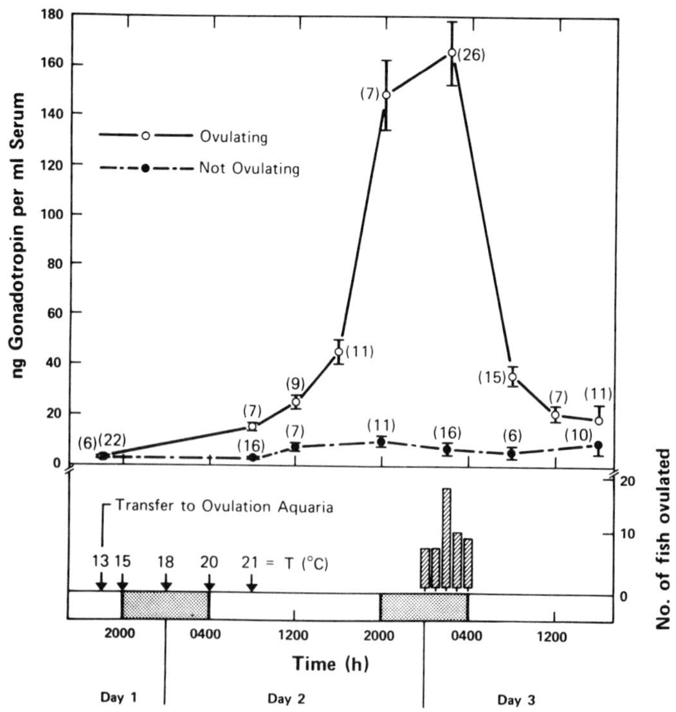

Figure 8. Time of ovulation and serum GtH (mean ± SEM) profiles of sexually mature goldfish kept on 16L:8D photoperiod and warmed from 13 ± 1 to 21 ± 1°. The number of fish ovulating each hour is shown in the lower graph. Numbers in parentheses = numbers of fish. (From Stacey *et al.*, 1979, Ovulatory surge of gonadotropin in the goldfish, *Carassius auratus*. *Gen. Comp. Endocrinol.* **37**:246–249.)

ii. Goldfish (Carassius auratus). In the goldfish, most of the events associated with oocyte maturation and ovulation take place over a period of 2–3 days, compared to 40 days in the rainbow trout. Ovulation can be induced in sexually mature females by increasing the water temperature from 13°C to 21°C and/or by placing aquatic vegetation in the aquaria. If fish kept on a 16L:8D photoperiod (to stimulate ovarian development) are warmed from 13°C to 21°C over a period of 10 hr, ovulation takes place after approximately 20 hr and always just before "dawn." As in the trout, this is accompanied by a preoocyte maturation surge of GtH (Figure 8), which then declines rapidly after ovulation.

iii. Sea Bass (Dicentrarchus labrax). At the time of oocyte maturation in the sea bass, a switchover occurs from the synthesis of androgen and estrogen biosynthesis to 11-deoxycorticosterone (DOC) and 11-deoxycortisol (substance S). These steroids are formed by 21-hydroxylation of progesterone and 17α-hydroxyprogesterone, respectively. 21-Hydroxylase activity has been demonstrated in the mature ovaries of at least five other teleost species (all saltwater fish of the order Perciformes). It is likely in these species that DOC and/or S perform the same role as the hydroxylated progestagens in the order Salmoniformes.

iv. Turbot. The turbot is a batch spawner; i.e., it does not ovulate all its eggs at once but in several batches and usually at fairly regular intervals. As can be seen in Figure 9, the ovulatory rhythm is accompanied by peaks in plasma testosterone, androstenedione, and 11-deoxycortisol which correspond to the absence of ova-fed eggs. There is no evidence for raised levels of progesterone, 17α-hydroxy progesterone, 17α-hydroxy,20β-dihydroprogesterone, or 11-deoxycorticosterone.

2.3.2j. Ovulation

Ovulation, which involves the active expulsion of oocytes from the follicles, is controlled by prostaglandins (PGs), which appear to act directly on the smooth muscle fibers which are found in the theca layers of the follicles and in the ovarian lamellae. A degree of species specificity in response to different prostaglandins has been found. For example, $PGF_{2\alpha}$, but not PGE_2, induces *in vitro* ovulation in rainbow trout oocytes. In the perch (*Perca fluviatilis*), however, PGE_2 can induce ovulation, and in the goldfish PGE_1, PGE_2, and $PGF_{2\alpha}$. In the brook trout, ovulation is enhanced *in vitro* by $PGF_{2\alpha}$ and inhibited by PGE_1, whereas PGE_2 sometimes appears to be stimulatory and sometimes inhibitory. The inhibitory effect of these latter prostaglandins is probably mediated through cyclic AMP, which is known to relax smooth muscle fibers. It inhibits ovulation in the brook trout (even in $PGF_{2\alpha}$-treated oocytes).

2.3.2k. Oviposition

The shedding of ovulated eggs (oviposition) generally requires the presence of a male, a suitable spawning substrate (e.g., vegetation for goldfish, gravel for

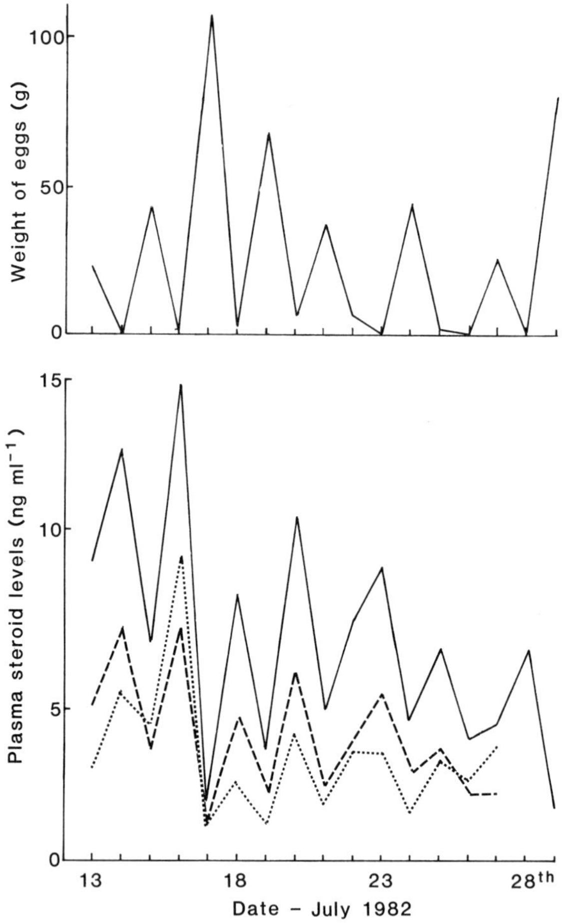

Figure 9. Ovulatory cycles in a single female turbot (*Scophthalmus maximus*). The top graph shows the weight of buoyant (i.e., freshly ovulated) eggs that could be stripped from the female; stripping was done once a day. The lower graph shows fluctuations in plasma levels of testosterone (—), androstenedione (- - -), and 11-deoxycortisol (······) as measured by direct radioimmunoassays.

trout), and appropriate environmental conditions (e.g., warm temperature for goldfish, monsoons for Indian catfish). It is usually accompanied by elaborate courtship behavior, involving chemical, visual, and tactile signaling. Studies on at least 19 oviparous species have shown that sexually mature females, which are ready to spawn, emit pheromones. These serve to attract males and stimulate their sexual activity. The significant event in the onset of pheromone production appears to be the occurrence of ovulation.

Ovulated eggs (especially in species that reproduce at fairly high temperatures and have small eggs) generally have an exceedingly short life-span (e.g., 1 hr in carp, 24 hr in turbot). This means that unless they are shed and fertilized very soon after ovulation they become inviable. Work on goldfish has established that spawning behavior in the female occurs almost immediately after ovulation. It appears to be triggered by the presence of ovulated eggs in the ovarian lumen. This has been shown by removing the ovulated eggs, which terminates spawning behavior, and injecting them back into the ovarian lumen, which restores spawning behavior. Several findings indicate that these actions are mediated by prostaglandins. Injections of $PGF_{2\alpha}$ elicit spawning behavior, whereas indomethacin (an inhibitor of prostaglandin synthesis) inhibits the onset of spawning behavior following ovulation. It is probable that the prostaglandins that are produced by the ovary during ovulation are the same ones that elicit spawning behavior, and they probably do so by acting (via the blood stream) directly on the brain.

2.3.2l. Involvement of Other Endocrine Glands in the Female Reproductive Cycle

Arginine vasotocin (AVT), one of the octapeptides secreted by the neurohypophysis, stimulates contractions in isolated oviducts and induces oviposition and spawning behavior in certain species.

Plasma levels of cortisol, which is secreted by the interrenal, are significantly elevated during the secondary growth stage in plaice, the preovulatory period in goldfish, and the postovulatory period in brown trout. The significance of this is uncertain.

There is evidence that in a number of species (e.g., Indian catfish) oocyte-maturation-inducing steroids (e.g., 11-DOC and 11-deoxycortisol) are derived from the interrenal rather than the ovary.

The thyroid appears to be especially active during vitellogenesis, and administration of thyroxine to maturing females has been shown to stimulate egg growth.

The pineal gland also appears to play an important role in the regulation of the timing of reproduction (see Sundararaj, B. I., 1981, *Reproductive Physiology of Teleost Fishes,* Aquaculture Development and Coordination Program, ADCP Rep. 81/16, The Food and Agriculture Organisation of the United Nations, Rome, 82 pp.). Morphological studies indicate that it has a dual photosensory and secretory function. It secretes a number of substances—among them melatonin and AVT—which are considered to be antigonadotropic.

2.3.3. The Male

2.3.3a. Control of Spermatogenesis

Like the ovaries, the testes of teleost fishes require the support of an intact and functioning pituitary gland for their maintenance.

Two stages of spermatogenesis have been pinpointed as being especially dependent on the presence of the pituitary: spermatogonial mitosis and the transformation of spermatogonia into spermatocytes. Once formed, spermatocytes can transform into spermatids and these into spermatozoa in the absence of a pituitary. Hypophysectomy inhibits mitotic spermatogonial divisions completely in most species studied but only slows them down in others. When testes are at an advanced stage of maturation, hypophysectomy usually results in postspermatogonial stages undergoing a normal transformation into spermatozoa. In some species, however, they may degenerate. Injections of mammalian LH or fish GtHs can overcome the effects of hypophysectomy and also stimulate full testicular development in immature fish. It is probable, however, that they act indirectly via steroids produced by the Leydig and/or Sertoli cells (hypophysectomy usually brings about regressive changes in these cells).

It has been found in hypophysectomized Indian catfish and goldfish that full spermatogenesis can be restored by injecting the fish with testosterone. Low doses of testosterone only maintain secondary spermatogonia in hypophysectomized goldfish, but high doses stimulate the formation of spermatocytes and spermatids.

2.3.3b. Control of Spermiation

There is some uncertainty as to the nature of the factor(s) controlling spermiation (i.e., hydration of the testes) in teleosts. In most fishes, hypophysectomy blocks spermiation, and it can be restored by injection of mammalian LH or teleost GtH preparations. In cyprinids, elevated levels of GtH in the plasma have been shown to be associated with spermiation. In rainbow trout, however, attempts to measure GtH in spermiating males and to elicit spermiation by GtH injection have given confusing results. Pituitary GtH levels are higher during spermiation than during the rest of spermatogenesis. However, there seems to be no discernible or consistent pattern of GtH plasma levels in relation to spermiation (or indeed spermatogenesis or steroid cycles) in male rainbow trout. Injections of GtH into males during the spermiation period cause only a temporary rise in the volume of milt produced, despite the fact that immunoreactive GtH levels remain elevated in the plasma and large numbers of spermatozoa are still present in the testes. Such results could be accounted for if, in males, GtH was, as in females, released from the pituitary in a pulsatile fashion.

2.3.3c. Changes in Steroid Hormone Levels during the Spermatogenetic Cycle in Teleosts

It can be seen that, in male rainbow trout (Figure 10), testosterone and 11-ketotestosterone levels in the plasma increase over the period when testicular weight is increasing and spermatogenesis is most active. The levels start to fall about the time that spermiation begins, but do not fall to basal levels until spermiation has been completed and the testes have regressed. 11-Ketotestosterone levels may sometimes peak at the same time as testosterone levels (as in Figure 10) or sometimes as much as 1 or 2 months later. The significance of this fact is not known. It is

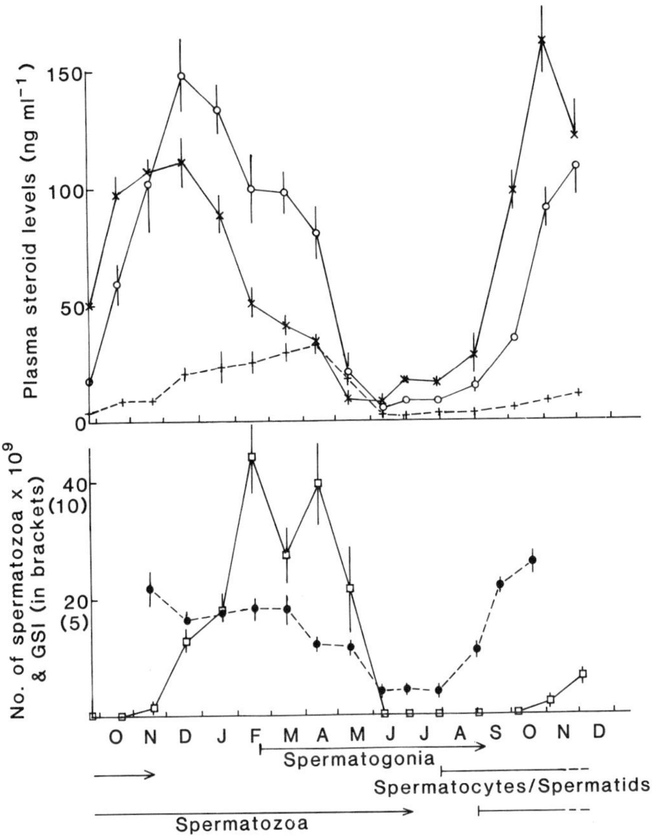

Figure 10. Plasma levels of 11-ketotestosterone (○), testosterone (X), and 17α-hydroxy,20β-dihydroprogesterone (+), GSI (●), and volume of sperm produced (□) during the annual reproductive cycle of male rainbow trout. Vertical bars represent SEM. Horizontal bars at bottom show period over which various germ cell stages were present in testes.

impossible to correlate either hormone with the volume, sperm number, or density of the milt obtained by hand stripping during the period of spermiation. There is a very close correspondence, however, between sperm production and plasma levels of 17α-hydroxy,20β-dihydroprogesterone. Levels of this steroid are significantly correlated with the volume of milt, total sperm count, K^+ content, and K^+/Na^+ ratio in seminal fluid. Injections of 17α-hydroxy,20β-dihydroprogesterone into intact spermiating males do not affect the volume of milt produced but significantly elevate the K^+ concentration of the seminal fluid. This suggests that 17α-hydroxy,20β-dihydroprogesterone found in the plasma of spermiating males may be involved in the regulation of the ionic composition of the intratubular fluid.

Levels of both 11-ketotestosterone and testosterone are elevated during spermatogenesis in Pacific salmon, brook trout, brown trout, and Atlantic salmon, and

17α-hydroxy,20β-dihydroprogesterone has been found in the plasma of mature male *Oncorhynchus nerka*. Plasma levels of 11β-OH testosterone glucuronide have been shown to be elevated in mature brown trout.

3. Class Osteichthyes: Superorder Chondrostei

The reproductive biology of the Chondrostei, the major living representatives of which are the sturgeons (*Acipenser* spp.), is similar to that of teleosts. Artificial induction of oocyte maturation/ovulation by injection of homologous pituitaries is practiced widely in the USSR. The active principle has been purified and characterized and is similar to the GtH isolated from teleost pituitaries.

4. Class Chondrichthyes (Cartilaginous Fishes)

Although the research on the endocrine control of reproduction in cartilaginous fishes, which comprise the Elasmobranchii (sharks, skates, rays, and dogfishes) and Holocephali (rabbit fish), is not nearly as extensive as that carried out on teleosts, it has mainly been concentrated on a single species, the dogfish, *Scyliorhinus canicula*. Thus, the information available (which has recently been comprehensively reviewed by J. M. Dodd and J. P. Sumpter [1984, Reproductive cycles of cyclostomes, elasmobranchs and bony fishes, in: *Marshall's Physiology of Reproduction,* 4th ed. (G. E. Lamming, ed.), Churchill Livingstone, London, pp. 1–126] is fairly consistent.

4.1. Gonadal Structure

Female dogfish are oviparous. They have an extended spawning season, the rate of egg laying being relatively constant throughout most of the year, but decreasing appreciably in the late summer. The ovary is single and contains oocytes varying in size from those filled with yolk, weighing 3 g, down to those which are small, translucent, and yolk-free. Dogfish exhibit, in other words, *asynchronous* oocyte development, a pattern that has not been very well investigated endocrinologically in teleosts. It is therefore difficult to make direct comparisons between the classes.

The testes in dogfish and other elasmobranchs have structures that are very different from those found in teleosts. They are based on a system of closed ampullae, rather than tubules, and these are arranged in concentric zones each consisting of ampullae at the same stage of spermatogenesis. Ampullogenesis takes place throughout maturity, divisions of the germ cells within the ampullae are synchronous, and each ampulla contains germ cells at the same stage of development, the dividing cells being embedded in Sertoli cell cytoplasm. Each ampulla contains some 500 Sertoli cells, each containing 64 sperm. A "zone of breakdown" gradually moves through the concentric zones over the course of the year.

4.2. Gonadal Steroidogenesis

Elasmobranch ovaries are capable of synthesizing most of the common sex steroids—e.g., progesterone, androstenedione, testosterone, and 17β-estradiol. The cellular source of these steroids is not known with certainty although 3β-hydroxysteroid dehydrogenase has been localized in the granulosa cells of the follicle in dogfish *Squalus acanthias*.

Testes can synthesize androstenedione, testosterone, 11-deoxycorticosterone, 11-deoxycortisol, and 17α,20β-dihydroxy-4-en-3-one as well as water-soluble conjugates. 11-Ketoandrogens have not been detected. The steroidogenic capacity of individual ampullae differs markedly depending on the stage of development of the germ cells. Because of this, and because Leydig cells are difficult to detect, it has been suggested that the Sertoli cells are the main sites of steroidogenesis in the testis.

4.3. The Pituitary Gland

The chondrichthyean pituitary is subdivided into separate lobes to an unusual degree. Directly beneath the brain are the rostral, median, and neurointermediata lobes, and at some distance from these and connected to the median lobe only by a thin stalk is the ventral lobe in elasmobranchs. This is the origin of the gonadotropin(s).

S. canicula is the only elasmobranch from which a GtH has been partially purified. It is a glycoprotein and similar in its physicochemical properties to LH.

4.4. The Hypothalamus

Hypothalamic control of GtH function in the Chondrichthyii appears to differ from that in virtually all the other vertebrates, except cyclostomes (see below), in that the GtH cells are entirely independent of a hypophyseal portal system—nor are they in contact with nerve fibers. However, a factor exists in the hypothalamus of dogfish that can elevate plasma sex steroid levels and stimulate ovulation. Also, synthetic LH-RH can stimulate ovulation in the dogfish and rabbit fish (*Hydrolagus collei*). Certain areas of the hypothalamus of dogfish, and the pineal complex, have also been shown to bind sex steroids, suggesting that they may be sites of feedback action by sex steroids on GtH synthesis. Dodd and Sumpter (1984) have suggested that the chondrichthyean hypothalamus produces a GnRH that reaches the ventral lobe via the systemic circulation (see Part III).

4.5. Endocrine Changes during the Reproductive Cycle

Despite having an extended egg-laying season, female dogfish exhibit marked cycles in pituitary GtH content and plasma levels of testosterone and 17β-estradiol—the levels of testosterone ranging between 1 and 8 ng ml^{-1} and of 17β-estradiol between 5 and 40 ng ml^{-1}. Vitellogenin is also present in the plasma of mature females although the levels are far lower than those typically found in teleosts. However, this may be related to the fact that oocyte development is asynchronous. As in other oviparous vertebrates, estrogen injections increase the levels of vitellogenin in the plasma.

Removing the ventral lobe of female dogfish results in complete inhibition of ovulation, and all eggs that have started vitellogenesis become atretic. However, plasma levels of 17β-estradiol, testosterone, and vitellogenin appear to be unaffected. On the other hand, testosterone, but not 17β-estradiol, levels fall in response to the injection of antisera raised against purified GtH.

In male dogfish, ventral lobectomy results in the disintegration of the spermatogonia undergoing their ultimate mitotic division but does not affect plasma testosterone levels. On the other hand, administration of a homogenate of ventral lobes to ventral lobectomized males prevents testis breakdown and results in a rapid rise in plasma androgen levels. These findings are somewhat difficult to interpret!

5. Class Cyclostomes

There are two groups of cyclostomes, the lampreys (Petromyzonids) and hagfishes (Myxinoids), whose reproductive biology differs markedly. Most of the lampreys, of which there are 38 living species, produce large numbers of small eggs. They spawn only once in their lives and then die (cf. eels). Spawning takes place in fresh water. Hagfishes, on the other hand, produce only a few and fairly large eggs (20–25 mm diameter), and, with the exception of *Eptatretus burgeri,* they live in a constant environment, 50–100 m deep in the oceans, and breed throughout the year. Oocyte development is synchronous in lampreys.

The fine structure of the ovarian follicles in cyclostomes is virtually indistinguishable from that of the other vertebrates, except that mature hagfish oocytes develop a unique tough shell.

Exogenous vitellogenesis has been demonstrated in both groups of cyclostomes. The blood of mature female *E. stouti* (a hagfish), for example, contains a phosphate-rich protein, which can also be induced in males and immature fish by injections of 17β-estradiol and estrone. It therefore appears that hepatic vitellogenesis is a universal phenomenon in oviparous vertebrates (see Chapter 8).

With regard to gonadal steroidogenesis, however, cyclostomes depart radically from the common vertebrate pattern. Several attempts have been made to measure the classical sex steroids, testosterone and 17β-estradiol, in the blood of maturing fish. No clear-cut pattern has emerged, however. Although measurable levels of these hormones have often been found, it has been impossible to correlate them with the stage of gonadal development or even with the sex of the fish. However, it has recently been shown that although cyclostome gonads are capable of synthesizing *in vitro* small amounts of the normal vertebrate steroids, they mainly produce androgens substituted with hydroxyl groups at unusual positions in the steroid molecule—the major products of *M. glutinosa* gonads being 6β-hydroxytestosterone and 5α-androstane-3β,7α,17β-triol, and the major product of *Petromyzon marinus* and *Lampetra fluviatilis* gonads being 15α-hydroxytestosterone. No attempts have yet been made to measure these steroids *in vivo*. None of the cyclostome gonads produce water-soluble conjugates.

The pituitary in lampreys is, as in higher vertebrates, divided into distinct regions, and the proximal region contains basophils that respond to gonadectomy. The hagfish pituitary consists mainly of undifferentiated cords of cells with no identifiable gonadotrops. In both groups, the pituitary is completely separated from the neurohypophysis by a layer of connective tissue, and there is no evidence for a pituitary portal system. Hypophysectomy in hagfish induces some degenerative changes in the gonads, but gametogenetic activity appears to be largely unaffected.

Hypophysectomy in lampreys slows down the rate at which sexual maturation proceeds but does not inhibit it. However, it prevents the appearance of secondary sexual characteristics and possibly also ovulation in females. Its effect on oogonial and spermatogonial mitoses has not been tested, as these events take place during larval life.

Although there is no direct connection between the hypothalamus and pituitary, immunoreactive LH-RH-like material has been found in neurons of the preoptic region of the brain in the Japanese lamprey, *Entosphenus japonica*. It is therefore possible, as in the chondrichthyeans, that a GnRH from the hypothalamus reaches the pituitary via the systemic circulation. Pituitaries transplanted to a different region of the body are capable of maintaining full reproductive development in lampreys.

6. Conclusions

Moving down the fish evolutionary scale there is evidence for less and less central control of the reproductive cycle. For example, in teleosts, gonad development is completely controlled by hormones secreted by the pituitary. The pituitary is linked to the hypothalamus via direct innervation and a blood portal system. Most of the endocrine secretions are virtually the same as those found in mammals. In elasmobranchs, on the other hand, gonad development appears to be only partially dependent on the pituitary, and the hypothalamus and pituitary are not intimately linked. The hormones that have been found, however, are similar to those of mammals. In the cyclostomes, much of gonad development is autonomous. Furthermore, the hypothalamus and pituitary are physically separated by a cartilaginous barrier, and the hormones secreted by the gonads are extremely unusual.

Among the teleosts I have dealt with, the majority are temperate species. This has been because I am familiar with some of these species, and not because there is a lack of information on reproduction in tropical and subtropical fishes. This latter subject has been covered by Sundararaj (1981), and the same basic mechanisms and principles I have already outlined are applicable.

7. Selected Readings

Dodd, J. M., and Sumpter, J. P., 1984, Reproductive cycles of cyclostomes, elasmobranchs and bony fishes, in: *Marshall's Physiology of Reproduction* (G. E. Lamming, ed.), 4th ed., Churchill Livingstone, London, pp. 1–126.

Ginzburg, A. S., 1972, *Fertilisation in Fishes and the Problem of Polyspermy,* Moscow Academiya Nauk SSSR. Institut Biologii Razvitiya (translated by Israel Program for Scientific Translations Ltd., Jerusalem, 1972). Available from Dept. of Commerce, National Technical Information Service, Springfield, VA 22151, 366 pp.

Hoar, W. S., Randall, D. J., and Donaldson, E. M., eds., 1983, *Fish Physiology,* Vol. IXA and IXB. *Reproduction.* Academic Press, New York and London.

Proceedings of the International Symposium on Reproductive Physiology of Fish, Paimpoint, France, 1977. *Ann. Biol. Anim. Bioch. Biophys.* **18**(4):759–1106, 1978.

Proceedings of the International Symposium on Reproductive Physiology of Fish, Wageningen, The Netherlands, August 1982 (C. J. J. Richter and H. J. Th. Goos, eds.), Pudoc, Wageningen, 1982, 256 pp.

Sundararaj, B. I., 1981, *Reproductive Physiology of Teleost Fishes,* Aquaculture Development and Coordination Programme, ADCP/Rep. 81/16, The Food and Agriculture Organisation of the United Nations, Rome, 82 pp.

8

Vitellogenesis and Viviparity

Ian P. Callard and S-m. Ho

1. General Introduction

In vertebrates, there are three principal modes of nutrition for developing embryos: (1) the laying down of food stores, synthesized in the liver, in the oocyte (vitellogenesis) (this is the most common mode in nonmammalian species); (2) the direct provision of nutrients to the developing young through the intermediary of the uterine wall by either secretory activity (histotroph) or (3) a direct maternal connection (placentation) (Table I). In egg-laying vertebrates, yolk is the only source of fetal nutrition; however, in live-bearing (viviparous) species, at least two of these mechanisms may exist in an individual. Thus, aplacental rays and torpedos have eggs with a very small amount of endogenous yolk, sufficient only for early development. The bulk of the nutrients are supplied as ''histotroph''* by specialized secretory elements of the reproductive tract. In ''placental'' elasmobranchs (some sharks), yolk is present and may be reduced in amount, but the yolk-sac placenta forms a site for exchange of nutrients from the mother to the fetus. This is also true of placental reptiles (certain small lizards and snakes). However, in reptiles both a yolk-sac placenta (choriovitelline or omphaloplacenta) and the phylogenetically newer chorioallantoic placenta exist and function sequentially depending on the quantity of endogenous yolk supplied with the egg. In eutherian mammals, no true yolk is present in the eggs, the developing young being dependent on histotroph from the secretory epithelium for variable times depending on the species (Table II).

*Histo = relationship to tissue; troph = nourishment.

Ian P. Callard • Department of Biology, Biological Science Center, Boston University, Boston, Massachusetts 02215. **S-m. Ho** • Department of Biology, Tufts University, Boston, Massachusetts 02155.

Table I
Modes of Embryofetal Nutrition in Vertebrates

I. Oviparity
II. Viviparity
 A. Aplacental viviparity
  1. Dependent solely on yolk reserves without a placental association (Elasmobranchii: *Squalus;* spiny dogfish)
  2. Oophagy/adelphophagy; intrauterine cannibalism (Elasmobranchii: *Lamna;* Porbeagle shark; *Carcharias;* sand shark)
  3. Placental analogues
   a. Uterine histotroph via uterine trophonemata (villi) (Elasmobranchii: *Torpedo,* electric ray; *Trygon,* sting ray; Amphibia: *Nectophrynoides;* Nimba toad)
   b. Ovarian follicular pseudoplacenta (Teleostei: *Poeciliidae* and *Anablepidae*)
   c. Ovarian cavity histotroph/branchial placenta or trophotaeniae (Teleostei: *Jenynsiidae* and *Goodeidae*)
   d. Cutaneous brood chambers (Amphibia: *Gastrotheca*)
 B. Placental viviparity
  1. Yolk sac (choriovitelline placenta—transition from yolk dependency to maternal nutrition (some elasmobranchs, all reptiles, mammals)
  2. Chorioallantoic placenta—follows yolk sac placenta ontogenically and phylogenically; maternal "on-demand" nutrition (reptiles, some metatherian and all eutherian mammals)

Then the chorioallantoic placenta follows, being complex in this group and the primary supporter of embryonic/fetal development. Here endocrine controls have been acquired in order to retain and maintain the embryo/fetus during the period of obligatory eutherian uterine development. In all viviparous vertebrate lines, there has been a gradual suppression of yolk protein synthesis in favor of alternative uterine means of fetal support once a species became committed to viviparity. The degree of inhibition of vitellogenesis and reduction of egg yolk deposits is directly correlated with the degree of complexity of the uterine mechanisms for nutrient supply, with yolk suppression being total in eutherian mammals in which the development of chorioallantoic placentation reaches its most complex. It thus appears that once a group of animals becomes committed to viviparity, the end result, operating through natural selection, is the suppression of hepatic yolk synthetic mechanisms in favor of *in utero* provision of nutrients. The probable advantages of viviparity versus oviparity as reproductive strategies are listed in Table III.

Study of elasmobranch and reptilian species in which yolk is reduced but not eliminated in favor of intrauterine nutritional mechanisms may allow us to understand the endocrine regulatory factors involved in the process of yolk elimination.

2. Viviparity and Placentation in Nonmammalian Species

Viviparity is most common in elasmobranchs and reptiles (Table IV) and will be emphasized in this chapter. Passing mention will be made of certain groups of

Table II
Origin and Relative Importance of Embryonic Nutrients in Viviparous Vertebrates[a,b]

Vertebrate group and species	Type/mode of nutrient supply	Source and relative contribution of embryonic nutrients		
		Autotrophic	Histotrophic	Hemotrophic
Elasmobranchs				
a. *Dilatias licha*	Aplacental; uterine villi	+++	+	−
b. *Trygon violacea*	Aplacental; uterine villi	+	+++	−
c. *Mustelus laevis*	Placental (choriovitelline, yolk sac)	+	++	+++
Teleosts				
a. *Heterandria formosa*	Intraovarian pseudoplacenta	+	++	+++
b. *Jenynsia lineata*	Intraovarian branchial placenta	+	++	+++
Amphibians				
a. *Gastrotheca ovifera*	Cutaneous brood pouch			
b. *Salamandra atra*	Aplacental, intrauterine	+++	+++	−
c. *Nectophrynoides occidentalis*	Aplacental, uterine villi	+	+++	−
Reptiles				
a. *Nerodia (Natrix) sipedon*	Choriovitelline → chorioallantoic placenta	++++	++	+
b. *Chalcides chalcides*	Choriovitelline → chorioallantoic placenta	++	+	+++
c. *Mbuya heathi*	Choriovitelline → chorioallantoic placenta	+	+	++++
Metatheria				
a. *Trichosurus vulpecula*	Choriovitelline placenta	−	+++	+
b. *Sminthopsis crassicuadata*	Choriovitelline placenta	−	+++	++
c. *Perameles nasuta*	Choriovitelline → chorioallantoc placenta	−	++	+++
Eutheria				
a. *Rattus norvegicus*	Choriovitelline → chorioallantoic placenta	−	++	+++
b. *Homo sapiens*	Choriovitelline → chorioallantoic placenta	−	+	++++

[a] Autotrophic, egg yolk source; histotrophic, oviduct secretory material; hemotrophic, nutrient supply across the maternal fetal vascular interface. Relative importance of nutrient sources: − negligible, + minor, ++ important, +++ major importance, ++++ almost exclusive.

[b] Mainly from Hughes, R. L., 1977, Egg membranes and ovarian function during pregnancy in monotremes and marsupials, in: *Reproduction and Evolution* (Calaby, H. and Tyndale-Biscoe, C. H., eds.) Australian Academy of Science, Canberra, Australia, pp. 281–291.

Table III
Advantages and Disadvantages of Viviparity

Advantages
1. Protection of embryos from predation
2. Protection from desiccation, soil molds, and bacteria
3. Predictability in the placement of newborn in relationship to the availability of seasonal resources
4. Maternal behavioral thermoregulation provides an optimal temperature range for development
5. Continuous supply of nutrients to the embryo by the mother during development, rather than making a one-time maternal commitment at ovulation

Disadvantages
1. Reduction in fecundity because retention of young suppresses reproductive cycles
2. Parent encumbered by enlarged eggs or embryos for prolonged periods; more vulnerable to predation near term. If maternal organism dies, total reproductive commitment is lost
3. Reduced genetic diversity due to reduced number of matings associated with single brood

teleosts and amphibians that have experimented with this reproductive strategy. A list of evolutionary innovations associated with viviparity is given in Table V.

In elasmobranchs and reptiles, there is a well-defined progression from oviparity through aplacental viviparity to placental viviparity with an associated reduction of egg yolk reserves. The reduction of yolk reserves correlates with an increase in the complexity and importance of maternal–fetal connections for supply of nutrients. *Aplacental viviparity takes on several forms.* (1) In both elasmobranchs and reptiles, large-yolked eggs may be retained in the oviduct without significant yolk reduction. In these species, the embryos are supplied with adequate yolk for the completion of development, though even so there is some close apposition of fetal and maternal epithelia suggesting the beginnings of placentation. (2) In some elasmobranchs placental analogs develop and form virtually the only source of maternal nutrients for the embryo. These nutrients—histotroph—are secreted by

Table IV
Occurence of Viviparity in Nonmammalian Vertebrates

Vertebrate group	No. families with viviparous species (total no. families)	No. genera with viviparous species (total no. genera)	No. species viviparous (total no. species)
Class Chondrichthyes	40	90	420
Subclass: Elasmobranchii	(98)	(145)	(600)
Class Osteichthyes	14	123	510
Subclass: Teleostei	(425)	(3900)	(18000)
Class Amphibia	3	8	11
	(27)	(278)	(3000)
Class Reptilia	21	120	275
Order Squamata	(27)	(887)	(6005)

Table V
Evolutionary Innovations Associated with Viviparity

1. A decrease in the number of eggs produced
2. Perfection of structural and behavioral adaptations necessary for internal fertilization
3. Utilization of the yolk sac for the absorption of maternally secreted nutrients (choriovitelline placenta)
4. Utilization of the allantois, when present, in the formation of the chorioallantoic placenta
5. Intrauterine care of relatively few young to an advanced stage of development
6. Participation of endocrine glands in the extension of gestation (e.g., corpus luteum), with the object of retaining eggs in the genital tract for periods beyond the limits of the normal ovarian cycle
7. Adaptation of endocrine control of lactation in preparation for early nutrition of the newborn
8. The gradual endocrine suppression of yolk synthesis and vitellogenesis as a mechanism for embryonic nutrient supply in favor of intrauterine supply of nutrients via maternal secretory activity or placentation

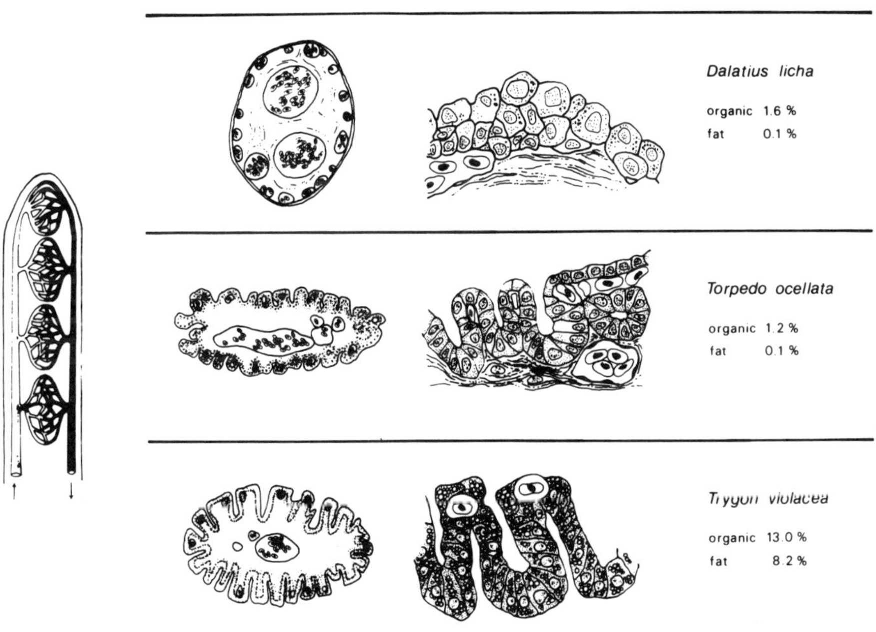

Figure 1. Villus structure in three elasmobranch species, shown in relationship to organic and fat content of uterine histotroph (at right). Left, generalized villus, arrows indicate vascular flow. Middle, cross-section of villus characteristic of each species shown at right. (×90.) Right, detail of surface epithelium for each species. (×600.) (Redrawn from Ranzi, S., 1934, *Le basi fisio-morfologiche dello svilluppo embrinale dei Selaci,* Part II et III, Vol. 13, Publicatione della Stazione Zoologica di Napoli, Naples, pp. 332–437.

Table VI
Changes in Weight and Composition of Selachian Embryos during Development in Oviparous, Aplacental Viviparous, and Placental Viviparous Species[a]

Characteristics in relation to embryo	*Scyliorhinus canicula* (dogfish)	*Dalatias (Scymnus)* licha	*Torpedo ocellata* (electric ray)	*Dasyatis (Trygon) violacea* (stingray)	*Mustelus laevis* (nursehound)
Reproductive mode	Oviparous	Aplacental viviparous	Aplacental viviparous	Aplacental viviparous	Placental viviparous
Number eggs/young		10–16	3–36	5–6	4–20
Gestation (months)	6	10	4	2	5
Weight of ovulated egg (g)	1.31	130	6.78	1.9	5.54
Weight of "completed" embryo (g)	2.69	188	13.37	118.0	189.6
Percent weight change in embryo	+105%	+45%	+97%	+6105%	+3326%
Percent change in water in embryo	+213%	+100%	+251%	+10412%	+5608%
Percent change ash in embryo	+292%	+30%	+157%	+10250%	+7609%
Percent change organic substances in embryo	−21%	−22%	−23%	+1628%	+1064%
Placental type or type of nutrition	Yolk	Yolk primary, uterine villi and histotroph secondary	Yolk initially then histotroph from uterine villi	Short-term yolk dependence; rich histotroph from uterine villi, which may enter gut	Intimate yolk sac

[a]Adapted from Table 8, p. 41 in Needham, J., 1950, *Biochemistry and Morphogenesis,* Cambridge University Press, Cambridge.

the maternal uterine epithelium by epithelial folds which reach their highest degree of development in the trophonemata of certain rays and torpedoes (Figure 1). In some species these trophonemata extend through the gill slits of the developing young into the upper portion of the embryonic gastrointestinal tract where the secretory material is directly absorbed. In some eutherian species, a histotrophic mode of early embryonic nutrition also occurs. This is best documented in the larger domestic animals, such as the horse, in which actual implantation and placental development is slow, and uterine secretions provide the critical early nutrient supply. (3) Other elasmobranchs (*Lamna, Sphryna*) have developed an alternative aplacental stategy which is based on continued yolk absorption and intrauterine cannibalism. In these species, ovulation continues throughout gestation, and the developing fetuses ingest the eggs as they pass through the reproductive tract. There

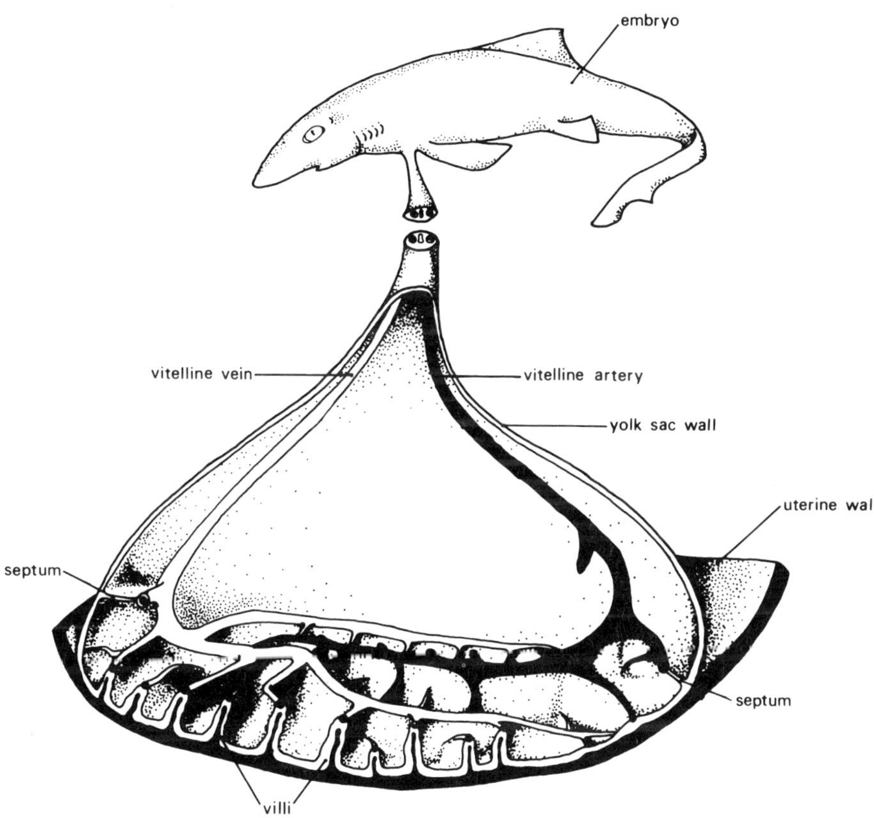

Figure 2. The relationship of the fetus, yolk sac, and choriovitelline placenta to the uterine wall in *Mustelus laevis*. In this species each embryo is contained in a uterine compartment which separates it from other embryos. [Redrawn from Ten Cate-Hoedemaker, N. J., 1933, Beitrage zur Kenntnis der Plazentation bei Haien und Reptilien. Der Bau der reifen plazenta von *Mustelus laevis* Risso und *Seps chalcides* Merr. (*Chalcides tridactylus* Laur.) *Z.f. Zellforschung u mikr. Anatomie* **18:**299–345.]

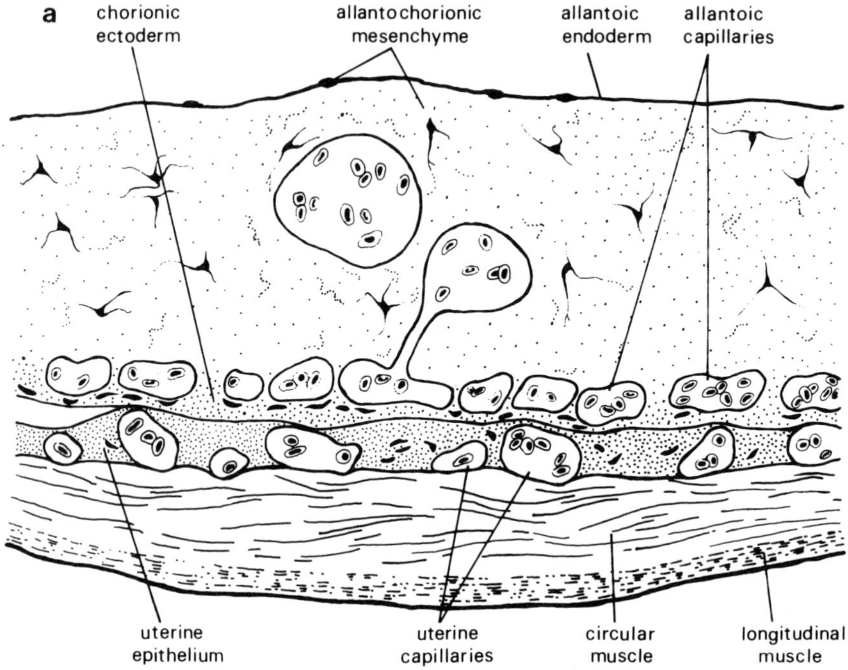

Figure 3. Chorioallantoic placental structure in lizards. (a) Section of mature chorioallantoic placenta of *Egernia cunninghami,* showing the folded maternal placental face. In the mature placenta maternal capillaries press against and often indent the bases of the overlying epithelial cells. The chorionic ectoderm cells are furrowed by bulging allantoic capillaries, so that often only the thinnest cytoplasmic membranes separate two circulations. (b) Section of the mature chorioallantoic placenta of *Lygosoma ocellatum,* showing apposition of maternal and fetal tissues. The uterine wall is extremely thin. In the fully developed placenta the intact maternal capillaries bulge at the surface of small uterine folds, the grooves between which are lined with glandular epithelium. The chorionic epithelial cells are enlarged and over most of the placenta interlock with the shallow maternal grooves, often extending cytoplasmic processes which penetrate for some distance between uterine epithelial cells. (c) Section of a young chorioallantoic placenta of *Egernia entrecasteauxi* showing folded maternal placental face. The folds are larger than in *Lygosoma ocellatum* shown in b. This is the most highly specialized type. The placental region is roughly eliptical in shape, and the folded maternal tissue immediately overlies the main longitudinal uterine blood vessels. The uterine folds are enlarged and complex and indent the fetal placental face. Each is filled with capillaries of various sizes and is lined by a single layer of enlarged, cuboidal, ciliated epithelial cells and immediately overlie the allantoic capillaries. (Redrawn from Weekes, H. C., 1935, A review of placentation among reptiles with particular regard to function and evolution of the placenta. *Proc. Zool. Soc. Lond.* **2**:625–646.)

is also a gradual reduction in the number of embryos as pregnancy progresses, due to intrauterine cannabalism, with a single, highly active, aggressive fetus remaining at term. The relative significance of the egg yolk supply, uterine histotroph, and placental association in a series of elasmobranchs is shown in Table VI.

Placental viviparity involves two different structures which involve the chorion, the choriovitelline placenta, and the chorioallantoic placenta. The yolk sac

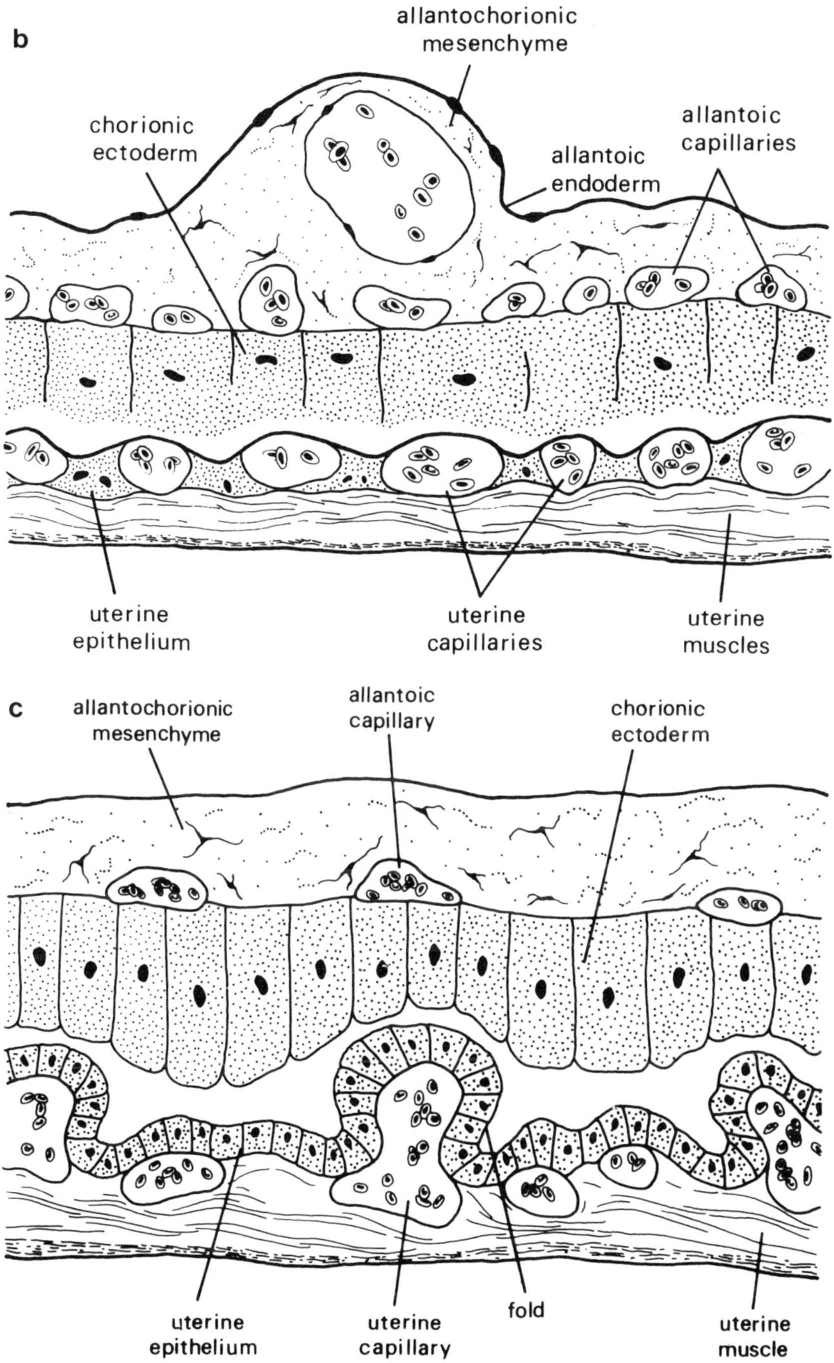

Figure 3 (*Continued*)

placenta is the only placental structure in elasmobranchs functioning to absorb yolk, but it also forms a significant fetal/maternal interface through which nutrients are absorbed as egg yolk supplies are reduced (e.g., *Mustelus canis,* the smooth dogfish; see Figure 2). In this species, the placenta resembles the epitheliochorial placenta of eutherian mammals, and the organic content (particularly fat) of uterine fluid remains as high as in some placental species. In amniote species, with the formation of the allantois, the chorioallantoic membrane develops into the definitive placenta of eutherian mammals and that of reptiles. This structure has variable degrees of complexity and intimacy of association with the maternal uterine wall (Figure 3a,b,c), which correlate with the associated reduction of egg yolk. In reptilian and mammalian species, the yolk sac placenta precedes the chorioallantoic placenta developmentally, and in addition to absorbing yolk supplied to the egg, this structure probably functions to absorb nutrients from the maternal circulation (as in placental sharks) in the manner of the chorioallantoic placenta (in the garter snake *Thamnophis,* for example), which later takes over this function.

It can be seen that viviparity as a reproductive strategy has evolved in parallel many times in the vertebrates. With this development has come an increasing direct dependence of the developing embryo on the maternal organism for nutrient supplies through either a histotrophic or placentotrophic mechanism as yolk is reduced and gestation extended. In some vertebrates ancestral to present-day mammals this trend was carried to the logical conclusion of total elimination of oocyte food stores in the form of yolk and the complete dependence on uterine mechanisms of embryo-fetal support. In some extant species of reptile, notably *Mabuya heathi* and *Chalcides chalcides,* the trend to yolk reduction is already extreme (see Table VII).

Since extant mammals evolved from reptilian ancestors, some discussion of the reptilian–therian transition might aid our understanding of the probable manner in which yolk was finally eliminated as a corollary of the development of the

Table VII
Volumes of Newly Ovulated Ova of Some Viviparous Reptiles[a]

Species	Ovum volume, mm^3
Mbuya heathi	0.52
Chalcides chalcides	12
Sceloporus aeneus bicanthalis	159
Hoplodactylus maculatus	231
Egernia whitei	524
Thamnophis sirtalis	567
Enhydris dussumieri	2356
Egernia cunninghami	4189

[a]From Blackburn, D. G., Vitt, J., and Beuchat, C. A., 1984, Eutherian-like reproductive specializations in a viviparous reptile. *Proc. Natl. Acad. Sci. USA* **81:**4860–4863.

definitive eutherian chorioallantoic placenta. Although the fossil record does not show reproductive modes of evolving reptiles, much may be gained by extrapolation from living forms.

3. Evolutionary Aspects of Viviparity and Vitellogenesis: The Reptilian–Therian Transition

Beginning with reptilian ancestors, mammalian evolution is considered to be diphyletic (Figure 4). The last common ancestor of the Theria and Prototheria existed in the Triassic (about 180 million years ago) and was probably a cynodont in

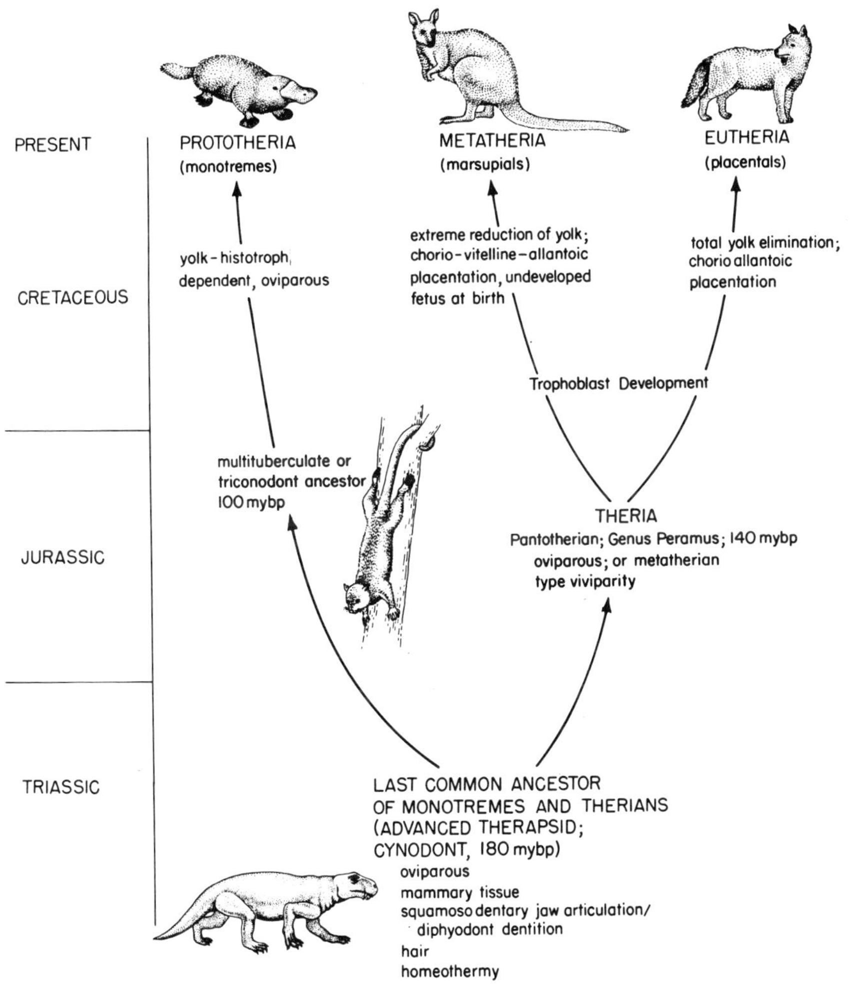

Figure 4. Plan of evolutionary changes from Triassic to the present showing the Theria and the emergence of monotremes, marsupials, and eutherian placentals.

which mammary tissue and the definitive mammalian squamosodentary articulation of the jaw were already present. One branch of this diphyletic tree led to the Prototheria, to give the modern-day Monotremata (*Platypus, Echidna*), which possibly evolved from a multituberculate or triconodont ancestor in the Cretaceous (about 100 m.y.b.p.). The monotremes retain many reptilian characteristics, but the one of most interest here is their egg-laying habit, suggesting derivation from an ancestral oviparous amniote (reptile?) with a large-yolked "cleidoic" egg.

The other branch of this diphyletic evolutionary tree leads to the Metatheria and Eutheria, whose immediate common ancestor is considered to be a pantotherian of the Jurassic era (about 140 m.y.b.p.). This ancestor might have been oviparous, necessitating independent evolution of viviparity, or viviparous, as in Metatheria (marsupials).

Whatever the reproductive mode in the last common ancestor of the Prototheria and Theria, on the one hand, and the Meta- and Eutheria, on the other, the modern-day Theria differ from their reptilian and prototherian ancestors in (1) total yolk suppression, and (2) the presence of the trophoblast. This latter structure develops from the extraembryonic tissues of the blastocyst and is thought to be responsible for the immunological protection of the fetus from maternal rejection. This is essential if the developing fetus is to attain the necessary close association with the maternal organism for essential nutrient supply. Thus, final yolk suppression and trophoblast development must have been closely concomitant events in order for the fetuses of the hypothetical ancestor to survive the close proximity to the mother and obtain necessary nutrients without a yolk supply.

Thus, the development of the trophoblast by metatherian and eutherian mammals is seen as essential for fetal protection demanded by the close apposition of fetal–maternal tissues and blood vessels, an essential corollary of yolk elimination. Once the "barrier" trophoblast was developed, this resulted in the evolution of a rich variety of reproductive adaptations in mammals.

4. Comparison of Reptilian and Mammalian Viviparity: Egg Membranes and Trophoblast

In all vertebrates the primary and secondary egg membranes are formed within the ovary, and the inner and outer tertiary membranes are added in the oviduct. These outer membranes remain as effective barriers between maternal and fetal tissues even in reptiles with the longest gestation periods and the highest degree of placental development and yolk reduction. It is these inert membranes which prevent activation of the maternal immune system during the intrauterine stay and thus allow a long period of intrauterine development dependent primarily, though not exclusively, on yolk deposits as in reptilian ancestors of mammals. However, while protecting the embryo from maternal rejection, these membranes also prevent that closeness of apposition of maternal and fetal tissues seen in Eutheria where there is total embryonic support through uterine mechanisms. Nevertheless, at the end of pregnancy, in viviparous lizards and snakes, young are born at an advanced state of development and are able to fend for themselves as in a number of mammalian species (e.g., horse, cow).

Thus, in our search for an understanding of the evolution of viviparity from an oviparous mode of reproduction, culminating in the eutherian condition, we must infer the need for the coincident elimination of tertiary egg membranes, the development of the trophoblast, and the elimination of yolk. We are able to identify nonmammalian vertebrate species in which one or two of these three conditions may be seen. Thus, in viviparous reptiles, tertiary egg membranes are clearly reduced when compared to oviparous relatives with shelled eggs. It should be possible by judicious selection of animals to examine possible endocrine mechanisms for tertiary egg membrane reduction. Similarly, it is possible to examine a species with reduced membranes and reduced yolk (e.g., the skinks, *Mabuya* and *Chalcides*), thus providing insight into the regulation of these mechanisms which were of importance for mammalian evolution and possibly future reptilian developments. Although all reptilian options for further elucidation of these questions have not yet been exhausted, it is likely that examination of the Metatheria and their reproductive adaptations may reveal species instructive for our understanding of the final stages of egg membrane elimination associated with trophoblast development.

In marsupials, various degrees of uterine erosion occur, a characteristic thus not confined to Eutheria. There is some degree of egg membrane dissolution. The consequent close association of mother and embryo allows rapid birth in 16-day pregnancy, without the additional factor of immunological rejection.

5. The Vertebrate Egg in Relation to Vitellogenesis and Viviparity

The large yolky oocyte of nonmammalian species is the product of the regulatory activities of the ovarian follicle wall, which produces hormones controlling vitellogenesis and through which vitellogenin molecules must pass before being sequestered in the oocyte. At maturity, oocytes are large compared to typical somatic cells, even in eutherian mammals with no yolk (in man, 60μm versus 20 μm; Figure 5). In nonmammalian species, oocytes are much larger (several millimeters in amphibians and teleost fishes and possibly several centimeters in elasmobranchs, reptiles, and birds). Once ovulated, in terrestrial egg-laying species facing a threat of dehydration, additional (tertiary) coats of aqueous, protective materials (albumin) are added to the yolk by the reproductive tract as the egg passes through. In addition, a shell, which may or may not be calcified, is added to the outside of the egg. In Figure 6, a typical oocyte, the structure of the follicular wall, and the final egg structure are shown for the two major reptilian subdivisions. In the evolution of viviparity, the development of the tertiary egg membrane is either delayed or suppressed, and the amount of yolk appears to be subsequently reduced as a corollary of alternate intrauterine methods of embryo support.

As a background for further understanding of oviparity, and the evolution of viviparity and the associated elimination of yolk, it is helpful to understand the physiology, biochemistry, and molecular biology of vitellogenesis.

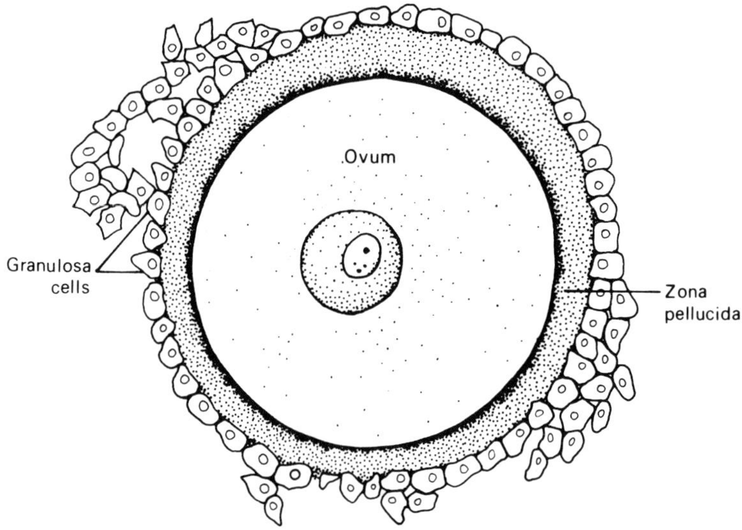

Figure 5. A generalized mammalian oocyte with surrounding granulosa cells (cumulus cells).

6. Structure, Biochemistry, and Molecular Biology of Vitellogenin (Yolk)

The yolk protein precursor is a phospholipoglycoprotein which exists as 400,000- to 500,000-dalton forms. The native circulating form is a dimer of two polypeptides, each of 200,000–250,000 daltons (Figure 7). This dimer may be composed of two dissimilar polypeptides (heterodimer) or two similar peptides (homodimer). There are multiple forms of the native molecule and the vitellogenin polypeptides in the plasma of the quail, the goldfish, the South African clawed frog, and the chicken, possibly owing to the expression of multiple vitellogenin genes.

Each vitellogenin polypeptide consists of a phosphate-rich and a lipid-rich portion, and once taken up by the oocyte these two portions are enzymatically cleaved to lipid-rich (lipovitellin) and phosphate-rich (phosvitin) proteins. Multiple forms of these proteins occur in the yolk, indicating variable enzymatic processing. The phosvitin consists of approximately 50% serine residues, of which the majority are phosphorylated in the hydroxyl side chains. The polypeptide also contains carbohydrate moieties (three mannose, three galactose, 5-*N*-acetylglucosamine, and two sialic acid residues) per molecule arranged in a branched structure attached to the protein via an asparagine residue. The total phosphate content in phosvitin is 10% and in lipovitellin less than 1%. Lipidation occurs mainly in the lipovitellin portion, which is about 10–20% lipid by weight. Vitellogenins bind calcium and ferric ions, the calcium improving the solubility of the complex in plasma. In *Xenopus laevis* (the South African clawed frog) vitellogenin binds two moles of bile

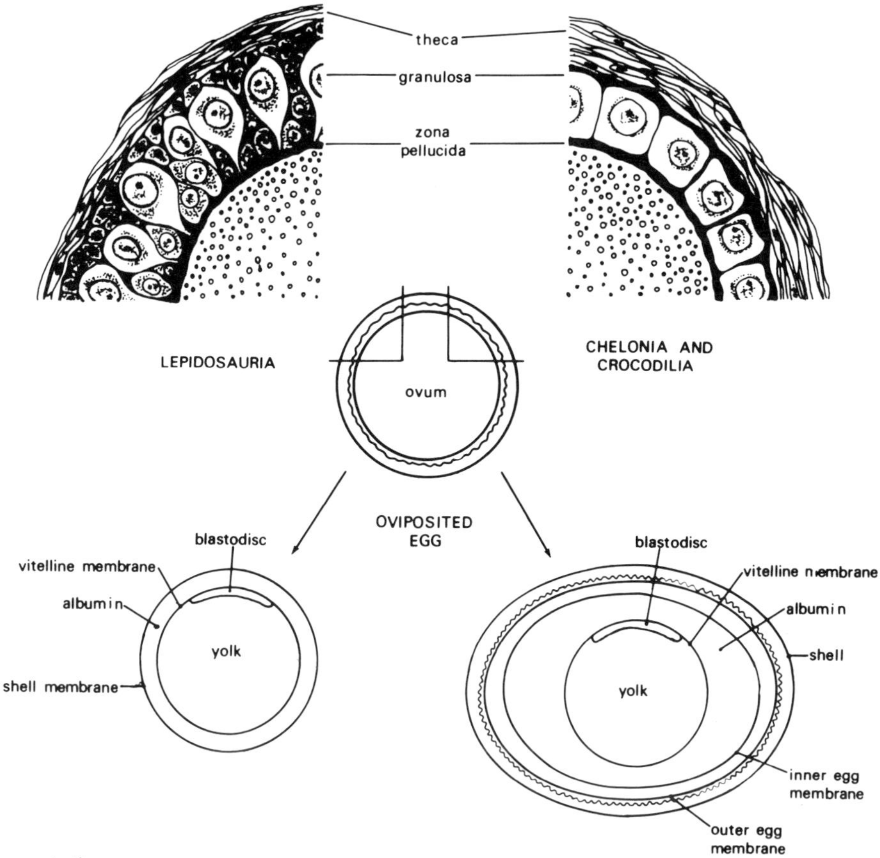

Figure 6. Follicular wall and mature egg differences (top) in the two major subdivisions of reptiles [lizards and snakes (squamate); turtles; crocodiles]. It can be seen that the squamate follicle wall is more highly differentiated than that of the other two groups in which the wall is lined by a simple cuboidal epithelium. The megalecithal yolky oocyte is similar in gross structure (center), but after ovulation some differences in the deposition of tertiary membranes occur. Eggs of crocodilians are virtually identical to those of birds, having a thick, hard, calcareous shell comprised of calcite crystals in a matrix of proteinaceous fibers. The crocodilian eggshell is porous, allowing for gas exchange. A pair of tertiary egg membranes separates the eggshell from a thick albumin layer. The eggshell of turtles varies from being hard and brittle, to soft and pliable (parchmentlike). The shell is penetrated by pores allowing gaseous exchange. Two tertiary egg membranes lie between the thick layer of albumin and the eggshell. The eggs of oviparous squamates (*Lepidosauria*) usually have a flexible, parchmentlike shell containing relatively small amounts of calcium carbonate. No other membranes seem to be present on the inner aspect of the *Lepidosaurian* shell, and thc albumin layer is exceedingly small at the time of oviposition. In live-bearing species, calcium carbonate is not deposited, but the shell membrane usually persists with varying degrees of reduction. (Drawn in part from Packard, G. C., Tracy, C. R., and Roth, J. J., 1977, The physiological ecology of reptilian eggs and embryos, and the evolution of viviparity within the class Reptilia. *Biol. Rev.* **52:**71–105).

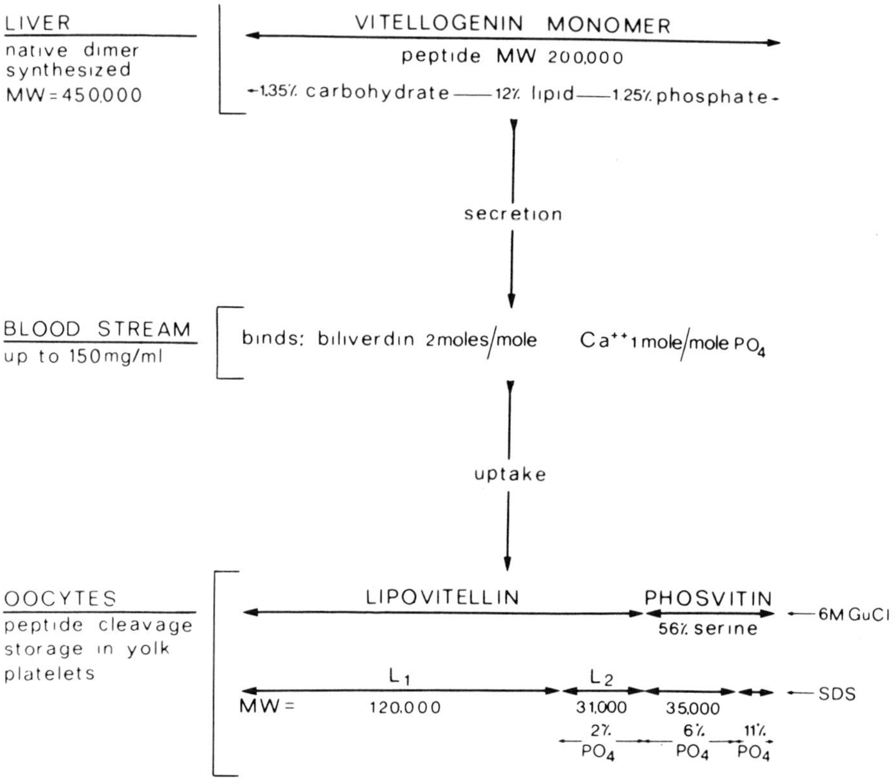

Figure 7. Relationship between hepatic, blood, and oocyte forms of yolk components. (From Wangh, L. J., Longthrone, R. F., and Knowland, J., 1976, Investigation of the mechanisms and consequences of steroid hormone action on vitellogenin synthesis in *Xenopus laevis,* in: *The Molecular Biology of Hormone Action,* Academic Press, New York.)

pigment (biliverdin) for every mole of vitellogenin, imparting a green color to the molecule.

7. Vitellogenin Genes and Vitellogenin Messenger mRNA Synthesis

Vitellogenin, like most abundant proteins of a differentiated cell is synthesized from genes that are present in single copies in the genome (Figures 8 and 9). The enzyme RNA polymerase II catalyzes the synthesis of numerous copies of a RNA nucleotide sequence that is complementary to the DNA sequence of the vitellogenin gene. The newly formed transcripts are then processed in the cell nucleus into mature messenger RNA molecules, which are exported to the cytoplasm to act as the templates for polypeptide synthesis. With advances in gene-cloning technology, it is now possible to excise specific regions of DNA and to determine the nucleotide sequences. Using this technology, the vitellogenin genes in both *Xenopus* and the

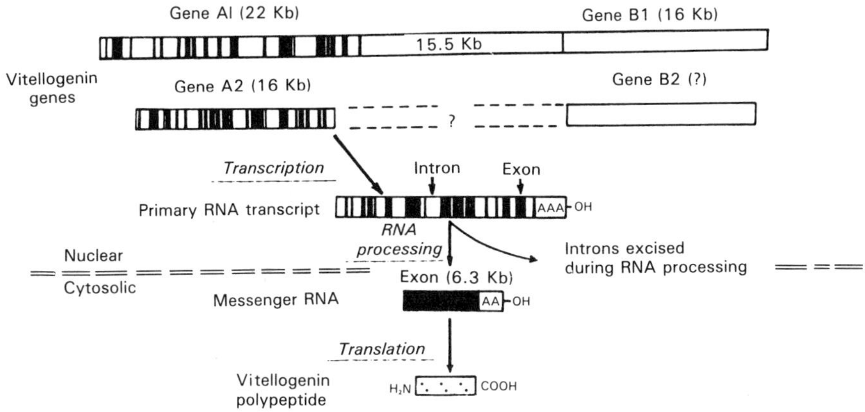

Figure 8. Relationships between the four vitellogenin genes in *Xenopus laevis* (the South African clawed toad). The four can be divided into two groups, A and B, according to the similarities in their coding sequences. A and B groups each compose two subgroups, A1 and A2 and B1 and B2. Groups A and B differ in approximately 20% of their coding sequences while the members of the subgroups (e.g., A1 and A2) differ only in about 5% of their coding sequences. The coding regions (exons) are of the same length in all four vitellogenin genes with a total exon length of 6.3 kilobases (kb). Nevertheless, the coding regions are interrupted by various lengths of intervening sequence thus resulting in vitellogenin genes of different total gene lengths (22 kb for A1, 16 kb for A2, and 16 kb for B1). Genes A1 and B1 are linked with a 15.5-kb-pair DNA between them. It is postulated that the four genes derived from two duplication events; first is a duplication of genes A and B from an ancestral vitellogenin gene followed by the second duplication, which involved duplication of the whole *Xenopus* genome.

chicken have been isolated and characterized. These studies reveal the expression of multiple vitellogenin genes that cannot be detected when vitellogenesis is studied at the protein synthetic level.

In *Xenopus,* vitellogenin is encoded by at least four related genes, all of which are under estrogen control. Analyses of the coding sequence of the four vitellogenin genes reveal that they can be divided into two major groups, A and B, which differ in approximately 20% of their coding sequences. The A and B groups are each composed of two subgroups (A1 and A2, B1 and B2), which differ in about 5% of their coding sequences. The portion of DNA that codes for the mRNA (coding regions, exons), are of the same lengths in all four vitellogenin genes. The total exon length is 6.3 kilobases (kb), coding for mRNA molecules of 6300 nucleotides. The coding regions are interrupted by various lengths and numbers of "introns" or intervening sequences that are not coded for the mature mRNA molecules. These different exon–intron arrangements thus result in vitellogenin genes of different total gene lengths: 22 kb for A1, 16 kb for gene A2, and 16 kb for B1. The two A genes have many structural similarities and are believed to have arisen from a duplication event. Thus, coding regions of A1 and A2 are 95% homologous, and both are interrupted 33 times by introns at homologous positions. Genes A1 and B1 are found to be linked with 15.5-kb pairs of DNA between them. Their coding regions are 80% homologous. It has been postulated that the *Xenopus* vitellogenin

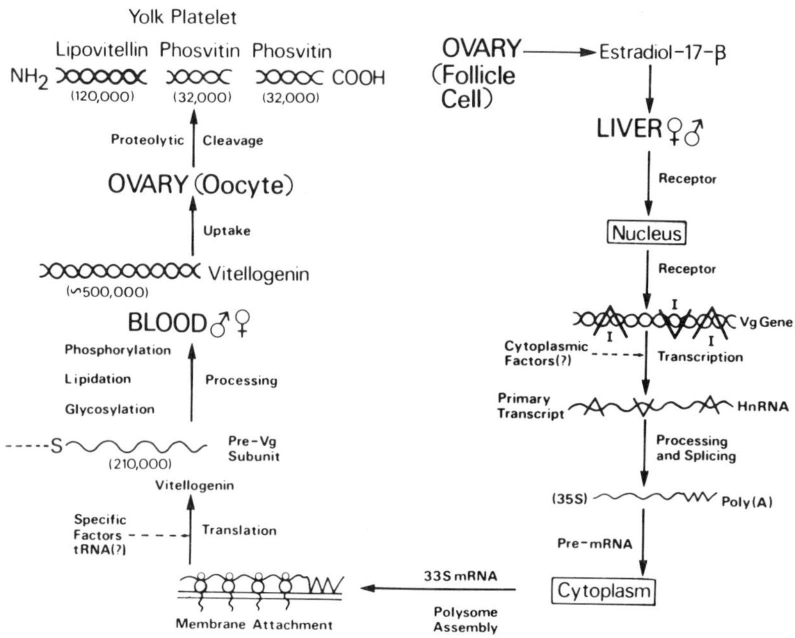

Figure 9. Scheme summarizing the sequence of major events triggered by estradiol and leading to the secretion of vitellogenin (Vg) into the blood of male or female *Xenopus,* and to the deposition of phosvitin and lipovitellin in egg yolk platelets. Numbers in parentheses refer to nominal size of RNA or protein. I, intervening sequences within the coding region of the gene; S, probable presence of the "signal" peptide in the primary translation product. (From Tata, J. R., and Smith, D. F., 1979, Vitellogenesis: A versatile model for hormone control of gene expression. *Rec. Prog. Horm. Res.* **35:** 47–95.)

gene family arose by two duplication events. The first involves the duplication of genes A and B from an ancestral vitellogenin gene; the two new genes remained linked on the same chromosome. Second, a whole genome duplication event occurred about 30 million years ago in *Xenopus,* thus resulting in one A gene and one B gene linked on one chromosome and the other A and B genes linked on a second chromosome.

During the transcription of the vitellogenin gene, the introns are transcribed along with the exons into nuclear precursors of vitellogenin mRNA. The long insertions of noncoding RNA sequences (introns) are spliced out from the precursor molecules, and the exon sequences are joined to each other to form the mature mRNA which is exported into the cytoplasm (Figure 5). Four distinct vitellogenin mRNA molecules, corresponding to genes A1, A2, B1, and B2, are found in estrogen-treated *Xenopus* hepatocyte cytoplasm. All can be translated *in vitro* into vitellogenin polypeptides of 200,000 daltons. The A and B proteins synthesized *in vitro* from mRNAs differ substantially in their primary structure, whereas the A1 and A2 and B1 and B2 polypeptides show only minor differences. In the plasma of estrogen-treated *Xenopus,* at least three vitellogenin polypeptides of slightly different molecular weights (197,000, 188,000, and 182,000 daltons) can be detected under denaturing conditions by high-resolution gel electrophoresis.

In contrast to *Xenopus,* only one vitellogenin gene and one vitellogenin mRNA have been discovered in the chicken. The intron–exon studies of the avian vitellogenin gene showed that the 23-kb gene is interrupted by at least 25 introns. The coding region of the avian gene is 6.7 kb in length. However, studies at the protein level clearly demonstrated two major vitellogenins, VTG I and VTG II, present in the plasma of estrogen-stimulated roosters. It has been suggested that the identified avian vitellogenin gene encodes VRT II, and that the gene coding for VTG I remains for future discovery.

Multiple forms of vitellogenin polypeptides have also been described in the goldfish, *Carassius auratus,* in the axolotl, *Ambystoma mexicanum,* and in the quail, *Coturnix.* It is unclear whether these polypeptides are products of different vitellogenin genes or results of posttranslational modifications.

The physiological basis for the apparent need for multiple vitellogenins is unclear, but from an evolutionary point of view it will be of great importance to characterize more vitellogenin genes and compare the structural differences in order to understand their evolutionary relationships. Furthermore, it will also be of interest to find out whether the different vitellogenin genes in the same cell are regulated by the same control mechanisms and expressed coordinately under different physiological conditions.

8. Posttranslational Modification of the Vitellogenin Polypeptide

As for other secretory proteins, the vitellogenin polypeptide is synthesized on ribosomes attached to the rough endoplasmic reticulum (ER) of stimulated hepatocytes. As the polypeptide is synthesized it threads its way across the rough ER membrane and ends inside the lumen of the rough ER. From there it is translocated through different membrane compartments of the cell before secretion into the circulation. The nascent polypeptide undergoes phosphorylation, glycosylation, and lipidation in an orderly manner as it passes through the various cellular compartments (Figure 9). In *Xenopus,* approximately 30% of the phosphate residues are covalently attached to the vitellogenin polypeptide in the rough ER after the termination of polypeptide synthesis, and the remainder of the phosphate is incorporated during intracellular translocation through membrane compartments. The partially phosphorylated vitellogenin polypeptide is then transferred to ER membrane compartments where the complex-type oligosaccharides are attached to the peptide backbone of vitellogenin. It has been suggested that both phosphorylation and glucosaminylation of the vitellogenin polypeptide can occur simultaneously at different regions of the vitellogenin molecule within the smooth membrane compartments. The glycosylated polypeptide is then transferred to the Golgi apparatus where trimming and modification of the core oligosaccharides may take place. The mature polypeptide destined for secretion is packaged into Golgi-associated secretory vesicles just prior to secretion. In the chicken, a different sequence of posttranslational modification seems to occur. The nascent polypeptide appears to be first glycosylated, followed by phosphorylation immediately prior to secretion. Little information is available on the site or mechanism of lipidation of the vitellogenin molecules. Nevertheless, it is known that vitellogenin is the richest source of lipid in most nonavian eggs, and in *Xenopus* at least 15% of lipid and cholesterol synthesized by the liver is incorporated into vitellogenin. No information on post-translational modification of vitellogenins is available for other vertebrate groups.

9. Plasma Vitellogenin Levels

Animals that have a seasonal breeding pattern generally also show a seasonal variation of vitellogenin levels in the blood stream. Some examples are given in Figure 10. In the U.S. eastern painted turtle, *Chrysemys picta,* which ovulates in June, plasma vitellogenin levels are highest in the months preceding ovulation. The rise in the autumn correlates with this period of follicular growth and is interrupted by a winter hibernation period before continuing in the spring. Rapid uptake of vitellogenin before ovulation causes the plasma levels to drop prior to ovulation. In a U.S. southern species, *Sternotherus odoratus,* which produces three clutches of eggs in spring and early summer, vitellogenin levels increase sharply in early spring before the first ovulation and decrease progressively at each ovulation. Plasma

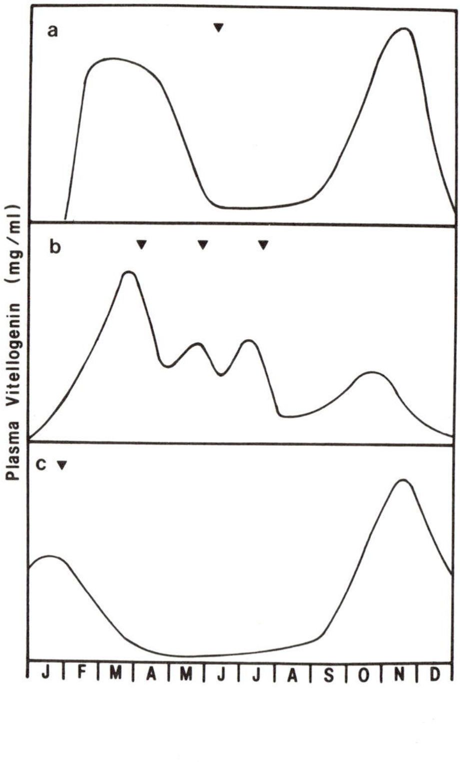

Figure 10. Seasonal changes in plasma vitellogenin levels in (a) *Chrysemys picta,* (b) *Sternotherus odoratus,* and (c) *Salmo gairdneri.* (Adapted from Ho, S.- m., *et al.*, 1982, Regulation of vitellogenesis in reptiles. *Herpetologica* **38:**40–50, and von Bohemen, C. H. G., and Lambent, J. G. D., 1981, Estrogen synthesis in relation to estradiol and vitellogenin plasma levels during the reproductive cycle of the female rainbow trout, *salmo gairdneri. Gen. Comp. Endocrinol.* **45:** 105–114.)

vitellogenin levels drop before each ovulation, probably reflecting an increase in uptake prior to ovulation. In the trout, *Salmo gairdneri,* which spawns in January/February, plasma vitellogenin reaches the highest level around November and December preceding the time of spawning.

A different pattern exists for the elasmobranch, *Scyliorhinus canicula,* in which oviposition continues for 10–12 months of the year. A very low vitellogenin level displaying little seasonal variation appears to reflect a steady secretion by the liver and a fairly constant uptake by the ovaries throughout the year. In contrast, in viviparous *Squalus acanthias,* which has a 22-month gestation period, follicular growth begins immediately after ovulation and continues throughout gestation. Ovarian follicles reach 4–5 cm in diameter at term, and plasma vitellogenin is undetectable in early pregnancy and increases toward ovulation. In laboratory-maintained *Xenopus,* a low level of vitellogenin exists at all times, apparently reflecting the reproductive habit of continual production of small groups of mature oocytes throughout the year. In the domestic hen, once maturity is achieved at 21–23 weeks of age, a fairly constant level of 20–25 mg/ml vitellogenin is found in females, which declines with age to 10–20 mg/ml.

10. Hormonal Requirements of Vitellogenesis

With few exceptions estrogen is the sole inducer of vitellogenesis. Of the natural estrogens, 17β-estradiol, the primary natural estrogen of vertebrates (see Part I), is the most potent, estrone being only 5–10% as effective in the rainbow trout, *Salmo gairdneri,* for example. The synthetic estrogen, diethylstilbestrol (DES), is as effective as estradiol in the domestic hen. In *Xenopus* hepatocyte cultures, however, DES is less potent than estradiol in vitellogenin mRNA induction. In the same system, 17-ethinylestradiol is less potent, and mestranol more potent, than estradiol. High doses of synthetic androgens (methyltestosterone, ethinyltestosterone, and methylandrostenediol) will induce vitellogenesis in the goldfish. In this case the hormones or their metabolites, when present in excess, may be binding to the estrogen receptor, or the androgens may be enzymatically converted to estrogenic hormones. However, in the teleost fish, *Gobius niger,* the black goby, a nonaromatizable androgen, dihydrotestosterone, has been reported to induce vitellogenin synthesis and to increase the binding of an estrogen receptor to the nuclear compartment of the hepatocyte.

Although estrogen is the primary inducer of vitellogenesis, other hormones may modify the response. In the iguanid lizard, *Dipsosaurus dorsalis,* gonadotropin and estrogen induce vitellogenesis and follicular growth. However, experimentally, gonadotropin in the absence of the pituitary does not so act and requires growth hormone to do so. Hepatic changes associated with estrogen induction of vitellogenesis did not occur in hypophysectomized animals in the absence of growth hormone. Thus, growth hormone is an important synergist for gonadotropin–estrogen induction of vitellogenesis in this species. Similar results have been found for the painted turtle (Figure 11).

Unlike these studies in reptiles, in *Xenopus,* vitellogenesis can be induced by estrogen in hypophysectomized animals. Nevertheless, even in this species, other hormones may modulate estrogen-induced vitellogenesis. Thus, metamorphic tadpoles will not synthesize vitellogenin in response to estrogen unless they are first exposed to thyroid hormones, which may be considered growth hormones in the amphibian at metamorphosis. In addition, cultured liver fragments from adult *Xenopus* lose their ability to respond to estrogen *in vitro* unless both thyroid hormone and an adrenal steroid analog (dexamethasone) are added to the culture medium. Recent studies in the domestic hen have suggested that this species may be similar to the reptile in that vitellogenesis is dependent on the pituitary and growth hormone.

There is evidence from reptilian species that other gonadal steroids may modify the estrogenic response. Thus, in the iguanid lizards *Dipsosaurus dorsalis* and *Sceloporus cyanogenys,* both progesterone and prolactin inhibit gonadotropin-induced ovarian growth and normal seasonal follicular growth associated with yolk protein uptake. Recent work with the turtle suggests that one way these hormones and testosterone work is by regulating the inductive action of estrogen on the liver. Thus, administration of progesterone or testosterone simultaneously with estrogen reduced and delayed the appearance of vitellogenin in the blood in a dose-dependent manner. The action of these two steroids in suppressing the estrogen effect may be in limiting the availability of estrogen receptor or its binding to the nucleus, both essential steps in the estrogenic induction of vitellogenesis. In the normal ovarian cycle both progesterone and androgen may be physiologically important modulators of estrogen action, since they are secreted in increasing amounts as the follicular cycle progresses. In addition, progesterone, at least, is an important hormone in the postovulatory phase of the cycle. The hormones may also influence follicular growth directly by inhibiting vitellogenin uptake by maturing follicles. Thus, the action of both hormones may be important in slowing the rate of follicular development, and in preventing vitellogenesis from occurring during gestation in certain species.

Adequate evidence exists to suggest that estrogen stimulates vitellogenin synthesis through a specific steroid receptor in hepatic cells of all nonmammalian groups. However, enough differences do exist to suggest that the mammalian model of steroid–receptor translocation from the cytosol to the nucleus cannot be applied to the induction of vitellogenin in birds, reptiles, amphibia, and fish without some modification. This will only be clarified by further work.

Once hepatic cells are exposed to estrogen for the first time, *in vivo* or *in vitro,* yolk protein synthesis is detected within a few hours. The lag period between exposure and detection of vitellogenin synthesis varies from one vertebrate group to another. Vitellogenin accumulates in plasma over the next few days and then gradually declines to zero over several weeks. This is called the primary response. If hepatic tissue is again exposed to estrogen, the lag period between hormone exposure and vitellogenin synthesis is shorter and the response more rapid and of greater magnitude. This is the secondary response. Because it can be observed for

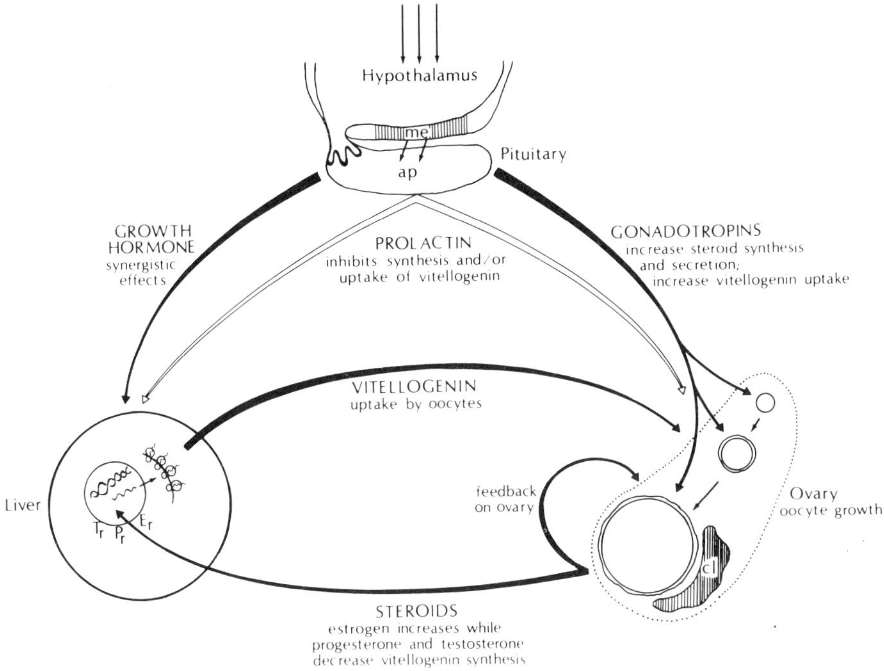

Figure 11. Summary diagram of the control of vitellogenesis in reptiles, derived primarily from work on squamates and turtles. See text for details.

up to several months after the primary induction, it is said to show a "memory effect." Recent studies suggest that the cellular events involve changes at both the genomic and translational level after the secondary exposure.

The major molecular effect of estrogen is to increase the number of copies of vitellogenin mRNA. Thus, in male *Xenopus,* a widely used model for these studies, vitellogenin mRNA increases from zero to 30,000 copies/hepatocyte. Current concepts of steroid hormone action suggest that enhanced transcription of the vitellogenin gene is a crucial part of the estrogen response. In addition, the primary exposure to estrogen may cause long-term changes in vitellogenin gene structure (decondensation of chromatin), which can be detected by enzymatic (DNase I) digestion.

In addition to its effects on vitellogenin gene transcription and structure, estrogen also markedly enhances the overall protein synthetic capacity of the liver as indicated by amino acid incorporation, increased numbers of ribosomes, and development of the endoplasmic reticulum and Golgi. All of these changes enhance the ability of the liver to synthesize vitellogenin and other proteins and probably contribute to both primary and secondary responses.

The amount of estrogen receptor in female turtle liver is at least partially dependent on the pituitary gland. Hypophysectomy reduces hepatic cytosol receptor levels to a very low level, and the receptor can be partially restored by growth hormone injection. It is likely that this effect of growth hormone on cytosol receptor levels partially explains the synergistic effect of growth hormone on vitellogenesis. Although direct evidence is not available, inhibitory effects of prolactin, progesterone, and testosterone on vitellogenesis are likely to be exerted, in part at least, through actions on the estrogen receptor, its synthesis, and movement to and from the nucleus. Probable hormonal interactions involved in the regulation of yolk protein synthesis and uptake in reptiles are diagrammed in Figure 11.

11. Uptake of Vitellogenin by the Growing Oocytes

In the vertebrate ovary, each oocyte is surrounded by a group of specialized ovarian cells that form its follicle wall, whose organization differs to some extent among the various vertebrate groups. The basic components are (Figure 8): (1) an acellular envelope, the zona pellucida or vitelline membrane, which forms the inner layer of the follicle wall; this is penetrated by microvilli from the oocytes; (2) external to the zona pellucida is the follicular epithelium or the granulosa layer; (3) the basal lamina; and (4) a thecal layer, which consists of bundles of collagen fibers, blood vessels, and fibroblasts or stromal cells (see Part I).

During the previtellogenic phase of follicle growth, the follicular epithelium develops into a multilayered structure and actively engages in the synthesis of macromolecules which are transported into the previtellogenic oocytes. As the oocytes enter the vitellogenic phase, the follicular epithelium becomes single-layered again and synthetic activity decreases. Vitellogenin molecules probably leave the perifollicular capillaries by intracellular (pinocytotic) transport across capillary endothelial cells and by passage between adjacent endothelial cells, across the thecal layer, and through the basal lamina and penetrate through the intracellular channels of the follicular epithelium to the zona pellucida and via its pore canals gain access to the growing oocyte. Vitellogenin is taken up by the oocyte by "receptor-mediated endocytosis." Special receptor proteins that bind vitellogenin with high affinity ($K_d = 6 \times 10^{-7}$ M) have been demonstrated on the plasma membrane of the oocytes. The receptors appear to recognize the phosvitin portion of the vitellogenin molecule. Studies on chicken oocyte vitellogenin receptors have shown that the receptor is of large size, approximately 500,000 daltons. In *Xenopus* oocytes, the number of vitellogenin receptors on the plasma membrane is between 2 and 28 $\times 10^{10}$ sites per oocyte. Once the vitellogenin molecule is bound to the vitellogenin receptor on the oocyte surface, the protein–receptor complex is internalized via coated pits at an accelerated rate. By this mechanism vitellogenin is selectively taken up by oocytes in the presence of other serum proteins. In culture, follicles take up vitellogenin 20–50 times more rapidly than bovine serum albumin. In *Xenopus,* vitellogenin receptors appear to be continuously internalized, regardless of the presence of vitellogenin.

Pituitary gonadotropins play a major role in the regulation of the uptake of vitellogenin by the oocytes. Hypophysectomy of vitellogenic females usually causes cessation of follicular growth, which can be reinduced by gonadotropic injection. Gonadotropin is believed to act on the follicular components of the follicle, regulating access of the vitellogenin to the oocyte. When oocytes above a certain size are removed from their follicular envelopes and provided with a saturating concentration of vitellogenin in culture, they can grow to sizes larger than normally found *in vivo*. However, vitellogenin uptake by oocytes without their follicular membranes cannot be regulated by gonadotropin. In the presence of gonadotropin *in vitro* uptake of vitellogenin by intact follicles increases to 350% of the control value. It is possible that the hormone increases the microcirculation or opens up special channels between follicular cells. Gonadotropin may also act indirectly by inducing a follicular factor that stimulates the rate of vitellogenin uptake by oocytes. Follicular steroid hormones may play a role in regulating vitellogenin uptake, since estrogen reduces endocytosis. Since gonadotropins stimulate the production of ovarian steroids, their action may be in part mediated by these steroids. Although little work has been done on the direct inhibitory effects of hormones on oocyte yolk protein uptake, progesterone reduces frog oocyte vitellogenin uptake *in vitro*, and both progesterone and prolactin inhibit gonadotropin-induced follicular growth in lizards.

12. Conclusion

The frequency of appearance of viviparity in different vertebrate groups suggests multiple origins and parallel evolution. Nevertheless, the reproductive substrate with which the process began was similar, and the process resulted in repetitive use of similar endocrine/reproductive adaptations involving the ovary, corpus luteum, reproductive tract, and hormones. In all groups, once viviparity has evolved, there is a trend toward increased complexity of the fetal–maternal interface, the placenta, or, in some species, reliance on other intrauterine mechanisms. These are correlated with an increased dependence of the embryo/fetus on a direct, maternal nutrient supply and a reduction in the amount of yolk supplied to the ovum prior to ovulation. These processes appear to have reached a natural culmination in the ancestors of present-day eutherian mammals in which yolk has been eliminated, and the placental association is close enough to allow the supply of nutrients directly through close apposition of maternal–fetal blood vessels.

As demonstrated in this chapter, a great deal is known about the hormonal regulation of yolk synthesis and its uptake by the oocyte. However, relatively little is known about mechanisms involved in the suppression of yolk synthesis, which must have played a role in the evolution of eutherian viviparity. It can be seen, however, that in certain reptiles, at least, other gonadal hormones (androgen, progesterone) probably act to inhibit vitellogenesis during the natural cycle. Extension of, and emphasis on, these processes in the evolution of viviparous species must be important in the final elimination of vitellogenesis in favor of placentation as an alternate mechanism for embryo/fetal nutrition. Presumably, final elimination of yolk came at some point in the early evolution of eutherian mammals from their pantotherian ancestors.

ACKNOWLEDGMENT. The author wishes to acknowledge the support of the National Science Foundation (NSF 81-04144) during the writing of these chapters and during research upon which some of the material is based.

13. Selected Readings

Amoroso, E. C., 1952, Placentation, in: *Marshalls Physiology of Reproduction,* Vol. 2 (A. S. Parkes, ed.), Longmans, London, pp. 127–311.

Ho. S-M., Kleis, S. M., McPherson, R., Heiserman, G. J., and Callard, I. P., 1982, Regulation of vitellogenesis in reptiles. *Herpetologica* **38:**40–50.

Lillegraven, J. A., Kielan-Jaworowska, Z., and Clemens, W. A., 1980, *Mesozoic Mammals,* Berkeley, University of California Press, pp. 1–311.

Needham, T., 1950, *Biochemistry and Morphogenesis.* Cambridge, Cambridge University Press, pp. 1–780.

Packard, G. C., Tracy, C. Richard, and Roth, J. J., 1977, The physiological ecology of reptilian eggs and embryos, and the evolution of viviparity within the class *Reptilia. Biol. Rev.* **52:**71–105.

Tata, J. R., and Smith, D. F., 1979, Vitellogenesis: A versatile model for hormonal regulation of gene expression. *Recent Prog. Horm. Res. V.* **35:**47–95.

Tinkle, D. W., and Whitfield Gibbons, J., 1977, *The Distribution and Evolution of Viviparity in Reptiles.* Museum of Zoology, University of Michigan.

Wallace, R. A., 1984, Vitellogenesis and oocyte growth: Non-mammalian vertebrates, in: *Developmental Biology: A Comprehensive Synthesis,* Vol. 1, *Oogenesis* (L. Browder, ed.), Plenum, New York.

Wourms, J. P., 1981, Viviparity: The maternal–fetal relationship in fishes. *Am. Zool.* **21:**473–515.

Part III

The Hypothalamus and Pituitary Gland

T. F. C. Batten and P. M. Ingleton

T. F. C. Batten • Department of Cardiovascular Studies, Medical School, University of Leeds, Leeds LS2 9JT, England. **P. M. Ingleton** • Department of Pathology, Medical School, University of Sheffield, Sheffield S10 2RX, England.

9

The Structure and Function of the Hypothalamus and Pituitary Gland

1. General Introduction

The name "pituitary" for the hormone-secreting gland beneath the brain originates from ideas of Aristotle that the brain secretes phlegm or pituita which cools the body. This fluid was thought to pass from the brain via the pituitary body and into the nasal passages. In the nineteenth century ideas about the function of the pituitary began to change when it was realized that certain pathological conditions were associated with tumors of the pituitary: for example, acromegaly, which is excessive growth of parts of the skeleton in adults. From a mixture of clinical observations and experiments on laboratory mammals, the relationships between the pituitary and other endocrine organs were established. Comparative studies have also made significant contributions to our understanding of pituitary hormone physiology. Early studies on tadpoles established the pituitary as an important controlling organ in development and function of the adrenals, gonads, and thyroid, and, moreover, that the pituitary was important for skin color changes in amphibia. The proliferative response of pigeon crop sac mucosa to prolactins from all tetrapods became a standard bioassay and hastened the isolation and characterization of this pituitary hormone. Because of the position of the pituitary beneath the brain, and its connections with it, it was realized that the pituitary must be a coordinating center for stimuli from the brain and, moreover, is probably dependent on signals from the brain for its function.

The important role of the hypothalamus in control of endocrine processes via the pituitary was first recognized in the 1930s, following experiments that demonstrated impaired gonadal function following disconnection of the pituitary from the brain, either by stalk section or by transplantation to a remote site. Pituitary pars intermedia cells were first shown to be innervated from the neural lobe, and it was supposed that the pars distalis cells must be similarly innervated; however, searches for these nerves proved fruitless. The idea that section of the pituitary stalk was interrupting vascular and not nervous supply to the gland did not gain momentum until experiments in the toad demonstrated the downward flow of blood in the portal vessels. Thereafter, over the next two decades, more critical experiments, involving hypophysectomy, pituitary grafts, and section of the portal vessels, made it clear that blood-borne factors from the median eminence and the hypothalamus were necessary for the elaboration and secretion of thc pituitary peptide hormones.

2. Anatomical Relationships of Brain and Pituitary

The forebrain (prosencephalon) of vertebrates can be divided into two parts, the anterior telencephalon (cerebral hemispheres) and, behind it, the diencephalon (''between-brain''). The lateral walls of the diencephalon may be subdivided into four horizontal zones, marked by furrows or ''sulci'' on the surface, which show the same relationships in all vertebrates. These are, passing ventrally:

1. The *epithalamus*—comprising various ganglia, the pineal, parapineal, and subcommissural organ.
2. The *dorsal thalamus*—a sensory integration center.
3. The *ventral thalamus*—a coordinating area with output to the lower brain areas.
4. The *hypothalamus*—responsible for integration of a whole range of inputs from other brain centers and for the control of visceral functions, in many cases via its effects on the secretions of the pituitary gland, which lies at its ventral surface (tuber cinereum). The central lumen of the diencephalon is the *third ventricle,* which often possesses lateral extensions into parts of the hypothalamus.

The pituitary lies beneath this central part of the brain and is connected to it by nervous and vascular links (Figure 1).

2.1. The Pituitary Gland

2.1.1. Embryonic Development and General Morphology

The pituitary is present in all vertebrates, and even in Protochordates the subneural gland (Urochordates) and Hatschek's pit (Cephalochordates) may be considered as homologs. The morphology of the gland is similar in all vertebrates, being composed of two principal parts, the neurohypophysis (NHP) and the adenohypophysis (AHP). Both parts are derived from embryonic ectoderm, the NHP as a downgrowth of neural ectoderm (the infundibular sac) from the floor of the diencephalon, and the AHP from Rathke's pouch, a dorsal outpushing of buccal ectoderm. Some recent evidence suggests that in avian (bird) embryos, cells that

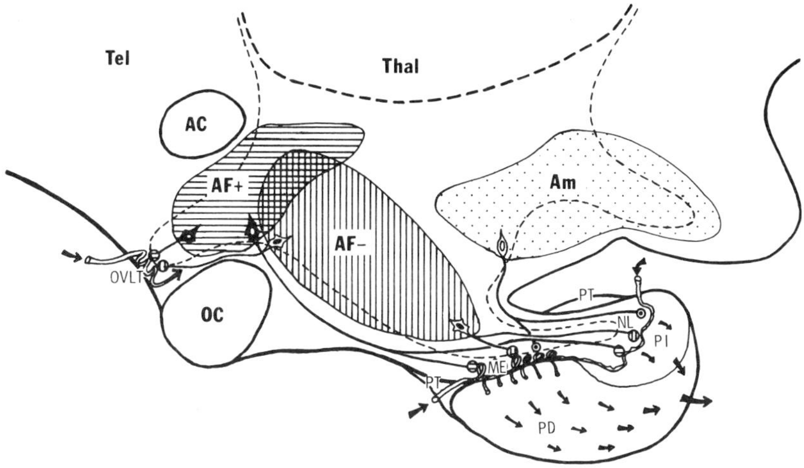

Figure 1. Principal relationships between pituitary and hypothalamic areas in a generalized vertebrate. AC, anterior commissure; AF+ or AF−, aldehyde fuchsin + or − staining neurons; Am, region of aminergic neurons; ME, median eminence; NL, neural lobe; OC, optic chiasma; OVLT, vascular organ of the lamina terminalis; PD, pars distalis; PI, pars intermedia; PT, pars tuberalis; Tel, telencephalon; Thal, thalamus. Arrows indicate the direction of blood flow; limits of the third ventricle are indicated by a dashed line. Anterior is to the left.

constitute Rathke's pouch may originate from ventral ridges of neural ectoderm and migrate to the pharyngeal wall and assume the form of Rathke's pouch: only further detailed studies will determine whether all chordate pituitaries develop in the same way. In general, it is accepted that the pituitary is a composite of neural and epithelial tissues, with vascular elements derived from the mesenchyme lying between the two areas (Figure 2) and enclosed in a membrane, sella turcica. Figure 3 shows a generalized diagrammatic representation of Rathke's pouch in the vertebrate embryo; all other amniote AHP rudiments are similar to this but with a larger oral lobe.

The NHP can be divided into three regions, distally the neural lobe (NL) (= pars nervosa), the neural stem, and proximally the median eminence (ME), the latter two zones together constituting the infundibulum (Table I). The AHP is formed of epithelial cells differentiated to secrete peptide hormones. It is regionalized, mainly in relation to the overlying brain, with the cells most closely apposed to neural tissue forming the *pars intermedia* (PI), which in many species is permanently divided from the rest of the AHP by the persistent lumen of Rathke's pouch, the hypophyseal cleft. Lateral outgrowths of Rathke's pouch fuse in the middorsal line to form the *pars tuberalis* (PT), and the anterior wall of the pouch thickens and gives rise to the *pars distalis* (PD), generally the largest and cytologically most complex region of the AHP (Figures 1–3).

In vertebrates, except teleosts, the vascular link between the floor of the brain and pituitary is particularly important because it carries factors from the central

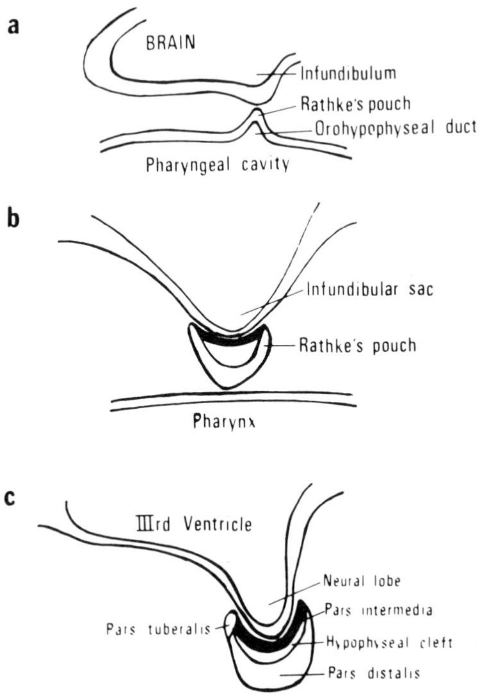

Figure 2. Embryonic development of the pituitary from Rathke's pouch and neural ectoderm.

nervous system (CNS) that control AHP cell activity. This vascular connection is via a system of portal vessels within the ME, so that virtually all blood reaching the AHP has passed through this specialized area. Within the AHP, vascular channels are extensive, and almost all cells contact a capillary or sinusoid, essential for transport of secreted peptide hormones and for receiving nutrients and controlling

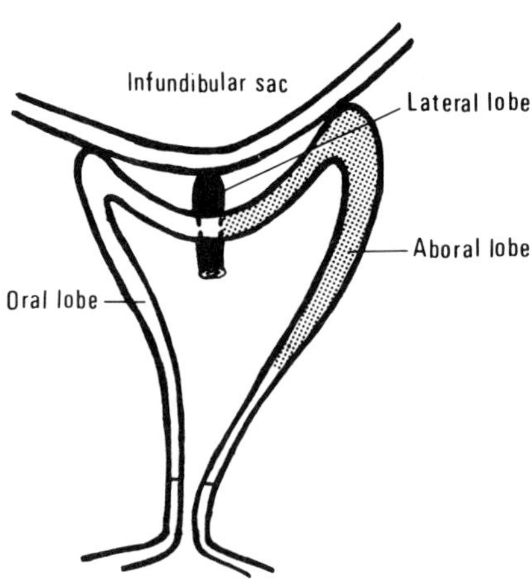

Figure 3. Rathke's pouch in a generalized vertebrate embryo. [Redrawn from Wingstrand, K. G., 1966, Comparative anatomy and evolution of the hypophysis, in: *The Pituitary Gland,* Vol. 1 (G. W. Harris and B. T. Donovan, eds.), Butterworths, London.]

Table I
Regions of the Hypophysis

Median eminence (ME)		
Adenohypophysis	Pars tuberalis (PT)	
	Pars distalis (PD)	Anterior lobe of the pituitary
	Pars intermedia (PI)	Posterior lobe of the pituitary
Neurohypophysis	Neural lobe (NL)	
	Neural stem	Infundibulum
	Median eminence (ME)	

signals. Capillaries are lined by endothelial cells, which are fenestrated and separated from pituitary cells by a double-layer basement membrane (BM) secreted by fibroblasts.

2.1.2. Hormones of the Adenohypophysis

Peptide hormones secreted by the AHP fall into three categories: (1) the tropic hormones, somatotropin (growth hormone, STH, GH) and prolactin (PRL, lactotropic hormone, LTH), (2) the glycoprotein hormones, thyrotropin (TSH), and gonadotropin(s) (GtH) of which there are two distinct hormones in most vertebrates, follicle-stimulating hormone (FSH) and luteinizing hormone (LH), and (3) corticomelanolipotropins, which include α- and β-melanotropin (MSH), adrenocorticotropin (ACTH), and β-lipotropin (LPH).

GH and PRL are secreted by distinct cells of the PD, GH by somatotrops and PRL by lactotrops or PRL cells. The term PRL cell is preferred because lactotrop refers only to the action of PRL on growth and differentiated activity of mammary glands, absent from all vertebrates except mammals. Thyrotrops secreting TSH and gonadotrops secreting GtH also occur exclusively in the PD, but ACTH is produced by cells of both the PD and PI, although only cells of the PD normally secrete the complete molecule. α-MSH, β-MSH, and β-LPH are produced by cells of the PI, by cleavage of a large molecule, proopiocortin.

2.1.3. Identification of Adenohypophyseal Cells

Identification of the cells of the AHP secreting the hormones noted earlier has been achieved gradually by several methods involving staining techniques, ultrastructural studies, and, more recently, immunological reactions with specific antisera. Staining and immune reactions depend on the chemical nature of the hormones, but there would be little definitive identification unless the hormones were stored in the cells that produce them, and it was by electron microscopy (EM) that the method of hormone storage was visualized.

2.1.3a. Staining Properties of AHP Cells

The details of staining techniques used for pituitaries are given in Section 16, but because cell types were originally identified by staining characteristics, we shall here give a brief outline of the techniques and their results. Before cells can be stained they must first be fixed, using materials that maintain the tissue in a condition as close to the living form as possible. For pituitary thc two best fixatives are Zenker–formol (Helly's fixative) and sublimated Bouin–Hollande; both of these include metal ions which enhance the uptake of stains and their reaction with tissue components. The most commonly used staining procedures are Cleveland–Wolfe trichrome, Herlant's tetrachrome, and alcian blue–PAS. The components of these react with particular cells according to the chemical nature of the stored hormone. Somatotrops and PRL cells are acidophilic, staining preferentially with orange G and erythrosin or azocarmine, respectively. Adrenocorticotrops also tend to be acidophilic and will stain with acid dyes including erythrosin and alizarin blue. Gonadotrops and thyrotrops were originally described as basophils, but are more accurately described as mucoid cells because they secrete glycoprotein hormones and in consequence the cells react with periodic acid Schiff reagent (PAS), which specifically identifies polysaccharide-containing molecules. Thyrotrops also react with aldehyde fuchsin, which stains acid mucopolysaccharide. A general identification of the cells is given in Table II, but it must be remembered that the various cell types do not always give exactly the same reaction in different animal species.

2.1.3b. Ultrastructure of AHP Cells

There is a remarkable similarity in the fine structure of the various pituitary cell types throughout the vertebrates. This being so, we shall describe their general features and illustrate them with electron micrographs.

i. General Ultrastructural Features of Peptide-Hormone-Secreting Cells. Peptide-hormone-secreting cells have the typical organization of glandular epithelial cells and store their product in secretory granules. Figure 4 shows a diagram of such a cell with cytoplasmic organelles concerned with peptide synthesis, including rough endoplasmic reticulum (RER), mitochondria, hormone storage granules, and smooth

Table II
Staining Properties of Adenohypophyseal Cells

Prolactin cells, η cells	Acidophilic, erythrosinophilic
Somatotrops, α cells	Acidophilic, orangeophilic (OG+)
Adrenocorticotrops, ϵ cells	Acidophilic, erythrosinophilic, alizarin blue+ (Aliz B+)
Gonadotrops, β and γ cells	Basophilic, PAS+
Thyrotrops, δ cells	Amphiphilic, PAS+, aldehyde fuchsin+ (AF+)

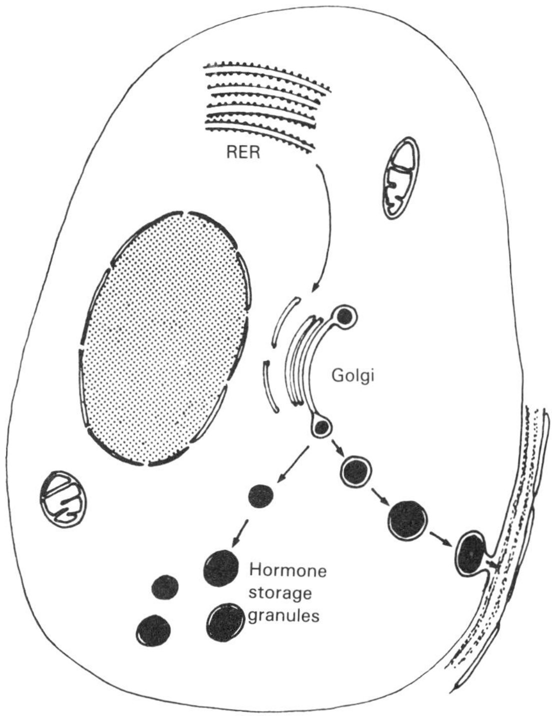

Figure 4. Intracellular secretory pathway of an endocrine cell. RER, rough endoplasmic reticulum. Amino acids assembled into peptides in the RER pass to the Golgi to be parceled into membrane-bound granules for storage or release.

endoplasmic reticulum forming the Golgi complex. The process of peptide synthesis occurs principally in the RER; messenger RNA (mRNA), encoded by nuclear DNA, passes to ribosomes attached to RER membranes, and here, by sequential processing of transfer RNAs (tRNA) with their attached amino acids, the specific peptide chain is assembled. It is believed that a specific signal sequence causes the membrane of the RER to recognize the beginning of the peptide and attach to it. This enables the growing peptide to pass into the lumen of the RER cisternae. The signal sequence is cleaved, and the particular peptide, e.g., growth hormone, passes along the RER lumen until a specialized area of membrane, without attached ribosomes, is reached. This is the Golgi complex, and it is here that peptides are parceled, and finally, by a pinching-off process at the membrane of the innermost cisternae of the Golgi stack, they are released into the cytoplasm as membrane-bound granules. The granules are then stored until the hormone is released by exocytosis. This process involves fusion of the granule membrane with the cell membrane so that the granule contents are released into the extracellular space and ultimately into the capillary lumen. Because peptide synthesis is very energy demanding, the cells usually have numerous mitochondria.

These are the basic components of peptide-hormone-secreting cells, but each of them is variable, and it is the spectrum of the forms of the various organelles, as well as the staining properties, location, and shape of AHP cells, that have identified them. The similarity in the cytological features of individual cell types among the vertebrates probably reflects conservation of the molecular structure of each hormone.

ii. Prolactin Cells. They are normally active cells, unique among pituitary cells in being under tonic inhibitory control, principally by dopamine (DA), a neurotransmitter. Generally the cells are well granulated, which means they take up stain avidly.

The cells are quite large, and when forming part of a follicle as in some teleosts, they are elongated, and in some species they may be polymorphic. They tend to be polarized, with the nucleus toward one pole of the cell and a concentration of granules at the opposite pole. The cells generally sit on a BM adjacent to a capillary, with hormone release occurring preferentially into the pericapillary space. Cytoplasm is moderately dense and RER is well developed, often with large concentrations of parallel arrays; there is at least one Golgi complex, which may be very large. Mitochondria are numerous and often have complex cristae giving "hank-of-wool" appearance. The nucleus is large, sometimes indented, and has at least one nucleolus. There are many hormone secretory granules, large when they are mature, oval, or spherical, but in some species they tend to be allantoid or pleomorphic; the mean size varies, but may be up to 800 nm across.

Figure 5 and 6 show the ultrastructure of PRL cells of a teleost (trout) and bird (zebra finch).

iii. Somatotrops. Although somatotrops stain preferentially with orange G in some species, in others, e.g., amphibia and rodents, it is difficult to distinguish them from PRL cells by staining alone. Ultrastructurally STH cells are normally well granulated, because when active, they secrete hormone episodically at few hourly intervals, but only an amount representing a small proportion of the stored hormone. However, in species exhibiting seasonal growth, there is marked variation between the active and regressed condition. The cells are generally polygonal and scattered throughout the PD, but confined to the proximal PD (PPD) in those pituitaries with distinct regionalization into rostral (RPD) and proximal zones. In lower vertebrates particularly, the cells tend to be arranged in cords, often interspersed with gonadotrops or thyrotrops, and adjacent to a capillary or sinusoid. The nucleus is generally large, rounded, and approximately central within the cell. The cytoplasm is moderately electron dense, mitochondria tend to be small, and RER is organized in groups of cisternae and tubules; only occasionally do large concentrations of parallel RER occur. Generally there is at least one Golgi complex close to the nucleus. Hormone granules are quite large and electron dense, with the contents closely apposed to the membrane; they are spherical and approximately 400–500 nm across (Figure 7).

iv. Corticotrops. ACTH-secreting cells generally show little affinity for histochemical stains, an observation that can be accounted for by the generally sparse

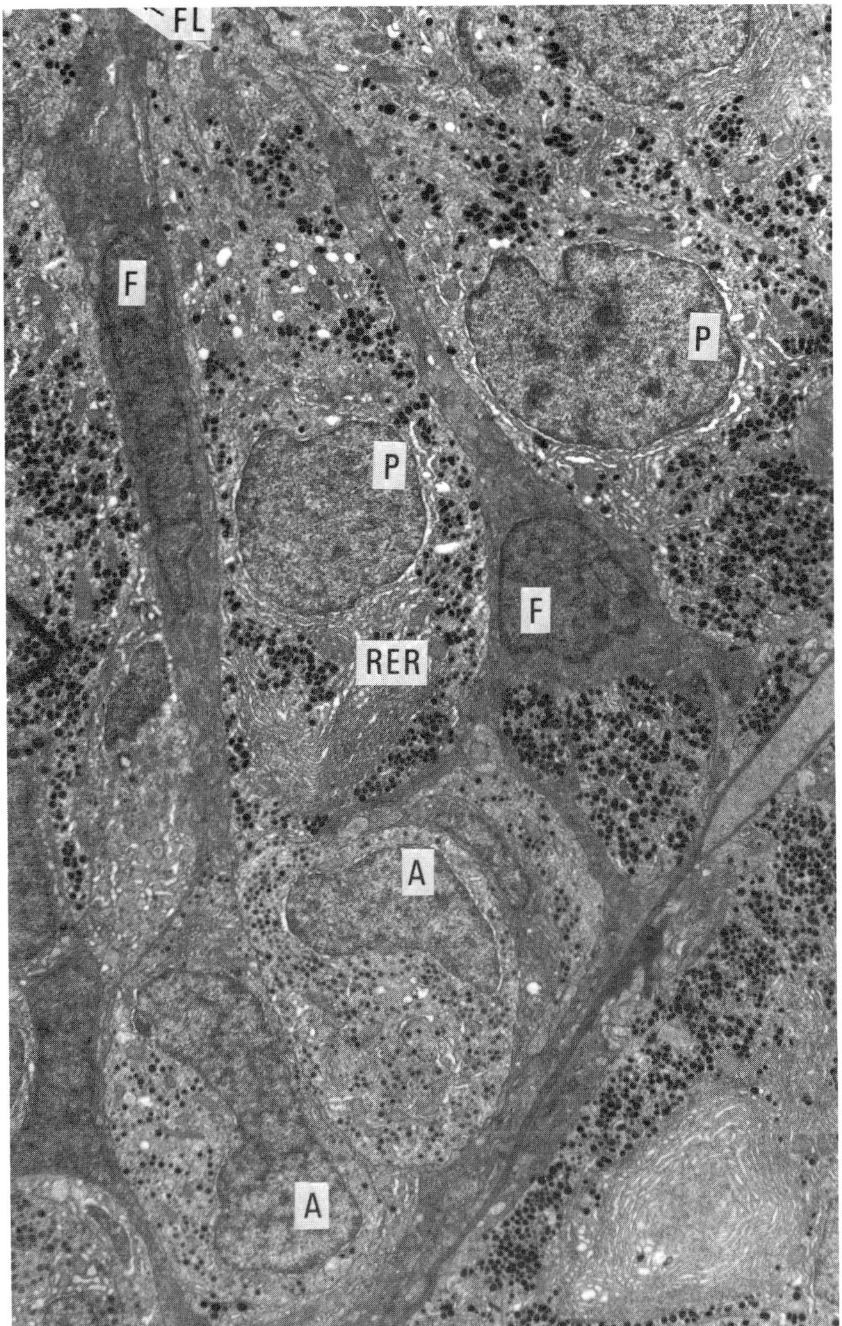

Figure 5. Cells of the rostral pars distalis of the trout, *Salmo gairdneri*. Note the elongated prolactin cells (P) with extensive RER at the pole of the cell opposite to the follicle lumen (FL→) and numerous pleomorphic granules, the smaller rounded adrenocorticotrop (A) and the follicular cells (F). (×4500.)

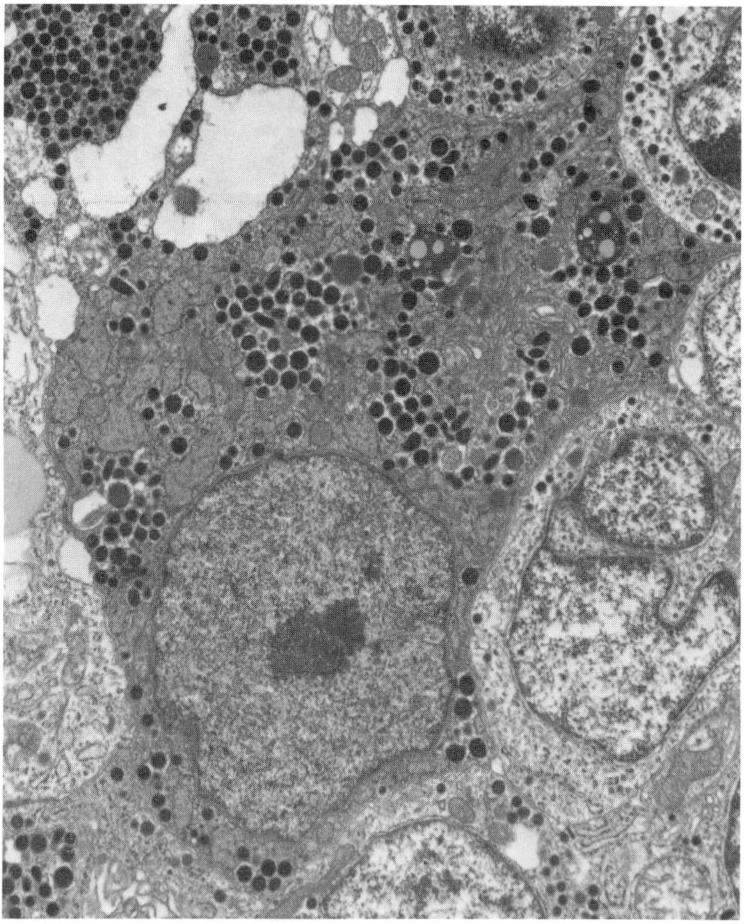

Figure 6. Prolactin cell of the zebra finch, *Poephila guttata castanotis*, with typical large pleomorphic granules, rounded nucleus, and dilated RER. (×10,000.)

granulation seen by EM, and because the granules are small, rarely larger than 250 nm and more frequently between 100 and 200 nm in diameter. The cells may be elongated, as in some teleosts, ovoid, or polymorphic. The Golgi complex is small, RER consists of scattered vesicles and short lamellae, mitochondria are small and slightly elongated, and the fine granulation is often arranged peripherally (Figure 5).

v. Glycoprotein-Hormone-Secreting Cells. Historically these have been referred to as basophils, but not all the cells react specifically with basic stains and a more acceptable term is mucoid cells.

Three glycoprotein hormones are secreted by the pituitary of all vertebrates so far studied, except species of squamate reptiles and fish; in these there appears to be a single gonadotropin with the properties of both luteinizing hormone (LH) and follicle-stimulating hormone (FSH) of other species. The third glycoprotein hormone is thyrotropin (TSH), which probably evolved from the gonadotropin molecule. The similar chemical structure of these hormones led to confusion in identifying the cells of origin. All the hormones are composed of two chains of amino acids, α and β, together with a carbohydrate moiety. The α peptide is identical in the three hormones of most species, and the β chains too have considerable homology. Thus, cells producing these hormones show similar staining properties, frequently basophilic, but not exclusively so, but all the cells react with PAS to give an intense magenta: this combination reacts specifically with the polysaccharide portion of the glycoprotein hormone molecule.

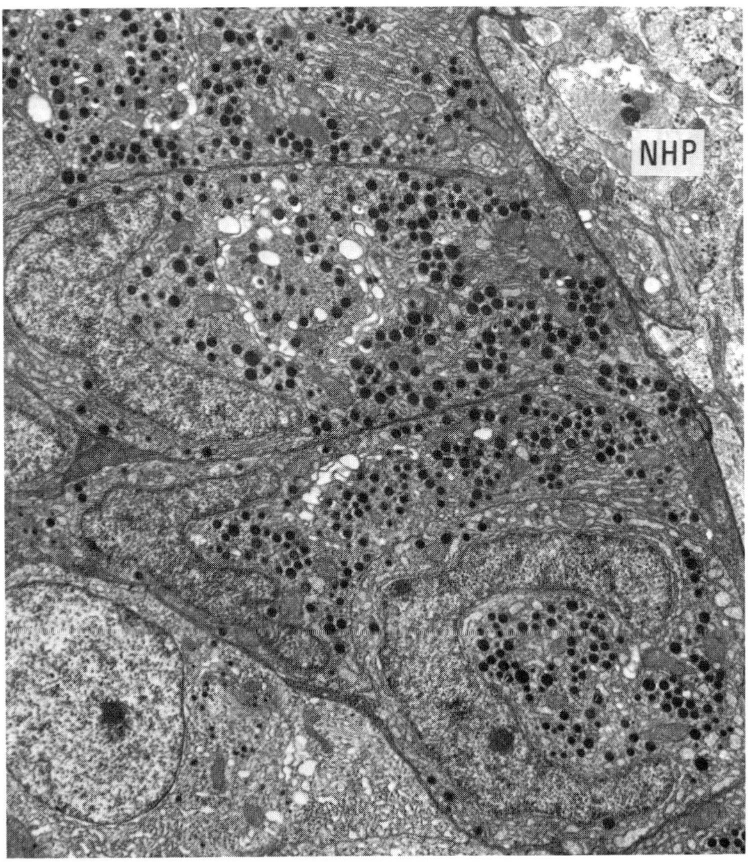

Figure 7. Growth hormone cells in the proximal pars distalis of the trout, *Salmo gairdneri*, bordering the neurohypophysis (NHP). Note the round, electron-dense hormone secretory granules and moderate amount of RER. (×4000.)

Thyrotrops. Thyrotrops are generally small rhomboidal or angular cells, often found in small groups or nests within the PD. However, the size and staining properties of the cells change with activity. The target organ for TSH is the thyroid, which produces thyroxine under the influence of TSH; in turn, thyroxine exerts an inhibitory feedback control on TSH production. If thyroxine secretion is reduced, for example by lack of iodine, then thyrotrops become more active; they enlarge and their staining properties change. Normally, TSH cells are AF+ and PAS+, but in the very active condition they show little affinity for AF, apparently owing to the absence of mature secretory granules. Ultrastructurally, TSH cells can be seen to have a moderately dense cytoplasm, mitochondria are small but quite numerous, RER is composed of scattered cisternae and vesicles, and there are several small Golgi and small, 100–200 nm, hormone storage granules (Figures 8 and 9).

Gonadotrops. Is there one or are there two distinct cell types secreting LH and FSH? In fish and squamate reptiles (lizards and snakes) only one gonadotropin has been identified, secreted by a single cell type, but in all other vertebrates examined, two cell types have been distinguished by staining techniques. Cells secreting FSH are AF− and PAS+, whereas LH-secreting cells are PAS+ but also stain with orange G (OG+); i.e., they are acidophilic. Gonadotrops are large cells but also show considerable increase in size and variability of ultrastructure depending on their activity, which is related to sexual cycles and activity of their target organs, the gonads. Gonadotrops can be recognized ultrastructurally by their vesicular RER, and the size heterogeneity of their granules, 100–300 nm, which also show variable

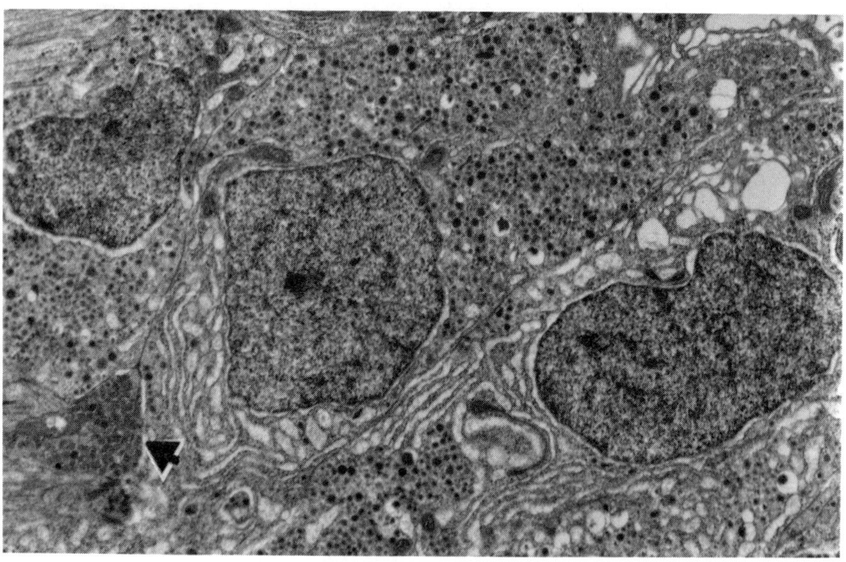

Figure 8. Thyrotrops in the proximal pars distalis of the molly, *Poecilia latipinna,* showing the small vesicular secretory granules and dilated RER characteristic of activated cells. Note the peptidergic-type neurosecretory fiber forming synaptoid connection with the cell (→) (×7000.)

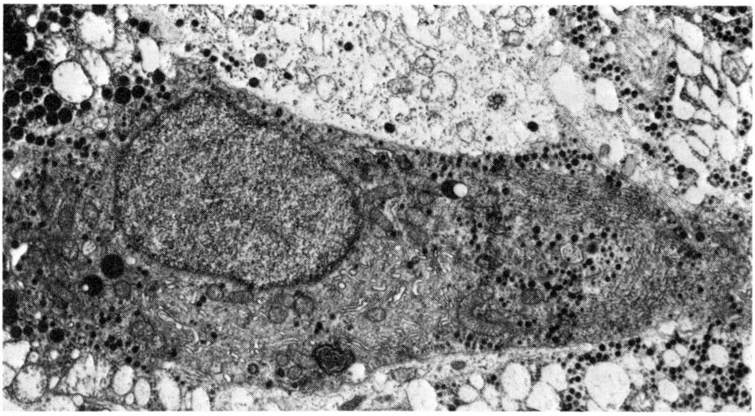

Figure 9. Thyrotrop in the rat pituitary. Note the tiny secretory granules, extensive Golgi, and organized RER typical of the euthyroid condition. (×5700.)

electron density; the nucleus tends to be polymorphic, and there are usually several small Golgi complexes (Figures 10 and 11). In very active cells the RER vesicles expand enormously and coalesce to form large vacuoles containing material of low electron density (Figure 10).

Antisera to specific β peptides of LH and FSH have been used to distinguish LH- and FSH-secreting cells, but the results have been less than definitive, in that immunoreactive LH and FSH were detected in the same cell, but separate cells also reacted with the individual antisera. Apparently both hormones can be produced by the same cell, but perhaps at different stages of reproductive activity some of the cells predominantly produce one hormone or the other.

vi. Cells of the Pars Intermedia. The PI is composed of, at most, two cell types, and in all vertebrates it is the source of melanocyte-stimulating hormone (MSH, melanotropin). The melanotrops stain with lead hematoxylin (PbH+) in all species; a second cell type in teleosts stains with PAS but does not secrete MSH. In tetrapods both PD corticotrops and PI melanotrops stain with both PbH and PAS. The origin of ACTH and α-MSH from a common precursor molecule, proopiocortin has been established. This is a large glycoprotein, which is cleaved by enzymes to give the various smaller hormone molecules (Section 13.3). The presence of the carbohydrate moiety explains the positive reaction with PAS, and the common peptide chain explains the similar staining reaction in both PI and PD cells and, moreover, why both react to antisera to ACTH and MSH. However, the PAS+ cells of teleost PI do not react with antisera to MSH or ACTH.

Thus, throughout the tetrapods the melanotrop is the predominant PI cell (Figure 12) and ultrastructurally appears to be active with organized RER, usually arranged in parallel arrays, small but quite numerous mitochondria, and small

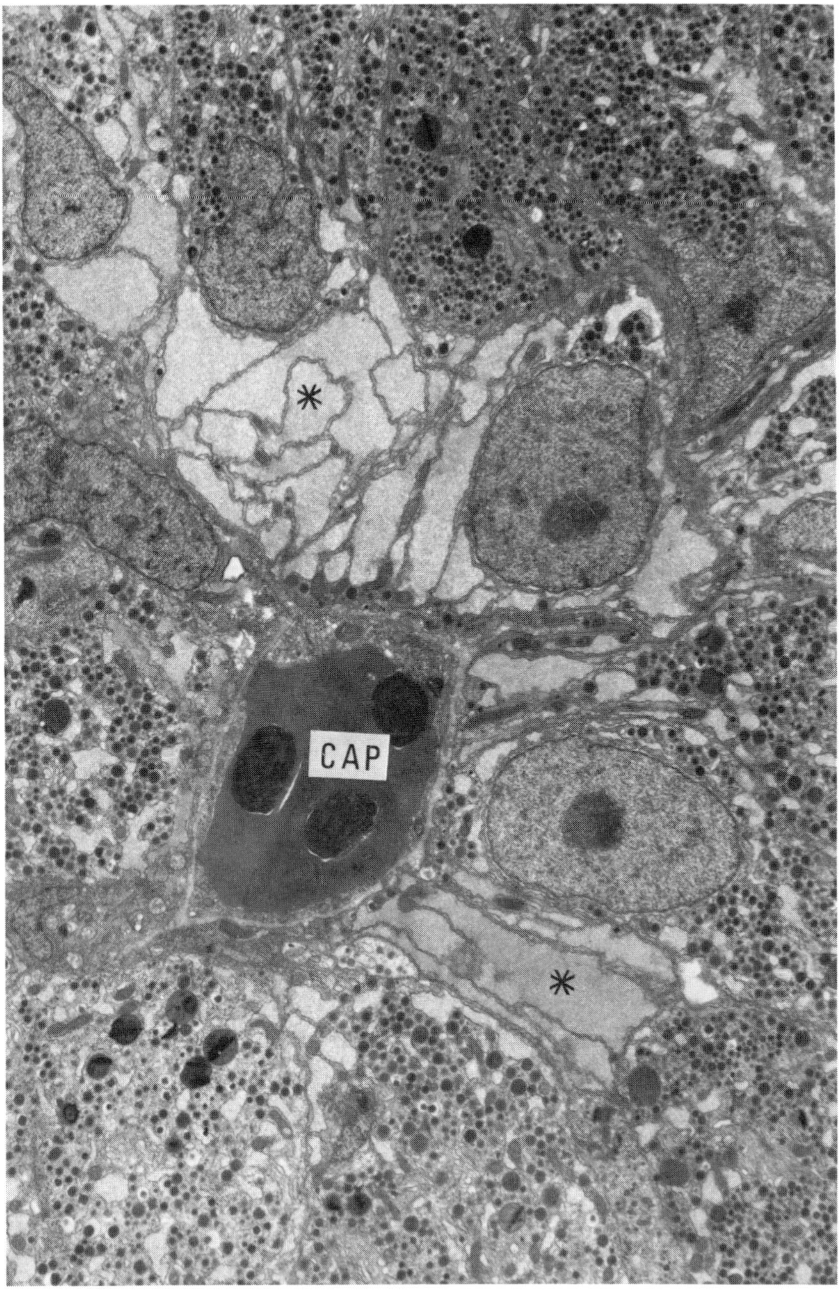

Figure 10. Gonadotrops surrounding a capillary (CAP) in the proximal pars distalis of the molly, *Poecilia latipinna*. Note the large distended RER cisternae (*) and pleomorphic secretory granules characteristic of vitellogenic fish. (×5000.)

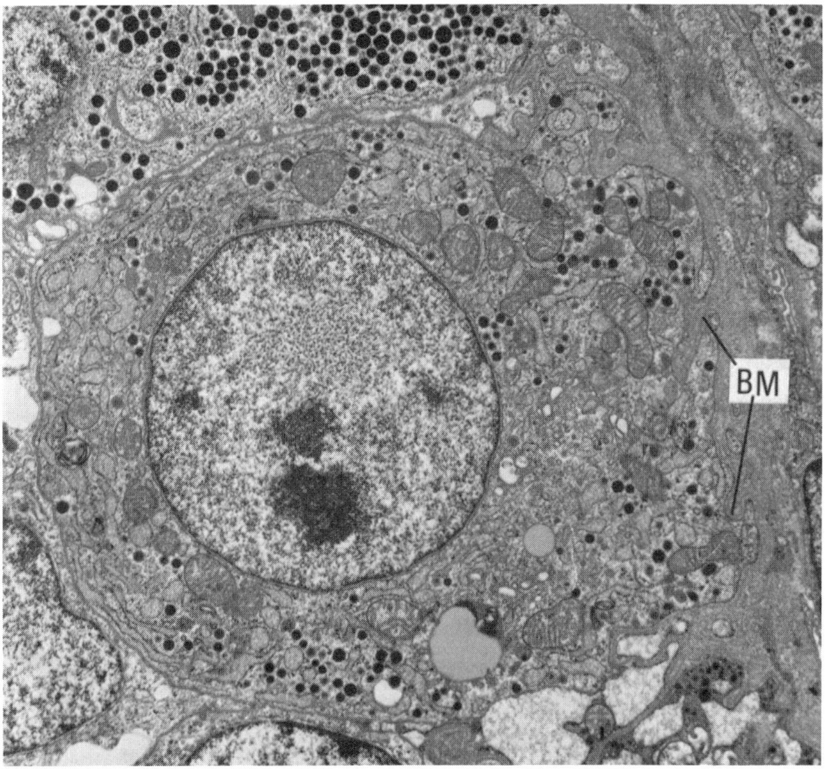

Figure 11. Gonadotrop of the zebra finch, *Poephila guttata castanotis*, containing typical small secretory granules, scattered dilated RER, and large mitochondria with simple cristae. Note, too, the intimate contact between the cell and basement membrane. (×8000.)

secretory granules, 100–200 nm, although in *Anolis carolinensis* (Reptilia) they fuse to form larger granules.

The function of PAS+ cells of teleost PI (Figure 12) has not yet been determined, but in some species, (e.g., *Poecilia, Blennius, Tilapia*) they do respond to background color changes being activated on black backgrounds, but in other teleosts (*Fundulus, Carassius*) they are activated in low ambient calcium concentrations.

vii. The Stellate–Follicular Cell System. These are the nonsecretory cells found throughout the pituitary PD and PI. The follicular or covering cells are most prominent in pituitaries with clear follicular organization of parenchymal cells, as in some teleosts, but even tiny follicles, as in rat PD, are lined by these cells, which have microvilli and sometimes cilia on their luminal surfaces. In their derivation they resemble ependymal cells, i.e., lining a lumen of invaginated ectoderm. Junc-

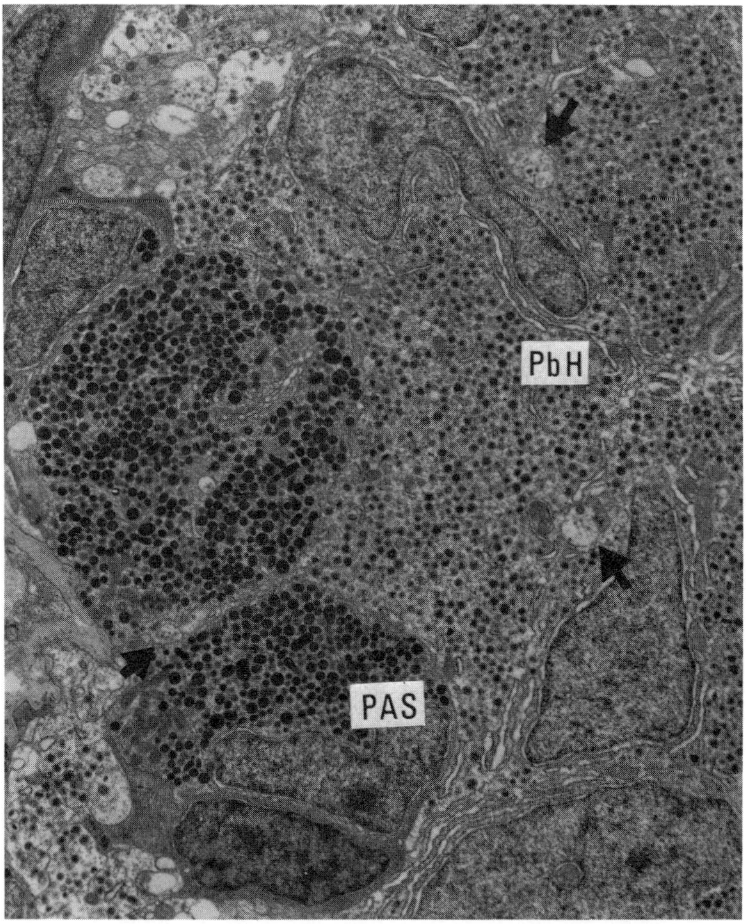

Figure 12. Pars intermedia of the molly, *Poecilia latipinna,* showing the relationship between aminergic-type nerve endings (↑) and the two types of secretory cells (PAS and PbH). (×6000.)

tional complexes join one cell to another, near the luminal surface, forming a tight but not impenetrable barrier, and fine cytoplasmic processes extend between the parenchymal cells, to contact a BM or process of another stellate cell. The stellate cells lie among the secretory cells with long processes lying between them creating a large interface (Figure 5). The cells contain no secretory granules and are chromophobic, but they frequently contain lipid droplets and glycogen, as well as lysosomes, a little RER, small Golgi, and few mitochondria. Their function is probably to act as scavengers of pituitary cell debris which results from death of secretory cells or lysosomal action within them.

2.1.3. The Hypothalamus

2.1.3a. Hypothalamic Nuclei and Neurosecretory Cells

The pituitary gland in all groups of vertebrates appears to be under the influence of specialized nerve cells within the brain known as neurosecretory (NS) cells or neurons. NS cells can be defined as nerve cells which show secretory activity to such an extent that they are comparable to gland cells, and which control other target cells via neurohormonal signals contained in NS granules (NSG). The subcellular organization of a NS cell is very similar to that of a typical protein-secreting cell (see Figure 4). Within the brain, NS cells are usually grouped together into discrete areas or nuclei, and those controlling pituitary activity are located in the hypothalamus (HT). The structure of hypothalamic nuclei, and the cells within them, shows a high degree of uniformity throughout the vertebrates; nevertheless some differences can be recognized which demonstrate an evolutionary progression from the primitive condition to an advanced condition (Figure 13). This may be seen not only in comparison of a lower vertebrate (e.g., lamprey) with a higher vertebrate (e.g., mammal), but within a group such as the fishes between a phy-

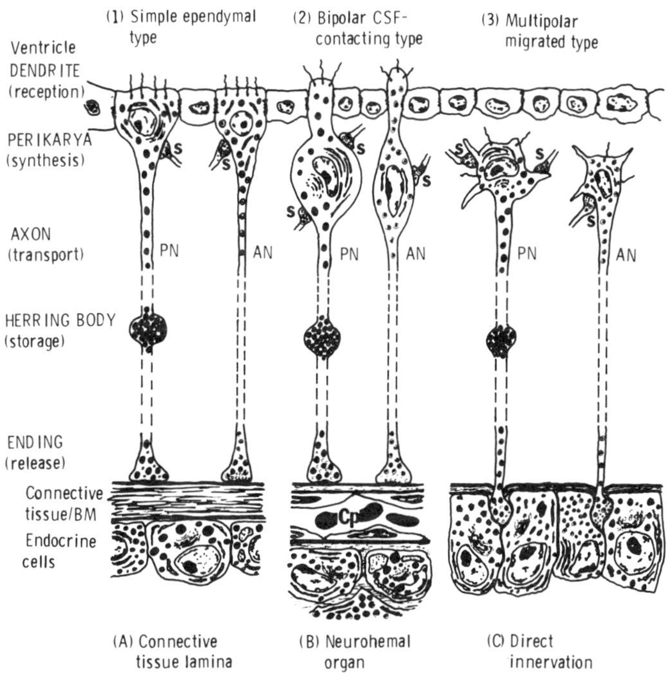

Figure 13. Classification of neurosecretory neurons and their relationships with target tissues. AN, aminergic neuron; Cp, capillary; PN, peptidergic neurons; S, synapse.

logenetically ancient species (e.g., sturgeon) and a phylogenetically younger teleost. In the primitive condition, the nuclei are composed of a few rows of cells lining the ventricle of the HT (third ventricle), but in subsequent stages of evolution more and more cells leave the periventricular position to aggregate into discrete functional cell clusters, and in the extreme condition they may "migrate" into the lateral areas of the HT to form nuclei remote from the ventricle. Concomitant with this, changes occur in the structure of the individual cells. In the primitive condition, these are of the simple "ependymal" type, with their cell bodies (perikarya) lying among the ependyma, i.e., cells that line the cavities of the CNS [Figure 13(1)]. Cilia and microvilli may project into the ventricle, perhaps indicating that the cell receives some form of higher control from factors within the cerebrospinal fluid (CSF). This type of NS cell is typical of those found in the nervous systems of invertebrates, protochordates, and the earliest stages of vertebrate embryonic development. In a more advanced condition, the NS perikarya lie in a subependymal position; however, contact with the ventricle is maintained by means of a ciliated CSF-contacting apical process, the dendrite [Figure 13(2)]. In addition to possible reception of signals from the CSF, these cells may also be controlled by synaptic contacts from nerve fibers originating from higher centers. Such bipolar NS cells form the bulk of most of the hypothalamic nuclei of lower vertebrates and also occur during mammalian development. The final stage of the evolution is the multipolar neuron, migrated away from the ventricle deep into the neuropil of the HT [Figure 13(3)]. The perikarya receive numerous synapses from other areas of the brain, but in addition bear some cilia, still possibly sensory, which project into the intercellular spaces. Multipolar perikarya are characteristic of the hypothalamic nuclei of adult birds and mammals.

From the NS perikarya, long, slender processes known as axons run down to the site of neurohormone release, where the NSG are expelled from the bulblike swellings at the end of the axon. Other swellings may occur along the axon, believed to be sites of granule storage; these are commonly observed in the peptidergic fibers of the NHP, where they are termed "Herring bodies" (Figure 13).

2.1.3b. Site of Neurohormone Release—Neurohemal Organs

In the most primitive condition the NS endings abut against a simple connective tissue lamina (Figure 13A) into which the neurohormone is released and subsequently diffuses or is transported to the local site of action or into the systemic circulation. Examples are the terminations on the floor of the brain in protochordates and cyclostomes. In more highly organized vertebrates, specialized neurohemal organs, devoid of a blood–brain barrier, are developed for release of the NS material into the blood (Figure 13B). The neurohemal areas associated with the HT are the NHP, comprising the NL and ME, and the vascular organ of the lamina terminalis (OVLT) (Figure 1). These areas lie around the midline ventricles of the brain and belong to a broader category of structures known as "circumventricular organs." Other circumventricular organs (the pineal, subfornical organ, subcom-

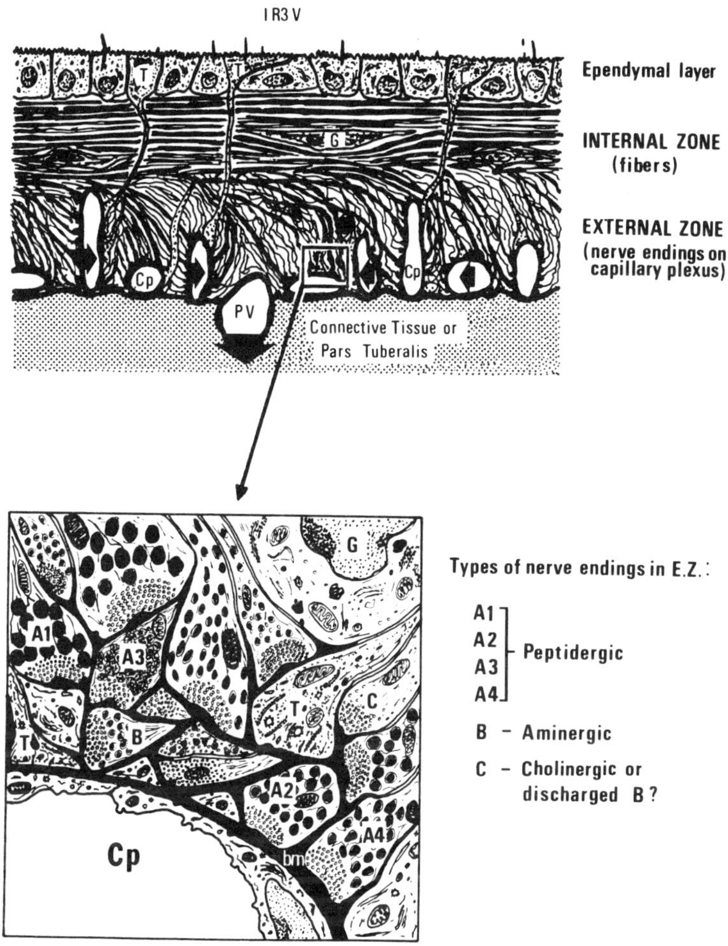

Figure 14. Detailed structure of the median eminence showing relationships between nerve endings and blood vessels. bm, Basement membrane; Cp, capillary; G, glial cell; IR3V, infundibular recess of the third ventricle; PV, portal vessel; T, tanycyte. Arrows indicate the direction of blood flow.

missural organ, area postrema) lie outside the boundaries of the HT. In addition to neurohormone release, these structures may also be associated with "sampling" the composition of the blood, for the regulation of neurohormone production/release. The neurohemal organs all have a typical three-layered structure (Figure 14): dorsally, ependymal cells lining the ventricle, centrally, a layer of axonal nerve fibers, and ventrally, an external palisade layer where the NS endings contact a narrow double BM, enclosing connective tissue, which surrounds the perivascular space of fenestrated capillaries. The area for neurovascular contact can be increased by two methods: first, by the elaboration of capillary loops into the layer of nerve endings;

second, by the ramification of the pericapillary BM away from the capillaries among the nerve endings. This is seen in the NHP, and in particular in the ME, of the most highly evolved groups of gnathostome (jawed) vertebrates. In the NL, material is released into the plexus intermedius and its derivatives (the series of blood vessels between the NL and PI) and is then conveyed into the general circulation or into capillary branches among the PI endocrine cells. The ME is situated on the base of the HT usually some distance away from the pituitary cells which the neurohormones are destined to control, so its capillaries (the primary plexus) converge to form portal veins which convey the blood to the gland. Within the PD the portal veins supply a secondary capillary plexus, which, often with the aid of intervascular channels, conveys the blood-borne material among the PD endocrine cells. It is unlikely that the OVLT is directly involved in pituitary control, since its capillaries drain not toward the gland but up into the HT. It is possible that material released at the OVLT reaches the general circulation or is destined to control neurons within the HT itself, perhaps exerting "feedback" effects (Section 14).

A third route of neurohormonal control is seen in the most highly developed groups within various vertebrate classes. This is direct synaptic contact between NS fiber endings and pituitary endocrine cells, bypassing the neurovascular route (Figure 13C). This form of control is usually restricted to the PI cells; however, in the most advanced fishes, particularly the teleosts, it also applies to some or all of the PD cell types.

2.1.3c. Classification of NS Neurons

The first of the "classical" stains for light microscopic (LM) study of NS neurons was chrome–alum–hematoxylin–phloxin (CAHP). This method has been superseded by the more effective aldehyde–fuchsin (AF) technique. These methods stain the same neurons, particularly those leading from the preoptic nucleus (NPO) to the NHP (the "classical" NS system). However, they work not by staining the neurohormone within the neuron, i.e., the octapeptide, but its carrier protein, the neurophysin (Section 13.4). The octapeptide can be stained, but more complex methods, such as performic acid–alcian blue (AB), must be used.

Stainability with AF and similar stains is not now considered to be a prerequisite for classification of neurons as neurosecretory. Large areas of the HT contain neurons that display cytological evidence of neurosecretion, but are AF negative, so other histochemical methods must be used for their classification. The most specific method is the Falck–Hillarp formaldehyde–fluorescence method, which, combined with microspectrofluorometry, can accurately identify neurons containing the biogenic amines dopamine, noradrenaline, and serotonin (5-hydroxytryptamine) (Section 16.4). At EM level these aminergic cells characteristically contain NS granules in the form of dense-cored vesicles (DCV) 60–100 nm in diameter, as opposed to the 150–250 nm NSG of peptidergic cells (Figure 14).

Not all the AF-negative neurons of the HT can be demonstrated to contain biogenic amines. It is possible that some of these nonaminergic AF-negative cells are peptidergic neurons producing the specific releasing (or inhibiting) hormones for the various pituitary cells. These peptides cannot be identified by straightforward staining or histochemical methods, but with the introduction of immunocytochemistry (ICC) (Section 16.3), some progress has been made, particularly in identification of neurons containing the probable releasing or inhibiting factors for gonadotropin (using antisera to luteinizing-hormone-releasing hormone, LH-RH) and for growth hormone (using antisomatostatin, SRIF). Where identified, the ultrastructure of these cells is similar to that of the AF-positive cell, with NSG possibly being smaller. NS fibers and endings have often been categorized as type A or type B, primarily on the basis of ultrastructure of the granular inclusions within them. The type B fibers are quite uniform throughout the vertebrate groups, containing DCV, 60–100 nm, together with clear synaptic vesicles, and are usually found to be aminergic. Type A fibers contain granules exceeding 100 nm and are believed to be peptidergic, but the precise size of the NS granules is variable, so that several authors have identified a number of subtypes: e.g., A1, A2, A3 (Figure 14). It is possible that this variation is due to the presence of different neurohormones within the granules; for instance, those containing NSG larger than 150 nm may be AF positive (octapeptide-containing) and those with granules smaller than 150 nm may be AF negative (releasing-factor-containing). An additional type of fiber, type C, has been recognized by several authors, characterized by the presence of clear vesicles only: this could be a separate non-NS nerve type or might represent a type B fiber that has released all its DCV, or at least has no DCV within the plane of section.

3. Protochordates

The neural complex of the tunicate (*Ciona intestinalis;* Urochordata) is composed of a neural structure, the cerebral ganglion, and an epithelial structure, the neural gland. The neural gland in Urochordates and Hatschek's pit in Cephalochordates (*Amphioxus*) are considered to be homologs of the pituitary of other chordates. Antisera to mammalian hormones have reacted with cells in the cerebral ganglion and gut epithelium of *Ciona;* these hormones include pancreatic polypeptide, substance P, somatostatin, secretin, and neurotensin. Other cells of the ganglion contain immunoreactive β-endorphin, vasoactive intestinal peptide (VIP), motilin, ACTH, and prolactin. These interesting findings highlight the dual origin of the pituitary gland from neural and stomodeal epithelia.

Before the clear development of the hypothalamus and pituitary, groups of NS cells in the primitive "brain" of ancestral chordates may have formed NS terminals on the ventral brain floor and thus discharged their secretions systemically. There is evidence for the presence of activity of NS cells in the "brains" of modern Protochordates. For example, during the breeding season cells in the cerebral ganglion of *Ciona* appear to control the rhythmical activity of the neural gland. In *Amphioxus* there are two pairs of peptidergic NS lying in the "brain bladder" with axons terminating on the ventral basement lamella. Perhaps related to this is the observation that there are immunoreactive LH cells in Hatschek's pit of *Amphioxus* which respond to exogenous LH-RH. Other groups of cells and nerve fibers of the cerebral ganglion contain immunoreactive LH-RH, SRIF, PRL, ACTH, and other neuropeptides.

The Protochordates are thought to have evolved as an offshoot of the main invertebrate–vertebrate line and, as such, represent an interesting stage of development. That there is a continuum of peptide and amine hormones and neurotransmitters throughout much of the metazoa is clear from studies using antisera to a variety of mammalian hormones and neurotransmitters which react, for example, with "brain" or gut epithelial cells of insects and polychaete worms.

4. Agnatha

In agnatha (Cyclostomes) there is only a very simple relationship between the brain and the adenohypophysis. Rudimentary vascular connections exist between the posterior NHP and the PI in lampreys (Petromyzonidae), and hagfishes (Myxinidae) have no distinct PI at all. The close apposition between the anterior NHP and the AHP suggests that hypophysiotropic substances of the anterior neurohypophysis (ANH) could diffuse across to the AHP.

Of the two groups of cyclostomes the lampreys lead more active lives, living in the shallower parts of the sea or in fresh water, whereas hagfishes live in a virtually constant environment, away from light and seasonal variations, in deep oceans. Thus, certain features of their brain/pituitary relationships are probably related to their life-style; for example, hagfishes appear to have no LH-RH, and may not be simply primitive features.

4.1. The Adenohypophysis

4.1.1. Embryonic Development

The AHP of lampreys (Petromyzonidae) arises from the nasopharyngeal duct, a backwardly directed elongated pouch that extends beneath the brain and, anteriorly, opens on the snout. The AHP anlage forms on the dorsal surface of the pouch as a solid structure, lying ventral to the infundibulum; the PI separates off early and aligns beneath the presumptive NHP. However, the origin of the AHP in hagfishes (Myxinidae) is apparently different not only from lampreys but also from all other vertebrates, if its derivation from endodermal placode is accepted. The adenohypophyseal placode is a thickened area on the dorsal wall of the archenteron, lying immediately below the posterior ventral tip of the infundibulum. Interpretation of the exact germ cell layer in which the AHP rudiment originates may not be possible; what is probably most important, in all vertebrates, is that cells in the region of the stomodeum/enteron which lie close to the infundibulum can be influenced to differentiate to the AHP.

4.1.2. Petromyzonidae

The AHP in lampreys is divided into two parts, the anterior pars distalis (PD), which can be further subdivided into rostral and proximal zones on the basis of cell types, and the PI, lying posterior to the pars distalis and separated from it by a connective tissue septum (Figure 17). The rostral zone is the homolog of the gnathostome PD (Figure 15); it is composed of cords of cells separated by connective tissue in which blood capillaries abound. Three secretory cell types have been described in the rostral zone, distinguished both by their staining properties as PAS+ and AF+ basophils and ultrastructurally by the size and density of their secretory granules. In larval stages, the ammocoetes, more chromophobes appear to

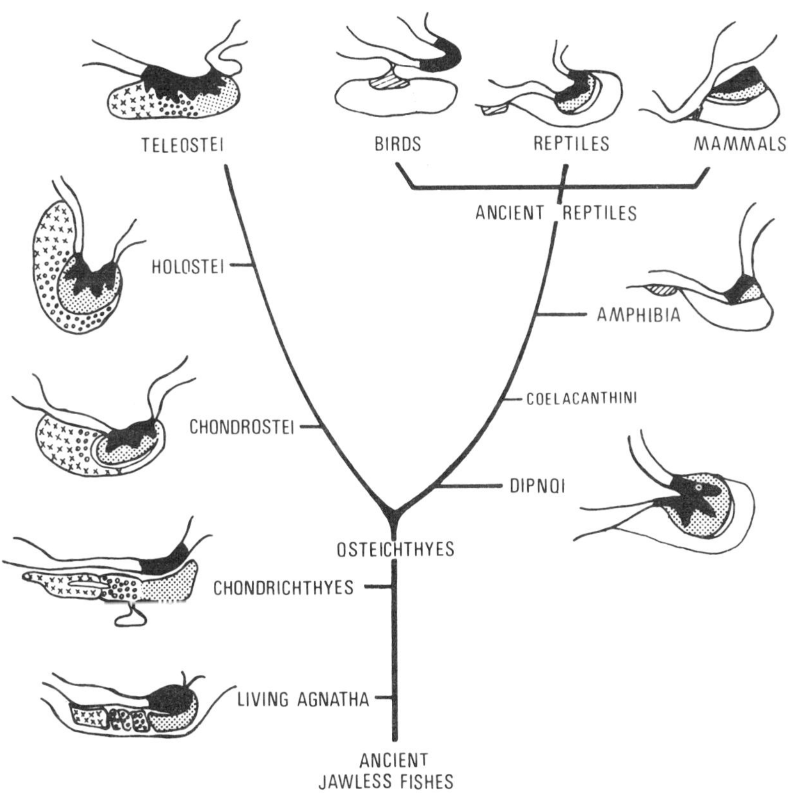

Figure 15. Hypothetical scheme for the evolution of pituitary glands. ■, Neural lobe; ⊞, pars intermedia; ⊞, proximal pars distalis, ⊠, rostral pars distalis; ▨, pars tuberalis. Pars distalis in tetrapods and dipnoi is a mosaic of cell types—□. (Redrawn from Gorbman, A., and Bern, H. A., 1962, *A Textbook of Comparative Endocrinology,* John Wiley & Sons, New York.)

be present, some of which may become granulated in the adult. Cells with cytoplasmic granules of 130–260 nm in the rostral zone and similar cells in the proximal zone may be gonadotropic. Other cells with small granules, 120–250 nm, may be adrenocorticotrops since ACTH-like activity has been shown in them in several species of lampreys, and ACTH-immunoreactivity also occurs in the majority of rostral zone cells. A third type of cell with more uniformly small granules of 100–140 nm occurs in the proximal zone; its function is unknown but it may be somatotropic. Recently, cells in the proximal zone have been shown to react specifically with antiserum to rat LHβ, but these are chromophobic and do not stain with PAS. Chromophobes persist throughout life, in both the rostral and proximal zones; some of these resemble and may serve the same purpose as stellate cells of other vertebrate pituitaries. However, there is no evidence of control of thyroid gland activity by a pituitary thyrotropic hormone (TSH), suggesting that none of the cells of the AHP is a thyrotrop.

The PI is closely apposed to the neural lobe, separated from it only by connective tissue containing a system of capillaries and penetrated by very few neurohypophyseal processes. There are neither blood capillaries nor nerve fibers running in the parenchyma of the PI, which is composed of two cell types, one of which is nongranulated and chromophobic, and the other generally AF+ and granulated. Ultrastructurally the granulated cells appear more varied; the most common type has moderately numerous secretory granules, 160–240 nm in diameter. Other granulated cells may contain large, electron-dense droplets, 1.2–2 μm in diameter. It is probable that the various cell "types" represent a single hormone secretory cell species in different stages of activity, and indeed, a recent study shows immunoreactive MSH in all the PI cells.

Processes of the secretory cell frequently contact the basal lamina of the pericapillary space at the PI/NHP junction. However, extensions of the nongranulated PI cells may lie between the granulated cells and the basal lamina. Granule release probably occurs in this region because the granulated PI cells are polarized and show marked accumulation of secretory granules at the basal lamina pole of the cell.

4.1.3. Myxinidae

The AHP in myxinids is simpler than in lampreys, probably a primitive condition, although possibly also related to life in an unvarying environment. The whole of the AHP consists of clusters of cells, sometimes in follicles, embedded in connective tissue lying below the diencephalon, and there is no division into PD and PI. Vascularization is poor, with only a few capillaries in the connective tissue and none at all in the follicles (Figure 16).

Several cell types have been described in the AHP of *Myxine*, some PAS+, AF+, and erythrosinophilic, as well as chromophobes. However, it is probable that two granulated cell types are basic, one of the proopiocortin cell line and the other of the GH/PRL cell line. Ultrastructurally one agranular and five granulated cell types have been described, of which the most abundant lies at the center of the cell clusters, closely bound together by desmosomes; granules are small, approximately 125 nm, and cytoplasmic organelles are sparse. A second cell type, away from the center of the cell clusters, has larger granules, 170–180 nm, with more developed RER, and cells with some granules, 110 nm, are situated peripherally. Two other cell types have intermediate-sized granules. Flat, polygonal, agranular cells form an epithelial net round the follicles; they interdigitate and are closely linked by desmosomes and their cytoplasm is rich in microfilaments. It seems probable that these are stellate cells concerned with transport of material to and from the granular cells, and they may be phagocytic. However, we know little either of the nature of the secretory material of myxinid pituitaries or of their physiological role, so that much more investigation is needed in these genera.

4.2. The Hypothalamus

The HT in both lampreys and hagfishes is a broad, flattened area at the base of the diencephalon and is composed mainly of small, poorly differentiated neurons. For hagfishes, the ventricles of the forebrain are much reduced, with apparently no communication between the rudimentary preoptic and infundibular recesses. Using classical AF staining methods, NS systems have been demonstrated in the HT of both hagfishes and lampreys. In the hagfish (Figure 16) the preoptic nucleus (NPO) lies dorsocaudal to the remnant of the preoptic recess. Two distinct parts can be recognized: the pars parvocellularis (NPOp), consisting of small cells that are nonstainable, and the more dorsocaudal pars magnocellularis (NPOm), with larger

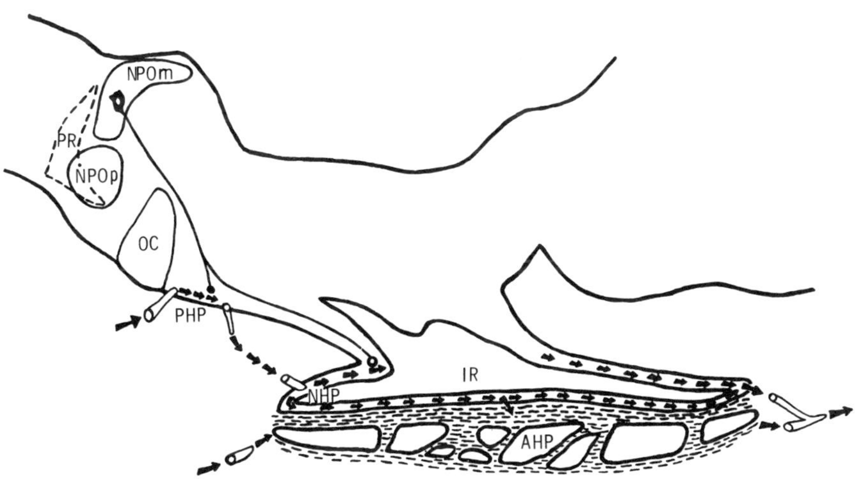

Figure 16. Hagfish hypothalamus and pituitary. Arrows represent the direction of blood flow. AHP, adenohypophysis; IR, infundibular recess; NHP, neurohypophysis; NPOm, NPOp, magnocellular and parvocellular preoptic nuclei; OC, optic chiasma; PHP, prehypophyseal plexus; PR, preoptic recess.

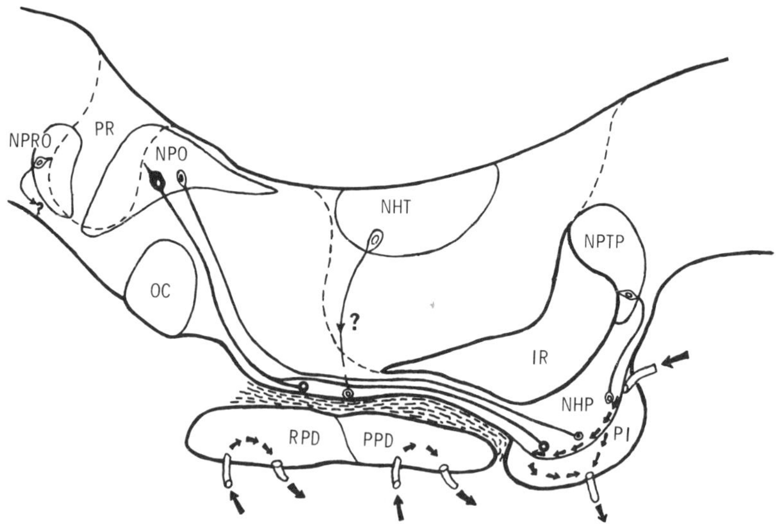

Figure 17. Hypothalamus and pituitary of the lamprey. IR, infundibular recess; NHT, dorsal hypothalamic/ventral thalamic nucleus; NPO, preoptic nucleus; NPRO, nucleus of the preoptic recess; NPTP, posterior paratubercular nucleus; PPD, proximal pars distalis (filled symbol), AF + neurons; (open symbol), AF − neurons. Arrows and other abbreviations as in Figure 16.

cells that stain well with AF. In several species of lampreys (Figure 17), the NPO lines the preoptic recess of the third ventricle, with some bipolar neuronal cell bodies interrupting the ependymal cell lining and projecting into the CSF and others buried more deeply among glial tissue. Most of these neurons have been shown by ICC to be vasotocinergic; i.e., they elaborate arginine vasotocin (AVT), although some neurons reacting with anti-LH-RH and anti-SRIF also occur in the subependymal layer. The axons of these neurons form diffuse tracts which eventually enter the NHP. AF-negative HT nuclei in cyclostomes are poorly differentiated. Monoaminergic systems have been demonstrated in a few species of lamprey, but not yet in the hagfishes. Scattered CSF-contacting catecholaminergic cells occur in front of the NPO, forming the *nucleus recessus preopticus* (NPRO), and a large diffuse subependymal sheet of monoaminergic neurons has been reported around the ventricles of the dorsal HT, but also extending into the posterior ventral HT and the ventral thalamus. The latter area has been termed the *n. hypothalamicus et thalamicus* (NHT). Scattered CSF-contacting monoaminergic cells also occur in a more caudal region, above the lateral and posterior recesses of the ventricle, the *n. paratubercularis posterioris* (NPTP)—a region that may correspond to the nucleus lateralis tuberis (NLT) or paraventricular organ of higher fishes.

4.3. Neurohypophysis

The cyclostomes generally appear to lack any neurohemal contact area on the hypothalamic floor directly comparable to the ME of other groups of vertebrates. In fact, both the hagfish and lamprey may illustrate an intermediate condition between the "brain bladder" of amphioxus and the vascular ME of the gnathostomes. The most striking feature of the "hypothalamohypophyseal" system in both groups is that the endocrine tissue of the PD (in lamprey) or AHP (in hagfish) is separated from the overlying infundibulum by a thick sheet of virtually avascular connective tissue.

In hagfishes, the fibers from the NPO follow two distinct routes (Figure 16). The first is to an area on the floor of the brain just behind the optic chiasma, where the fibers terminate on the capillaries of the prehypophyseal plexus. The fine structure of this neurohemal contact area is very similar to that of the tetrapod ME. From the prehypophyseal plexus blood is supplied via long portal vessels to the NHP. The second, more usual, route for NPO fibers is directly to the NHP itself, where most end in the dorsal wall, on capillaries of the secondary plexus: a dense network of vessels that drains into the general circulation. There are several different types of fiber ending, based on size of NS granules, and some of these may correspond to the peptidergic type A and aminergic type B fibers of higher vertebrates. The ventral floor and part of the anterodorsal wall of the hagfish NHP also contain fiber endings on capillaries, but in addition, directly above the AHP, there are both peptidergic and aminergic endings on the processes of ependymal cells which lie against the connective tissue layer between the AHP and NHP. The ependymal cell bodies lie in the wall of the infundibular cavity and appear to be secretory. This relationship between nerve endings and ependymal cells is reminiscent of that in the ME of tetrapods. If, indeed, this part of the NHP is a rudimentary ME, the question remains how does any hypophysiotropic material released here reach the AHP? In general there are no significant vascular connections between the two parts of the pituitary, which are supplied with blood from separate branches of the carotid artery. There are, however, species differences in this respect: in *Eptatretus burgeri* there are about 20 small blood vessels traversing the connective tissue layer (rudimentary portal vessels?), and in *Myxine* the AHP receives some blood from the prehypophyseal plexus. In other species with no such vascular links to the AHP, one must speculate that hypophysiotropic material released in the NHP reaches the endocrine cells either via the systemic circulation or by avascular transport across the connective tissue layer. There is now good experimental evidence that the latter can occur: small molecules such as horseradish peroxidase are absorbed by the NHP (or "ME") ependymal cells and transported along their processes across the connective tissue to the AHP.

The two separate lobes of the lamprey pituitary, the PD and the neurointermedia, have separate vascular supply and drainage (Figure 17). The majority of the nerve fibers from the NPO end in the NHP, which is only a slight thickening of the infundibular floor in contact with the PI. The NHP is a three-layered structure as in all vertebrates: dorsally ependymal cells, an inner fiber layer, and a ventral palisade layer, in which nerve endings abut the perivascular membrane of the blood vessels of the plexus intermedius, which completely separates the NHP from the PI. Some fibers also end on ependymal cell processes, which in turn form expanded end-feet against the perivascular membrane and a few pass beyond to end among the PI cells.

Most of the fibers appear to fall into two categories: type A, with 150–180 nm NSG, possibly containing the neurohypophyseal octapeptide arginine vasotocin (AVT), and the more rare type B, containing dense-cored vesicles (DCV), possibly aminergic. There is no neurohemal contact zone in lampreys corresponding to the prehypophyseal plexus of hagfish. Some fibers from the NPO appear to terminate on the floor of the HT anterior to the NHP, directly above the pituitary PD. This region has a remarkably ME-like structure, with both aminergic and peptidergic endings on ependymal cell processes and on the BM of the connective tissue layer overlying the PD. However, no blood vessels have been found linking this putative ME with the PD.

Thus in lampreys, as in hagfishes, hypophysiotropic material (if it exists!) would have to diffuse across the connective tissue in order to reach the AHP cells. An alternative hypothesis is that hypophysiotropic factors travel to the AHP via the systemic circulation. Indeed, this seems to be supported by the recent findings that the NPO of lamprey contains immunoreactive (ir-) LH-RH, which is distributed down axons to nerve endings, not in the putative ME, but in the NHP, where it is presumably released into the systemic circulation.

5. Elasmobranchs

The Elasmobranchs are jawed fishes whose skeleton is almost wholly cartilaginous and which evolved at an early stage, the Ordovician period, from the main evolutionary stem of jawed fishes. Today they form a very successful group and hence demonstrate both primitive and specialized characteristics. They can be divided into two main groups: the Selachii, including skates and rays, sharks and dogfishes, and the numerically much smaller group, the Holocephali, which includes the chimaeras, rat, and rabbit fishes.

5.1. Pituitary Morphology

The pituitary in all elasmobranchs is lobular, consisting of a rostral (RPD), proximal (or median) pars distalis (PPD), and neurointermediate lobe (NL + PI), together with an additional lobe, called the ventral lobe in selachians and the buccal lobe (rachendachhypophyse) in holocephalians (Figure 18). The ventral lobe of selachians is attached to the PD, but the buccal lobe is completely surrounded by cartilage and lymphoid tissue and has no attachment to the rest of the pituitary.

The AHP develops from Rathke's pouch, which remains hollow during development. The homologies of the various lobes have been the source of much controversy, but the development of the RPD from the anterior process of Rathke's pouch and the PPD from the aboral process indicates their correspondence to the cephalic and caudal lobes of the avian PD (Figure 15). However, within the elasmobranchs, the ventral and buccal lobes do not appear to be homologous because the ventral lobe develops from paired lateral outgrowths of Rathke's pouch, which bend toward

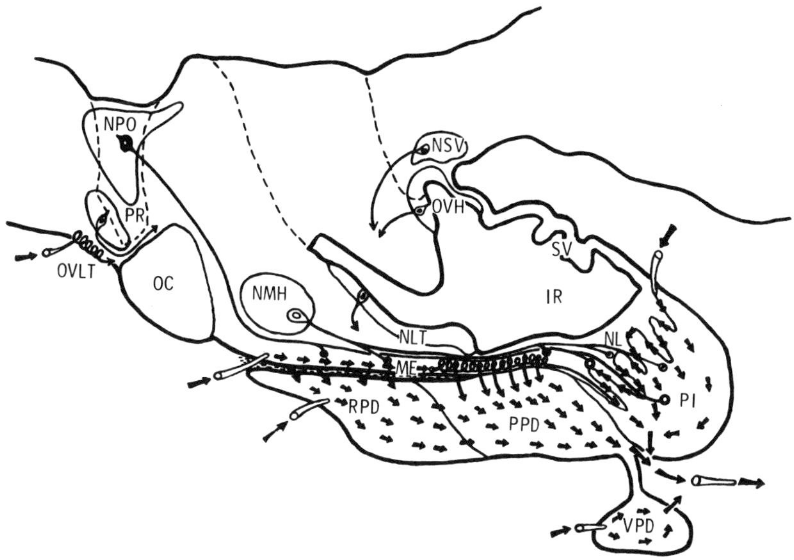

Figure 18. Hypothalamus and pituitary of the dogfish. ME, median eminence; NL, neural lobe; NLT, lateral tuberal nucleus; NMH, medial hypothalamic nucleus; NPO, preoptic nucleus; NSV, saccus vasculosus nucleus; OVH, hypothalamic vascular organ; OVLT, vascular organ of the lamina terminalis; PR, preoptic recess; SV, saccus vasculosus; VPD, ventral lobe of the pars distalis.

each other and fuse beneath the gland, whereas the buccal lobe apparently forms as a single entity from the oral region of the pouch. The ventral lobe appears to be a more likely homolog of the tetrapod pars tuberalis, and the buccal lobe is related to the buccal part of the pituitary of *Latimeria* (Crossopterygii, Section 8.2).

In sharks and dogfishes the PD is hollow (squaloid type); in the rostral region extensions of the hypophyseal cavity form vesicles and tubules, and posteriorly the proximal zone consists of folds of epithelial tissue around a large cavity. Typically in the skates and rays the PD is batoid; that is, the hypophyseal cavity is virtually obliterated and the entire PD is a mass of cords and clusters of cells. The PD of holocephalians is always hollow. The selachian ventral lobe too is hollow, and the remnant of the hypophyseal cleft in the holocephalian buccal lobe occurs as cavities in follicles formed by secretory cells. These follicular cavities contain colloid, which is PAS+, AF+, and AB+ and which is apparently secreted by chromophobic cells lining the follicles. These cells resemble stomodeal epithelial mucus-secreting cells and may be homologous with stellate cells of the AHP of most vertebrates and ductal and covering cells of primitive actinopterygian fishes.

Definitive identification of all the cell types of the elasmobranch pituitary has not been made, as yet. The standard procedures of ablation and replacement by tissue extracts, together with bioassays in heterologous systems, have indicated that gonadotropic activity occurs in the ventral lobe of selachian pituitaries and in the buccal lobe of holocephalians. Both ACTH and PRL activity have been demonstrated in extracts of RPD, and TSH activity in the ventral lobe of selachians. Immunocytochemistry confirmed the presence of ACTH in the RPD and gonadotropin in the ventral lobe. Ultrastructurally the GtH cells have very few granules, 100–200 nm diameter, and correspond to cells described as chromophobes by LM in *Torpedo*. Detailed EM studies of the spotted dogfish (*Scyllium canicula*) show two granulated cell types in the RPD, the more numerous having large secretory granules, 100–350 nm, concentrated near the capillary pole of the cell, and cells with small granules, 100 $\pm$ 40 nm, scattered evenly throughout the cell. Agranular supporting cells with high nuclear/cytoplasm ratio are moderately numerous. The PPD also contains two granulated cells; the major one shows extensive development of RER and Golgi, with large granules, 200–430 nm. The cells with small granules are similar to those in the RPD, and there are also agranular, stellate cells. In the ventral lobe the most abundant cells have small granules, 80–180 nm, and sparse RER; they probably secrete TSH. Cells with large granules, 180–250 nm, are much less numerous and are generally found near the pericapillary space. An unusual feature in newly hatched dogfish is the presence of giant cells which appear undifferentiated and stellate. The cavity of the lobe is bordered by either granular or stellate cells.

Tissue from the neural lobe of the NHP interdigitates with PI cells to form a composite neurointermedia (Figure 18). The degree of penetration of the PI cells by neural tissue is variable, the development of cords of intermedia cells is slight in some sharks (*Squalus* and *Etmopterus*) but extensive in the dogfish (*Scyliorhinus*) and rays (*Torpedo* and *Raia*). The cords of PI cells may be arranged in distinct lobules, separated by highly vascular connective tissue, or they may form a parenchymatous mass penetrated by an irregular blood plexus. The PI cells secrete MSH; the size and intimate association between the neural lobe and the PI indicate the importance of MSH in elasmobranch physiology. By LM there appears to be a single cell type which is weakly PAS+ but which also stains with OG and azocarmine. In the newly hatched dogfish these cells have been shown to contain large secretory granules, 320–340 nm, but there are a few other cells with small, 100-nm granules evenly distributed throughout the cell. The lobe has a cavity in larval dogfish, but this has been obliterated in newly hatched fish, and there are apparently no stellate cells. Thus, cells secreting PRL and GH have still to be identified.

5.2. Hypothalamus

Surprisingly little is known of the structure of the HT of either the selachians or holocephalians in view of the evolutionary importance of this group. Most studies have concentrated on the histology of the classical NS system and the vascularization of the hypothalamohypophyseal system. The NPO is a triangular sheet of bipolar AF+ neurons lying above the optic chiasma (Figure 18), and there appears to be no differentiation into magnocellular and parvocellular regions. Using the *in situ* AF staining technique the axons from the NPO can be seen to take a posterior path through the HT and then to turn ventromedially to form a single tract along the floor of the third ventricle which expands to form the NHP. There appear to be two types of axons from the NPO, on the basis of size of the NS granules: type A1 (NSG 180 nm), which pass through the NHP and into the PI, and type A2 (130 nm), which form endings mainly in the ME. Several AF-negative nuclei have also been recognized in the elasmobranch HT. The *organum vasculosum preopticum* consists of CSF-contacting catecholaminergic neurons in the wall of the preoptic recess. The *n. lateralis tuberis* (NLT) lies in the wall of the infundibular recess directly above the ME. Originally it was believed to be aminergic, but more recent work suggests that the NLT is an AF-ve peptidergic nucleus, supplying fibers to the ME. Another tuberal nucleus anterior to the NLT, the *n. medius hypothalami* (NMH), is serotonergic in dogfish, and this supplies fibers to the PI. Other aminergic centers are found in the posterior hypothalamus (*n. lobi inferioris* (NLI), *n. sacci vasculosi* NSV), and *organum vasculosum hypothalami* (OVH) and are probably homologous to the paraventricular organ (PVO) of tetrapods.

5.3. Neurohypophysis and Portal Blood System

The bulk of the AF+ fibers from the NPO terminate in the neural lobe of the NHP; however, this is an extremely variable structure in elasmobranchs. The most primitive condition is seen in the sharks, which have a discrete neural lobe interdigitating with the PI with the usual three-layered arrangement (ependyma, fibers, palisade). In the ventral palisade zone the type A fibers, of which there seem to be two subtypes with NSG 180 nm and 130 nm, and aminergic type B fibers (DCV $<$ 100 nm) end in contact with the septum enclosing the plexus intermedius or pass beyond and innervate the PI cells. The PI receives blood via capillary branches from the plexus intermedius. At the other extreme, there can be total fusion of neural and intermediate tissue (the neurointermediate lobe) as seen in skates and rays and the holocephalians, with the cords of PI cells being extensively penetrated by blood vessels and both type A and type B nerve endings. Directly above the elasmobranch neural lobe, the dorsal wall of the infundibulum forms the saccus vasculosus (Figure 18), a structure of unknown significance found in association with the NHP in many other groups of fishes.

The elasmobranch ME is differentiated into anterior and posterior parts supplying the rostral and medial lobes of the PD, respectively (Figure 18). In the posterior part, at least, there are the usual three layers. Dorsally the infundibular recess is lined by ependyma and tanycytes (secretory ependymal cells with "stretched-out" processes that run down to the capillary loops). The middle layer consists of AF+ fibers en route to the NHP, and the ventral palisade layer of nerve endings (AF+, AF−, and aminergic) on fenestrated capillaries, surrounded by supporting glial cells. Nevertheless, several features of the elasmobranch ME may be considered as "primitive." The ME is separated from the PD by a connective tissue lamina, although this is thin and discontinuous, and in the rostral part of the ME the capillaries are continuous with those of the rostral PD (i.e., no intervening portal vessels as in the caudal ME), and the capillary loops mostly do not penetrate into the palisade layer. Additional to those portal vessels supplying the PD dorsal lobe, a few also convey blood from the ME to the neurointermedia in several species. The dorsal lobe also receives a separate arterial supply from the carotid, and the relative importance of portal/arterial blood supply in different areas of the dorsal PD varies in different species.

An anomaly of the vascular system in elasmobranchs is that there is no portal blood supply to the ventral lobe of the pituitary, or to its counterpart, the "rachendachhypophyse" in holocephalians. As there is no convincing evidence of direct innervation to any part of the PD, the only possible route for HT control of this lobe is via the systemic circulation.

6. Primitive Bony Fishes

These bony fishes, often referred to as "ganoids," are the surviving remnants of the earliest groups that developed jaws. It is likely that they display both primitive and specialized characteristics. All belong to the class Actinopterygii, the ray-finned fishes, divided into the Chondrostei, which includes the Paleoniscoidei and Acipenseroidei, and the more recently evolved Holostei. Very little is known of the structure or function of the hypothalamohypophyseal system in any of these species, which is surprising in view of their evolutionary importance. Nevertheless, much of what is known is very interesting and reflects the evolutionary origins of the group.

6.1. Chondrostei

6.1.1. Adenohypophysis

The two surviving genera of Paleoniscoidei, the African freshwater species of *Polypterus* and *Calamoichthys,* have been extensively studied. The pituitary of these primitive fishes retains the cavity of Rathke's pouch in the anterior region of the gland (Figure 15), moreover linked by the persistent duct, the orohypophyseal duct, to the buccal cavity. Embryologically the AHP anlage develops as a solid bud of cells, with the cavity and duct developing later. The cavity ramifies through the anteroventral part of the pituitary, forming follicles; this area of the AHP is probably the homolog of the RPD of teleosts. The whole of the cavity and the duct are lined by mucus-secreting cells, and it seems probable that these are related to the stellate cell system of higher vertebrates. The colloid filling the cavities is PAS+ and AB+ as in all fishes. The AHP in *Polypterus* and *Calamoichthys* is unusual in that the orohypophyseal duct is retained in the adult, but the anterior PD, homologous to the rostral pars distalis of teleosts, is not follicular in organization.

The cell types of these pituitaries have been identified by standard staining techniques as well as ICC. It is interesting that in *Calamoichthys* the PRL cells, identified by immunostaining with antiserum to mammalian PRL, are not concentrated in the RPD but lie scattered throughout the PD. Identification of PRL cells by this technique showed them to be distinct from the PAS+ cells of the duct epithelium, which does not support the suggestion that this is their origin. Basophilic cells, which are PAS+, AB+, but AF−, also occur in the rostral region. The PD contains a small acidophil and two basophilic cells, one of which is strongly PAS+ and AF−, the other being weakly PAS+ but also AF+. All the acidophils react with antiserum to mammalian GH, and some of them also react with antimammalian PRL serum. The type I basophil is probably a thyrotrop and reacts with antiserum to the β subunit of mammalian TSH. Ultrastructurally, the second type of basophil resembles gonadotrops of teleosts and reacts with antisera to β subunits of mammalian LH and FSH (Section 13.1). A third weakly basophilic, PbH+ cell, lying close to the lumen of the AHP cavity, reacts with antiserum to ACTH. The PI contains only one secretory cell type, which is PAS+ and PbH+. These cells are arranged radially round the extensions of the neural lobe, but separated from its cells by a double BM enclosing an extravascular space.

Two genera of Acipenseroidei are extant, *Polyodon* and *Acipenser,* the sturgeon, of which only *Acipenser* has been studied in detail, mainly because of its commercial importance as the source of caviar. The AHP develops from a solid anlage, and by splitting, so forming a schizocoel, a large hypophyseal cleft is developed, which almost completely separates the PD from the PI and also divides the PD into dorsal and ventral regions (Figure 15). There is no orohypophyseal duct, but the hypophyseal cavity ramifies throughout the PD, probably all interconnected in the embryo, but ducts close in the adult leaving isolated vesicles. These vesicles are probably homologous with the follicles of holostean and primitive teleostean pituitaries. The cells surrounding the vesicles are acidophilic and appear to secrete PRL, detected by bioassay and by immunostaining with antiserum to ovine PRL. It is interesting that PRL cells have also been located, by ICC, in the PPD, scattered among the GH cells and basophils. This dual location of PRL cells, in both the RPD and PPD, is an intermediated condition between the organization found in teleosts and chondrosteans. Bio- and immunoassayable GH has been extracted from sturgeon pituitaries, probably originating from the acidophils of the dorsal region of the PPD, but in *A. güldstaeti* the amphiphilic cells of this region stain with antiserum to bovine GH. Basophils of the dorsal and ventral regions of the PPD become active coincidentally with increased gonadal growth. ICC demonstrates that the ventral basophils secrete gonadotropins, either both FSH and LH or LH alone, whereas TSH cells are scattered throughout the PPD.

ACTH activity and immunoreactivity have been detected in the RPD, in cells corresponding to the amphiphils or chromophobes.

The PI is large, consisting of basophils mainly, with few chromophobes and acidophils, arranged in cords. Cells of the PI that stain with PbH are associated with corticotropic activity. There is close association between the neural lobe and PI, as deep hollow diverticula of the NL penetrate between the cords of PI cells. A highly vascular double BM separates NL and PI cells, and the regional arrangement of saccus vasculosus, infundibular recess, NL cavity, and PI is similar to that seen in polypterides. The PD receives no direct innervation or processes of the NL.

6.1.2. Hypothalamus and Neurohypophysis

In the Paleoniscoidei, HT nuclei take the form of primitive periventricular plates of cells, although a distinct NPO is present, with rather diffuse AF+ tracts passing via the ME region into the NHP. The ME has a typical tetrapod structure with a tendency to rostrocaudal subdivision, as in all ganoids. Nearly all the nerve endings on the primary portal capillaries are aminergic (type B). No distinct NLT is present, and no aminergic nucleus that could be the source of the type B fibers in the ME has yet been found. Both an NPO (AF+) and a compact NLT (AF− and some AF+) lie in the usual positions in sturgeon (Figure 19), and in the absence of direct innervation of the adenohypophysis, the peptidergic type A fibers (two categories) and the aminergic type B fibers form endings in either the neural lobe or the ME. The ME of sturgeon is peculiar in that the neurohemal contact zone lies outside the main nervous bulk. The capillary loops do not penetrate the palisade layer but lie superficially within the flat sheet of connective tissue that separates the pars distalis from the infundibular floor. Here the primary capillary network receives numerous nerve endings, mostly of type B, which are clearly synaptic on the pericapillary BM and separated from the fenestrated capillary endothelium by a wide, collagenous pericapillary space. From this primary capillary plexus very short portal vessels pass down into the underlying PD, so that, in effect, primary and secondary plexi are in direct continuity.

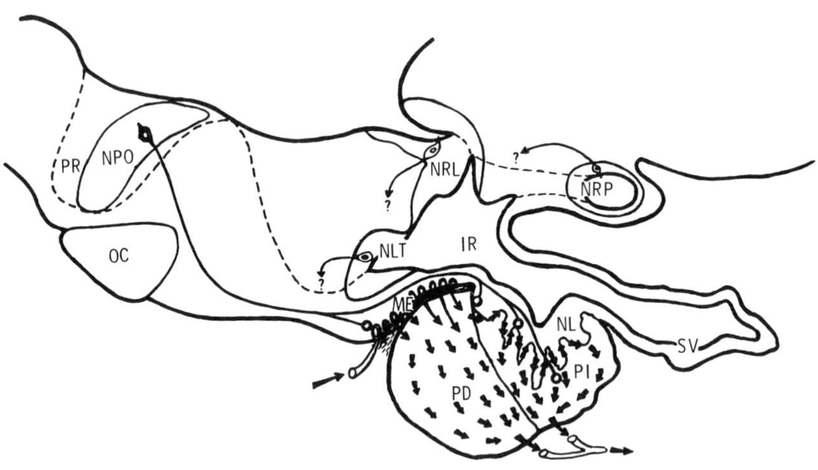

Figure 19. Generalized hypothalamus and pituitary of ganoid fishes. NRL, NRP, nuclei of the lateral and posterior recesses of the third ventricle; IR, infundibular recess; NPO, preoptic nucleus; NLT, nucleus lateralis tuberis; NL, neural lobe; ME, median eminence; OC, optic chiasma; PR, preoptic recess; PI, pars intermedia; PD, pars distalis; SV, saccus vasculosus.

The neural lobe has the same basic structure in all the ganoid fishes and has complex relationships with the PI and the saccus vasculosus (SV), which is always well developed (Figure 19). The thin floor of the infundibular recess (IR) has become extensively folded posteriorly to produce long neurohypophyseal processes down into the PI. The processes are hollow, penetrated by diverticuli of the IR, but become solid toward their distal ends. The diverticuli are lined by ependymal cells and, in *Polypterus* at least, by some cells that resemble the secretory "crown cells" which line the SV. Beneath the lining lie ependymal processes and nerve fibers of both type B (catecholaminergic) and type A (from the NPO tract). These fibers may end in contact with ependymal processes or on perivascular spaces of the capillaries. Extending all around the NHP processes and separating them from the PI cells is a double BM, which is highly vascular and often greatly distended, containing the plexus intermedius. This is equivalent to the simple neurointermediate septum of, on one hand, the cyclostomes and, on the other, the tetrapods. Type A and type B endings believed to control PI activity contact this BM opposite the PI cells.

6.2. Holostei

The holosteans were the dominant fishes of the Jurassic and Cretaceous periods and from them the teleosts have evolved. Only two representatives survive today, both inhabitants of fresh water in North America; *Amia,* the bowfin, and *Lepisosteus,* the gar-pike. As with *Polypterus* and *Calamoichthys,* it is probable that both these holosteans have always inhabited fresh water, although the majority of the group became marine. Because of their relationship to teleosts and their primitive characteristics they are particularly interesting.

6.2.1. Cytology of the Holostean Adenohypophysis

The AHP develops as a solid anlage; a schizocoel forms only transiently, so that in the adult there is neither hypophyseal cavity nor duct. The pituitary is attached along most of its length to the floor of the infundibulum with vascular connections between. Posteriorly the cavity of the infundibular recess connects with the cavities of the NL and the remarkably long saccus vasculosus. A rostral region of the PD can be distinguished from the proximal zone by the follicular arrangement of the cells around vesicles, which represent remnants of the diverticula from the hypophyseal cleft. In *Amia* the most abundant cell type is an acidophil which is columnar, containing 250–300 nm granules, and shows other features of very active cells—numerous mitochondria, well-developed RER, and large Golgi. PRL activity has been found in this region of the pituitary, and antisera to mammalian and teleostean PRL bind to these cells. Chromophobic cells, fewer in number, are found closely associated with the BM of the follicles, but ultrastructurally they can be seen to have some granules, 200–250 nm, and whorls of microfibrils in the cytoplasm and probably correspond to the ACTH cells identified by ICC. There are also a few basophils in the RPD, with larger (600 nm) granules, which are similar to the gonadotrops of the PPD. Nongranulated cells line the follicles, joined together by desmosomal junctions and sometimes extending down between the secretory cells to the BM.

Cells of the PPD are arranged in cords vertically, with acidophils, immunoreactive to anti-GH serum, predominating in the dorsal region; basophils are most abundant ventrally. The GH cells contain numerous moderately large, 270–300 nm granules, and of the two types of basophils, one resembles that found in the RPD, probably a gonadotrop, and the other is sparsely granulated by small, 130–170 nm granules found near the cell periphery: this is probably the thyrotrop. Nongranulated stellate cells form a network between the secretory cells. Extensive branching processes of the neural lobe ramify throughout the PI. There are two cell types in the PI, the most numerous is club-shaped and PbH+, with processes contacting the PI/PN membrane; the second cell is small, rounded, and PAS+. The pituitary of *Lepisosteus* is essentially similar.

6.2.2. Hypothalamus and Neurohypophysis

In the holosteans *Lepisosteus* and *Amia* the NPO is divisible into NPOp and NPOm, and the rather diffuse NLT is also composed of several distinct regions.

The posterior and lateral recesses of the third ventricle are lined by AF− neurons (c.f. PVO of higher vertebrates), but it is not known whether these are aminergic in nature.

AF+ and AF− fibers from the HT pass down into the NHP, which interdigitates with the PI, and a few fibers may penetrate into the PI and directly contact the endocrine cells. It also appears that in *Amia* a few processes from the NHP penetrate into the PPD, where the type A and type B fibers occasionally form endings on the intervascular space adjacent to the PPD endocrine cells. Here we see the beginnings of neurohypophyseal penetration of the pars distalis and direct innervation of the PD cells so characteristic of the teleosts. In spite of this, and in contrast to the teleosts, a well-developed ME is retained in the modern holosteans. The vascularization of the ME is typical; the capillary loops are just as elaborate as in mammals, and from them short portal vessels pass down into the rostral PD, and longer portal vessels into the proximal PD. The thick connective tissue sheet present between the PD and ME in other ganoids is not seen in *Amia,* and with the EM, each fenestrated capillary within the ME is seen to be surrounded by a narrow perivascular space which receives numerous synaptic contacts from exclusively type B nerve endings. The structure of the neural lobe is essentially the same as in the Chondrostei.

7. Teleosts

Teleosts, the bony fishes, evolved from holostean stock and have proved to be extremely successful, so that by now there are more than 20,000 species. During evolution of all fishes the tendency has been to produce forms lighter and more agile, by reduction of bony plates to small, thin scales, by reduction of the tail, and, in teleosts, by the development of the swim bladder. Metabolically these features have enabled fish to live more economically, and as a result, a greater proportion of their energy could be devoted to reproduction (see Chapter 7). In the relatively unchanging, nutritionally rich seawater environment such developments must have given impetus to successful radiation of early teleosts. The early forms were all marine, and obviously physiological modifications were essential for successful reinvasion of freshwater, although fossil evidence, or rather its lack, suggests that Salmonids, for example, may have evolved in freshwater. Some of the necessary physiological modifications occurred in pituitary hormone utilization and will be discussed in Section 14.1.

The most primitive living teleost is considered to be the tarpon (Elopidae), but the herrings (Clupeids) are close to the basic stock. The salmon–trout group is relatively primitive, and the eels (Apoda) diverged early from the primitive stock and, although specialized, have retained primitive features. The most advanced group of teleosts have evolved spiny fins (Acanthopterygii); the majority of these are marine, but some are freshwater inhabitants. In spite of adaptations to various aquatic, and partially terrestrial, habitats, pituitary hormone physiology and the basic morphology of the pituitary are remarkably similar in all species. Unlike all the other vertebrates, pituitary cells in teleosts receive direct innervation from the hypothalamus, and there is no portal vascular system. These developments are related to the absence of a distinct median eminence, which appears to have been incorporated into the anterior neurohypophysis. The concentration of cell types in zones of the AHP has helped considerably in studies of the physiology and function of the various cells. Detailed studies have only been made on relatively few species, and it is intended to refer to two broad categories, the primitive type of morphology of eels and the salmon/trout group and the more advanced form exemplified by *Poecilia latipinna,* the molly, *Leuciscus rutilis,* the roach, and *Oreochromis (Sarotherodon) mossambicus,* tilapia.

7.1. Morphology and Cytology of the Adenohypophysis

The gland is in two main portions. Posteriorly the PI makes intimate contact with the NL, usually by numerous complex interdigitations, and in some species, e.g., the trout, may constitute half or two-thirds of the total gland; however, in *Poecilia* and other cyprinodonts it is the smallest region. Anteriorly the PD is divided clearly into a rostral (RPD) and proximal (PPD) zone, both of which contact the NHP (Figure 21).

7.1.1. Pars Distalis

The main mass of cells in the RPD is composed of PRL-secreting cells, which in primitive species are arranged in follicles (Figure 5). Intercalated between the PRL cells and lining the follicle lumen are chromophobic cells, which have no secretory granules. These follicular cells are usually bound tightly together by complex desmosomal junctions permitting only processes of PRL cells to contact the luminal fluid. The lumen of the follicle may contain granules, mitochondria, ER, and Golgi membranes probably derived from breakdown of PRL cells. In the Isospondyli each follicle lumen is an extension of the orohypophyseal duct, which in larval stages opens into the pharynx, but is closed in adults. In salmonids the lumen is derived *de novo* as a schizocoel during development, and in some clupeids the orohypophyseal duct may remain patent in adult fish. The PRL cells are generally columnar, and their basal pole sits on a double-layered BM, which may contact capillaries and nerve fibers of the NHP. In the primitive species nerve fibers do not penetrate the BM into the RPD follicles. In species of teleosts considered to be more advanced, the PRL cells form a compact mass, as in *Poecilia, Leuciscus,* and *Sarotherodon,* never follicles. The cells also sit on a BM, and sometimes nerve fibers from the NHP run between the cells and synapse directly on the PRL cells but not directly with the ACTH cells (Figure 20).

Prolactin cells have been identified by their erythrosinophilia and by cross-reaction with antisera to mammalian PRL. Ultrastructurally PRL cells have a large nucleus, moderately dense cytoplasm with well-developed, organized RER, elongated mitochondria (''hank of wool'' form), cytoplasmic microtubules and microfilaments, and numerous electron-dense hormone secretory granules, which are generally round or slightly elongated, 200–300 nm (Figure 5). Follicular cells have a dense nucleus and long cytoplasmic extensions running between the PRL cells. At the border of the lumen in primitive teleost RPD, both follicle cells and PRL cells are bound together by complex tight junctions and both have microvilli and cilia projecting into the follicle lumen.

ACTH cells are usually located in the RPD too, close to the NHP border (Figures 5 and 20). The cells are small and may stain with AB or azocarmine, but in some species, e.g., *Poecilia,* they are chromophobic. In the eel they are well granulated and take up AB intensely. At EM level the cells can be seen to have cytoplasm and nuclei of even density, usually relatively sparse RER, poorly developed Golgi, with small, 150-nm granules.

In some teleosts (eels, clupeids, and cyprinids), TSH cells are also located in the RPD, usually between the follicles of PRL cells but sometimes between PRL cells of the follicle wall. In cichlids and salmonoids, however, TSH cells lie between the RPD and PPD, and in *Poecilia* and other cyprinodonts they are located within the PPD. Though their position may vary, their morphology and staining properties are uniform; the cells are small and angular with basophilic, PAS+ cytoplasm. Granule size is very variable even within the same pituitary in some species, e.g., trout, and they are between 100 and 800 nm (Figure 8).

Within the PPD of all teleosts GH and GtH cells are arranged in cords. GH cells are polygonal with acidophilic cytoplasm and stain selectively with orange G (OG+). In immature teleosts GH cells are by far the most numerous cell type, but with onset of sexual maturity GtH cells increase in size and number. Ultrastructurally GH cells are well granulated, with mature secretory granules, 400 nm, well-

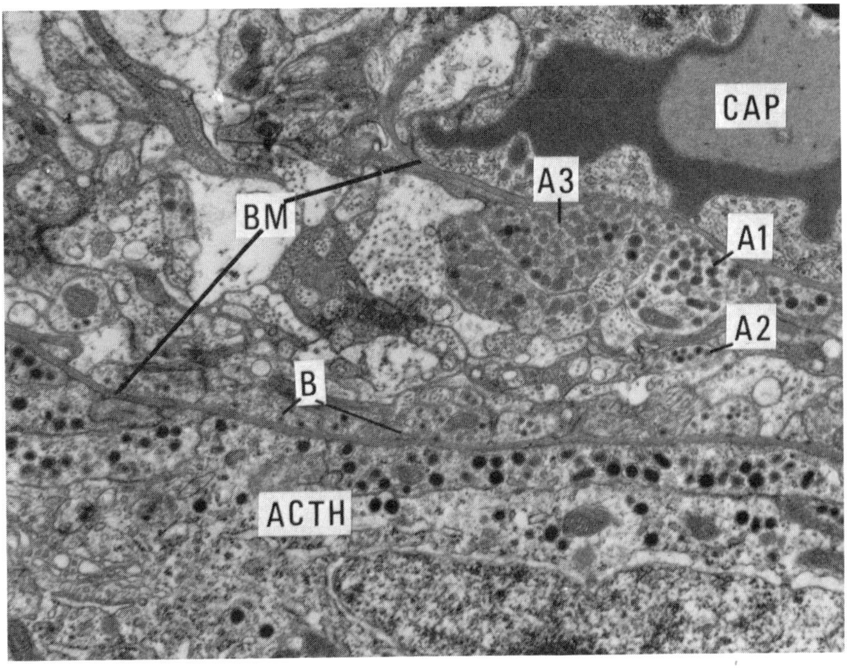

Figure 20. Relationships between different peptidergic (A1, A1, A3) and aminergic (B) neurosecretory fiber types, blood capillaries (CAP), basement membrane (BM) of the neurohypophysis, and secretory cells of the pars distalis in the molly, *Poecilia latipinna*. ACTH, adrenocorticotrop.

organized RER, and a nucleus frequently elongated or indented (Figure 7). GtH cells in immature teleosts show relatively sparse granulation; the granules themselves are small, 150 nm, and there is little organized RER and only a small Golgi. During sexual maturation in both male and female teleosts the GtH cells show considerable change in structure; granulation and granule size increase considerably, there is extensive development of RER, which becomes organized into parallel arrays, and the Golgi enlarges. Frequently the RER cisternae become greatly dilated and filled with flocculent material (Figure 10). GtH cells are strongly basophilic and PAS+ owing to the glycoprotein hormone stored in the secretory granules. The problem of whether or not there are two types of GtH cell, as in mammals, has not been satisfactorily decided, but it is more likely that there is only one cell type secreting a single gonadotropin, and that differences between cells reflect their states of activity (see Section 14.2.5 and Chapter 7). Ramifying between the GH and GtH cells are extensions of stellate cells forming a network separating the secretory cells from each other. Projections of the NHP and fibers of the NHP abut the BM of the PPD cells and make synaptoid connections, but only penetrate between the secretory cells to a limited extent in a few species.

7.1.2. Pars Intermedia

In all species the PI is closely apposed to the posterior region of the NHP and is usually invaded by neurohypophyseal tracts. There are generally two cell types distinguished by their staining properties; one of them is PAS+ and normally abuts the NHP, and the other is PbH+. Sometimes by LM the cells appear chromophobic, as in the trout, in which there appear to be no PAS+ cells; however, by EM two cell types are clearly distinguishable on the basis of their granulation: the few chromophobic (PAS+) cells have few tiny granules, 100 nm, well-organized RER, and polymorphic nuclei. This contrasts with the heavily granulated appearance of the PAS+ cells in some other species (Figure 12). The PbH+ cells have larger granules with more diffuse cores shrunken away from the membranes; the nucleus is frequently a bizarre shape and the abundant RER often dilated (Figure 12).

7.2. Hypothalamus

Many different hypothalamic nuclei have been described in a multitude of teleost species; however, only three (NPO, NLT, PVO) have been studied in detail in a representative range of species and have been implicated in hypothalamic control of the pituitary. The NPO is an arc-shaped nucleus lining the third ventricle above the optic chiasma (Figure 21). Dorsally it consists of large AF+ neurons (pars magnocellularis) and ventrally much smaller, less densely staining or AF− neurons. Many of these neurons are bipolar or multipolar, with short processes, often ciliated, projecting between the ependymal cell lining into the ventricle. Observed with the EM, the magnocellular NPO neurons contain an extensive system of parallel RER, usually in the form of a "polar" cap between the nucleus and the ventricle, and numerous dense NSG (Figure 22). The cell bodies receive numerous synaptic contacts from fibers that appear to be aminergic. In some species of teleosts the NPOm neurons have been classified into two categories: one type with NSG of ca. 200 nm, sometimes described as "dark cells," and a second type with NSG ca. 150 nm, or "light cells." There is now some evidence that these correspond to the separate vasotocin (AVT)- and isotocin (IT)-producing cells identified with ICC methods. The axonal processes of the NPO neurons run, at first laterally and then ventrally, away from the nucleus in diffuse tracts, before converging in the midline into a single tract which runs back through the NLT region and then expands to form a large part of the NHP.

Other AF− nuclei have been described in the region of the preoptic recess (Figure 21). The *n. anterioris recessus preopticus* (NAR) lies in front of the NPO in the floor of the recess, and in some teleosts this area contains monoaminergic cells and is well vascularized, showing structural resemblances to the vascular organ of the *lamina terminalis* (OVLT) of tetrapods and the *organum vasculosum preopticum* of cyclostomes and elasmobranchs. Caudal to the preoptic recess is the anterior periventricular nucleus (NAP), consisting of both periventricular and lateral AF− neurons: the latter part receives monoaminergic innervation and may corre-

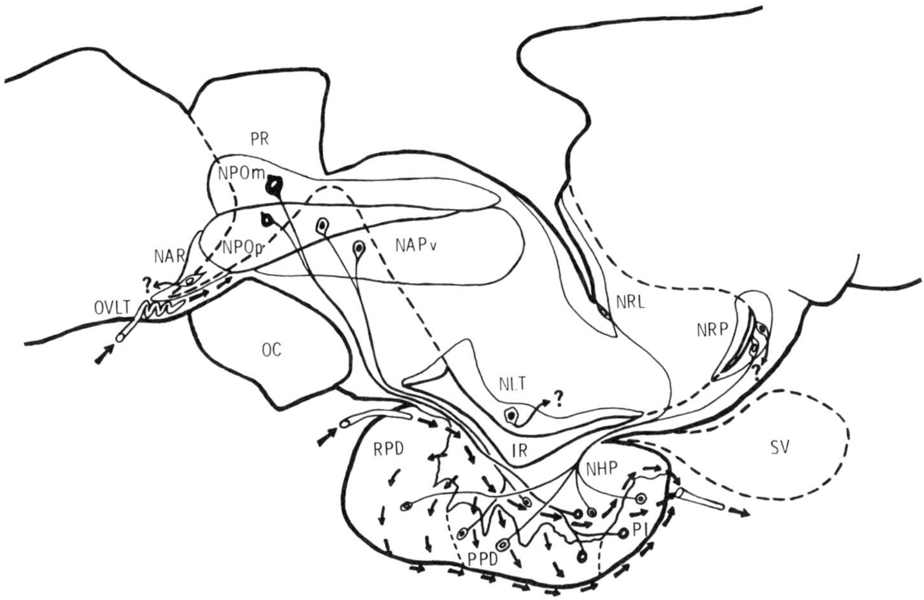

Figure 21. Hypothalamus and pituitary of a teleost. NAPv, ventral part of the anterior periventricular nucleus; NAR, nucleus of the anterior preoptic recess.

spond to the *n. medius hypothalami* of elasmobranchs. ICC studies on several species show that some of the AF− neurons of the preoptic region (NPOp, NAP) may produce pituitary releasing factors; so far LH-RH, CRF, and SRIF have been localized in this area.

The tuberal region of the teleost HT contains many small AF− neurons, which are usually considered to be organized into three nuclear areas (Figure 21). The anterior tuberal nucleus and posterior tuberal nucleus are separated by the lateral recess of the third ventricle and are "migrated" nuclei (Section 2.2). The NLT lies in the basal tuberal HT, directly above the pituitary, and consists of mainly periventricular neurons around the anterior and lateral walls of the IR. In most species the NLT has been described as AF− throughout; however, in a few species at least some cells appear to be AF+. The NLT can be divided, on the basis of neuron ultrastructure, into several distinct regions, although again these seem to vary in different species. A fairly typical regionalization of the NLT occurs in the molly, *Poecilia latipinna*. The NLT pars anterioris (NLTa) comprises several rows of small neurons lining the dorsolateral walls of the anterior infundibular recess, above the pituitary PD. More posteriorly, above the PI, the NLT widens into the pars posterioris (NLTp), and behind the PI it narrows again into the pars inferioris (NLTi). The appearance of the neurons in each of these three regions is broadly similar; small, unipolar, with RER and Golgi apparatus usually poorly developed. Granular inclusions are rarely present, but a few dense-cored vesicles (DCV) ca. 80 nm are

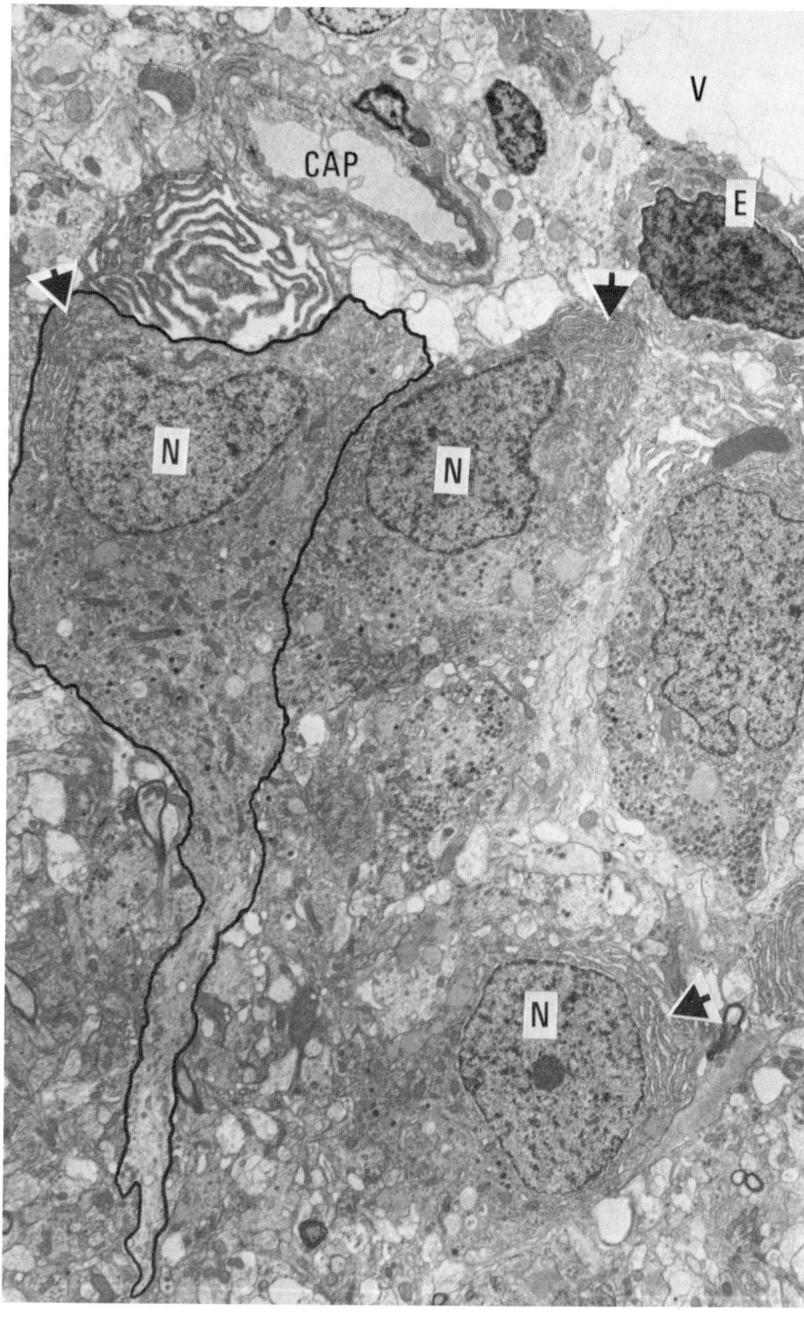

Figure 22. Perikarya in the preoptic nucleus of the teleost, *Poecilia latipinna,* showing the typical ultrastructure of peptidergic neurons. Note the polar cap of RER (➧) around the nucleus (N) and the long axonal process extending out from one cell (outlined) into the neuropil; V, ventricle; E, ependymal cell; CAP, capillary. (×4800.)

sometimes seen in the NLTp cells. Lateral to the NLTa is the NLT pars lateralis (NLTl), composed of scattered "migrated" groups of large cell bodies in the infundibular floor, above the sides of the PD, and often projecting into the submeningeal space. The NLTl neurons resemble those of the NPOm, with extensive RER and ca. 150 nm NSG. A few similar migrated cell bodies also occur anterior to the NLTa, among the neuropil of the horizontal commissure, this area being defined as the NLT pars rostralis (NLTr). Early work on the NLT suggested that the neurons were aminergic; however, this has now been disproved for several species, any aminergic content in this area being attributed to fibers from other parts of the brain passing between and ending on the NLT cells. It seems likely that the NLTl at least contains NSG of a peptidergic nature, although because of the poor stainability of these neurons, tracts from here into the pituitary cannot be followed. Peptide localization by ICC has been reported in the NLT of a few species: SRIF in the NLTa, LH-RH in the NLTp, and melanin-concentrating hormone in the NLTl.

Proven aminergic nuclei have been demonstrated in various parts of the teleost HT; however, because of the different nomenclature adopted by different authors, it is difficult to make comparisons. Most of these areas lie in the posterior HT around the recesses of the third ventricle and probably correspond to the paraventricular organ (PVO) of tetrapods. In the molly there are two paired nuclei showing strong monoamine fluorescence: the *n. recessus posterioris* (NRP) and part of the *n. recessus lateralis*. The latter areas have been termed the *pars ventralis* of the posterior periventricular nucleus (nPPv) in other species. Both the NRP and NRL give rise to fluorescent tracts in the direction of the NLT and pituitary. EM observations show that the NRL and NRP are composed of spindle-shaped bipolar neurons, with small cell bodies, ciliated apical dendrites projecting into the ventricles, and long axonal processes leading away from the nucleus in an anteroventral direction. Cell body, dendrite, and axon each contain numerous DCV ca. 80 nm. In some species, spectral analysis of the fluorescence (Section 16.4) and immunocytochemistry, have detected dopamine, noradrenaline, and 5-hydroxytryptamine (serotonin) in separate cells of the PVO.

7.3. Vascular Supply

Despite the sporadic appearance of reports to the contrary, no teleost yet studied has any vascular area in the infundibular floor comparable to the median eminence–portal system of other vertebrates. The blood to the PD is supplied by one (or two) hypophyseal artery(ies), and although this runs along the floor of the HT for a variable distance, forming arteriolar loops in the meninx in some species, no NS terminations on the vessels in this region have ever been observed. After entering the pituitary in the anterior NHP, the arteries give rise to the primary longitudinal capillary plexus, which lies within the NHP and its interdigitations into the PD, close to the neuro-adenohypophysis interface. From the primary longitudinal plexus, capillaries pass into the PD, forming an elaborate network of sinusoidal vessels and their perivascular spaces among the endocrine cells, the secondary centrifugal plexus. At the interface of the NHP with the PI is the plexus intermedius: in some species this is in continuity with the primary longitudinal plexus, but in others it receives a separate supply from the caudal hypothalamic artery. There is little vascular penetration of the PI glandular tissue. Blood is collected into a venous plexus on the surface of the gland and then drains via posterior vessels into the anterior cardinal veins.

7.4. Neurohypophysis

The teleost NHP can usually be subdivided on the basis of structure into an anterior part, which ramifies into the PD, and posterior part, which ramifies into the PI. The posterior part is largely composed of peptidergic fibers (type A), of which several distinct categories based on granule size can usually be recognized. The fibers often form huge axonal swellings packed with NSG, known as Herring bodies, in the interior of the NHP. The infundibular wall of the NHP is lined by ependymal cells, and in the primitive teleosts (e.g., eel) this forms hollow, fingerlike projections into the PI. Modified glial cells, termed pituicytes, are scattered throughout the NHP and may be secretory. The majority of the type A fibers terminate within the posterior NHP, either on pituicytes or on the capillaries of the plexus intermedius, or its intervascular channels enclosed by the double BM, which forms the boundary between the NHP and the PI. Such sites are believed to represent the sites of release of the neurohypophyseal octapeptide hormones into the blood stream.

The main cateogries of type A fibers, perhaps originating from different classes of NPO neurons, may contain different neuropeptides. AVT, IT, and SRIF- and CRF-staining fibers have recently been demonstrated in the teleostean NHP. In some species a proportion of the type A fibers penetrate the NHP/PI boundary and form endings on the endocrine cells, suggesting that at least some of these fibers may play a part in the control of PI secretion. A few aminergic type B fibers are also observed within the posterior NHP, and these form endings, often synaptic, on capillaries, BM, or pituicytes, or directly on the PI cells.

In the anterior part of the NHP, peptidergic type A fibers, which by ICC have been shown to contain SRIF, LH-RH, CRF, AVT, and IT, are much fewer in number and mainly terminate on blood vessels in the interior (Figure 20), rather than in the interdigitations into the PD. A few, however, do penetrate into the proximal PD, ending on capillaries or BM, and in some species form endings directly on the PPD endocrine cells. The vast majority of fibers within the anterior NHP are aminergic type B containing small (ca. 80 nm) DCV. These form endings, frequently synaptic, or capillaries, on pituicytes (less numerous than in the posterior NHP), and particularly on the almost continuous BM which marks the boundary between the NHP and RPD (Figure 20). In most species, type B fibers penetrate the BM in places and form endings on some or all of the PD endocrine cells, i.e., direct innervation. With the Falck–Hillarp technique (Section 16.4), small fluorescent varicosities can sometimes be seen among the PD cells, corresponding to the aminergic type B endings. In addition to the pituicytes, there are in the anterior PD darker glial processes. It is possible that these originate from ependymal cells lining the infundibular recess and therefore represent tanycytes. There is now a large amount of evidence supporting the concept that, on functional, anatomical, and embryological grounds, the teleostean anterior NHP is equivalent to the ME of other vertebrates, as if the teleost ME has become encapsulated within the pituitary. The fine structure of the anterior NHP shows striking similarities to that of the typical tetrapod (e.g., anuran amphibian) ME (Figure 14), in the relationships between nerve endings, capillaries, and BM, especially in the primitive teleosts (e.g., salmonids). Hypophysiotropic material appears to be released into the primary capillary plexus within the NHP instead of within the ME and conveyed to the endocrine cells via the short secondary centrifugal capillaries rather than by long portal vessels. The evolution of the system seems to have been taken a stage further in the more "advanced" species which have developed direct innervation of the endocrine cells. Embryologically, the teleost NHP arises from an area of the infundibular floor which in tetrapods develops into the ME, whereas the NHP of tetrapods is derived from the infundibular sac, which in most teleosts gives rise to the *saccus vasculosus*. This structure, which is an outgrowth of the infundibular floor behind the pituitary, has been studied rather more in teleosts than in elasmobranchs or ganoids, but appears to have no endocrine connections with the gland. It is lined by ependymal cells, some of which are specialized into "crown cells": these are believed to secrete polysaccharides into the CSF and may also be involved in maintenance of the osmotic and ionic balance of the CSF.

8. Crossopterygian Fishes

The Crossopterygii (lobe-finned fishes) form a distinct group that first appeared in the Devonian era in three separate forms, the Osteolepids (all now extinct), Coelacanthini, and Dipnoi, which were established by the middle Devonian. The tetrapods evolved from the basic Osteolepid stock in the latter part of the Devonian; thus, the surviving genus *Latimeria* (Coelacanthini) and the six species of lungfishes (Dipnoi) are closer than any other living fishes to the ancestors of the tetrapods. The structure of the pituitary gland is of special interest in that it may show both tetrapod and fishlike characteristics. The most detailed studies have been made on the African lungfish, *Protopterus*.

8.1. Dipnoi

Basically the pituitary organization is more amphibian than fishlike (Figures 15, 23, 24); the neural lobe is well marked, there is a ME, no saccus vasculosus, and the PD cells are arranged as a mosaic. However, not all features are tetrapodlike: the neural lobe does make intimate contact with the PI, PD cells are directly innervated, there is a large hypophyseal cleft between the PD and PI, which

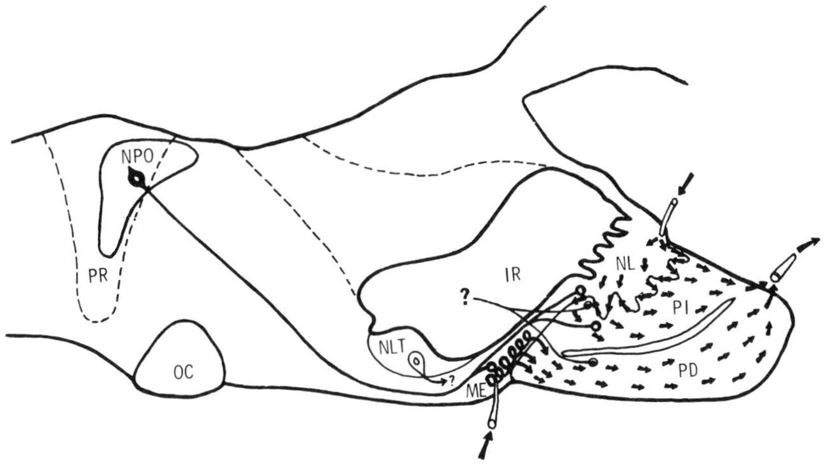

Figure 23. Hypothalamus and pituitary of the lungfish. IR, infundibular recess; ME, median eminence; NL, neural lobe; NLT, nucleus lateralis tuberis; NPO, preoptic nucleus; OC, optic chiasma; PD, pars distalis; PI, pars intermedia; PR, preoptic recess.

appears as a schizocoel in the solid AHP anlage, and there is no pars tuberalis (PT). In *Protopterus* the PD lies behind the neural lobe, an amphibian feature.

8.1.1. Cytology of the Adenohypophysis

Cells of the PD are arranged around capillaries, which usually form sinusoids; sometimes, instead of being solid and compact, the cords of cells have a central lumen containing colloid (a feature seen also in turtles, birds, and mammals). Five cell types have been distinguished on the basis of their staining properties, but lack of physiological studies means that their function can only be inferred by comparison with other vertebrate pituitaries, particularly amphibians. In *Protopterus* the two acidophils stain differentially with erythrosin and OG, one presumed to secrete PRL and the other GH. There are three basophils: type 1, which appears early in development and is presumed to secrete TSH; type 2, found only in adult fish and presumed to secrete GtH; and type 3, presumed to secrete ACTH. These cells are particularly located round the point of entry of the portal vessels and are amphiphilic as in amphibians.

Immunocytochemical studies of the South American lungfish, *Lepidosiren paradoxa,* AHP show that the acidophils react with antiserum to ovine PRL and bovine GH. GH cells are more abundant near the hypophyseal cleft and PRL cells scattered throughout the PD; however, immune cross-reactions occur between both cell types indicating molecular structures common to both hormones. At EM level the PRL cells show abundant 200–300 nm secretory granules, whereas the GH cells have only few, unusually small, 80–160 nm granules. Antisera to synthetic 1-24

ACTH react with cells resembling type 3 basophils, which are also PbH+, as well as with a few cells of the PI.

8.1.2. Vascular Supply

In common with amniotes and amphibia the dipnoan PD is supplied with portal blood which passes through the ME before entering the anterior PD (Figure 23). Primary portal capillaries, supplied by branches of the hypothalamic arteries, penetrate into the ventral, palisade, layer of the ME and connect extensively with capillaries of the PD. The neurointermediate lobe is vascularized separately from the PD, a tetrapodlike feature. NS nerve endings can be seen in the pericapillary spaces. Short portal vessels run from the primary capillary bed to the rostral tip of the PD. The portal system appears to be a specialized area of the ''mantle plexus'' between the infundibular floor and the PD (as in sturgeons).

8.1.3. Hypothalamus

The lungfishes closely resemble amphibians in the organization of the hypothalamohypophyseal system (Figures 23 and 24). The only hypothalamic nucleus yet described in detail is the NPO, which is relatively far anterior and not very long. The neurons stain with the classical NS stains and send short dendrites into the ventricle, terminating in small globules. With the EM, only one type of perikarya has been observed, containing 140–190 nm NSG. However, ICC studies have identified separate neurons, not preferentially located, which produce the two octapeptides vasotocin and mesotocin. In *Lepidosiren* there are small AF− neurons lying above the ME, containing 90–100 nm DCV and resembling the NLT perikarya of teleosts. No aminergic nucleus corresponding to the PVO of amphibians has yet been described. From the NPO the AF+ axons form two tracts, one ventral supplying the ME (mainly vasotocinergic) and one dorsal supplying the neural lobe (mainly mesotocinergic).

8.1.4. Neurohypophysis

The ultrastructure of the ME has been studied in *Lepidosiren*. The anterior region shows the typical arrangement (Figure 14), with NS fibers forming endings on the fenestrated primary capillaries, which penetrate into the palisade layer and drain into the portal vessels to the rostral PD. There appear to be three types of type A ending, based on NSG morphology, but no type B fibers are observed. The arrangement in the posterior ME, which lies just above the PD, is different and perhaps represents the persistence of the primitive condition exemplified in cyclostomes. Here the NS fibers, again exclusively type A, end on a thick layer of avascular connective tissue interposed between the ME and the PD.

The neural lobe (NL) of lungfishes is composed of tubules or lobules, which may have a closed core of ependymal cells or contain extensions of the infundibular cavity, lined by ependymal cells. These ependymal cells may be secretory, as they contain large spherical granules and produce long tanycytelike processes into the interior of the NL. Beneath the ependyma is a fibrous layer of axons, ependymal processes, and scattered pituicytes. Next is the palisade layer, composed of much larger NS fibers, packed with granules, with endings on a connective tissue sheath, which separates the NHP from the PI. This sheath encloses the elaborately folded capillaries of the plexus intermedius. The endings on the capillaries in this region are mainly type A (two types perhaps corresponding to AVT and mesotocin-secreting fibers identified by ICC). Between the nerve endings there are also end-feet from the ependymal (tanycyte?) processes. Aminergic type B fibers do occur in the NL but are restricted to the areas close to the PI and therefore appear to be concerned only with control of the intermediate cells. Some of these pass from the NHP into the PI and innervate the cells directly.

An unexpected feature of the pituitary innervation in lungfishes is that scattered aminergic fibers are found within the PD, ending on the endocrine cells, although no synapses have yet been found. This type of direct innervation of the PD, similar to, but less extensive than, that of teleosts, is puzzling from an evolutionary point of view, since it is combined with a well-developed ME, although the latter itself seems to lack aminergic innervation. The explanation could be that teleosts and lungfishes both developed direct innervation independently, from ancestors that had the conventional ME–portal system.

8.2. Coelacanthini

This group of Crossopterygii had been considered to be extinct until a specimen was caught in 1939. This was *Latimeria chalumnae,* and since the group occupies a position between lungfishes and tetrapods, it was hoped that the pituitary of *Latimeria* would show both primitive and advanced features, to indicate the direction of evolutionary changes. In fact, in gross anatomy, the pituitary of *Latimeria* resembles that of selachians most closely. but some features may be neotenous.

8.2.1. Cytology of the Adenohypophysis

The AHP is very elongated, consisting of a broad PI that interdigitates with the NL, and the PD is narrower, commencing dorsally to the PI and extending anteriorly beneath the telencephalon. A curious feature is the presence of islands of PD tissue enclosed in a tube of fibrous material anterior to the PD, the rostral islets. These may be the equivalent of the rachendachhypophyse of holocephalians and ventral lobe of selachians. The islet cells are basophilic and resemble GtH cells of other fish. The rest of the PD can be divided into a rostral and proximal zone; the cells of the PD are follicular in the RPD and stain with acid dyes, suggesting a resemblance to PRL cells. There are three basophils of the RPD, which may be comprised of mature and immature GtH cells and thyrotrops. The follicle lumen contains a colloid AB+, PAS+, and AF+, and the lumen of some follicles is continuous with the hypophyseal cleft. Two cell types in the PPD are acidophilic, one resembling ACTH cells in morphology and location, and a second acidophil, a putative GH cell. One secretory cell type is obvious in the PI, which has no affinity for PbH, but which nevertheless has large flocculent secretory granules; a second, infrequent cell type has small, dense granules.

8.2.2. Hypothalamus and Neurohypophysis

The fragmentary studies on the HT of *Latimeria chalumnae* revealed a NPO with magnocellular and parvocellular areas. The ME receives mainly aminergic fibers, with primary capillaries embedded in dense connective tissue and a portal system supplying the main part of the PD, but probably not the most distal islets of buccal PD. The NHP interdigitates to some extent with the cerebral PD, and LM studies have revealed scattered AF+ nerve fibers within the PD. This has been interpreted, as in the ganoid *Amia,* as representing "a meagerly developed non-synaptic accessory innervation" of the PD. A poorly developed saccus vasculosus has been reported.

9. Amphibians

In the adult form the pituitary of each of the three groups of amphibians, Anura (frogs and toads), Urodela (salamanders and newts), and Apoda (limbless amphibians), all differ from one another, but nevertheless homologies with the Dipnoi and Amniotes are clear. The AHP develops as a solid anlage and remains compact in anurans and urodeles, but in apodans there is a transient schizocoel. The aboral surface of the epithelial bud contacts the infundibulum and is induced to differentiate into the PI. In anurans and urodeles the PI remains dorsal, but it lies posterior to the PD in apodans. The PT arises as two anteriorly directed outgrowths from the AHP anlage; these maintain contact with the PD in urodeles and apodans, but become isolated in anurans as patches of tissue in the tuber cinereum. The PI is separated from the NL by a network of capillaries, the plexus intermedius, in anurans, but in those urodeles which are neotenous and largely aquatic (*Necturus, Amphiuma*), there are projections of the cavity of the NHP into the PI, so resembling dipnoan organization. There is no recess separating the PD from the PI, because there is no infundibular stem between the ME and NL.

The apodan AHP is flattened and elongated anteroposteriorly with the PD anterior to the PI (absent in *Typhlonectes*), which encircles the NL. A groove runs along the dorsal surface of the PD widening at the posterior end, in which lies the infundibular stem and NL.

9.1. Cytology of the Amphibian Adenohypophysis

Five cell types have been distinguished in amphibian pituitaries on the basis of location, staining properties, ultrastructure, and immunological reactivity. Studies have been made on several anurans, *Rana* sp., *Xenopus, Bufo, Bombina,* and

Nectophrynoides, and among the urodeles *Triton, Triturus,* and *Ambystoma*. There are two acidophils and three basophils (mucoid cells), recognized by staining, although differentiation of acidophils into erythrosinophils and orangeophils can be rarely achieved. Of the acidophils A1 cells have been identified as PRL cells; they stain preferentially with erythrosin in some species and react with antisera to mammalian PRL, almost exclusively, although in *Rana temporaria* there is cross-reaction with antiserum to GH. Ultrastructurally PRL cells have an abundance of large spherical secretory granules ranging in size between 200 and 500 nm (mean 350 nm), and the RER is mainly in the form of small cisternae. The cells are generally large and contact the capillary via a BM. The activity of these cells diminishes during metamorphosis and migration to land. The A2 acidophils are classified as GH cells, and in anurans and urodeles there is a tendency for them to be concentrated in the dorsocaudal region of the gland, particularly in bufonids (toads), with the PRL cells scattered more evenly throughout the PD. The A2 cells react with antiserum to mammalian GH, but there is sometimes (*Rana temporaria*) cross-reaction with antiserum to PRL. These cross-reactions show that there are similarities between amphibian GH and PRL molecules, which can be detected by their antigenicity. The hormone storage granules of GH cells are irregularly shaped and quite small, 120–250 nm, mitochondria are small with prominent cristae, and there are sometimes lipid droplets in the cytoplasm.

The three basophils are PAS+ and AB+; B1 basophils are identified as thyrotrops, because they change with thyroid activity. They are small cells and, in some species, tend to be concentrated toward the dorsal aspect of the central PD. Their hormone secretory granules are unusual for thyrotrops in showing considerable size heterogeneity varying between 120 and 500 nm in anurans and 180 and 250 nm in urodeles (*Triturus*); similar cells occur in apodans. The RER consists mainly of small dilated cisternae, and lysosomes are frequent. The cell morphology changes as the cells become more active during metamorphosis, and in relation to PRL cells at that time. The second basophil, B2, is a typical gonadotrop, generally a large cell with a tendency to lie ventrally within the central PD. Hormone granules are numerous, pleomorphic, 150–400 nm in diameter, and the cells often include large lysosomal bodies, up to 300 nm in diameter. Although both FSH and LH have been detected by bioassay in some amphibian pituitaries (*Ambystoma, Rana pipiens* and *catesbeiana*), the stored hormone reacts only with antiserum to β-LH, and no reaction with antiserum to mammalian FSH has been demonstrated so far; with onset of reproductive activity the RER shows considerable dilation in these cells. The B3 adrenocorticotrops are characterized by abundant small secretory granules, 100–250 nm, prominent mitochondria with complex cristae, small groups of parallel RER membranes, and few lysosomes. The adrenocorticotrops are found in the anterior region of the anuran and urodele PD, nearest to the ME, and generally close to the secondary capillaries of the hypophyseal portal system; in the apodan, *Chthonorpeton,* they occupy the entire anterior region of the PD. In common with ACTH cells of other vertebrates they have an affinity for acid stains as well as being PAS+ and AB−.

Only one cell type in the PI of amphibia secretes MSH. The cells become more active in animals maintained on dark backgrounds. The secretory granules are small, 150–200 nm, irregularly shaped, and with an indistinct limiting membrane; mitochondria are small and dense, RER is arranged in parallel arrays, and there are occasional lysosomes.

9.2. Hypothalamus

In amphibians the NPO lies in the anterior HT along either side of the preoptic recess (Figure 24) and, in the toad *Bufo* at least, appears to be comprised of two parts. The ventral region is AF+ and PAS+ and gives rise to a tract of fibers terminating mainly in the NL. The dorsal region is AF+ but PAS− and produces fibers that project mainly to the ME. The perikarya are large, often CSF-contacting, and with similar ultrastructure to those of fishes. ICC studies on the NPO of frogs demonstrate the presence of separate vasotocin- and mesotocin-containing neurons, together with some containing somatostatin, but with little regionalization. Neurons reactive with anti-LH-RH antiserum occur anterior to the NPO in an area termed the medial septal nucleus. AF− aminergic cells are found in some parts of the NPO; these also form two groups of cells in a more ventral position, the preoptic recess organ (PRO) and the OVLT. The sites of termination of axons from these two areas are not known.

Various parvocellular AF− nuclei have been identified in the basal HT (Figure 24): the infundibular nucleus, the periventricular nucleus, and the *pars ventralis tuberis* or "gonadotropic center." It now seems likely that these are all part of the

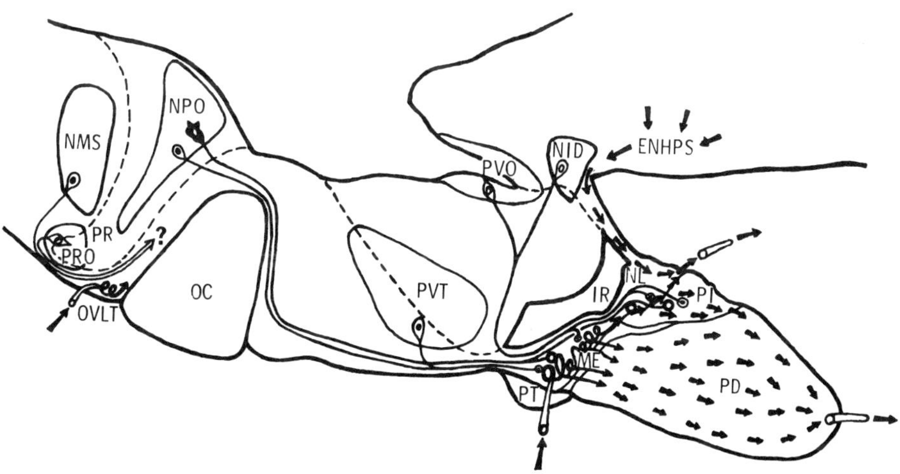

Figure 24. Hypothalamus and pituitary of an anuran amphibian. ENHPS, encephaloneurohypophyseal portal system; NID, dorsal infundibular nucleus; PRO, preoptic recess organ; PT, pars tuberalis; PVO, paraventricular organ; PVT, pars ventralis tuberis.

same system, homologous with the infundibular nucleus of reptiles and birds and are arcuate nucleus of mammals. Groups of LH-RH and somatostatin immunoreactive neurons have been demonstrated within this area, which may contribute to the AF− innervation of the ME. The main AF− aminergic region is the paraventricular organ (PVO) of the caudal HT, which often appears to embrace several subdivisions that may be categorized as separate nuclei. Tracts of fibers can be traced from the PVO through the internal zone of the ME and into the NL, where some terminate, but most pass on into the PI. The cell bodies, as in fishes, are the bipolar CSF-contacting type. Similar aminergic neurons also occur caudal to the PVO in an area designated the *n. infundibularis dorsalis* (NID).

9.3. Neurohypophysis

The degree of development of the NL seems to be related to osmoregulatory demands and is therefore greater in the more terrestrial species. It is clearly separated from the PI by a connective tissue sheet, enclosing the plexus intermedius, except in the case of some aquatic urodeles, which have a lungfishlike arrangement of hollow neural tubules into the PI. The bulk of the NL is composed of numerous AF+ nerve fibers, capillaries, and rather few pituicytes. At least two kinds of type A fiber have been identified by ICC to be vasotocin-containing and mesotocin-containing, and in the toad five fiber types have been described on the basis of ultrastructure. Aminergic type B fibers are rare in the interior of the NL, being restricted to the area close to the neurointermedia septum, from which many penetrate into the PI and innervate the intermedia cells. Some mesotocinergic AF+ fibers also cross the septum into the PI and innervate the "globular" intermedia cells of the internal zone. Within the NL the A fibers terminate, often synaptically, on capillaries and their pericapillary spaces without the interposition of glial processes. Some fibers make endings on other axons or on pituicytes. Two types of pituicytes have been described, both with long processes among the nerve fibers and some terminating on the capillaries. The ependymal cells lining the infundibular recess are similar, but some cells are apparently specialized for transport. Near their apical surfaces and in their basal processes, which end on the subependymal capillaries, they have small vesicles which may be pinocytotic (involved in uptake of fluid from the CSF).

The amphibian ME, again most highly developed in anurans, has the characteristic three-layered structure and many detailed similarities to the mammalian ME (Figure 14). The cells of the ependymal layer line the floor of the infundibular recess, and some (tanycytes) have basal processes that end on capillaries in the external zone. The internal zone contains numerous nerve fibers and scattered glial cells. The bulk of these fibers belong to the NPO–NHP tract (mesotocinergic and vasotocinergic AF+ fibers), together with the type B fibers passing through to the PI. Some of the AF+ fibers and the PD-controlling AF− and type B fibers leave the tracts here and pass down into the external zone. This zone has the palisade arrangement and carries the primary portal capillaries on its ventral surface. The structure of this layer is particularly elaborate in toads, with the primary plexus forming numerous short capillary loops, together with a few long capillary loops that reach up as far as the ependymal layer. Both types of loop receive endings from nerve fibers and ependymal processes, and the structural relationship between these endings and the capillaries is remarkable. The pericapillary BM bears elaborate ramifications that form a three-dimensional connective tissue network around the blood vessel. This network greatly increases the surface area available to receive nerve endings and probably provides a system of channels for transport of material from the nerve endings into the capillary. In frogs, the extensions of the pericapillary BM are nowhere near as elaborate. The types of nerve endings in this contact zone have been analyzed in several species of anuran and urodele, with a good general agreement of accounts. Two type A endings are usually described, possibly corresponding to vasotocinergic and mesotocinergic fibers from the NPO, together with the usual type B aminergic endings of fibers thought to originate from the PVO. In addition, other endings similar to the type B but with larger granules have been found; these may represent the AF-negative innervation from the infundibular nucleus and preoptic recess and may correspond to the LH-RH and SRIF reactive endings demonstrated in the external zone by ICC.

The urodele ME is less well differentiated. The infundibular floor is hardly thickened, and the primary capillary plexus is superficial and embedded in a connective tissue layer which indents into the external zone, where nerve terminations are received. The ME capillaries are directly connected to the PD vasculature by short capillaries, not by portal veins, as in the anterior ME of elasmobranchs and in ganoid fishes. The urodele ME varies considerably with habit, being most elaborate in the terrestrial salamanders and very simplified in the cave-dwelling and neotenic species—possibly this shows a degenerate condition.

The endocrine cells of the amphibian PD are never directly innervated in the adult, but a sparse aminergic innervation of the PD has been described recently in *Rana temporaria* tadpole and *Bufo* tadpole, which disappear at the onset of climax of metamorphosis. This is obviously a point of great phylogenetic significance.

9.4. Vascular Supply

The blood supply to the ME and pituitary in anurans follows the usual tetrapod pattern. The primary capillary plexus of the ME is supplied by branches of the hypophyseal artery via a series of tortuous arteriolar loops which may serve to slow blood flow thereby facilitating neurohemal exchange in the ME. Blood from the ME capillaries passes into the portal veins, which drain into the secondary capillary plexus of the PD. The plexus intermedius is supplied by a separate branch from the hypophyseal vein, but also appears to receive some blood from the ME capillaries and from an extraordinary second portal system, the encephaloneurohypophyseal portal system (Figure 24). The latter is a plexus spread widely throughout the mesencephalon of the brain and the preoptic region of the HT, which drains into one or two large vessels in the roof of the infundibular recess. These supply the NL capillaries, which in turn drain into the plexus intermedius. The PI receives blood from the NL containing high concentrations of octapeptides, as in fishes, reptiles, and mammals, and from the ME and wide areas of the HT and mesencephalon, a curiously comprehensive supply.

10. Reptiles

The reptiles occupy a central position in the evolution of the tetrapods because birds, mammals, and modern reptiles all evolved from early reptile stock (Figure 15). Extant forms are the sparse remnants of the great numbers and variety of forms dominant in the Cretaceous period. By development of the cleidoic egg the reptiles were the first group to become completely emancipated from aquatic environments, with young hatching from the egg already equipped to live on dry land. "Stem reptiles" of the Carboniferous diversified throughout the period; one group separated early to ultimately give rise to the mammals and a later offshoot, the birds. The birds arose from a group related to the dinosaurs, of which the only living examples are the crocodiles. Snakes (Ophidia) and lizards (Lacertilia) evolved from a line that first gave rise to the Rhynchocephalia, represented now by only one living species, *Sphenodon,* the tuatara of New Zealand, but the group believed to show the most primitive characteristics is the Chelonia, the tortoises and turtles. The wide variation displayed by the hypothalamohypophyseal system of reptiles reflects the phylogenetic position of the groups: the arrangement in chelonians strongly resembles that in amphibians, whereas that in crocodilians is similar to the arrangement in birds. The pituitary organization in *Sphenodon,* crocodiles, and turtles is thought to show the most primitive condition; the PI and PT are well developed with the PD divided into cephalic (rostral) and caudal lobes, as in birds. Snakes have no PT, and it is often reduced in lizards; hence the portal vessels pass from the ME to the PD in connective tissue, not via PT tissue.

10.1. General Morphology of the Adenohypophysis

10.1.1. Rhynchocephalia (*Sphenodon*)

Similarities with lungfishes and amphibia are evident in *Sphenodon* AHP; the PD is divided into caudal and cephalic regions, linking posteriorly with the PI and anteriorly with the ME via the PT, so forming a portotuberal tract with associated portal vessels, as in birds. The PT in *Sphenodon,* as in chelonians, has a region apposed to the ME, the juxta PT, and a region in contact with the PD, the PT interna.

10.1.2. Chelonia

Compared with *Sphenodon* the AHP of chelonians is generally more closely applied to the infundibular floor and the PT interna is much more developed. Recent studies show extensive development of PT interna dorsally, in the PD, forming a cortical layer, as well as connections with the juxta PT. The PT forms as a pair of lateral outgrowths from the anterior aspect of the pituitary anlage (Figure 25) and is largest in embryo and young turtles. There is indistinct division of the PD into caudal and cephalic regions histologically; the PI is well developed, surrounding the structurally simple NL. The PD commonly contains colloid-filled vesicles, sometimes very large, and the hypophyseal cleft persists but is obliterated in other reptiles.

10.1.3. Crocodilia

The AHP is relatively short and deep with the PD sharply divided into cephalic and caudal lobes. The embryonic outgrowths forming the PT attach to the ME and remain as thin plates, the juxta PT, with a large portotuberal zone attached to the PD. The hypophyseal cleft is present only in young animals leaving direct attachment of the PI to the PD in adults. The PI is well developed with tissue penetrating between the lobules of the NL.

10.1.4. Squamata

In lizards there is considerable variation in the anatomy of the AHP, and, although generally symmetrical, occasional species show marked asymmetry (*Xantusia*). In most species a deep cleft, filled with connective tissue, separates the PD from the infundibular floor. The PD is clearly divided into cephalic and caudal lobes. In adults the PT persists only as plaques beneath the ME and on the PD, so that there is no portal PT linking the interna and juxta PT; ultimately, the neural juxta PT invades ME and hypothalamus because the BM and pia separating the

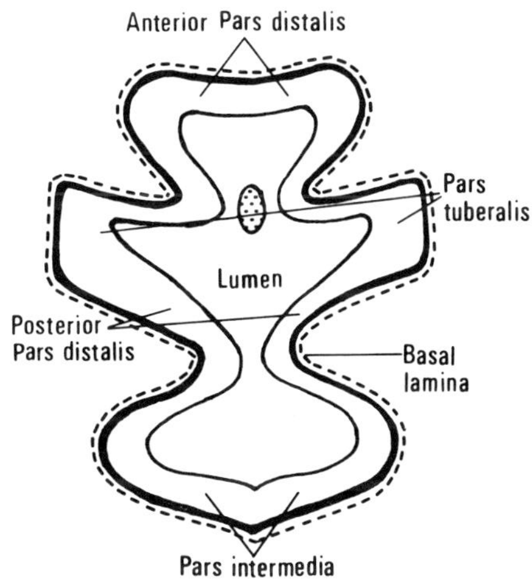

Figure 25. Horizontal section of the embryonic reptile pituitary. Development of a large lumen in the anlage determines the spatial arrangement of the various cell types. (Redrawn from Licht, P., and Pearson, A. K., 1978, Cytophysiology of the reptilian pituitary gland. *Int. Rev. Cytol.* Suppl. **1:** 239–286.)

tissues disappear. The PI generally surrounds the NL, except dorsally, but is very variable in size, only a few cells thick (*Lacerta, Anguis, Xantusia*) or very large (*Anolis*), and completely atrophied in blind, burrowing lizards, amphibaenids, and trogonophids. A hypophyseal cleft may be present and occasionally (*Varanus niloticus*) sends diverticuli into the caudal PD as well as the PI (cf. *Acipenser*).

The AHP in snakes is markedly flattened dorsoventrally, a characteristic feature of serpentine vertebrates, and in adult snakes it is left/right asymmetrical, a development that occurs just before hatching. In some primitive, burrowing snakes the AHP remains orientated anteroposteriorly, but in most snakes the PD is to one side and the PI to the other. There is clear histological differentiation of the PD into caudal and cephalic lobes; the PT is completely absent from adult snakes, only vestigial lateral lobes being produced from Rathke's pouch in the embryo. Thus, the portal vessels pass from ME to PD in connective tissue only. In most snakes the PI is very large, but is missing from burrowing snakes, and there are no scattered PI cells in the PD, as seen in bird and mammal AHP without a discrete PI. However, when present, the PI normally forms a thick cup investing the NL, which is at the end of a long, generally asymmetrical, infundibular stem.

10.2. Cytology of the Adenohypophysis

The distribution and morphology of reptile AHP cells have been studied by standard techniques, LM with histochemical staining, EM, ICC, and changes produced by target organ ablation or stimulation. In a number of earlier studies the cell types were designated by alphabet, but it seems more sensible now to recognize the cells on the basis of the chemical nature of their secretions; Table III shows the various classifications of the cells.

Five secretory cell types have been recognized in the PD of species of all families of reptiles, but the most detailed studies have been on the lizard, *Anolis carolinensis*. The most convincing evidence shows that PRL cells are exclusive to the rostral (cephalic) region of the PD, although in one species of *Trionyx* they appear to be caudal. They are acidophilic, staining with erythrosin, but are PAS, AF, and AB negative; they also react with antiserum to mammalian PRL. The cells vary in shape depending on where they are in the PD, but tend to be elongated and arranged in cords around sinusoids. They have an abundance of large, round secretory granules up to 800 nm, mainly 300–400 nm in diameter. The cell nucleus is

Table III
Cells of Reptile Adenohypophysis

A_1 B B X	Acidophil, erythrosinophil, PAS-, AB-, AF-, PRL cell
A_2 D X	Acidophil, orangeophil, PAS-, AB-, AF-, GH cell
B_1 E	Basophil, PAS+, AB+, AF+, Thyrotrop
B_2 A B	Basophil, PAS+, AB+, AF+, Gonadotrop
B_3 C	Amphiphil, PbH+, PAS±, Adrenocorticotrop

irregular, generally indented, and situated basally; RER consists of parallel membranes scattered among secretory granules, and the Golgi is well developed, usually found close to the nucleus. GH cells are OG+, do not stain with PAS, AF, or AB, and are situated predominantly in the caudal lobe of the PD (except perhaps in *Trionyx*). They are polygonal or cuboidal with a basal, round nucleus. In turtles, snakes, and lizards GH cells have large, 200–500 nm electron-dense granules. The GH cells constitute the largest cell population in reptile PD, as they do in mammals. Adrenocorticotrops are most usually concentrated in the rostral PD, but some cells are scattered caudally. They are ovoid and stain strongly with PbH and both acid and basic stains. In *Anolis* granule size is variable, between 150 and 500 nm, and particularly in young *Tiliqua* (Scincidae) and sexually inactive females the cells are heavily granulated; in other reptiles too, these cells change with reproductive cycles. Thyrotrops are scattered throughout the PD, with the largest concentrations laterally; nevertheless they constitute the smallest cell population of the PD, probably developing from the proximal remains of the lateral lobes of Rathke's pouch. The cells are characteristically basophilic, PAS+, AB+, AF+, with numerous, small (200 nm) secretory granules, and an irregular nucleus, which may be deeply indented. Gonadotrops are scattered throughout the PD and make up about 20% of the total cell population. They have characteristic staining properties, basophilic, PAS+, AB+, AF+, but morphologically show considerable modifications with the onset of the breeding cycles. In the unstimulated condition they show sparse, scattered cisternae of RER, granules, moderately numerous, but varying in size from 100 to 500 nm, and lipid droplets in the cytoplasm. All gonadotrops react with antiserum to LH, and in several chelonian pituitaries this reaction is specific with antiserum to the β subunit. However, some of the gonadotrops also react with antiserum to FSH, but no exclusive reaction to β-FSH antiserum has been reported. Thus, gonadotrops are a mixed population which may synthesize immunoreactive LH exclusively or both FSH and LH, but never FSH alone.

Recently PT cells of several chelonian species have been examined by ICC for glycoprotein hormones as well as GH and PRL. Paradoxically, although high concentrations of gonadotrops were seen in the peripheral PT interna of several turtles, none of the cells stained with PAS. LH cells are also present in the proximal juxta PT, but not in the neural end of the juxta PT. Thyrotrops, distinct from LH cells, also occur in the peripheral PT interna (*Pseudemys scripta*), together with a few GH cells. These secretory cells probably occur in the PT because during embryonic development a large lumen forms in the anlage, creating peripheral regionalization, so that the lateral lobes from which the PT derives are composed of cells similar to those forming the PD (Figure 25).

The PI is very greatly developed in some reptiles, notably *Anolis carolinensis*, but may be absent from some snakes and burrowing lizards. Generally only one cell type is present which secretes MSH. Staining properties are variable among reptile species: they may be acidophilic, amphilic, or chromophobic. The MSH cells are large and columnar, in the simplest organization (*Klauberina* sp.) forming the only layer of cells lining the hypophyseal cleft. The primitive tuatara (*Sphenodon*) has only one PI cell type, which ultrastructurally has large secretory granules, generally lining the cell membrane adjacent to the capillary, with nucleus and RER toward the opposite pole. Secretory granules of *Anolis* are large because small ones fuse together, becoming less electron dense, and RER in active cells is organized into parallel, lamellar arrays.

10.3. Hypothalamus

The classical AF+ neurosecretory system has been well described in a number of reptilian groups, and recently a few extensive cytoarchitectural studies have demonstrated various AF− nuclei too. From this fragmentary information there appears to be a considerable anatomical difference between Chelonia, Crocodilia, Lacertilia, and Ophidia. As yet it is difficult to draw a generalized picture of the reptilian HT; the variations may demonstrate an increasing complexity of ''sub-unit'' organization ranging from the diffuse amphibian type to the well-demarcated bird/mammalian type.

One universal feature of the reptilian HT (Figure 26) is the division of the magnocellular NS system into a supraoptic nucleus (NSO) and a paraventricular nucleus (NPV), as in birds and mammals. The magnocellular neurons of the two nuclei are bipolar with CSF-contacting dendrites and axons that converge to form intermingled tracts projecting to both the ME and the NL. Both the NSO and NPV have been shown to comprise separate vasotocin- and mesotocin-producing cells with little or no preferential distribution. The relative size and positional relationship between NSO or NPV are variable, and several additional AF+ nuclei, which may in fact be extensions of the NSO and NPV, have been reported. AF+ ''bridge cells'' occur between the two nuclei in some species (mainly lizards), and

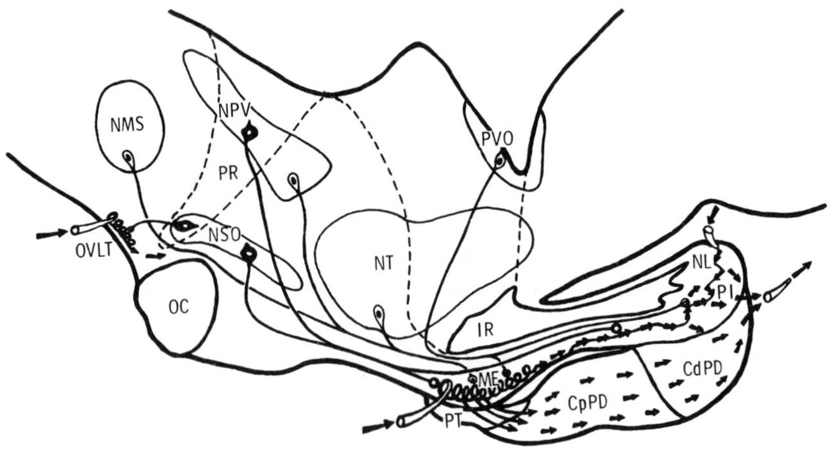

Figure 26. Hypothalamus and pituitary of a reptile. CdPD, caudal pars distalis; CpPD, cephalic pars distalis, NPV, paraventricular nucleus; NSO, supraoptic nucleus; NT, tuberal nucleus.

in the cobra two groups of cells, the *n. circularis* and the *n. retrochiasmaticus,* may be ventrocaudal divisions of the NSO. In some lizards AF+ cells occur in the tuberal region, and there are also vasotocin- and mesotocin-containing cells in the anterior preoptic area, which project not to the NHP but to the OVLT.

Numerous parvocellular AF− nuclei have been described in various reptiles, 22 for example in the HT of cobra. It is difficult to generalize, but some seem to correspond to those established in mammals: the arcuate nucleus, dorsomedial and ventromedial nuclei (together comprising the periventricular nucleus), and the mamillary nucleus. Several of the nuclei of the caudal hypothalamus may be part of the PVO system. The PVO is the only reptilian nucleus yet identified as being aminergic, as in other vertebrates. Catecholaminergic fibers radiate from the PVO to innervate the ME and other nuclei throughout the HT, so the PVO may serve as a coordinating area for HT function. ICC has indicated that some of the cells in AF− areas may contain releasing hormones: cells reactive for LH-RH have been found in several preoptic areas including the medial septal nucleus (NMS) and somatostatin-positive cells in the ventral NPV and in the tuberal region, with both LH-RH and somatostatin reactive fibers traceable to the ME.

10.4. Neurohypophysis

The NHP in *Sphenodon,* chelonians, and some lizards has a primitive structure, forming only a thin lining of the IR. This simply consists of three layers: ependyma, fibers, and a palisade layer, with very few glial cells or pituicytes. In the burrowing lizards and crocodilians the NL retains a narrow cavity continuous with the IR, but the lobule walls are thicker and contain pituicytes. A more advanced condition exists in the snakes: the NL is pierced by few tubules and forms solid, pituicyte-containing lobules bounded by vascular connective septa, penetrating into the PI. In the most primitive condition, exhibited by *Sphenodon,* the ependymal cells lining the IR form long basal processes that expand into end-feet on the BM of the plexus intermedius, forming an almost complete barrier between neural tissue and pericapillary space. All neurovascular contacts occur on the BM of the plexus intermedius, since there are no capillary branches into the interior of the NL. In the fibrous layer of the NHP two categories of A fiber are numerous, type A1 and type A2, which may correspond to the vasotocin and mesotocin fibers demonstrated by ICC, together with a few type B fibers. In the neurohemal contact zone type A1 and type B endings predominate; type A2 contacts are rare. The fine structure of the NHP appears to be basically similar in all lizards so far studied, although there are some differences in the extent of ependymal end-feet interposition between nerve endings and pericapillary BM and the extent of capillary penetration into the bulk of the NL. One universal feature, unique among vertebrates, is the total lack of innervation of the PI, which has been demonstrated by both ultrastructural and cytochemical studies. In all reptiles the vascular system between the NL and the PI is continuous, so a neurovascular link must be the sole method of control of the PI cells.

All reptiles possess a well-differentiated ME, but again there are some important phylogenetic differences within the class. The palisade layer shows considerable structural variation. In *Sphenodon* the primary portal capillaries do not penetrate up into the ME; instead the palisade layer sends down projections into the PT, where the portal capillaries run between the tuberalis cells. The palisade projections abut the capillaries, and their nerve fibers form endings on the narrow BM separating the ME projections from the blood vessels. This is reminiscent of the arrangement within the rostral ME of elasmobranchs. In other reptiles the primary portal capillaries may form a superficial network over the ME (in chelonians, small lizards) or form loops embedded within the palisade layer (crocodilians, large lizards, snakes). Anterior and posterior divisions of the ME can be recognized, with a suggestion of "point-to-point" relationship between certain HT nuclei and groups of primary capillaries (as in birds). Type A1 and A2 nerve endings mainly occur in the AF+ anterior region and may correspond to the numerous vasotocinergic and few mesotocinergic endings of fibers from the NSO and NPV demonstrated in the external zone by ICC. The type B endings, probably aminergic and from the PVO, predominate in the posterior ME. Somatostatin and LH-RH-reactive fibers have been found in the external zone of a few species, but the endings of these fibers have not yet been identified at the ultrastructural level.

10.5. Vascular Supply

As mentioned in Section 9.1, there are important phylogenetic differences in the vasculature of the hypothalamohypophyseal system of reptiles. The primitive condition is seen in some lizards, where blood from the infundibular arteries supplies the superficial primary plexus, which has no loops penetrating up into the ME. This plexus is continuous with a capillary network on the surface of the NL, which also receives some blood directly from the infundibular arteries (Figure 26). The superficial plexus of the NL forms the plexus intermedius along the interface with the PI, but there is no vascular penetration of either NL or PI. From the ME plexus, four to six portal vessels supply a secondary plexus within the PD. All the blood received by the PD is portal; there is no separate arterial supply. In more "advanced" lizards, the ME plexus is mainly superficial, but some loops do penetrate the palisade layer. The portal vessels are larger and fewer in number (one or two). The rest of the NHP receives only a few direct connections from the ME, the plexus intermedius being mainly formed from branches of the infundibular arteries. Capillary branches from the plexus intermedius penetrate the NL and also run between the lobules of the PI. In the most advanced condition, seen in snakes, the ME primary plexus is very elaborate, with capillary loops penetrating deep into the palisade layer and seven or eight portal vessels supplying the cephalic PD. The NL receives little blood from the ME plexus, being largely supplied by the hypophyseal arteries, and both the NL and PI contain a network of capillary branches from the plexus intermedius. The PD of the snake also has a blood supply independent from the portal system, via small arterial branches from the internal carotids.

11. Birds

Birds are considered to have evolved from Archosaurian reptiles, which also gave rise to the dinosaurs. Two groups probably diverged early, the Paleognathae or ratites, which are flightless, and the Impennae, the penquins, specialized for an aquatic life; but most living birds belong to the Neognathae, the flying birds, which show very uniform skeletal morphology and general anatomy reflecting the physical restraints inherent in their aerial life-style.

The general morphology of the hypothalamohypophyseal system is similar to that of reptiles. As in other vertebrates the AHP of birds develops from Rathke's pouch, which associates with the infundibular sac of the diencephalon. Two areas of the pouch enlarge to form the oral and aboral lobes; the oral lobe reaches up to the region of the ME and forms part of the PD. Lateral outgrowths of the oral lobe develop into the PT, which, as in mammals and some reptiles (see Figure 25), at their proximal ends become included in the PD and may form a PT interna. The PD derives from both the oral and aboral lobes, which give rise to the cephalic (rostral) and caudal (proximal) lobes, respectively. Externally the PD may be divided by a horizontal sulcus, but this does not correspond to the histological division of the two lobes. A remnant of the oral part of the pouch may persist as an epithelial stalk. The PT forms a bridge between the ME and PD and in many species forms a portotuberal tract of portal vessels, connective tissue, and PT elements. In many species the tract is not closely associated with neural tissue forming the neural stalk. A PT is present in all birds and in some species invades the ME, as in squamate reptiles. The division of the ME into anterior and posterior regions supplying the cephalic and caudal lobes of the PD is also reminiscent of squamate reptiles. Unlike the other living descendants of the dinosaurs, the crocodile, the birds have no PI (Figure 15).

11.1. Cytology of the Pars Distalis

Secretory cells of the PD are arranged in cords and acini, limited by a BM and interwoven by a network of capillaries. The PD secretory cells make direct contact with the perivascular space, but small chromophobic cells lie toward the center of the cell groups, away from the capillaries. The center of the acini appear to be filled with colloid, which may be chromophobic or PAS+; the "colloid," however, is fibrous and arises from the cytoplasm of the apical zones of several gland cells. Detailed studies have been made on only a few species of bird pituitaries, including duck, chicken, pigeon, quail, and white-crowned sparrow, but there is general agreement on the form and location of the cell types.

PRL cells are found in the cephalic lobe mainly, with occasional incursions into the caudal. As in the AHP of other vertebrates, they are erythrosinophilic and will stain purple with AB and AF. They have been identified by changes in ultrastructure related to the onset of broodiness or "lactation," particularly in pigeons, hens, and turkeys. Antisera to mammalian (ovine) PRL specifically react with the

PRL cells confirming previous interpretations of histochemical observations. The cells, situated around sinusoids, have a generous complement of large polymorphic, electron-dense granules, the longest axis of which may be up to 800 nm (Figure 6). RER is well developed in PRL cells, and arranged in parallel arrays, Golgi is prominent, and mitochondria are large and elongated. Somatotrops can be distinguished from PRL acidophils immediately by location as they are confined to the caudal lobe. These cells are round or oval, generally with a large nucleus, and cytoplasm with many electron-dense, spherical secretory granules, 250–300 nm in diameter; RER is more sparse than in PRL cells and may consist of dilated sacs or threadlike profiles; the Golgi areas are small and mitochondria quite large, spherical or elongated. Corticotrops are large oval or columnar; AF+ cells are found in the cephalic lobe, arranged in cords around the sinusoids. They have numerous, electron-dense secretory granules, 250–300 nm, well-developed RER, and small, round mitochondria. Thyrotrops are confined to the cephalic lobe; they are normally infrequent AB+ cells, adjacent to sinusoids, with most intracellular structures small in dimension—granules 100–150 nm, mitochondria small and round, and RER composed of sparse, small cisternae.

Much discussion has concerned the possible dual location of TSH cells in both the cephalic and caudal lobes, reported to be the case in several species including

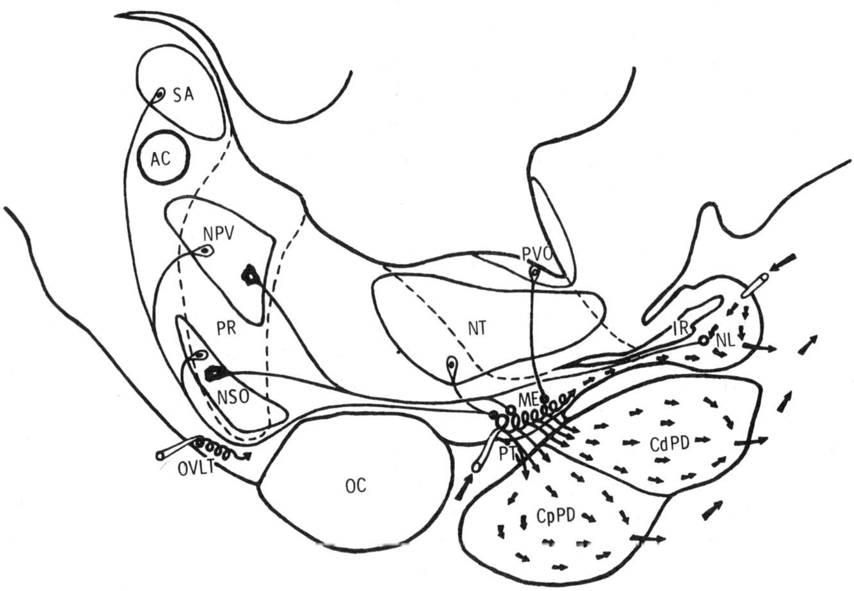

Figure 27. Hypothalamus and pituitary of a bird. AC, anterior commissure; CdPD, caudal pars distalis, CpPD, cephalic pars distalis; IR, infundibular recess; ME, median eminence; NL, neural lobe; NPV, paraventricular nucleus; NSO, supraoptic nucleus; NT, tuberal nucleus; OC, optic chiasma, OVLT, vascular organ of the lamina terminalis; PR, preoptic recess; PT, pars tuberalis; PVO, paraventricular organ; SA, septal area.

duck, Japanese quail, guinea fowl, pigeon, and red bishop. However, thyroidectomy only affects cephalic lobe AB+ cells in most species and, as yet, there has been no reaction with antiserum to mammalian TSH. However, antiserum to chicken LH reacts with both gonadotrops and thyrotrops, but this may be due to the similarity between the α-amino acid chains of the two hormones. Gonadotrops are basophilic cells found in both cephalic and caudal lobes; they are generally large round or polygonal cells, situated near sinusoids (Figure 11). There are some morphological differences between the gonadotrops of the two lobes; generally the secretory granules of the cephalic lobe cells are more variable in size and electron density than those of the caudal lobe. FSH activity has been associated with the cephalic lobe and LH with the caudal in chicks, but the hormone-releasing factor LH-RH stimulates both gonadotrops. Two separate regions of the HT control GtH cell activity—an anterior infundibular nucleus complex controls FSH secretion and a posterior complex controls LH secretion (see Section 14.2.5). Further evidence indicates that in the Japanese quail, LH is preferentially secreted by caudal lobe GtH cells, which react strongly with antiserum to chicken LH, whereas cephalic GtH cells only react weakly. It is clear that more pituitaries need to be examined with antisera to the specific α- and β-amino acid chains of LH, FSH, and TSH to differentiate the GtH and TSH cells definitively.

The PT has not been extensively studied, but in the chicken secretory cells containing granules of variable size, mainly between 180 and 250 nm, have been described as well as nongranulated cells; but the nature of the secretory material is unknown. The cells also contain stored glycogen.

11.2. Hypothalamus

The structure of the avian HT (Figure 27) has been analyzed with varying degrees of completeness in about a dozen different species; however, because of the very diffuse nature of the HT nuclei of birds, there are inconsistences in nomenclature. The AF+ neurosecretory system basically comprises two nuclei, the NSO and the NPV, as in reptiles, but many authors have found it necessary to subdivide the two areas further. The NSO usually consists of a median cluster of neurons around the preoptic recess, together with lateral extensions along the chiasma and optic nerves. The NPV lies posterodorsally to the NSO, is extremely diffuse, and varies in size in different groups of birds. It consists of several groups of cells close to the third ventricle, plus other more lateral groups within the AF+ tract, known as "endopeduncular" divisions. Very few studies have examined the neurons of the NSO and NPV at the EM level. In the sparrow, two classes of neurons were identified in the NSO, one with 150–250 nm NS granules believed to supply the NL, and one with 100 nm granules possibly supplying the anterior ME. On the other hand, in the fowl both the NSO and the NPV appear to contain only one multipolar type of magnocellular neuron with 100–200 nm granules. In contrast to lower vertebrates, the dendrites of the magnocellular neurons of birds do not appear to be CSF-contacting; instead they are embedded within a rich plexus of aminergic fibers and terminals from other parts of the brain. The AF+ fibers from the various divisions of the NSO and NPV run laterally into the entopenduncular region and then run posteriorly and ventromedially to join in the midline. The tract then follows the caudal edge of the optic chiasma into the ME, where it divides into two: one branch into the anterior ME and one to the NL. ICC studies on several avian species have demonstrated that vasotocin and mesotocin are produced by separate cells throughout all parts of the NSO/NPV, with axons of both types of neurons joining a common tract. In addition, perikarya reactive with antisomatostatin antiserum are scattered throughout the preoptic region, including parts of both the NSO and the NPV.

The AF− neurons of the avian HT form a number of poorly defined nuclei in the tuberoinfundibular region. Again, there are difficulties with terminology, since probably homologous areas have been variously termed infundibular nucleus, tubernal n., arcuate n., inferior hypothalamic n., posterior medial hypothalamic n., and ventromedial n. In the sparrow, the ''tuberal nucleus'' has been shown to consist of three zones of mosaically arranged cell clusters, believed to consist of aminergic neurons and neurons producing releasing factors, projecting to the anterior ME with a point-to-point relationship between groups of neurons and specific areas of portal vessels. Although the tuberoinfundibular region has a rich supply of aminergic fibers and terminals throughout, the main aggregation of aminergic cell bodies is the PVO, as in other vertebrates. Ultrastructural studies on the tuberoinfundibular region have revealed perikarya with DCV of ca. 100 nm, which might represent releasing-factor-containing neurons. ICC studies on the duck HT have demonstrated scattered LH-RH-positive cells, but in most species examined the tuberoinfundibular region, although containing reactive fibers, appears not to contain reactive perikarya, these being located mainly in the preoptic region and among the NSO/NPV neurons and also more anterodorsally in the septal area.

11.3. Neurohypophysis

The avian NL is quite separate from the anterior pituitary and exhibits wide structural variations, ranging from a thin-walled infundibular sac to a compact lobe around a small lumen (Figures 15 and 27). It is composed mainly of structurally uniform fibers from the AF+ tract, containing 150–250 nm NSG, and aminergic fibers are rarely seen. ICC studies on a variety of species have shown that mesotocinergic fibers occur mainly around the lumen, whereas vasotocinergic fibers occupy the deeper and more peripheral parts of the lobe. The fibers do not appear to terminate directly on the capillaries: rather they are separated from the capillary BM by a ''cuff'' of glial cell processes.

The ME is composed of the usual three layers—ependymal, fibrous, and palisade—found in other vertebrates, and with the LM, anterior and posterior divisions of the palisade zone can be recognized, the anterior region being strongly AF+ but the posterior region containing little AF+ material. At the EM level the distinction is much less clear, most of the endings throughout containing NSG of

120–250 nm, slightly smaller than those in the profiles of the NL. ICC studies on a range of species show that these probably correspond to vasotocinergic terminals, since mesotocin reactivity is restricted to the internal fiber layer. A second type of type A ending has also been distinguished in the palisade zone of several species, containing 100–120 nm NSG, and in the duck and the pigeon at least, the predominance of this type of fiber in the anterior ME may account for the regional differences in staining seen with the LM. A third category of ending occurs in the avian ME, with 70–100 nm DCV, presumably corresponding to the aminergic type B fibers of other vertebrates. However, the evidence for the aminergic nature of these type B terminals is far from conclusive: although the ME of the fowl and the gull show strong aminergic fluorescence, little fluorescence is seen in the pigeon even though the type B termination predominates throughout the posterior ME. The question of which type of nerve endings contain the pituitary releasing factors is still unresolved. LH-RH-reactive endings appear to be scattered throughout the palisade zone, whereas somatostatin-reactive fibers are restricted to the anterior part. The relationship between the nerve endings and the capillaries in the palisade zone is similar to that in the NL, with a collar of ependymal processes usually intervening. It has been estimated in sparrows that depending on species, between 10% and 25% of the nerve endings make direct contact with the portal capillaries. In addition to the endings on the capillaries, the type B fibers also make contacts on other nerve fibers and on glial or ependymal processes (tanycytes?) which may themselves secrete into the portal capillaries.

11.4. Vascular Supply

The vascular system of the avian ME is considerably more complex than in the lower vertebrates. The primary portal plexus is divided into two areas, supplied independently from the hypophyseal arteries, covering respectively the surfaces of the anterior and posterior division of the ME (Figure 27). Capillary loops into the palisade zones are very highly developed. From the two capillary plexi, blood drains into the portal vessels, which supply specific areas of the PD. Thus, blood from the anterior ME enters the PD cephalic lobe, and from the posterior ME enters the caudal lobe. The anatomical evidence, therefore, suggests a point-to-point relationship between specific hypothalamic neurons, specific portal capillaries, and specific areas of the pituitary, to a greater extent than in reptiles. Blood supply to the NL is direct, not via the portal system, and is independent from the supply to the PD.

12. Mammals

There are three orders of mammals, of which the eutherians, the placental mammals, are the most numerous and have been studied most extensively and in greatest detail. Here we shall only give a brief account of the pituitary anatomy and cytology.

The PD in mammals differs from that in other amniotes in being mainly formed from the aboral lobe of Rathke's pouch; the oral lobe, although present in embryo, fails to grow, remaining as a vestige at the anterior end of the gland close to the PT (Figure 3).

12.1. General Morphology of the Adenohypophysis

12.1.1. Prototheria

The duck-billed platypus and spiney anteaters, which constitute this group of egg-laying mammals, have pituitaries with features common to reptiles as well as higher mammals. They have a portotuberal tract similar to birds and many reptiles, but the PT and remnant of the oral lobe are localized in the anterior end of the PD, typical of mammals. Development of the PI is variable, depending on the degree of contact in embryo between the glandular and neural epithelial primordia; most spiney anteaters have small PI, but the platypus has a large PI surrounding the NL.

12.1.2. Metatheria

In the Metatheria (Marsupials), the pituitary shows similarities with both Prototheria and Eutheria. The PT consists of a prominent PT externa and small PT interna, with strings of cells connecting them. The PT interna lies anteriorly within the PD, which is large and separated from the PI along most of the interface by the hypophyseal cleft. The PI contacts the NL only along the ventral and lateral aspects.

12.1.3. Eutheria

Only in this group of mammals does the true placenta develop, and moreover the brain shows much more differentiation and development than in lower mammals.

The PT, when present, often forms a complete, or near-complete, collar round the infundibular stem; however, it is quite absent from some genera including the pangolins and edentates. The PD is generally large and simple in form; the PI is variable; most frequently it does not extend dorsally and caudally around the NL, but in camels and llamas a large PI develops that completely surrounds the NL and is many cell layers thick. Similar, disproportionately large, PIs are found in other mammals adapted to arid environments; it is absent from the human AHP.

12.2. Cytology of the Adenohypophysis

There are five cell types of the mammalian PD, with a single MSH-secreting cell in the PI. The PD cells form a mosaic, sometimes with regional concentrations of particular cell types. Somatotrops are similar in form and size to their counterparts in lower vertebrates showing similar staining properties; they are acidophilic and OG+. They are normally well granulated with spherical secretory granules, approximately 400 nm in diameter; the maintenance of this population of stored granules depends on thyroid hormones. In primates particularly, but generally in most mammals, somatotrops are the most numerous cell type in nonpregnant animals, and there is a tendency for the cells to be large and more numerous in males than females. PRL cells are erythrosinophilic and distributed uniformly within the PD. They are larger than somatotrops and more irregularly shaped with pleomorphic secretory granules, up to 800 nm. The cells increase in size and number during pregnancy and lactation, particularly in primates, and show increased secretory activity, with considerable proliferation of highly organized RER, large accumulations of RNA, and enlargement of the Golgi apparatus. TSH cells of lower mammals (Figure 9) resemble those of lower vertebrates, and there may be a tendency for cells to be concentrated in the centromedial areas of the wings of the pituitaries, as in the rat pituitary for example. In the primates these cells are characterized by affinity for AB when used with PAS after oxidation by performic acid; they also stain with AF after chromation. Gonadotrops are scattered throughout the PD, but may tend to be more numerous in the cephalic region near the infundibular attachment; in larger mammals, including primates, basophilic cells predominate in the median zone, called the mucoid wedge, which is an extension of the PT interna. Immunocytochemical evidence indicates that both gonadotropins, LH and FSH, are secreted by the same cell. Adrenocorticotrops, as in other vertebrates, are small polyhedral cells, with long processes; they make up only about 4% of the total cell population and are generally distributed throughout the PD except for a narrow region beside the PI. In rats ACTH cells are AB+, PAS+, and AF+, and melanotrops of the PI are AB−, PAS+, and AF+.

12.3. Hypothalamus

The HT of the adult eutherian mammal comprises a diffuse array of neurons, many of which are aggregated into distinct nuclei (e.g., the magnocellular AF-positive nuclei), but others of which lie within ill-defined zones that are better described as "areas" rather than nuclei (Figure 28). In the preoptic region of the anterior HT there are two paired magnocellular nuclei, stainable with the classical NS stains, the supraoptic (NSO) and paraventricular (NPV), as in reptiles and birds. In addition, scattered "accessory" neurons may occur between the two nuclei. The perikarya are multipolar but are located away from the ventricle, without CSF-contacting dendrites, and are embedded in rich innervation. They are often found close to capillaries, since this area of the HT is much more highly vascular than in other vertebrates. The ultrastructure of these magnocellular neurons is very similar to that in other vertebrate groups, with a polar region of RER, a large lobed nucleus and 150–200 nm NSG, which are scattered down the basal process of the cell. Variations do occur, but are believed to represent different secretory phases rather

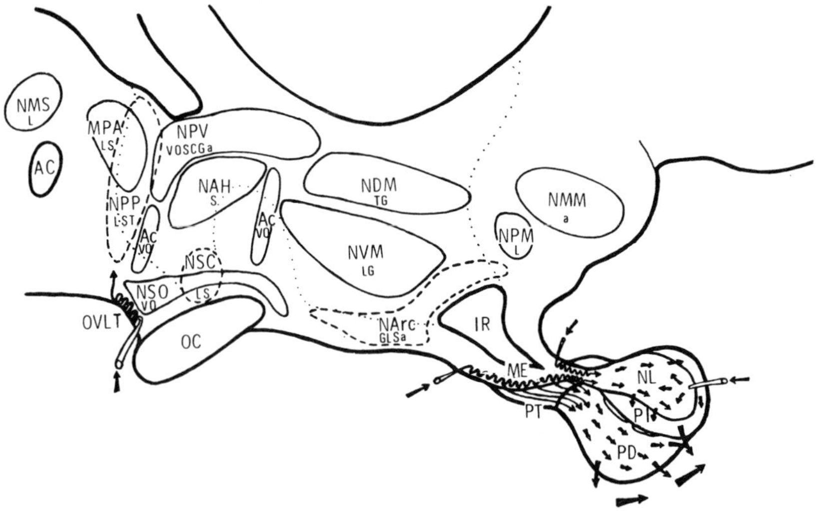

Figure 28. Hypothalamus and pituitary of a mammal. AC, anterior commissure; Ac, accessory preoptic nuclei; MPA, medial preoptic area; NAH, anterior hypothalamic area; NArc, arcuate nucleus; NDM, dorsal medial nucleus; NMM, mammillary nucleus; NMS, medial septal nucleus; NPM, premammillary nucleus; NPP, periventricular preoptic area; NPV, paraventricular nucleus; NSC, suprachiasmatic nucleus; NSO, supraoptic nucleus; NVM, ventromedial nucleus. Small type indicates neurosecretory substances detected in each nuclear region: a, amines; C, corticotropin-releasing hormone; G, growth-hormone-releasing hormone; L, luteinizing-hormone-releasing hormone; O, oxytocin; S, somatostatin; T, thyrotropin-releasing hormone; V, vasopressin.

than different classes of neuron. ICC studies, using antisera against the two mammalian octapeptide hormones, have demonstrated that oxytocin and vasopressin are produced by separate populations of neurons, but that both types of neurons are dispersed throughout both the NSO and NPV. Furthermore, oxytocin and vasopressin reactivity are also found in some parvocellular areas of the preoptic region, particularly the suprachiasmatic nucleus (NSC) in perikarya containing NSG smaller than those of the NSO and NPV (only 80–100 nm). The axons from these preoptic neurons converge to form tracts that pass down to the floor of the HT and mostly enter the infundibular stem and expand to form the NL. A proportion of the fibers, however, mainly from the NPV, enter the ME.

As mentioned earlier, division of the parvocellular (AF−) areas of the HT into definite nuclei is by no means clear cut; however, the major nuclear divisions (shown in Figure 28) are: in the periventricular zone of the HT: the NSC, the periventricular preoptic area (NPP), arcuate nucleus (NArc), or infundibular nucleus, and in the medial zone of the HT: the medial preoptic area (MPA), the anterior hypothalamic area (NAH), the dorsomedial and ventromedial nuclei (NDM and NVM), and the premammillary and mammillary nuclei (NPM and NMM). Much of the anterior periventricular area of the HT, including the NArc, is regarded as being the "hypophysiotropic area" of the HT, the presumptive site of neurons producing the releasing and inhibiting hormones controlling pituitary PD function. Surgical manipulation and stimulation of this area does, in most mammals studied, affect various pituitary functions, but verification of the sites of projection of the parvocellular neurons is made difficult by the fact that they are not normally stainable with any of the neurosecretory stains. Nevertheless, this has been done in some cases using the techniques of silver impregnation and retrograde peroxidase uptake, demonstrating that the axons do indeed terminate in the ME. Recently ICC has been used extensively in an attempt to localize the areas of the HT producing the various hypophysiotropic hormones. Generally results have proved difficult to interpret in that reactive perikarya and fibers are often observed scattered throughout the HT rather than being located in definite functional groups, and there is quite a wide variation in the results obtained by different workers and with different species. LH-RH immunoreactive perikarya have been reported scattered throughout the NPP, NSC, NVM, and MPA, but the main systems appear to be in the NArc, with fibers to the ME, and the MPA, with separate tracts of fibers to the ME, the OVLT, and extrahypothalamic areas, particularly the septal area, where the medial septal nucleus (NMS) also contains LH-RH-positive cells. ICC studies at the EM level have shown that the reaction is located in NSG of ca. 90–130 nm. Somatostatin immunoreactive perikarya are scattered subependymally in the walls of the third ventricle from the NPP through the NPV, NSC, and NArc down to the ME. Short fibers project to other neurons in most of the hypothalamic nuclei, especially the NArc, and longer fibers to extrahypothalamic regions, the ME, and possibly the OVLT. TRH-positive cells occur in the NDM and scattered through the NPP, with fibers to other tuberal nuclei and the ME. Recently antisera to CRH and GRH have been shown to stain perikarya in the NPV and nerve endings in the external ME. Furthermore, GRH-immunoreactive cells are found in the NArc, NVM, and NDM. Besides the known hypophysiotropic hormones, other peptides can be detected in hypothalamic neurons, for example, the gut-related peptides gastrin, cholecystokinin, and substance P and the ACTH-related peptides ACTH, LPH, and β-MSH, α- and β-endorphin, and enkephalin. Evidence is now accumulating that several of these peptides may play a physiological role as neurotransmitters or neuromodulators in pituitary hormone release—for example, endorphin in modulation of dopamine inhibition of prolactin release. It is particularly relevant, therefore, that this peptide has been localized in NArc neurons which project both to the ME and to other NArc neurons.

Aminergic cells of the bipolar CSF-contacting type seen in lower vertebrates are not found in the HT of mammals, but multipolar neuronal aminergic cells do occur, although they are difficult to demonstrate. Using the Falck–Hillarp fluorescence method (Section 16.4), and ICC using antisera to amines or to the enzymes involved in amine synthesis, aminergic fibers and varicosities can be observed throughout the HT, but cell bodies are restricted to a few, scattered in the periventricular HT from the preoptic region to the NMM. The highest concentration occurs in the NArc, from which fibers pass down to the ME.

The vascularization of the mammalian pituitary is highly complex compared to lower vertebrates, particularly in the primates and in man. There is a wide range of variation in the details of the system between different mammalian groups, but the general layout is usually the same. The arterial supply to the region is from branches of the internal carotid arteries. The superior hypophyseal branch supplies blood to the upper part of the NHP, the ME, and infundibular stem, and the inferior hypophyseal branch supplies the lower part of the NHP, the NL. There is usually no direct arterial supply to the PD, or to the largely avascular PI if present. On entering the ME, the arterioles are organized into a series of elaborate capillary formations. These may be either simple loops penetrating into the neural tissue before returning to the surface and emptying into the portal veins or larger, more complicated, "tufted" vessels, which are best developed in primates. Blood drains into the PD from the ME via several long portal vessels, and also from the capillary bed of the NL via a few short portal vessels (Figure 28). The significance of the latter supply varies, being less in those mammals retaining a well-developed cleft between the PD and the PI. Venous blood from all parts of the pituitary drains into the systemic circulation via the intracranial venous sinuses.

12.4. Neurohypophysis

The fine structure of the mammalian ME, primates excepted, is quite simple, the arrangement being similar to that found in the ME (or NHP) of lower vertebrates. The usual three layers exist: ependymal, fibrous, and palisade (Figure 14). The wall of the infundibular recess is lined by ciliated ependymal cells, with tanycytes interspersed among them. The tanycytes are distinguishable from the normal ependymal cells at the EM level since their surfaces bear microvilli, not cilia, their cytoplasm has a more active appearance, and they are "stretched out" into long basal processes that end on capillaries in the palisade layer. These characteristics have led to the hypothesis, still controversial, that hypophysiotropic substances may be released into the CSF and transferred by tanycytes across the ME from the third ventricle to the primary capillary plexus. The fibrous layer consists of various categories of NS fibers passing through to terminate more distally in the NHP, together with tanycyte processes and glial elements. In the external (palisade) layer of the ME, the several types of nerve fibers and tanycyte processes form endings on the perivascular BM of the fenestrated capillaries of the primary plexus.

The rest of the NHP consists of an infundibular stem of variable length, together with a bulbous NL. The complex is generally solid, but in a few groups of mammals the infundibular cavity extends into the stem, or even partially into the NL, in which case it is lined by ependymal cells similar to those of the ME. The bulk of the NL is equivalent to the fiber layer of the ME, consisting of axonal nerve fibers, often expanded into swellings packed with NSG ("Herring bodies"), and glial elements, including the large pituicyte cells. In the neurohemal contact areas of the NL the fiber endings, which again can be classified into several types, show a similar relationship with the capillaries as those in the ME. The capillaries of the NL, as in most vertebrates, are surrounded by a quite extensive network of perivascular spaces bounded by BM, which appear to be continuous with the fine intercellular spaces running between the NS endings. This is believed to provide a route by which NS material released from the more distant endings is conveyed to the capillary.

The nerve fibers and endings can be classified into several different categories, based on granule size and morphology, which appear to be similar in all parts of the NHP, for any particular species. Those with the smallest granules, DCV of 60–80 nm, similar to the type B fibers of lower vertebrates, probably correspond to the aminergic (dopaminergic and serotonergic) fibers, which have been demonstrated throughout the ME and NL by a combination of fluorescence histochemistry, autoradiography, and ICC. The remaining fibers, with NSG in excess of 100 nm (type A), are believed to contain the two neurohypophyseal octapeptide hormones and the peptide pituitary releasing/inhibiting hormones. In a few cases this has been proved using EM–ICC. In the rat, for example, vasotocin and oxytocin have been localized in similar fibers containing 150–200 nm NSG. Both fiber types occur intermingled throughout the fibrous layer of the ME and the interior of the NL, with fewer numbers in the ME external zone. Somatostatin has been localized in scattered fibers in the external zone containing granules of 90–110 nm, and LH-RH-positive fibers show a similar distribution and contain slightly smaller granules (75–95 nm). Several other peptides have been localized in the ME and NL including TRH, CRH, and the ACTH-related peptides, but cannot yet be assigned to any particular category of fiber.

The mammalian PI, when present, appears to receive a dual innervation by both aminergic and vasotocinergic fibers, which penetrate the BM between NL and PI and make contact with the PI cells.

13. Chemistry of Pituitary and Hypothalamic Hormones

13.1. Glycoprotein Hormones

The pituitary glycoprotein hormones of all vertebrates, LH, FSH, and TSH, are composed of two peptide subunits, α and β, linked by noncovalent bonds, plus a carbohydrate moiety, giving a total molecular weight of approximately 30,000 daltons. The α subunits of the three hormones in a given species are identical or virtually so, but the β subunits are different, although often sharing homologies in amino acid sequence. It is the β subunit which confers biological specificity on the molecule, but full biological activity only resides in the intact glycoprotein. Amino acid sequence of both subunits and analysis of carbohydrate content have been determined for gonadotropins and thyrotropins of representatives of all vertebrate classes; the degree of homology between them indicates the ancient lineage of the molecules and conservation of their composition and structure. Moreover, there are significant sequence homologies between GH and α subunits of glycoprotein hormones, which suggest that they both evolved from a common ancestral molecule of

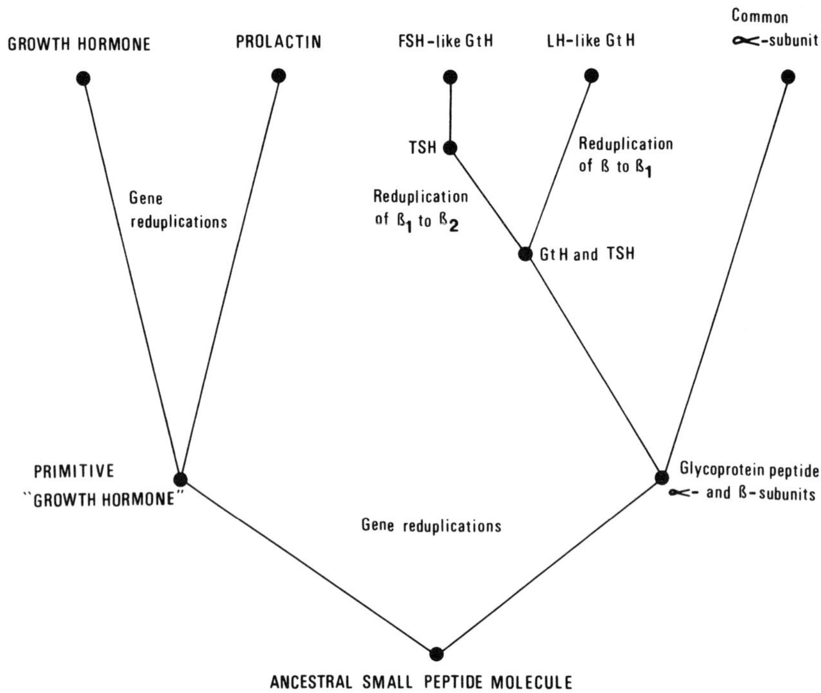

Figure 29. Hypothetical evolution of pituitary hormones.

29–35 residues. Gonadotropins are considered to be the most ancient of the glycoprotein hormones; TSH was probably derived by gene duplication of the β subunit, and from this the β-subunit characteristic of FSH was later derived. Figure 29 gives a résumé of these proposed evolutionary trends.

13.1.1. Gonadotropins

In fish and squamate reptiles it appears that probably there is only one gonadotropin which serves the purpose of both LH and FSH of other vertebrates, although there may be a second hormone, low in carbohydrate, concerned with vitellogenin incorporation in fish oocytes. Fish GtH is composed of α and β subunits and a carbohydrate moiety. Amino acid composition of the α and β subunits of carp GtH show greatest similarity with tetrapod LH subunits, and in general, fish GtHs have high proline content, again similar to LH and different from FSHs, which are poor in proline. Table IV shows amino acid contents of α and β subunits of some vertebrates.

Carbohydrates of gonadotropins are a variable proportion of the total molecular weight, most frequently between 10 and 15%. They include hexose, hexosamine, and sialic acid, which is only a very small part of the total carbohydrate in LH and fish GtH, but a significantly higher proportion in FSH. Carbohydrate is attached to both peptide subunits. The presence of carbohydrate is important for biological activity of the molecule, because removal by enzyme action, for example, sialic acid by neuraminidase, virtually destroys biological activity of FSH. Table V shows the carbohydrate content of LH and FSH of several vertebrate species.

13.1.2. Thyrotropin

Analysis of TSH α and β subunits of submammalian vertebrates has been considerably hampered by the difficulty in gathering enough material, since TSH is found only in low concentrations, and the pituitaries are generally small. TSH has been prepared from bullfrog (Amphibia) pituitary, but only TSH of sea turtle (Reptiles) has been analyzed. Table VI shows the amino acid composition of sea turtle and ox TSH. They are obviously very similar, the major differences being much lower content of threonine and higher leucine in sea turtle. Sequence homologies must exist between the various tetrapod TSH molecules because they all show some degree of stimulating activity in mammalian or chick thyroid cell assays. However, extracts of teleost pituitaries do not stimulate mammalian thyroid cells;

Table IV
Amino Acid Composition of α and β Subunits from Various LH Molecules and Carp GtH

	α Subunits				β Subunits			
Amino acids	Sheep	Turkey	Sea turtle	Carp	Sheep	Turkey	Sea turtle	Carp
Lys	10	10.1	12	10	2	2.1	3	4
His	3	3.0	4	3	3	0.4	2	3
Arg	3	5.1	4	4	8	7.8	4	5
Asp	6	7.1	7	11	5	6.2	6	10
Thr	9	9.4	9	6	7	7.7	7	10
Ser	6	5.2	5	6	8	6.5	10	7
Glu	8	7.2	7	6	6	9.4	7	11
Pro	7	6.5	7	6	20	15.3	13	13
Gly	4	4.6	5	3	7	17.5	10	4
Ala	7	6.6	4	5	8	10.8	14	2
1/2Cys	10	8.8	9	10	12	8.1	12	12
Val	5	3.4	4	8	8	9.9	11	12
Met	4	2.6	3	3	3	0.0	1	2
Ile	2	4.5	4	2	5	4.0	3	3
Leu	2	4.4	4	5	12	8.5	6	9
Tyr	5	2.9	4	5	2	3.3	4	6
Phe	5	4.5	5	3	3	2.0	5	3
Trp	—	—	—	0	—	—	—	0
Total	96	96	96	96	119	119	119	119

Table V
Carbohydrate Composition[a] of Some Vertebrate Gonadotropins

	Species	Hexose	Hexosamine	Sialic acid
GtH	Carp	12.5	8.7	0.35
	Tilapia	5.9	5.8	0.7
LH	Bullfrog	3.6	5.5	0.2
	Sea turtle	6.4	8.9	1.9
	Alligator	3.9	7.3	0.4
	Chicken	5.2	7.1	1.4
	Sheep	7.2	9.1	0.4
FSH	Bullfrog	7.2	9.8	3.0
	Sea turtle	4.9	6.6	0.6
	Chicken	3.1	3.2	0.8
	Sheep	5.7	4.5	2.8

[a]g/100 g glycoprotein.

nevertheless antiserum to human TSH β subunit specifically stains thyrotrops of several teleost species indicating antigenic similarities between the molecules. Biologically specificity of the molecule resides in the β subunit, but this too, at least in mammals, shows common sequence homologies with FSH and LH, and with chorionic gonadotropin, the placental gonadotropin of the same species.

Table VI
Amino Acid Composition of Sea Turtle TSH Compared to Bovine TSH

Amino acid	Sea turtle	Bovine TSH
Lysine	17.9	19
Histidine	4.5	6
Arginine	8.0	7
Aspartic	19.2	15
Threonine	10.4	20
Serine	12.6	11
Glutamic	16.6	15
Proline	15.2	14
Glycine	11.4	8
Alanine	12.7	13
1/2 Cystine	11.4	22
Valine	11.4	11
Methionine	5.3	9
Isoleucine	9.5	8
Leucine	11.1	6
Tyrosine	17.6	16
Phenylalanine	9.7	9

The carbohydrate of TSH includes hexoses, hexosamines, and sometimes a very small proportion of sialic acid, which in this case does not seem to be essential for biological activity, as it is for LH and FSH.

13.2. Tropic Hormones—Growth Hormone and Prolactin

Growth hormones and prolactins belong to the same family of peptides derived probably from a small ancestral molecule, approximately 30 amino acid residues, by gene reduplication and point mutation (see Figure 29). Whether PRL or GH was the primogenitor of the other is still a matter for conjecture, but their close chemical relatedness is clear. Both molecules are simple peptide chains; in mammals GH is characterized by two disulfide bridges and a single tryptophan residue, whereas PRL has three disulfide bonds and two tryptophan residues. The molecules are composed of between 190 and 199 residues and weigh about 21,000–23,000 daltons; in solutions of high salt concentration, GH particularly tends to aggregate and form dimers. The close chemical relationship between mammalian GH and PRL is shown in Figure 30.

Immunologically the molecules are similar, showing cross-reactions with antiserum to rat GH to a degree broadly in line with their phyletic distance from mammals. Similar relatedness is shown by their activity in stimulating epiphyseal growth in rat tibia (standard bioassay), except for bullfrog (Amphibia) GH, which is unexpectedly potent. Until recently there was no evidence that GH of modern teleosts was active in this assay; however, purified tilapia GH has low, but detectable, growth-stimulating activity. Pituitary extracts of shark (Elasmobranch), sturgeon (Chondrostei), and lungfish (Dipnoi) are all stimulatory. Table VII summarizes results of tibia tests of various vertebrate growth hormones.

Table VIII illustrates the remarkable uniformity of amino acid composition of

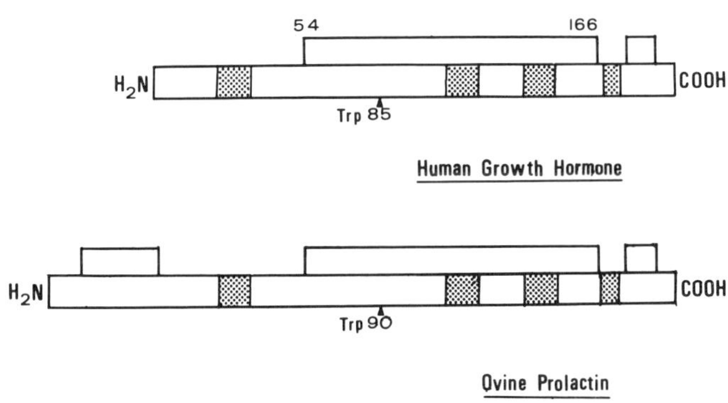

Figure 30. Schematic drawing of human growth hormone and ovine prolactin. Square brackets show disulfide bridges, shaded blocks show internal-sequence homologies; prolactin also differs by the additional bond at the extended amino-terminal end.

Table VII
Potencies of Growth Hormone Determined in the Rat Tibia Assay

Species	Potency[a]
Sturgeon	0.40
Tilapia	0.05
Bullfrog	1.15
Leopard frog	0.36
Snapping turtle	0.24
Sea turtle	0.12
Ostrich	0.94
Duck	0.16
Marsupials	0.03
Mammals	1.00

[a]Bovine GH standard = 1.0. Marsupial GHs were nonparallel.

GH from various vertebrate species, only the content of aspartic acid, leucine, and phenylalanine showing much variation.

The NH_2-terminal amino acid sequences too are similar (Table IX). However the N terminus of sturgeon GH is tyrosine, with isoleucine, leucine, and valine in tilapia differing from the phenylalanine found in most other vertebrates so far examined.

Few prolactins of nonmammalian vertebrates have been isolated and analyzed, but there is at least one from each class. The general picture is that there are only two disulfide bridges in PRL of lower vertebrates, compared with three in mammals, but in this respect too, bullfrog PRL more closely resembles mammalian PRL, in having three disulfide bridges. Tilapia and salmon (Teleostei) PRL more closely resemble mammalian GH in having only a single tryptophan residue as well as only two disulfide bridges. Table X shows the amino acid composition of PRL of some vertebrate species. It is tempting to infer that the presence of two tryptophans and three disulfide bridges is related to the lactotropic and lactogenic properties of mammalian PRL, but recent studies show that a teleost (Tilapia) PRL can stimulate mammalian casein gene expression.

13.3. Melanocorticotropins

MSH secreted by the cells of the pars intermedia (PI) and ACTH secreted by pars distalis (PD) corticotrops are both derived from a common precursor molecule, proopiocortin, by a cascade of enzyme cleavages. The precursor molecule is a glycoprotein of 31,000 daltons. In the corticotrops of the PD proopiocortin is cleaved to produce ACTH and β-lipotropin, but further splitting in the PI cells produces α-MSH, corticotropic intermediate lobe peptide (CLIP), β-endorphin, β-

Table VIII
Amino Acid Composition of Various Species of Avian, Reptilian, Amphibian, and Piscine GH[a]

Amino acid	Sheep	Ostrich	Snapping turtle	Bullfrog	Tilapia	Sturgeon
Lysine	11	14	12	11	8	15
Histidine	3	5	4	6	5	4
Arginine	13	13	13	17	11	11
Aspartic	16	18	20	30	19	20
Threonine	12	10	9	12	12	10
Serine	13	13	14	12	21	18
Glutamic acid	24	25	24	18	29	22
Proline	6	9	7	6	7	7
Glycine	10	8	9	7	7	8
Alanine	15	12	11	6	8	10
1/2 Cystine	4	4	4	4	4	5
Valine	6	8	8	11	6	9
Methionine	4	3	4	4	1	5
Isoleucine	7	6	7	8	9	6
Leucine	27	27	16	18	27	14
Tyrosine	6	7	7	10	7	6
Phenylalanine	13	10	12	11	7	11
Tryptophan	1	1	1	1	1	1

[a]Calculated on the basis of 191 residues per mole; values rounded to nearest integer. [From Papkoff, H., 1982, A comparative view of the chemistry and function of the anterior pituitary hormones, in: *Pituitary Hormones and Related Peptides* (M. Motta, M. Zanisi, and F. Piva eds.), Academic Press, London and New York, pp. 17–30.]

MSH, and α-endorphin. The structural relationships of these molecules are shown in Figure 31.

The proteolytic enzyme action invariably splits the precursor molecule at a point between paired basic amino acids (e.g., Arg-Arg or Lys-Lys). Production of proopiocortin by the PD and PI cells, therefore, is similar, but enzyme complements

Table IX
N-Terminal Amino Acids of Some Vertebrate Growth Hormones

		1				5
Sheep	H_2N -	Ala -	Phe -	Pro -	Ala -	Met
Duck and pigeon	H_2N	———	Phe -	Pro -	Ala -	Met
Snapping turtle	H_2N -	Ala -	Phe -	Pro -		
Bullfrog	H_2N	———	Phe -	Pro -	Phe	

Table X
Amino Acid Composition of Some Prolactins

	Ostrich	Bullfrog	Sea turtle	Chum Salmon	Sheep
Lys	11	9	12	9	9
His	8	7.5	6	5	8
Arg	13	13	9	7	11
Asp	23	22	24	16	22
Thr	6	8	8	9	9
Ser	14	15	15	26	15
Glu	26	26	32	17	22
Pro	7	11	13	13	11
Gly	12	11	14	9	11
Ala	11	8	15	6	9
1/2 Cys	4	6	3	4	6
Val	10	8	8	7	10
Met	2	3	2	9	7
Ile	10	13	7	10	11
Leu	28	25	20	26	23
Tyr	5	4	4	2	7
Phe	7	7	7	7	6
Trp	2	ND	ND	1	2

of the two cell types result in secretion of different hormones. Thus, ACTH is 39 amino acids; α-MSH corresponds to residues 1–13 of ACTH; β-LPH is 91 amino acids and includes β-MSH within its core, residues 41–58, as well as β-endorphin, corresponding to residues 61–91. ACTH, α-MSH, β-MSH, and LPH are further related by a common heptapeptide core, Met-Glu-His-Phe-Arg-Phe-Gly.

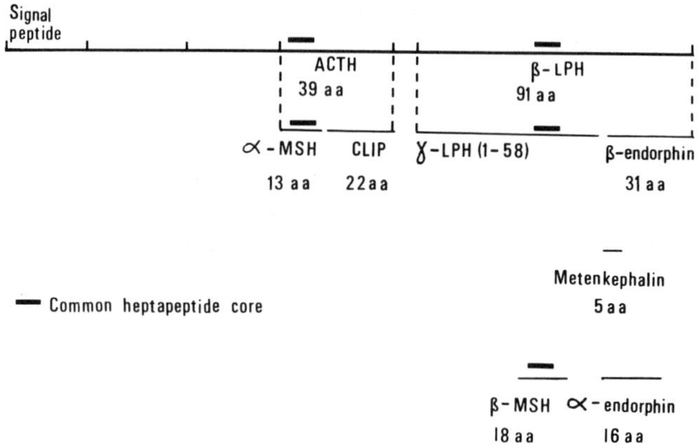

Figure 31. Enzyme cleavage products of proopiomelanocortin.

13.3.1. Adrenocorticotropin

The molecule is 4500 daltons, composed of 39 amino acid residues, with the N-terminal 24 amino acids constant in mammals. This is regarded as being the biologically active end of the molecule, but residues 25–33 are variable between mammalian species. The complete sequence for dogfish ACTH (*Squalus acanthias*), the only nonmammalian ACTH completely analyzed, shows different residues at positions 13 and 15, but 11 amino acids at the C terminal are variable, indicating relative conservation of the biologically active portion of the molecule. Composition of bird ACTH (chicken and ostrich) also shows only minor differences from the mammalian molecule. Cross-reactions between pituitary extracts of species of all vertebrate classes with antiserum to ACTH residues 1–24 show that the molecule has antigenic similarities throughout. Similarly, the biological activity has been preserved because extracts of pituitaries of teleosts, amphibians, reptiles, and birds all stimulate steroidogenesis by isolated mammalian adrenal cells; thus, the molecule has been highly preserved during evolution, probably reflecting the importance of adrenal steroid production for successful life.

13.3.2. α-Melanocyte-Stimulating Hormone (α-MSH)

This peptide of 13 amino acid residues corresponds to positions 1–13 of ACTH. It is highly conserved, the sequence being identical in all mammals examined, and in dogfish α-MSH the only difference is at residue 13, which is methionine instead of the valine of mammal α-MSH. Following enzymatic cleavage of proopiocortin, α-MSH is generally acetylated at the N-terminal end and amidated at the C terminal to protect it from rapid degradation. Immunoreactive α-MSH occurs in amphibian, reptilian, and teleostean pituitary extracts, again indicating conservation of the molecular composition, which is probably related to the essential steroidogenic activity of ACTH. Figure 32 shows the composition of mammalian and dogfish α-MSH.

13.3.3. Lipotropin, β-MSH, and β-Endorphin

Cleavage of β-LPH produces β-MSH and β-endorphin. These molecules show much more variation in their amino acid composition, even within the mammals, and β-MSH from dogfish (*Squalus* and *Scyliorhinus*) is considerably different from the mammalian molecule (See Figure 33).

Mammalian	CH_3CO-Ser-Tyr-Ser-Met-Glu-His-Phe-Arg-Trp-Gly-Lys-Pro-Val-NH_2
Dogfish	——————————————————-Met-NH_2

Figure 32. Amino acid composition of mammalian and dogfish α-MSH.

Bovine	H-Asp-Ser-Gly-Pro-Tyr-Lys-Met-Glu-His-Phe-Arg-Trp-Gly-Ser-Pro-Pro-Lys-Asp-OH
Squalus	H-Asp-Gly-Asp-Asp-Tyr-Lys-Phe-Gly-His-Phe-Arg-Trp-Ser-Val-Leu-OH

Figure 33. Amino acid composition of mammalian and dogfish β-MSH.

The full amino acid sequence of β-LPH has only been worked out for some mammals and in all of them is composed of 91 amino acids. The other cleavage product of β-LPH, β-endorphin, shows much more uniformity, at least in the mammals so far examined, suggesting it has an important biological function.

13.4. Neurohypophyseal Hormones

All the NHP hormones are nine-residue peptides (nonapeptides), but distinct structural variations occur between different vertebrate groups. In most vertebrates two hormones are present, a basic "vasopressor" principle and a neutral "oxytocic" principle. The NHP peptides are produced not, as the name might imply, within the NHP, but in the NS cells of hypothalamic nuclei. Synthesis and storage is in the form of a larger prohormone, with the peptide bound to a specific carrier protein, neurophysin, together with a glycopeptide. During transport of the hormone, packaged in NS granules, down axons to the sites of release (median eminence, pituitary stalk, neural lobe), it is believed to dissociate from its neurophysin carrier. It is, however, the neurophysin, not the active peptide, which gives the neurohypophyseal tract its characteristic staining properties (Section 2.2).

13.4.1. Composition of Neurohypophyseal Hormones

The first NHP hormones to be extracted from the pituitary, purified, and chemically characterized were the mammalian hormones vasopressin and oxytocin. Since then a whole range of different, but structurally similar, peptides has been discovered in the various vertebrate groups. All are composed of nine amino acids, the first and sixth being cysteine residues, linked by their sulfur atoms to form one cystine residue (hence the molecule is usually referred to as an octapeptide rather than a nonapeptide). This disulfide bridge results in a six-amino-acid ring, to which a tail chain of three amino acids is attached (Figure 34). In all the octapeptides so far characterized the amino acids at positions 1, 5, 6, 7, and 9 are invariable, whereas those at 3, 4, and 8 have apparently been liable to evolutionary change.

13.4.2. Structure and Occurrence in Vertebrate Groups

The amino acid sequences of the usual mammalian hormones oxytocin (OT) and arginine vasopressin (AVP) are shown in Figure 34. There are differences between the two at only two positions: position 3 (in the ring) is isoleucine in OT

	Neutral Hormone	Basic Hormone
MAMMALS *Placentals* All except pigs	1 2 3 4 5 6 7 8 9 Cys-Tyr-*Ile*-Gln-Asn-Cys-Pro-*Leu*-Gly-NH_2 Oxytocin	1 2 3 4 5 6 7 8 9 Cys-Tyr-*Phe*-Gln-Asn-Cys-Pro-*Arg*-Gly-NH_2 Arginine vasopressin
Domestic pigs	Oxytocin	1 – 2 – 3 – 4 – 5 – 6 – 7 - Lys - 9 Lysine vasopressin
Marsupials Macropodidae	Oxytocin	Lysine vasopressin + 1 - Phe – 3 – 4 – 5 – 6 – 7 – 8 – 9 Phenypressin
Phalangeridae	1 – 2 – 3 – 4 – 5 – 6 – 7 - Ile - 9 Mesotocin	Arginine vasopressin
BIRDS, REPTILES, AMPHIBIANS AND LUNGFISHES	Mesotocin	1 – 2 – Ile – 4 – 5 – 6 – 7 – 8 – 9 Arginine vasotocin
TELEOSTEAN AND GANOID FISHES	1 – 2 – 3 - Ser - 5 – 6 – 7 - Ile - 9 Isotocin	Arginine vasotocin
ELASMOBRANCHS *Selachii* Skates and Rays	1 – 2 – 3 - Ser - 5 – 6 – 7 - Gln - 9 Glumitocin	Arginine vasotocin
Sharks and Dogfishes	1 – 2 – 3 – 4 – 5 – 6 – 7 - Val - 9 Valitocin + 1 – 2 – 3 - Asn - 5 – 6 – 7 – 8 – 9 Aspar(g)tocin	Arginine vasotocin
Holocephali CYCLOSTOMES	Oxytocin ------	Arginine vasotocin Arginine vasotocin

Figure 34. Structure and phylogeny of vertebrate neurohypophyseal hormones. Residue differences between the two mammalian molecules are underlined; variable residues throughout each series (basic or neutral) are boxed in the mammalian molecule. Numbered residues correspond to those of the mammalian molecule.

but phenylalanine in AVP, and position 8 (in the tail) is leucine in OT but arginine in AVP. Hence AVP can be referred to as (Phe^3, Arg^8)-oxytocin. In some members of the pig family a different vasopressin occurs, lysine vasopressin (LVP) or (Phe^3,Lys^8)-OT (see Figure 34), in addition to or in place of AVP. The presence of LVP has also been reported in some strains of mice. In the marsupials it is difficult to generalize because of the few species examined and the variable results obtained, but the kangaroos and wallabies of family *Macropodidae* have OT, LVP, and an additional vasopressor principle phenypressin (Phe^2,Phe^3,Arg^8)-OT (Figure 34). In another marsupial family *Phalangeridae,* however, none of these peptides occur; instead AVP and mesotocin (Ile^8)-OT (Figure 34) are present. All species of nonmammalian tetrapods (birds, reptiles, and amphibians) so far studied possess mesotocin as the neutral principle together with arginine vasotocin (AVT), or (Arg^8)-OT as the basic principle. In lungfish, OT may occur in addition to mesotocin and AVT. In the teleost fishes and in the ganoid fish *Polypterus,* AVT also occurs, but the neutral principle is isotocin (Ser^4,Ile^8)-OT (Figure 34). In contrast to the other vertebrate classes, the elasmobranchs show a marked heterogeneity in these hormones. AVT appears to occur throughout in very small amounts, but the neutral principle is different in each subclass. Of the selachii, rays have glumitocin (Ser^4,Gln^8)-OT, whereas sharks and dogfishes have both valitocin (Val^8)-OT and/or aspargtocin (Asn^4)-OT, and the ratfish (Holocephali) has OT itself. In the cyclostomes AVT is the only octapeptide so far detected.

13.4.3. Phylogenetic Conclusions

In view of the small number of species examined within each vertebrate class, it is still rather premature to draw conclusions about the evolution of these peptides. It is, however, tempting to propose AVT as the ancestral NHP hormone, since it appears to occur in all lower vertebrates, with a single gene duplication giving rise to two lines of peptides, the basic vasotocinlike line and the neutral oxytocinlike line. In the basic line the gene appears to have remained stable until the evolution of mammals, when the change to AVP could be explained by a single base change in the amino acid codons. A duplication of the AVP gene, followed by a single base change, in the evolution of the pigs may have resulted in coproduction of LVP. The whole scheme has now, however, been confused by two factors: the finding of AVT in fetal mammals, and the discovery of LVP and a second basic octapeptide in marsupials. Further gene duplications would have to be postulated to account for this. The evolution of the peptides in the oxytocinlike series is even more obscure. The presence of OT in the ''ancient'' chimaeriformes would tend to suggest that this was the first neutral peptide to evolve. The amino acid substitution at position 8 to give mesotocin can be explained by a single base change in the codon at the time of evolution of the modern nonmammalian tetrapods. A similar change would also have been necessary in the evolution of the Phalangeridae marsupials. Alternatively, the mammals (marsupials included) may have evolved from mesotocin-producing tetrapods and undergone a ''back mutation'' of the gene after the branching off of one marsupial line. A single base change would also account for the evolution of valitocin in sharks. Unfortunately, the substitutions at position 4 to give aspargtocin, glumitocin, and isotocin cannot simply be explained by single base changes in the genetic code: hypothetical intermediate peptides would be necessary in the scheme.

13.4.4. Neurophysins

These NHP hormone carrier proteins have now been characterized in seven species of mammal (man, rat, cattle, sheep, pig, horse, whale), and two or three distinct components are found in each. Originally these were classified as neurophysins I, II, (and III) based on their electrophoretic migration characteristics, but more recently they have been classified as two phylogenetic lines named MESL- and VLDV-neurophysins according to the initial letters of the amino acid residues at positions 2, 3, 6, and 7. All MESL-neurophysins so far sequenced are polypeptides of 93–95 residues, molecular weight approximately 10,000, with an unusually high content of proline and glutamic acid, and seven disulfide bridges. The sequence is highly conserved in the seven species so far studied: 80% of the chain is identical, and the few substitutions occur mainly at the C-terminal end of the molecule. Analysis of the VLDV-neurophysins has proceeded more slowly than for the MESL series, but, in the pig at least, there are considerable differences in primary structure from the MESL sequence. There is immunocytochemical evidence of similar proteins in the preopticoneurohypophyseal tract in all classes of vertebrates.

Although both MESL- and VLDV-neurophysins bind the two mammalian octapeptides, and those of other vertebrates, equally well *in vitro,* in the preopticoneurohypophyseal neurons MESL-neurophysin is associated with AVP and VLDV with OT. It is now known that both octapeptide hormone and specific neurophysin originate from a common precursor molecule. Gene analysis in the ox has predicted a 147-residue "preprovasopressin," containing AVP, MESL-neurophysin, and a glycopeptide (copeptin), and a 106-residue "preprooxytocin," containing OT, VLDV-neurophysin, and histidine. Neurophysins and copeptin appear to be secreted into the blood in the NHP, along with the octapeptide, but there is no good evidence that they have any physiological function.

13.5. Hypophysiotropic Hormones

13.5.1. Thyrotropin-Releasing Hormone (TRH)

TRH was the first releasing factor to be characterized in the late 1960s. Purified extracts of pig and sheep hypothalami with TSH-releasing activity were found to contain the amino acids glutamic acid, histidine, and proline in equimolar ratio. When the tripeptide Glu-His-Pro was synthesized, it proved to be chromatographically identical to natural TRH and to have the same biological activity.

Thyrotropin-releasing Hormone (TRH)

pyr-Glu-His-Pro-NH_2

Somatostatin (SRIF)

H-Ala-Gly-Cys-Lys-Asn-Phe-Phe-Trp-Lys-Thr-Phe-Thr-Ser-Cys-OH

Growth Hormone-releasing Hormone (GH-RH)

H-Tyr-Ala-Asp-Ala-Ile-Phe-Thr-Asn-Ser-Tyr-Arg-Lys-Val-Leu-Gly-Gln-Leu-Ser-Ala-Arg-Lys-Leu---

--Leu-Glu-Asp-Ile-Met-Ser-Arg-Gln-Gln-Gly-Glu-Arg-Asn-Gln-Glu-Gln-Gly-Ala-Arg-Val-Arg-Leu-NH_2

Luteinizing Hormone-releasing Hormone (LH-RH)

pyr-Glu-His-Trp-Ser-Tyr-Gly-Leu-Arg-Pro-Gly-NH_2

Corticotropin-releasing Hormone (CRH)

H-Ser-Gln-Glu-Pro-Pro-Ile-Ser-Leu-Asp-Leu-Thr-Phe-His-Leu-Leu-Arg-Glu-Val-Leu-Glu-Met-Thr--

--Lys-Ala-Asp-Gln-Leu-Ala-Gln-Gln-Ala-His-Ser-Asn-Arg-Lys-Leu-Leu-Asp-Ile-Ala-NH_2

Figure 35. Chemistry of hypothalamic peptidergic releasing hormones.

A full structural analysis of natural TRH was later carried out, confirming a cyclical tripeptide with protected N- and C-terminal amino acids (see Figure 35).

Synthetic TRH has been tested for biological activity in various vertebrate groups. In the lower vertebrates it is usually found to be ineffective in stimulation of TSH release, but appears to have other, often unrelated, functions, such as a role in control of prolactin secretion (Section 14.2.1). Using RIA and ICC, immunoreactive TRH has been demonstrated in high concentrations not only in the hypothalamus, but in other areas of the central nervous system, the pineal gland, and the skin of many vertebrates, and in neural tissues of amphioxus and even snails. Extracts from these extrahypothalamic tissues have in some cases been shown to contain truly biologically active TRH, although no characterization of the molecule from a submammalian vertebrate has yet been done, because of the requirement for such large quantities of hypothalami for extraction. It seems unlikely, however, considering the simple tripeptide structure, that there is much scope for evolutionary modification of the molecule without drastically affecting its biological activity and immunoreactivity.

13.5.2. Somatostatin (SRIF)

The presence of two antagonistic principles in mammalian hypothalamic extracts, one promoting growth hormone (GH) release and one inhibiting release, has been known since the late 1960s. The first of these to be chemically characterized was the inhibiting factor, named somatostatin or somatotropin release-inhibiting factor (SRIF), which was isolated and purified in 1973, from 500,000 ovine hypothalamic fragments and had potent GH-inhibiting activity in pituitary cell cultures.

Structural analysis showed that the active principle was a tetradecapeptide (sequence shown in Figure 35) with a disulfide bridge between cysteine residues at positions 3 and 14. Unlike all other HT hormones so far characterized, there is no pyroglutamyl residue at the N terminal. Porcine SRIF has now also been sequenced and is identical to ovine SRIF. Native SRIF is a cyclic tetradecapeptide, which may originate from a 28-residue precursor. Linear SRIF is marginally less active in bioassays than the cyclical hormone: cyclization of the linear form appears to take place both *in vivo* and *in vitro*. Neither the disulfide bridge nor the N-terminal dipeptide H-Ala-Gly is essential for biological activity.

RIA and ICC have shown that immunoreactive SRIF is located not only in the mammalian HT and NHP but in extrahypothalamic brain tissue, particularly the pineal, distributed throughout the CNS, and in cells and nerve fibers in many parts of the alimentary tract, including the pancreas, stomach, and salivary glands. In the submammalian vertebrates, immunoreactive SRIF has been reported in both the brain (HT included) and gastrointestinal tract in all classes, and also in the esophagus and stomach mucosa of a tunicate and the CNS of invertebrates. In all cases, when tested, the material extracted has been found to be indistinguishable from mammalian SRIF in both biological activity and immunological properties. This suggests that SRIF is a phylogenetically ancient molecule, which perhaps originally had a general depressant neurotransmitter or neuromodulator function on secretory neurons and later became more specifically involved in regulation of GH secretion.

13.5.3. Growth-Hormone-Releasing Hormone (GRH)

The presence of a factor stimulating GH release in mammalian hypothalamic extracts has been known since the late 1960s; however, isolation and characterization have only recently proved successful. GRH from porcine bovine and ovine hypothalami all have 44 amino acid residues with -COOH terminal amidated peptides; human hypothalamic GRH also has 44 amino acids and that of the rat has 43 with a free C terminus. A similar peptide had earlier been extracted from pancreatic tumors of two humans showing acromegaly. The compound, which was 44 residues from one tumor, and only 40 from the other, was given the trivial name somatocrinin. Studies with these and other synthetic molecules showed that residues 1–29 were essential for biological activity. The sequence of GRH is shown in Figure 35. Chemically GRH is related to the secretin, glucagon, porcine intestinal peptide (PHI), and vasoactive intestinal peptide (VIP) group of peptides. Immunocytochemical studies show that most GRH neurons are located in the arcuate nucleus of the infundibulum with few scattered neurons in the NPV, NVM, and NDM. Both the hypothalamic and tumor-derived releasing hormones are potent stimulators of GH secretion *in vivo* and *in vitro*.

Physiological evidence points to the existence of GRH in several groups of lower vertebrates, but there is no evidence yet about the chemistry of such factors. To date there is very little evidence for any action of mammalian GRH or somatocrinin on GH release in nonmammalian vertebrates.

13.5.4. Luteinizing-Hormone-Releasing Hormone (LH-RH)

A decapeptide (Figure 35) isolated and characterized from porcine, ovine, and bovine hypothalamic extracts was shown to be luteinizing-hormone-releasing hormone (LH-RH or LRH) in 1971. Subsequently the decapeptide was synthesized, and surprisingly it proved effective in releasing not only LH from the pituitary but to a lesser extent FSH. Because of this the decapeptide has frequently been referred to as LH/FSH-RH or simply gonadotropin-releasing hormone (GnRH). There is, however, some physiological evidence for the existence of a separate releasing factor for FSH (Section 14.2.6), and partial purifications of a FSH-RH have occasionally been reported, so the matter of whether there are one or two releasing factors for the gonadotropins still remains controversial. Numerous synthetic analogs of LH-RH have been synthesized and studied, some with reduced and some with enhanced biological activity. It appears that the residues that are most crucial for stimulating activity are $pyrGlu^{1}$, His^{2}, Trp^{3}, Arg^{8}, and Pro^{9}.

Extractable gonadotropin-releasing activity has been found in the HT of birds, amphibians, and fishes, and the synthetic decapeptide is active in stimulating gonadotropin release in these groups. Immunoreactive LH-RH has been demonstrated by either ICC or RIA in all classes. Recent immunochemical investigations, however, show that the LH-RH molecule is likely to differ in various vertebrates. Reptilian and piscine LH-RH have an identical composition but differ at two positions (Trp7, Leu8) from the amphibian and mammalian molecules (Leu7, Arg8). Chicken LH-RH appears to differ from mammalian LH-RH only at position 8, which is Gln. In spite of these differences in the sequence of the molecule, LH-RH of all vertebrates binds to receptors of all lower vertebrates as evidenced by biological activity. However, mammalian receptors appear to specifically respond only to mammalian LH-RH.

13.5.5. Corticotropin-Releasing Hormone (CRH)

Corticotropin-releasing hormone (CRH) was the first hypothalamic releasing factor to be proposed, but it is only very recently that the structure of a true CRH (Figure 35) has been elucidated. In the mid-1950s it was shown that hypothalamic extracts caused the release of ACTH from cultured pituitary cells; however, the presence of other substances in the extracts, particularly vasopressin, potentiated ACTH release. There is now good evidence that a peptide CRH distinct from vasopressin exists; in the sheep this has been characterized and is a 41-residue peptide with the sequence shown in Figure 35. This peptide has been synthesized and shown to be active in release of ACTH and the structurally related peptide β-endorphin from the pituitary both *in vivo* and *in vitro*. The sequences of rat and human CRH appear to be identical, but differ from sheep CRH at 7 out of 41 residues.

There is some evidence for the stimulation of ACTH release by synthetic CRH in amphibians and teleost fishes. Immunocytochemical studies have demonstrated

CRH immunoreactivity in the parvocellular NPO and NHP of these groups. It is interesting to note that a peptide of 40 residues isolated from the skin of a South American frog, sauvagine, has several sequence homologies with the ovine CRH, as has urotensin I, a 41-residue peptide from the teleostean urophysis. These molecules may be members of a family of neuropeptides distributed widely throughout vertebrate tissues, as are TRH and somatostatin.

13.5.6. MSH-Controlling Factors

Mammalian hypothalamic extracts appear to contain peptides with both melanocyte-inhibiting activity (MIF or melanostatin) and melanocyte-stimulating activity (MRF or melanoliberin); however, the identification and assay of these are made difficult by the presence in the extract of neurotransmitters that themselves modulate MSH secretion (Section 14.2.4). Several small peptides that have been isolated from hypothalamic extract and have stimulatory or inhibitory effects on MSH secretion *in vitro* and *in vivo* have been identified as enzymatic cleavage products of neurohypophyseal hormones. These have been proposed by some workers as the physiological MSH-controlling factors, but the matter is still controversial.

The first of these peptides to be isolated was the tripeptide Pro-Leu-Gly-NH_2, which is the side chain of oxytocin (see Figure 34), which has MIF activity. The tripeptide sidechains of arginine and lysine vasopressin (Figure 34) also inhibit MSH secretion, as does tocinoic acid, the ring structure of oxytocin, in some circumstances. A pentapeptide MIF has been reported in bovine hypothalamic extract, with the structure Pro-His-Arg-Phe-Gly-NH_2. A pentapeptide derived from the opened N-terminal ring of oxytocin, H-Cys-Tyr-Ile-Gln-Asn-Cys-OH, has been proposed as the MRF.

MSH secretion is under predominantly inhibitory control in amphibians and fishes, but little is known of the nature of the MIF. Catecholamines appear to be involved (Section 14.2.4), and several peptide derivatives of neurohypophyseal hormones have been reported as having MSH-stimulating or -inhibiting activity.

13.5.7. Prolactin-Controlling Factor

The presence of a prolactin-release-inhibiting factor (PIF) in mammalian hypothalamic extracts is well established; however, its isolation and purification have been hampered by a lack of suitable assay systems. Some workers have reported that the entire PIF activity of hypothalamic extracts can be accounted for by the presence in them of the catecholamine neurotransmitters noradrenaline and dopamine, and these substances certainly have physiological effects on PRL secretion both *in vitro* and *in vivo* (see Section 14.2.1). Others, however, have claimed inhibition of PRL secretion by extracts of low catecholamine content, suggesting the presence of an as yet uncharacterized peptide PIF. Whether this material is distinct from SRIF, which is now known to modulate PRL secretion (Section 14.2.1), has not yet been established.

The existence of a PRF in mammals has been proposed, and there is good evidence for this in birds. TRH is able to increase PRL secretion *in vitro,* but there is evidence to suggest that this is not the physiological PRF. Several other neuropeptides and neurotransmitters have at least a pharmacological PRF activity, including vasoactive intestinal peptide (VIP), neurotensin, substance P and endorphin, vasopressin and vasotocin, and serotonin (Section 14.2.1) although these might act via a hypothalamic PRF rather than directly on the pituitary.

The bulk of the experimental evidence from amphibians and teleost fishes shows PRL secretion to be predominantly under inhibitory control, probably by a catecholamine PIF.

13.5.8. Other Substances Implicated in Pituitary Control

In addition to the specific peptide-releasing hormones, it is now known that the neurons in the brain, including the HT, produce other substances that modify the secretion of one or more pituitary hormones, either directly or by effect on the release of releasing hormones. These substances can be divided for convenience into two broad categories: neurotransmitters and neuropeptides, although the neuropeptides may in fact function as neurotransmitters.

13.5.8a. Neurotransmitters

These small molecules are the chemical mediators involved in transfer of information from one neuron to the next, via synapses. Since the peptidergic neurons of the HT receive numerous synapses from nerve fibers originating in other parts of the brain, it is not surprising that neurotransmitters have been found to affect their activity. The "classical" neurotransmitters are amines or amino acids

Dopamine

Noradrenaline (Norepinephrine)

5-Hydroxytryptamine (Serotonin)

Acetylcholine

γ-Aminobutyric acid (GABA)

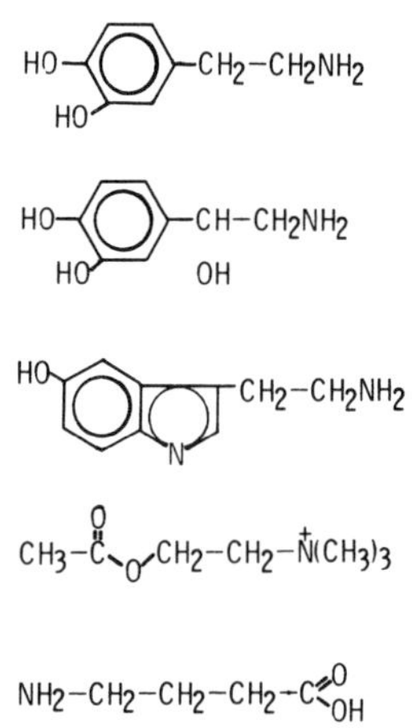

Figure 36. Principal nonpeptidergic neurotransmitters.

Table XI
Brain Peptides and Pituitary Hormone Release: CNS Action[a]

Peptide	ACTH	Prolactin	GH	TSH	FSH	LH
CCK	+	+	+	−	0	−
Gastrin	nt	−	+	−	0	−
VIP	nt	+	+	0	0	+
SP	nt	?+	+	0	0	+
NT	nt	−	+	0	0	−
Opioids	+	+	+	−	−	−
Bradykinin	nt	−	0	0	−	0
Angiotensin II	+	−	−	0	−	0
Bombesin	nt	+	+	nt	nt	nt
Somatostatin	nt	0	+	−	−	−
Vasotocin	nt	−	nt	nt	−	0

[a]nt, Not tested; CCK, cholecystokinin; VIP, vasoactive intestinal peptide; SP, substance P; NT, neurotensin.

and include adrenaline, noradrenaline (NA), serotonin (or 5-hydroxytryptamine, 5-HT), dopamine (DA), acetylcholine (ACh), and γ-aminobutyric acid (GABA) (for structure see Figure 36). DA, NA, and 5-HT are classed as monoamines, DA and NA as catecholamines. There is evidence for the involvement of DA and ACh in release of oxytocin and vasopressin; ACh, DA, NA, GABA, and 5-HT in gonadotropin release; DA (via PIF?) and 5-HT, and possibly also ACh and NA, in PRL release; NA and 5-HT in TSH release; 5-HT, and possibly NA and ACh, in ACTH release; and NA, DA, and 5-HT in GH release (see Section 14).

Table XII
Brain Peptides and Pituitary Hormone Release: Pituitary Action

Peptide	Prolactin	GH	TSH	FSH	LH
CCK	0	0	0	0	0
Gastrin	0	0	−	0	0
VIP	+	0	0	0	0
SP	+	0	0	0	0
NT	+	0	+	0	0
Opioids	0	0	0	0	0
Bradykinin	0	?−	0	−?	0
Angiotensin II	+[a]	+[a]	+[a]	0	+[a]
Bombesin	0	0	nt	nt	nt
Vasotocin	+	nt	nt	+	+
Inhibin	nt	nt	nt	−	0[b]

[a]Only high doses effective (2–25 μg ml^{-1}).
[b]Higher doses also suppress LH.

13.5.8b. Neuropeptides

A whole range of neuropeptides have now been extracted from and localized within the HT and extrahypothalamic brain, and some also occur in other tissues of the body. The list now includes the proopiomelanocortins (POMC peptides) (for structure see Figure 31), substance P (SP) and neurotensin (NT), bombesin and the gastrointestinal peptides: vasoactive intestinal peptide (VIP), cholecystokinin (CCK), gastrin, angiotensins, bradykinin, and glucagon. Several of these have now been detected in the portal blood and have been found to have effects on the secretion of particular pituitary hormones (Tables XI and XII). These effects may be similar to those of the neurotransmitters, modifying the activity of hypothalamic neurons producing releasing hormones, or at the pituitary cell level as "neuromodulators" (see Section 14).

14. Hypothalamic Control of Pituitary Functions

14.1. General Considerations

14.1.1. Mechanism of Action of Hypothalamic Hormones on Pituitary Cells

The hypothalamic hormones ("releasing and inhibiting" factors) exert their effects on pituitary cells through binding to specific receptors on the cell membrane. Each receptor is usually highly specific for a particular factor (secretagogue): even a few amino acid changes in the secretagogue result in a loss of affinity. For some factors (e.g., TRH, LH-RH) pituitary cells appear to possess two types of receptor: a low-affinity, high-capacity type and a high-affinity, low-capacity type. In some cells, two or more substances may have additive or opposing effects on hormone secretions, which results from either blockade of a single receptor or from separate and independent binding at different receptors. In addition, the number and concentration of receptors vary under the influence of local and systemic "modulating" factors (e.g., neurotransmitters, steroids, hormonal feedback), so that the response of the cell to its primary secretagogue is regulated by the availability of its receptors.

The exact intracellular pathway leading to increased or decreased hormone secretion following secretagogue binding to a pituitary cell is still not clear. Two mechanisms have been implicated involving either or both the adenyl cyclase enzyme system and changes in intracellular cAMP concentration, and changes in intracellular calcium ion concentration due to influx or mobilization of intracellular stores. Possible pathways are shown in Figure 37. Changes in secretion by the cell may be effected by action at three points in the intracellular secretory pathway: synthesis of new hormone by the RER, release of hormone from granules at the cell membrane, and degradation of old hormone by lysosomes. In some cases the processes of synthesis and release appear to be activated independently; for in-

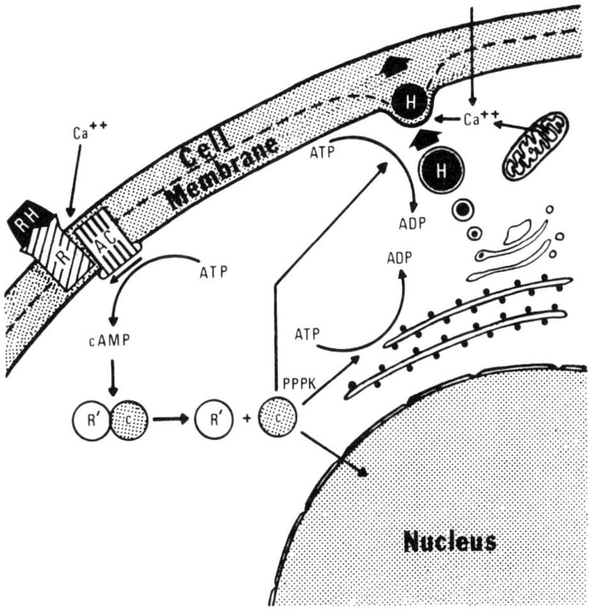

Figure 37. Intracellular pathways and biochemical events following stimulation of hormone release. AC, adenylate cyclase; ADP, adenosine diphosphate; ATP, adenosine triphosphate; C, catalytic component; H, hormone; PPPK, protein kinase; R, cell membrane receptor; R′, regulatory protein; RH, releasing hormone.

stance, LH-RH stimulation of LH secretion from the mammalian gonadotrop is not accompanied by *de novo* protein synthesis unless the cell has received a prior exposure (''priming'') to LH-RH. However, in all cells the process of exocytosis is known to be calcium dependent.

14.1.2. Higher Control of Hypothalamic Hormone Release

The hypophysiotropic NS neurons within the hypothalamus are capable of a considerable degree of autonomous function. If the HT is (effectively) isolated from the rest of the brain by lesions, in many cases the basal secretion of pituitary hormones is unaltered; however, changes in secretion normally associated with stimuli from the internal and external environments are frequently lost. This implies a second, ''higher'' level of control over the production of the hypothalamic hormones. This may be by inhibitory or stimulatory input from other extrahypothalamic areas of the brain, directly or indirectly from the special sense organs (vision, hearing, smell), exteroreceptors (pain, temperature, reproductive stimuli), chemoreceptors, stress, or inbuilt rhythms. Alternatively, control may be through changes in the levels of hormones, ions, and other substances in the blood supply to the HT and in the CSF.

Although many areas of the brain affect activity of the HT indirectly, the following centers are the major sources of neural input to the HT: stria terminalis, amygdala, and hippocampus (together forming the limbic system); anterior thalamus; piriform cortex; midbrain. In several cases one or more of these centers has been implicated in the higher control of particular pituitary functions.

Extrahypothalamic inputs to the hypophysiotropic neurons are mediated by a variety of chemical neurotransmitters. These are produced in the cell bodies of the higher brain centers, are transported down the axons to nerve terminals, where they are stored in vesicles, and are released at synapses onto other neurons to produce a local effect. Using fluorescence histochemistry and immunocytochemistry (see Section 16) it has been possible to demonstrate the DA, NA, and 5-HT containing fibers innervating the HT from the areas of the brain listed earlier. It is difficult to allocate specific roles to neurotransmitters in the control of pituitary function. A single neurotransmitter may be involved in the control of a whole range of pituitary cells, and the nature of its effect may depend on its point of action in a multisynaptic pathway. Nevertheless, definite functions for these substances in the higher control of specific hormones have in some cases been proposed.

When a neurotransmitter is released at a synapse on a hypothalamic neuron, it is believed to cause an action potential that passes down the axon, resulting in depolarization and calcium ion influx, leading to exocytosis of the NS granules at the ending. Neurotransmitters may also cause changes in cAMP levels in the neuron, by a similar mechanism to that described above for pituitary endocrine cells (Figure 37). Prostaglandins have also been implicated in the regulation of hypophysiotropic hormone release, possibly by modifying or modulating the effect of the neurotransmitter. This may also be the way in which steroids exert controlling effects at the hypothalamic level.

14.1.3. Feedback Mechanisms in the Control of Pituitary Function

Feedback mechanisms are of great importance in regulating the levels of pituitary secretion and are best exemplified by the pituitary–target gland model (Figure 38) in which the pituitary, under the influence of a hypothalamic factor, secretes a tropic hormone that stimulates a target gland to release its hormone. As the level of the target hormone in the blood rises, it inhibits the release of the tropic hormone, and as it falls again, the inhibition is removed and tropic hormone secretion recommences. This is known as *long-loop negative feedback*. In rare instances the feedback may be positive rather than negative, when the target hormone stimulates release of the tropic hormone: this is an inherently unstable system leading to a rapid surge in secretion. Inhibitory feedback mechanisms occur in the pituitary–gonad, pituitary–thyroid, and pituitary–adrenal axes, but similar processes appear to regulate the secretion of the pituitary hormones that have no target gland. In these cases the feedback effects are mediated by metabolites and hormones from other target glands. The feedback action of the target hormone or metabolite may take place at various levels, not just the anterior pituitary, for instance, the basal hypothalamus,

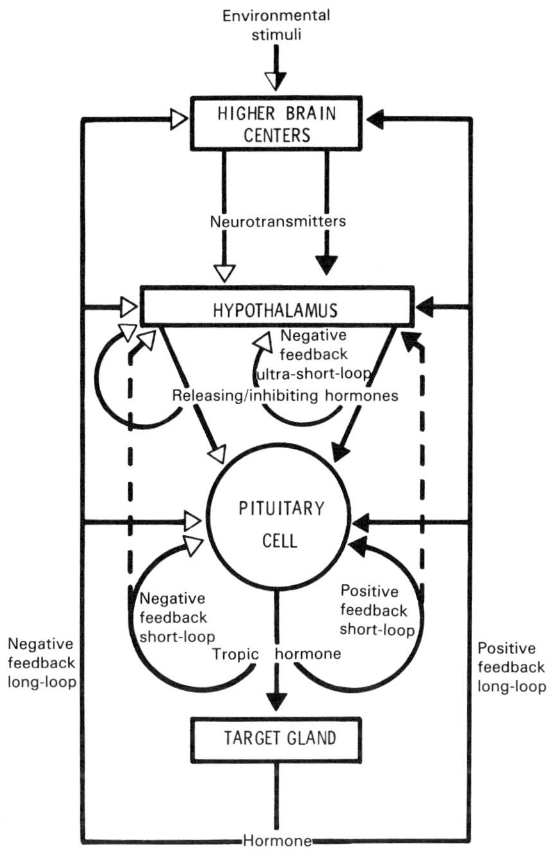

Figure 38. Generalized diagram to show the factors and feedback pathways in the control of pituitary hormone secretion.

preoptic area, and extrahypothalamic sites, which control the activity of the hypothalamus.

There is also evidence that some pituitary hormones can exert positive or negative regulation of their own secretion by *short-loop feedback* (Figure 38). This may be at pituitary level, for example, the increasing levels of the hormone within the pituitary capillaries "switch off" further release, or at the basal hypothalamic level, possibly via retrograde flow of blood up the portal vessels to the median eminence, thus inhibiting the release of hypothalamic factors. It is also possible that some hypothalamic hormones might control their own release by *ultra-short-loop feedback* via either a vascular route or a system of axosomatic synapses.

14.2. Control of Adenohypophyseal Hormone Secretion

The majority of studies on control of adenohypophyseal hormone secretion have been made in mammals, particularly rodents, so that in this chapter, inevitably, the systems discussed will mainly refer to mammals, but, when possible, differences and similarities in nonmammalian vertebrates will be noted.

14.2.1. Prolactin

Prolactin cells are unique among pituitary PD cells in being principally under tonic inhibitory control. They generally exhibit autonomous secretion, so that when removed from hypothalamic connections and autotransplanted they will continue to secrete hormone, suggesting that this is an ancient property of PRL cells. PRL cells of autotransplanted pituitaries of some teleosts, including the molly (*Poecilia latipinna*), goby (*Gillichthys mirabilis*), and tilapia (*Sarotherodon mossambicus*), of amphibians (larval tadpole and toads), as well as mammalian species, remain active. However, circumstantial evidence indicates that autotransplanted PRL cells of birds (pigeon and duck) and some teleosts (eel, goldfish, and rainbow trout) become inactive and may atrophy. Correspondingly, pituitaries of some teleosts, amphibians, mammals, and the terrapin (the only reptile studied) continue to secrete PRL when cultured *in vitro*. However, two species of bird, the quail and blackbird, do not release PRL spontaneously from cultured pituitaries, but Peking duck and mallard pituitaries do; so that even among the birds, PRL secretion may be controlled by tonic inhibition in some species.

In mammals the hypothalamic inhibitory factor is considered to be DA, so that *in vivo* treatments, including drugs such as chlorpromazine and reserpine, which deplete hypothalamic catecholamine stores, increase PRL secretion; conversely, increased hypothalamic catecholamine concentrations reduce PRL secretion. In several species of teleosts reserpine treatment increases PRL cell activity, but in the eel (*Anguilla anguilla*) reserpine has little effect, although PRL cells are activated by dopamine receptor antagonists. The situation in euryhaline teleosts is complicated by the responsiveness of PRL cells to osmotic pressure (OP), so that in media of high OP (seawater) DA has little effect on PRL secretion, but in low-OP media (freshwater) PRL secretion is much reduced by increased hypothalamic DA concentrations. In the amphibian *Notophthalmus* there is circumstantial evidence that drug-induced decreases in hypothalamic DA concentrations increase PRL secretion. In birds the situation remains unclear; reserpine may increase PRL secretion or have no effect. The evidence that catecholamines may be involved in control of PRL secretion in reptiles is fragmentary and contradictory; acid extracts of rat and turtle ME (presumably containing catecholamines) stimulate PRL secretion by cultured turtle and lizard pituitaries. However, earlier studies had indicated an inhibitory control of PRL secretion in reptiles, and indeed turtle pituitaries will secrete PRL at low levels *in vitro* autonomously. Reptiles show a spectrum of responses similar to those of bird pituitaries, but more data are needed to evaluate the evolutionary significance of PRL control in reptiles. The important point about control of PRL secretion by DA is that in all the *in vitro* studies of teleosts, amphibian, and mammalian pituitaries, only release of PRL is inhibited; DA does not appear to affect hormone synthesis.

Other factors that may be concerned with control of PRL secretion include vasoactive intestinal peptide (VIP), TRH, steroid hormones, and neurotransmitters. TRH has been shown to stimulate PRL secretion by pituitaries from several vertebrate species both *in vivo* and *in vitro*. It will partially antagonize the action of DA, and its effectiveness appears to be enhanced by estrogens and following suckling of the mammary gland in rats. Release of prolactin *in vivo* occurs episodically, showing a circadian rhythm in most vertebrates studied, but also varying in relation to steroid hormone secretion, particularly estrogens in females. In mammals PRL levels rise after parturition owing to the fall in plasma progesterone. Circadian rhythms are entrained by environmental stimuli, particularly light, which enters via both the eye and the pineal gland, are integrated in the CNS, and exert their influence via the hypothalamus. Steroids, particularly estrogens, act directly on the PRL cells, on hypothalamic nuclei, and on neurons of the CNS, to stimulate PRL secretion. In mammals this action is opposed by progestins and androgens. Serotonin, NA, and GABA as well as endorphins, enkephalins, and VIP probably have a modulating influence *in vivo* on other factors controlling PRL synthesis and release. A potent stimulant for PRL release, peculiar to mammals, is suckling of the mammary gland, an example of a nervous stimulus prompting an endocrine response. The mechanisms involved in control of PRL secretion are shown in Figure 39.

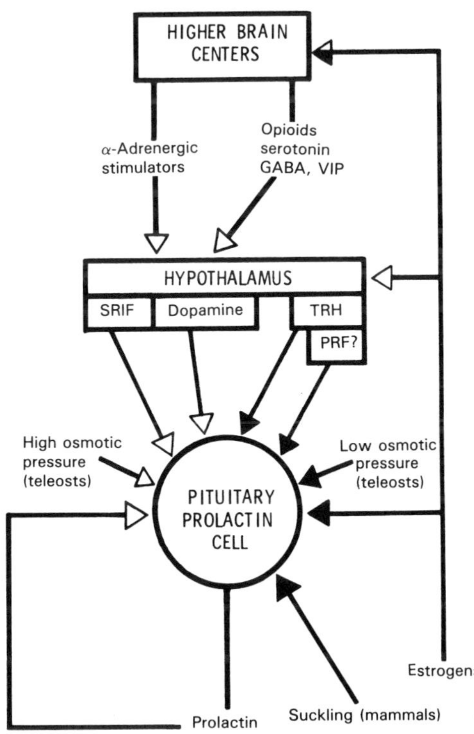

Figure 39. Factors involved in control of prolactin secretion. GABA, γ-amino butyric acid; PRF, prolactin-releasing factor; SRIF, somatostatin; TRH, thyrotropin-releasing hormone; VIP, vasoactive intestinal peptide.

14.2.2. Growth Hormone

Growth hormone is released from the pituitary episodically, but in vertebrates with seasonal growth, release is greater during the growing period. The most detailed studies on control of GH secretion have been made in experimental mammals, mainly rats and mice, largely because sensitive assays have not been available until recently for GH of lower vertebrates.

Hypothalamic control of GH secretion involves both a releasing factor (GRH) and an inhibitory factor (SRIF, somatostatin). There is a large concentration of SRIF-containing nerve endings in the median eminence, but cell bodies are located in various parts of the hypothalamus, particularly the preoptic region (see Section 12.3, and Figure 28). GRH-containing neurons have been detected recently in the ventromedial and arcuate nuclei of the human hypothalamus. Studies in rats show a dynamic, antagonistic interplay between GRH and SRIF in controlling GH secretion: *in vivo* responses to injected GRH can only be seen in rats with SRIF concentrations reduced by treatment with specific antibody. However, SRIF appears to play only a minor role in the generation of troughs of serum GH concentrations during normal rhythmical release of GH and in stress-induced suppression, because lesions in the region of the NPO in rats produce increased frequency of pulsatile release of GH, but no change in the levels at the low point between pulses. GH release can be obtained by several other factors including stress, starvation, low blood glucose (in rats), high blood glucose (in humans), anesthesia, and short-term direct feedback. The CNS plays a coordinating role via neurotransmitters; DA, NA, and 5-HT generally all enhance GH secretion. Amino acids, such as arginine, also stimulate GH release, and thyroxine stimulates synthesis.

A target organ for GH is the liver, which it induces to produce somatomedin(s), and most recent results show that somatomedin C will inhibit secretion of GH stimulated by GRH.

There is relatively little information on control of GH secretion in nonmammalian vertebrates. In birds the hypothalamus apparently contains both stimulatory and inhibitory substances, and mammalian GRH stimulates GH secretion in the chicken. Some pharmacological evidence indicates that 5-HT, A, and NA slightly inhibit GH secretion, probably by action at β-adrenergic receptors. *In vitro* release of GH from a teleost pituitary can be inhibited by hypothalamic extracts and SRIF, suggesting they may be involved *in vivo*. Urodele pituitaries show a considerable degree of autonomous GH secretion, but anuran pituitaries appear to depend on a hypothalamic releasing factor; there are, however, no physiological studies on control of GH secretion by amphibian pituitaries.

Factors controlling GH secretion exert their action on somatotrops via calcium uptake and by the increase of intracellular cAMP in the case of GRH, but SRIF may act by decreasing cAMP or in the mechanisms beyond intracellular cAMP conversion.

Secretion of GH is then a rhythmical process controlled through the CNS and hypothalamic factors SRIF and GRH and which can be modified by metabolic

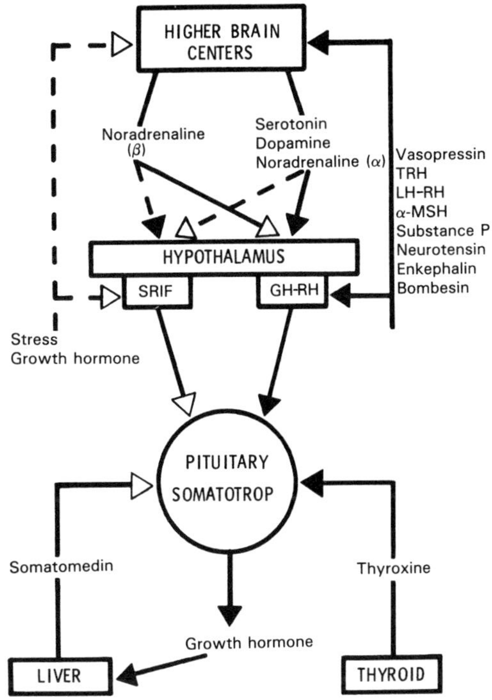

Figure 40. Factors involved in control of growth hormone secretion. GH-RH, growth-hormone-releasing hormone; LH-RH, luteinizing-hormone-releasing hormone: α-MSH, melanocyte-stimulating hormone; SRIF, somatostatin; TRH, thyrotropin-stimulating hormone.

products and stress (Figure 40). Research on control in nonmammalian vertebrates could contribute considerably to our understanding of the evolution of these complex mechanisms.

14.2.3. Adrenocorticotropin

It has long been known that secretion of ACTH is dependent on a hypothalamic releasing factor, but the molecule was only recently characterized (Section 13.5.5, and Figure 35). Extracts of hypothalami of all classes of vertebrates contain CRF activity, which is presumed to stimulate ACTH secretion *in vivo*. Secretion of CRF is controlled by neurotransmitters from the CNS, where external cues are integrated. The primary action of ACTH is to stimulate adrenal cortex or interrenal cells to secrete glucocorticoids, mainly cortisol and corticosterone, which exercise feedback inhibition on release of ACTH by pituitary corticotrops. The adrenal steroids in the circulation act directly on both corticotrops and hypothalamic nuclei.

In mammals, at least, basal secretion of ACTH is episodic, with bursts of hormone release every 1–3 hr. These occur rhythmically and in relation to dark/light and activity cycles, so that the highest circulating levels of ACTH and glucocorticoids occur just before start of activity (waking) and reach their nadir just before cessation of activity (sleeping). Circadian changes in hypothalamic CRF concentrations are also related to activity cycles. Stress, of various kinds, is a potent stimulator of ACTH release and overrides the normal episodic bursts; however, the magnitude of response varies and is greatest when resting serum glucocorticoid levels are at their lowest. Among nonmammalian vertebrates circadian rhythms of secretion of cortisol have been noted in teleost fishes, the goldfish and rainbow trout, suggesting that ACTH is also similarly released. The problem in lower vertebrates is that the secretion of cortisol by interrenal tissue can be directly stimulated by salt and by renin/angiotensin; however, stress-induced increases in serum cortisol appear to be mediated through increases in ACTH secretion. In teleosts this may be due, in part, to release from the NHP of arginine vasotocin or isotocin, which have CRF activity. In amphibians too, these NHP octapeptides may act as a CRF, but a distinct CRH is also secreted by other hypothalamic neurons: a similar situation occurs in reptiles.

The general pattern of control is that hypothalamic CRH acts on the corticotrops of the pituitary to stimulate ACTH release. Release of CRH is stimulated by neurotransmitters of the CNS; 5-HT has been most strongly implicated. CRH release may be inhibited by other neurotransmitters, such as NA and GABA. Stress overrides any inhibitory controls and prompts release of CRH leading to increased secretion of ACTH and ultimately corticosteroids, which act as feedback controls inhibiting release of both CRH and ACTH (Figure 41).

14.2.4. Melanocyte-Stimulating Hormone

The vascular supply to the PI of most vertebrates is generally poor, so that compared with cells of the PD, blood-borne factors play a relatively less significant role in controlling MSH secretion. Instead, PI cells are innervated directly, or indirectly via adjacent basement membranes, by both peptidergic and aminergic nerve fibers from the hypothalamus. There are two hypothalamic peptides controlling melanotrop activity, one inhibitory (MIF) and the other stimulatory (MRF) (Section 13.5.5). The neural amines that have been implicated in control of MSH secretion are DA, NA, 5-HT, and GABA, and acetylcholine may also have an effect. In addition, MSH exerts a direct feedback inhibition on the melanotrops. It is the interrelationships between these factors which control MSH release and synthesis, and which mediate the action of external stimuli such as light, stress, suckling, sound, touch, and hypertonic saline.

The weight of evidence shows that in all vertebrates, melanotrop activity is predominantly under tonic inhibitory control. The major inhibitory factors are DA, acting via the DA receptors, and NA, acting at the α-adrenergic receptors of the melanotrop, as well as the peptide MIF. In one species, at least, *Xenopus laevis* (Anura), both synthesis and release of MSH are inhibited. Stimuli that increase MSH release do so by interrupting the catecholaminergic inhibitory pathway: GABA and 5-HT are probably transmitters involved in this process. There is another stimulatory pathway via cholinergic nerves; acetylcholine stimulates MSH release from frog (*Rana pipens*) and mouse pituitaries *in vitro* and has a similar

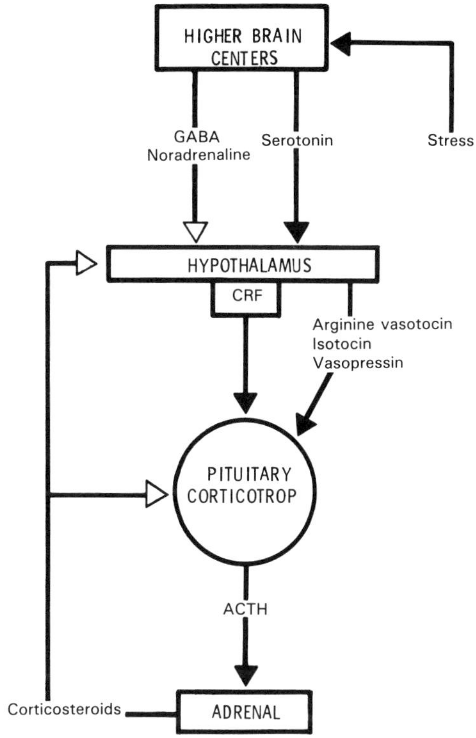

Figure 41. Factors involved in the control of adrenocorticotropin secretion. CRF, corticotropin-releasing factor; GABA, γ-aminobutyric acid.

effect *in vivo,* perhaps by action at the hypothalamus as well as the pituitary. Although the stimulatory peptide MRF is found in the hypothalamus and prompts MSH release when injected, its physiological role in control of MSH secretion is not clear at present. The major external stimulus to MSH release in nonmammalian vertebrates is the absence of light or a dark background, which results in inhibition of DA release and hence increased MSH secretion, which causes skin melanophores to expand. In mammals, however, MSH is not concerned with changing skin color but may play a part in natriuresis (excretion of sodium), and a very large PI is found in some desert mammals. In experimental rats hypertonic saline induces release of MSH, and in some varanid lizards the PI is most developed in those species which suffer periods of drought. However, the converse is true in some species of amphibia, in which adaptation to saline medium causes atrophy of the PI, and in some teleost fishes PI cells become active in media of low osmotic pressure. In mammals suckling and vaginal stimulation at estrus cause MSH release from the pituitary. These responses to salinity and suckling may be related to the close apposition between the PI and NHP; the neurohypophyseal hormone oxytocin is released during suckling, as is prolactin, the latter owing to inhibition of DA release. Such an inhibition may also result in increased MSH secretion, and oxytocin, released as the complete molecule, may reduce MIF formation. The natriuresis during salt loading may be due to synergism between vasopressin, the other NHP hormone, and MSH.

Another controlling factor in mammals is estrogen, which stimulates both MSH secretion and melanotrop proliferation and moreover depletes hypothalamic MIF stores. The parallelism between the control of PRL cells and melanotrops by DA, α-adrenergic agonists, and estrogen, as well as their responses to suckling in mammals, are remarkable, but whether this is significant in pituitary hormone physiology is not known. Figure 42 outlines the control mechanisms of MSH secretion.

14.2.5. Gonadotropins

Secretion of gonadotropin by the pituitary, and ultimately the initiation of physiological processes leading to reproduction, is controlled by a hypothalamic releasing hormone, GnRH, or LH-RH. This appears to be true for all vertebrates, although the GnRH may vary chemically (Section 13.5.4). However, a host of other factors is involved, including a possible peptide inhibiting factor, gonadal steroids, adrenal steroids, thyroid hormones, inhibin, prostaglandins, and neurotransmitters (Figure 43). Through a balance of these factors, reproductive activity can be regulated in relation to physiological and environmental conditions (via the CNS), such as food availability, stress, visual and tactile stimuli, olfactory stimuli (pheromones), daylength, and temperature.

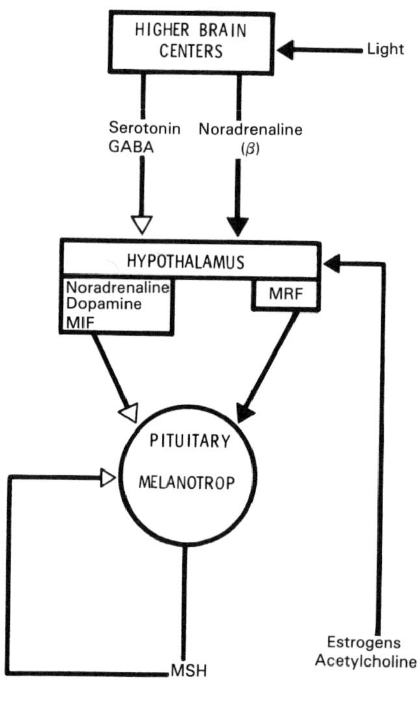

Figure 42. Factors involved in the control of melanotropin secretion. MIF, melanocyte-inhibiting factor; MRF, melanocyte-releasing factor.

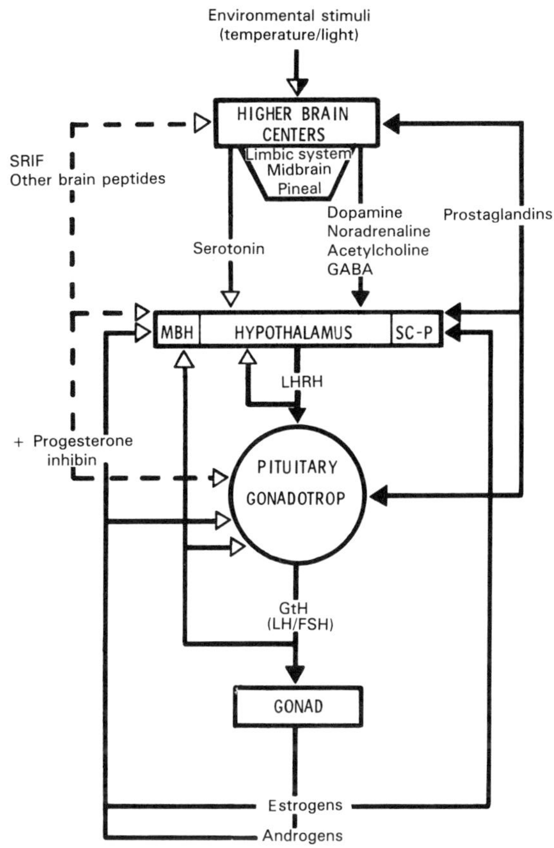

Figure 43. Factors involved in the control of secretion of gonadotropins. GABA, γ-amino butyric acid, GtH, gonadotropin; LH-RH, luteinizing-hormone-releasing hormone; MBH, medial basal hypothalamus; SC–P, suprachiasmatic–preoptic region.

Two gonadotropins, LH and FSH, are secreted in mammals, birds, most reptiles, and amphibians, but the balance of evidence is now in favor of one releasing factor that controls the release of both hormones; however, a second releasing hormone, GnRH II, has recently been isolated from chicken hypothalami. Much of the evidence indicating dissociated patterns of LH and FSH release under certain physiological conditions can be explained by a differential response to LH-RH of the GtH cell, possibly secreting LH and FSH simultaneously, as a result of modulation by other factors (e.g., steroids, inhibin) to which they are exposed. In mammals, the basal levels of LH and FSH secretion necessary for gonadal integrity are maintained by a succession of episodic discharges, regulated by negative feedback of the gonadal steroids. In the breeding female, however, the cyclical preovulatory release of gonadotropins is triggered by positive feedback of steroid from the developing follicle. For details of the patterns of LH and FSH secretion and their control in the cyclical or seasonal breeding female mammal the reader is referred to Part II.

There is evidence from lesioning studies on the rat that the sites of this "dual-control" mechanism are two separate areas of the HT. The basal control appears to reside in the medial basal HT (comprising NAH, NVM, and NArc; see Section 12.3), whereas the cyclical control center lies more anteriorly in the suprachiasmatic–preoptic (NSC–NPO) region. The latter region has extensive connections with the higher brain centers and integration areas that may supply information from the environment. Both control regions could contain LH-RH-producing neurons sending axons to the ME (or OVLT), with differing sensitivity to steroids; alternatively, the NSC–NPO region could consist of non-LH-RH neurons that influence the LH-RH-neurons of the medial basal HT. Results from ICC studies tend to support the first hypothesis, as LH-RH-immunoreactive neurons have been observed in both regions. Dual control of cyclical and basal GtH release may be a feature of all vertebrate groups. Evidence for this among birds and reptiles is only tentative, but in amphibians and teleost fishes is quite good, both from lesioning studies and from ICC. In amphibians (see Section 9.2) an anterior region, the medial septal area of the posterior telencephalon, appears to be responsible for induction of ovulation, with a ventral infundibular nucleus probably controlling basal GtH levels. In teleosts (see Section 7.2) the anterior control center may be the posteroventral telencephalon–preoptic area, and the tuberal center the NLT.

The control mechanisms for cyclical GtH release probably reside not in the NSC–NPO region of the HT itself, but in a higher center of the brain. The limbic system (Section 14.1.2.) is particularly implicated, since it has extensive connections to the NSC–NPO region and contains neurons with sex steroid binding sites. Depending on which part of the limbic system is stimulated electrically, GtH secretion can be either facilitated or inhibited. The same is true for certain parts of the midbrain, which is linked by a complex network of tracts to the limbic system, to the NSC–NPO area, and also to the medial basal HT. These extrahypothalamic areas may affect GtH secretion by regulation of LH-RH release from the two controlling areas of the HT, or by direct modulating effects of their neurotransmitters on the pituitary. Some neurotransmitter effects on GtH secretion have been characterized. NA stimulates GtH release, probably via the HT by regulating the LH-RH neurons of the preoptic area. DA also generally stimulates GtH secretion via the HT, the hypothesis being that DA neurons in the NArc regulate the release of LH-RH from terminals in the ME. In certain steroid environments, however, DA can directly inhibit release of GtH from the pituitary. ACh too can stimulate LH release, possibly through excitation of the tuberoinfundibular DA neurons, and there is also indirect evidence for a stimulatory effect of GABA. Inhibition of GtH secretion appears to be mediated by a serotonergic pathway, since injections of 5-HT into the HT suppress both tonic and cyclical GtH release. The pineal is another part of the brain believed to influence GtH secretion, particularly in seasonally breeding animals. Several pineal "hormones" (indolamine neurotransmitters) including melatonin show various GtH-stimulating and -inhibiting effects at both HT and pituitary levels.

Various brain peptides (Table XI and XII) have recently been found to modulate GtH release, usually, it seems, by affecting central neurotransmitter levels, although effects on LH-RH neurons and directly on the pituitary may also occur.

Little is known of the influence of the extrahypothalamic brain and neurotransmitters in submammalian vertebrates; however, in teleosts an aminergic GnIF has been proposed by some workers.

As mentioned earlier, gonadal steroids exert an important controlling influence on GtH secretion by feedback on the HT, either negatively in the case of basal GtH secretion or positively in the case of cyclical release in the female. Both hypothalamic control centers contain neurons with estrogen binding sites: estrogen implanted in the medial basal HT inhibits GtH secretion and in the NSC–NPO area stimulates secretion. Androgens are less potent than estrogens in inhibiting LH release, and progesterone is normally ineffective. Suppression of FSH secretion requires the synergistic action of estrogen, testosterone, and a protein, inhibin, secreted by the germinal epithelium of the ovary. Testosterone may effect a positive feedback for the preovulatory release of FSH. In addition to feedback at the HT, the same steroids may also act at the pituitary level.

Effects of steroids on LH-RH-induced GtH secretion from the *in vitro* incubated pituitary are difficult to interpret: the same steroid can inhibit or facilitate secretion depending on the conditions, and synergistic action may be important. The gonadotropins to some extent appear capable of regulating their own release by short-loop feedback, either directly on the gonadotrops or on hypothalamic LH-RH secretion. Ultra-short-loop feedback of LH-RH on its own secretion is also a possibility.

There is similar, but often fragmentary, evidence of steroid feedbacks on GtH secretion in birds, reptiles, amphibians, and fishes, and steroid binding sites have, in several cases, been located in the presumptive controlling areas of the HT.

14.2.6. Thyrotropin

In mammals TSH is secreted by the pituitary thyrotrops, under the influence of two main regulating factors: hypothalamic TRH, which stimulates synthesis and release of TSH, and circulating levels of thyroid hormones, which inhibit TSH, secretion by negative feedback. This simple model is sufficient to account for most of the changes observed in the pituitary–thyroid axis, but not for thyroid responses to factors such as temperature and stress, or for the circadian changes in TSH secretion observed in some mammals. Steroids, some neurotransmitters, and neuropeptides (particularly SRIF) have also been implicated in control of TSH secretion (Figure 44). Relatively little is known about the controlling mechanisms in lower vertebrates; nevertheless, it is clear that there are some important differences from the mammalian model.

Disconnection of the pituitary from the HT by stalk section, or autotransplantation, leads to decreased TSH secretion in mammals, implying tonic stimulation by a releasing factor. TRH has been characterized (Figure 35) and in its synthetic form is potent in releasing TSH from the pituitary both *in vitro* and *in vivo*. The precise area

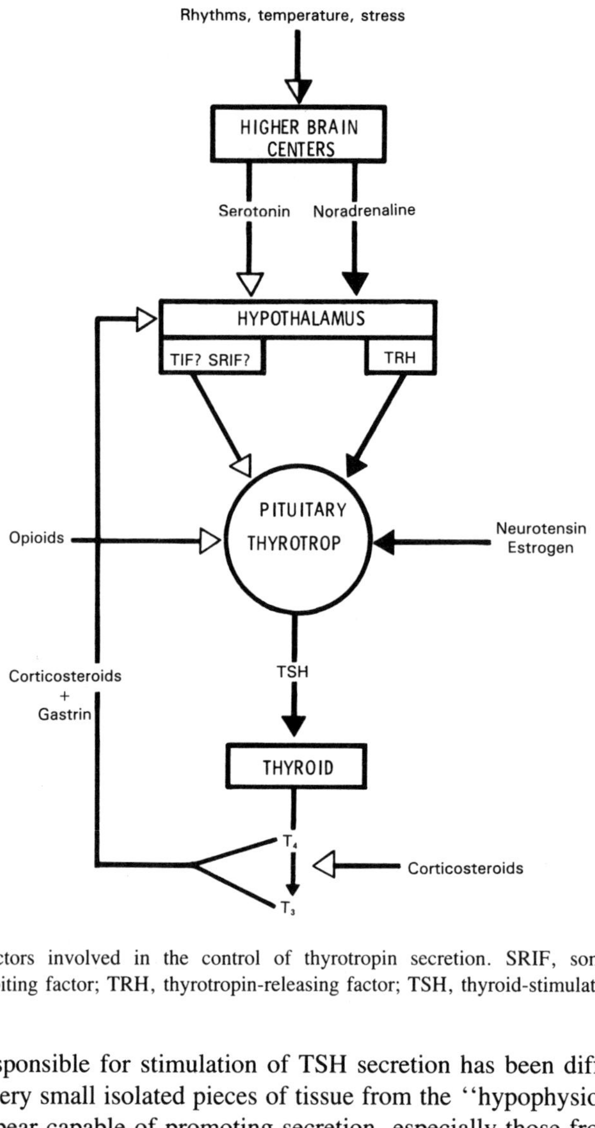

Figure 44. Factors involved in the control of thyrotropin secretion. SRIF, somatostatin; TIF, thyrotropin-inhibiting factor; TRH, thyrotropin-releasing factor; TSH, thyroid-stimulating factor.

of the HT responsible for stimulation of TSH secretion has been difficult to pinpoint. Even very small isolated pieces of tissue from the "hypophysiotropic" area of the HT appear capable of promoting secretion, especially those from the NVM and NArc regions (see Section 12.3), in the absence of inputs from other parts of the brain. Lesions in the anterior HT, around the NPV, cause a reduction in basal TSH secretion, but do not interfere with changes in secretion in response to thyroid hormone levels. This could be explained by a direct responsiveness of the thyrotrops to thyroid hormones, with TRH merely setting the basal level of TSH secretion, the "set point," at which the negative feedback switches off the thyrotrop activity. The concentration of TRH within the HT, assessed by RIA and ICC, is greatest in the ME and ventral tuberal region, with lesser amounts throughout the hypophysiotropic area. These results must, however, be interpreted with caution, as some TRH immunoreactivity also occurs throughout the extrahypothalamic brain and spinal cord.

In birds, lesions of the anterior HT again reduce TSH secretion, but on the whole, thyrotrop activity is considerably more autonomous than in mammals. Synthetic TRH is effective in releasing TSH from the avian pituitary. Conversely, in reptiles, a variety of lesions in the HT cause increased thyroid function, suggesting the presence of an inhibiting factor (TIF). Although immunoreactive TRH has been detected in the snake brain, there is no evidence that the tripeptide has any action on the reptilian pituitary. Among amphibians, the dependence of TSH secretion on the HT is very variable, but the majority of experimental studies indicate some degree of stimulation by a TRF. In the adult frog a TSH-controlling center has been located in the NIV (cf. GtH control), but in larvae control appears to lie more anteriorly in the dorsal NPO. Immunoreactive TRH occurs in high concentrations throughout the brain, in the blood, and in the skin, and extracts from these tissues release TSH from the mammalian pituitary. It now seems that amphibian TRF is *not* identical to mammalian TRH, which appears to play no part in the control of TSH secretion in amphibians; possibly it is involved in control of PRL secretion or has some general neurotransmitter function. In teleost fishes, as in reptiles, the balance of evidence is in favor of inhibitory control of TSH secretion; the TIF probably originating in the NLT region of the HT. In some species, however, dual control by a TIF and a TRF has been suggested. Synthetic TRH has no consistent effect on the thyrotrops of teleosts, but syn-SRIF may have an inhibitory action. Immunoreactive TRH has been detected in the brains of fishes, but, as in amphibians, the tripeptide may be more concerned with control of PRL secretion than TSH secretion.

In some mammals phasic patterns of TSH secretion can occur, due, for example, to circadian rhythms, and TSH release can often be induced by stimuli such as cold exposure and severe stress. It is probable that these changes are mediated by inputs from higher brain centers, but virtually nothing is known about the pathways involved. A few lesioning and electrical stimulation studies have suggested mainly inhibitory inputs from the habenula, the limbic system, and the pineal body. Again, there is little conclusive evidence for the identity of the central neurotransmitters or neuropeptides that might be involved in these inputs. The catecholamines NA and 5-HT seem to have opposing stimulatory (NA) and inhibitory (5-HT) actions on TRH release by the HT, and melatonin may have an inhibitory effect. Of the neuropeptides so far tested, gastrin inhibits TSH secretion at both hypothalamic and pituitary levels, opioids inhibit directly at pituitary level, and neurotensin stimulates at pituitary level. SRIF may have a physiological role in inhibition of TSH secretion, particularly in lower vertebrates.

In mammals the primary factor determining the rate of TSH secretion is the circulating concentration of the thyroid hormones, thyroxine (T_4) and triiodothyronine (T_3). By "negative feedback," high levels of T_3 and T_4 decrease the responsiveness of the thyrotrop to TRH, by reducing the number of TRH-binding sites on the cell membrane. Although both thyroid hormones have been shown to do this, T_3 is much more effective than T_4 in suppressing TSH secretion. Monodeiodination of T_4 to T_3 is unlikely to occur within the pituitary, but conversion of T_4 to T_3 elsewhere in the body will affect the magnitude of the feedback, as will the levels of T_3- and T_4-binding proteins, since the thyrotrops are only responsive to free circulating T_3 and T_4. There is also some evidence to indicate feedback by thyroid hormones on the HT. The preoptic area is the most sensitive to hormone implants, reducing TSH secretion, but whether this is by inhibition of TRH neurons or activation of an inhibitory (somatostatinergic?) pathway is not known.

Thyroid hormones also appear to effect a negative feedback suppression of TSH secretion in amphibians, possibly by reducing TRF release from the HT. In teleosts, thyroxine lowers TSH secretion, apparently acting at both pituitary level and hypothalamic level (by release of TIF?).

Certain steroids may also be involved in control of TSH secretion by similar mechanisms. Cortisol can further reduce TSH release after maximal suppression with T_3, by acting at pituitary and hypothalamic levels. Adrenocorticoids can also affect the pituitary–thyroid axis by reducing peripheral conversion of T_4 to T_3. Estrogen has a stimulatory effect on TSH secretion, by increasing the responsiveness to TRH (i.e., number of TRH-binding sites) of the thyrotrop, in direct opposition to the actions of T_4 and T_3. The activity of the pituitary–gonad and pituitary–thyroid axes in lower vertebrates often appears to be closely related, but whether this is due to the effect of gonadal steroids on TSH secretion is not known.

14.3. Control of Neurohypophyseal Hormone Secretion

In mammals, oxytocin release in response to suckling is through a typical neuroendocrine reflex, the "milk letdown" reflex. Sensory receptors in the breast send neural inputs to the midbrain, from where other fibers send impulses to the NSO and NPV. Dilation of the cervix at parturition also stimulates OT release, and stress inhibits OT release, by similar pathways.

Vasopressin release is controlled through a reflex, via the brain stem, initiated by stimulation of blood pressure/volume sensors (baroreceptors) in the heart, aorta, and carotid sinus. Plasma osmolality is also important in the control of secretion of AVP, and to a lesser extent OT. As this falls, AVP secretion rises and causes an antidiuretic response (see Part IV). Osmoreception is probably not directly by the AVP perikarya of the NSO and NPV, but by other neurons located nearby in the anterior HT. The AVP-secreting NS cells may be functionally heterogeneous, some responding to osmoregulatory afferents, some to baroregulatory afferents, acting independently.

Excitatory inputs to the AVP and OT perikarya are probably mediated via cholinergic afferents, and inhibitory inputs via noradrenergic afferents. These neurotransmitters have been demonstrated in synapses on the perikarya by a combination of EM autoradiographic and immunocytochemical methods. Sex steroid receptors have also been shown on AVP and OT cells by similar colocalization techniques, and estrogen stimulates OT release. Both populations of NS neurons exhibit synchronized activity: this may be due to control by interneurons or direct coupling through collaterals and contacts between adjacent perikarya.

It is premature to speculate whether similar control mechanisms apply to lower vertebrate groups.

15. Actions of Pituitary Hormones

15.1. Prolactin

Prolactin is uniquely versatile in the range of its actions within the vertebrates, apparently interacting with most tissues derived embryonically from endoderm and ectoderm. Although numerous (ca. 100) functions have been identified, they fall into three broad categories: (1) growth and development, (2) osmoregulation, and (3) reproduction, and in many of these PRL acts in concert with steroid hormones of the gonads or adrenals. Probably the most primitive and fundamental actions are concerned with growth and osmoregulation.

15.1.1. Growth and Development

The action of PRL in promoting growth usually involves cell proliferation as well as hypertrophy, but such activity demonstrated in nonmammalian vertebrates has usually depended on mammalian PRL. However, recent work has shown that a teleost (tilapia) PRL promotes growth of its own juveniles, and bullfrog PRL has much more potent growth-promoting activity in tadpoles than has ovine PRL. Adult frogs do not respond to mammalian PRL, but young pigeons and a lizard, *Lacerta s. sicula,* show enhanced growth. In both anuran and urodele tadpoles PRL is antimetamorphic and prompts return to water (water drive) and morphological changes associated with reproduction in some urodeles. The most familiar proliferative action of PRL in stimulating pigeon crop sac mucosa (forming crop "milk") has been widely adopted as a bioassay for tetrapod PRL, but teleost PRL does not produce a typical response. In mammals PRL promotes growth by stimulating the liver to produce the intermediary growth factor(s) somatomedin and increasing proliferation of mammary epithelium, although this may not be due to direct action on the epithelial cells themselves. Table XIII lists some growth-promoting actions of PRL.

15.1.2. Osmoregulation

Prolactin plays a role in water and electrolyte balance in all classes of vertebrates, but in some euryhaline teleosts it is vital for survival in freshwater. PRL acts on the epithelia of the gut, skin, urinary bladder, and kidney of all vertebrates and also on the gills of fish and larval amphibians. In freshwater fish and amphibia PRL

Table XIII
Actions of Prolactin Involving Growth Promotion: Stimulation of Somatic Growth Observed in Adult Reptiles, Birds, and Mammals and in Larval and Adult Amphibians

Teleosts
1. Proliferation of melanocytes
2. Growth of seminal vesicles
3. Renal glomerular growth, tubule stimulation and proliferation

Amphibians
1. Tail and gill growth
2. Limb regeneration
3. Proliferation of melanocytes
4. Structural changes accompanying water drive
5. Brain growth in tadpoles

Reptiles
1. Tail regeneration
2. Skin sloughing

Birds
1. Proliferation of pigeon crop sac mucosa
2. Epidermal hyperplasia in brood patch
3. Feather growth
4. Development of female reproductive tract

Mammals
1. Mammary development
2. Sebaceous and preputial gland growth
3. Hair growth
4. Renotropic

Growth-hormone-like metabolic actions
1. Effects on lipid deposition and/or mobilization reported in teleosts, amphibians, reptiles, birds, and mammals
2. Hyperglycemic–diabetogenic action reported in amphibians, birds, and mammals

maintains internal sodium (Na^+) levels by a variety of mechanisms, including reduction in permeability of the gills to sodium efflux and water influx, increased mucus production by skin, gills, and buccal epithelium, increased Na^+ reabsorption from the urinary bladder, and stimulation of kidney epithelia to form more and larger glomeruli. There is no clear definition of the role of PRL in osmoregulation in reptiles, but it apparently stimulates the rate of plasma filtration by the kidney and increases ion reabsorption from urine. In birds PRL stimulates salt excretion by nasal glands, a process probably aided by the antidiuretic action of PRL together with stimulation of food and water ingestion. In the chick embryo PRL acts on the allantoic membrane to help control the volume of fluid space during desiccation. This action of PRL on the extraembryonic membranes of chick is perhaps related to the role of PRL in water and ion transport by amniotic membranes of the mammalian fetus. In mammals PRL also acts on the kidney, intestine, and mammary epithelium in lactating rats, to help control internal electrolyte balance, by promoting reabsorption of Na^+, Cl^-, and water. The intestine absorbs more glucose, amino acids, and Ca^{2+} under PRL stimulation. Table XIV lists some actions of PRL on hydromineral balance among the vertebrates (see also Part IV).

15.1.3. Reproduction

The most familiar role of PRL in reproduction is stimulation of mammary gland growth and milk production. PRL increases mitosis of mammary epithelial cells during development, but in addition, during pregnancy and lactation, promotes synthesis of the milk proteins casein and α-lactalbumin. In cichlid fish, PRL also prompts skin mucus-secreting cells to produce copious amounts of mucus on which the young fish feed.

Mammalian PRL affects the activity of the ovary in all vertebrates, but whether the homologous hormone acts similarly has not yet been studied. In mammals it can be luteolytic and luteotropic, in birds and the few fishes studied it stimulates ovarian growth, but in reptiles there is no clear action. High circulating levels of PRL in primates cause amenorrhea, due in part to direct action on the ovary, but PRL also interferes with pituitary GtH production, particularly episodic LH secretion, which

Table XIV
Actions of Prolactin Affecting Water and Electrolyte Balance

Cyclostomes
1. Electrolyte metabolism in hagfish (ACTH-like)
Teleosts
1. Survival of hypophysectomized euryhaline freshwater species
2. Restoration of water turnover in hypophysectomized *Fundulus kansae*
3. Skin, buccal, and gill mucus secretion
4. Reduced gill Na^+ efflux (reduced permeability)
5. Reduced gill permeability to water
6. Inhibition of gill Na^+,K^+-ATPase
7. Increased urinary water elimination and decreased salt excretion
8. Stimulation of renal Na^+,K^+-ATPase
Amphibians
1. Skin and electrolyte changes associated with water drive
2. Sodium and water transport across toad bladder
3. Restoration of plasma Na^+ in hypophysectomized newts
Reptiles
1. Restoration of plasma Na^+ levels in hypophysectomized lizard
Birds
1. Stimulation of nasal (orbital) salt gland secretion
Mammals
1. Lactation
2. Increased Na^+ retention at renal level

would normally initiate and maintain the corpus luteum. In males PRL acts on the testes and accessory sex glands, causing increases in seminal vesicle weight, and in newts it will restore growth of regressed nuptial pads.

There is evidence that PRL may play an important role in coordinating physiological and behavioral aspects of reproduction in nonmammalian vertebrates. In birds it is involved in onset of broodiness and in egg incubation. Broodiness behavior in a cichlid fish depends on PRL, and in the paradise fish the hormone stimulates secretion of mucus bubbles which form a nest. Some of the effects of PRL on aspects of reproductive behavior and physiology are listed in Table XV.

The wide range of PRL actions in vertebrate physiology suggests that the molecule is very ancient and has played a vital part in vertebrate survival and evolutionary radiation.

15.2. Growth Hormone

In all adult tetrapods GH is the major growth-promoting hormone of the pituitary; in larval amphibians, juvenile reptiles, and teleost fishes PRL also has significant growth-stimulating activity. GH acts by a variety of mechanisms to

Table XV
Actions of Prolactin Related to Reproduction

Teleosts
1. Skin mucus secretion (e.g., discus ''milk'')
2. Growth and secretion of seminal vesicles
3. Parental behavior (nest building, fin fanning, buccal incubation of eggs)
4. Maintenance of brood pouch in male seahorse
5. Gonadotropic

Amphibians
1. Water drive (prior to reproduction)
2. Antispermatogenic
3. Stimulation of cloacal gland development

Reptiles
1. Antigonadotropic

Birds
1. Production of crop ''milk''
2. Formation of brood patch
3. Antigonadal
4. Premigratory restlessness
5. Synergism with steroids on female reproductive tract
6. Suppression of sexual phase of reproductive cycle

Mammals
1. Mammary development and lactation
2. Synergism with androgen on male sex accessory glands
3. Luteotropic
4. Luteolytic action in rats
5. Vaginal mucification in rats

increase body length and carcass weight. The majority of studies, even in nonmammalian vertebrates, have used mammalian hormones, but the few experiments with homologous hormones have shown similar actions, generally with higher potency. The episodic release of GH ensures a constant supply of hormone for interaction with other hormones, particularly from the adrenal and thyroid. Circulating levels of GH do not differ greatly throughout life in mammals, but the greatest rate of linear growth occurs before puberty, mainly due to enhanced tissue sensitivity at this stage. In teleosts, GH increases body length of fish held in short photoperiod, but has little additional effect on fish in long photoperiod, due probably to high concentrations of endogenous GH and receptor saturation within sensitive tissue.

15.2.1. GH and Somatomedins (SM)

Serum of all vertebrates except teleosts contains the growth-promoting factor(s) somatomedin(s) (SM). In mammals it is established that SM is produced by the liver under the influence of GH (PRL also stimulates SM production), but as yet there is no direct evidence for this action in other vertebrates. SM acts on various tissues, inducing cell replication, but acts particularly on the epiphyseal plate of the long bones to stimulate mitosis and sulfate uptake into cartilage. The liver then may be considered to be a major target organ for GH.

15.2.2. Metabolic Action of GH

GH is also involved in growth by its effects on metabolism of lipid, carbohydrate, and protein, and through stimulation of food consumption. As a result of its lipolytic action, fat stores are mobilized leading to a rise in serum free fatty acids, which are used as an energy source, particularly for protein synthesis, causing protein accumulation in liver and muscle. The direct effects of GH on protein synthesis are twofold: uptake of amino acids into cells is enhanced as well as their intracellular incorporation into proteins, although the two processes are not interdependent. Thyroid hormones also synergize with GH to stimulate amino acid uptake into some tissues (e.g., liver). Protein accumulation is also protected by the antagonistic effect of GH to the promotion of gluconeogenesis from amino acids by glucocorticoids. The result of these several actions is positive nitrogen balance and elevated plasma glucose through stimulation of glucose formation from liver glycogen, but not from muscle carbohydrate stores. This glycogenolytic action of GH is due partly to potentiation of glucagon and adrenaline action with release of glucose from the cells due to antagonism to the action of insulin (diabetogenic effect). In times of stress the antagonistic action of GH to the effects of glucocorticoids prevents excessive protein catabolism.

In species of reptiles and postmetamorphic amphibia, as well as in mammals, GH stimulates linear growth and body weight increase in animals maintained on a controlled diet; if food is freely available, then both food consumption and body weight increase considerably. Thus, GH has multiple actions that increase body size; the metabolic effects appear to be direct, but increase in linear growth may be mediated in tetrapods through SM.

15.3. Adrenocorticotropin

Pituitary control of interrenal cells is a secondary evolutionary development, since production of steroids predates pituitary organization and hormone secretion. Although in mammals the primary function of ACTH is to maintain secretory integrity of all adrenal cortical cells and to stimulate zona fasciculata and reticularis cells to produce glucocorticoids (cortisol, cortisone, and corticosterone), the situation in nonmammalian vertebrates is different because interrenal tissue is not divided into zones secreting either glucocorticoids or mineralocorticoid (aldosterone). Determining the role of ACTH in control of steroidogenesis in nonmammalian vertebrates is complicated by the responsiveness of the cells to extrapituitary factors including ions, osmotic pressure, and renin. However, it is apparent that whatever the environmental ionic situation, ACTH maintains basal secretion of steroids in all vertebrates. The greatest difference from mammals is shown by amphibians, in which ACTH is concerned mainly with aldosterone secretion and the adrenocorticotrops of the pituitary respond to inhibitory feedback control of this steroid. Adrenal tissue from duck and turtle will also secrete aldosterone under stimulation by ACTH *in vitro,* but whether this is a significant physiological interaction *in vivo* is not known. In teleosts, cortisol controls mineral ion metabolism, and in several species serum cortisol levels rise following ACTH treatment. Similarly, in species of reptiles and birds hypophysectomy causes a fall in serum corticosterone and ACTH restores normal levels (see also Part I).

The biological activity of ACTH in all species resides in the first 1–24 amino acids; residues 25–39 are important for maintaining molecule stability in serum and reducing the rate of breakdown.

ACTH may also share the intrinsic function of other hormones derived from the proopiocortin molecule, β-LPH, α- and β-MSH, and opioid peptides, including lipolytic action to mobilize fat, induction of insulin release, influences on memory and learning, and particularly melanin dispersion. These actions are probably due to common amino acid chains shared by these hormones.

15.4. Melanotropins (α- and β-MSH)

The melanotropins in vertebrates, except birds and mammals, are concerned with melanogenesis and melanin dispersion; in mammals MSH will also induce melanogenesis, but this is not a very significant physiological effect and here α-MSH may be more concerned with fetal growth. In cyclostomes (lampreys), fishes, and amphibians MSH exerts an important physiological effect in causing dispersion of melanin granules within melanophores, enabling animals to escape predators by camouflage. Change in pituitary MSH content accompanies adaptation to background in reptiles, and the response of melanophores of *Anolis carolinensis*

(Reptilia, Lacertilia) is a bioassay for MSH. Some birds and mammals that change their pelage color according to season will respond to MSH treatment provided they are maintained on appropriate daylength, so it seems that other factors besides MSH are involved in these responses.

15.5. Glycoprotein Hormones

In contrast to the prolactin–growth hormone family of molecules, the glycoprotein hormones have actions that are almost entirely restricted to single target organs.

15.5.1. Thyrotropin

A few extrathyroidal actions of TSH have been claimed in mammals, for example, lipolysis in isolated adipose tissues and the induction of exopthalamic edema, but, as far as has been tested, the principal action in all vertebrate groups is stimulation of the thyroid gland. TSH appears to influence many aspects of thyroid function, including uptake of ionic iodine into the cells, thyroglobulin synthesis and proteolysis, oxidation of the iodide and iodination of the tyrosyl residues of thyroglobulin to produce iodotyrosine (hormonally inactive), coupling of iodotyrosines to produce the iodothyronines T_3 and T_4, and release of the active T_3 and T_4 into the blood. Histologically, this stimulation of thyroid metabolism by TSH can be seen as a hypertrophy (increase in cellular size) and hyperplasia (increase in cell numbers) of the follicles, by an increased vascularization and a decrease in colloid within the follicles. At the EM level, RER, Golgi, and secretory granules in the follicular epithelial cells are increased after TSH treatment, as are the numbers of microvilli on the cell surface.

Mammalian TSH is active in other groups of vertebrates; however, in fishes mammalian LH and FSH also have some thyrotropic activity. On the other hand, fish TSH appears to have little or no activity in mammals. These findings emphasize the close structural similarities between the three glycoprotein hormones (Section 13.1) and suggest that there has been considerable evolutionary modification both in the TSH molecule and in the thyroid cell receptors for it.

15.5.2. Follicle-Stimulating Hormone and Luteinizing Hormone

The gonadotropic hormones have an effect on only one target gland, the gonad; no other tissue appears to possess receptors for them. It is, however, far from easy to generalize about the precise role played by each gonadotropin in reproductive processes, which follow diverse adaptive patterns both within and between different vertebrate groups (see Part II). Basically, the gonad of any animal, male or female, performs two separate functions: production of gametes (gametogenesis) and of steroid hormones (steroidogenesis). FSH appears mainly involved with gametogenesis and LH with steroidogenesis, although there is often a high degree of interaction (synergism) between the two gonadotropins, and with steroid hormones, in control of these processes. The distinct roles of FSH and LH are seen most clearly in the mammalian testis. The targets for FSH, judged by the distribution of receptors, are the Sertoli cells of the seminiferous tubules, but for spermatogenesis the synergistic action of LH and testosterone is necessary; LH acts primarily on the interstitial (Leydig) cells, which produce testosterone. The development of follicles within the ovary is much more complex, and the response of the tissues to LH and FSH depends to a large extent on the phase of the ovarian cycle (see Part II). Nevertheless, FSH is the hormone again mainly concerned with gametogenesis (follicle development), and LH mainly with steroid, especially progesterone, secretion. Ovulation appears to be induced primarily by the surge in LH secretion, and the same hormone has an important function in maintenance of the corpus luteum (luteinization).

The manner in which the gonadotropins interact to control reproduction in lower vertebrates is much less clear, and it is difficult to make comparisons with the mammalian model, since both the reproductive processes, and the nature of the hormones themselves, are quite different in some groups. For example, in fishes there is almost certainly only a single gonadotropin, which, although having some activity in amphibians, has none in mammals. Other hormones, such as steroids and prolactin, appear to have much more importance in the regulation of reproductive events than in mammals; however, gonadotropin(s) may be involved indirectly through their steroidogenic action.

15.6. Neurohypophyseal Hormones

The octapeptide hormones (Figure 34) of the neurohypophysis have received considerable attention in comparative research, providing a fairly complete picture of structure and function. The systemic effects of these hormones include control of water balance, reproductive tract and vascular smooth muscle contraction, and mammary gland milk ejection. Phylogenetically these systemic actions are probably recent events, and it may be that neurohypophyseal hormones were originally neurotransmitters.

Evolutionary changes in the molecular structure of these peptides may be associated with a change of action. Arginine vasotocin, found in all nonmammalian vertebrates, is effective in all target tissues mentioned earlier, but its pressor action in fish is postulated to be its most primitive function. AVT will act on mammary gland and reproductive tract but is much less potent than the later-evolving vasopressin and oxytocin. The pressor effect in fish is opposed by isotocin and mesotocin. Birds differ from all other vertebrates in responding to neurohypophyseal hormones with a drop in blood pressure. The action of basic neurohypophyseal hormones on vascular smooth muscle is an integral part of their role in regulation of urine flow. In tetrapods the action is antidiuretic, but in some freshwater fish (e.g., lungfish) AVT causes diuresis. In mammals the appearance of two hormones replacing AVT has led to clear differentiation in their function, with vasopressins being the more potent regulators of water balance and oxytocin concerned in reproductive physiology by acting on the uterus and mammary gland. For further discussion see Part IV.

16. Techniques for Identifying Pituitary and Hypothalamic Cells

16.1. Staining Techniques for Pituitary Cells

16.1.1. Introduction

Differential staining of adenohypophyseal cells depends both on the chemical nature of the hormones and on the fact that the hormone is stored in granules within the cell cytoplasm, so that the activity of the cells and amount of stored hormone can affect the sensitivity and intensity of staining. Before cells can be stained at all they must first be fixed, in a condition as near as possible to the living state. The pituitary must be removed rapidly from the animal for prompt fixation, but for accurate orientation, particularly of very small pituitaries, it is helpful to leave it attached to the whole or part of the brain.

16.1.2. Fixation

The two most generally successful fixatives for pituitary tissue are Zenker–formol and sublimated Bouin–Hollande.

16.1.2a. Zenker–Formol

Stock solution:

Distilled water	100 ml
Potassium dichromate	2.5 g
Mercuric chloride	4.0–5.0 g
Sodium sulfate	1.0 g

Formalin added just before use, 5.0 ml/100 ml of stock solution.
Fixation time: 6–24 hr.

16.1.2b. Sublimated Bouin–Hollande

Stock solution:

Distilled water	100 ml
Picric acid	4.0 g
Cupric acetate	2.5 g

Formalin, 10 ml, and saturated mercuric chloride, 10 ml, added per 100 ml of stock solution just before use.
Fixation time: 2–7 days.

16.1.2c. Decalcification

If the pituitary was fixed surrounded by bone, then this must be removed before further processing, by immersion in a solution of 5% formic acid in 5% formalin, changed daily for 14 days.

After several hours' washing in water, the tissue can then be dehydrated through graded alcohols, cleared, and embedded as routinely.

16.1.3. Staining Protocols

16.1.3a. Cleveland–Wolfe Trichrome

Tissue sections are immersed sequentially as follows:

1. Ehrlich's hematoxylin for 5 min, then wash in alkaline alcohol to stain nuclei blue.
2. Rinse in tap water (TW) and distilled water (DW) briefly.
3. 1% aqueous erythrosin for 5 min.
4. Rinse in TW and DW briefly.
5. 2% orange G in 1% phosphomolybdic acid for 8–10 sec.
6. Rinse in TW and DW rapidly.
7. 0.25% aniline blue in water with acetic acid added to pH 3–4 for 90 sec.
8. Rinse very rapidly in two changes of absolute alcohol.

16.1.3b. Herlant's Alizarin Blue Tetrachrome

Three solutions containing stains are used sequentially in this technique.

1. Erythrosin, 1% in distilled water
2. Mallory 2, containing:
   aniline blue, 0.5 g
   orange G, 2.0 g
   distilled water, 100 ml
   acetic acid, 1.0 ml
3. Acid alizarin blue, containing:
   acid alizarin blue, 0.5 g
   aluminium sulfate, 10.0 g
   warm distilled water, 100 ml

The solution is boiled for 5 min, volume corrected to 100 ml, and then filtered.

Procedure. Tissue sections are immersed sequentially as follows:

1. Erythrosin for 10 min.
2. Rinse rapidly in TW and DW.

3. Mallory 2 solution for 2 min.
4. Rinse rapidly in TW and then in 5% phosphomolybdic acid.
5. Alizarin blue for 5 min.
6. Rinse rapidly in TW then in 5% phosphomolybdic acid for 5 min.
7. Rinse rapidly in 95% alcohol and dehydrate quickly in absolute alcohol.

16.1.3c. Alcian Blue–Periodic Acid Schiff Technique

The principle of this technique depends on the oxidation of tissue carbohydrates and carbohydrate-containing molecules to aldehydes, which then form a magenta color with Schiff's reagent, which is decolorized basic fuchsin. Three solutions are used in this technique:

1. For oxidation:
    - 2.5% potassium permanganate
    - 5.0% sulfuric acid

   For use mix 10 ml of each solution with 60 ml distilled water.
2. Schiff's reagent:
    - basic fuchsin, 0.5–1.0 g
    - sodium metabisulfite, 1.9 g
    - 1 N hydrochloric acid, 15 ml
    - distilled water, 85 ml

The solution must be shaken at intervals for at least 2 hr, then decolorized by addition of 200 mg activated charcoal, allowed to stand for 2 min, and then filtered. The solution must be stored in an airtight bottle in the refrigerator.

3. Alcian blue:
    - For pH 3.0:
        - alcian blue, 1.0 g
        - distilled water, 100 ml
        - glacial acetic acid, 1.0 ml
    - For pH 0.2:
        - alcian blue, 1.0 g
        - 10% sulfuric acid, 100 ml

Procedure

1. Oxidize in permanganate/sulfuric acid for 5 min.
2. Rinse in running tap water.
3. Bleach in 5% sodium bisulfate.
4. Rinse in running TW and then DW.
5. Immerse in 1% alcian blue for 30 min.
6. Rinse in running TW.

7. Oxidize again in 1% periodic acid for 5 min.
8. Rinse in DW.
9. Immerse in Schiff's reagent for 30 min.
10. Wash in running tap water for at least 15 min, but preferably overnight to intensify the Schiff coloration.
11. Immerse in Ehrlich's hematoxylin for 5 min.
12. Blue in TW.
13. Counterstain in orange G for 8–10 sec.
14. Wash rapidly in TW and DW.
15. Dehydrate rapidly, clear, and mount.

Table XVI shows the tinctorial differentiation of pituitary cells possible following these various staining procedures.

16.2. Electron Microscopy

16.2.1. The Instrument

Whereas the light microscope (LM) gives a magnified image by focusing light waves from a specimen, the transmission electron microscope (EM) does so by focusing waves of invisible electrons. The electrons are produced from a filament made of tungsten—a metal that emits electrons when heated by the passage of an electric current—and then accelerated down the column due to the large potential difference (40–100 kv) between the cathode (immediately surrounding the filament) and the anode (further down the column). After passing through the hole in the anode, the electrons are converged into a narrow beam on to the specimen by a series of condenser lenses. Electrons passing through the specimen are imaged by an objective lens, and the ''electron image'' is magnified by a series of intermediate and projector lenses. These lenses are not made from glass as in the LM, but are electromagnets, with variable focal lengths dependent on the current flowing

Table XVI
Staining Properties of Adenohypophyseal Cells

Cell type	Cleveland–Wolfe	Herlant tetrachrome	Alcian blue–PAS
Prolactin cells	Dull red	Red	Yellow
Somatotrops	Bright red/orange	Yellow/orange	Yellow
Corticotrops	Colorless/faint black	Pale red/pink	Yellow
Thyrotrops	Pale gray/blue	Dark blue	Dark blue
FSH cells	Clear blue	Pale blue	Violet
LH cells	Clear blue	Purple/violet	Brick red
Melanotrops[a]	Purple	Pale blue/purple	

[a]Melanotrops will also stain with lead hematoxylin (PbH+).

through them—in this way focusing of the electron image can be achieved. The electron image itself is invisible, so it cannot be viewed directly by ocular lenses; however, a visible image can be formed on a phosphor-coated screen, since electrons hitting the coating of the screen cause it to fluoresce. A photographic image can be obtained when desired by tilting the viewing screen out of the way and opening the shutter through to the camera, where the electrons expose the photographic plates in a manner similar to visible light. To ensure a continuous passage of electrons down the length of the column, the instrument operates at a high internal vacuum (at least 10^{-5} torr); thus, the probability of any of the electrons colliding with gas molecules and being scattered from the beam is very small. In order to maintain the vacuum during operation, insertion of the specimen and photographic plates into the column must be done via airlocks.

16.2.2. Fixation and Specimen Preparation

As soon as cells are removed from their normal physiological environment, disruption of intracellular organization and autolysis (breakdown of structural components) begin. Fixation is a procedure that aims to prevent this, so that ultrastructure the same as, or as near as possible to, that found in the living cell is maintained throughout the subsequent harsh environments of dehydration and embedding. Most procedures for EM fixation, as for LM, utilize aldehydes (formaldehyde and/or glutaraldehyde), which stabilize the tissue components by cross-linking the free amino groups of proteins. Osmium tetroxide is often used too, as it cross-links unsaturated fatty acids, so preserving membranes. Osmium also has the advantage of being a heavy metal, and so enhances the electron density of membranes.

The transmission EM relies on the electron beam passing through the specimen in order to produce an image. Because of the limited penetration of electrons through tissue at accelerating voltages up to 100 kv, it is important for adequate resolution that sections no thicker than 100 nm be used. In order to produce such thin sections, special methods of supporting (embedding) and cutting the tissue have been developed. The most widely used embedding media are epoxy resins (e.g., Araldite, Epon, Spurr's resin). Fixed tissue cannot be placed directly into these resins, since they are immiscible with water—dehydration and infiltration is necessary first. Dehydration is simply achieved by fairly rapid transfer of the tissue through a graded series of ethanol/water mixtures (e.g., 50%, 70%, 90%, 95%, 100% $\times$ 3: approximately 10 min in each). Infiltration takes longer, usually several days, in order to ensure full penetration of the resin through the interstices of the tissue and into the densest components of the cells. If the resin used is soluble in ethanol (e.g., Spurr's resin), tissue can be infiltrated in a 50% resin:50% ethanol mixture; otherwise (using Epon, Araldite) it must be transferred through graded mixtures of ethanol:propylene oxide and proplylene oxide:resin. In either case, infiltration is completed in several changes of fresh resin—it is advantageous to mechanically rotate the pots containing the tissue and resin during this whole process. After infiltration the tissue is placed in molds containing fresh resin, which is polymerized (at 60–70°C, for 8–48 hr) to form hard blocks from which thin sections can be cut. This is done on an ultramicrotome, using a glass or diamond knife. As the sections are cut, they float out on to water in ribbons, which are picked up on fine grids (usually copper) for examination in the EM. It is usually necessary to improve the contrast of the sections prior to examination by staining with heavy metals. This is done by immersing in a solution of a suitable salt of the metal: a combination of uranyl acetate, which stains nucleic acids and proteins, and lead citrate, which stains lipids, is frequently used. The heavy metal atoms thus introduced to the section act by "knocking out" electrons from the beam, therefore darkening the image given by the stained areas.

16.3. Immunocytochemistry

The technique of immunocytochemistry depends on the detection of tissue components by specific antibody and subsequent visualization by fluorescence or histochemistry. Most biological molecules, under the right conditions, are antigenic (immunogenic) in foreign species and prompt the formation of circulating antibodies (Ab); these antibodies can then be used to detect the antigen in tissues of the species of origin. There are three stages in the process of immunocytochemistry:

1. Production of specific antibody
2. Preparation of tissue containing active antigen
3. Binding of Ab to antigen (Ag) and visualization of Ab/Ag complex.

16.3.1. Production of Antibody

Large protein or peptide molecules are immunogenic in foreign species, but small molecules, such as steroids, catecholamines, and small peptides, must be conjugated to large peptides (e.g., bovine serum albumin) before they will cause formation of specific antibody. The antigen should be as pure as possible because impurities will also prompt antibody formation and lead to nonspecific reactions in tissue tests. Thus the antigen, solubilized in physiological saline and emulsified in "adjuvant," is injected intramuscularly or intradermally into a recipient animal (usually rabbit, guinea pig, or goat). After two or more injections, and when the antibody titer is sufficiently high, serum is collected. Specific antibody may be separated from whole serum by chromotography (Con A-Sepharose) as a component of the immunoglobulins, or by absorption on media coated with the specific antigen and subsequent splitting of the Ag/Ab complex. Because peptide hormones from a wide variety of species are similar in amino acid composition, it is possible to use antibody against hormone of one species (e.g., sheep prolactin) to detect the related hormone in a different species (e.g., teleost), a heterologous system. However, it is preferable to use a homologous antiserum that has been carefully screened for cross-reactivity with other tissue antigens so that positive reactions are definitive. It must be borne in mind that in heterologous systems there may be more false-reactions than in the homologous, and that negative reactions are more likely to be due to differences in molecular configurations between the native hormone molecule and that from the original species than to the absence of the particular hormone. In the case of some hypothalamic peptides there is complete molecular identity throughout the vertebrates, and a single antibody will detect, for example, TRH in all of them, but because the molecule is only a short sequence of amino acids, then these may be detected adventitiously in totally unrelated peptides. It is essential then, in immunochemical reactions, to have both negative and positive controls.

16.3.2. Preparation of Tissue

Unless ICC localization is to be applied to living cells or to frozen sections, some method of fixation of the tissue will have been previously carried out: this may cause chemical modification of the tissue Ag. The action of fixatives usually involves cross-linking and denaturation of tissue components, and if this occurs in the critical region (''antigenic determinants'') of the Ag molecule, binding with the Ab may be reduced or even completely abolished. If this happens, the tissue is said to have lost its ''antigenicity'' for a particular Ab. The effect of fixatives on antigenicity is far from uniform; some antigens are disrupted by mild fixatives, while others seem to be able to withstand even the harshest fixation procedures.

Loss of antigenicity due to fixation is not often encountered in ICC at the LM level, as the fixatives used, e.g., Bouin's, are formaldehyde-based and quite mild in their effect. Problems do, however, arise in EM immunocytochemistry. Osmium fixation and epoxy resin embedding sometimes destroy antigenicity completely, and some antigens are also very sensitive to glutaraldehyde. Usually one has to reach a compromise between some reduction in antigenicity and reasonably good morphological tissue preservation. To this end, several special fixatives and embedding media for EM immunocytochemistry have been developed.

16.3.3. Detection and Visualization of Ag/Ab Complex

Detection of tissue antigen by immunocytochemistry is by ''direct'' or ''indirect'' methods (Figure 45). In the direct method a fluorochrome or enzyme complex (2) is coupled to the primary antibody (1), which then, after the antibody has bound to tissue antigen, can be visualized by examination in a fluorescence microscope or by formation of a colored or electron-dense precipitate by enzyme/substrate reaction. The fluorochromes generally used are rhodamine (red) or fluorescein isothiocyanate (yellow/green). One enzyme substrate reaction involves horseradish peroxidase, which, in the presence of peroxide, will cause precipitation of a colored product from diaminobenzidine (brown) or chloronaphthol (blue). This technique can be used for EM–ICC and enhanced by reaçtion with osmium, but colloidal gold and ferritin are also used as electron-dense markers. The indirect method is more sensitive and easily controlled than the direct method, involving intermediate steps of second antibody to the species of the primary antibody and of the antihorseradish

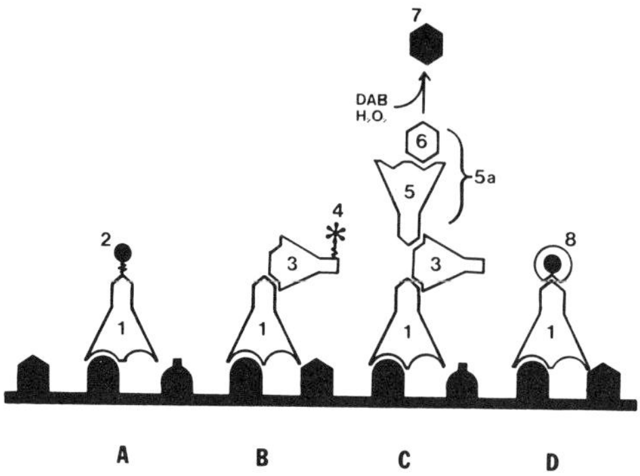

Figure 45. Diagram showing steps in immunocytochemical staining techniques. A, direct method; B, indirect immunofluorescence; C, immunoperoxidase; D, immunogold. 1, Primary antibody; 2, marker molecule (e.g., ferritin, chromogen); 3, second antibody (e.g., goat antirabbit IgG); 4, fluorochrome (e.g., fluoroscein isothiocyanate); 5, antiperoxidase; 5a, peroxidase/antiperoxidase complex; 6, horseradish peroxidase; 7, reaction product of diaminobenzidine (DAB); 8, colloidal gold bound to second antibody or to protein A.

peroxidase, which can be prebound to peroxidase as PAP complex. The steps are outlined below:

16.3.3a. Examples of Procedure (Referring to Figure 45)

i. Indirect Immunofluorescence Method for LM (B)

1. Dewax and rehydrate slides with tissue sections.
2. Preincubate in normal goat serum (NGS), 1:10, 30 min (see note 2).
3. Incubate in primary antibody (1), 1:100–1:10000, 1–24 hr (see note 3).
4. Incubate in goat antirabbit IgG (3) labeled with fluorescein isothiocyanate (4), 1:10, 1 hr.
5. Mount slides in glycerine/buffer and examine on fluorescence microscope.

ii. Indirect Immunoperoxidase Method for LM or EM (C)

1. Dewax and rehydrate slides, or "etch" thin EM sections on nickel grids (see note 4), 20 min in 10% H_2O_2, or 2% NaOH in 50% ethanol.
2. Preincubate in NGS, as in *i*.
3. Primary antibody, as in *i*.
4. Goat antirabbit IgG (3), 1:20–1:100, 30 min.
5. Peroxidase–antiperoxidase (5) 1:50–1:200, 30 min.
6. Diaminobenzidine (50 mg/200 ml), freshly prepared, filtered, and 0.003% H_2O_2 added, 5–15 min.
7. Improve contrast of reaction product (7) if necessary in osmium tetroxide vapor, 30 min.
8. Dehydrate and mount slides for LM. Stain EM sections if necessary in uranyl acetate/lead citrate and blot dry.

iii. Indirect Immunogold Methods for EM (D)

1. Etch sections on nickel grids.
2. Preincubate in NGS, as in *i*.

3. Primary antibody, as in *i*.
4. Protein A or goat antirabbit coated colloidal gold (8) 1:10 1 hr.
5. Rinse grids, stain if necessary, blot dry, and examine on EM.

16.3.3b. Notes

1. Between each step slides or grids are thoroughly rinsed with buffer, usually tris or phosphate buffered saline, 0.05 M, pH 7.6. One drop of Triton-X100 may be added per liter to improve wetting. The same buffer is used for dilution of antisera, etc. Buffer must be millipore-filtered for EM technique.
2. NGS preincubation "blocks" sites in tissue that would otherwise bind goat antirabbit nonspecifically.
3. If procedure is done at room temperature, slides must be in sealed humidity chamber.
4. To remove resin from the section surface and thereby expose antigenic groups.

16.4. Localization of Amines by Formaldehyde Fluorescence

Biogenic amines which act as neurotransmitters in the brain–hypothalamus–pituitary system, and which ultimately control release of AHP hormones, have been demonstrated most successfully by the induction of fluorescence by paraformaldehyde in the vapor phase.

All the biogenic amines, i.e., catecholamines—adrenaline (A), noradrenaline (NA), indoleamines—serotonin (5-hydroxytryptamine, 5-HT) and dopamine (DA) and 5-hydroxytryptophan (5-HTP) all diffuse readily in aqueous fixatives, but under dry conditions they react with formaldehyde vapor to form highly fluorescent trihydroxyindoles. Primary amines (DA, NA) produce maximum fluorescence at 410 nm and secondary amines (A) at 480 nm, whereas the fluorescent product of 5-HT has peak emission at 530 nm. Hence, catecholamines can be distinguished by their green fluorescence from the yellow of indoleamines.

To demonstrate amines in brain and pituitary the tissues must be rapidly excised and cooled immediately to below −50° in a mixture of propane/propylene (9:1) at the temperature of liquid nitrogen, to prevent ice crystal formation in the tissue. All water is then removed by lyophilization, and, when dried, tissues are reacted with paraformaldehyde vapor. For successful reaction with tissue amines the relative humidity (RH) of the paraformaldehyde is very important; RH between 50 and 80% gives strong reactions. To achieve this, 6 g of paraformaldehyde should be stored in a 1-liter closed vessel containing sulfuric acid (density 1.30, 40% solution) for 5–7 days before use. When dry the tissue is transferred, in a small nylon mesh container, to a 1-liter glass vessel containing 5 g paraformaldehyde, which is then sealed and heated to 80° for 1 hr; however, for maximal fluorescence of adrenaline the reaction must continue for 3 hr. Tissues are rapidly embedded in degassed paraffin wax. Sections are mounted on dry slides in liquid paraffin for studies on adrenaline, but for some other amines they may be dewaxed with xylene before temporary mounting in a liquid-paraffin-based mountant.

16.4.1. Microscopic Examination

Sections are examined in a microscope equipped with a high-pressure mercury vapor lamp to provide ultraviolet light (UV), a blue excitor filter (e.g., Schott BG 12, 3–7 mm thick), and barrier filters to absorb the blue background.

16.4.2. Differentiation of Specific from Nonspecific Fluorescence

Specific amine fluorescence fades with exposure to UV light whereas nonspecific (e.g., autofluorescence) is generally stable. Specific fluorescence can be rapidly quenched by the reducing action of sodium borohydride. Tissue sections must be immersed in a 0.1% solution of sodium borohydride in 90% isopropanol for 1–2 min.

17. Selected Readings

Anatomy, Morphology, and Cytology of the Hypothalamohypophyseal System

Doerr-Schott, J., 1976, Immunohistochemical detection, by light and electron microscopy, of pituitary hormones in cold-blooded vertebrates. I. Fish and amphibians. II. Reptiles. *Gen. Comp. Endocrinol.* **28:**487–512, 513–529.

Gorbman, A., 1983, *Comparative Endocrinology,* J. Wiley & Sons, New York.

Hansen, G. N., 1983, Cell types in the adenohypophysis of the primitive Actinopterygia, with special reference to immunohistochemical identification of pituitary hormone producing cells in the distal lobe. *Acat Zool.* Suppl. 1–87.

Harris, G. W., and Donovan, B. T., eds., 1966, *The Pituitary Gland,* Vol. 1–3, Butterworth, London.

Holmes, R. L., and Ball, J. N., 1974, *The Pituitary Gland: A Comparative Account,* Cambridge University Press, Cambridge.

Licth, P., and Pearson, A. K., 1978, Cytophysiology of the reptilian pituitary gland, *Int. Rev. Cytol.* Suppl. **7:**239–286.

Tixier-Vidal, A., and Follett, B. K., 1973, The adenohypophysis, in: *Avian Biology,* Vol. III (D. S. Farner and J. R. King, eds.), Academic Press, New York, London, pp. 109–182.

Vila-Porcile, E., and Olivier, E., 1984, The problem of the folliculo-stellate cells in the pituitary gland, in: *Ultrastructure of Endocrine Cells and Tissues* (P. M. Motta, ed.), Martinus Nijhoff Publishers, Boston, The Hague.

Control of Pituitary Function

Ball, J. N., 1981, Hypothalamic control of the pars distalis in fishes, amphibians and reptiles, *Gen. Comp. Endocrinol.* **44:**135–137.

Collu, R., 1981, Neuroendocrine control of pituitary hormone secretion, in: *Pediatric Endocrinology* (R. Collu, J. R. Ducharme, and H. Guyda, eds.), Raven Press, New York, pp. 1–28.

Nistico, G., and Bolis, L., eds., 1983, *Progress in Non-mammalian Brain Research,* Vol. III, CRC Press, Boca Raton, FL.

Sowers, J. R., 1980, *Hypothalamic Hormones,* Vol. 14, *Benchmark Papers in Human Physiology* (L. L. Langley, series ed.), Academic Press, New York.

Hormone Chemistry and Actions

Acher, R., 1980, Molecular evolution of biologically active polypeptides, *Proc. Roy. Soc. (London) B* **210:**21–43.

Barrington, E. J. W., ed. 1979, *Hormones and Evolution,* Vol. 2, Academic Press, London, New York.

Ganong, W. F., and Martini, L., eds., 1978, *Frontiers in Neuroendocrinology,* Vol. 5, Raven Press, New York.

Guillemin, R., Brazeau, P., Böhlen, P., Esch, F., Ling, N., Wehrenberg, W. B., Bloch, B., Mougin, C., Zeytin, F., and Baird, A., 1984, Somatocrinin, the growth hormone releasing factor. *Rec. Prog. Horm. Res.* **40:**233–299.

Jeffcoate, S. L., and Hutchinson, J. S. M., 1978, *The Endocrine Hypothalamus,* Academic Press, London.

Li, C. H., ed., 1973–1983, *Hormonal Proteins and Peptides,* Vol. I–XI, Academic Press, New York, London.

Vale, W., Spiess, J., Rivier, C., and Rivier, J., 1981, Characterisation of a 41-residue ovine hypothalamic peptide that stimulates secretion of corticotrophin and β-endorphin. *Science* **213:**1394–1396.

Neurosecretion

Knowles, F., and Vollrath, L., eds. 1974, *Neurosecretion—The Final Common Pathway,* Springer Verlag, Berlin.

Kobayashi, H., Bern, H. A., and Urano, A., eds., 1985, *Neurosecretion and the Biology of Neuropeptides,* Japan Scientific Society Press, Tokyo/Springer Verlag, Berlin.

Techniques

Björklund, A., and Hökfelt, T., eds., 1983, *Handbook of Chemical Neuroanatomy,* Vol. I, *Methods in Chemical Neuroanatomy,* Elsevier, Amsterdam.

Marchalonis, J. J., and Warr, G. W., eds., 1982, *Antibody as a Tool,* J. Wiley & Sons, Chichester, England.

Polak, J. M., and Van Noorden, S., 1983, *Immunocytochemistry: Practical Applications in Pathology and Biology,* Wright, P. S. G., Bristol, England.

Polak, J. M., and Varndell, I. M., 1984, *Immunolabelling for Electron Microscopy,* Elsevier, Amsterdam.

Part IV

Secretion of Endocrine Glands and Their Relationship to Osmoregulation

R. J. Balment and I. W. Henderson

R. J. Balment • Department of Zoology, University of Manchester, Manchester M13 9PL, England. **I. W. Henderson** • Department of Zoology, University of Sheffield, Sheffield S10 2TN, England.

10

Environmental Constraints and Adaptive Mechanisms

1. Introduction

In all higher animals extracellular fluid (plasma, interstitial, and transcellular fluids) provides an environment in which cells function separated from the external medium. This internal environment, as termed by the nineteenth-century French physiologist Claude Bernard, is the stable medium necessary for normal cell activity, possessing the correct levels of oxygen, carbon dioxide, and metabolic substrates. Maintenance of the optimum composition of this extracellular fluid compartment is ensured through the activity of several organ systems. Thus, the lungs, liver, and kidneys in mammals regulate levels of respiratory gases, metabolic substrates, and waste products, the ionic composition and volume of extracellular fluid, and, indirectly, the same variables within the cell. This part of the book examines current understanding of these latter aspects of homeostasis, principally the mechanisms controlling volume and specific ion and osmotic concentrations of body fluids—osmoregulation.

Body fluid composition and volume are sustained in all vertebrates by a precise balance of water and electrolyte uptake and excretion. The compositional profile of extracellular fluid, as determined by plasma analysis (Table I), is similar in most modern vertebrates despite wide variation in their external osmotic environments. The pattern is only marred by the myxinids (hagfish), which are an ancient group of jawless fish, and the elasmobranchs, which were separated at an early evolutionary stage from the main vertebrate series.

A variety of mechanisms are directed toward the maintenance of the optimum internal compartment. Deviations in one direction bring about opposing reactions in such mechanisms, in accord with Cannon's principle of negative feedback for homeostasis. Changes in the external environment are thus buffered by this system,

Table I
Extracellular Fluid Composition[a]

		Osm	Na	K	Ca	Mg	Cl	Urea
Seawater		1000	459	10	10	53	538	—
Cyclostomata								
Hagfish	SW	1002	454	7	9	23	532	3
Lamprey	FW	248	120	3	2	2	96	0.4
Elasmobranchii								
Dogfish	SW	1075	269	4	3	1	258	376
Stingray	FW	308	150	6	7	4	149	1
Teleostei								
Flounder	SW	337	180	4	3	1	160	
Goldfish	FW	293	142	2	6	3	107	
Crossopterygii								
Coelocanth	SW	1181	181	51	7	29	199	355
Amphibia								
Frog	FW	210	92	3	2	2	70	2
Toad	TER	458	184	4			122	148
Crab-eating	FW	290	125	9			98	40
Frog	SW	830	252	14			227	350
Reptilia								
Alligator	FW	278	148	4	5	3	111	
Iguana	SW	310	159	3	3	1	133	
Aves								
Duck	FW	294	139	3	2		103	
	SW	313	147	3	3		112	
Mammalia								
Man	TER	310	142	4	3	2	103	6
Rat	TER	300	140	6	3	2	120	7

[a]SW, seawater; FW, freshwater; TER, terrestial. Concentrations are mmoles/liter and osmotic concentration (Osm) mOsmoles/liter.

allowing intracellular activities to proceed normally. As Bernard suggested, maintaining the constancy of the internal environment is a requirement for a free and independent life.

Marked differences exist in the availability of water and salts in the various osmotic niches occupied by vertebrates; variability within a particular environment may also be pronounced. Terrestrial species suffer high evaporative water loss from the skin and respiratory surfaces, which is exacerbated in homeothermic groups (birds and mammals) by very high rates of respiratory exchange, panting, and sweating. Egg laying and lactation place further demands on the fluid economy of such species. The osmotic problems encountered by aquatic vertebrates differ markedly in freshwater (FW) and seawater (SW) species. FW is hypoosmotic (less concentrated) to body fluids; animals in this medium tend to gain water osmotically and lose salts by passive diffusion across permeable areas of the integument. Apart from amphibians, the skin of most vertebrates is relatively impermeable to water, although more free exchange may occur across the fish gill. SW is hyperosmotic (more concentrated) to typical vertebrate body fluid (see Table I); marine species, therefore, tend to suffer osmotic loss of body water and passive diffusional gain of salt. Exceptions are the myxinids, which are isoosmotic (equal concentration) with SW, and the elasmobranchs, which achieve slight hypertonicity to SW by urea retention and high plasma sodium and chloride levels.

The terms isotonic, hypertonic, and hypotonic are often used as synonyms for isoosmotic, etc. This is not strictly correct as tonicity is not defined physically, as is osmotic pressure, but is defined biologically, according to the response of living cells immersed in solution. Cells swell in hypotonic solution, shrink in hypertonic solution, and maintain their volume in isotonic medium. We shall normally define the osmotic concentration of a solution in mOsmoles/liter. Within the physiological range of concentrations, 1 mmoles/liter of most ions or compounds in biological fluids is equivalent to 1 mOsmoles/liter.

The body fluids of all living organisms are dilute salt solutions, perhaps reflecting the origin of life in the sea, though they also contain organic solutes. The main inorganic salts contain the restricted array of elements shown in Figure 1, which represents the levels present in the average 70-kg human. Most of the cal-

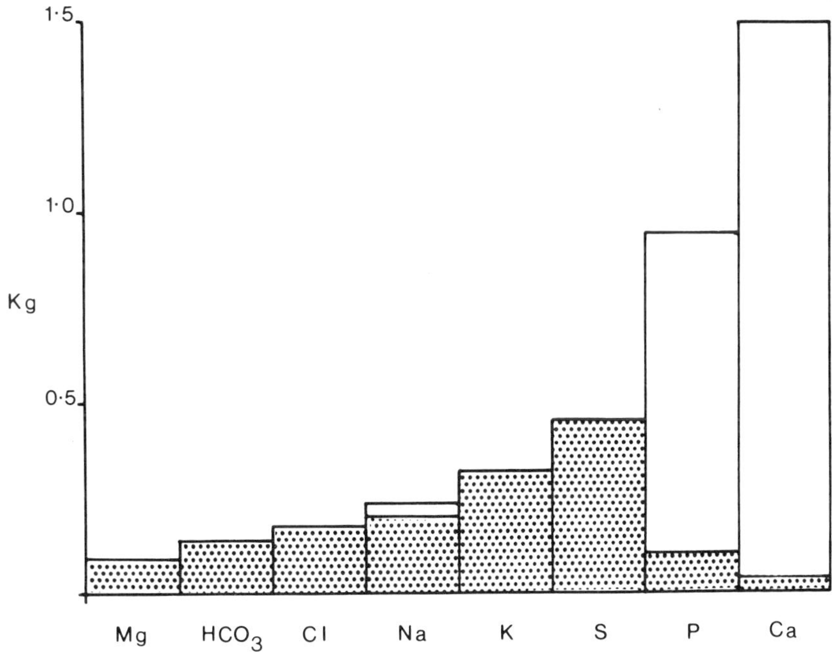

Figure 1. Typical inorganic composition of human body. Clear areas indicate bone content. (Modified from Rankin, J. C., and Davenport, J. A., 1981, *Animal Osmoregulation*, Blackie, Glasgow and London.)

cium and phosphorus and a large part of the sodium are present in bone (clear region of columns, Figure 1). Much of the remaining calcium and phosphorus, sulfur, and magnesium are not freely ionized in solution but are bound in organic compounds. The major salts that therefore contribute to the osmotic concentration of body fluids are chlorides, phosphates, and bicarbonates of sodium and potassium. Osmoregulatory processes are accordingly directed toward the management of plasma levels of these ions. It is not always clear whether animals regulate the volumc and osmotic concentration of their body fluids per se, or whether control of these parameters results from primary regulation of the concentration of the major individual solutes present. The hagfish, for example, may not be an osmoregulator but does control levels of specific ions in the face of changes in external medium composition.

Aquatic animals can be broadly divided on the basis of their ability to survive in environments of varying osmotic concentration. Irrespective of the absolute

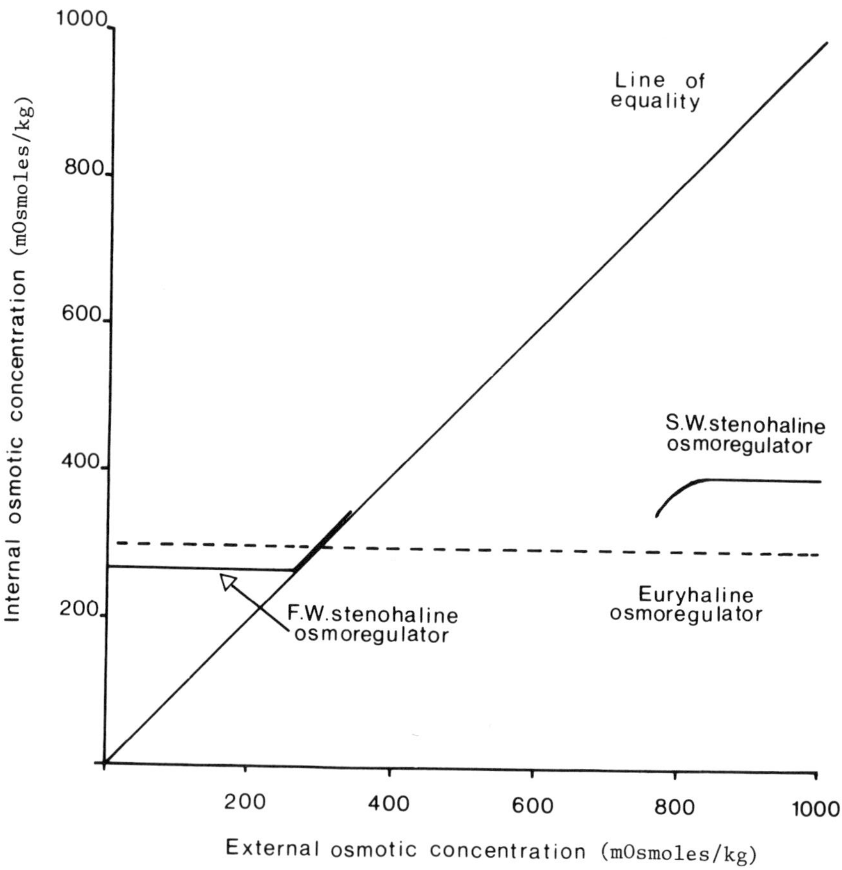

Figure 2. Patterns of plasma osmotic concentration regulation in aquatic vertebrates.

concentration range, animals with a limited tolerance to changes in the environment are called stenohaline. In fact, relatively few species can withstand large variations in environmental osmotic concentration, and these are referred to as euryhaline. The large stenohaline group of aquatic vertebrates can be further divided according to the change in body fluid osmotic concentration following variation in the external media. The body fluid composition of osmoconformers varies more or less in parallel with changes in the external medium (Figure 2). By contrast, the body fluids of osmoregulators remain fairly constant in the face of external change. Euryhaline groups are often good osmoregulators over a limited range of environmental salinity and are osmoconformers over another range.

2. The Osmoregulatory Process

Osmoregulatory mechanisms are based on (1) membranes that have regulated permeabilities to water and specific ions, and (2) the actions of active "pumps" which can transport ions against concentration gradients. These types of mechanisms are to be found in the integument, gut, kidney, and other organs of excretion. Passive influx and efflux of water and electrolytes may occur across the integument, but it is the sites of active salt uptake and excretion which are important for regulation (see Figure 3). A wide variety of specialized tissues and organs are responsible for active salt transport in vertebrates, including fish gill, elasmobranch rectal gland, amphibian ventral skin, reptilian and avian cephalic "salt" glands, and the ubiquitous kidney. The major movements of water inevitably follow the osmotic gradients created at these sites. In spite of the many attempts, no molecular mechanism for the active transport of water across cells has yet been demonstrated. All water flux results from molecular diffusion down thermodynamic activity gradients, mainly osmotic gradients. Water movement associated with transmembrane ion transport often occurs in the opposite direction to the gross osmotic gradient that exists or, in the absence of an osmotic gradient, between the water source and body fluids. This latter isoosmotic transport of water is important in many osmoregulatory tissues, relying on the local osmotic effects of active solute transport. This creates small regions of higher solute concentration on one side of the membrane than on the other, providing a local osmotic force for water transport. The bulk reabsorption of water filtered in the kidney relies on such a mechanism operating in the proximal tubule segment.

Much of our present understanding of the mechanism of salt transport across epithelia stems from experiments on frog skin and toad urinary bladder. This work, pioneered by Ussing, utilizes an isolated area of epithelia clamped between two Ringer-filled chambers. Active transport of sodium in a mucosal (external surface of skin or bladder) to serosal (blood side of skin or bladder) direction is the driving force. This "sodium pump" is based on the enzyme Na^+, K^+-activated ATPase, which is specifically blocked by the drug ouabain. All animals' cell membranes incorporate such a sodium pump, which exchanges three molecules of intracellular

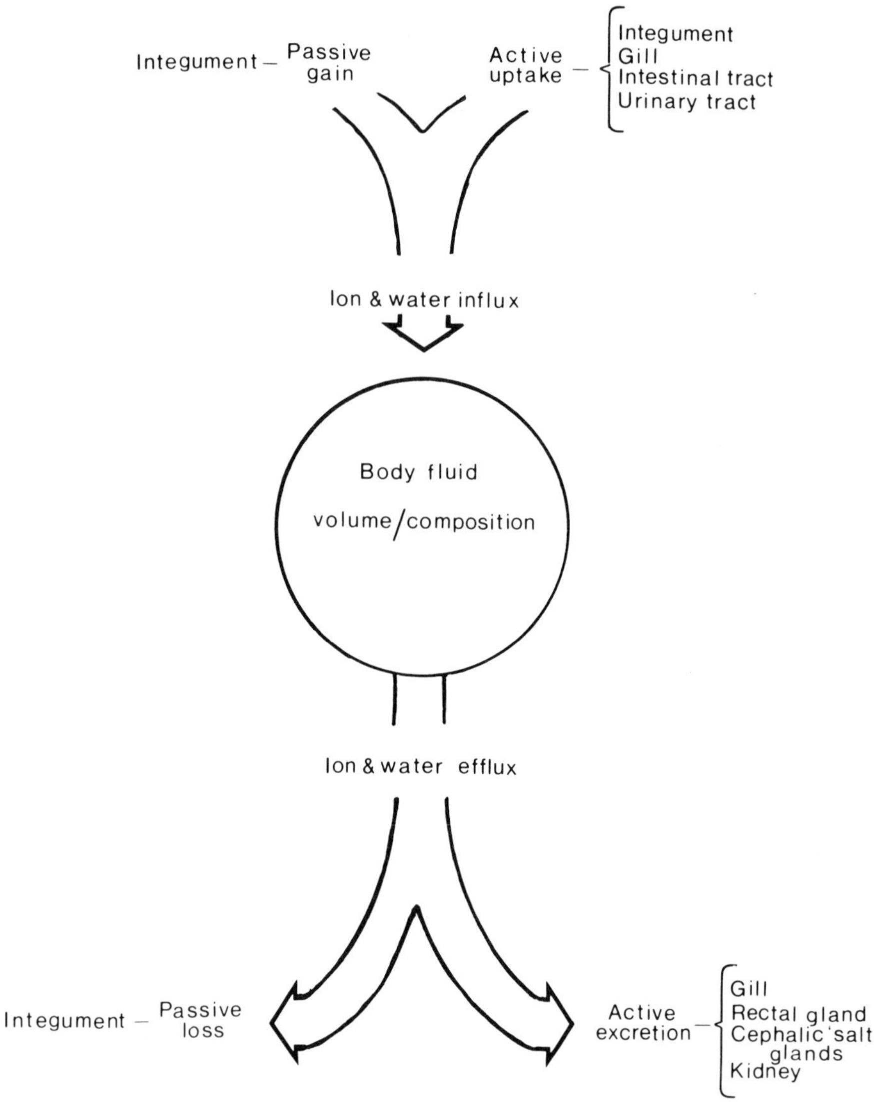

Figure 3. Major sites of osmotic exchange.

Na for two molecules of extracellular K. This pump and differential membrane permeability to Na and K are responsible for maintaining the high intracellular K concentration (160 mmoles/liter) and low Na concentration (10 mmoles/liter) relative to extracellular fluid (compare with Table I). Sodium pump activity on the serosal cell surface in epithelia supports active influx of sodium (Figure 4). A transepithelial potential will build up across the epithelia with a positive charge on the serosal surface. The continual buildup of potential is prevented by the consequent

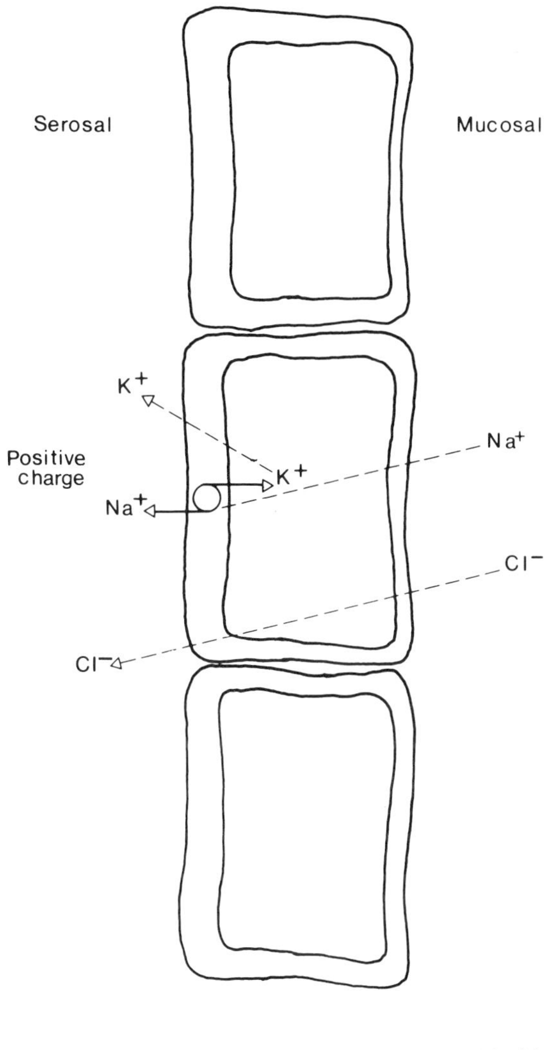

Figure 4. Active sodium transport. Dashed lines indicate diffusion down electrochemical gradients established by the sodium pump activity.

increase in positive charge on the serosal surface, which will encourage chloride ions to move down their electrochemical gradient and cancel part of this charge. The potential difference developed (positive charge on the serosal relative to mucosal side) will depend on the balance between sodium and chloride diffusion. This may be as high as 100 mV for amphibian epithelia.

3. The Vertebrate Series

A bewildering variety of organs serve an osmoregulatory role in the vertebrate subphylum. The separate functions of these organs reflect the prevailing osmotic environment, the manner of nitrogenous waste excretion, and, to some extent, the functional capacity of the kidney in each of the major animal groups. We shall see later, in Chapters 11 and 12, that, despite variation in osmoregulatory demand and detailed manner of body fluid management, the endocrine control of water and electrolyte economy is consistently dominated by the hormones of the pituitary, adrenal cortex (interrenal), and kidney. In order to understand the significance of the actions of these hormones in different groups of vertebrates, it is first essential to examine in some detail the separate osmoregulatory mechanisms they control.

3.1. Ancestral Vertebrates

The earliest true vertebrates to emerge were the jawless agnathans, which are represented today by two distinct lines, the lampreys and myxinids (hagfish). These two groups of animals differ markedly in their osmoregulatory physiology. Lampreys show considerable specialization which permits migration between SW and FW at precise stages in their life-cycle. By contrast, the exclusively marine hagfish present a more simplistic osmoregulatory capacity, perhaps closer to that of ancestral vertebrates (see Chapter 13 for discussion). Their body fluids are isoosmotic to SW, and as a consequence, intracellular ion levels in the hagfish differ greatly from those in other vertebrates (Table II), resembling more closely those in marine invertebrates. Notably, the relative proportions of intracellular sodium and potassium are nearer unity in hagfish than in other vertebrates.

If the myxinids do present an ancestral form of osmoregulatory physiology, it is interesting to note that, although body fluids are isoosmotic to SW, the concentrations of particular ions in plasma are maintained somewhat different from those in SW (Table II). Potassium, calcium, and magnesium levels are lower than in surrounding SW, and sodium levels are elevated in compensation. A similar composition profile is found in many marine invertebrates. These small differences between ionic composition of plasma and SW are critical and are sustained largely through the activities of the kidney and numerous mucous glands of the integument and gills. Urine and mucous gland secretions are rich in potassium, calcium, and magnesium and are low in sodium (see Table III).

Table II
Major Intracellular Ion Concentrations in Marine Vertebrates and Invertebrates[a,b]

Invertebrates	Hagfish	Other vertebrates	Seawater
Sodium 54–325	122	8–45	475
Potassium 48–175	117	83–185	10
Calcium 3–89	2	2–9	10
Magnesium 8–96	13	7–11	53
Chloride 54–380	107	11–30	550

[a]Concentrations all mmoles/liter.
[b](After McFarland, W. N., Pough, F. H., Cade, T. J., and Heiser, J. B., 1979, *Vertebrate Life*, Macmillan Publishing Co., Inc., New York.)

Table III
Composition of Myxinid Plasma, Urine, and Mucus[a]

	Plasma	Urine	Mucus	Seawater
Sodium	549	481	95	470
Potassium	11	13	207	12
Calcium	5	10	11	8
Magnesium	20	28	39	50
Chloride	563	536		550

[a]Concentrations all mmoles/liter.

Hagfish undoubtedly possess the most primitive of all vertebrate kidneys, the organ being based on 15–20 pairs of nephrons arranged segmentally along the length of the body cavity. The glomerulus and short tubule produce a urine almost isoosmotic to plasma. Sodium and water do not appear to be reabsorbed in the nephron, but glucose is conserved, and potassium and divalent ions are actively secreted into the forming urine. The bile may provide an additional avenue of magnesium excretion to maintain the low plasma levels observed.

Hagfish osmoregulatory mechanisms appear to be restricted to management of plasma levels of specific ions; water conservation is not a problem in these animals as they are isoosmotic to their environment. However, when provided with the challenge of being placed in dilute SW, body fluid tonicity falls to match that of the external medium. The hagfish is an osmoconformer. The simplistic hagfish capacity for management of water and electrolyte economy is similar to that of most marine invertebrates but differs markedly from that in all other vertebrates.

Recent experiments by Reigel (1978) suggest that, even in the hagfish, the kidney may perform a volume regulatory role. As the plasma osmotic pressure tends to be slightly higher than surrounding SW, these fish tend to gain water osmotically. The kidney has the potential capacity to remove water accumulated in this way. Experiments in which individual glomeruli have been perfused indicate that filtration rate (approximately equal to excretion rate) rises steeply with an increase in arterial blood pressure. Since expansion of blood volume, as a result of osmotic accumulation of water, would tend to increase blood pressure, this provides an automatic volume-regulatory mechanism. Indeed, the Pacific hagfish, *Eptatretus stoutii,* maintains constant blood volume even in SW diluted to 80% of its original concentration. This type of volume-regulatory function, mediated by changes in systemic blood pressure altering rates of glomerular filtration, is common to most nonmammalian vertebrates.

3.2. Fish

The fossil record indicates that early fish colonizing FW gave rise to modern fish groups and subsequently to tetrapods. The modern bony (teleosts) and cartilaginous (elasmobranchs) fishes are considered to be derived from this FW stage. It is, therefore, pertinent at this stage to examine the osmoregulatory physiology of extant FW fish to gain insight into this important stage in phylogenetic development.

3.2.1. FW Fish

Very few elasmobranchs are found in FW, though this medium has been successfully exploited by teleosts. FW fish maintain their body fluids at much lower ionic and osmotic concentrations than do the marine myxinids, an essential adaptation to life in such a dilute medium. All higher vertebrate groups have inherited this more dilute condition of their body fluids, reflecting the FW stage in their ancestry. FW fish face the problem of continual passive osmotic gain of water and diffusional loss of ions (see Figure 5).

Most osmotic exchange of water occurs across the gills, which have a higher permeability and total surface area than the integument. Gill water permeability and thus water influx are influenced by the level of calcium in the external medium. Accumulated water is excreted by the kidney, which has greatly increased numbers of nephroi that are no longer segmentally arranged. The brown trout kidney stretches the length of the body cavity, with the bulk of its 40,000 nephroi distributed in the middle and most caudal portions. High urine flow is achieved by high rates of glomerular filtration and low water reabsorption in the renal tubule, usually less than 80% of filtrate. Variations in the rate of urine production are largely achieved by altered rates of glomerular filtration.

Although the kidney is important in regulating body fluid volume in FW fish, it

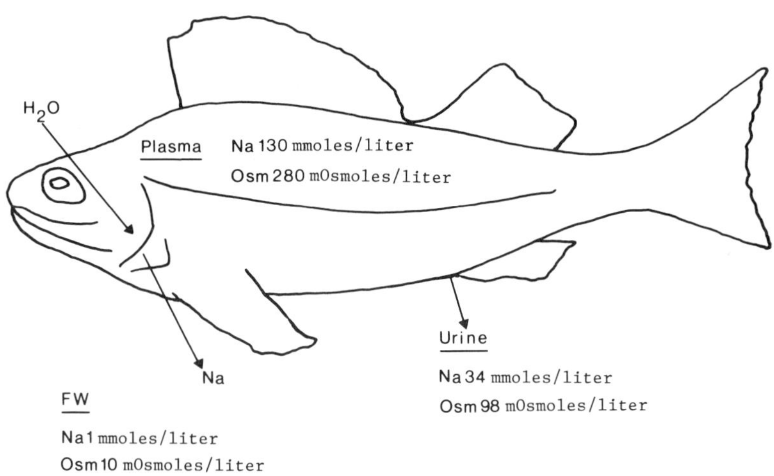

Figure 5. Water and ion exchange in freshwater fish. Typical plasma, urine, and water compositions are shown.

makes little contribution toward the management of the osmotically important ions like sodium, potassium, and chloride. Indeed, despite more than 90% of the filtered sodium and chloride being reabsorbed along the renal tubule, the copious urine flow in these fish ensures considerable renal depletion of body salt. This may account for 8% of total body content each day in the goldfish, greatly exceeding the diffusional loss of ions across the gills. Electrolyte lost in this way is replaced through active uptake across the gills though, depending on the diet, this may not be necessary in feeding fish. Active uptake of ions at the gills involves exchange with ions in the body, for example, Na^+/NH_4^+ and Cl^-/HCO_3^-. Changes in the rate of exchange to meet the animals' needs are largely achieved by altered rates of blood flow through the gill epithelia.

The kidney in FW fish may be an embarrassment in monovalent ion management, but it has a vital role to play in divalent ion economy as it does in the myxinids. Calcium, magnesium, and sulfate are actively secreted into the forming urine by the renal tubule. The increased number of nephroi and the addition of a proximal tubular segment to ensure reabsorption of most of the filtered monovalent ions are the critical innovations in the renal apparatus of FW fish. This basic renal unit, responsible for volume regulation and divalent ion excretion, has been inherited by all subsequent groups of vertebrates. The distal addition of extratubular segments has, however, enabled considerable further modification of the nephrons' functional capacity.

3.2.2. SW Fish

With the exception of the myxinids, extant marine fish are generally assumed to be derived from a FW stage in vertebrate evolution. Movement to the sea from FW posed considerable osmotic problems. The inherited low osmotic concentration of body fluids ensured that such fish entering SW would suffer osmotic loss of body water and diffusional gain of ions. Two osmoregulatory strategies have appeared in marine fish to overcome these problems. The first involves the accumulation of organic solutes, mainly urea and trimethylamine oxide (TMAO), to raise the osmotic concentration of extracellular fluid to equal or surpass that of SW. This is characteristic of the cartilaginous elasmobranchs (sharks and rays) and crossopterygians (only extant species is the coelacanth, *Latimeria chalumnae*). By contrast, marine teleosts have adopted the strategy of maintaining their body fluids hypoosmotic to SW and have evolved mechanisms to overcome the resulting loss of body water and accumulation of salts.

3.2.3. Elasmobranchs

Although urea and TMAO provide the essential osmotic ballast in elasmobranchs, plasma sodium and chloride levels are also maintained at higher levels than observed in FW fish (Figure 6). Most elasmobranchs sustain their body fluids slightly hypertonic to SW, achieving a small osmotic gain of water from their

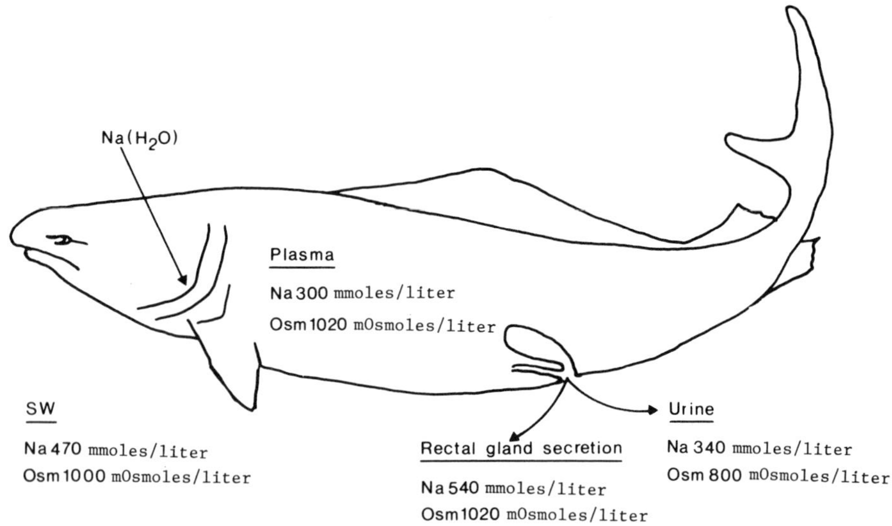

Figure 6. Water and ion exchange in marine elasmobranchs. Typical plasma, urine, rectal gland secretion, and seawater compositions are shown. (Modified from Rankin, J. C., and Davenport, J. A., 1981, *Animal Osmoregulation.* Blackie, Glasgow and London.)

environment. This gain of osmotically free water is sufficient to replace fluid lost in urine and rectal gland secretion. In most aquatic animals, nitrogenous waste is excreted as ammonia; its high solubility ensures easy disposal, despite its high toxicity. In elasmobranchs, ammonia, derived from protein catabolism, is converted to urea [$CO(NH_2)_2$] in the liver by the enzymes of the ornithine cycle. Though urea is less toxic than ammonia, levels well below those in elasmobranchs would be fatal in most other vertebrates. Elasmobranchs not only have developed a tolerance to urea, but many cellular enzyme systems will not function optimally in the absence of urea. Urea rapidly crosses cell membranes, so it exerts no net osmotic effect on cells. Maintaining the high levels of urea in body fluids relies not only on its constant synthesis but also on low gill permeability and exhaustive renal tubular reabsorption to restrict losses across the gill and in the urine. The renal tubule of elasmobranchs characteristically has an exceptionally extensive intermediate segment which is responsible for the reabsorption of urea.

Urea and TMAO ameliorate problems of water economy in SW, but elasmobranchs are still faced with the diffusional gain of sodium and chloride across the gills. The turnover of sodium chloride is relatively low by comparison with marine teleosts (see Table IV), less than 1% of body content each hour, but the accumulated salt must be excreted. This is achieved by a combination of efflux across the gills, rectal gland secretion, and renal excretion. Gill losses probably involve active secretion but the exact location is unclear. Renal excretion is important in the maintenance of ionic balance, despite the production of, at best, an isoosmotic urine. Elasmobranchs are able to sustain a much higher rate of urine production than marine teleosts. Rectal gland secretion may also account for a significant proportion of the sodium excreted, perhaps up to a third of the total. This gland, a characteristic of the elasmobranchs, is small, weighing only 1–3 g in sharks weighing more than 10 kg. Histologically it consists of masses of tubules, like the avian and reptilian "salt glands," containing very high levels of Na^+, K^+-activated ATPase. A ducting system drains the secretion into the rectal region of the gut. The secretion is isoosmotic to plasma, but contains almost pure sodium chloride solution at twice the concentration of plasma (see Figure 6). Normally the rectal gland may only contribute 5–10% of the total sodium efflux, and its surgical removal may not impair normal ionic balance. The importance of the gland may lie in its capacity to excrete abnormal salt loads which follow excessive dietary intake as, for example, results from ingesting marine invertebrates that have body fluid sodium levels similar to SW.

3.2.4. Teleosts

Hypoosmotic regulation of body fluids in teleosts is arguably more demanding in energy than the isoosmotic regulation practiced by elasmobranchs. Marine teleosts suffer diffusional gain of sodium and chloride and osmotic loss of water (Figure 7). Exchange of sodium with the environment may be some hundred times greater than that in FW teleosts and 50 times greater than in marine elasmobranchs (see Table IV). The principal avenue of monovalent ion excretion in teleosts is the gill. In the 1930s, Keys and Willmer (1932) demonstrated active extrusion of chloride-based salts using the isolated head (heart/gill) preparation. Large, prominent epithelial "chloride" cells at the base of the gill leaflets are considered to be the site of

Table IV
Total Sodium Fluxes, Measured as Outflux of Injected ^{24}Na, in Marine and Fresh Water Teleost and Elasmobranch Fish[a,b]

		Total Na flux (μM/100 g per hr)
Teleosts		
Goldfish	FW	27
Blenny	SW	2700
Flounder	FW	43
	SW	2600
Eel	FW	4
	SW	1321
Elasmobranchs		
Dogfish	SW	59
Lip shark	SW	75

[a]FW, freshwater; SW, seawater.
[b](Data from Bentley, P. J., 1971, *Endocrines and Osmoregulation. A Comparative Account of the Regulation of Water and Salt in Vertebrates,* Springer Verlag, New York, Berlin, Heidelberg.)

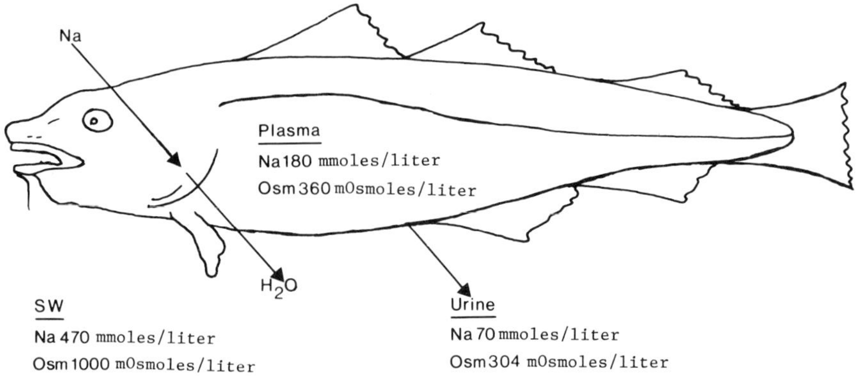

Figure 7. Water and ion exchange in marine bony fish. Typical plasma, urine, and seawater compositions are shown.

the active ion extrusion, largely because of their absence in FW fish. Similar "chloride cells" are also found on the epithelia lining the operculum in some fish like the killifish, *Fundulus heteroclitus*. Experimentation with isolated opercular epithelia indicates that the active excretion of chloride salts involves a chloride pump, though Na^+,K^+-activated ATPase may also be involved. Although the precise mechanism remains unclear, there is no dispute of the preeminent role of gills in the excretion of accumulated salt. The kidney of marine teleosts makes a relatively minor contribution, urine flow being much lower than in FW animals. The glomerular size and number are reduced in marine species, and in a few, like the angler fish, *Lophius piscatorius,* the kidney may be completely devoid of glomeruli. Blind-ended renal tubules in this fish receive only venous blood from the renal portal vein. The secretion of magnesium by the renal tubular cells appears to be one of the major driving forces of urine production in such fish. Indeed, the main function of kidney in marine teleosts is probably the excretion of divalent ions.

Any renal losses of water in the marine teleost exacerbate the steady osmotic depletion of body water to the medium. It is, therefore, important that urine flow in these fish is restricted. Marine teleosts necessarily must drink large volumes of SW to replace the renal and osmotic fluid losses that occur. Elasmobranchs and FW teleosts, it should be noted, drink very little water. Imbibition of the hypertonic SW initially adds further to the osmotic problems facing marine teleosts. Low water permeability of the esophagus ensures that much of the ingested salt diffuses into the body, reducing the osmotic concentration of fluid entering the stomach. Active salt absorption in the intestine rapidly reverses the osmotic gradient, facilitating at last the gut absorption of water. Such drinking and gut absorption are clearly essential to alleviate water losses and restore body fluid volume, but this process adds further to the salt load that must be excreted at the gill. Sodium and chloride taken up in this way may account for 10–20% of the total gill efflux.

3.2.5. Euryhaline Fish

Body fluid management in purely FW or SW fish is geared to the relative stability of these contrasting osmotic media. Constant osmotic and ionic gradients between the animal and its environment are overcome by rigid patterns of osmoregulatory organ function. By contrast, one might expect a more sensitive and plastic osmoregulatory system to prevail in euryhaline species, which survive the fluctuations in external salinity in estuaries or migrate between SW and FW. A few teleost species, such as the salmon and eel, make anadromous (SW to FW) or catadromous (FW to SW) spawning migrations. Such euryhaline species have attracted a disproportionally large amount of research attention, as they provide an opportunity to examine dynamic changes in osmoregulatory function to meet controlled changes in osmotic demands. Much of our understanding of osmoregulatory physiology in stenohaline (purely FW or SW) species has been deduced from observations of euryhaline fish. However, it should be remembered that euryhalinity requires considerable physiological specialization, which has been achieved by relatively few species. There may well be detailed differences in the adaptive physiology of stenohaline and euryhaline fish.

The three major sites of water and electrolyte management (gill, gut, and kidney) exhibit marked changes in function to accommodate variation in external salinity. The transfer of euryhaline fish to FW may not pose large difficulties in the essential gill uptake of sodium and chloride; similar sodium/ammonium and chloride/bicarbonate exchanges are observed in marine animals. However, the potential diffusional loss of ions to the FW medium does pose a serious problem. Transfer of the eel to FW results in an immediate reduction in sodium outflux (instantaneous regulation), which may involve changes in the transbranchial electrical potential. This is followed 30 min later by a further sustained fall in outflux (secondary regulation), probably resulting from a reduction in the ion permeability of the gill epithelia. Stenohaline fish do not show this reduction in ion outflux and rapidly lose a large proportion of their body sodium and die. The reverse transfer of eels from FW to SW is accompanied by an increase in the numbers of "chloride cells" in gills and a rise in the level of branchial Na^{+},K^{+}-activated ATPase activity.

Movement of animals between SW and FW also clearly necessitates altered patterns of drinking behavior (Figure 8). The increased drinking observed in animals held in a hypertonic medium is also associated with adaptive changes in the absorptive capacity of the alimentary canal. This largely involves increased permeability of the esophagus to sodium and chloride, to allow desalting of the ingested SW, and an increased rate of fluid and electrolyte uptake in the intestine (see Figure 8).

Renal function in FW- and SW-adapted euryhaline fish also differs markedly. The large volumes of dilute urine produced in FW animals would, if unchecked, greatly increase the water conservation problems of fish in SW. Transfer to SW is normally associated with a dramatic fall in urine flow, consequent to a reduction in

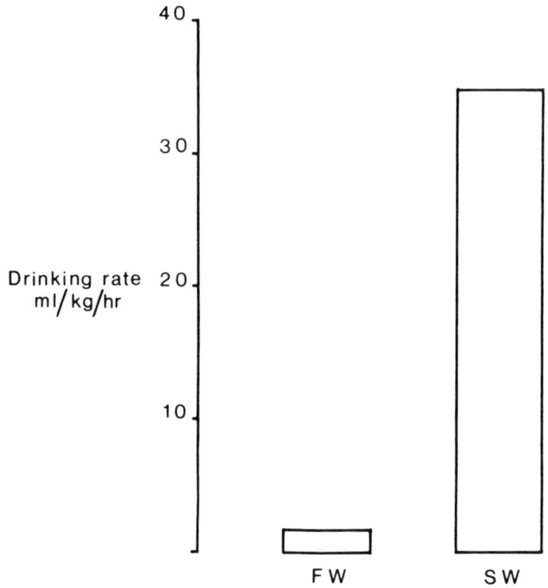

In vitro Isolated Gut	FW	SW
* Fluid uptake ml/hr	0·28	0·55
* Sodium uptake µM/hr	52	109
* Chloride uptake µM/hr	59	103

Figure 8. Drinking behavior, gut fluid, and ion uptake in seawater (SW) and freshwater (FW) adapted flounder. Data represent mean values for at least eight animals. *Gut uptake values are expressed per gram of wet tissue weight.

the rate of glomerular filtration (Figure 9). This decline in glomerular filtration rate (GFR) may not simply rely on a general reduction in the GFR of individual nephroi, but may rather result from a decline in the number of glomeruli filtering (glomerular derecruitment). An infusion of ferrocyanide is freely filtered through glomeruli and can be visualized within the nephron as Prussian blue. Recent experiments employing this technique indicate that not all glomeruli in the fish kidney are filtering, as Prussian blue is absent from the tubular component of these nonfiltering nephroi. Indeed, glomeruli of some nephroi contain no Prussian blue and may not receive arterial perfusion (nonperfused). The relative proportions of nephroi being perfused and filtering change dramatically on transfer of fish between SW and FW (see Figure 10). Only 5% of the nephroi may contribute to the whole kidney rate of glomerular filtration in SW-adapted animals. The rise in glomerular filtration upon return to FW appears to arise from a large increase in the proportion of glomeruli that are filtering (glomerular recruitment). Such intermittency of glomerular function is probably not restricted to euryhaline fish, having also been implied in the control of urine production in nonmammalian tetrapods.

In spite of the dramatic changes in osmoregulatory organ function consequent to transfer between SW and FW, the body fluid composition of fish maintained in these respective media differs significantly (Figure 11). Movement from SW to FW is associated with a fall in plasma concentration of the main solutes (sodium and chloride) and consequent decline in plasma osmolality. It appears that the set point of body fluid composition, toward which the homeostatic osmoregulatory mechanisms work, differs in these extreme osmotic environments.

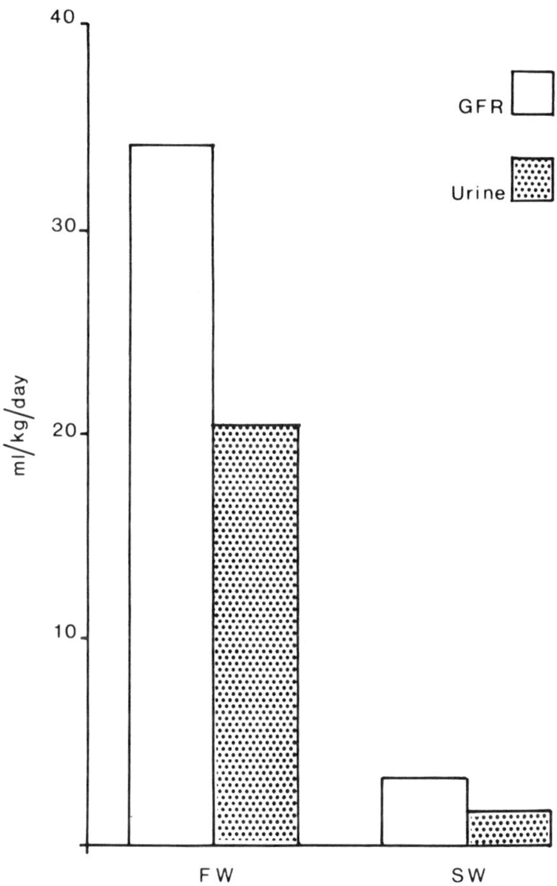

Figure 9. Urine production and glomerular filtration rate (GFR) in freshwater (FW) and seawater (SW) adapted flounders. Data represent mean values for six animals.

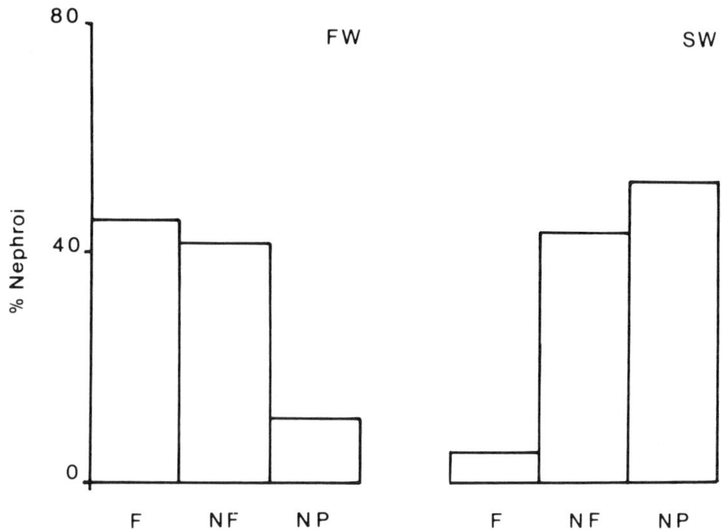

Figure 10. Distribution of functional types of nephron in seawater (SW) and freshwater (FW) adapted trout. Data represent mean values for five animals. F, filtering; NF, nonfiltering; NP, nonperfused. (Redrawn from Brown, J. A., *et al.*, 1980, Angiotensin and single nephron glomerular filtration in the trout, *Salmo gairdneri*. *Am. J. Physiol.* **239:**R509–R514.)

3.3. Amphibians

The terrestrial tetrapod groups (amphibians, reptiles, birds, and mammals) owe their origins to FW fish, as do modern lungfish (Dipnoi). Phylogenetically the Amphibia are the first group to exhibit a largely terrestrial life-style, though no species are fully independent of the aquatic environment. Egg laying and initial larval development are dependent on access to water, though adult forms may be wholly terrestrial. The contemporary Amphibia are represented by two main orders, the Anura (frogs and toads) and Urodela (newts and salamanders). Although frogs and toads air-breathe, in many species most oxygen uptake occurs across the skin, which is kept moist by the watery secretions of the numerous integumental mucous glands. This high integumental permeability by comparison with piscine groups can potentially result in enormous evaporative water loss, and many amphibians appear to alternate between periods of dehydration and rehydration. Skin permeability to water and ions does vary between species, largely in accordance with their prevailing osmotic environment, and may be subject to physiological regulation.

The amphibian group as a whole exhibits, uncharacteristically for vertebrates, great tolerance of fluctuations in body fluid volume and composition. Many species may lose up to 50% of their body water before death occurs. This group also differs from other vertebrates in the fully hydrated state in having a higher water content, some 80% rather than a more normal 70% of body weight. This extra

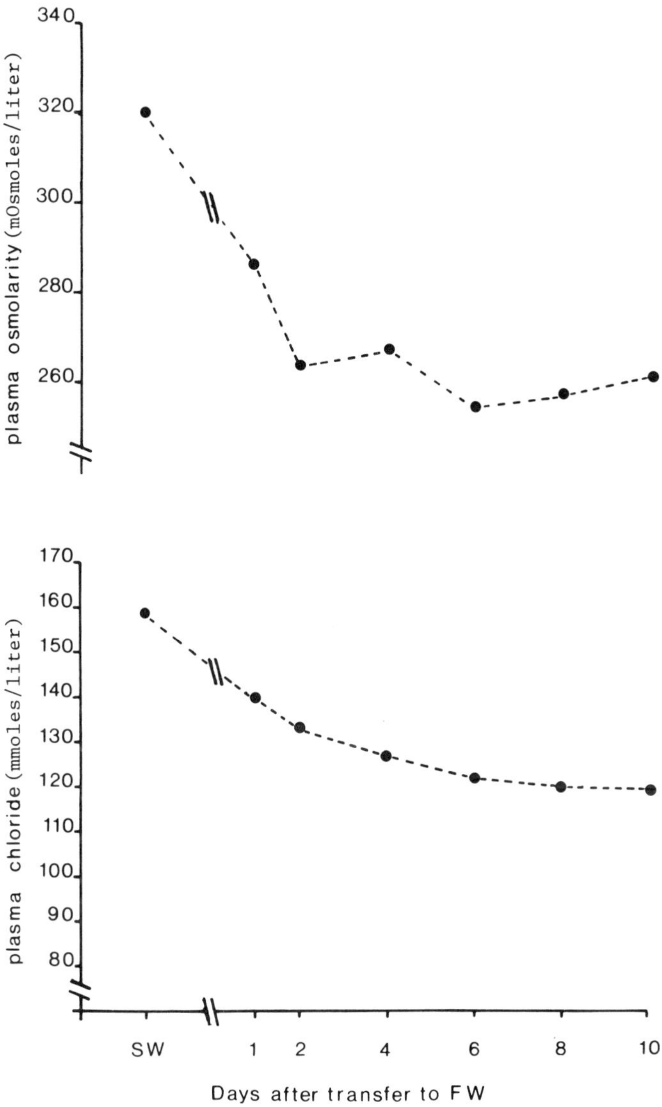

Figure 11. Altered plasma composition following transfer of flounders from seawater (SW) to freshwater (FW). Data represent mean values for at least seven animals.

water is large extracellular and is responsible for the low plasma osmotic (200 mOsmoles/liter) and sodium (100 mmoles/liter) concentrations (see Table I). In most species, behavioral adaptations ensure that exposure to arid evaporative water loss in dry air is minimized. As a tetrapod innovation, a urinary bladder, derived embryologically from the cloaca, makes its phylogenetic debut in the Amphibia.

Table V
Plasma Composition in Salt Water and Estivating Amphibians[a,b]

	Sodium (mmoles/liter)	Urea (mmoles/liter)	Osmolarity (mOsmoles/liter)
Spadefoot toad			
In FW	159	39	305
During estivation	228	286	630
Crab-eating frog			
In FW	125	40	290
In 80% SW	252	350	830

[a]FW, freshwater; SW, seawater.
[b](Data from Bentley, P. J., 1971, *Endocrines and Osmoregulation. A Comparative Account of the Regulation of Water and Salt in Vertebrates,* Spinger Verlag, New York, Berlin, Heidelberg.)

This may serve an important function as a fluid reserve in some species, perhaps representing 50% of the body weight of animals from arid environments, though in purely aquatic species it is nearer 1% body weight.

The North American spadefoot toad, *Scaphiopus couchi,* can survive long periods of drought in deserts through estivation in burrows. The same phenomenon is observed in lungfish stranded in dried-up river beds. The concentration of body fluids rises considerably during the period of estivation as a result of the accumulation of urea and dehydration (Table V). Body fluid concentration is also raised in the semimarine, crab-eating frog, *Rana cancrivora,* as a result of urea retention, but here, as in elasmobranchs, the urea is retained specifically to serve as an osmotic ballast to reduce the osmotic gradient and consequent water loss to its hyperosmotic environment.

Amphibians do not appear to drink, in contrast to other vertebrates. Rehydration is achieved by uptake of water from the bladder urine and across the ventral skin surface in animals in water or resting on moist soil. In predominantly aquatic species, osmotic accumulation of water across the skin poses a significant excretory problem, representing 30–40% of body weight each day. Urine production is accordingly high in such animals. The rate of urine production is adjusted to the hydrational state of the animal and may cease altogether in dehydrated frogs to limit further water loss. These changes in urine flow result from altered rates of glomerular filtration, perhaps involving variation in the proportion of nephroi contributing to the renal filtrate (glomerular intermittency). Like the fish groups, amphibians are unable to produce a urine hypertonic to plasma but, unlike fish, some adjustments to the final osmotic concentration of urine can be achieved in the most distal portions of the nephron tubule.

The excreted urine contains little salt, sodium and chloride having been absorbed along the renal tubule and later across the wall of the urinary bladder. The copious urine flow in well-hydrated animals, however, may lead to significant depletion of these ions. Those animals immersed in FW for long periods suffer further diffusional loss of ions. These renal and diffusional losses are balanced to a large extent in many species by the active uptake of sodium and chloride by the integument. Active ion transport across the ventral skin surface has been demonstrated in many anurans and some urodeles. The active transport of sodium and chloride across the isolated, *in vitro* preparation of amphibian skin has proved a valuable experimental model of ion pump mechanisms. The uptake of sodium from the environment responds to the prevailing osmotic demands. Bathing frogs in deionized water enhances their ability to accumulate sodium and chloride across the integument.

3.4. Reptiles

Reptiles emerged from amphibianlike ancestors in the early Mesozoic period and are considered phylogenetically to be the first truly terrestrial animals. They are much less dependent on free access to water than amphibians. Their skin is relatively impermeable to water, reducing cutaneous water loss to less than 2% of that in amphibians. Reptilian production of the amniotic and cleidoic egg, which is normally laid on land, further reduces dependence on water for reproduction. The water economy of many reptiles, particularly those in arid areas, is further eased by their ability to produce uric acid (rather than urea or ammonia) as the principal end product of nitrogen metabolism. This greatly reduces the amount of water required for renal excretion of catabolic nitrogen. It has been calculated that 1 g of nitrogen requires 300–500 ml of water for excretion by ammoniotelic species like the alligator, but only 10 ml of water in uricotelic species. Uricotelism, production of the amniotic egg and a relatively impermeable skin, distinguishes reptiles from their amphibian ancestors and has contributed to their wide geographical distribution.

The composition of reptilian body fluids conforms to the usual vertebrate pattern. However, variations do occur; marine reptiles have slightly higher plasma ionic and osmotic concentrations than terrestrial species. Individuals may also tolerate quite large fluctuations in body fluid composition, though not to the same degree as amphibians.

On land, reptiles lose water by evaporation across the integument and lungs, though this is small by comparison with amphibians, birds, and mammals. Cutaneous loss, which is not controlled in relation to osmotic demands, may represent 50–60% of the total, though this varies from species to species. The evaporative loss of water normally greatly exceeds urinary and fecal losses, though the exact proportion will vary according to ambient temperature.

Reptiles possess a metanephric kidney, as do birds and mammals, differing in its embryological origin from the mesonephros of fish and amphibians. The metanephric kidney arises from presumptive mesoderm caudad to the mesonephros. In contrast to the metanephric kidney of birds and mammals, the reptilian renal apparatus is unable to produce a urine more concentrated than (hypertonic to) plasma. Potentially, as in amphibia, renal excretion of solute in the reptiles necessitates excretion of a volume of water equal to the volume in which the solutes were

Table VI
Renal Function following Water Loading and Salt Loading in the Lizard, *Varanus Gouldii*[a]

	Control	Water-loaded	Salt-loaded
Glomerular filtration rate (ml/k/g per hr)	6.8	12.8	8.4
Proportion of filtrate reabsorbed (%)	53	38	79
Urine			
Volume (ml/kg/per hr)	3.1	7.7	1.7
Osmolarity (mOsmoles/liter)	146	114	299
Plasma			
Sodium (mmoles/liter)	151	148	202
Osmolarity (mOsmoles/liter)	328	287	421

[a](Data taken from Green, B., 1972, Aspects of renal function in the lizard *Varanus gouldii*. *Comp. Biochem. Physiol.* **43A:**747–756.)

distributed in body fluids. Water cannot be conserved during excretion by increasing the osmotic concentration of urine above that of plasma, as is achieved in avian and mammalian kidneys. However, in those reptiles which are capable of converting increasing amounts of waste nitrogen to uric acid, some water economy can be achieved as uric acid has lower osmotic activity than the equivalent amounts of ammonia or urea. Complexing of a significant proportion of the cations present in urine in the form of urates further reduces the osmotic activity of total urine solutes and reduces further the volume of water required for their excretion.

Urine flow in reptiles shows considerable lability in accordance with the state of hydration, dehydrated animals becoming anuric (see Table VI). Urine volume is largely controlled through variation in the rate of glomerular filtration, which may vary by a factor of 10 in the crocodile. The changes in GFR result from a combination of altered filtration rates across individual nephroi and from altered total numbers of nephroi filtering (glomerular intermittency). At low rates of urine flow, reabsorption of water from the forming urine in the distal tubule may significantly alter the volume of urine excreted. At high flow rates, the contribution of such tubular adjustments in water content will be relatively small.

Urine is not voided directly from the reptilian kidney, but passes initially into the cloaca and, in some species, from there into a urinary bladder. Of the four major groups of extant reptiles, Crocodilia, Ophidia (snakes), Chelonia (turtles and tortoises), and Lacertalia (lizards), urinary bladders are present only in the latter two. Urinary and fecal pellets contain little water in most reptiles, which suggests that, as in birds, water reabsorption occurs across the cloaca and large intestine prior to pellet expulsion. Water reabsorption in association with active salt absorption may similarly occur across the urinary bladder. Indeed, the urinary bladder of Chelonia and Lacertalia may serve as an important reservoir, representing perhaps 20% of the animals' body weight.

The kidney is normally the major route for excretion of solute accumulated through the diet. In FW species with access to ample osmotically free water, this is easily achieved. However, in many reptiles, especially marine species, large amounts of solute are accumulated which are beyond the excretory capacity of the kidney. Their excretion in isoosmotic urine would lead to an unacceptable depletion of body water. Such renal "incompetence" is apparently overcome in two ways. Many reptiles exposed to high salt intakes possess cephalic "salt glands," which permit excretion of sodium and potassium chloride solutions considerably hypertonic to plasma. Other species that lack such extrarenal avenues of salt excretion often exhibit an ability to withstand considerable increase in body fluid osmotic concentration. Plasma sodium levels may rise to 300 mmoles/liter during the summer in the lizard, *Amphibolurus ornatus*. With the return of rain, these animals drink the available water, providing the essential fluid for renal excretion of this accumulated salt.

Cephalic "salt glands" in marine and terrestrial reptiles are based on modified tear (supraorbital or Harderian) glands or, as in birds, nasal glands (see Figure 12). The secretion produced is hypertonic to body fluids, with electrolyte levels in excess of those in SW. More than 90% of the salt excretion in these animals may occur through the "salt glands." Such efficient salt-excreting devices may permit some marine species to drink SW to repair any deficits in body water content.

3.5. Birds

Avian and mammalian skins are largely impermeable to water, and apart from sweat production in mammals, these groups suffer little cutaneous water loss. However, the endothermic groups are disadvantaged by comparison with poikilothermic tetrapod (amphibians and reptiles) ancestors by their high resting oxygen requirements which lead to high respiratory water losses. High oxygen requirements in birds and mammals are associated with the development of a completely divided heart, so that blood is always pumped to the lungs at exactly the same rate as to the rest of the body. Reptiles can reduce respiratory water losses by diverting blood in their semidivided heart away from the lungs between breaths. Thus, they do not expose blood to surfaces from which evaporation will occur unless this is necessary to meet respiratory demands.

Birds and mammals occupy the same geographical regions and habitats and accordingly face similar osmotic problems. Birds, however, sustain a higher (3–4°C) body temperature than mammals with an associated higher metabolic rate, resulting in elevated respiratory water loss. Both homeothermic groups have a renal capacity to produce a urine that is hyperosmotic to body fluids. This is based on the function of the intermediate tubular segment (loop of Henle), which acts as a countercurrent multiplier system, generating a standing hyperosmotic gradient within the renal medulla (see Section 3.6 for detailed mechanism). This osmotic gradient provides the driving force for the osmotic reabsorption of water in the most distal tubular segment. The maximum urine concentration achieved by most birds is two to three times that of body fluids, much less than that in many mammals.

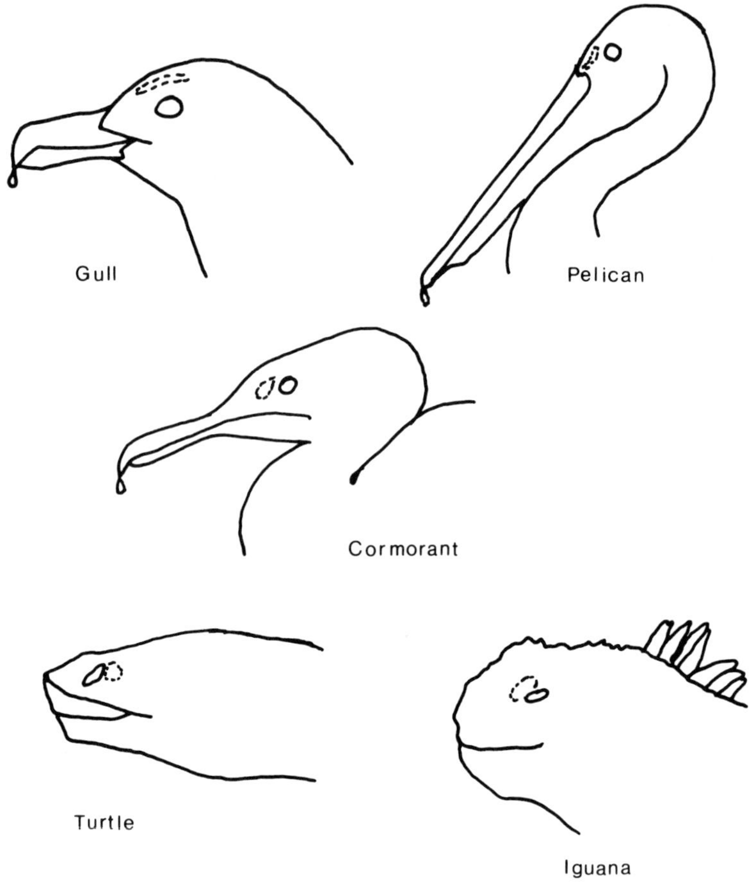

Figure 12. Salt glands in marine birds and reptiles. Hatched line indicates salt gland location. (Modified from Schmidt-Nielsen, K., 1959, *Salt Glands: Scientific American,* W. H. Freeman and Co., San Francisco.)

However, retention from their reptilian ancestors of the ability to convert metabolic nitrogen to uric acid allows additional renal water conservation in birds that is not available for ureotelic mammals.

Salt excretion is principally based on the kidney in birds, as it is in mammals, but in those marine and terrestrial species with very high dietary salt intakes extrarenal cephalic "salt glands" may also be present. These "salt glands" are analogous to those in reptiles (see Figure 12), providing an avenue of highly hypertonic saline excretion. Figure 13 illustrates the main routes of ion and water exchange in marine birds.

The avian kidney is unusual in that it contains two morphologically and functionally distinct populations of nephroi (Figure 14). Mammalian-type nephroi, which have loops of Henle, form a medullary cone at the base of a broader expanse of renal tissue comprised of reptilian-type nephroi, which lack the loop of Henle. To this extent, the avian kidney is an admixture of the renal organization present in mammals and reptiles. Variation in the rate of urine production can be achieved by altered tubular water reabsorption or by changes in the GFR. Changes in GFR may involve adjustments in the relative proportions of reptilian- and mammalian-type nephroi contributing (see Table VII). Reptilian-type nephroi may cease to filter during periods of water shortage, allowing the production of reduced volumes of more concentrated urine from the mammalian-type nephroi.

Birds lack a urinary bladder; urine passes from the ureters into the cloaca and then retrogradely into the colon and rectum. Ureteral urine volume and composition can be considerably modified in this region prior to final excretion. This is an important consideration when attempting to assess avian and reptilian renal function from examination of the excreted urinary product. Gut absorption of monovalent

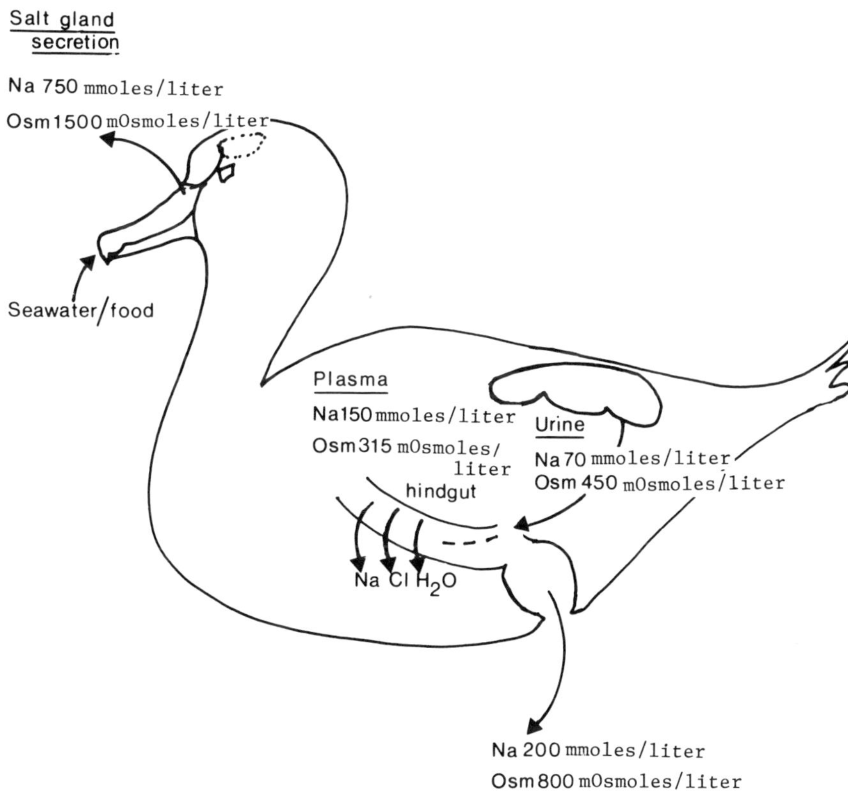

Figure 13. Main avenues of salt and water exchange in marine birds. Typical plasma, urine, excreted feces, and salt gland fluid composition are shown.

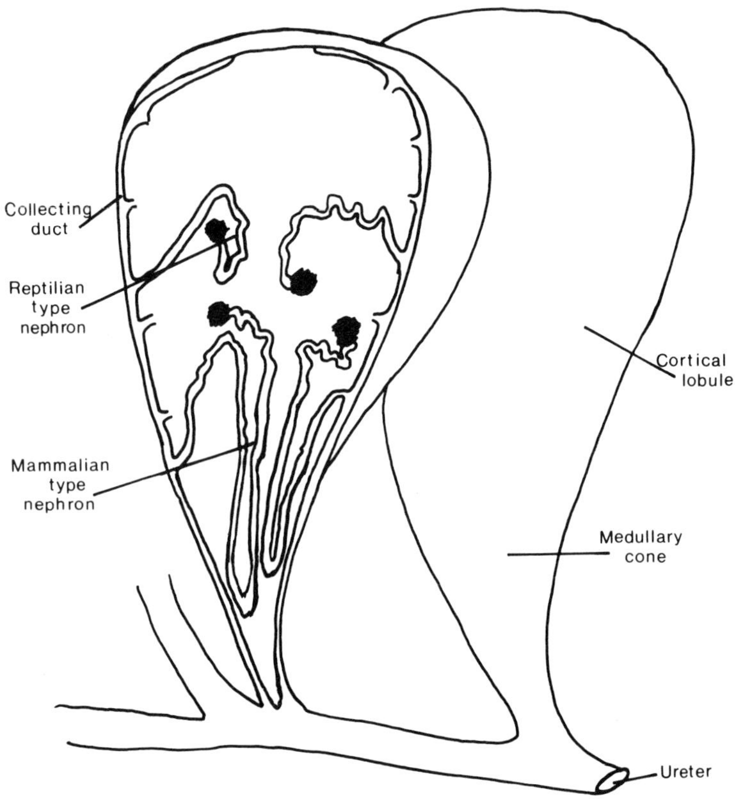

Figure 14. Renal lobules of the avian kidney. The relationship between the reptilian-type and mammalian-type nephroi and between the nephroi and collecting duct system is indicated.

ions is accompanied by osmotic equivalent volumes of water, dramatically reducing the potential urinary losses of water and electrolyte. In species with cephalic "salt glands," the gut absorption of salt and water, coupled with "salt gland" excretion of the electrolyte load in hypertonic solution, provides an integrated mechanism for generating osmotically free water. Such a mechanism clearly parallels the generation of osmotically free water in marine teleost fish, which is achieved through gut absorption of imbibed SW and gill excretion of the salt load.

The production of uric acid not only reduces the volume of urine necessary for renal solute excretion but, as Rankin and Davenport (1981) have recently pointed out, this also facilitates water conservation when urine enters the cloaca. Most of the uric acid is present in urine in the form of urate salts of sodium and potassium. These urates are relatively insoluble, and their precipitation essentially prevents much of the nitrogen and monovalent cations from exerting osmotic activity. Such urates are carried in suspension. Therefore, a milky ureteral urine of birds and uricotelic reptiles may remain hypoosmotic to body fluids but does, in fact, carry

Table VII
Single Nephron Glomerular Filtration Rates in the Desert Quail following Salt and Water Loading[a]

	Mammalian type (nl/min)	Reptilian type (nl/min)	Reptilian type filtering (%)
Control	14.6	6.4	71
Salt-loaded	12.7	—	0
Water-loaded	33.2	11.4	100

[a](Data from Braun, E. J., and Dantzler, W. H., 1972, Functions of mammalian-type and reptilian-type nephrons in the kidney of the desert quail. *Am. J. Physiol.* **222:**617–629.)

large quantities of solute essentially out of solution. Provided that further urinary water is reabsorbed postrenally, in the cloaca and large intestine, large amounts of solute may be excreted in insoluble form with little water loss or the necessity for the energy expensive production of a highly hypertonic urine. Indeed, if renal urine was highly hypertonic, urate precipitates would condense and block the renal tubules. True hard crystals of urate are normally formed only following water reabsorption in the hind gut distal to the ureters.

Despite the water conservation afforded by the gut and kidney, most birds drink to maintain positive water balance; marine forms may even drink SW in order to repair renal and respiratory losses. More than 90% of the salt load absorbed from the imbibed SW may be excreted by the cephalic "salt glands." The partition of electrolyte elimination between cloaca and "salt gland" in the cormorant, *Phalacrocorax auritus,* is shown in Table VIII. A few mammals also drink SW to replenish body water losses, though here the associated salt load must be excreted renally. Only those mammalian species which can produce very highly concentrated urine are capable of excreting the accumulated salt load in a smaller volume of urine

Table VIII
Contributions of Cloacal Excretion and Nasal Secretion to the Elimination of a Salt Load in the Cormorant[a]

	Cloacal excretion	Nasal secretion
Sodium	52%	48%
Potassium	90%	10%
Chloride	51%	49%
Water	68%	32%

[a]Seawater was given by stomach tube and collections made over the following 8 hr. (Data from Schmidt-Nielsen, K., Jorgensen, C. B., and Osaki, H., 1958, Extrarenal salt excretion in birds. *Am. J. Physiol.* **193:**101–107.)

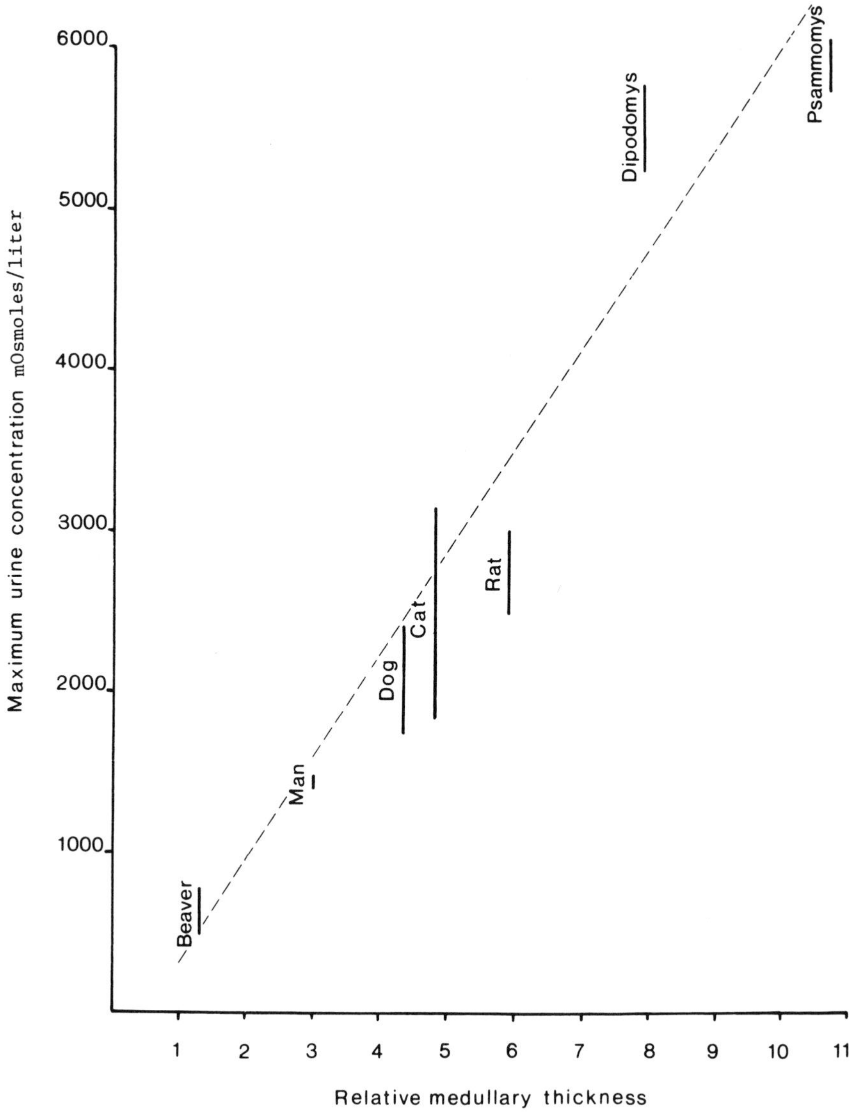

Figure 15. Variation in urine-concentrating ability and length of loop of Henle (relative medullary thickness). The range in maximum urine concentration is compared with the relative thickness of the medullary region of the kidney. (Drawn from data presented by Schmidt-Nielsen, B., and O'Dell, R., 1961, Structure and concentrating mechanism in the mammalian kidney. *Am. J. Physiol.* **200:**1119–1124.)

than the original volume of SW drunk. Dehydrated mammals with insuffient renal-concentrating ability, or birds without "salt glands," become further dehydrated after drinking SW, as a result of the enhanced renal fluid loss.

3.6. Mammals: Eutheria

In addition to respiratory water loss, evaporation of sweat may serve an important cooling function in many mammals. Fluid loss from sweat may be large; a man working in the desert can lose up to 15 liters of water a day. Sweating can be reduced in dehydrated animals; thus, the camel, after 24 hr without water, perspires at 60–70% of the rate in normally hydrated animals. The salt content, particularly sodium and potassium chloride levels, in the sweat secreted may also be modified in accordance with the animal's needs.

The secretion of saliva in herbivores can also place an enormous strain on fluid and salt economy, though ultimately most of the water and electrolyte may be reabsorbed in the gut. The gut fluid content in herbivores may represent 10–15% of the body weight and be largely derived from saliva secretion. Salt-depleted animals can reduce salivary salt content and increase the gut uptake of ions.

Renal loss of water, to excrete dietary excess or as an unavoidable consequence of essential solute excretion, varies enormously between mammalian species. The size of such losses depends on dietary composition, environmental conditions, and the degree to which the animal may concentrate urine. Renal excretion usually represents less than 50% of total daily water loss. The ability to concentrate urine is clearly related to the osmotic demands of the prevailing habitat of each species. The maximum urine concentration achieved by the mountain beaver, *Aplodontia rufa,* is about 500 mOsmoles/liter, which contrasts with the extreme concentration (9000 mOsmoles/liter) of urine produced by the desert-living Australian hopping mouse, *Notomys alexis*. There is an apparent direct relationship between morphological length of the loop of Henle (as measured by the relative medullary thickness) and maximum concentration of urine produced in different species (Figure 15).

The loops of Henle contain the basic mechanism that enables the production of a urine hypertonic to body fluids. Fluid entering the loop is isoosmotic to plasma but emerges, after its passage through the loop, into the distal tubule considerably hypoosmotic. It is somewhat paradoxical that the nephron segment which "dilutes" tubular fluid also determines the maximum concentration of the urine excreted. This, of course, relies on the countercurrent multiplier system of the loop to generate a highly hypertonic interstitial fluid in the renal medulla. The concentration of urine is achieved by the osmotic abstraction of water in the collecting duct against this medullary gradient.

Establishment of hypertonicity within the renal medulla relies initially on the active extrusion of sodium chloride from the thick ascending limb of the loop of Henle (Figure 16). Although the positive charge of tubular fluid relative to the interstitium supports active chloride followed by passive sodium transport, recent

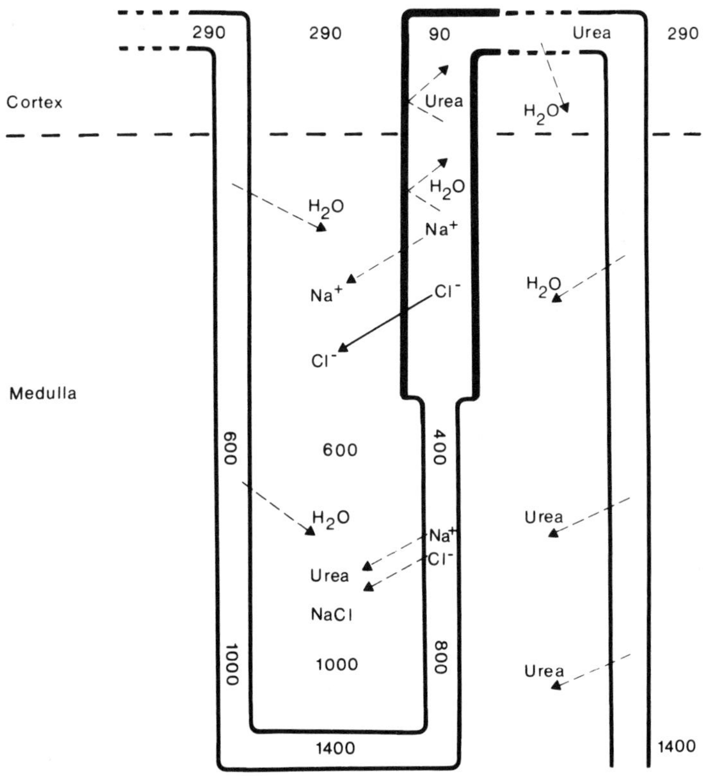

Figure 16. Basic mechanisms generating renal medullary hypertonicity. Active transport is indicated by solid line, passive diffusion down electrochemical gradient by broken line. Figures represent typical values of tubular fluid and interstitial fluid osmolality, mOsmoles/kg.

studies indicate that Na–Cl cotransport is also involved. Nonetheless, it is active extrusion of these ions in the water-impermeable thick ascending limb which raises interstitial fluid tonicity. An additional passive component of the medullary osmotic gradient has recently become evident. This is based on the passive recycling of urea between the cortical and medullary tubules (Figure 16). Salt reabsorption in the urea-impermeable ascending limb, coupled with water abstraction in the distal tubule, produces a high urea concentration in tubular fluid entering the collecting duct. Urea passively diffuses out of the more permeable medullary part of the collecting duct, thus raising the urea content and osmotic concentration of the medullary interstitium. The elevated interstitial fluid osmolality (now urea and NaCl) effectively removes water from the salt-impermeable descending limb to increase luminal salt concentration. As the tubular fluid rounds the tip of the loop and enters the salt-permeable thin ascending limb, salt may now passively diffuse into the interstitium, adding further to the medullary osmolality. Urea uptake in the thin ascending limb of the loop will be recycled via the distal tubule and collecting duct. A similar osmotic concentration persists at any transverse level of the loop. Movement of tubular fluid through the loop ensures that the gradient between ascending and descending limbs is multiplied to create a standing osmotic gradient through the medulla. The collecting ducts, like the ascending limb, are relatively impermeable to water, but here permeability may be increased in the presence of pituitary antidiuretic hormone (ADH). Urine concentration is effectively raised in the presence of ADH as a result of its permissive effect in allowing reabsorption of water in the collecting duct against the medullary osmotic gradient. Urine in the collecting duct can theoretically achieve the same osmotic concentration as the medullary interstitium. As indicated in Figure 16, collecting duct fluid entering the medulla is normally isotonic with body fluids, as a result of the osmotic equilibration and further solute-linked fluid reabsorption that occurs in the distal tubule and cortical portion of the collecting duct.

Management of body salt content is centered on the kidney in mammals, which lack the variety of extrarenal organs of salt excretion and uptake to be found in nonmammalian groups. Mammals characteristically have a high rate of glomerular filtration, which may represent the equivalent of the whole blood volume 40 to 50 times each day. The bulk of the filtered solute and water is reabsorbed, perhaps up to 99%. Rapid recycling of body fluid and water in this way enables quick correction of imbalances in body fluid volume and composition simply by adjusting the rates of tubular reabsorption. The capacity of the mammalian kidney to produce a hypertonic urine also ensures that an increase in renal excretion of solute does not inevitably lead to an osmotic equivalent increase in renal loss of water, as normally occurs in most nonmammalian vertebrates. Equally, in the advent of body fluid dilution, adjustments in renal water excretion can be achieved without undue salt loss. Variations in water and salt excretion result from adjustments in specific tubular reabsorptive activities, largely under the control of pituitary and adrenal hormones. Glomerular filtration rates are remarkably stable in mammals, ensuring a steady rate of filtration and excretion of metabolic waste. With control largely based on tubular rather than glomerular mechanisms, the management of salt and water can be largely achieved independently of each other within the mammalian kidney.

There is no reliance on extrarenal means of salt management, even among mammals living in extreme osmotic environments, though marine species do maintain their blood osmotic concentration higher (dolphin, 420 mOsmoles/liter) than most terrestrial mammals (ca. 300 mOsmoles/liter). Respiratory water losses are small in these animals, as they breathe air of relatively high humidity. Urine and feces are the major avenues of water loss (up to 90% of total). A positive water balance can be achieved with the osmotically free water generated in the kidney and the metabolic water gained from food, obviating the need for marine mammals to drink SW.

11

Endocrine Glands and Their Secretions

1. Introduction

Integration and coordination of the complex osmoregulatory processes described in Chapter 10 appear to be largely under endocrine control. As we saw earlier, management of body fluids is based on two principal mechanisms: (1) alteration of membrane permeability to water and particular ions; and (2) active ion pumps transporting ions against their concentration gradient. Adequate functioning of osmoregulatory tissues that incorporate these properties is ultimately dependent on blood supply. Hormones may thus influence the activity of osmoregulatory organs in several ways:

1. By altering the active transport of ions (Na^+, K^+, H^+, Cl^-, and HCO_3^-) through direct actions on the cells involved. Adrenal steroids, posterior pituitary peptides, and catecholamines influence active sodium and potassium transport in renal and extrarenal tissues in most vertebrates.
2. By changing osmotic movement of water and ion diffusion across epithelia by adjusting membrane permeability. Pituitary prolactin alters sodium permeability of the teleost gill, limiting diffusional loss in freshwater. Posterior pituitary hormones may similarly alter the water permeability of a number of tissues.
3. By altering blood flow to and within osmoregulatory organs by adjusting vascular smooth muscle tone and thus blood vessel diameter. In this way catecholamines, angiotensins, and neurohypophyseal peptides may modify the rate of glomerular filtration and, accordingly, urine flow in the nonmammalian kidney. Variation in regional blood flow distribution can also influence the secretory activity of tissues like the teleost gill.

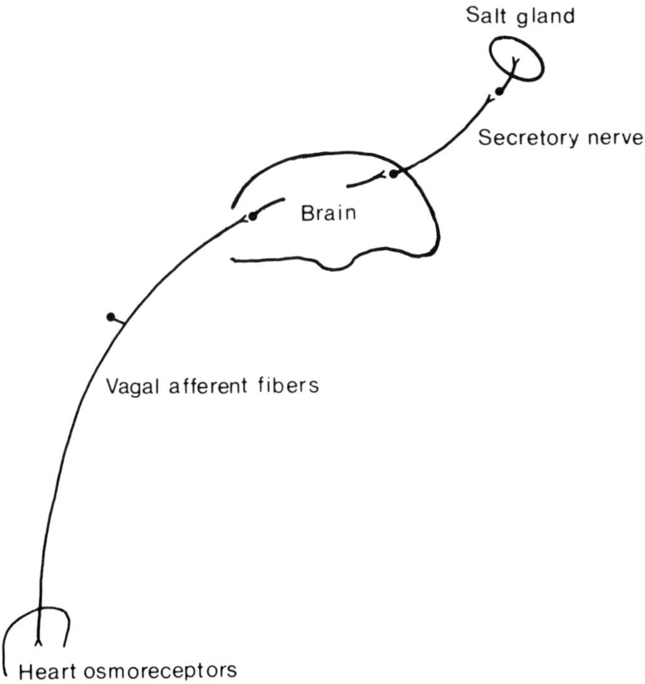

Figure 1. Reflex arc for stimulating avian salt gland secretion. [Redrawn from Hanwell, A., Linzell, J. L., and Peaker, M., 1972, Nature and location of the receptors for salt-gland secretion in the goose. *J. Physiol. (Lond.)* **226:**453–472.]

Although we have dismissed consideration of neural integration of osmoregulatory mechanisms as uncommon, we should be aware that it may play a role. Neural mechanisms mediate the initial reception of altered blood volume, pressure, and composition, though the effector system evoking adaptive changes in osmoregulatory organ function is usually hormonally based. The cephalic "salt glands" of birds and reptiles are exceptional in that they are stimulated to secrete by autonomic cholinergic neurons, though the final form of secretion may be altered by adrenal steroids (see Figure 1). The influence of adrenergic neurons on the tone of vascular smooth muscle may also indirectly lead to altered function of tissues like the fish gill and the kidney.

Analysis of the vertebrate series reveals that particular endocrine centers are regularly involved in the control of osmoregulatory processes. Prolactin of the anterior pituitary, posterior pituitary peptides, adrenal (interrenal) steroids, and the renin–angiotensin system are most prominent, though their exact roles vary from group to group. Before we examine the detailed actions of these hormones (Chapter 12), it is profitable to consider their phylogenetic distribution. We shall review variations in the homologous glands that produce them, principal stimuli for their secretion, and general features of their actions at target organs.

2. Pituitary Hormones (see also Part III)

Hormones of the pituitary gland play a central role in controlling aspects of vertebrate physiology, including osmoregulation. As a functional unit, the hypothalamic–pituitary complex plays an important integrative role for signals from both the brain and peripheral tissues. There are two main areas of the pituitary: the adenohypophysis, derived from stomodeal ectoderm, and the neurohypophysis, originating from neural ectoderm. The connection between the two main areas is variable among the vertebrates, but is achieved principally by development of neurohemal organs. The median eminence of the neurohypophysis is the most important area for transmission of factors controlling the cells of the pars distalis of the adenohypophysis, whereas the pars intermedia cells are controlled by direct nervous links. In advanced teleost fishes, however, some or all pars distalis cells are directly innervated. In addition to their local (hypophysiotropic) function, neurosecretions also enter the general circulation to influence peripheral organs (Figure 2).

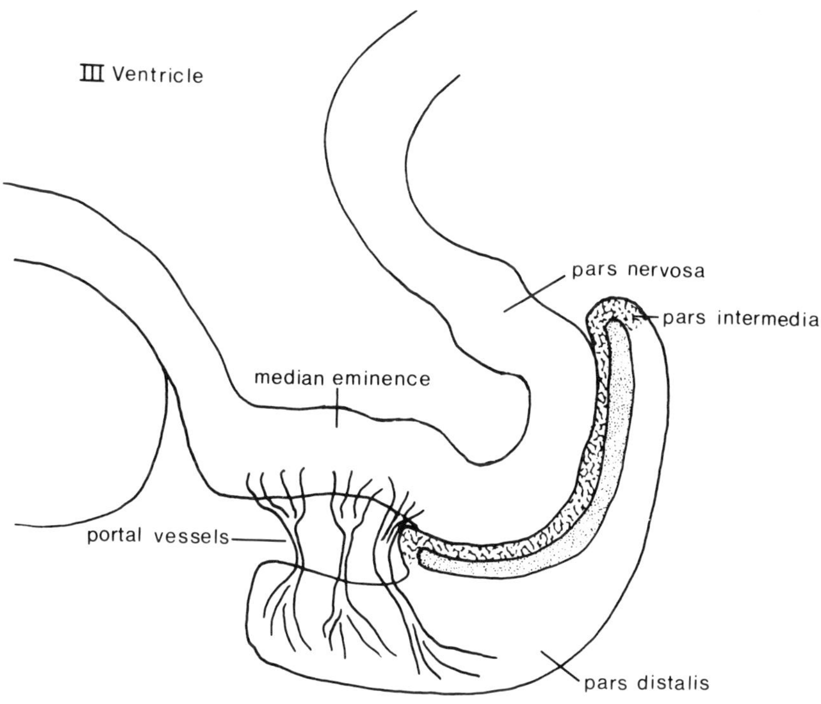

Figure 2. Idealized amniote pituitary gland. Neurohypophyseal tissue (pars nervosa and median eminence) is closely apposed to the adenohypophysis (pars intermedia and pars distalis), though the extent of vascular and neuronal link varies.

2.1. Neurohypophyseal Hormones

The systemic effects of neurohypophyseal hormones include control of water balance, reproductive tract and vascular smooth muscle contraction, and mammary gland milk ejection. These systemic functions of the pars nervosa are probably more recent, in a phylogenetic sense, than the hypophysiotropic role of the median eminence. Clearly, the regulation of milk ejection in mammals is a very recent phylogenetic development. All vertebrates, however, have a neurohemal link for the systemic release of neurohypophyseal hormones; that in hagfish (Agnatha) is particularly well organized, though the physiological significance of the peptide hormone secreted (arginine vasotocin) is unclear.

All neurohypophyseal hormones are peptides with nine amino acid residues, including two cysteines linked by a disulfide bond to form a single cystine residue (Table I). Conventionally, the hormones are considered to be octapeptides with a common structure based on a six-membered ring with a three-amino-acid side chain. The linkage of cysteines at positions 1 and 6 forms the ring structure. Since the composition of oxytocin, isolated from pig and beef neurohypophyses, was first identified in 1953 by du Vigneaud and colleagues, 10 distinct neurohypophyseal peptides have been identified from the vertebrate subphyla. These can be divided into two groups: (1) "basic" antidiuretic/vasoactive peptides, and (2) "neutral" oxytocinlike peptides. The four basic peptides—arginine vasotocin, arginine vasopressin, lysine vasopressin, and phenypressin—all contain a basic amino acid (arginine or lysine) in position 8. The remainder possess neutral amino acids (leucine, isoleucine, valine, or glutamine) in position 8. Contrasting the structures of the 10 peptides, it is noteworthy that amino acid substitutions into the common structure occur only at positions 3, 4, and 8. The phenypressin, recently discovered

Table I
Amino Acid Sequences of Vertebrate Neurohypophyseal Hormones[a]

Common structure	1 Cys	2 —	3 —	4 —	5 Asn	6 Cys	7 Pro	8 —	9 Gly
Basic peptides									
Arginine vasotocin		Tyr	Ileu	Gln				Arg	
Arginine vasopressin		Tyr	Phe	Gln				Arg	
Lysine vasopressin		Tyr	Phe	Gln				Lys	
Phenypressin		Phe	Phe	Gln				Arg	
Neutral peptides									
Oxytocin		Tyr	Ileu	Gln				Leu	
Mesotocin		Tyr	Ileu	Gln				Ileu	
Valitocin		Tyr	Ileu	Gln				Val	
Isotocin		Tyr	Ileu	Ser				Ileu	
Glumitocin		Tyr	Ileu	Ser				Gln	
Aspartocin		Tyr	Ileu	Asn				Val	

[a]Cysteine amino acids at positions 1 and 6 are connected by a disulfide bond.

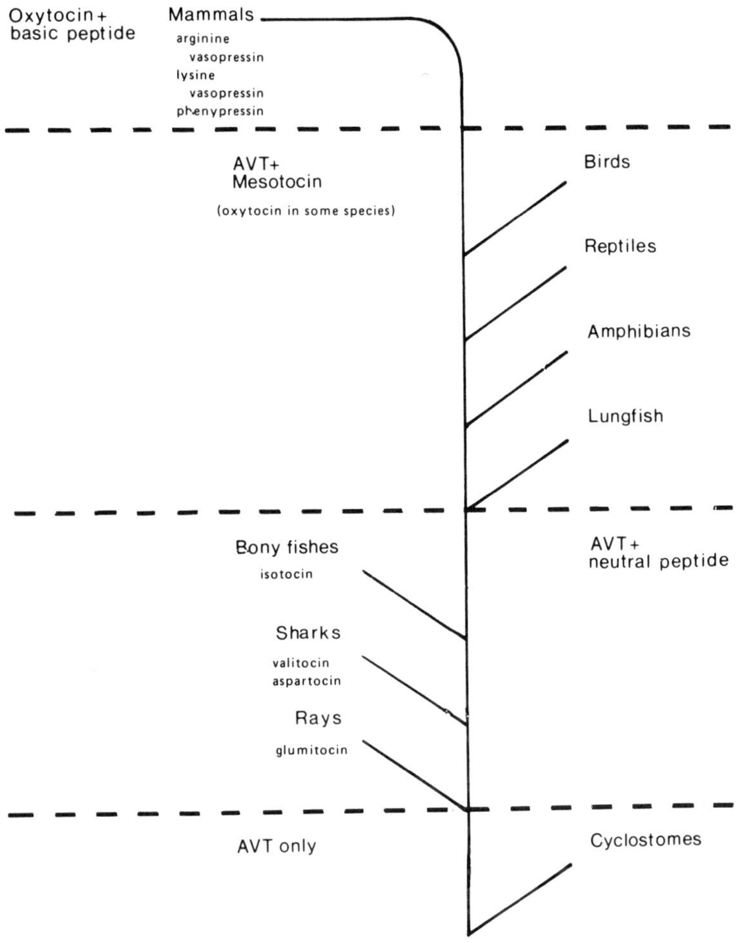

Figure 3. Distribution of vertebrate neurohypophyseal hormones. All vertebrates except cyclostomes produce a neutral peptide and a basic peptide. The basic peptide in all groups other than mammals is arginine vasotocin (AVT). (Modified from Gorbman, A., *et al.*, *Comparative Endocrinology*, John Wiley and Sons, New York.)

in marsupial mammals, differs from all other peptides in possessing phenylalanine rather than tyrosine at position 2.

The neurohypophyses of all jawed vertebrates contain at least one basic and one neutral peptide (Table I), though the ancient cyclostome neurohypophysis apparently contains only arginine vasotocin (AVT). Analysis of neurohypophyseal peptide distribution through the vertebrate series (Figure 3) indicates that AVT is probably the most primitive and certainly the most commonly occurring peptide. It is interesting that the sarcopterygian fish (lungfish) and nonmammalian tetrapods are united in their possession of AVT and mesotocin. These groups are, of course, conventionally linked as representing the transition in the vertebrate series from aquatic to terrestrial life-style.

Mammals are exceptional in having no AVT in their adult neurohypophyses, though it may be present in embryonic or neonatal stages. Instead, most mammals have arginine vasopressin as the basic peptide, with lysine vasopressin restricted to members of the suborder Suina (pig family), some marsupials, and the Peru strain of mice. Phenypressin appears to be restricted to the isolated marsupial group of mammals.

Neurohypophyseal hormones are synthesized in large nerve cell bodies of the anterior hypothalamus. These so-called magnocellular neurons are concentrated in the preoptic nucleus in lower vertebrates, whereas in the amniotes (birds, reptiles, and mammals) cell bodies are found in two paired structures, the supraoptic and paraventricular nuclei. At synthesis and during axonal transport to the neurohypophysis (see Figure 4), hormones are linked to a large carrier peptide (99 amino acids), neurophysin. Hormone and neurophysin are cleaved from a common precursor peptide molecule. Neurosecretory granules containing the hormone are transported down the axons and stored at the nerve endings in the neurohypophysis. Upon receipt of an oppropriate stimulus, leading to depolarization of the magnocellular neurons, granules are released by exocytosis. This so-called stimulus–secretion coupling involves the transduction of a neural signal (nerve depolarization) to a hormonal signal (hormone secretion). The electrical properties of the magnocellular neurons appear to be similar to those of other neurons. The arrival of an action potential at the nerve terminal leads to an influx of sodium and calcium ions, which triggers the process of stimulus–secretion coupling. In particular, it appears that the rise in intracellular free calcium ions is a prerequisite for the secretion of neurohormone, as it is also for neurotransmitter secretion. There is also evidence that the cyclic nucleotide, cyclic AMP, may also be involved. Neurohypophyseal hormone, now separated from neurophysin, is discharged into the circulation as neurosecretory granule. The sequence of events involved in neurohormone secretion is summarized in Figure 4.

Although the experimental evidence is based almost exclusively on mammalian studies, a universal concept has emerged that each neurosecretory neuron produces only one type of hormone. Thus, the mammalian magnocellular neurons are considered to be either vasopressinergic or oxytocinergic. Immunocytochemical techniques have been employed to trace fiber tracts reacting with either antivasopressin or antioxytocin antisera. These studies indicate that such fibers extend not only to the neurohypophysis, but also to other regions of the brain and central nervous system. Vasopressin-containing nerve endings are present in the median eminence, the region of releasing factor/hormone secretion, and oxytocinergic and vasopressinergic fibers have been identified in the spinal cord. Indeed, vasopressin has been specifically implicated in memory and learning processes. Such observations have led to the speculation that molecules like the neurohypophyseal peptides may serve as both systemic hormones and neurotransmitters or neuromodulators in

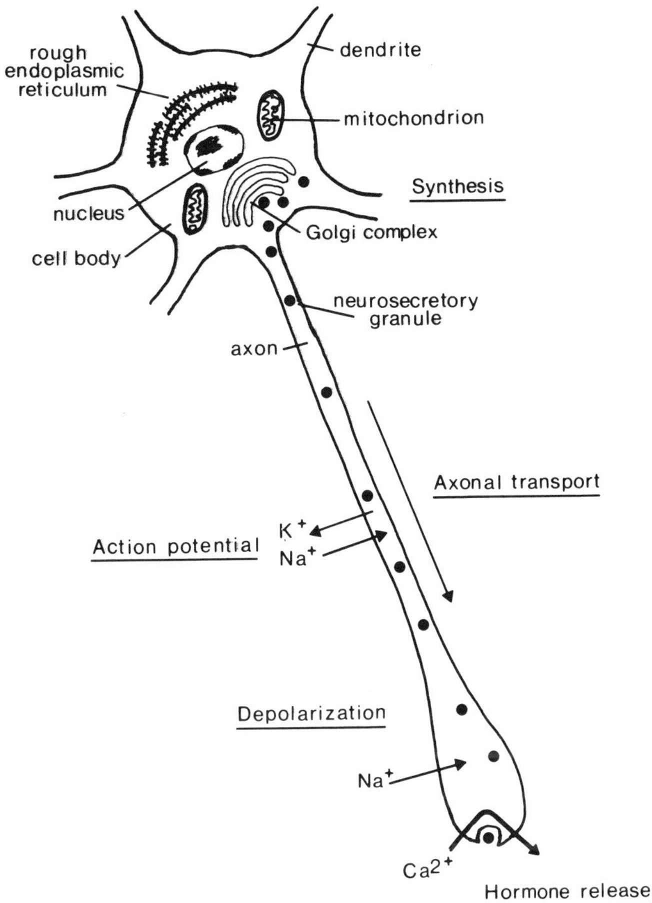

Figure 4. Principal components of neurosecretion. (After Bennett, G. W., and Whitehead, S. A., 1983. See also Figure 9)

the central nervous system. The relative importance of these two functions may vary between species and perhaps between the major vertebrate groups. Of course, the dual properties of these peptides allow the attractive possibility that one molecular species may be a common mediator (and, therefore, integrator) of neural and endocrine responses to specific physiological disturbances.

Circulating vasopressin levels in normally hydrated mammals vary between 10^{-11} and 10^{-12} M. Hormone levels tend to be higher in small animals like the mouse (2×10^{-11} M) than in larger species like man (4×10^{-12} M). Hormone concentrations may increase 25-fold as a result of increased secretion during dehydration. The relationship between plasma osmotic concentration and plasma vasopressin levels in the rat is shown in Figure 5. Radioimmunoassays have re-

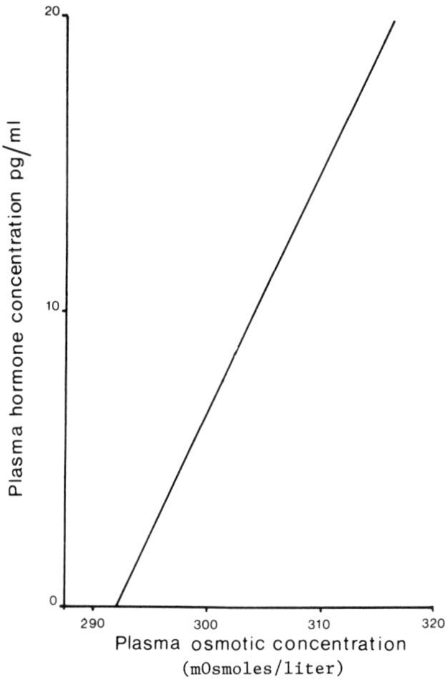

Figure 5. Relation between plasma vasopressin and osmotic concentrations. The linear relationship has been drawn from plasma hormone measures in the rat following salt loading. (From Dunn, F. L., *et al.*, The role of blood osmolality and volume in regulating vasopressin in the rat. *J. Clin. Invest.* **52:** 3212–3219.)

cently become available for arginine vasotocin, and a similar relationship between AVT levels and plasma osmotic concentration appears to hold for a few nonmammalian species studied.

Neurohypophyseal hormones are carried in free solution in the circulation and accordingly are readily taken up by receptors at target tissues. The peptides are rapidly excreted by the kidney and metabolized in the liver, resulting in a characteristically short half-life. For vasopressin, the half-life is about 2 min in the rat, but is nearer 15 min in man; the difference largely reflects the influence of body size on metabolic rate. In poikilothermic groups, the lower metabolic rate results in longer half-lives; for example, in the toad, *Bufo marinus,* the half-life of AVT at 20° is 33 min, but is only 18 min in the domestic duck at 43°.

An increase in systemic blood pressure was the earliest observed action of mammalian neurohypophyseal extracts. On this basis the mammalian peptide was termed vasopressin, though the physiological importance of the vascular action is now doubted. It is generally considered that the plasma levels of hormone induced by exogenous administration in order to demonstrate a vascular response are beyond the range normally achieved by endogenous secretion. As in any endocrine study, it is essential to separate the physiological from pharmacological actions of a hormone. Accordingly, the short-term vasopressor action of vasopressin in the mammal is thought to be pharmacological, though there is continuing clinical interest in a possible contribution of vasopressin to the development of some forms of chronically high blood pressure (hypertension).

Basic neurohypophyseal peptides are more characteristically associated with regulation of urine flow and are sometimes loosely termed antidiuretic hormones, on the basis of their inhibition of urine production in tetrapods. Such an antidiuretic action may be absent in fish; indeed, a diuresis (increase in urine flow) is a regular response to AVT in freshwater fish including the lungfish. The hormone-induced changes in urine flow appear to be based on effects on the renal vasculature in lower vertebrates, though more specific actions on renal tubular water permeability replace these to varying degrees in tetrapod groups. Neurohypophyseal hormones similarly influence electrolyte management, both renal and extrarenal, through a combination of vascular effects and specific actions on transport mechanisms. The precise roles of neurohypophyseal peptides in water and electrolyte economy are discussed in detail in Chapter 12.

The secretion of vasopressin, as measured by magnocellular neuron firing rate or plasma hormone concentration, follows a linear relationship over a limited range with increase in plasma osmolarity (see Figure 5). Vasopressin neurons are highly sensitive to an increase in plasma osmolarity above a threshold of 290 mOsmoles/liter. Changes in blood osmolarity are detected by hypothalamic osmoreceptors. These may represent specialized cells neurally connected to the magnocellular neurons, or the magnocellular neurons themselves may fulfill this function. Sodium and its associated anions contribute over 95% of the osmotically active constitutes of plasma, and it has been suggested that these hypothalamic osmoreceptors are primarily sensitive to changes in sodium rather than osmotic concentration per se. Experimental evidence to date has failed to resolve this issue satisfactorily.

Changes in body fluid volume can also alter vasopressin secretion rate. Two main groups of stretch receptors have been identified in the cardiovascular system, which project neurally to the hypothalamus. These are located in low-pressure (left atrium wall) and high-pressure (carotid sinuses and aortic arch) sides of the vascular system. Other volume receptors may also be present in the abdomen, detecting gut fluid uptake. A fall in blood volume results in a decrease in the tonic neural inhibition to the hypothalamus, whereas a rise in fluid volume further increases this inhibitory neural influence to suppress vasopressin secretion. A 9–10% reduction in blood volume appears to be the threshold for vasopressin secretion, which rises proportionally with subsequent hemorrhage over the range 10–50% blood volume. In the mammal, a fall in blood volume may be a more potent stimulus for hormone secretion, in terms of the plasma peptide concentration achieved, than an increase in plasma osmotic concentration. Clearly, the firing rate of magnocellular neurons, and therefore the rate of hormone secretion, are determined by the effective balance of stimulatory and inhibitory neuronal inputs from the osmoreceptors and volume receptors (see Figure 6).

Control of vasotocin secretion is considered to be similar to that for vasopressin. However, direct measurements of plasma AVT or recordings of magnocellular neuron activity in nonmammalian vertebrates are scanty. Recent studies of circulating AVT levels in frogs indicate that hemorrhage is associated with elevation of plasma AVT, though infusion of hypertonic saline produced less consistent effects. Plasma AVT levels show a linear relationship with plasma osmolarity in the euryhaline eel (Figure 7), similar, in principle, to the relationship shown in Figure 5

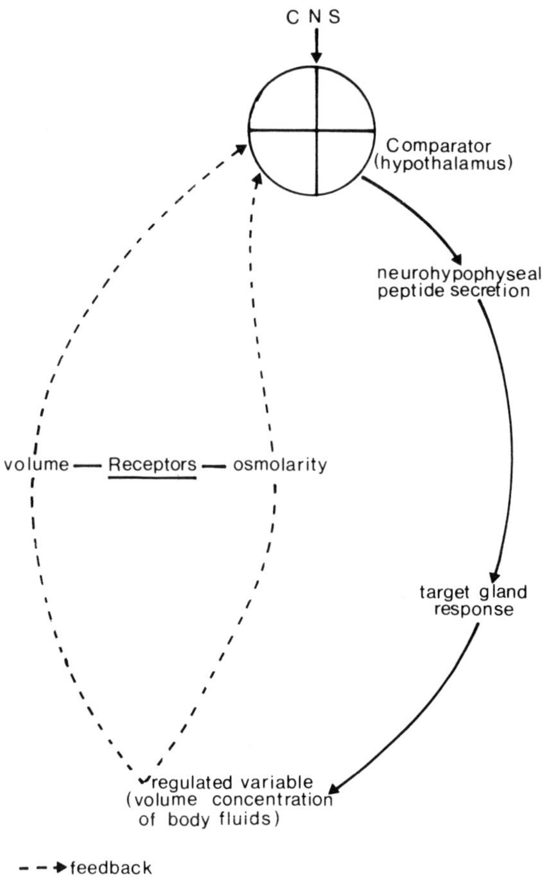

Figure 6. Feedback mechanisms controlling neurohypophyseal secretion. Primary feedback information is related to the volume and osmotic concentration of body fluids.

for the rat. This supports the concept that, in general, control of AVT secretion is based on a scheme similar to that outlined in Figure 6.

Regulation of neurohypophyseal hormone secretion is, in common with other endocrines, based on a negative feedback system. Information concerning changes in the controlled variables (body fluid volume and osmotic concentration) is fed back to the hypothalamus to influence subsequent secretory patterns. The control system depends on a comparison of estimates of existing body fluid volume and osmolarity with preset optima. This comparator function is essentially fulfilled by the balanced response of the magnocellular neurons to the neural inputs from osmotic and volume receptors. For example, in the mammal, vasopressin administration will lead to renal water retention, with subsequent rise in body fluid volume and fall in osmolarity. These changes will effectively increase inhibition of the immediate further secretion of endogenous hormone.

Peptide hormones like the neurohypophyseal secretions generally effect their response at target tissues by interaction initially with specific receptors on the surface membranes of target cells. The only vasopressin receptors clearly identified to date are in the collecting duct of the renal tubule. Binding of radioactively labeled vasopressin occurs in membrane preparations of the renal medulla, receptors showing a dissociation constant in the range 10^{-9} to 10^{-10} M (dependent on species), with maximal binding capacity in the order of pM/mg protein. In other words, the receptors are highly sensitive to very low concentrations of vasopressin. It has been estimated that occupancy of less than 5% of the available vasopressin receptors results in a maximum antidiuresis, indicating a very large receptor reserve. This

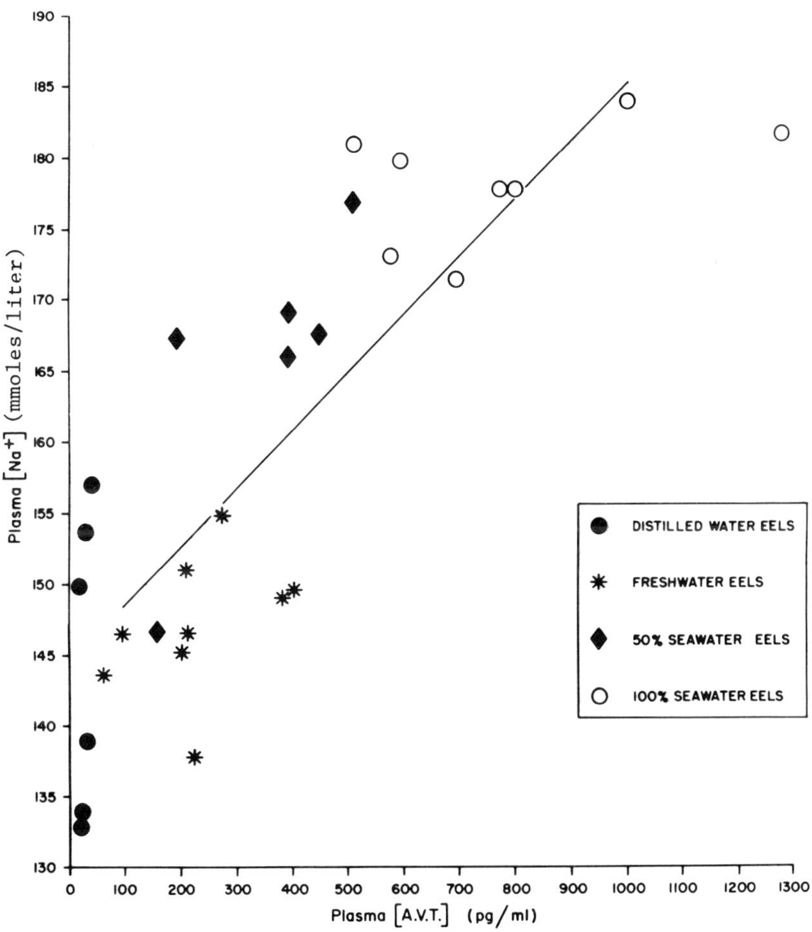

Figure 7. Relationship between plasma AVT and sodium (osmotic) concentrations in the eel. (Courtesy K. M. Hughes.)

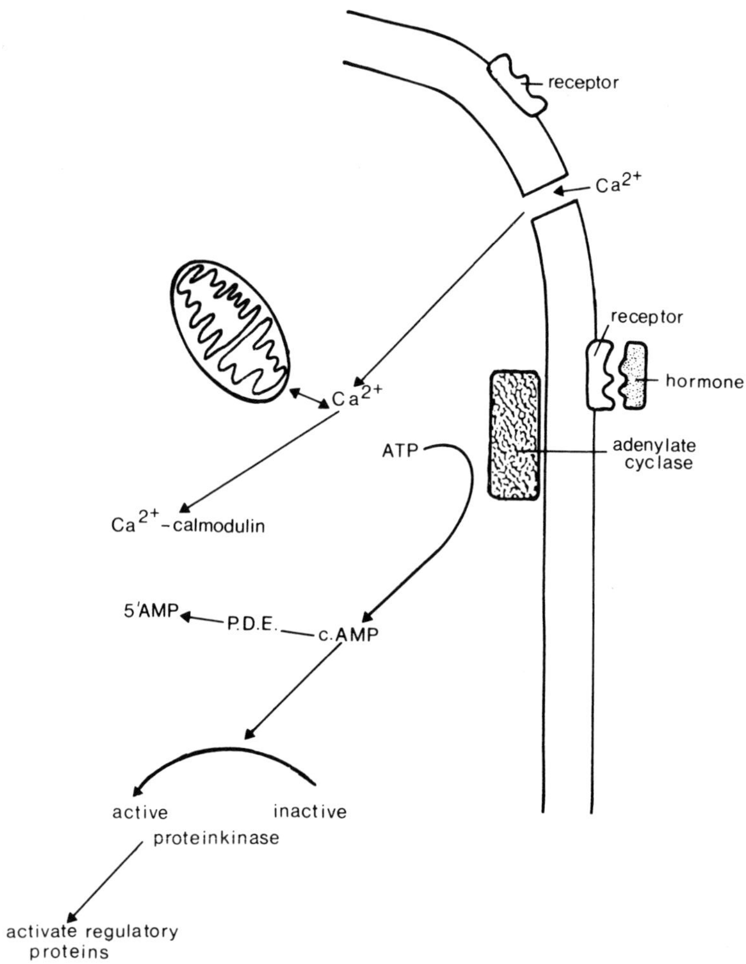

Figure 8. Mechanism of peptide hormone action. Cell membrane receptors may mediate hormone actions be changes in cyclic AMP (cAMP) and calcium mobilization. P.D.E., phosphodiesterase.

picture underlines the general concept of hormone-mediated regulation being based on the secretion of very small quantities of messenger molecules in association with high-affinity uptake by specific receptors at target cells.

Cyclic AMP production mediates the postreceptor binding steps in hormone action (Figure 8). Cyclic AMP serves this intracellular second messenger role for many of the peptide-based hormones. Vasopressin-sensitive adenylate cyclase (enzyme converting ATP to cyclic AMP) is located in the basal membrane (blood side) of cells in the collecting duct of the renal tubule. Vasopressin binding at the receptor activates the adenylate cyclase, producing an increase in intracellular cyclic AMP which promotes a cascade of reactions, ultimately leading to an increase in the water permeability of the luminal cell membrane (in contact with forming urine). This results in an antidiuresis as water is reabsorbed from urine following the osmotic gradient between tubular fluid and the renal medullary interstitium (see Chapter 10). Hormone binding to target receptors is also associated with increased intracellular free calcium levels which, like cyclic AMP, are thought to be involved in the intracellular response to the hormone. Largely from studies of the isolated frog skin or toad bladder, it is generally assumed that AVT-induced effects at target tissues are achieved in a similar manner. It is interesting to note, however, that frog skin exhibits more than one type of adenylate cyclase-mediated response to AVT. The hormone promotes both sodium and water movement across amphibian skin, effects that are linked to separate hormone receptors. It seems likely that two adenylate cyclases and/or two "pools" of cyclic AMP exist in the tissue, each mediating a distinct response.

2.2. Adenohypophyseal Hormones

Hormones of the adenohypophysis are involved in a wide range of physiological functions, from reproduction and growth to intermediary metabolism and behavior. Such effects emanate from both direct action of anterior pituitary hormones and their tropic activation of other endocrine centers. The direct actions of prolactin and activation of adrenal (interrenal) steroidogenesis by adrenocorticotropin (ACTH) are particularly important in osmoregulatory physiology.

2.2.1. Prolactin

Prolactin was a term first applied in 1928 to an extract of the pituitary that promoted lactation in the pseudopregnant rat, though subsequent research has revealed this hormone to have a wide range of physiological actions. Most of the known vertebrate prolactins are just under 200 amino acid residues in length, with a molecular weight around 23,000 daltons. Characteristically, the spiral arms of the peptide are linked by at least three disulfide bridges between adjacent cysteine residues (Figure 9). Pituitary extracts contain, in addition to the 23,000-dalton prolactin, a "big" prolactin of 56,000 daltons. This may represent a dimer of the prolactin molecule, though it is more likely to be a prohormone from which prolactin is enzymatically cleaved. Up to 20% of the prolactin molecule can be degraded enzymatically without serious loss of biological activity. Thus, as for other peptide hormones, it seems that certain amino acid sequences, including disulfide bridges in this case, are unnecessary for biological activity. Recognition by, and subsequent activation of, receptors on target tissues only appear to involve part of the hormone molecule.

Prolactin is produced by a specific cell type of the adenohypophysis which stains preferentially with acidic dyes, particularly erythrosin. The staining properties are due to storage of the hormone in cytoplasmic secretory granules which, at

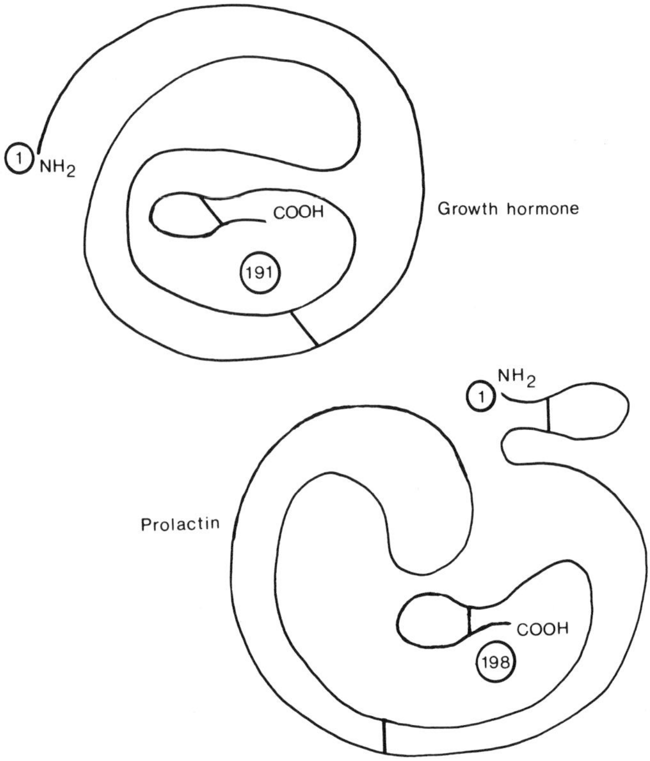

Figure 9. Structures of growth hormone and prolactin indicating positions of the disulfide bridges. (After Bennett, G. W., and Whitehead, S. A., 1983, *Mammalian Neuroendocrinology,* Croom Helm, London.)

electron microscope level, in prolactin cells are characteristically large (up to 700–800 nm) and pleomorphic.

The molecular structure of prolactin is remarkably similar to that of growth hormone (somatotropin) (see Part III). Their primary structures are identical at approximately 25% of all amino acid locations (Figure 9). This inevitably leads to some overlap in biological action between the two hormones. It has been suggested that growth hormone and prolactin arose for a common ancestral molecular structure in early vertebrates, perhaps in a similar way to the development of neurohypophyseal peptides. Unfortunately, there is little information of growth hormone and prolactin in agnathous (jawless) vertebrates.

Over the last 50 years, it has become clear that prolactin is easily the most biologically versatile of the anterior pituitary hormones. A review by Nicoll in 1974 stated that, at the time, there were 85 clearly established functions known for prolactin within the vertebrate series. These were listed under five broad headings: somatotropic effects, reproduction-related effects, actions on the integument and its derivatives, steroid hormone synergism, and osmoregulatory effects. Our consideration of prolactin's role in osmoregulation represents only a part of the whole gamut of actions attributed to this hormone.

The importance of prolactin in osmoregulation was first established in certain euryhaline teleosts; injections of mammalian prolactin enabled their survival in freshwater following experimental removal of the pituitary. The hormone restricted ion loss (particularly sodium) via the gill and kidney and also promoted renal water excretion. There is considerable species variability in the specific ions (Na, K, Ca, Cl) influenced by prolactin, and some fish, like the eel and goldfish, may osmoregulate largely independently of prolactin. The lack of sensitivity of goldfish and eels to hypophysectomy seems to result from the restricted permeability of their bodies, particularly gills, to sodium. Even in these species, however, prolactin can be shown to decrease branchial sodium loss (efflux). Prolactin also restricts the water permeability of the renal tubule and urinary bladder, reducing water reabsorption, and accordingly facilitates renal water excretion. Prolactin is therefore important in freshwater fish to prevent overhydration. As we shall see in Section 2.2.2 these actions of prolactin in fish are very largely counterpoised by those of the adrenal steroid cortisol, which is an important survival factor for animals in seawater.

Although prolactin was first considered only to be involved in teleostean osmoregulation, it is now clear that this hormone has similar actions to restrict salt loss in osmotically sensitive tissues in all vertebrate groups. In mammals, it restricts renal sodium losses, interacting with a similar function of the adrenal steroid aldosterone. Interestingly, the amniotic fluid also contains high concentrations of prolactin, which may play a role in salt and water balance of the fetus.

Control of prolactin secretion from the glandular cells of the adenohypophysis relies on neurosecretory factors from the hypothalamus. In mammals and most vertebrates studied, hypophysiotropic influence is largely inhibitory. Hypothalamic lesions (median basal hypothalamus in the rat) lead to a reduction in this inhibitory activity and a dramatic rise in prolactin secretion. The prolactin inhibitory factor is antagonized, in mammals, by estrogens and stimulated by thyrotropin-releasing factor and vasoactive intestinal peptides. There is considerable evidence to support the conclusion that dopamine (a neurotransmitter) is the principal inhibitory factor.

Like all other anterior pituitary hormones, the secretion of prolactin is pulsatile in nature. Plasma levels may also vary between 2 and 50 ng/ml in mammals in a diurnal rhythm. Prolactin is not significantly bound to plasma proteins and accordingly has a short half-life in the order of 15 min. Apart from the feedback effects of other hormones (particularly gonadal) and plasma ionic and osmotic composition, there is good evidence in the mammal that prolactin may regulate its own secretion (autoregulation) via a short negative feedback loop (Figure 10). When circulating levels of prolactin are raised, the turnover rate of hypothalamic dopamine is increased, as well as the concentration of this amine in hypophyseal portal blood. It is proposed that prolactin may stimulate the release of dopamine from the nerve terminals in the median eminence, which regulates further prolactin secretion.

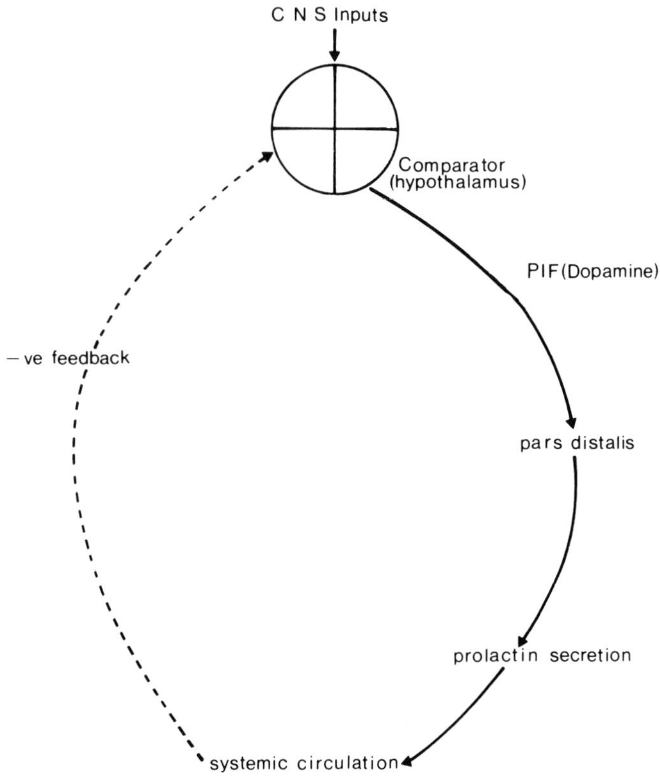

Figure 10. Control of prolactin secretion. Negative feedback is largely directed to enhance prolactin inhibitory factor (PIF) secretion from the hypothalamus, though a prolactin-releasing factor may exist.

Recent evidence of retrograde blood flow in some hypophyseal portal blood vessels has added credence to this short negative feedback loop. However, such feedback could be mediated by prolactin in the general circulation.

The stimuli for prolactin secretion are not always clear, but may result in large changes in circulating hormone levels. There is a 12-fold increase in prolactin concentration following transfer of the teleost *Poecilia latipinna* from seawater to freshwater. In this case, it is probably the lowered blood osmotic pressure which is the primary stimulus for enhanced prolactin secretion. Both fish and mammalian pituitaries in culture respond directly to a lowering of the osmotic concentration of the culture medium. Prolactin secretion is also sensitive to calcium levels in the environment and is raised in low calcium media like freshwater. Indeed, prolactin has been shown to alter calcium metabolism in teleosts, tending to raise blood calcium levels. The importance of calcium in limiting cell membrane permeability, its involvement in ion transport, hormone release, and hormone action, has led to the speculation that some prolactin effects may stem from its primary influence on calcium metabolism.

In animals that undergo a seasonal variation in osmotic environment, external cues such as changes in daylength may also be important in triggering appropriate changes in prolactin secretion. For the stickleback, *Gasterosteus aculeatus,* it appears that lengthening of the day in spring is an important cue to trigger increased activity of prolactin-secreting cells. This prepares the fish for their annual migration from the sea into freshwater.

Binding studies with ^{125}I-labeled prolactin indicate that receptors to the hormone are located on the cell surface of target tissues, which appear to be widely distributed through the body. Unlike other peptide hormones, there is no substantive evidence that prolactin acts via cyclic AMP. In some tissues, prolactin enters cells to bring about its biological action. The prolactin–receptor complex may be taken up by liver cells (internalization) and combine with the Golgi apparatus. Whether this internalization is important for the initiation of the response or represents a stage in the degradation of the hormone remains to be established. Evidence is also accumulating that some peptide hormones become slowly incorporated within the nuclei of target cells. This has the implication that such hormones may induce their effects intracellularly in a manner analogous to the action of steroid hormones (see Section 2.3.5) altering the pattern of nuclear DNA transcription.

2.2.2. Adrenocorticotropin

Anterior pituitary production of ACTH is essential for the maintenance of steroid secretion by the mammalian, avian, and reptilian adrenal cortex and the anamniote interrenal. Adrenal/interrenal steroids regulate the balance of carbohydrate, protein, and fat metabolism. In this homeostatic capacity and their permissive role in the action of many other hormones, they help animals face a variety of stressful situations. However, it is the role of these steroid hormones in water and electrolyte management which is their more vital function. Indeed, removal of the steroid-producing tissue will lead to the death of the animal unless salt supplements are provided. The pituitary–adrenal axis is, therefore, a key part of the complex of endocrines involved in body fluid homeostasis.

Corticotrops of the pars distalis, which synthesize and secrete ACTH, contain relatively few, small secretory granules. ACTH itself is a 39-amino-acid peptide (4500 daltons molecular weight), derived by enzymatic cleavage from a large precursor molecule, proopiocortin. Biological activity is restricted to the N-terminal 1–24 amino acids. The C-terminal sequence (25–39) is not essential for full steroidogenic activity at the adrenal. Variations in the ACTH structure among mammalian species indicate that amino acid substitutions are restricted to this "inactive" (25–39) portion of the molecule. It is, therefore, interesting that the recently sequenced dogfish ACTH showed substitutions at positions 13, 15, and 20 in the "active" portion of the molecule. Dogfish ACTH was only 15% as effective in stimulating steroidogenesis as human ACTH in the dogfish.

Much of our understanding of the synthesis, secretion, and biological function of ACTH has been reevaluated recently since the discovery that this hormone is produced from part of a macromolecular precursor protein or glycoprotein—pro-

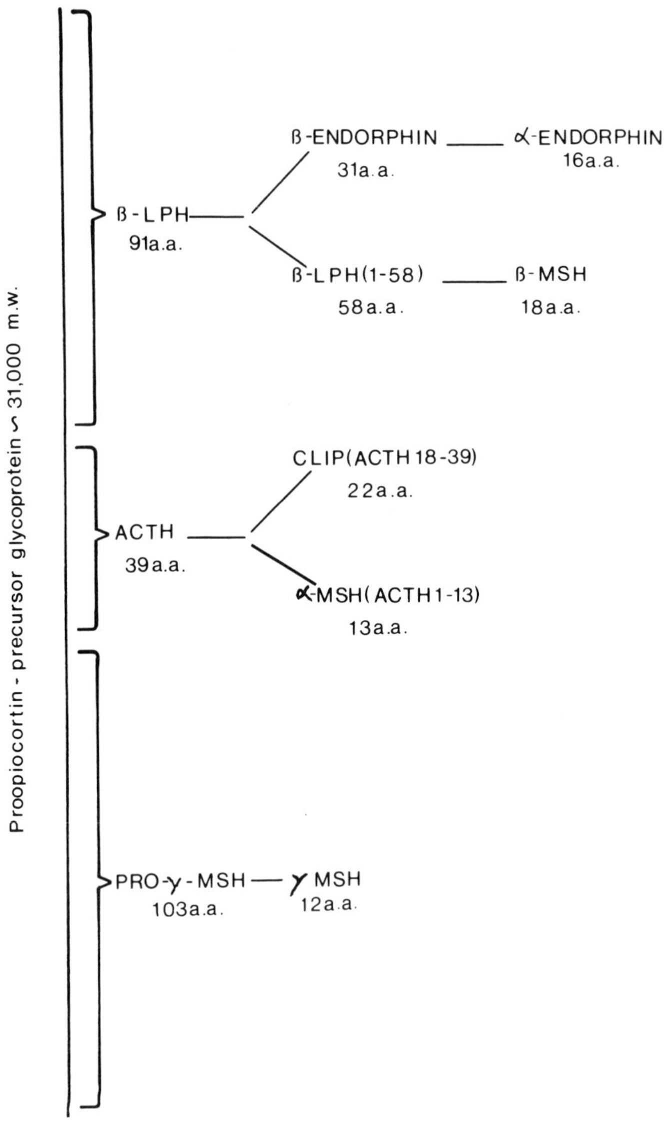

Figure 11. Sequence homologies and possible derivations of ACTH and related peptides. a.a., Amino acid residues.

opiocortin. This precursor contains not only the ACTH sequence of amino acids, but also a larger polypeptide component, β-lipotropin (β-LPH), that is cosecreted from identical granules and may function as a hormone in its own right. In addition, as Figure 11 indicates, both ACTH and β-LPH may serve as precursors for other biologically active peptides or hormones, including β-endorphin, α-melanophore-stimulating hormone (α-MSH), and corticotropinlike intermediate lobe peptide (CLIP). This has led to the speculation that peptides cosecreted with ACTH may act as cohormones regulating related functions in the adrenal and other sites.

The various peptide fragments are processed from their precursors by the action of specific proteolytic enzymes (peptidases) in different cells. ACTH and β-LPH are elaborated in the pars distalis corticotrops, whereas α-MSH and CLIP are processed from ACTH precursors in the pars intermedia. β-Endorphin is a powerful opiate which, along with enkephalin, competes with morphine for receptor binding in the brain. This has been found in both pars distalis and intermedia. Proopiocortin, ACTH, β-LPH, and β-endorphin have all recently been identified in lower vertebrates. It may therefore soon be possible to begin evaluation of the evolutionary development of the proopiocortin molecule, its complex of products, and their physiological functions.

It has been known for more than 25 years that ACTH secretion is under neurohormone control by the hypothalamus. The structure of corticotropin-releasing factor (CRF) has, however, remained elusive because of its lability and lack of specificity of its release. A 41-amino-acid linear polypeptide has recently been isolated from sheep hypothalami which may represent CRF, though a number of other substances, including vasopressin, have been shown to induce ACTH secretion. The recently isolated CRF has unusual amino acid sequencing, which it shares with a neuropeptide (urotensin I) of the teleost caudal neurosecretory system.

CRF stimulates the secretion of both ACTH and β-endorphin from rat pituitaries. As for other hypophysiotropic factors, this effect is thought to be mediated by the intracellular messenger, cyclic AMP. Vasopressin potentiates secretion induced by CRF, and stimulation of vasopressin secretion by angiotensin (see Section 2.4 on the renin–angiotensin system) may be the basis of the CRF activity of this peptide.

The basal secretion of ACTH is episodic, periodic bursts of secretion occurring every 1–3 hr, depending on the time of day. This gives rise to a pronounced circadian rhythm in plasma ACTH and thus corticosteroid levels, with a peak just prior to the most active period of the day. A variety of stimuli, broadly described as stressful, may superimpose changes on this diurnal rhythm of hypothalamic–pituitary–adrenal axis activity. In the rat, activation of the neuronal inputs to the CRF neurons leads to a release of CRF and, within 3–4 min, an increase in plasma ACTH. A four- to fivefold increase in plasma ACTH levels is translated within 10 min into an 8- to 10-fold increase in plasma corticosterone levels. These rapid changes indicate the highly tuned nature of the hypothalamic–pituitary–adrenal axis and the large amplification of response that occurs from brain to adrenal.

The secretion of CRF and ACTH is under negative feedback control from corticosteroid secretion (see Figure 12); steroids exert their inhibitory influence either directly on corticotrop cells or indirectly via CRF neurons. There is also an apparent short-loop feedback of the pituitary hormone inhibiting hypothalamic activity. This latter mechanism may be particularly important in times of severe stress to prevent overstimulation of the adrenal.

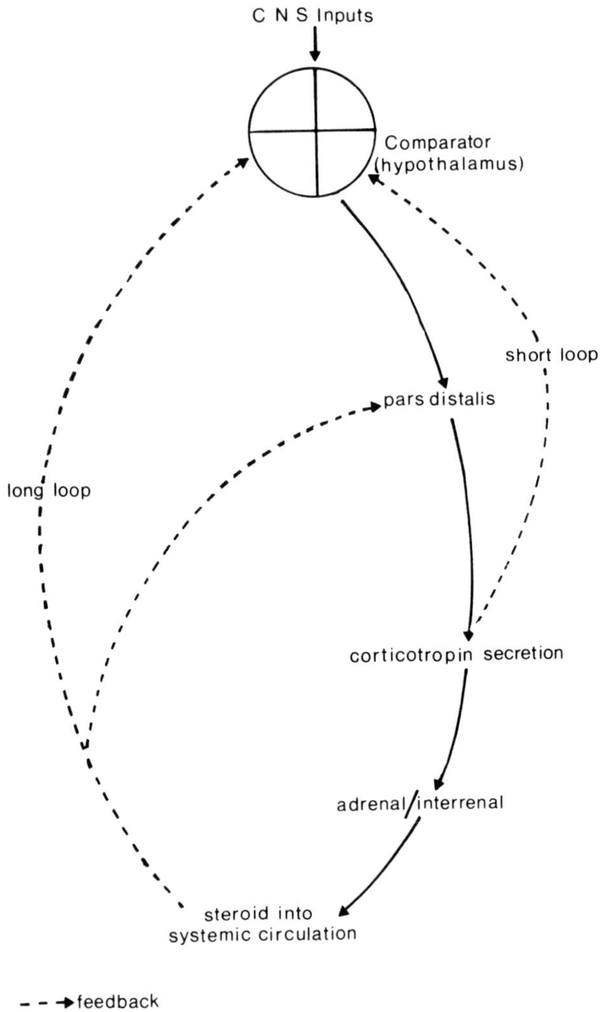

Figure 12. Control of adrenal/interrenal secretion. Negative feedback is provided by circulating steroid on both hypothalamus and pituitary secretory activity. A short-loop feedback is also evident between corticotropin secretion and the hypothalamus.

Mammalian-type basic hypothalamic–pituitary–adrenal interrelationships have been demonstrated in representatives of all the major vertebrate groups, except perhaps the cyclostomes and elasmobranchs. Hypophysectomy is generally observed to result in atrophy of adrenal/interrenal cells and to reduce steroid secretion, whereas mammalian ACTH can activate steroidogenesis. Treatment of animals with corticosteroids or their analogs, such as dexamethasone, causes adrenal inactivity and atrophy as a result of the inhibition of ACTH secretion. Although adrenal function in nonmammalian groups enjoys a similar mode of regulation to that in mammals, there appears to be less dependence on an intact hypothalamic–pituitary axis.

2.3. Steroids from the Adrenal/Interrenal—The Corticosteroids

The structure of steroids and the pathways of their formation (steroidogenesis) are discussed in Chapter 1. The morphology of the gonadal and adrenal and interrenal tissues that produce the hormonal steroids is fully covered in Chapters 2 and 3.

Certain steroids made by the adrenal/interrenal glands predominate, and these are called corticosteroids (Figures 13 and 14). For eutherian mammals, they are

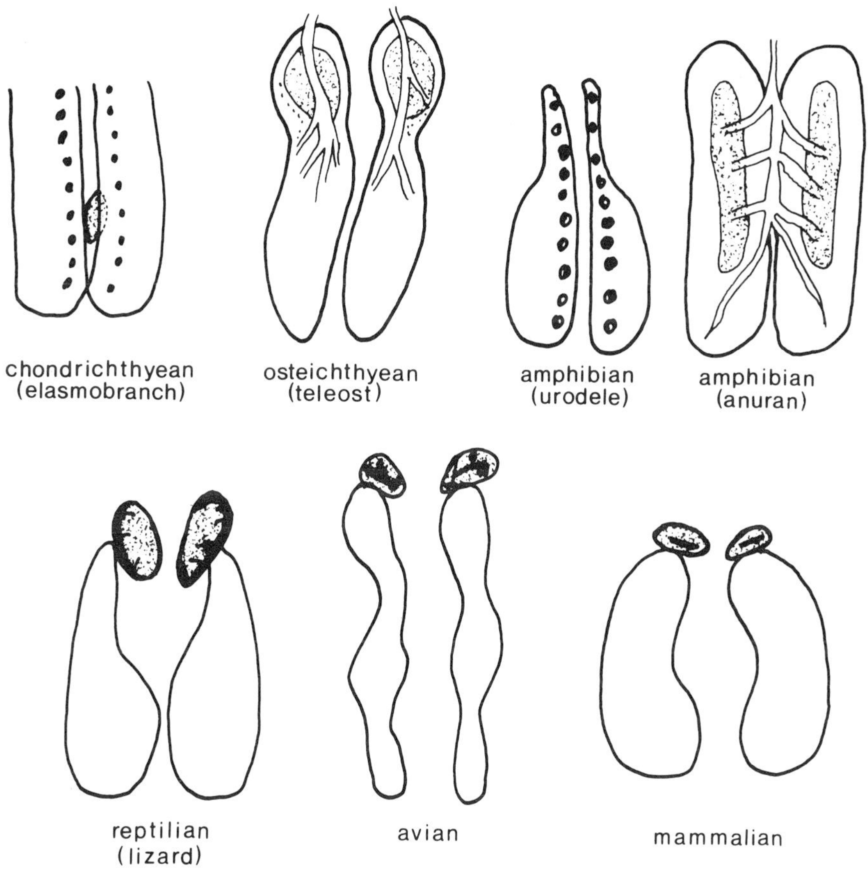

Figure 13. Types of vertebrate adrenals. The close association with renal tissue is evident; stippled area indicates steroidogenic tissue and solid black indicates catecholaminergic tissue. (Redrawn from Gorbman, A., *et al.*, *Comparative Endocrinology*, John Wiley and Sons, New York.)

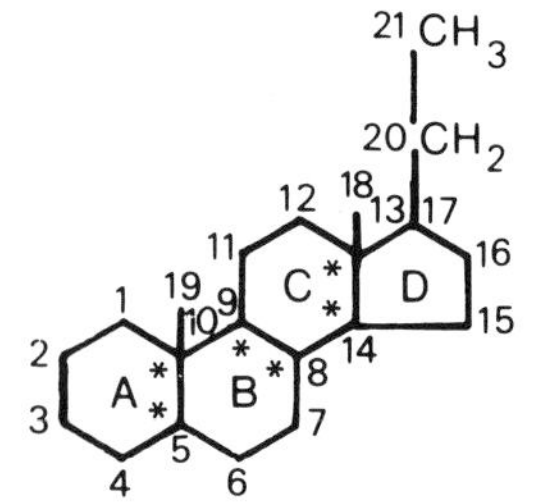

Numbering of carbon atoms in steroid nucleus

Mammalian glucocorticoids

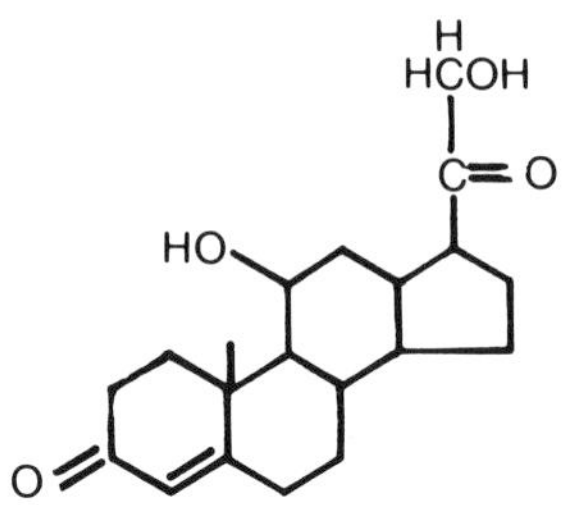

Corticosterone(Compound B) 11β:21-dihydroxypregn-4-en -3:20-dione

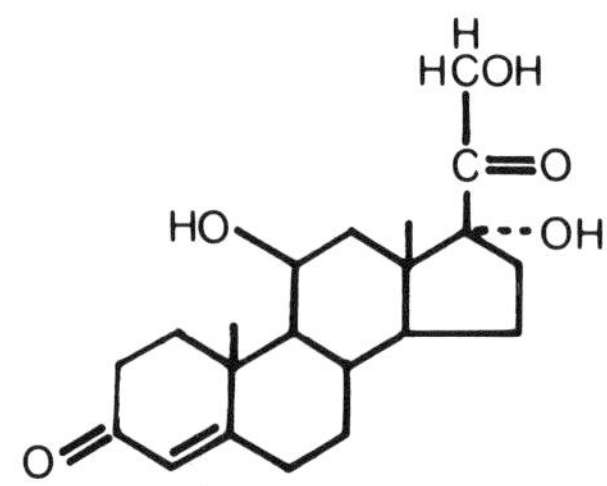

Cortisol(Hydrocortisone, compound F) 11β:17 :21-trihydroxypregn-4-en-3:20-dione

Mammalian mineralocorticoids

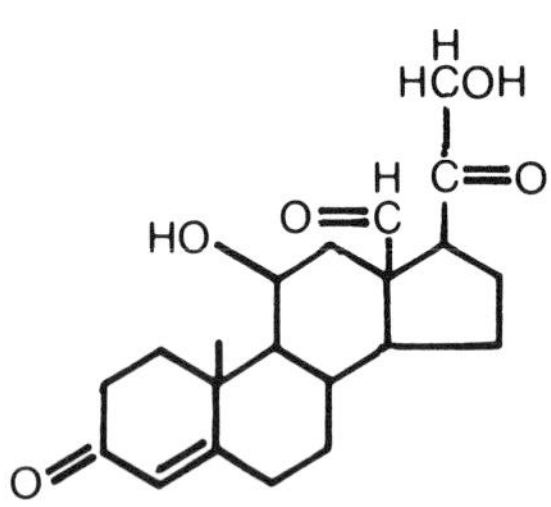

Aldosterone-aldehyde form 11β:21-dihydroxy-3:20-dioxopregn-4en-18-al

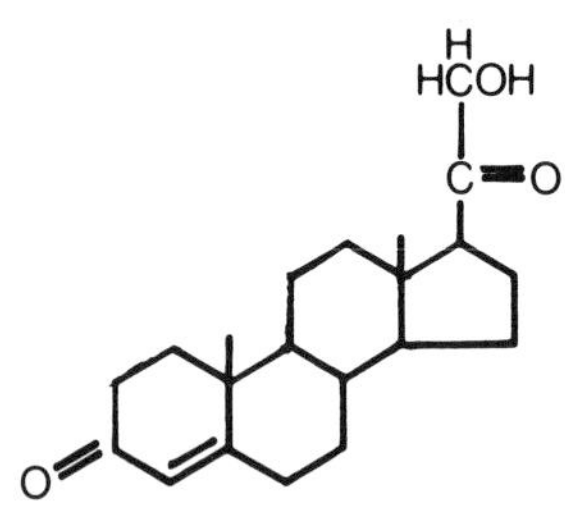

Deoxycorticosterone(DOC , desoxycorticosterone; if 21-acetate, DCA or DOCA) 21-hydroxypregn-4-en-3-20-dione

Figure 14. Some important corticosteroids.

aldosterone and cortisol, though rodents are generally corticosterone secretors; for Amphibia, reptiles and birds, aldosterone and corticosterone; for bony fish, teleosts, cortisol; and for cartilaginous fish, elasmobranchs, a special compound—the unusual 1α-hydroxycorticosterone. In fish, Pisces, that is, all groups of fish, the occurrence of aldosterone is dubious (Table II). All these hormones are formed in steroidogenic cells which are gathered together and make up the adrenal and interrenal (Table II).

Table II
Distribution of Major Corticosteroids[a]

Group	Corticosteroids in circulation
Elasmobranchs	*1-Hydrocorticosterone*, corticosterone, 11-deoxycorticosterone, 11-deoxycortisol
Bony fish	*Cortisol*, cortisone, 11-deoxycortisol, corticosterone
Lungfish	Cortisol, aldosterone, 11-deoxycortisol, 11-deoxycorticosterone, corticosterone
Amphibians	*Corticosterone*, 18-hydroxycorticosterone, aldosterone, 11-deoxycorticosterone, cortisol
Reptiles	*Corticosterone*, 18-hydroxycorticosterone, aldosterone
Birds	*Corticosterone*, aldosterone, 11-deoxycorticosterone
Mammals	*Cortisol/corticosterone*, aldosterone, cortisone, 18-hydroxycorticosterone, 11-deoxycorticosterone

[a]Italicized steroids are the major secretory product.

The term adrenal gland is used when the organ is near, but separate from, the kidney on each side of the body. This is characteristic of mammals, reptiles, and birds. These are jawed vertebrates (Gnathostomata) with fetal membranes—Amniota. Cells of the interrenal gland intermingle, to varying degrees, with kidney tissue, represented by amphibians and fish. The Agnatha (without jaws) possess living members of the order Cyclostomata (roundmouths) and may have groups of cells representing the interrenal gland. The designation suprarenal gland is confined to medical literature (Figure 13).

2.3.1. Mammalia: Eutheria

Much of our knowledge comes from man and laboratory and agricultural animals and spreads over to help comparative studies (see Chapter 3). The adrenal glands are paired structures lying cephalad to the anterior pole of each kidney. They are vascularized by branches from the main abdominal vessels, chiefly the aorta, phrenic, and renal arteries. The numerous arterioles run to the surface of the gland, coming to lie in the loose external connective tissue whence their twigs, in an ill-defined plexus, vascularize the gland. For venous drainage, one vein emerges from the hilus of each gland to open into the inferior cava or the left renal vein (some species, e.g., the dog, are unusual in having up to four emergent veins).

The adrenal is, in fact, made up of two glands of diverse embryological origin.

The center of the gland, the medulla, comprises coalesced chromaffin cells—so called because of their coloration by potassium bichromate, the "chromaffin reaction" (Figure 15). They secrete adrenaline (epinephrine) and noradrenaline (norepinephrine), whereas the surrounding adrenal cortex produces steroids. Initially chromaffin cells arise from neighboring paraganglion cells of the neural crest. The adrenal cortex arises from mesoderm, and the chromaffin cells pass through its anlage.

In the fully formed gland, the cortex is divided, on histological criteria, into three zones: the zona glomerulosa, zona fasciculata, and zona reticularis.

The cells of the zona glomerulosa consist of ovoid groups or "balls" immediately inside the outer connective tissue capsule. This region may be particularly obvious, as in sheep, and difficult to discern in small rodents (e.g., mouse, gerbil). The nuclei vary in shape from a sausage to an oval appearance, with one or two nucleoli. The cytoplasm contains a varying number of lipid droplets (liposomes), depending on the species and its physiological status, and these lie between the nucleus and that side of the cells which abuts on its local capillary.

The zona fasciculata appears histologically in radial lines when the sections are cut in appropriate orientations. The cells of the zone are polyhedral with round nuclei and one or two nucleoli. They are packed with cytoplasmic liposomes in some species (e.g., the laboratory rat, which provides so great a basis for endocrinology) and not in others (e.g., the hamster).

The zona reticularis presents a variety of aspects. Generally, it comprises anastomosing networks of reticular cell cords surrounding the large blood sinuses. The cells are smaller than those of the rest of the cortex. At different ages and in a range of species, the reticular cells appear degenerative. Sometimes a zona reticularis is not apparent and at others a histologically "healthy" zone is to be seen.

The vasculature, in general in the Eutheria, comprises arterioles that (1) form a network to embrace the balls of cells of the zona glomerulosa; (2) proceed radially concomitantly with the columns of the zona fasciculata and enter the zona reticularis to splay out into sinuses; (3) pass through the adrenal cortex to vascularize the medulla in addition to the blood seeping in from reticularis sinuses; and (4) occasionally loop back to the outer connective tissue capsule (e.g., man). Blood drains to the major medullary vein which carries away formed steroid hormones and the catecholamines.

2.3.2. Aves (Birds)

The paired adrenal glands are found, each by a kidney, as separate encapsulated organs, composed of adrenocortical and chromaffin tissues. These are intermingled, to varying degrees, depending on the species. There is no coalescence of chromaffin tissue to form a medulla (characteristic of the Eutheria and Metatheria). The glands lie, often wholly or partly covered by the gonads, at the anterior end of the kidneys (metanephric in embryological origin) and just behind the lungs.

The basic unit of avian adrenocortical tissue is a cord composed of a double

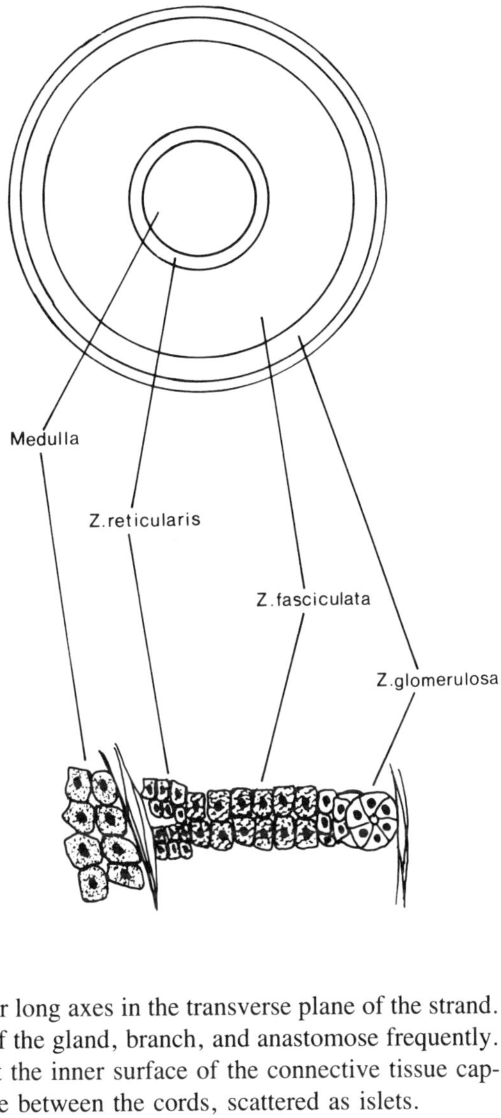

Figure 15. Mammalian adrenal gland. The cortical tissue surrounding the central medulla is organized into three distinct zones, the inner zona reticularis, the zona fasciculata, and the outer zona glomerulosa.

row of parenchymal cells, with their long axes in the transverse plane of the strand. The cords radiate from the center of the gland, branch, and anastomose frequently. At the periphery, they loop against the inner surface of the connective tissue capsule. Groups of chromaffin cells lie between the cords, scattered as islets.

2.3.3. Reptilia

Reptiles are the third tetrapod group with an amnion—Amniota. It characteristically has the "new" kidney, the metanephros in the evolutionary sequence. As a consequence, the mesodermal cells that form the adrenal gland come to lie, as in birds and mammals, near the kidney. It is a discrete encapsulated organ and, basically, it does not differ substantially from the varying degrees of intermingling of adrenocortical and chromaffin tissues observed in birds.

2.3.4. Amphibia

Amphibians belong to the Tetrapoda, and the major interrenal secretions are aldosterone and corticosterone. However, in this group the functional kidney is the mesonephros. The same region of presumptive mesoderm gives rise to both the kidney and interrenal tissue, which, therefore, are intermingled in their embryology. As a representative of the Anamniota, the interrenal lies ventrally on the kidney (mesonephros) and intermingles with the renal tubules and interstitium. It is thus vascularized by common interrenal and renal blood vessels. Though not a discrete encapsulated gland, so characteristic of mammals, birds, and reptiles, its basic microstructure displays the general vertebrate pattern. It can be seen that the interrenal, the adrenocortical homolog, comprises discrete or scattered tissues set about the ventral surface of the functional opisthonephros (a term to include the mesonephros and its posterior parts). Bilateral branches of the dorsal aorta supply both the gland and the kidneys. The blood exits through venules into renal veins to enter the posterior vena cava. Venous blood from the renal portal vein enters numerous small vessels to reach the outer edges of the kidneys and is also carried to the vena cava by the renal veins.

2.3.5. Pisces (Fish)

This is a general term for a mixed group of the "fish type." Among the main examples are bony fish, Teleostei, cartilaginous fish, Elasmobranchii, and jawless Cyclostomata.

The teleost interrenal consists of groups, cords, or strands of cells set about the posterior cardinal veins and their branches, vascularized by twigs of the dorsal aorta. The interrenal is usually confined to the anterior part of the kidney, named the "head kidney," commencing posteriorly to the last gill slit. Chromaffin cells occur in isolated groups near the interrenal cells.

The elasmobranch interrenal lies posteriorly between or on the lobes of the kidney. It is completely separate from chromaffin tissue, which is set segmentally arranged.

The cyclostome interrenal tissue is in some doubt. In Petromyzonida, it presumably lies along the walls of the cardinal veins. Even more doubtful is the situation for the hagfish (Myxinidae).

2.3.6. Steroid Secretion and Action

Steroids produced by the adrenocortical homolog can be categorized on a functional basis. The principal actions of corticosteroids are either broadly metabolic (glucocorticoid) or impinge on electrolyte management (mineralocorticoid). In mammals and probably other amniotes (reptiles and birds), these mineralocorticoid and glucocorticoid properties are associated with specific steroid molecular structures. Thus, aldosterone is the dominant mineralocorticoid hormone in these groups, whereas cortisol and corticosterone have predominantly glucocorticoid properties. In fish and amphibians, interrenal steroids may not be so precisely classified according to their actions. Each hormone may exhibit similar potency in mineralocorticoid and glucocorticoid activities. Cortisol in bony fish regulates gill sodium exchange and is important in the control of intermediary metabolism.

In the amniote groups (reptiles, birds, and mammals), where mineralocorticoid and glucocorticoid properties are associated with particular steroid structure, there is some separation in the mechanisms controlling the rates of secretion of the two hormone types. Accordingly, synthesis and secretion of glucocorticoids is largely under pituitary ACTH control, whereas mineralocorticoid secretion is regulated to a very large extent directly by blood potassium levels and indirectly by the renin–angiotensin system. The renin–angiotensin system is discussed in detail in Section 2.4. From the few direct studies undertaken, it appears that steroidogenic actions of ACTH and angiotensin in fish and amphibians are less discriminatory.

The distinction between steroids with predominantly mineralocorticoid and glucocorticoid properties is clearest in mammals. It is evident that their contrasting activities in this group rely on particular structural components of the steroid molecule. Glucocorticoids characteristically have an oxygen (=O or —OH) group on carbon 11, whereas mineralocorticoids are 11-deoxy compounds (no oxygen at carbon 11). The structures of the major mammalian mineralocorticoids and glucocorticoids are given in Figures 14 and 16 and in Chapter 1.

The zona glomerulosa cells produce the major mineralcocorticoid, aldosterone, whereas the zonae fasciculata and reticularis cells produce glucocorticoid hormones and some sex steroids. The cells of the inner two zones lack the enzyme required for oxidation of steroids at carbon 18 and therefore cannot convert precursor steroids (corticosterone) to aldosterone. Thus, the secretory activity of the inner two zones is largely under pituitary ACTH control, whereas steroidogenesis in the zona glomerulosa is more sensitive to plasma potassium levels and the activity of the renin–angiotensin system.

The influence of ACTH on the adrenal can be considered as either tropic or steroidogenic. The tropic action refers to the ability of ACTH to maintain the size and differentiated state of the gland. Pituitary removal (hypophysectomy) in the rat leads to a marked reduction in the volume of zonae reticularis and fasciculata but little change in the zona glomerulosa. Clearly, the inner zones require the supportive influence of ACTH, which stimulates cell division, protein synthesis, and mitochondrial multiplication. The steroidogenic action of ACTH involves increasing the levels of precursors of steroid synthesis and activating specific enzymes in the steroidogenic pathways. These effects of ACTH are mediated by the intracellular secondary messenger, cyclic AMP, following hormone binding to surface receptors on the adrenal cells. Removal of pituitary ACTH through hypophysectomy reduces aldosterone secretion from the zona glomerulosa (see Table III), but to a much lesser extent than inner zone steroid production. The secretory capacity of the zona glomerulosa shows considerable independence of the pituitary. Angiotensin and high plasma potassium concentration are more potent stimuli than ACTH for aldosterone secretion. Raising the potassium concentration in the medium bathing isolated adrenal cells *in vitro* is very effective in increasing aldosterone secretion. This type of dual pituitary and pituitary-independent regulation of adrenal/interrenal

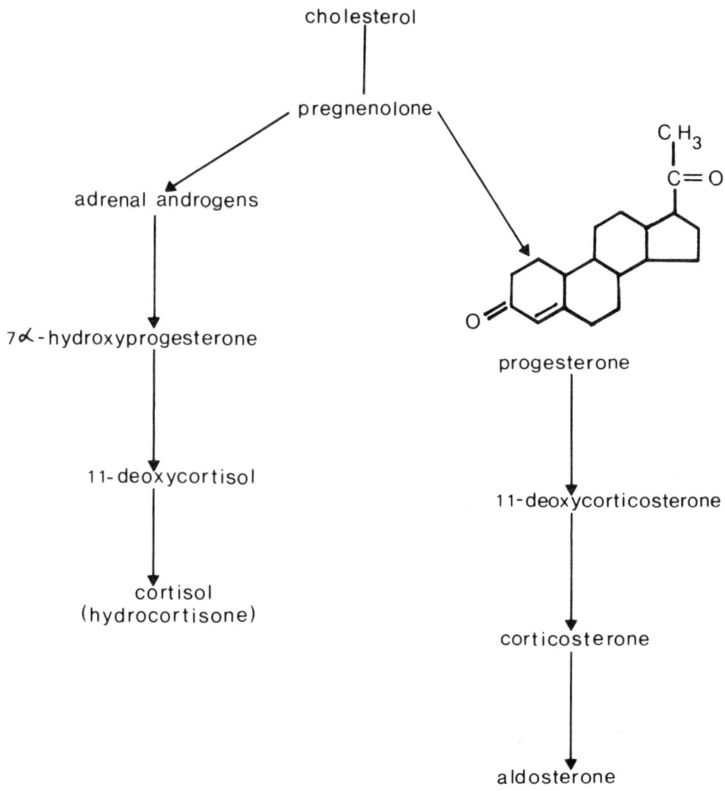

Figure 16. Synthesis of adrenal steroid hormones. The major pathways to cortisol and aldosterone synthesis are shown.

steroid secretion is less well understood in nonmammalian vertebrates, though it is evident that potassium balance and the activity of the renin–angiotensin system can influence interrenal secretion. In the absence of a clear separation of mineralocorticoid and glucocorticoid functions in different steroid molecular structures, such differential control of steroidogenesis may be less important in lower vertebrates.

There is some variation through the vertebrate series in the major steroid product of the adrenocortical homolog (Table II). Either cortisol or corticosterone is the major product in almost all groups. Aldosterone, the important mineralocorticoid of amniotes, has not been reliably detected in any of the fish groups other than lungfish, which, of course, are most closely related to tetrapods. Corticosteroid levels in the blood of agnathans are generally low, and their physiological significance is unclear. The elasmobranchs, which were separated from the main vertebrate line at an early stage, characteristically produce a unique steroid, 1α-hydroxycorticosterone. Other vertebrates appear to lack the 1α-hydroxylase enzyme in their adrenal tissue, essential for the production of this steroid. The 1α-hydroxylase

Table III
Effects of Hypophysectomy on Adrenal Function[a]

	Adrenal weight (mg)	Aldosterone secretion (ng/min)	Corticosterone secretion (ng/min)
Intact	37	1.4	699
Hypophysectomized	10	0.7	14

[a]Thirteen-week-old rats: 3 weeks after sham operation (intact) or hypophysectomy.

enzyme is usually found only in the kidney in other vertebrates, where it is important for the metabolism of vitamin D. Interestingly, 1α-hydroxycorticosterone is not apparently present in the ratfish, which belongs to the other major subclass of the chondrichthyean (cartilaginous) fishes, Holocephali.

Unlike the peptide hormones we considered earlier, corticosteroids are not secreted from granular stores; there is little hormone storage in adrenocortical cells. Steroid secretion is dependent on the rapid release of *de novo* synthesized hormone. Control of steroid secretion by ACTH or other mechanisms involves stimulation of biosynthetic pathways rather than promoting exocytosis of hormone storage granules.

Corticosteroids entering the circulation are rapidly bound to plasma proteins. Only 6% of the circulating steroid in mammals will be in a free unbound state; this is the biologically active form. More than 80% of the binding is to a specific cortisol-binding globulin (transcortin) and about 14% is bound to albumin. Transcortin binds cortisol, corticosterone, and progesterone, but does not bind aldosterone. A dynamic equilibrium exists between the bound and free hormone. The bound corticosteroid therefore provides a large reserve circulating pool. Binding to plasma proteins confers some protection against the normal degradation of hormone in the liver and kidney. Accordingly, corticosterone and cortisol are removed at a much slower rate than aldosterone. In humans, the half-life of cortisol is 90 min and that for aldosterone only 8 min. The latter is, of course, closer to that for peptide hormones.

Steroids, in contrast to peptides, are lipophilic and readily pass through cell membranes. It is perhaps not surprising, then, that receptors for corticosteroid hormones are present in the cytosol rather than cell surface. Adrenal steroids appear to interact with these intracellular receptors without prior metabolic transformation of molecular structure, which often seems to be a prerequisite to sex steroid binding to their receptors. The receptor-bound hormone is translocated to a nuclear binding site where this complex induces changes in nuclear DNA transcription, leading to altered RNA production and ultimately new cytosol protein synthesis (see Figure 17). The new cytosol protein may augment existing cell activities or radically alter target cell function to express the hormone action. Clearly, the process of target cell

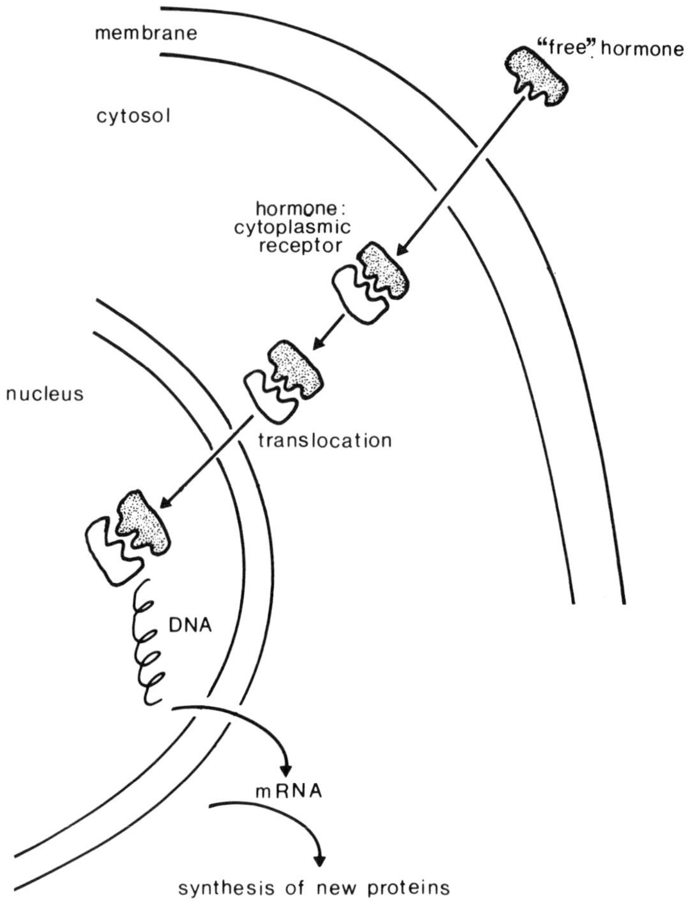

Figure 17. Mechanism of steroid hormone action. Cytoplasmic hormone receptors may effectively concentrate hormone within the cytoplasm and aid in its translocation to the nucleus. Only hormone "free" of carrier protein binding is available to enter the cell.

stimulation is considerably longer than the simple activation of intracellular messengers (cyclic AMP) associated with peptide hormones. Responses to corticosteroids in mammals are characteristically delayed for 60 min or more after initial hormone administration. This contrasts with the 2- to 3-min delay in response to the neurohypophyseal peptide vasopressin. However, the steroid response will normally persist for 1–2 hr by comparison with the 5- to 10-min response to vasopressin.

The physiological actions of corticosteroids impinge on a range of functions, from metabolism and osmoregulation to behavior. Many of the effects are apparently achieved through permissive or synergistic interactions with other hormones, particularly neurohypophyseal peptides and prolactin. Thus, the reduction in water permeability of the skin, gut, gills, and urinary bladder induced by prolactin during freshwater adaptation of the eel is dependent on the presence of cortisol. Cortisol also stimulates sodium uptake in these tissues. The coordinated actions of cortisol and prolactin allow uptake of sodium without the diluting effect of associated osmotic uptake of water. Metabolic actions of corticosteroids appear to be fairly uniform through the vertebrates, elevating blood sugar level and increasing liver and muscle glycogen stores. Adrenal steroids have been implicated in the reproduction and migratory behavior of birds. The influence of corticosteroids on osmoregulatory physiology (mineralocorticoid effects) is focused on the epithelia involved in active sodium and potassium transport. Target tissues therefore vary between the major vertebrate groups and include the gill, urinary bladder, skin, cephalic "salt glands," and, in tetrapods, the kidney.

2.4. Renin–Angiotensin System

The renin–angiotensin system (RAS) is considered to be a "diffuse" hormonal system, as its peptide substrate and enzyme components are located in the blood

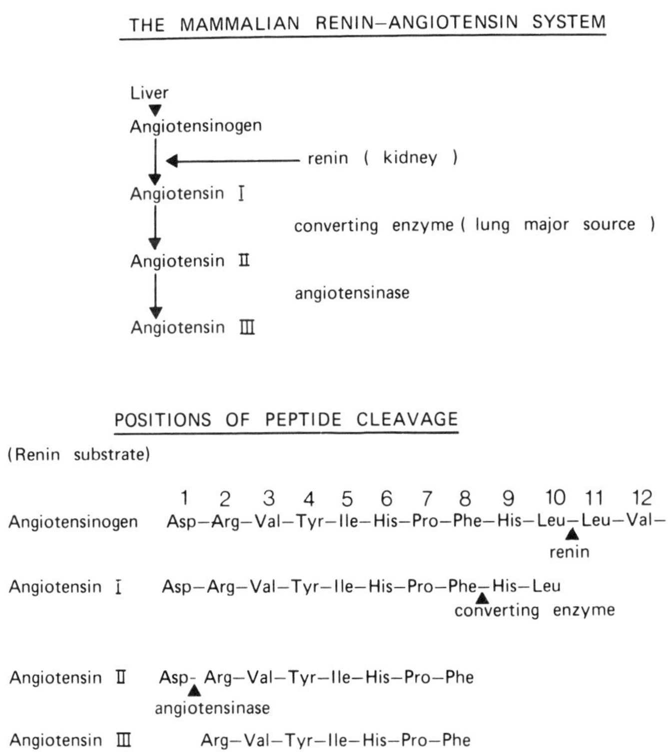

Figure 18. The mammalian renin–angiotensin system.

and in tissues like the liver, lung, and kidney. Production of the major biologically active component, angiotensin II (AII), relies on the concerted action of several enzymes in a variety of tissues (Figure 18). The starting point of the process could be considered the production of a large glycoprotein, angiotensinogen (50,000–100,000 daltons molecular weight), by the liver. An enzyme secreted by the kidney, renin, is capable of cleaving the terminal 10 amino acids from angiotensinogen in the circulation to form angiotensin I (AI) (Table IV). This decapeptide is converted to the biologically active AII or AIII by peptidase enzymes of the lung, liver, and general vascular bed. A converting enzyme removes two amino acids from the C terminal of AI to yield the most biologically potent peptide, AII. The heptapeptide form, AIII, is less potent than AII.

The level of activity of the RAS is set essentially by the rate of renin secretion from the kidney, as renin substrate and other enzymes are normally present in excess. The rate of renin release effectively controls the rate of AII formation. Renin is produced by cells lining the blood vessels supplying the glomerulus, the afferent and sometimes efferent arterioles. These so-called juxtaglomerular cells appear to have developed from smooth muscle cells that line the arterioles. Renin release by the juxtaglomerular cells is controlled by a variety of factors, including blood pressure in the renal arterioles, renal sodium excretion, activity of the sympathetic nervous system, and other humoral factors, such as adrenaline, vasopressin, potassium, and some steroids. Changes in blood pressure are probably detected directly by the juxtaglomerular cells as altered stretch of arteriolar walls. A fall in blood pressure stimulates renin release in all vertebrate groups examined thus far. The AII ultimately generated as a result of the renin secretion is potently vasopressor and will contribute directly to the resulting rise in blood pressure. In mammals, the RAS is thought to play a particularly important role in the vascular

Table IV
Angiotensin I Amino Acid Sequences from Various Vertebrates

	Amino acid sequence									
Common structure	1 —	2 Arg	3 Val	4 Tyr	5 —	6 His	7 Pro	8 Phe	9 —	10 Leu
Species variation										
Eel	Asp				Val				Gly	
Salmon	Asn				Val				Asn	
Goosefish	Asn				Val				His	
Bullfrog	Asp				Val				Asn	
Snake	Asp/Asn				Val				Tyr	
Fowl	Asp				Val				Ser	
Bovine	Asp				Val				His	
Human, pig, rat, dog, horse	Asp				Ile				His	

tone adjustments required to cope with postural changes. The RAS is activated following standing up from the lying position in order to sustain normal blood pressure.

On a day-to-day basis, adjustments in plasma renin activity in accordance with renal sodium excretion are probably the major influence controlling aldosterone secretion. In the mammalian kidney, a specialized region of the distal convoluted tubule, the macula densa, is considered to be responsible for monitoring the rate of sodium excretion. The epithelial cells of this tubular section are closely applied to the juxtaglomerular cells around the glomerulus of the same nephron. Lowered tubular sodium gives rise to increased renin secretion, and this, through AII formation and aldosterone stimulation, will ultimately lead to a compensatory increase in renal tubular sodium conservation. This complex of juxtaglomerular cells and macula densa is often referred to as the juxtaglomerular apparatus.

AII and other secretagogues for aldosterone (potassium and ACTH) appear to act through a common mechanism in the mammalian zona glomerulosa cells. All contribute to elevation in intracellular cyclic AMP levels and increased activity of cholesterol side-chain cleavage enzymes. Potassium and AII, the more potent secretagogues, specifically activate the 18-hydroxylase enzyme responsible for conversion of corticosterone to aldosterone.

AII appears to influence mammalian renal function additionally through aldosterone-independent mechanisms. The local, very high concentrations of AII generated within the kidney may alter urinary excretion directly through changes in glomerular filtration rate or tubular electrolyte reabsorption and/or indirectly through altered patterns of intrarenal blood flow distribution. Renal function in nonmammalian vertebrates is very sensitive to variations in glomerular filtration rate consequent of changes in systemic blood pressure. In these groups, the wider systemic vascular effects of AII may also impinge on renal excretory patterns.

In recent years it has become clear that the renin–angiotensin system influences water management not only through effects on renal excretion, but also through central dipsogenic actions of AII. Application of AII to the preoptic area of the mammalian brain induces drinking behavior even in water-replete animals. The RAS may be involved in normal thirst mechanisms. It is not clear, however, whether the AII involved enters the brain from the systemic circulation or is produced locally within the brain. All the major components of the RAS have been identified in brain tissue; thus, thirst and related functions may be regulated by this so-called iso–renin–angiotensin system. AII has also been shown to stimulate neurohypophyseal hormone secretion. Thus, the RAS affords a balanced response to dehydration through stimulation of compensatory drinking and of neurohypophyseal hormone-mediated renal water conservation.

We have in the RAS the potential for a control system capable of coordinating the body's regulation of both salt and water economy, an essential requirement to maintain overall body fluid volume and composition. The physiological actions of the RAS, however, range beyond salt and water management to include blood pressure regulation, though of course these physiological phenomena are closely interlinked. In mammals, at least, the exciting concept has emerged that the RAS may play a central integrative role in body fluid homeostasis. This type of concept is at present difficult to uphold in nonmammalian species, as experimental information is rather scanty.

The RAS appears to be present in all jawed vertebrates, though its demonstration in elasmobranchs is rather controversial. Attempts to establish its presence in cyclostomes have proved unsuccessful. From thc few species in which the amino acid sequence of AI has been established (Table IV), it appears that there is some variation in structure between the major groups. Substitutions in the basic molecular structure are restricted to amino acids at positions 1, 5, and 9. It is interesting that variations in positions 1 and 5 are more conservative than at position 9. The former are, of course, more critical to the molecule's biological activity, as they are retained in the AII peptide after converting enzyme cleavage.

The macula densa type of structure, seen closely applied to the juxtaglomerular cells in mammals, is absent in most nonmammalian groups and has only been clearly demonstrated in the mammalian-type nephroi of the mixed bird kidney (see Chapter 10). Absence of this structure in lower vertebrates suggests that renin release in these groups is not closely related to renal sodium excretion. This may have evolved late in the vertebrate series, as the kidney took on a more dominant role in monovalent ion management. What, then, is the function of the RAS in nonmammalian vertebrates? An increase in blood pressure appears to be a universal response to AII in jawed vertebrates, even elasmobranchs. Effects of AII on renal function and drinking behavior have also been demonstrated in teleost fish, reptiles, and birds. AII has also been shown to be steroidogenic in teleosts and amphibians, but perhaps not birds. These observations may, however, reflect the pharmacological actions of high doses of AII, the molecular structure of which may also vary from the experimental animal's native peptide. Demonstration of a biological action of angiotensin does not necessarily endorse its normal physiological function. Nishimura (1980) has speculated that the RAS evolved primarily as a blood pressure-regulating mechanism and that other functions, such as steroidogenesis, appeared secondarily. The influence of AII on mammalian blood pressure may, therefore, represent a relic of the primitive RAS function.

2.5. Putative Endocrine Tissues

Hormones of the pituitary, the adrenal gland, and the RAS consistently feature in the control of vertebrate osmoregulatory mechanisms. However, other tissues that elaborate secretions may play a role in osmoregulatory physiology. Two such tissues in fish are the corpuscles of Stannius and the caudal neurosecretory system. Though their endocrine status is not fully established, they are worthy of brief consideration.

2.5.1. Corpuscles of Stannius (see Chapter 12)

Unique to bony fish are the small "white bodies" associated with the renal mass, the corpuscles of Stannius, initially described by Stannius in 1839. Their number varies considerably; in most teleosts there are two to six, whereas holostean fish like the bowfin, *Amia calva,* may have 40–50. The corpuscle cells have the appearance of peptide-secreting cells and contain secretory granules. The granule content, a possible measure of secretory activity, varies according to the stage of the animal's breeding cycle, life-cycle, and osmotic environment. Although the tissue is considered to have an endocrine role, probably linked to osmoregulation, this has yet to be clearly defined. Among the better supported effects of the tissues is the hypercalcemia (high blood calcium) that results from its removal and a pressor action of extracts which is similar to that for angiotensin. The precise nature of the secretory products of this gland and their relevance to osmoregulatory physiology await definition.

2.5.2. Caudal Neurosecretory System

The urophyseal or caudal neurosecretory system is found in the tail of bony fish, and although it was initially described in the last century (1813), its functional significance remains obscure. Like the more anterior neurohypophysis, the urophysis is composed entirely of neuronal tissue. Nerve cell bodies in the most caudal region of the spinal cord project axons to a vascularized swelling ventral to the terminal vertebra (see Figure 19). This swelling, the urophysis, is essentially a

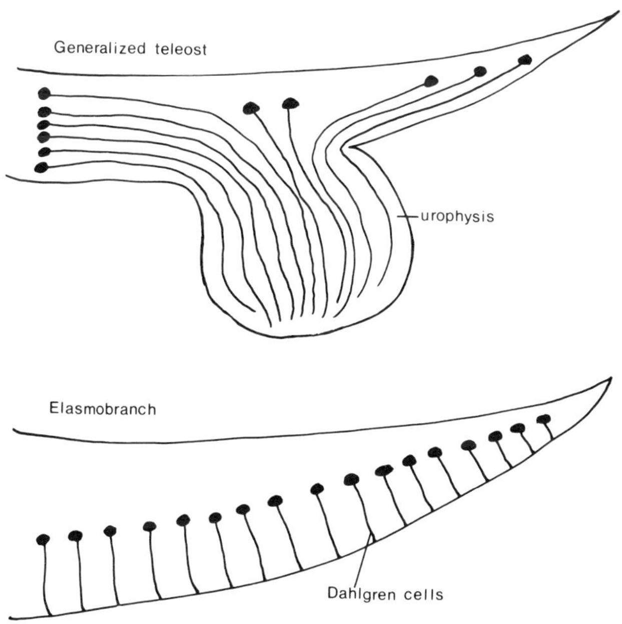

Figure 19. Caudal neurosecretory systems. Idealized structures seen in bony fish (teleosts) and elasmobranchs are shown. The neurosecretory cell bodies in the caudal spinal cord project axons to a well-defined neurohemal organ in bony fish—urophysis.

neurohemal organ enabling the link between neurosecretory nerve terminals and the vascular system. The neurosecretory granule content of the gland varies according to the osmotic concentration of the external environment and the animal's reproductive activity. At least two biologically active peptides have been isolated from the urophysis, urotensin I and II. These peptides, like neurohypophyseal hormones, are linked to a large carrier peptide (urophysin) during axonal transport and granule storage.

Urotensins have a variety of described biological actions, including effects on renal, gut, and gill handling of fluid and electrolyte. However, much further work is required to characterize fully this system and to establish its physiological significance.

The chondrichthyean fish possess a similar caudal neurosecretory system, though it involves fewer, giant neurosecretory cells (Dahlgren cells) scattered through a large portion of the caudal spinal cord. There is no focal neurohemal organ; the axons of Dahlgren cells separately make contact with blood vessels ventral to the spinal cord. Comparable tissues appear to be present in holostean fishes, but not in cyclostomes.

12

Integration of Hormonal Actions to Regulate Extracellular Fluid Volume and Composition

1. Introduction

Chapters 10 and 11 have outlined organ systems concerned with general homeostatic processes and have described the hormones that primarily or secondarily influence the effector mechanisms. It is axiomatic, although often unstated, that comparative physiology aims to develop all-embracing generalizations to describe the regulation and function of particular effectors, notwithstanding perceived differences in environmental adaptations. Thus, a common hormone may be identified and a common effector recognized in disparate species, and the problem is then to arrive at an understanding of the two in the same terms. Sometimes this is possible at molecular and cellular levels of activity, but rarely with respect to the whole animal, since environmental constraints may differ so markedly.

As noted in Chapter 10, the extracellular fluid is regulated in terms of both volume and composition. In the case of the former, water is the primary agent under control, whereas for the latter the position is extremely complex, since so many substances are involved and each may be under almost separate government. For example, the total solute of the extracellular fluid comprises sodium chloride in abundance, yet other substances—quite essential for normal life—are present in relatively trace amounts—calcium, phosphate, and magnesium, for example—but are nevertheless carefully controlled. In addition, it is worth noting again that the regulation of water, sodium chloride, and other substances results from the concerted integration of a whole series of delicately linked processes, some yet to be identified. Moreover, the secretion and actions of a hormone that perhaps overtly regulates water metabolism are significantly governed by appropriate concentrations of ions such as calcium in the extra- and intracellular milieus. This chapter aims to integrate the various controlling factors for water (volume), sodium chloride, and calcium (composition) metabolism. Neurohypophyseal hormones, adrenocorticosteroids, and the renin–angiotensin system are identified as key factors, but they are inextricably linked with a spectrum of other materials, including parathyroid hormone, adenohypophyseal factors (notably prolactin, growth hormone, TSH), calcitonin, thyroid hormones, the vitamin D_3 (cholecalciferol) endocrine system, and, in teleost fishes at least, factor(s) from the corpuscles of Stannius. The material drawn upon includes much current literature, some of which is preliminary or unpublished, so that the chapter contains large speculative elements which will be carefully identified.

2. Calcium

Calcium is the most abundant cation in the vertebrate body, and its concentration in both extra- and intracellular compartments is carefully controlled by many hormones that ultimately have the purpose of retaining body stores or excreting excess to maintain optimal distributions within and between cells. The role of calcium in the body is widespread. The spectrum of its involvement includes cellular secretion (including hormone release), skeletal and smooth muscle contraction, integrity of neural function, blood coagulation, and many membrane and cellular processes (including hormone action). It is in many ways hardly surprising that hormones and calcium are closely linked, the one equally affecting the other. Such generalities have produced notions of ''procalcium'' and ''anticalcium'' hormones, which respectively promote or inhibit calcium entry into the vertebrate body. In this context, it is of considerable interest that the view is currently being promoted of a general disruption of extracellular regulation of calcium associated with tumor development—the so-called ''hypercalcemia of malignancy.''

Procalcium hormones may be seen to be parathyroid hormone, 1,25-dihydroxycholecalciferol, and prolactin; anticalcium hormones include calcitonin, somatostatin, and factor(s) from the corpuscles of Stannius. Clearly, phylogenetic variations are apparent—for example, the lack of parathyroid glands in nontetrapod vertebrates and the uniqueness of the corpuscles of Stannius to teleost fishes—but this general scheme is conceptually very important in understanding calcium homeostasis in vertebrates generally.

As an alkaline metal, earth calcium is found in abundance as deposits as phosphates, carbonates, and sulphates, which are essentially inert biologically in that they are unavailable to both animals and plants. When dissolved in weak acids, ionic forms of calcium appear, and these are of biological significance. Calcium is thus available in varying amounts in the three broad environments of seawater, freshwater, and land: in seawater, calcium is present in abundance, but poorly so in freshwater or on land. In freshwater, the fecundity (= numbers of species and their biomass) of lakes is closely related to the calcium content. Those with a low content harbor few species in small numbers; those with higher amounts support rich varieties of both plant and animal species in large numbers.

Terrestrial vertebrates, like those in freshwaters, may have inconstant supplies of calcium. Dietary calcium, in food and drinking water, varies in amount, and there is some evidence of a specific calcium appetite or drive among a few mammals and birds. Carnivores are adequately supplied from the calcified skeletons of their prey, but herbivores and some omnivores may become calcium-deficient under certain circumstances.

Given the necessity for calcium for normal structure and function in vertebrates, and the variable supply of the material, it is not surprising that some hormones act to limit entry under high calcium regimens, whereas in calcium-deficient environments, others promote its entry and conservation. Once calcium is in the body, the same hormones participate in governing its distribution and release from stores in times of need. One example is the laying bird, in which there is considerable accumulation and storage of calcium in medullary bone prior to the onset of laying eggs, but when eggshells are to be laid down, large quantities are released. A gamut of regulatory factors is involved, and the regulation of calcium must take place against the background of sexual maturation, ovulation, and oviposition.

In blood plasma calcium occurs in three major forms—a free biologically active ion, bound to plasma proteins, and complexed with phosphates, sulfates, and citrates. Outside the extracellular fluid, the bony skeletons, ear ossicles, lime sacs (of amphibians), and cells generally contain large amounts of calcium, whose physical form is carefully monitored and controlled. Within cells themselves, very low concentrations of calcium are present in cytoplasm, with the majority being held in structures such as the mitochondria. The phenomena associated with calcium homeostasis with respect to calcified tissues and skeletal structures are beyond the scope of this chapter, which only considers the regulation of extracellular concentrations. It is, however, pertinent to note that among vertebrates today, some have fully ossified skeletons, others, such as cyclostomes, have at most ossified portions of cartilage, and elasmobranchs have almost wholly cartilaginous skeletons. This may be a secondary phylogenetic acquisition, since certain highly specialized areas within their skeletons resemble typical vertebrate bone. This raises the interesting possibility that, during their evolution, these fishes were equipped with hormonal systems that essentially prevented ossification. Furthermore, among teleost fishes, two types of skeleton have been recognized—the so-called acellular bone and the cellular bone. The former is thought to be the "more advanced." The basic difference between the two types is that acellular bone has compacted areas lacking cells and which presumably do not have the ability to reform and remodel, so not contributing to "exchangeable pools" of calcium within the body. This varying skeletal function impinges on the mechanisms for extracellular regulation of calcium.

Many hormones influence calcium homeostasis. In mammals and probably other tetrapods, three major regulators are viewed as central: parathyroid hormone, vitamin D_3 (cholecalciferol) and its metabolites, and calcitonin. In addition, prolactin, "hypocalcin" (= teleocalcin?), gonadal hormones, adrenal and interrenal corticosteroids, and growth hormone have secondary effects on calcium regulation.

There is a marked lack of uniformity among vertebrates as regards some of these factors and their putative actions. This appears to be related to aspects such as environment, the normal plasma concentrations, the type of respiratory surface, the type of skeleton, the production of shelled eggs, and oogenesis itself. The latter, vitellogenesis, which involves the production of large quantities of calcium-binding protein (vitellogenin, phosvitin), is a key feature under estrogenic control (see Chapter 8).

The general calcium homeostatic mechanisms show broad similarities in mammals, birds, reptiles, and amphibians, with differences usually associated with the formation or otherwise of cleidoic eggs and specialized storage structures such as the vertebral lime sacs of amphibians. The latter show remarkable seasonal cycles with large stores being laid down at one phase and then subsequently released at other times. However, within these types, the maintenance of extracellular calcium concentrations reflects the conjoint activities of parathyroid hormone, active metabolites of cholecalciferol, and, under some circumstances, calcitonin.

Parathyroid glands are found in all tetrapod vertebrates, although, interestingly, they are not found in premetamorphic urodele amphibians, perhaps suggesting that their evolutionary appearance was in some way associated with the colonization of the terrestrial environment with concomitant demands for increased skeletal support. Embryologically, parathyroid glands are derivatives of branchial pouches, although neuroectoderm may also be incorporated. It is thus significant that parathyroid hormone and hypophyseal ACTH share certain amino acid sequences and that, in some species, the adenohypophysis contains parathyroid hormonelike biological and immunological activities.

Parathyroid hormone itself is an 84-amino-acid polypeptide, of which the first 34 units are essential for biological activity—the production of hypercalcemia. The control of the hormone's release seems to be predominantly by plasma calcium concentrations, with a short-loop negative feedback; hypocalcemia induces release, and hypercalcemia inhibits. There is increasingly persuasive evidence that the vitamin D_3 endocrine system also regulates parathyroid function. The effects of the hormone in bone are complex and probably involve actions on osteoclastic activity to give overall increases in bone remodeling, with a release of calcium into the extracellular fluid. Two other actions of parathyroid hormone are of primary physiological significance: the hormone reduces renal calcium excretion while increasing that of sodium, and it also stimulates the 1-hydroxylase for 25-hydroxycholecalciferol to produce the hypercalcemic 1,25-dihydroxycholecalciferol. In one sense, therefore, parathyroid hormone may be viewed as a renotropic hormone, although it may also have direct actions on intestinal calcium absorption. A final, and only recently suggested, action, although not verified, is the hypotensive effects of parathyroid hormone. Such actions are of great physiological significance, since calcium is becoming increasingly implicated in the etiology of hypertension in man.

The vitamin D_3 endocrine system is a relatively recent concept in the vertebrate endocrine repertoire and has yet to be fully delineated. The basic components are outlined in Figure 1. Briefly, vitamin D_3 may be obtained in the diet or may be

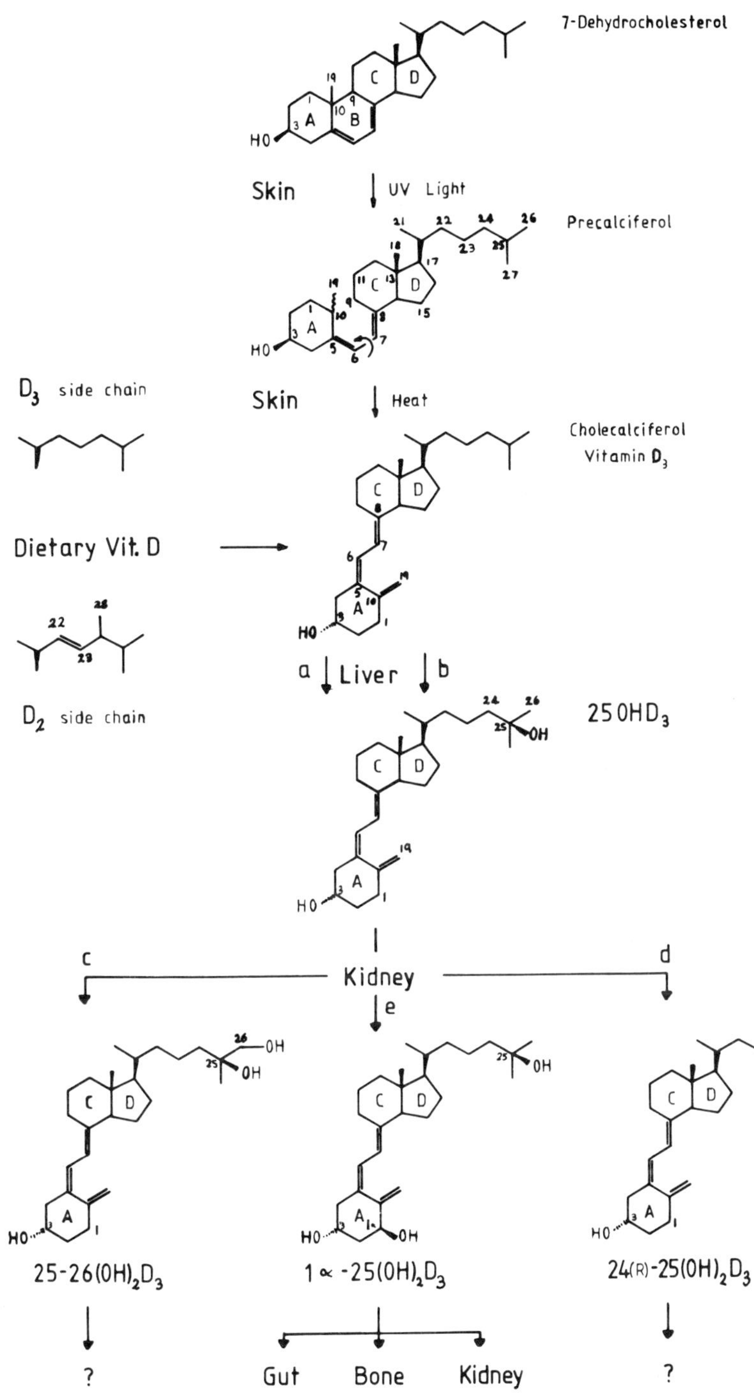

Figure 1. Outline of vitamin D_3 metabolism in mammals and birds. (a) Microsomal vitamin D_3 25-hydroxylase, (b) mitochondrial vitamin D_3 25-hydroxylase, (c) ? 25-hydroxy D_3 26-hydroxylase, (d) mitochondrial 25-hydroxy D_3 24-hydroxylase, (e) mitochondrial 25-hydroxy D_3, 1α-hydroxylase.

produced, at least in mammals and birds, in the skin, with ultraviolet (UV) radiation and heat producing the necessary cleavage and isomerization. Anecdotal evidence suggests that 7-dehydrocholesterol may be secreted onto fur. UV radiation would give the hormone precursor, ingested by licking. This might apply to birds when the feathers are preened, though bare areas of skin—legs, for example—are also rich in 7-dehydrocholesterol.

Once in the blood, transported there by specific intestinal binding proteins, probably synthesized under the control of parathyroid hormone, cholecalciferol undergoes a well-ordered sequence of hydroxylations to biologically active metabolites or metabolites with as yet undetermined actions. The liver is the primary site for 25-hydroxylation, and the kidney is the major one for 1-hydroxylation.

1,25-Dihydroxycholecalciferol is the metabolite well documented for activities in calcium homeostasis. Its production is influenced by the availability of vitamin D_3, either preformed or produced *de novo* by the organism via calcium status, parathyroid hormone, and phosphate concentrations. It also appears that during periods of "calcium stress"—growth, reproduction, lactation, or laying of shelled eggs, for example—prolactin, cortisol, and growth hormone may stimulate 1,25-dihydroxycholecalciferol production by acting directly to stimulate the renal 1-hydroxylase.

Bone resorption, muscle function, intestinal calcium absorption, the reticuloendothelial system, and renal excretory patterns of calcium are all influenced by vitamin D_3 metabolites, whereas specific receptors for the secosteroids are present in the hypophysis, pancreas, and placenta. Some of the actions are overtly associated with calcium homeostasis as such; of the others, some may be related to the key role that calcium plays in regulating so many bodily functions.

Little is known about the physiology of the vitamin D_3 metabolites in vertebrates other than mammals and birds, and there are many enigmatic aspects when fishes are examined. For example, teleost fishes have large stores of vitamin D_3 in their livers (hence the long-known efficacy of fish liver oils in treating selected bone disorders). It is of interest, but as yet unexplained, that marine teleosts contain greater hepatic stores of vitamin D_3 than freshwater species. There is still some doubt whether these fishes, or indeed elasmobranchs and cyclostomes, contain the full complement of hydroxylating systems, and certainly the physiological role of vitamin D_3 is unknown other than as a crude pharmacological extrapolation.

The third apparent calcium-regulating hormone—calcitonin—has a much more widespread occurrence among vertebrates, having been described either immunologically or by bioassay in all groups except cyclostomes. The classical action of calcitonin is its hypocalcemic activity, although there is an element of doubt about this being its general and genuine *physiological* action. In nonmammalian vertebrates, the source of calcitonin is the ultimobranchial bodies and, in mammals, the homologous C cells embedded in the thyroid gland. The secretory cells are derived from the sixth branchial pouch and associated neuroectodermal amine precursor uptake and deamination (APUD) cells. The 32-amino-acid sequence of calcitonin has been elucidated from a number of species, including salmon, eel, pig, sheep, and man. The hypocalcemic activities of calcitonins are greatly affected by single amino acid substitutions, and it is known, for example, that in mammalian bioassays, including clinical studies, salmon calcitonin is much more active than the homologous hormone.

Like parathyroid hormone, calcitonin secretion is regulated by a short-loop negative feedback system with plasma calcium concentrations, such that hypercalcemia increases, and hypocalcemia inhibits, secretion. Many other factors influence the release of calcitonin, including gastrin, progesterone, and glucagon, but the physiological relevance of these observations remains unestablished.

The hypocalcemic actions of calcitonin are incompletely understood and often controversial. However, it appears that effects on bone resorption, with increased mineral accretion, are major components. Renal effects, wherein the kidney contains specific calcitonin receptors, are probably also important, and actions on ileal calcium absorption may also be involved in the overall response. There are, however, a number of inconsistent observations. In particular, changing calcium regimes do not always alter calcitonin secretion, and it appears that, in mammals, the young are much more sensitive to the hypocalcemic actions than the adult. It has thus been suggested that actions of calcitonin on calcium homeostasis are only relevant to the physiology of the sexually immature animal.

In teleost and elasmobranch fishes, both containing large ultimobranchial stores of calcitonin, functional roles for the hormone have not been definitively demonstrated, although current evidence suggests an association with reproduction and, in euryhaline teleosts, with anadromous migration. Hypocalcemic actions of calcitonin in amphibians and birds have been demonstrated and influence lime sac calcium uptake and preovulatory calcium retention, respectively.

The three hormones thus far examined with respect to calcium homeostasis—parathyroid hormone, calcitonin, and vitamin D_3 metabolites—plausibly explain, albeit only in general and imprecise terms, the maintenance of plasma calcium concentrations. Antagonism, agonism, and synergism at sites from bone to the intestine and kidney, as well as other exchange surfaces, reasonably account for the regulation of the output, uptake, retention, and elimination of calcium. However, it must be emphasized that the exact interrelationships are far from being comprehensively understood, and many other hormones must be interpolated. Pituitary factor(s) especially may be involved. Prolactin, for example, is hypercalcemic in most vertebrate types examined, and in some cases this is a direct effect of the hormone, whereas in others, secondary actions consequent on the stimulation of 1,25-dihydroxycholecalciferol production are apparent.

Growth hormone, by affecting skeletal growth, intestinal absorption of calcium, and 1,25-dihydroxycholecalciferol production, corticosteroids, by affecting intestinal calcium absorption and vitamin D_3 turnover, and estrogens, with effects of the production of calcium-binding proteins as well as on the vitamin D_3 endocrine system, are additional hormones that directly impinge on calcium homeostasis.

Finally, the corpuscles of Stannius, which were referred to in Chapter 11, apparently have hypocalcemic activity, since their surgical removal in some species produces a significant hypercalcemia. The specific nature of the calcium response is in some doubt, however, since other homeostatic indices, such as sodium, potassium, and chloride, are also affected. Cell types vary within the corpuscles, according to environmental calcium concentrations, and their activities are said to result in the production of agents that influence skeletal metabolism, calcium uptake mechanisms in the gill, and possibly intestinal calcium transport. In addition, a relationship with the onset of sexual maturity has been suggested. The nature of the secretory products is not known with certainty, but three names have been coined, "teleocalcin," "hypocalcin," and "parathrin," and a further material resembling an angiotensin has also been suggested. Which of these isolated materials—some characterized either in terms of biological activity, immunological reactivity, or chemically—is responsible for physiological actions is not known.

3. Water

Water is the major constituent of life, comprising up to 85% of the body weight of vertebrates. It is, in effect, the solvent for most biological reactions and provides the environment for the vast majority of species. Since, for each vertebrate, more than 70 g of each 100 g body weight is water, it is inevitable that intake and output is regulated by mechanisms that are precisely controlled.

Gain of water may take place as a result of active, neuroendocrine measures or as a result of passive, osmotic flow of water from freshwater across branchial epithelia or integuments. Similar or equivalent "active" and "passive" mechanisms govern water losses, and to maintain total body water constant, sensors or receptors must be responsive to deviations from optimum and provoke appropriate measures at the effector sites to balance gains against losses. The afferent and efferent pathways of the feedback systems to monitor and, if necessary, correct alterations in body water may be broadly defined in terms of intracellular content and extracellular volume. Central nervous and peripheral receptors detecting what has been termed cellular "swelling" or "shrinkage" and volume are involved, and their stimulation (or inhibition) sets in train a cascade of responses involving neural, neuroendocrine, behavioral, and effector systems themselves.

Given the relative uniformity of body water content, it is clear that the primary environments—seawater, freshwater, and land—impose differing demands on the homeostatic processes. In seawater and under most terrestrial conditions, the purely passive gradients between the animal and its environment are such that acquisition and retention of water are principal activities; in other words, the vertebrates are likely to dehydrate unless specific measures are undertaken. In freshwater environments, largely opposite features are presented; mechanisms have developed to eliminate water and to minimize its passive gain.

Vertebrates may gain or lose water passively, and the major sites involved are the integument and respiratory epithelia (both gills and lungs). The sites where water loss or gain is demonstrably under active control are renal and gastrointestinal. In addition, specialized secretions from "salt glands," sweat, sebaceous, lacrimal, and mammary glands all contain significant amounts of water of relevance to the overall economy of the organism. A final general feature of water management is the association that apparently exists between water availability and the metabolism of nitrogen-containing compounds, in particular whether the major end products are ammonia, urea, or uric acid. It is necessary to emphasize that to eliminate 1 g of nitrogen as ammonia requires several-hundred-fold more water than if the end product is uric acid or urea.

The manipulation of patterns of nitrogenous waste excretion to adjust to particular water regimes has a very ancient heritage, and there have almost certainly been convergent mechanisms to solve equivalent problems. The classical "physiological uremia" of marine elasmobranchs has its counterparts in certain amphibia, both adult and premetamorphic, and even within the mammalian kidney (see Chapter 10). It may be that the underlying metabolic mechanisms are present in all vertebrates but are latent until water shortage is extreme, when neuroendocrine influences induce their appearance. Before specific examples are discussed, it is important to contemplate the phenomena wherein large quantities of urea are present in individual tissues (e.g., mammalian kidney) or throughout the body (amphibia and elasmobranchs), because it is a relatively toxic material as a potent protein-denaturing agent. Certainly, the concentrations of urea that exist, for example, in elasmobranch plasma interfere with the structure and function of nonelasmobranch proteins and enzymes; indeed, some elasmobranch proteins display urea tolerance whereas others do not. The problem of how they function *in vivo* was uncertain until the methylamine compounds, in particular trimethylamine oxide, were found to act as "biochemical antagonists" to the disruptive actions of urea. *In vitro* there appears to be an optimal trimethylamine-oxide-to-urea ratio of approximately 1:2 to achieve protection of proteins against denaturation by urea. Such ratios are observed in the intracellular compartment of some elasmobranch species. Mechanisms must exist in other vertebrates that contain especially high either systemic or local concentrations of urea, but they have not been described.

Received evolutionary theory states that elasmobranch fishes adopted iso-osmocity between their extracellular fluids and the environmental seawater, whereas teleosts regulated their extracellular fluid osmolarity against large osmotic gradients in both seawater and freshwater environments. In the case of elasmobranchs, there is essentially an equilibrium for water between the animal and its environment so that "passive" fluxes will be minimal. However, the fish maintain total water volumes, and losses from the kidney and from the gut must be compensated for, either by drinking or by altering the permeability of the gills. The endocrine mechanisms are not known, and the most that can be said is that renal water output and branchial water fluxes are altered by hypophysectomy. In the few studies carried out, the usual dipsogenic stimuli for most other vertebrates, such as angiotensin II (see below), do not affect drinking in elasmobranch fishes. A final difference that appears to exist between elasmobranchs and teleosts is that the renal tubular permeability to water may be the functional site for modifying the output of fluid, since, under a wide range of urine flows, glomerular filtration rates are remarkably unchanged; this is reminiscent of the functional state of the mammalian kidney.

To summarize water homeostasis in elasmobranchs: it appears that these animals minimize overall fluxes across the integument and gills, and volumes are regulated, as in mammals, by differential renal tubular water reabsorption. The roles of the idiosyncratic spectrum of interrenal and neurohypophyseal hormones are largely unexplored in this context.

Amphibia can be conceived as spanning the phylogenetic step between purely aquatic vertebrates and their terrestrial successors. The contemporary groups contain those that are indeed purely aquatic (the urodeles) and those that are terrestrial as adults (the apodans) and those that essentially dwell either in water or on land (the anurans). The larval stages are all aquatic, and it is during metamorphosis that one sees the physiological, biochemical, and morphological changes that might be perceived as mirroring those changes that were prerequisite when vertebrates emerged to colonize land. Two examples are apt: First, during ontogenesis, anuran tadpoles make the biochemical switch from ammonotelism to ureotelism, whereas when some anuran amphibians are placed in environments hyperosmotic to their extracellular fluid, or when chronically dehydrated, they too turn to ureotelism. There is one species of anuran that, in times of water shortage, becomes uricotelic. The underlying biochemical aspects of these physiologically apt changes are incompletely understood, although the adrenocortical steroids appear to influence the hepatic and possibly other enzymes involved in the sequential degradation/synthesis of nitrogen-containing substances and their end products for elimination. It is also likely that the interplay between thyroid hormones and prolactin is a vital component in some of these changes in the light of their known and well-established actions in metamorphosis, in which prolactin is considered to promote larval characteristics and thryoid hormones adult.

One marine species of frog—the crab-eating *Rana cancrivora*—has given rise to speculation concerning the convergence of the ''uremic adaptation'' between this species and elasmobranch fishes. The endocrine repertoire of *Rana cancrivora* is, however, that of a typical amphibian, although the precise actions of the hormones with regard to regulating fluid balance are poorly understood. The existence of this species provides a platform from which to embark on a more general discussion of the contrasting pressures on water economy of marine, freshwater, and terrestrial vertebrates and the general regulatory mechanisms. The latter, in many, involve adrenocortical and neurohypophyseal hormones, prolactin, and the renin–angiotensin system.

Some vertebrates—the elasmobranch fishes and *Rana cancrivora*—have mechanisms that minimize water gradients, whereas others acquired systems to eliminate an excess of osmotically free water (in freshwater) or actively to imbibe fluid to gain osmotically free water (in the sea or on land) by developing dipsogenic ''drives'' which are under very positive neuroendocrine control. For dipsogenic activity, the renin–angiotensin system is of central importance, although a notable exception seems to be the anuran amphibians, which fail to respond to angiotensin injections with drinking responses. In other terrestrial vertebrates, there is evidence that the sensitivity of the dipsogenic response to angiotensin varies with the ecology of the species; those which live in environments with scarce supplies of water are extremely sensitive, whereas those rich in water are relatively insensitive.

A second hormone that is involved with water balance in not altogether clear ways is prolactin or its homolog. Classically, prolactin is the "water drive" material in animals such as the red eft, and it may be that, in these animals, prolactin takes on the role of angiotensin, and it is perhaps significant that recent studies suggest that angiotensin may be a regulator of prolactin secretion, at least in mammals.

Having elicited the drive to take in water, either by drinking or passively by uptake across integumentary–environmental gradients, the next phase of water regulation is its retention and distribution, followed by feedback modification of the acquisitional phases. Two principal hormonal systems are involved: the neurohypophyseal, antidiuretic peptides (arginine vasotocin in all vertebrates except mammals and elasmobranchs) and the adrenocortical steroids. It is likely that these two factors regulate *inter alia* the fractional distributions of water across all cellular membranes and either primarily or secondarily affect the relative amounts of water that leave the animals, either via the kidney or passively across extrarenal surfaces. By definition, antidiuretic hormones reduce water excretion, and it is hardly surprising, therefore, that there are feedback relationships between dipsogenic actions of angiotensin and the secretory patterns of antidiuretic hormone. The usually accepted scheme is that angiotensin promotes water acquisition, whereas antidiuretic hormone promotes retention of the acquired water. Once appropriate amounts of water have been retained under the influence of antidiuretic hormone, this same hormone inhibits the release of renin to temper water drive and drinking. Equally, angiotensin stimulates the release of antidiuretic hormone.

The ability of angiotensin to stimulate adrenocortical function suggests that the corticosteroids act on water metabolism, but there may be more obscure relationships in the evolutionary sense. For example, there is a very close association between sodium chloride retention and elimination and thirst, but it is very difficult to separate the involvement of angiotensins. Moreover, it is well known that the ability of the mammalian kidney to concentrate urine, in particular the essential "diluting" steps in the distal convolutions, depends on corticosteroids. It is thus self-evident that the actions of antidiuretic hormone in retaining osmotically free water depend on corticosteroids, which themselves apparently maintain renal tubular water impermeability in certain regions and, at the same time, promote solute removal.

4. Sodium

Sodium, as the major cation circulating in blood, makes a major contribution to the osmolality of extracellular fluid, and in addition, there are major reserves of this ion in tissues such as bone. It is hardly surprising, therefore, that the regulation of the intake, elimination, and, importantly, its distribution within the milieu intérieur come under close neuroendocrine scrutiny. In addition, there is now strong evidence that many, if not all, vertebrates possess exquisitely sensitive monitoring systems for assaying total body sodium, and that there are specific coded appetites for this substance. A now classic example is that of sheep, in which it is possible to deplete an individual of accurately known amounts of sodium by implanting a catheter into the parotid salivary gland and allowing loss of saliva, which contains large amounts of sodium. Such an animal will, when presented with a solution of sodium bicarbonate, drink, within 5 min, an amount almost identical to the imposed deficit, and this before the substance would have been completely absorbed from the gut! Other familiar examples of sodium sensitivity include the "salt sense" of migratory teleost fish, the migration of many mammals to "salt licks," and the so-called "salinity preferences" of poorly euryhaline brackish water teleosts.

Among vertebrates generally there is a continuum of sodium availability with respect to environmental sodium and the regulation of a relatively constant total body sodium content. In other words, there are environments that are especially rich in sodium, either by virtue of the environment itself (seawater for example) or because the species consumes large amounts of sodium in the diet. Other environments have little sodium; freshwater is an obvious example. Those species which eat food scarce in sodium (many herbivorous species) provide other examples.

There is considerable controversy, especially within the medical literature, as to whether any one *individual* of a species has a particular "setting" for its total body sodium, but it is axiomatic that each species will have a set point, with perhaps an error of 5% and no more. The important question to which this section addresses itself is whether one can identify hormones and hormonal systems throughout the vertebrate series that regulate this setting of total body sodium, the appetite for this ion, and its elimination.

Within mammals, the key hormones and hormonal systems that primarily or secondarily influence one or all of these phenomena are unquestionably the adrenocortical steroids (in particular, aldosterone), the renin-angiotensin system, antidiuretic hormone, and putative natriuretic materials. The latter speculatively have their origins within the hypothalamus, the kidney itself, and, on recent evidence, the atria of the heart. Based on some limited clinical and laboratory studies, thyroid hormones, prolactin, and parathyroid hormone have also been suggested as participating in this overall sodium homeostasis.

Aldosterone, by virtue of its antinatriuretic action, produces sodium retention, but at the same time, to exaggerate the sodium retention, it may institute a sodium drive; aldosterone, of course, is controlled by the renin–angiotensin system, so that angiotensin has been invoked as a secondary influence on sodium acquisition, although a number of studies suggest that there are primary CNS actions on sodium appetite. The role of antidiuretic hormone is probably secondary with respect to sodium homeostasis, in that actions on water primarily influence the concentrations of sodium within the extracellular fluid, and in addition, it may physiologically inhibit renin release, with consequences on both adrenocortical function and renal excretion.

Such a general, relatively coherent scheme cannot, however, be applied to nonmammalian vertebrates. In teleost fishes, for example, it appears that the renin–angiotensin system is more active in sodium-rich environments (the sea) than in freshwater; it has been suggested that the stimulation of angiotensin in the hyperosmotic seawater produces the necessary relative polydipsia and there is stimulation of hypothetical sodium excretory mechanisms (corticosteroids + others?). In anuran amphibians, there are also enigmatic observations in that sodium depletion of frogs fails to alter plasma renin significantly, although there is strong stimulation of aldosterone production. The position is further complicated by the fact that, in these animals, aldosterone does not produce clear-cut actions on renal tubular sodium reabsorption; rather, its primary sodium-retentive actions are on the integument, the colon, and the urinary bladder. Indeed, under circumstances by stimulating extrarenal sodium accumulation, the steroid perhaps secondarily will induce a renal naturiuresis.

The presence of "salt glands" in elasmobranch fishes, reptiles, and birds and the gills of marine teleosts—clearly not homologous in morphological terms—provide clues to understanding the evolution of the endocrine control of sodium appetites, aversions, and general sodium homeostasis. In a sense they originally provided a simple direct route for sodium elimination, unencumbered by the many functions of the kidney in terms of its overall contribution to individual solutes such as calcium and phosphate and to extracellular volume regulation. Thus particular hormones and their receptors evolved within these structures to allow sodium excretion *per se,* so that renal homeostatic mechanisms, themselves perhaps governed by the same hormones with their receptors sited on the vasculature, were free to undertake nonsodium regulatory processes.

5. General Conclusions

This chapter has dealt with three elements of fluid and solute homeostasis—calcium, sodium, and water. These have been considered rather superficially and separately, and clearly this is not the way in which homeostasis functions. Figure 2 attempts to place some of the hormones mentioned into a general scheme such that directional changes in the particular hormone are indicated in relation to environmental and dietary availability of water, calcium, and sodium. It should be emphasized that much of the scheme is derived from the mammalian literature, and it must be borne in mind that, for example, water deprivation may be associated with concurrent sodium deprivation in some animals, whereas in others incipient sodium deficits accompany an excess of either dietary or environmental water supply.

However, it is pertinent to note that the selection of sodium, water, and calcium for particular discussion and speculation is not random. The importance of these three elements in terms of human physiology and pathology may be briefly discussed. There are certainly many reasons to suppose that, for example, cardiovascular diseases, hypertension in particular, are associated with abnormalities in each or all of sodium, calcium, and water, and often this is associated with

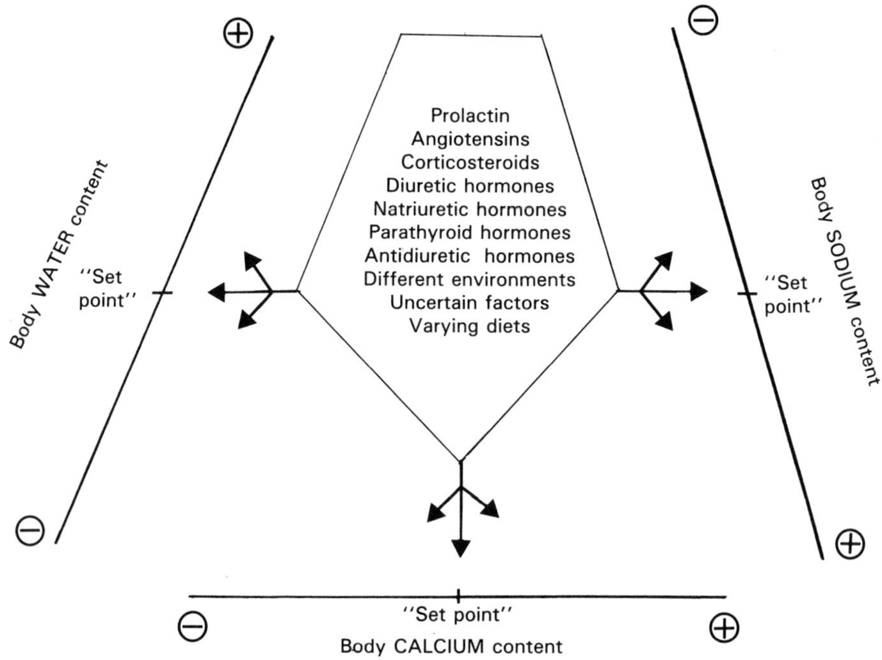

Figure 2. Homeostatic factors in maintaining balance of body water, sodium, and calcium content at "set points" for any one species, each with varying diets and environments. ⊕ Gains, ⊖ losses.

inappropriate profiles of the hormones outlined in Figure 2. General mechanisms of hypertension may be briefly outlined: this condition, in which blood pressure rests above a stated "norm," may result from an increased volume contained within the cardiovasculature and/or (usually "or") an increased resistance within the vasculature itself.

Sodium has a long, if somewhat controversial, history of being involved in some way with the development of hypertension, and a failure of the kidney to excrete an excess of volume (in extreme cases, anuria or renal failure) will clearly increase blood volume and thence pressure. Two endocrine consequences of an excess of sodium intake alongside compromised renal excretory function with renal ischemia are elevated antidiuretic hormone and elevated angiotensin; both these agents are potently hypertensive.

A further complication when considering the putatively hypertensive action of high dietary intake is that some studies have shown calcium to be protective, whereas others have shown that calcium antagonists relieve the hypertension. Moreover, currently there is much interest in the possible hypotensive activity of parathyroid hormone and in the vascular actions of calcitonin-related peptides. Finally, there have been studies that sought to reconcile the calcium–sodium interactions by attempting to describe an optimal calcium:sodium ratio that governed overall blood pressure. This must be in some way related to the setting of total body calcium and total body sodium. Thus, on the scheme shown, one can imagine a series of lines drawn between each of sodium, calcium, and water that give the particular individual complete homeostasis with regard to the cardiovascular system. It requires little imagination to see the complexity of the homeostatic process when one considers just three variables (calcium, sodium, and water balance) in relation to just six hormones and their consequent effects on one homeostatic function, that of arterial blood pressure! (Figure 2.)

13

A Phylogenetic Perspective

1. Introduction

There is considerable variety in vertebrate osmoregulatory mechanisms and in their mode of endocrine control. Clearly, such variation between major groups of animals reflects specialization and adaptation to particular environmental niches. However, it is possible to look beyond these features and discern some progressive steps in the evolutionary development of vertebrate osmoregulatory physiology. This is the task we set ourselves in this chapter.

Inevitably, when assessing the physiology of extant species we are assuming that basic features common to a particular group are truly more or less well developed. This, of course, may be a dangerous assumption and probably oversimplifies the problem. One inevitably has to categorize features as either being basic to the group or representing a particular adaptation and specialization of modern-day members. For example, the low number (or absence) of glomeruli in the kidney of marine bony fish is regarded as a secondary development in this line. The kidney of modern freshwater bony fish is considered to be more closely akin to the renal apparatus possessed by the ancestral freshwater fish stock that gave rise to early amphibians. Clearly, in making such distinctions, one relies on not only physiological, but anatomical and fossil evidence. The reader should be aware of the basis of these assumptions and generalizations and be particularly aware that the problem is exacerbated by the paucity of physiological studies in nonmammalian groups. Despite these shortcomings, a phylogenetic perspective is valuable, especially in the understanding of many aspects of osmoregulatory physiology in advanced vertebrates, including man.

2. The Renal Apparatus

The kidney plays a central role in the management of body fluid volume and divalent ion (Mg and SO_4) content throughout the vertebrate series, perhaps reflecting its primitive function in the earliest ancestors. Management of the major monovalent ions (Na, K, and Cl) is, by contrast, largely performed by extrarenal organs (gills, rectal gland, skin, bladder, salt glands) in lower vertebrate groups. Only in the Amniota (reptiles, birds, and mammals) does the kidney assume a more prominent role in this latter, vital regulatory function. Following the presumed lines of vertebrate evolution (Figure 1), it is possible to trace a gradual elaboration of renal function and a diminution of reliance on extrarenal devices. A linking thread in the evolutionary development of vertebrate osmoregulatory mechanisms is, therefore, the development of the renal apparatus.

Our analysis of the vertebrate subphylum must begin with a consideration of

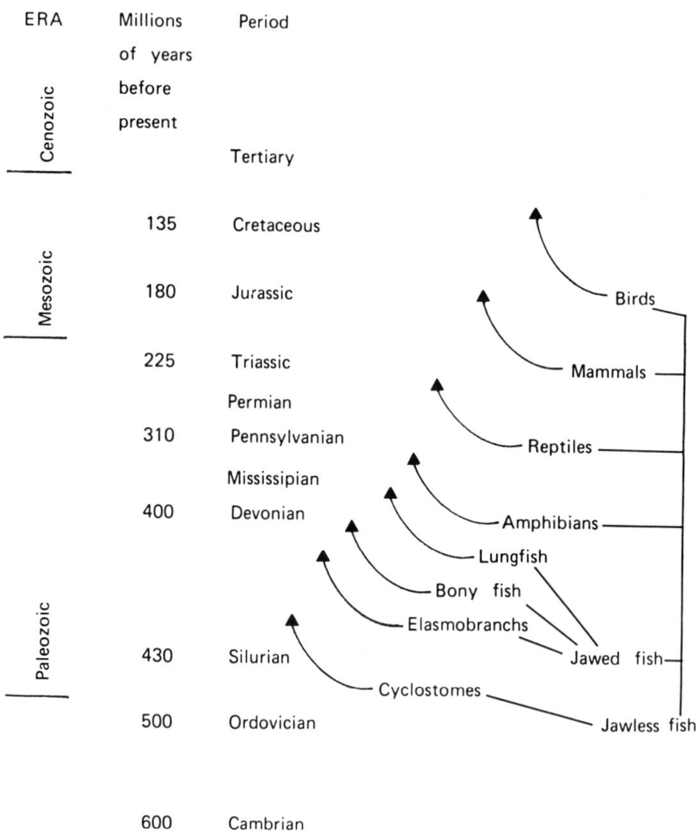

Figure 1. Initial appearance of the major vertebrate groups.

ancestral osmoregulatory problems and how these might have been overcome. Undoubtedly, subsequent developments in later vertebrates have been greatly influenced by the osmoregulatory apparatus inherited from these earliest forms.

Earliest vertebrates, the jawless agnathans, are represented today by two groups, the myxinids (hagfish) and lampreys. Lampreys exhibit a highly specialized life-cycle, migrating between the sea and freshwater. The more simplistic osmoregulatory physiology of the hagfish, which are exclusively marine, is probably closer to that of the protovertebrates. The hagfish contrast with other vertebrates in maintaining their body fluids isoosmotic with seawater.

The first agnathans presumably shared many characters with contemporary chordates (hemi-, uro-, and cephalochordates), the extant representatives of which are all marine and are also isoosmotic with seawater. Indeed, a marine origin of the vertebrates is suggested by the earliest fossil remains, which date from the late Cambrian and Ordovician periods (500–600 million years ago). These fragmentary remains of ostracoderms were found on the east coast of the Baltic Sea and in a large deposit that includes the Harding formation, extending from Arkansas to Montana. X-ray diffraction, boron analysis, and stratigraphic techniques have led to the conclusion that these earliest fish died and were deposited in a marine environment.

Hagfish osmoregulatory mechanisms appear to be restricted to a simple, segmentally arranged, glomerular kidney and the secretory activity of integumental mucous glands. These enable management of specific ion levels and allow small adjustments to body fluid volume. If this type of osmoregulatory physiology is similar to that of the earliest vertebrates, why should the vital development of the glomerular kidney have occurred in the marine environment? Homer Smith, who has molded much of our present-day understanding of renal physiology, pointed out that the glomerular kidney is an excellent device, through filtration and solute reabsorption, for the excretion of water. Such a device is essential for life to be sustained in a freshwater environment. Smith suggested that the ubiquitous presence of the glomerular kidney in vertebrates provided strong physiological evidence for a freshwater origin of the subphylum. Smith could also point to the tendency for glomeruli size and number to be reduced in modern marine fish. In support of this concept, the paleontological evidence at the time (1930s) also pointed toward a freshwater origin for vertebrates. Although current interpretation of the fossil record places vertebrate origins back in the sea, this still leaves the question of why the glomerular kidney did develop in the marine environment. Clearly, its inheritance in subsequent groups was vital as a preadaptation to movement into freshwater.

The hypothetical ancestral form of nephron consisted of a coelomoduct which drained coelomic fluid. A tubular component allowed modification of the excreted fluid through reabsorption and secretion. This type of excretory system is common to many invertebrate groups. The insertion of a tuft of capillaries (glomerulus) into the head of the coelomoduct (Figure 2) might then be considered to provide the basis of the vertebrate nephron. Such a structure is similar to the nephron of larval hagfish, though adult myxinids possess a more usual glomerular structure and simple tubule. What advantage did this glomerular apparatus convey over the sim-

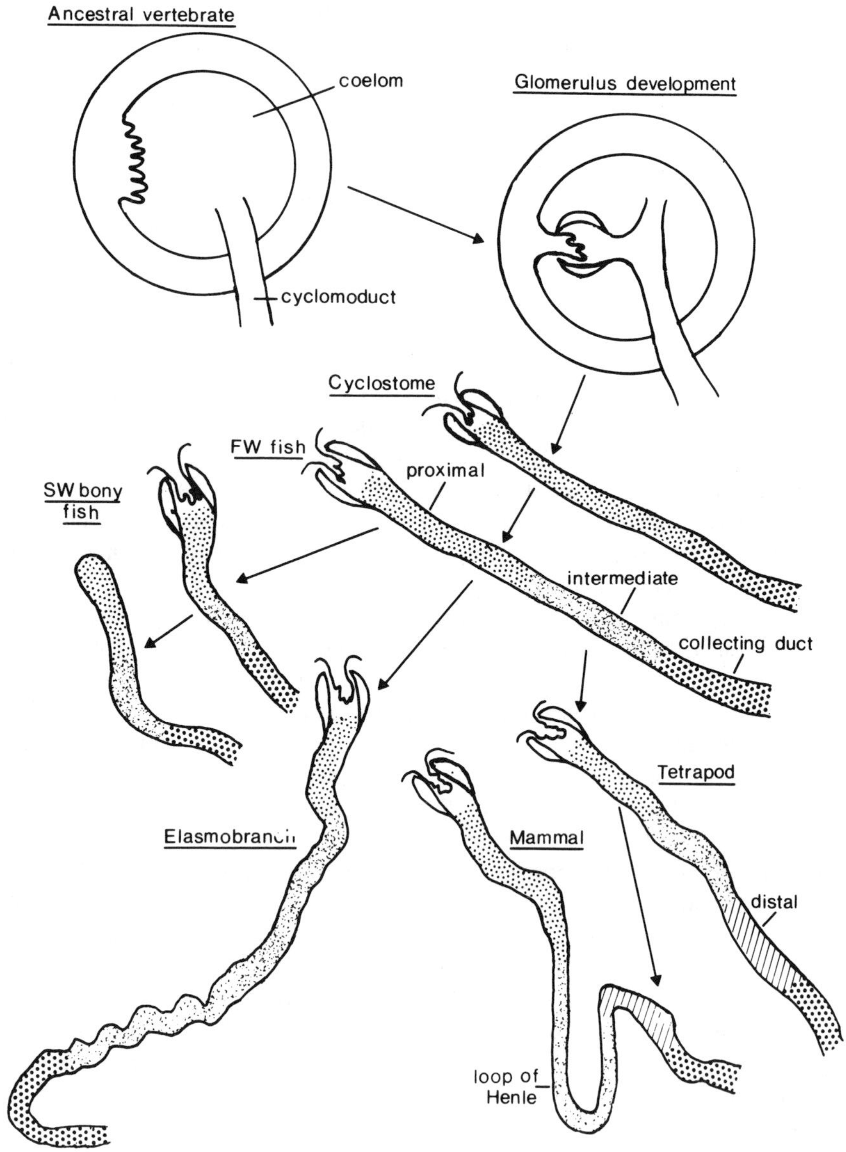

Figure 2. Evolutionary development of the vertebrate nephron. The idealized scheme indicates the major changes in glomerular and tubular components through the vertebrate series. SW, seawater; FW, freshwater.

ple excretory duct for animals living in a stable marine environment? Why should this marvelous preadaptation to life in freshwater have emerged in marine animals?

Examination of advanced invertebrate osmoregulatory systems may provide some insight into these questions. The "kidney" of decapod Crustacea, though structurally different, functions in much the same way as the vertebrate nephron, that is, as a filtration, reabsorption, and secretion system. The adaptive advantage of this type of system is considered to arise from the increased rate of body fluid turnover that it affords, enabling a more rapid and effective management of body fluid composition than simple excretory ducts. The renal apparatus of Crustacea has emerged in a line of animals that are exclusively marine. Those species of Crustacea which have entered estuaries or freshwater have exploited this device, by the addition of major solute reabsorptive segments, for the excretion of osmotically accumulated water. In marine forms, like *Cancer* and *Carcinus,* the main function of the kidney appears to be ion excretion (largely divalent) and adjustment of body fluid volume. This, of course, is comparable to the observed renal function of myxinids. If one may use the Crustacea as an evolutionary precedent, it is possible to propose that the vertebrate glomerular kidney developed primitively as an organ for metabolic waste excretion and for divalent ion management. The kidney has secondarily assumed a prominent role in body water economy in the subsequent groups that entered estuaries and freshwater. Indeed, it has been argued that, like the Crustacea, without prior development of the glomerular kidney vertebrates would have been unable to colonize freshwater and to move eventually onto dry land.

3. The Radiation of Vertebrates

The kidney of the earliest fish moving into freshwater had a capacity for divalent ion excretion and a limited capacity for fluid elimination. A large increase in nephron number and the addition of tubule segments (proximal) for bulk reabsorption of filtered solute produced the renal mechanism for water excretion so essential in a hypoosmotic environment. This initial stock of fish in freshwater, of course, gave rise to all modern fish groups (including marine forms) and to the terrestrial tetrapods (see Figure 3). It is noteworthy, therefore, that the general features of these freshwater ancestors, including body fluid composition profile and renal apparatus (large nephron number, tubule segments for bulk solute reabsorption, and divalent ion secretion), have been inherited by all subsequent groups in the vertebrate series.

The basic vertebrate nephron structure has proved very plastic to evolutionary change, enabling the contrasting renal function exhibited by marine, freshwater, and terrestrial species. The return of fish to seawater is associated with a reduction in glomerular filtration capacity (glomerulus size and number) in hypoosmotic species (teleosts) and with the development of a large intermediate tubular segment for exhaustive urea reabsorption in the hyperosmoregulatory elasmobranchs. Movement to land, and consequent pressure for water conservation, is coincident with the

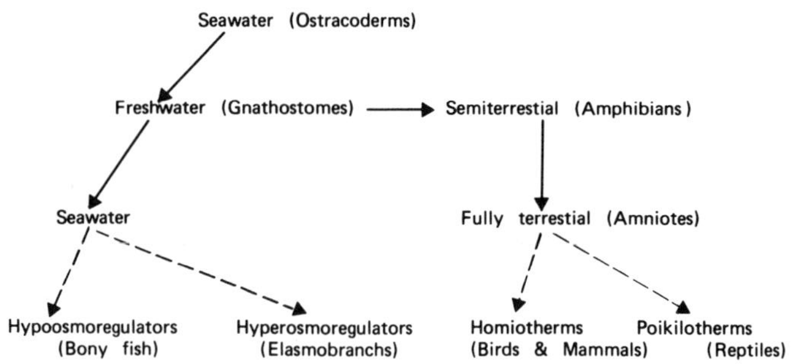

Figure 3. Major changes in vertebrate osmoregulatory physiology.

appearance of an ability for the most distal portions of the nephron to adjust monovalent ion and water content of excreted urine. Although these features are poorly developed in the Amphibia, they make a more significant contribution to the salt and water economy of amniote (reptiles, birds, and mammals) vertebrates. This depends on the lower urine volumes in these groups following the adoption of uricotelism (reptiles and birds) and/or the development of the ability to produce a urine hypertonic to plasma (birds and mammals). In birds and mammals, the kidney permits monovalent salt excretion without excessive depletion of body water and, when necessary, can allow water excretion in dilute urine without undue loss of important electrolyte. Thus, a degree of separate management of salt and water is afforded by the kidney in these groups. The greatly elevated blood pressure and circulation rate associated with higher metabolic rates in these homeothermic groups further enhance the effectiveness of the renal apparatus in osmoregulatory management. The very high rate of blood filtration, tubular reabsorption, and secretion ensures rapid adjustments to water and electrolyte excretion and enables maintenance of a very precise composition and volume of body fluids.

In summary, the major factors that have influenced the development of vertebrate osmoregulatory mechanisms appear to be

1. Prevailing osmotic environment.
2. Mode of nitrogen excretion.
3. Mode of body temperature regulation (poikilothermy versus homeothermy).

These factors have arguably provided the driving force for evolutionary change in renal structure and function. Developments of the renal apparatus have largely entailed distal addition of tubular segments to the basic ancestral nephron form (Figure 2). The gradual elaboration of the functional capacity of the vertebrate kidney has, of course, been associated with a reciprocal decline in the dependence on extrarenal mechanisms, particularly for monovalent ion management.

4. Endocrine Controls

Changes in the effector mechanisms of osmoregulation must be paralleled by alterations in their manner of endocrine control. Although we may point to a consistent set of endocrines that are involved (pituitary, adrenal/interrenal, and renal hormones), their target tissues and specific sites of action vary considerably through the vertebrate series. There is some variation in the major stimuli for secretion of these endocrines. As in all aspects of evolutionary change, these developments are inevitably associated with increasing specialization of function. Thus, for example, the hormones of the posterior pituitary and adrenal cortex are separately associated with water and salt management, respectively, in homeotherms. Such a distinction in hormone function is less clear, or not evident, in lower vertebrate groups, in which there is often considerable overlap in the actions of the two sets of hormones.

Contrasting sites of action and functions of particular hormones in different groups of vertebrates are also often associated with alteration in the chemical structure of a homologous series of endocrines. There is considerable species variation in the amino acid substitutions in pituitary peptide hormones, like prolactin and adrenocorticotropin. The neurohypophyseal hormones and angiotensins similarly vary in amino substitutions from group to group. By contrast, the important adrenal/interrenal steroid hormones are conservative in structure. The cortisol of fish is identical to that produced in man. However, the major secretory products of the adrenocortical homolog do vary within the vertebrate series. The characteristic product of amphibians and reptiles is corticosterone, rather than the cortisol of bony fish and mammals, whereas elasmobranchs elaborate their unique steroid, 1-hydroxycorticosterone. The major mineralocorticoid of mammals, aldosterone, has not been consistently found in fish and, although found in amphibians, may be without physiological function in this group.

The appearance of new hormone structures in the course of vertebrate evolution must have been matched by the appearance of appropriate receptors at target tissues. One of the major dilemmas facing the comparative endocrinologist must be the manner by which this coordinated emergence of new hormones and their receptors has been achieved. The variation from group to group in hormone structure and also target tissue receptor specificity greatly complicates the interpretation of comparative physiological studies. Very often, hormone preparations applied to lower vertebrates are either derived from mammalian material (pig, sheep, cow) or synthesized to mammalian hormone structure. Clearly, interpretation of the effect of such preparations in nonmammalian vertebrates at best must be cautious and may tell us little of the physiological function of the native hormone in these animals, the structure of which is often not known.

The octapeptide hormones of the posterior pituitary have probably received the greatest attention in comparative research and provide a relatively complete picture of distribution of hormone structure and function. It is, therefore, possible to speculate on the evolutionary pathways leading to the observed pattern of distribution of neurohypophyseal peptides. Clearly, this type of speculation not only should consider variations in hormone structure, but should include assessment of phylogenetic changes in hormone function. The structural changes in a hormone would be expected to confer some adaptational advantage in its physiological function to the evolving species. The parallel emergence of vasopressin and oxytocin as novel peptides in mammals illustrates this concept. Oxytocin has much greater potency in causing uterine contraction and milk ejection than in promoting water retention by the mammalian kidney. On the other hand, the vasopressins (arginine and lysine) are potently antidiuretic, but have little effect on uterine or mammary gland smooth muscle. Arginine vasotocin, which is the ubiquitous peptide of nonmammalian groups, is effective at all three target tissues, but is less potent than oxytocin for milk ejection and uterine contraction and than vasopressin for renal antidiuresis (see Table I). In mammals, then, the replacement of vasotocin by vasopressin and the appearance of oxytocin as the neutral peptide is associated with a clear separation in neurohypophyseal regulation of water economy (vasopressin) and reproductive physiology (oxytocin). Presumably, this separation of octapeptide functions has offered considerable adaptive advantage to the mammalian group. Paucity of information concerning the biological activities of homologous peptides in lower vertebrates unfortunately prevents similar speculation on their origins.

Although a homologous series of hormones, such as the neurohypophyseal peptides, may exhibit some consistent biological actions, these effects may vary considerably in physiological importance from group to group. Thus, this series of hormones has been implicated in systemic blood pressure regulation since their initial discovery. An increase in blood pressure was the earliest observed effect of mammalian neurohypophyseal extracts. As Table II indicates, this type of pressor response to neurohypophyseal hormones is common to all vertebrates except birds, which characteristically exhibit a depressor response. The pressor action of arginine vasotocin (AVT) in fish has been suggested by Sawyer (1977) to be the most primitive or earliest function of the peptide. Most of the osmoregulatory effects of

Table I
Biological Activities of Some Neurohypophyseal Hormones[a]

Hormone	Rat antidiuresis	Rat uterine contractions	Rabbit milk ejection
Arginine vasotocin	260	120	220
Arginine vasopressin	435	17	70
Oxytocin	5	450	450

[a]Activity measured in International U/μmole. (Data from Berde, B., and Boissonnas, R. A., 1968, *Handbook of Experimental Pharmacology,* Vol. XXIII, *Neurohypophysial Hormones and Similar Peptides,* Springer Verlag, New York.)

Table II
Neurohypophyseal Hormone Effects on Systemic Blood Pressure

Group	Systemic blood pressure response	
	Pressor	Depressor
Mammals	Vasopressin and vasotocin	—
Birds	(Vasotocin in pigeon only)	Vasotocin and mesotocin
Reptiles	Vasotocin and mesotocin	Vasotocin and mesotocin
Amphibians	Vasotocin and mesotocin	Mesotocin
Teleosts	Vasotocin	Mesotocin and isotocin
Elasmobranchs	Vasotocin	
Cyclostomes	Vasotocin	

this hormone can be understood in terms of such a vascular action, and it is interesting that the pressor action of AVT is opposed by the depressor effects of the neutral neurohypophyseal hormones found in fish (isotocin and mesotocin). Vasoactive effects of neurohypophyseal hormones, altering gill and renal function in fish, probably predate more specific and direct actions of these peptides on the permeability and transport processes of epithelial membranes so characteristic in tetrapods. It is interesting, however, that in recent years there has been a renewal of interest by mammalian renal and cardiovascular physiologists in the possible function of vasopressin in influencing cardiovascular reactivity and blood flow distribution within the kidney. Vasopressin has often been implicated in the development of certain types of hypertension. The archaic vascular actions of neurohypophyseal hormones may, therefore, have lingering physiological consequences in the mammals, including man.

Let us move our attention from a consideration of hormone action to examine the stimuli for hormone secretion. The differing roles of vasotocin in eliciting an antidiuresis in tetrapods and a diuresis, or antidiuresis, in fish demonstrate the necessity for some variation in the mechanisms regulating hormone secretion. In freshwater fish, body fluid dilution and/or volume expansion would be appropriate stimuli for the secretion of vasotocin as a diuretic agent. By contrast, its antidiuretic role in tetrapods would be better served if the hormone were secreted in response to concentration and contraction of body fluids. There is considerable evidence to support the latter model in tetrapods, but there is little experimental work from which to assess the former. There is a precipitous fall in the stored pituitary content of AVT following the transfer of the euryhaline flounder from freshwater to seawater (Figure 4). This suggests that AVT secretion was elevated on entering the hypertonic medium, which is known to be associated with hemoconcentration. Thus, in the flounder at least, the model of increased AVT secretion following dilution of body fluids does not appear to hold. The altered pituitary AVT levels in the flounder are more easily understood in terms of the tetrapod scheme. It is hoped

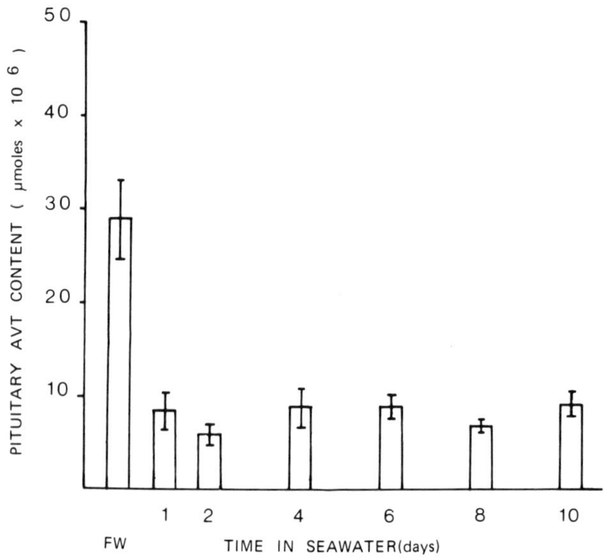

Figure 4. Changes in flounder pituitary AVT content. Data represent the mean ± standard error for at least five animals. FW, freshwater.

that the recently developed radioimmunoassays for AVT will provide direct measurement of plasma hormone levels and clarify this question.

Systematic differences in the use of homologous endocrines necessarily require altered responsiveness of the endocrine gland to excitatory stimuli (Table III). A facile example of this is the release of prolactin during suckling in the mammal. As other vertebrates lack mammary tissue, this represents a phylogenetically unique and novel stimulus. Prolactin secretion in euryhaline fish migrating from seawater to freshwater is, by contrast, apparently dependent on the low calcium concentration of the freshwater medium. In such fish, the reverse migration (freshwater to seawater) is associated with accumulation of salt and an elevation of plasma sodium

Table III
Effect of Prolactins from Different Vertebrate Groups on a Variety of Physiological Functions

Fish	Amphibians, reptiles, and birds	Mammals	Response
+	?	+	Teleost osmoregulation
–	+	+	Pigeon crop sac hyperplasia
–	?	+	Rat tibial plate growth

levels. This hypernatremia leads to raised interrenal secretion and gill salt elimination. Interrenal secretion in tetrapods is also sensitive to changes in sodium status, but in these groups it is a fall in plasma sodium levels that stimulates steroidogenesis. This is, of course, in line with the sodium-conserving role that such hormones play in tetrapods. Clearly, phylogenetic differences in the capacity of endocrine glands to respond to particular stimuli are an integral part of evolutionary change to hormone function.

5. Selected Readings

Balment, R. J., 1982, Evolutionary consideration of vertebrate osmoregulatory mechanisms, in: *Aspects of Avian Endocrinology: Practical and Theoretical Implications, Vol. 26* (C. G. Scanes, M. A. Ottinger, A. D. Kenny, J. Balthazart, J. Cronshaw and I. Chester-Jones eds.), Texas Tech University Press, Lubbock, pp. 297–307.

Balment, R. J., Henderson, I. W. and Chester-Jones, I., 1980, The adrenal cortex and its homologues in vertebrates; evolutionary considerations, in: *General, Comparative and Clinical Endocrinology of the Adrenal Cortex,* Vol. 3 (I. Chester-Jones and I. W. Henderson, eds.), Academic Press, London and New York, pp. 525–562.

Bentley, P. J., 1971, *Endocrines and Osmoregulation. A Comparative Account of the Regulation of Water and Salt in Vertebrates,* Springer Verlag, New York, Berlin, Heidelberg.

Bentley, P. J., 1982, *Comparative Vertebrate Endocrinology,* 2nd Edition, Cambridge University Press, Cambridge.

Dacke, C. G., 1979, *Calcium Regulation in Sub-mammalian Vertebrates,* Academic Press, London.

du Vigneaud, V., Ressler, C., Swan, J. M., Roberts, C. W., Katsoyannis, P. G., and Gordon, S., 1953, The synthesis of an octapeptide amide with the hormonal activity of oxytocin. *J. Am. Chem. Soc.* **75:**4870.

Gilles, R. (ed), 1979, *Mechanisms of Osmoregulation in Animals: Maintenance of Cell Volume,* John Wiley and Sons, Chichester.

Gorbman, A., Dickhoff, W. W., Vigna, S. R., Clark, N. B., and Ralph, C. L., 1983, *Comparative Endocrinology,* John Wiley and Sons, Chichester.

Gupta, B. L., Moreton, R. B., Oschman, J. L., and Wall, B. J. (eds.), 1977, *Transport of Ions and Water in Animals,* Academic Press, London.

Keys, A. B., and Willmer, E. N., 1932, ''Chloride-secreting cells'' in the gills of fishes with special reference to the common eel. *J. Physiol (Lond.)* **76:**368–378.

Lahlou, B. (ed.), 1980, *Epithelial Transport in the Lower Vertebrates,* Cambridge University Press, Cambridge.

Maloiy, G. M. O. (ed.), 1979, *Comparative Physiology of Osmoregulation in Animals,* Academic Press, London.

Nicoll, C. S., 1974, Physiological actions of prolactin, in: *Handbook of Physiology,* Section 7, Endocrinology, Vol. IV, Part 2 (E. Knobil and W. H. Sawyer eds.), American Physiological Society, Bethesda, MD, pp. 253–292.

Nicoll, C. S., 1981, Role of prolactin in water and electrolyte balance in vertebrates, in: *Current Endocrinology* (R. Jaffe, ed.), Elsevier, New York, pp. 127–166.

Nishimura, H., 1980, Comparative endocrinology of renin and angiotensin, in: *The Renin–Angiotensin System* (J. A. Johnson and R. R. Anderson, eds.), Plenum Press, New York, pp. 29–77.

Oguro, C., and Pang, P. K. T., 1982, *Comparative Endocrinology of Calcium Regulation,* Japanese Science Society Press, Tokyo.

Pang, P. K. T., and Epple, A. (eds.), 1980, *Evolution of Vertebrate Endocrine Systems,* Texas Tech. University Press, Lubbock.

Rankin, J. C., and Davenport, J. A., 1981, *Animal Osmoregulation,* Blackie, Glasgow and London.

Riegel, J. A., 1978, Factors affecting glomerular functions in the Pacific hagfish *Eptatretus stoutii* (Lockington). *J. Exp. Biol.* **73:**261–277.

Sawyer, W. H., 1977, Evolution of active neurohypophysial principles among the vertebrates. *Am. Zool.* **17:**727–737.

Part V

Endocrine and Related Factors in the Control of Metabolism in Nonmammalian Vertebrates

D. A. Gapp

D. A. Gapp • Biology Department, Hamilton College, Clinton, New York 13323. This part is dedicated to I.P.C., I.C.J., and I.A.M., three who led me into the field of endocrinology through their interest, teaching, and enthusiasm.

14

Introduction

Part V attempts to provide a comprehensive overview of the structure and function of the gastroenteropancreatic (GEP) hormone system in submammalian vertebrates. This system is responsible for regulating and coordinating the functions of the gastrointestinal tract and for the appropriate distribution of the products of digestion as well as the maintenance of appropriate levels of circulating metabolites essential for cellular metabolic activities. The origin, structure, and functions of the various hormones in the vertebrates are presented, and, where appropriate, the GEP hormone activities are integrated with the hormonal background provided by other essential hormones, such as the thyroid, adrenocortical, and adrenomedullary secretions. It is hoped that evolutionary trends in the regulation of the processing and distribution of a nutrient load can be identified and brought together in a cohesive structure. In this undertaking there are excellent reviews which serve as important sources and entry points into the original literature. A few of the more recent papers have been listed in the selected readings section at the end of Chapter 18.

Ingestion and utilization of the nutrient materials needed to provide cells and tissues with the appropriate substrate for energy generation and anabolic processes is essential for any vertebrate organism. This is a complex activity and involves physiological systems responsible for food ingestion and digestion, absorption and distribution of the nutrient load, and the appropriate utilization of nutrients. The digestion, distribution, and utilization of a nutrient load are under intricate hormonal regulation, much of it quite complex involving the interplay of hormones from the gastrointestinal tract and the endocrine pancreas which act on a hormonal background provided by adrenocortical and thyroid hormones.

Much of what we know concerning hormonal controls of nutrient processing and utilization has been derived from studies of mammalian systems and is being applied to the study and treatment of diseases in humans. Advances in radioimmunoassay and immunocytochemical techniques have greatly helped in unraveling interplay among the neural, gastrointestinal, and pancreatic endocrine controls that integrate activities associated with nutrient processing and utilization. Moreover, many peptide hormones identified and associated originally with the GEP system also have been found in the central and peripheral nervous systems. Here, then, is an expansion of the frontiers of neurosciences and neuroendocrinology, integrating them with endocrinology and the general field of chemical transmission.

It has been unfortunate, however, that much of our knowledge concerning mammalian GEP systems has yet to be applied systematically to the study of submammalian endocrine systems. In fact, our approach to the study of the lower vertebrate endocrine systems is couched in a mammalian bias. The roles and regulation of hormones in mammals are passed on, often by default, to lower vertebrates with the assumption that a hormone triggers the same cellular response or is released by the same stimulus in a reptile or amphibian as in the common laboratory rat or inbred strain of mouse. This is, and continues to be, an unsettling assumption. Availability of purified hormones and antisera to these hormones from mammals, but not from lower vertebrates, has led to a situation where, for example, a purified mammalian peptide hormone is used to assess gastrointestinal secretory function in a submammalian species when that hormone may be entirely inappropriate to that animal. For many years the presence of gastrin in the avian gastrointestinal tract had been equivocal until a gastrin 17-like component was demonstrated in a narrow zone of the chicken gizzard/duodenum juction. The function of gastrin in avian digestion, however, may be a minor one given both its relatively restricted distribution and that the widely distributed avian pancreatic polypeptide possesses all the attributes ascribed to mammalian gastrin when evaluated for effects on chicken crop, proventricular, and gizzard secretory function. Analogously, pancreatic B cell secretory function in submammalian vertebrates is routinely evaluated employing a glucose challenge, often without considering that another metabolite (e.g., an amino acid) may be the more physiological insulinotropic agent. Such might be the case in the lizard, *Anolis,* where insulin release in response to glucose occurs only when levels are very elevated.

Increased awareness of the benefits of comparative studies coupled with some of the flexibility in radioimmunoassay and immunocytochemical techniques has permitted a resurgence in comparative studies of the GEP system. Moreover, many investigators are trying to reduce the mammalian bias when elucidating lower vertebrate endocrine systems. Ultimately, availability of purified hormones and corresponding species-specific antisera for submammalian vertebrates will permit critical studies to be completed. Until that time, however, we must bear in mind when investigating nonmammalian endocrine systems using analytical techniques based on mammalian hormones that "immunostaining and radioimmunoassay are not infallible. Species specificity and evolutionary alteration of the structure of a peptide may prevent a reaction so that a negative result means very little . . . . It is obviously better if the animal's own hormone can be extracted and purified, but if cross-reactivity occurs it is more convenient to use an available antibody" [see Van

Noorden, S., *et al.*, 1980, Evolution of the gastrointestinal endocrine system, in: *Evolution of Vertebrate Endocrine Systems* (P. K. T. Pang and A. Epple, eds.), Texas Tech Press, Lubbock, TX]. Thus, if studies are well controlled and analyzed in light of the appropriate limitations and restrictions, accurate information on the GEP system of lower vertebrates will be forthcoming.

15

Gastroenteropancreatic Cells

1. Characteristics and Origins

The many gastroenteropancreatic (GEP) cells that are involved in the processing, distribution, and utilization of nutrients exhibit common morphological and functional characteristics which provide a unifying framework for their study. A single classification system for these cells, however, and for the broad spectrum of peptide and protein hormones they produce is far from complete, even for the intensively studied human GEP system. Although gut endocrinology in particular is a rapidly expanding field in medicine, systematic morphological studies coupled with chemical and secretory studies are not abundant for submammalian vertebrates. Likewise, the endocrine pancreas is a field of intense investigation, but again the efforts have been directed primarily toward the study of mammalian systems with a focus on the human diabetic condition. Nevertheless, comparative studies are on the increase, and evolutionary trends are being delineated as GEP cells are identified and classified in all major vertebrate groups, in protochordates, and in invertebrates including the Protozoa.

1.1. Morphological Considerations

GEP cells produce peptides or protein endocrine factors and possibly amines too. They are (1) distributed as single cells scattered throughout a tissue, as are the majority of gut endocrine cells, thus constituting a diffuse endocrine system (e.g., secretin-containing cells in the duodenal mucosa epithelium); (2) aggregated into discrete masses of cells exhibiting various degrees of organization (e.g., cells of the pancreatic islet); or (3) found in close association with other GEP cells, but in low numbers (e.g., somatostatin-containing cells found associated with gastrin-contain-

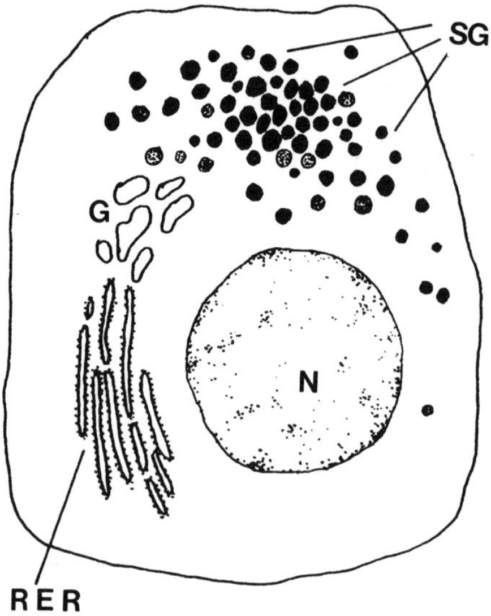

Figure 1. Diagram of a generalized peptide-secreting cell. N, nucleus; RER, rough endoplasmic reticulum; G, Golgi apparatus; SG, secretory granules.

ing cells of the antral mucosa). Morphologically, GEP cells are typical protein-producing cells characterized by extensive rough endoplasmic reticulum, prominent Golgi apparatus, and the presence of numerous electron-dense, membrane-bound secretory granules (Figure 1). Granule size and morphology have been the principal characteristics by which these cells have been identified. In general, the granules are small (400 nm or less), compared to zymogen granules in the acinar cells of the exocrine pancreas or salivary glands, which are much greater in diameter. GEP cell granules are generally regular in shape with individual cells characterized either by granules with the membrane tightly apposed to the granule contents, or by granules with a halo between the membrane and granule contents. Granule heterogeneity

Table I
Secretory Granule Diameter in Pancreatic Islet Cells (mean ± SEM)[a]

Species	Insulin	Glucagon	Somatostatin	Pancreatic polypeptide
Cynops hongkongiensis	407 ± 19	229 ± 8	131 ± 6	—
Hyla arborea	110 ± 4	315 ± 11	161 ± 7	—
Chelydra serpentina	333 ± 12	210 ± 5	236 ± 5	225 ± 8
Agkistrodon piscivora	495 ± 27	553 ± 27	325 ± 13	—
Alligator mississippiensis	187 ± 6	224 ± 11	296 ± 11	190 ± 7
Human	270 ± 9	250 ± 6	400 ± 24	210 ± 4

[a](From Buchan, A. M. J., 1980, Phylogeny of the gut hormones from amphibians to man, Doctoral thesis, University of London.)

among species (see, for example, Table I), developmentally, within a species, or even within a single cell has made universal application of granule morphology somewhat difficult, even when coupled with informative cytochemical characteristics such as reactions to various silver stains. Expanded classification systems now rely on granule morphology and immunostaining for specific hormones to describe not only granule structure but its content as well; hence, a functional classification for the GEP cells is possible.

GEP cells are included in a category of cells called paraneurons by Fujita in order to reinforce the idea that these endocrine cells have characteristics relating them to neurons. This concept of neurons and paraneurons builds on Pearse's amine precursor uptake and decarboxylation (APUD) concept, which postulates a common neuroectodermal origin for protein-secreting endocrine cells based on content of monoamines and the processing of amine precursors (see Section 1.3). The paraneuron concept originated from observations on gut endocrine cells and has been expanded to include many cells such as pancreatic endocrine cells, basal-granulated cells in the bronchial and alveolar epithelium, pituitary cells, pinealocytes, and even sensory cells in the epidermis and gastrodermis of *Hydra*. Paraneurons, then, are receptosecretory cells (as are neurons) which produce biologically active peptides, which may or may not produce amines as neurotransmitters, and for which the collected characteristics overlap with neurons to the extent where discrimination between neurons and paraneurons is impossible as the two are considered to be part of a continuum.

Paraneurons are classified as being open-type or closed-type, where the open-type is an epithelial cell that has its apical pole in contact with the lumen or cavity of a hollow organ while the basal-granulated pole rests on the basement membrane of the epithelium and is in close proximity to the vasculature. The open-type paraneuron is considered to be intermediate between the neuron and the endocrine paraneuron by virtue of its sensory apical pole and its seretory basal pole. The closed-type paraneuron is located outside the epithelium with no direct contact with the lumen but receiving input via mechanical, local, neural, or blood-borne stimuli and secreting into the extracellular space. The closed-type represents the typical endocrine cell and could be considered a modified open-type paraneuron, thus representing a later evolutionary form. Figure 2 illustrates how this classification is applied to gut endocrine cells where the open-type lies in the mucosal epithelium, the apically located microvilli serving as the sensory or chemoreceptive portion of the cell. This type is readily illustrated by cells secreting hormones such as gastrin, secretin, or cholecystokinin (Figure 3A). The closed-type, on the other hand, lies outside the mucosal epithelium and has no direct contact with the gut lumen; the somatostatin cells found in the mammalian stomach, intestine, or avian proventriculus are examples of closed-type paraneurons (Figure 3B). The endocrine pancreatic cells of vertebrates have closed-type paraneuron configuration whereas islet cell precursors in protochordates and invertebrates are open-type paraneurons (Figure 4). Apparently during the early course of vertebrate evolution cells producing insulin, glucagon, somatostatin, and pancreatic polypeptide left the gut epithelium,

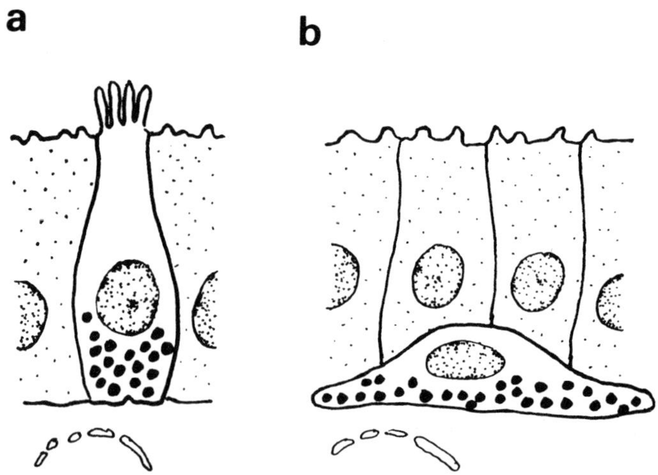

Figure 2. Gastrointestinal endocrine cells of open (a) and closed (b) types. [Redrawn from Fujita, T., Kobayashi, S., Yul, R., Hirano, T., and Wada, M., 1980, Evolution of neurons and paraneurons, in: *Hormones, Adaptation and Evolution* (S. Ishii, T. Hirano, and M. Wada, eds.), Japan Scientific Society Press, Tokyo/Springer-Verlag, Berlin, pp. 35–43.]

possibly before the appearance of the exocrine pancreas, to form a system of closed-type paraneurons.

1.2. Classification of GEP Cells

Attempting a complete classification of vertebrate GEP cells is impractical because of the great heterogeneity in granule morphology among species (Table I) and because of great gaps in our knowledge concerning localization and identification of GEP cells in many important vertebrate groups. Nevertheless, schemes for some individual species are available and are informative in providing an overview of the GEP system especially when the scheme is based on localization of the GEP cell, its granule morphology, and its hormone content. As might be expected, a system for human GEP cells is probably the most complete at present and will serve as a point of reference for classification of GEP cells. Human GEP cells can be placed into three groups based on the extent of their characterization: (1) GEP cells with established products (Table II); (2) candidate GEP cells for which peptide products are identified but not at present internationally accepted (Table III); and (3) unclassified GEP cells which have no known peptide product (Table IV).

Established GEP cells include the pancreatic B and A cells, which produce insulin and glucagon, respectively. The former cell type is characterized by large pleomorphic granules with a distinct halo around the granule contents. The pancreatic A cell has round granules with a strong electron-dense core and a small clear halo between the core and the argyrophilic granule membrane. Electron micro-

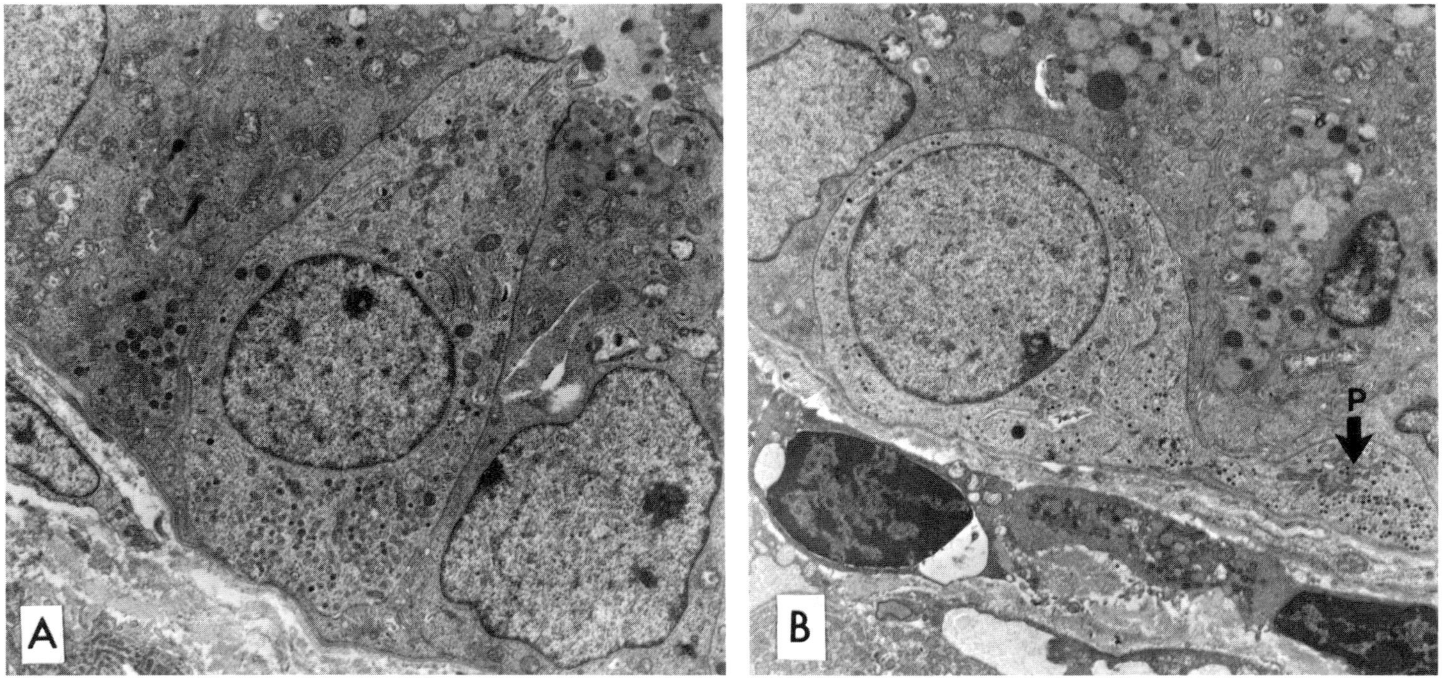

Figure 3. (A) Electron micrograph of a gastrin-containing open-type GEP cell from cat antral mucosa. (B) Electron micrograph of a somatostatin-containing, closed-type GEP cell from cat duodenum. Note the granule-filled process (P) which runs along the basement membrane. (Micrographs courtesy of Ian Varndell and Julia M. Polak.)

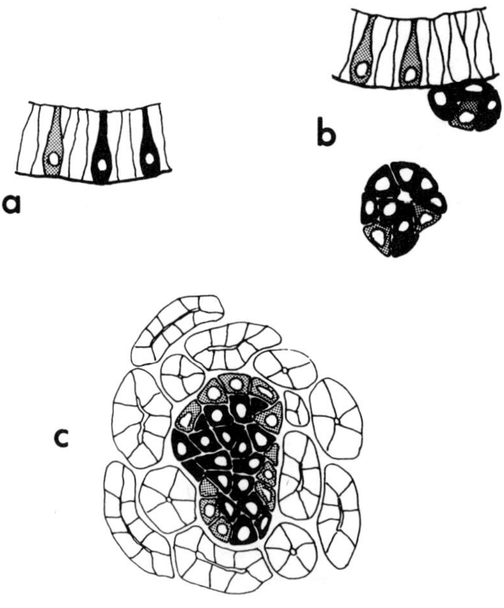

Figure 4. Diagrammatic representation of the evolution of the vertebrate islet arising from open-type cells in the mucosa of invertebrates and protochordates (a), to the formation of islet lobules in the cyclostomes (b), and, finally, to the formation of the true vertebrate islet associated with the exocrine pancreas (c). Dark shaded cells represent the insulin-containing cells, and stippled cells represent the non-insulin-containing cell populations. [Redrawn from Fujita, T., Kobayashi, S., Yui, R., Hirano, T., and Iwanaga, T., 1980, in: *Hormones, Adaptation and Evolution* (S. Ishii, T. Hirano, and M. Wada, eds.), Japan Scientific Society Press, Tokyo/Springer-Verlag, Berlin, pp. 35–43.]

scopic immunocytochemistry reveals immunoreactivity for proglucagon localized in the halo of the A granule. The enteroglucagon (EG) cells of the lower small intestine and of the large intestine produce enteroglucagon (gut glucagon, glicentin), a large form of glucagon (see Section 2.2). EG cells exhibit two morphological forms; one, common to the ileum, is an open-type cell with apical luminal contact and granulated basal pole; the other EG cell is a closed-type, with well-developed, long basal processes, and is common to the colon.

Somatostatin cells are found throughout the GI tract and exhibit similar granule morphology in the various regions of the gut. The somatostatin-containing pancreatic D cell is arrayed around the islet periphery and exhibits a rather irregular morphology. Antral and small-intestine somatostatin cells are open-type cells with the apical pole at the gut lumen and the basal portion exhibiting long processes running parallel to the basal membrane. Fundic somatostatin cells, on the other hand, are like pancreatic D cells in that they are a closed-cell-type GEP cell and have no connection to the gut lumen.

Pancreatic polypeptide (PP) cells found in the pancreas and large intestine are characterized by small, electron-dense granules with closely applied membranes. In the human pancreas they are closed-type GEP cells which are restricted to the islet whereas in the large intestine they are open-type cells with apical contact with the lumen and the basally located secretory granules. Interestingly, localization of this cell type within the pancreas of other mammals is quite variable. In the mouse, rat, and sheep pancreas it is found in the islets exclusively, whereas in the cat and dog PP cells are exclusively or predominantly found in the exocrine parenchyma. In the

Table II
Functional Classification (Human): Established Cell Types[a]

Area	Cell type	Product	Granule diameter (nm)	Characteristics
Pancreas	I	Insulin		Polymorphic
	PG	Glucagon		Grimelius positive halo
	SOM	Somatostatin	260–370	Relatively osmiophilic
	PP	Pancreatic polypeptide	150–170	Round, osmiophilic
Stomach	SOM	Somatostatin	260–370	Relatively osmiophilic
	AG	Antral gastrin	200–400	Some electron-lucent
Small intestine (upper)	S	Secretin	240	Irregular, electron-dense
	SOM	Somatostatin	260–370	Relatively osmiophilic
	GIP	Gastric inhibitory polypeptide	350	Round to irregular, dense osmiophilic core
	CCK	Cholecystokinin	250	Round, electron-dense
Small intestine (lower)	N	Neurotensin	300	Round, electron-dense
	EG	Enteroglucagon	210	Round, electron-dense
	SOM	Somatostatin	260–370	Relatively osmiophilic
Large intestine	EG	Enteroglucagon	210	Round, electron-dense
	SOM	Somatostatin	260–370	Relatively osmiophilic

[a](From Buchan, A. M. J., and Polak, J. M., 1980, The classification of the human gastroenteropancreatic endocrine cells. *Invest. Cell Pathol.* **3**:51–71.)

guinea pig and hamster PP cells are found evenly distributed between the islets and exocrine tissue.

Gastrin cells appear at several sites in the human GI tract. Antral gastrin (AG) cells are open-type GEP cells concentrated in the midthird of the antral mucosa and with secretory granules varying in size and electron density. Intestinal gastrin (IG) cells are also of open type but do not resemble AG cells, which are seldom found in

Table III
Functional Classification (Human): Candidate Cells[a]

Area	Cell type	Product	Granule diameter (nm)	Characteristics
Small intestine (upper)	IG	Intestinal gastrin	190	Round, electron-dense
	M }	Motilin	160–190	Round, electron-dense
	EC_M }		—	Argentaffin positive
	EC_{SP}	Substance P	—	Argentaffin positive
Large intestine	PP	Pancreatic polypeptide	150–170	Round, electron-dense

[a](From Buchan, A. M. J., and Polak, J. M., 1980, The classification of the human gastroenteropancreatic endocrine cells. *Invest. Cell Pathol.* **3**:51–71.)

Table IV
Functional Classification (Human): Unclassified Cells[a]

Area	Cell type	Product	Granule diameter (nm)	Characteristics
Pancreas	D_1	?	180–200	Round, electron-dense
Stomach	X	?	250	Round, electron-dense
	ECL	?	—	Polymorphic, argyrophilic
	D_1	?	180–200	Round, electron-dense
	P	?	90–100	Round, electron-dense
	EC	? + amine	—	Polymorphic, argentaffin
Small and large intestine	P (upper)	?	90–100	Round, electron-dense
	EC	? + amine	—	Polymorphic, argentaffin
	D_1	?	180–200	Round, electron-dense

[a](Buchan, A. M. J., and Polak, J. M., 1980, The classification of the human gastroenteropancreatic endocrine cells. *Invest. Cell Pathol.* **3**:51–71.)

the intestine. IG cells have small, uniformly round, electron-dense granules and are thought to produce gastrin-34, the large form of gastrin typically predominating in intestinal extracts. AG cells appear to store mainly the smaller form of the hormone, gastrin-17.

Cholecystokinin (CCK) is produced by open-type GEP cells found in the duodenal and jejunal mucosa. CCK cells are characterized by round, intermediate-size, electron-dense granules but do not have argentaffin (ability to reduce silver salts to metallic silver without a reducing agent) or argyrophil (impregnated silver salts require reducing agent to produce metallic silver) properties. Secretin cells are also found in the duodenal and jejunal mucosa with intermediate-size irregular granules. Gastric inhibitory polypeptide is localized to GEP cells of the duodenum and jejunum exhibiting large round to irregular-shaped granules. Neurotensin appears to be restricted to open-type cells in the mucosa of the ileum characterized by large, round, electron-dense granules.

Motilin (M) cells in the human are candidate GEP cells which exhibit two populations. M cells represent about 30% of the motilin-positive cells and are characterized by small, round, relatively electron-dense granules. They are found in both the villi and crypts of the upper small intestine, but only villus M cells exhibit apparent connections with the lumen. The EC_M cells, on the other hand, are typical enterochromaffin (EC) cells exhibiting large polymorphic granules which are argentaffin-positive. [EC cells in general contain serotonin (5-hydroxytryptamine) and belong to a cell type found abundantly throughout the gastrointestinal (GI) tract.] In addition, there is a population of substance P-containing enterochromaffin cells which are sparsely distributed in the small intestine.

The list of unclassified GEP cells (Table IV) is composed of those cells which as yet have no associated peptide. And, since there is a greater number of cells that immunostain for neuron-specific enolase (see Section 1.3) than with any of the known peptides, one would assume that many more peptides remain to be found.

Table V
Peptides Localized to Enteric Neurons

Vasoactive intestinal polypeptide (VIP)
Substance P
Enkephalin
Bombesin
Somatostatin
Neurotensin
Thyrotropin releasing hormone (TRH)
Cholecystokinin octapeptide (CCK-8)
Neuropeptide tyrosine (NPY)
Peptide histidine isoleucine (PHI)

For example, the newly characterized peptides* [peptide histidine isoleucine (PHI), peptide tyrosine tyrosine, and neuropeptide tyrosine (NPY)] may ultimately be ascribed to some of the unclassified cells appearing in Table IV although at present NPY and PHI appear to be restricted to neurons and not found in endocrine cells. It is certain that this list and the others will change significantly in the future.

As a conclusion to the discussion of the classification and distribution of the GEP hormones in the GI tract, it is essential to note that many, if not all, of the hormones mentioned have counterparts in the central and peripheral nervous systems. The pancreas and gut are richly innervated by nerves that release peptides along with, or in addition to, the usual adrenergic or cholinergic neurotransmitter substances. Many of the peptides found in these peptidergic nerve endings are also found in the diffuse GEP endocrine system. The peptidergic system of the gastrointestinal tract is an important counterpart to the cholinergic and adrenergic nervous system and plays a role in integrating GI responses to neural, hormonal, and other stimuli. The list of peptide-containing nerves is substantial (Table V), but it is not yet clear whether the peptidergic enteric neurons are totally distinct from the other autonomic neurons. It is probable that there is overlap since, for example, it has been estimated that 60–70% of adrenergic neurons of the inferior mesenteric ganglion contain somatostatin, and enkephalin has been localized to some of the adrenergic cells in the rat superior cervical ganglion. The distribution of these peptides in the enteric neurons is uniform throughout the gut, not as in the GEP cells which are distributed differentially in the various regions of the GI tract.

1.3. Developmental Aspects

The GI tract is of endodermal derivation, the straight-tube primitive gut giving rise to the stomach and intestines with the many variations found throughout the

*Newly isolated peptides are now being named using the single letter designation for amino acids (e.g., F = phenylalanine, M = methionine, R = arginine) employing the N-terminal and C-terminal residues. For example, PHI refers to the 27-amino-acid peptide with an N-terminal histidine (H) and a C-terminal isoleucine (I).

vertebrates. The simplest GI system is found in the adult hagfish, which has no stomach but only a straight-tube intestine, whereas in mammals, especially some ruminants, the GI tract is much more complex. The pancreas is included as an integral part of the GI tract as it is derived from the primitive gut as dorsal and ventral buds which subsequently fuse. The dorsal bud (or anlage) gives rise to the body, tail, and most of the head portion, and the ventral anlage gives rise to a specific portion of the head region (cf. Chapter 17). A discrete exocrine pancreas is not found in cyclostomes, but appears to have first developed in the cartilaginous and bony fishes. Cyclostomes do, however, possess an islet organ located around and derived from the bile duct epithelium; the zymogen (eventually pancreatic exocrine) cells are found in the intestinal mucosa (cf. Figure 2, Chapter 17).

The ontogenic appearance of the endocrine cells of the stomach, intestine, and pancreas is best known in the rat, mouse, and chick embryo, with more data being available on the developmental appearance of the pancreatic endocrine cells. Pancreatic endocrine cells are presumed to be derived from gut ductal epithelium, possibly from the same embryonic pool that gives rise to the gut endocrine cells. In the mouse, cells exhibiting cytochemical properties characteristic of GEP cells are observed at 8–9 days of gestation in the presumptive duodenum. In general, however, gut endocrine cells identifiable by immunostaining do not appear until late in gestational life, often postnatally, and usually later than most of the pancreatic endocrine cells.

The ontogenic appearance of endocrine cells in the stomach and duodenum of the rat is well documented. In the corpus of the stomach, enteroglucagon and somatostatin cells first appear around the twentieth day of gestation, both cell types increasing somewhat, early postnatally, until 10–16 days of age when the frequency of somatostatin cells increases markedly. Antral gastrin and somatostatin cells are first observed around 19–20 days of gestation, but not regularly until day 1 postnatally. During the following 3 weeks both cell types proliferate rapidly with gastrin cells being observed at 2 or 3 times the frequency of somatostatin cells. In the duodenum, gastrin and somatostatin cells are present by 18 days of gestation. Again, gastrin cells are the more frequent of the two. Both undergo marked proliferation during the first 4 postnatal days to reach their maximal frequency, and thereafter, proliferation declines so that by 3 weeks of age the adult frequencies for the duodenal gastrin and somatostatin cells have been established.

CCK cells are observed only occasionally in the late fetal or early postnatal rat duodenum and are generally scarce until 9 or 10 days of age. From days 10 through 15 CCK cell frequency increases dramatically followed by a decrease in frequency, although the numbers are still elevated at 3 weeks compared to the adult frequency. Secretin cells are scarce at the end of gestation but reach a maximum frequency between 1 and 4 days postnatally. Adult numbers of secretin cells are demonstrable by 3 weeks of age. Glucagon-containing duodenal cells appear early, being present by the eighteenth day of gestation and reaching maximal frequency at several days of age. Most of these cells stain for enteroglucagon, but a small portion stain for pancreatic glucagon as well.

Gut hormone cells in the rat appear ontogenically in a parallel or nonparallel manner. For example, gastrin and somatostatin cells arise simultaneously in the gastric antrum, and glucagon and somatostatin cells appear simultaneously in the corpus of the stomach. On the other hand, the appearance of gastrin and somatostatin cells in the duodenum is not synchronized (nonparallel) with the appearance of the antral gastrin and somatostatin cells, which are demonstrable only several days after the appearance of their duodenal counterparts.

Some GEP cells will appear only transiently in some regions during the development of the gut and pancreas. PP cells, for example, appear briefly in the gastric antrum during the perinatal period, and a few insulin-containing cells can be observed in the duodenum between the eighteenth and nineteenth days of gestation. Gastrin-containing D cells appear early in pancreas development and outnumber for a while the somatostatin-containing D cells. Pancreatic gastrin cell frequency increases until 1–4 days postnatally and declines thereafter such that by day 21 the pancreas is essentially devoid of gastrin cells. The role of these transient pancreatic gastrin cells is unclear, but given the tropic action of gastrin on the stomach, it is possible these cells may be responsible for some developmental processes in the pancreas.

Identifiable pancreatic endocrine cells are first observed somewhat prior to or about the same time as the appearance of the pancreatic diverticula in the primitive gut. In the rat the dorsal diverticulum appears around the eleventh day of gestation (20–25 somites), although A cells, but not B cells, are identifiable at the ultrastructural level prior to this. Both insulin and glucagon, however, are measurable by radioimmunoassay at 11 days with glucagon present in 100-fold excess over insulin. By 11½ days immunostainable glucagon is evident in the dorsal and ventral walls of the duodenum which will give rise to the pancreas, and by day 12 of gestation immunostainable B cells are present. Primitive islets appear shortly thereafter, and immunostainable D cells appear around day 15. Between days 18 and 20 of gestation islets become separated from the exocrine tissue by a thin connective tissue capsule, and the core of B cells with its A and PP cell mantle becomes established. By this time the concentration of insulin is 25-fold that of glucagon, reflecting the predominance of B over A cells and the reversal of the early embryonic condition. PP cells are not present in the rat pancreas prior to birth, but at parturition a few scattered cells can be found in the exocrine parenchyma or islets. PP cell frequency increases markedly between 5 and 7 days of age, and by 2 weeks PP cells are observed only in the islets and with adult frequencies.

The mouse pancreas exhibits a similar developmental pattern. The dorsal diverticulum develops around day 11, but endocrine cell types with polymorphic granules and APUD characteristics appear in the presumptive duodenum before day 11. Immunostainable cells are absent until day 14 of gestation when islet formation commences. A, B, and D cells are clearly demonstrable by day 16, and thereafter normal distribution of cell types within the islet is apparent.

Glucagon-containing A cells are the first cell type to appear in the dorsal pancreatic bud in the 3-day (30-somite) chick embryo. (The ventral bud develops somewhat later.) B cells appear at the 40–43 somite (3½ day) stage, and only after day 13 do small insulin islets appear in the central lobe, which is devoid of glucagon islets. Somatostatin-containing D cells appear shortly after the B cells first appear, and PP cells become evident around 9 days of incubation.

Pancreatic anlage in the frog, *Rana pipiens,* appears at, or shortly before, embryonic stage 22 (8 mm). B cells are present at this time occurring singly or in groups of 3–4 cells. In stage 24–25 embryos identifiable glucagon and pancreatic polypeptide cells occur singly or in clusters of two to three cells. B cells appear in clusters of four to five cells by larval stages I and II, and it is apparent at this time that glucagon and PP are localized to the same cell type. Scattered somatostatin-containing D cells first appear in stage II larvae. From stages V through IX, B cells begin to form large clusters (later cordlike arrangements) with the glucagon/PP cells and the D cells arrayed around the periphery, typical of the adult islet. During metamorphosis, however, the islet undergoes further changes; B cells increase progressively until stage XVIII, when a precipitous drop in B-cell mass follows during stages XIX and XX, just after the onset of exocrine cell degeneration. Decreased B-cell stainability for insulin may also occur at this time, coincident with the reorganization and redifferentiation of the exocrine tissue during this period of metamorphic climax.

The original embryonic source of GEP cells is a topic of considerable controversy centered around the APUD concept of Pearse, which proposed a neural crest* origin for peptide- and protein-producing endocrine cells. According to this hypothesis, GEP cells are derived from neural crest cells which migrate early from sites adjacent to the developing neural tube and colonize the gut epithelium and are considered, therefore, to be of ectodermal origin. Alternative proposals suggest that the GEP cells are derived from gut epithelial cells that did not originate from other germ layers and therefore can be considered as endodermal derivatives. This dilemma aside, the histogenesis of GEP cells which occurs after postulated migratory and/or differential events involves production of cells from the gut epithelium or a derivative such as the pancreatic duct. Gut endocrine cells are thought to arise from stem cells in the base of villus crypts or glands, and the pancreatic endocrine cells are thought to arise from pancreatic duct epithelium. In adult animals production of islet cells from the ductal epithelium or from mitosis of preexisting endocrine cells within the islet normally occurs at a low rate. A, B, and PP are observed, albeit infrequently, in the ductal epithelium and probably represent cells destined to become associated with islets. (PP cells are also observed in the ductal epithelium, but in many species this is a normal occurrence as this pancreatic cell may not be restricted to the islets in other species.) Mitotic figures are seen infrequently in islets

*The neural crest arises from the the junction of neural and nonneural ectoderm as a result of the neural folds coming together to form the neural tube. These neural crest cells which overlie the neural tube and extend along down the trunk are highly migratory and colonize a wide variety of tissues within the embryo. Neural crest cells give rise to many different tissues, including the sympathetic and parasympathetic ganglia, the calcitonin-producing cells, pigment cells, the dorsal root ganglion cells of the sensory nerves, and the chromaffin cells of the adrenal medulla.

from adult animals (other than in pathological conditions), and therefore preexisting islet cells do not represent a likely source for new islet cells. With respect to the B cells, once islet formation is complete late postnatally, there is little regenerative or replicative potential since destruction of the B-cell mass by a cytotoxic agent such as streptozotocin produces a permanent diabetic condition. If, however, streptozotocin is administered, for example, to newborn rats, only a transient diabetes is manifested; normoglycemia is regained, and much of the lost B-cell mass is recovered by events which include budding and formation of new islets from the small pancreatic ducts and by mitosis of new and remnant B cells. This replicative potential of B cells is apparently characteristic of fetal and neonatal animals only. A, D, and PP cells, however, appear to retain some ability to divide. Human diabetes, chemical diabetes in animals, and some forms of the murine genetic diabetes/obesity syndromes are characterized by proliferation of A, D, and PP cells to varying degrees apparently in response to a changing B-cell mass (cf. Chapter 17). In addition, acinar cells containing endocrine secretory granules have been observed in representatives from all vertebrate classes. This acinoinsular transformation has been proposed as a mechanism for islet cell histogenesis, but given the infrequency of occurrence and the possibility of preparation artifact, it has not been considered an important contributory factor in islet cell formation.

The APUD theory of Pearse and colleagues proposes a common embryonic origin for peptide-producing cells. Cells in the APUD series are characterized by the ability to produce peptide hormones (or hormonelike substances) and by the ability to produce and store amines. Early studies with embryological allografting techniques producing chick–quail chimeras, and with cytochemical marker experiments, validated that ultimobranchial C cells, thyroid C cells, type I cells of the carotid body, melanocytes, and adrenomedullary cells belonged to this series. APUD cells were assumed to be derived from the neural crest of the embryo which migrate at various states of development to colonize specific sites. Considerable evidence, however, has been marshalled against neural crest origin for gut and pancreatic endocrine cells. Results from culture of isolated chick embryo gut, removed presumably prior to neural crest cell migration, from neural crest ablation studies, and from allografts of quail neural tube to isochronic chick embryo hosts have argued against a neural crest origin for GEP cells. Similarly, presumptive rat embryo gut removed and cultured prior to presumed neural crest colonization showed normal development of pancreatic B cells in the developing pancreatic diverticula. These and other experiments have been used to support an endodermal origin for GEP cells. This theory supposes that endodermally derived presumptive gut or pancreatic endocrine cells originally elaborated gut peptides which evolved from an ancestral enzymelike protein produced by zymogen cells of the gut. These enzymes would have been secreted into the gut lumen, where enzymatic activity would have broken down these ancestral molecules into smaller fragments which could have been reabsorbed by the gut mucosa and passed into the blood stream, there to act upon presumptive "receptor" sites of peripheral tissues. Ultimately, enzymatic processing of the ancestral molecule would occur intracellularly and the active fragment(s) secreted directly into the blood, similar to cleavage of prohormone precursors of the peptide hormones. Ultimately, some of these cells would migrate from contact with the gut lumen to sites elsewhere such as the pancreas.

Modifications of the APUD concept now suggest that the source of these cells is neuroectoderm, not necessarily neural crest per se, and that migration of cells from the neural tube or associated neuroectoderm may occur very early in embryogenesis and in a pulsatile fashion such that operative interference in neural crest formation and dispersion may not be possible. Further evidence supporting a neural origin for GEP cells now comes from the localization of many GEP hormones such as CCK or gastrin to neurons in the hypothalamus or higher brain centers and to skin glands of some amphibians, all cells of known neuroectodermal or ectodermal origin. More recently, neuron-specific enolase, an enzyme localized exclusively in neurons, is also localized in pancreatic islet, adrenomedullary, thyroid C, and gastric G and somatostatin cells. In fact, any cell that produces peptide hormones stains for neuron-specific enolase. This perhaps is one of the strongest pieces of evidence pointing to a neural origin for peptide-producing endocrine cells, or at least points to a common precursor both ontogenically and possibly phylogenetically.

2. The GEP Hormones

The GEP hormones may be categorized according to different criteria such as molecular structure, a dual localization in the nervous system and the gut, counterparts in amphibian skin, or by lack of affinity to any particular grouping (see Table VI). The first of these categorizations provides perhaps the strongest scheme since it is based on relatedness of molecular structure and therefore has implications for molecular evolution and for the specificity of action of these hormones. The other categories do, however, provide alternative ways of viewing the GEP hormones. Eventually gray areas among these groupings will disappear and the interrelatedness of all GEP hormones will be clarified.

The gastrin and glucagon families of hormones are the best elucidated of the GEP hormone relationships although recently the insulin and pancreatic polypeptide families have been established and become well delineated. The gastrin family includes gastrin, CCK, and cerulein, and the glucagon family includes glucagon, secretin, vasoactive intestinal polypeptide (VIP), PHI, and gastric inhibitory polypeptide (GIP). Within each family members show striking overlap in sequence structure among them and are assumed to be derived from common ancestral molecules. Presumably each group arose from a single ancestral gene which duplicated one or more times, each daughter gene then proceeding to evolve at its own rate.

2.1. The Gastrin Family

The similarity among the members of the gastrin family (Figure 5) is restricted to the identical carboxy-terminal pentapeptide sequence and the tyrosine residues at

Table VI
GEP Peptide Classification

	GEP	Nervous system	Amphibian skin
Hormone families			
1. Gastrin	Gastrin		
	CCK	CCK-like	Cerulein
2. Glucagon	Glucagon		
	Secretin		
	VIP	VIP	
	GIP		
		PHI	
3. Pancreatic polypeptide	PP		
	PYY		
		NPY	
4. Insulin	Insulin	Insulinlike	
	IGF-I[a]		
	IGF-II[a]		
	Relaxin[a]		
	NGF[a]		
Peptides with counterparts in amphibian skin and the nervous system	Bombesin	Bombesin	Bombesin
	Substance P	Substance P	Physalaemin
	Neurotensin	Neurotensin	Xenopsin
	(Gastrin)		
	(CCK)	(CCK-like)	(Cerulein)
Peptides with dual location in nervous system and gut but not amphibian skin	Enkephalin	Enkephalin	
	Somatostatin	Somatostatin	
	(VIP)	(VIP)	
Hormones with no clear affinity to other groups	Motilin		

[a]IGF-I, IGF-II, relaxin, and NGF are not localized to cells in the GEP system but have been included because of their membership in the particular family of peptides. [Modified from Van Noorden, S., and Polak, J. M., 1980, Hormones of the alimentary tract, in: *Hormones and Evolution,* Vol. 2 (E. J. W. Barrington, ed.), Academic Press, New York, pp. 791–828.]

position 6 (gastrin) or position 7 (CCK, cerulein). The tyrosine is sulfated in CCK, cerulein, and gastrin II but unsulfated in gastrin I. The carboxy-terminal tetrapeptide for both CCK and gastrin is the minimal essential fragment required for biological activity of gastrin and CCK, and the position of the tyrosine determines whether the whole molecule produces a CCK or gastrin effect. Both gastrin and CCK exhibit multiple molecular forms. The principal forms of gastrin occurring in tissues and in circulation in mammals are gastrin-17 (mol. wt. 2100, little gastrin) and gastrin-34 (mol. wt. 3800, big gastrin), the former produced and secreted by antral gastrin cells, the latter produced by intestinal gastrin cells. The 17-amino-acid carboxy-terminal sequence of gastrin-34 is identical to that of gastrin-17. The amino-terminal pentadecapeptide of gastrin-34 is linked by two lysine residues to gastrin-17. This linkage, of course, is highly significant in that it represents the type of sequence

GASTRIN-17	
CHOLECYSTOKININ	LYS - ALA - PRO - SER - GLY - ARG - VAL - SER - MET - ILE - LYS - ASN...
CAERULEIN	
GASTRIN-17	PGL - GLY - PRO - TRP - LEU - GLU - GLU...
CHOLECYSTOKININ	...LEU - GLN - SER - LEU - ASP - PRO - SER - HIS - ARG - ILE - SER...
CAERULEIN	
	SO_3H (on TYR)
GASTRIN-17	...GLU - GLU - GLU - ALA - TYR - GLY - TRP - MET - ASP - PHE - NH_2
CHOLECYSTOKININ	...ASP - ARG - ASP - TYR - MET - GLY - TRP - MET - ASP - PHE - NH_2
	SO_3H (on TYR)
CAERULEIN	PGL - GLN - ASP - TYR - THR - GLY - TRP - MET - ASP - PHE - NH_2
	SO_3H (on TYR)

Figure 5. Amino acid sequences for members of the gastrin family. pGL, pyroglutamate.

found in prohormone molecules and represents a trypsin-sensitive site that is important in the proteolytic processing of precursor molecules. Cerulein is localized only in amphibian skin glands and appears not to be found in any other sources. It is most closely related to CCK (Figure 5) which is isolated from the duodenum and has 33 amino acid residues (CCK-33). A biologically active 39-residue variant has also been isolated (CCK-39) from the gut, and in the nervous system CCK-8 and CCK-4 variants are found both centrally and peripherally.

2.2. The Glucagon Family

Amino acid similarities among glucagon (mol. wt. 3485), secretin (mol. wt. 3073), VIP (mol. wt. 3326), and GIP (mol. wt. 5105) become apparent when the polypeptides are aligned from the amino acid terminus (Figure 6). Secretin, VIP, and glucagon are 27, 28, and 29 amino acid residues, respectively, and GIP (43 amino acids) is about half again as large. Although there is only one amino acid common to all four peptides (phenylalanine at position 6), there are 13 sites where three of four of the peptides have identical amino acids. PHI is a recently identified member of the glucagon family and is composed of 27 amino acid residues (Figure 6). The overlap of PHI with other family members is extensive as it has 13 residues identical with VIP, 11 with secretin, 10 with glucagon, and seven with GIP. Much of the conservation of structure among these molecules occurs in the amino-terminal region, possibly indicating the functional importance and/or the antiquity of the region. Amino acid sequences determined for a number of mammalian glucagons are identical, and avian glucagons differ only at one (chicken, turkey) or two (duck) positions when compared to the mammalian sequence. Thus, only 3–7% of the residues have been altered since the last ancestor to birds and mammals. These are conservative changes when compared to avian (chicken) or porcine insulin with 12% differing residues or to avian and bovine pancreatic polypeptide with 44%

differing residues. VIP also shows a fair degree of conservatism in that avian (chicken) and porcine differ in only 4 of 28 residues (15%).

GIP (mol. wt. 5105) is the largest member of this family, but little is known yet concerning its molecular size variants or differences in amino acid sequence, even among mammalian species. Glicentin, also known as enteroglucagon (gut glucagon, glucagonlike immunoreactivity), is also a member of this family. Glicentin is identical to pancreatic proglucagon and is found in endocrine cells of the ileum, jejunum, and colon. The glicentin molecule (mol. wt. 11,625) is composed of a glucagon-related polypeptide which is 63 amino acid residues in length followed by the 29-amino-acid residue sequence for pancreatic glucagon plus a C-terminal octapeptide.

2.3. The Insulin Family

Insulin is one of the best known hormones because of its central role in human diabetes. Localized primarily to the B cells of the pancreatic islets, insulin has also been found in the brain, identifying it with most of the gut peptides, which also have distributions in the nervous system. Insulin consists of a 21-amino-acid A chain and a 30-amino-acid B chain, which are linked by two disulfide bridges between cysteine residues at A_7 and B_7 and at A_{20} and B_{19}. There is also a disulfide bridge connecting cysteines at positions 6 and 11 within the A chain. The insulin molecule (mol. wt. 5734) is synthesized as a single-chain precursor, proinsulin (mol. wt. 9000), which is cleaved proteolytically by trypsinlike and carboxypeptidase B-like enzymes. This processing removes the C, or connecting, peptide, which appears to have no biological activity. Presumably the C peptide serves as a spacer region permitting appropriate folding of the proinsulin molecule and alignment for disulfide bridge formation. C-peptide sequences that have been determined show a high degree of variability, which supports the concept of a purely spacer role for the

					5					10				
Glucagon	His	Ser	Gln	Gly	Gly	Phe	Thr	Ser	Asp	Tyr	Ser	Lys	Tyr	Leu...
VIP	His	Ser	Asp	Ala	Val	Phe	Thr	Asp	Asn	Tyr	Thr	Arg	Leu	Arg...
Secretin	His	Ser	Asp	Gly	Thr	Phe	Thr	Ser	Glu	Leu	Ser	Arg	Leu	Arg...
PHI	His	Ala	Asp	Gly	Val	Phe	Thr	Ser	Asp	Phe	Ser	Arg	Leu	Leu...
GIP	Tyr	Ala	Glu	Gly	Thr	Phe	Ile	Ser	Asp	Tyr	Ser	Ile	Ala	Met...

	15					20					25				
Glucagon	...Asp	Ser	Arg	Arg	Ala	Gln	Asp	Phe	Val	Gln	Trp	Leu	Met	Asp	Thr
VIP	...Lys	Gln	Met	Ala	Val	Lys	Lys	Tyr	Leu	Asn	Ser	Ile	Leu	Asn	NH_2
Secretin	...Asp	Ser	Ala	Arg	Leu	Gln	Arg	Leu	Leu	Gln	Gly	Leu	Val	NH_2	
PHI	...Gly	Gln	Leu	Ser	Ala	Lys	Lys	Tyr	Leu	Glu	Ser	Leu	Ile	NH_2	
GIP	...Asp	Lys	Ile	Arg	Gln	Gln	Asp	Phe	Val	Asn	Trp	Leu	Leu	Ala	Gln...

Figure 6. Amino acid sequences for members of the glucagon family.

C peptide. Preproinsulin has been isolated and characterized as being a 24-amino-acid extension of the proinsulin molecule. The sequences for a number of insulin peptides from submammalian vertebrates are known, including the preproinsulin molecule from the hagfish, *Myxine glutinosa*. Hagfish preproinsulin is 115 amino acids in length (two amino acids longer than the mammalian counterparts) with a presequence of 26 amino acids, a 31-amino-acid B chain, a 37-amino-acid C peptide, and a 21-amino-acid A chain. The presequence shows considerable difference from mammalian presequences although it contains topological features common to the mammalian counterparts, such as a strong hydrophobic central region, which is important in facilitating secretory protein segregation and cleavage of the presequence. Although the overall homology of hagfish insulin with other known insulins is low (Figure 7), features such as conservation of cysteine residue location, presence of invariant positions common to all insulins, and location of basic amino acid residues in the prohormone molecule which are essential for proteolytic processing point to a conservation and retention of certain molecular characteristics essential to the biosynthetic process and, of course, to biological activity. Similarly, comparison of the carp insulin sequence to those of hagfish and mammalian insulins shows substitutions of 16 to 21 amino acids, respectively, indicating an evolutionary distance between the primitive hagfish and the carp comparable to the distance between the teleost and the mammal. Nevertheless, the important molecular features permitting biological activity are conserved.

Insulin in some cases exhibits several variants within a species as in some fish (tuna and toadfish) and the insulin I and insulin II in the rat and mouse. As yet, variation between the pancreatic insulin and insulin from the nervous system is not known. A number of other peptide hormones have been shown to bear some structural homologies with insulin, and this suggests that there may be a family of

A-CHAINS — 1, 5, 10, 15

HAGFISH: GLY - ILE - VAL - GLU - GLN - CYS - CYS - HIS - LYS - ARG - CYS - SER - ILE - TYR - ASN...

HUMAN: GLY - ILE - VAL - GLU - GLN - CYS - CYS - THR - SER - ILE - CYS - SER - LEU - TYR - GLN...

20

...LEU - GLN - ASN - TYR - CYS - ASN

...LEU - GLU - ASN - TYR - CYS - ASN

B-CHAINS — 1, 5, 10, 15

HAGFISH: ARG - THR - THR - GLY - HIS - LEU - CYS - GLY - LYS - ASP - LEU - VAL - ASN - ALA - LEU...

HUMAN: PHE - VAL - ASN - GLN - HIS - LEU - CYS - GLY - SER - HIS - LEU - VAL - GLU - ALA - LEU...

20, 25, 30

...TYR - ILE - ALA - CYS - GLY - VAL - ARG - GLY - PHE - PHE - TYR - ASP - PRO - THR - LYS - MET

...TYR - LEU - VAL - CYS - GLY - GLU - ARG - GLY - PHE - PHE - TYR - THR - PRO - LYS - THR

Figure 7. Amino acid sequences for hagfish and human insulins. Underlined amino acid residues are known to be invariant among hagfish, anglerfish, chicken, rat I and II, mouse I and II, pig, and human insulins.

insulinlike peptides comparable to the families of peptides described earlier. Nerve growth factor is found in the granular convoluted tubules of the submandibular gland and has significant homologies with proinsulin. The ovarian hormone relaxin consists of an A and B chain of 22 and 30 amino acids, respectively, which are linked by disulfide bridges in the same positions as insulin. The homologies in amino acid sequences are, however, somewhat limited. Finally, one of the insulinlike growth factors found in serum and nonsuppressible by insulin antibodies is a single-chain peptide of 70 residues cross-linked by three disulfide bridges. This molecule, insulinlike growth factor I, is classified a somatomedin (a growth-hormone-dependent group of hormones) and exhibits considerable insulinlike biological activity.

2.4. The Pancreatic Polypeptide Family

PP is localized almost exclusively to the pancreas in the mammalian GEP system although it may be distributed in other parts of the alimentary tract in submammalian vertebrates. Isolated originally as a contaminant of chicken insulin preparations, PP is a straight-chain polypeptide of 36 amino acids and a molecular weight of approximately 4200 daltons. The amino acid sequences for bovine, avian, and alligator PP are shown in Figure 8. Ovine, porcine, canine, and human pancreatic polypeptides differ from bovine PP in only one or two residues whereas avian PP differs from bovine PP at 44% of the positions. Alligator PP is unique among the PP molecules known in that it lacks the C-terminal amidated tyrosine and has instead a C-terminal amidated phenylalanine. This feature makes it very similar to the molluscan cardioexcitatory hormone FMRF-amide (phe-met-arg-phe-NH_2). As yet, no molecular variants within a species have been reported.

Two recently isolated and characterized peptides, PYY and NPY, show remarkable similarities to the pancreatic polypeptides (Figure 8). PYY is a 36-amino-acid straight-chain peptide isolated from porcine intestine and can be localized to endocrine cells in the distal part of the small intestine. NPY, on the other hand, is a 36-amino-acid peptide isolated from porcine brain with a 70% sequence homology with PYY and a 50% homology with PP. NPY is restricted in its localization to nervous tissue whereas PYY has been localized only to gut endocrine cells. The presence of NPY in the brain may explain previous demonstrations of PP-like material in this tissue. In addition to the close homologies with pancreatic polypeptide, these new peptides are remarkable in their identical number of amino acids and the unique carboxy-terminal tyrosines. The spectra of biological activities are similar as well, and it appears that PYY and NPY will help constitute a new, previously unrecognized family of peptides.

2.5. Substance P and Bombesin-Related Peptides

This grouping of peptides is characterized by a common carboxy-terminal-Leu-Met-NH_2 sequence for all but three of the structures shown in Figure 9A,

	5	10	15

BPP Ala - Pro - Leu - Glu - Pro - Glu - Tyr - Pro - Gly - Asp - Asn - Ala - Thr - Pro - Glu - Gln - Met - Ala...
APP Gly - Pro - Ser - Gln - Pro - Thr - Tyr - Pro - Gly - Asp - Asp - Ala - Pro - Val - Glu - Asp - Leu - Ile...
AMPP Thr - Pro - Leu - Gln - Pro - Lys - Tyr - Pro - Gly - Asp - Gly - Ala - Pro - Val - Glu - Asp - Leu - Ile...
NPY Tyr - Pro - Ser - Lys - Pro - Asp - Asn - Pro - Gly - Glu - Asp - Ala - Pro - Ala - Glu - Asp - Leu - Ala...
PYY Tyr - Pro - Ala - Lys - Pro - Glu - Ala - Pro - Gly - Glu - Asp - Ala - Ser - Pro - Glu - Glu - Leu - Ser...

20 25 30 35

...Gln - Tyr - Ala - Ala - Glu - Leu - Arg - Arg - Tyr - Ile - Asn - Met - Leu - Thr - Arg - Pro - Arg - Tyr - NH_2
...Arg - Phe - Tyr - Asp - Asn - Leu - Gln - Gln - Tyr - Leu - Asn - Val - Val - Thr - Arg - His - Arg - Tyr - NH_2
...Gln - Phe - Tyr - Asn - Asp - Leu - Gln - Gln - Tyr - Leu - Asn - Val - Val - Thr - Arg - Pro - Arg - Phe - NH_2
...Arg - Tyr - Tyr - Ser - Ala - Leu - Arg - His - Tyr - Ile - Asn - Leu - Ile - Thr - Arg - Gln - Arg - Tyr - NH_2
...Arg - Tyr - Tyr - Ala - Ser - Leu - Arg - His - Tyr - Leu - Asn - Leu - Val - Thr - Arg - Gln - Arg - Tyr - NH_2

Figure 8. Amino acid sequence of bovine pancreatic polypeptide (BPP), avian pancreatic polypeptide (APP), alligator pancreatic polypeptide (AMPP), neuropeptide Y (NPY), and peptide YY (PYY). Ovine PP differs from BPP by a serine at position 2; human PP differs from BPP by valine and aspartate at positions 6 and 23, respectively; and canine and porcine PP both differ from BPP by valine and aspartate at positions 6 and 11, respectively. Alligator PP is unique among the members of the PP family in that it possesses an Arg-Phe amide instead of the Arg-Tyr amide carboxy terminus.

although two of these show a -Phe-Met-NH_2 sequence. Moreover, the many other similarities include (1) identical phenylalanine and glycine residues at two positions in substance P, physalaemin, eledoisin, and kassinin, (2) a common -Gln-Trp-Ala-Val-Gly-His- sequence in bombesin, alytesin, ranatensin, and litorin, and (3) a near-complete identity, except positions 2 and 6, in the bombesin and alytesin molecules.

Mammalian representatives of this group include substance P, bombesin, and neurotensin, the latter being a somewhat atypical member of the group but with sufficient similarities to be included. Isolated originally from the hypothalamus, neurotensin (NT) has been localized to a number of tissues, including endocrine cells in the small intestine, and is structurally similar to xenopsin, a peptide isolated from the skin of *Xenopus laevis*. There appears to be a family of NT peptides with biological activity that is being conserved in the C-terminal portion of the molecule. Chickens possess two NT molecules, one which differs from bovine NT at only two amino acid residues, and a six amino acid NT variant called LANT-6. The latter NT has the four C-terminal residues identical to bovine and chicken NT but has Asn and Lys residues at positions 5 and 6, respectively, from the C-terminus.

Bombesin was isolated originally from amphibian (*Bombina bombina*) skin but has been localized to endocrine cells of the mammalian gut and lung and to parts of the nervous system. A separate peptide with characteristics very similar to those of bombesin has now been isolated from mammalian and avian tissues and is known as gastrin releasing peptide (Figure 9B). Substance P was isolated originally from mammalian brain and gut whereas its close molecular counterparts were isolated

A

	1 2 3 4 5 6 7 8 9 10 11 12 13 14
SUBSTANCE P	ARG - PRO - LYS - PRO - GLN - GLN - PHE - PHE - GLY - LEU - MET - NH_2
PHYSALAEMIN	PGL - ALA - ASP - PRO - ASN - LYS - PHE - TYR - GLY - LEU - MET - NH_2
ELEDOISIN	PGL - PRO - SER - LYS - ASP - ALA - PHE - ILE - GLY - LEU - MET - NH_2
KASSININ	ASP - VAL - PRO - LYS - SER - ASP - GLN - PHE - VAL - GLY - LEU - MET - NH_2
BOMBESIN	PGL - GLN - ARG - LEU - GLY - ASN - GLN - TRP - ALA - VAL - GLY - HIS - LEU - MET - NH_2
ALYTESIN	PGL - GLY - ARG - LEU - GLY - THR - GLN - TRP - ALA - VAL - GLY - HIS - LEU - MET - NH_2
RANATENSIN	PGL - VAL - PRO - GLN - TRP - ALA - VAL - GLY - HIS - PHE - MET - NH_2
LITORIN	PGL - GLN - TRP - ALA - VAL - GLY - HIS - PHE - MET - NH_2
NEUROTENSIN	PGL - LEU - TYR - GLU - ASN - LYS - PRO - ARG - ARG - PRO - TYR - ILE - LEU

B

PORCINE GRP	ALA - PRO - VAL - SER - VAL - GLY - GLY - GLY - THR - VAL - LEU - ALA - LYS...
CHICKEN GRP	ALA - PRO - LEU - GLN - PRO - GLY - GLY - SER - PRO - ALA - LEU - THR - LYS...

BOMBESIN	PGL - GLN - ARG - LEU - GLY - ASN - GLN - TRP - ALA - VAL - GLY - HIS - LEU - MET - NH_2
PORCINE GRP	...MET - TYR - PRO - ARG - GLY - ASN - HIS - TRP - ALA - VAL - GLY - HIS - LEU - MET - NH_2
CHICKEN GRP	...ILE - TYR - PRO - ARG - GLY - SER - HIS - TRP - ALA - VAL - GLY - HIS - LEU - MET - NH_2

Figure 9. (A) Amino acid sequences for substance P, bombesin, and related peptides. (B) Comparison of the amino acid sequence of bombesin with porcine and chicken gastrin-releasing peptides. pGL, pyroglutamate.

from amphibian skin (physalaemin, kassinin) and mollusk salivary gland (eledoisin). Of the peptides closely related to bombesin (alytesin, ranatensin, litorin), all were isolated from amphibian skin.

2.6. The β-Lipotropin/β-Endorphin Family

Methionine-enkephalin (Tyr-Gly-Gly-Phe-Met-NH_2) and leucine-enkephalin (Tyr-Gly-Gly-Phe-Leu-NH_2) are pentapeptides occurring in a ratio of about 4 to 1 in mammals, respectively, and are related to the β-lipotropin/β-endorphin group of peptides. Both the enkephalins and endorphins possess endogenous opiate activity. The endorphins are derived from a large biosynthetic precursor molecule, proopiomelanocortin, which, depending on the tissue, is processed to produce various complements of peptides including α- and β-melanotropin (α- and β-MSH), β-lipotropin (β-LPH), β-endorphin, adrenocorticotropin, and corticotropinlike peptide. The sequence for β-endorphin is found as $\beta\text{-LPH}_{61-91}$, and the sequence for the enkephalin ($\beta\text{-LPH}_{61-65}$) is found in β-endorphin and β-LPH. Although the enkephalin sequence is embedded in the proopiomelanocortin molecule, separate precursor molecules have been identified for these pentapeptides. Moreover, enkephalins have a different distribution in the body than do the endorphin derivatives of proopiomelanocortin. Enkephalins are localized in the antral gastrin cells and in the GEP cells in the duodenum and pancreas whereas the endorphins are localized to the pituitary and to regions of the brain different from the enkephalins.

GEP peptides without clear family associations include somatostatin and motilin. Somatostatin is a tetradecapeptide (Figure 10) isolated originally from the hypothalamus but with a distribution throughout the nervous system and the gastrointestinal tract. A number of somatostatinlike peptides, some up to 10 times the size of somatostatin, have been reported and presumed to represent precursor forms of the hormone. One of these, somatostatin-28, isolated from duodenum and proposed to be prosomatostatin, has been found to be secreted and possess biological activity and therefore may not be the precursor peptide. Cloning of the somatostatin gene from anglerfish (*Lophius americanus*) islets has demonstrated two genes coding for somatostatin in this species. Preprosomatostatin I and preprosomatostatin II are 121 amino acids (mol. wt. 13,300) and 125 amino acids (mol. wt. 14,100), respectively, with corresponding prosomatostatins of 97 amino acids (mol. wt. 10,600) and 101 amino acids (mol. wt. 11,600). These values are comparable to the 12,500-dalton precursor reported for rat prosomatostatin. Figure 10 shows the relationship of the carboxy-terminal portions of the anglerfish preprosomatostatins to mammalian somatostatins-14 and -28. The tetradecapeptides corresponding to anglerfish somatostatins I and II differ only at positions 7 and 10 and, moreover, anglerfish somatostatin I is identical to the corresponding portions of the mammalian somatostatin molecules. Considerable homology also exists in parts of the precursor regions of the anglerfish preprosomatostatins and somatostatin-28, indicating some degree of conservation of these molecules during the course of evolution.

Motilin is a peptide of 22 amino acids (mol. wt. 2700, Figure 11) found in the

	-10
Anglerfish I	... Leu - Glu - Arg - Ala - Ala - Ser - Gly - Gly - Pro - Leu - Leu - Ala - Pro...
Anglerfish II	... Leu - Glu - Arg - Ser - Val - Asp - Ser - Thr - Asn - Asn - Leu - Pro - Pro...
Somatostatin-28	Ser - Ala - Asn - Ser - Asn - Pro - Ala - Met - Ala - Pro...
Somatostatin-14	

-1 1 10
...Arg - Glu - Arg - Lys - Ala - Gly - Cys - Lys - Asn - Phe - Phe - Trp - Lys - Thr - Phe - Thr - Ser - Cys
...Arg - Glu - Arg - Lys - Ala - Gly - Cys - Lys - Asn - Phe - Tyr - Trp - Lys - Gly - Phe - Thr - Ser - Cys
...Arg - Glu - Arg - Lys - Ala - Gly - Cys - Lys - Asn - Phe - Phe - Trp - Lys - Thr - Phe - Thr - Ser - Cys
Ala - Gly - Cys - Lys - Asn - Phe - Phe - Trp - Lys - Thr - Phe - Thr - Ser - Cys

Figure 10. Amino acid sequences for somatostatin-14 and -28 and partial sequences for anglerfish preprosomatostatins I and II. (From Hobart, P., Crawford, R., Shen, L., Pictet, R., and Rutter, W. J., 1980, Cloning and sequence analysis of cDNAs encoding two distinct somatostatin precursors found in the endocrine pancreas of anglerfish. *Nature* **288:**137–141.)

PHE - VAL - PRO - ILE - PHE - THR - TYR - GLY - GLU - LEU - GLN...

...ARG - MET - GLN - GLU - LYS - GLU - ARG - ASN - LYS - GLY - GLN

Figure 11. Amino acid sequence for motilin.

M (nonenterochromaffin) cells of the upper small intestine. A larger form has been isolated and localized to the EC_M enterochromaffin cell of the small intestine.

As mentioned previously, many, if not all, of the peptides identified in GEP cells are found to have molecular counterparts within the nervous system. This peptidergic branch of the autonomic nervous system functions to control gut motility, to regulate endocrine and exocrine secretion, and is possibly involved in tropic maintenance of the gut. Table VII shows a partial list of the peptides and their chemical characteristics which have been localized to enteric neurons. Peptidergic nerves are found along the entire length of the gastrointestinal tract and within each layer of the gut wall (Figure 12). The distribution is more uniform than that of the endocrine cells, which are often restricted to particular regions of the gastrointestinal tract. Moreover, some peptides, such as VIP, enkephalin, and CCK-8, are more likely restricted to or predominate in the neural rather than the endocrine elements of the gut. In some cases, the neuropeptide is quite different from its counterpart in the endocrine cell; e.g., CCK-33 is the predominant form of cholecystokinin found in the intestinal endocrine cells, whereas CCK-12, CCK-8, and CCK-4 are generally restricted to nerves.

Table VII
Chemical Characteristics of Peptides Localized in Enteric Neurons[a]

Peptide	Molecular variant	No. of amino acids	Molecular weight	Sequence similarities
Bombesin	Amphibian	14	1620	Substance P
	Porcine	27	2806	Ranatensin
	Chicken	27	2841	Litorin
CCK-8	Octa-	8	1143	Gastrin/CCK family
Enkephalin	Leu^5	5	574	
	Met^5	5	556	Endorphin
Motilin	Porcine	22	2700	2 forms in gut and plasma
Neurotensin	Human	13	1673	Xenopsin
Substance P	Bovine	11	1348	Physalaemin
				Eledoisin
TRH	Ovine	3	362	Not known
VIP	Porcine	28	3326	Secretin
				Glucagon
				PHI

[a](From Bishop, A. E., Ferri, G-L., Probert, L., Bloom, S. R., and Polak, J. M., 1982, Peptidergic nerves. *Scand. J. Gastroenterol.* **17**(suppl. 71):43–59.)

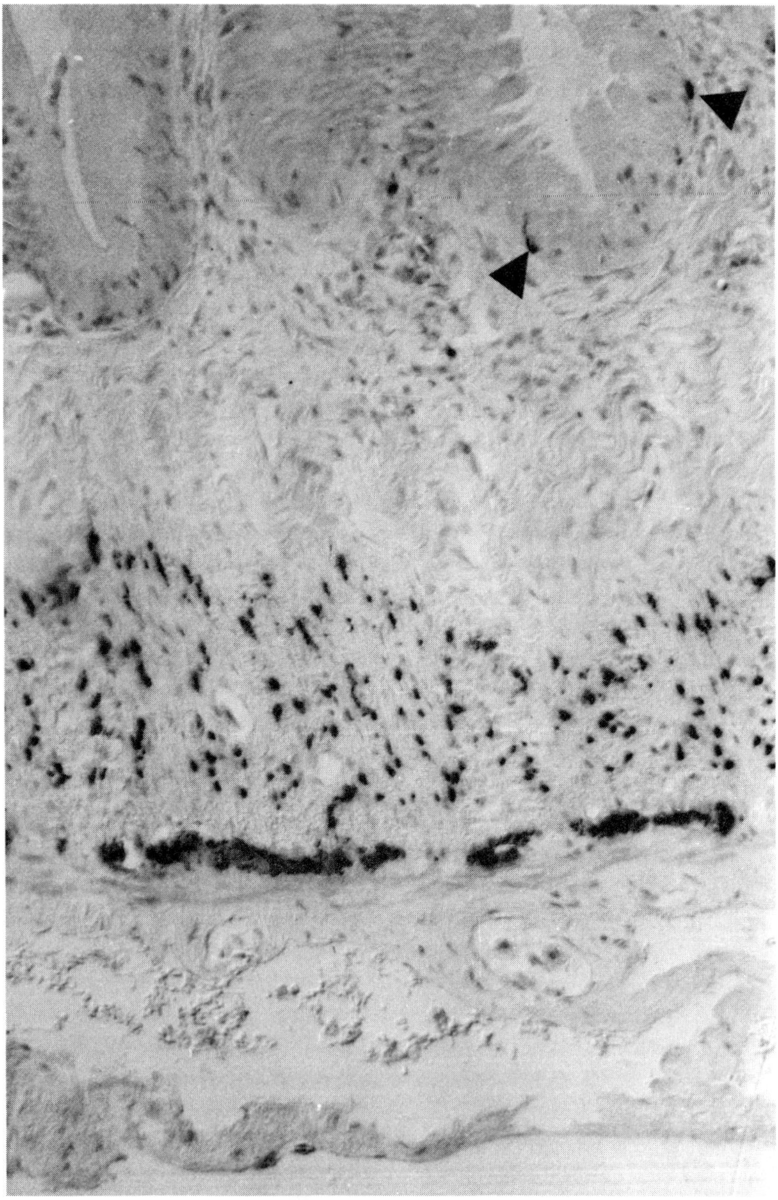

Figure 12. Section of eel (*Anguilla anguilla*) intestine immunostained by the peroxidase antiperoxidase method for VIP. Cells and fibers immunoreactive for VIP are seen in the middle portion of the field associated with the myenteric plexus. GEP cells immunostaining for VIP (arrows) are seen in the upper portion of the section localized within the duodenal mucosa epithelium. VIP tends to be more restricted to the enteric nervous system in higher vertebrates and is found less frequently in GEP cells. (Micrograph courtesy of S. Van Noorden.)

3. Modes of GEP Hormone Delivery

Delivery of GEP peptides to their target cells may occur by several different modes (Figure 13). Most of these, neurotransmitter, neuroendocrine, and endocrine, are familiar modes of delivery. Neurotransmitters such as acetylcholine and norepinephrine transmit information between elements of the nervous system or between the nervous system and effector cells. Some of the neuropeptides mentioned previously and localized to enteric neurons also belong to this category and excite, inhibit, or in some way modulate neuron or effector cell activity. The distance over which neurotransmitters act is very short, usually just the distance across a synaptic cleft. The neuroendocrine mode of delivery involves neurons that release their hormones into the circulation, the hormones acting on target tissues distally to the site of release. Examples of neuroendocrine systems include the well-known hormones of the posterior pituitary, oxytocin and vasopressin, as well as the hypothalamic regulatory hormones, which control secretion from the anterior and intermediate lobes of the pituitary. The endocrine cells, of course, are localized to glands or to diffuse systems that release their products into the blood stream for transport to target tissues distal to the site of hormone production and release.

Alternatively to the neurotransmitter, neuroendocrine, and endocrine regulation of cell functions, paracrine control is a concept which assumes that peptide hormones may be employed as regulatory substances at the local level. Paracrine

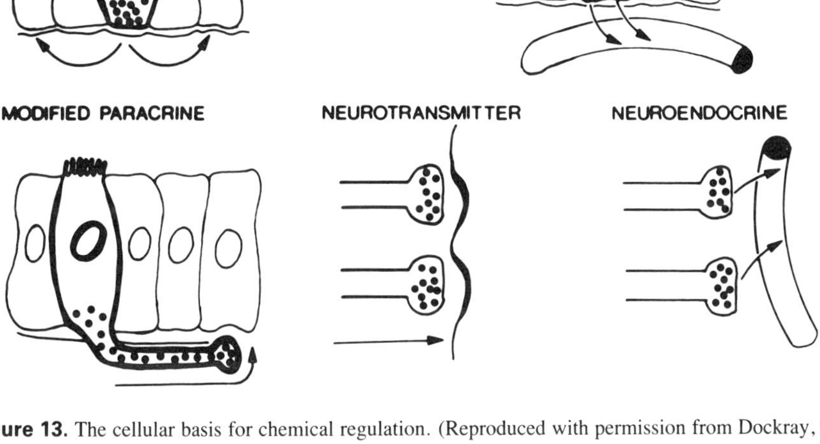

Figure 13. The cellular basis for chemical regulation. (Reproduced with permission from Dockray, G. J., 1981, Brain-gut peptides. *Viewpoints Dig. Dis.* **13**:5–8.)

control involves the release of peptides from open- or closed-type paraneurons and diffusion of the peptides through the extracellular spaces to affect activities of adjacent or neighboring cells. Some of the cells in the gut that are suspected of participating in paracrine control are modified to have long, granule-filled processes, which increases the number of contacts with other cells (see, for example, Figure 3B). Paracrine regulation permits a control of cellular activities in response to local conditions and allows for compartmentalization of potent regulatory substances without involving the systemic circulation.

Somatostatin plays a central role in the concept of paracrine control. This tetradecapeptide is found throughout the gut and in the D cells of the endocrine pancreas. Somatostatin inhibits release of all of the pancreatic hormones from the islets, and it can inhibit the release of all the known circulating gastrointestinal hormones. Moreover, it can block gastrin-induced acid secretion, CCK-induced gall bladder contraction, and secretin-induced bicarbonate secretion. Cells containing somatostatin differ somewhat in morphology, the fundic and pancreatic somatostatin cells being closed-typed paraneurons whereas antral and small-intestine somatostatin cells are open-type paraneurons. These morphological differences probably relate to the information the cells can process, the open-type cells having the advantage of being able to monitor the luminal contents of the gut.

The possibilities of paracrine controls are extensive given the many different peptides produced by the GEP cells. In the pancreatic islets, somatostatin does hold a central role in presumed paracrine regulation of islet function, but the other islet hormones also appear to affect secretory activity of other cell types (cf. Chapter 17). As yet, paracrine control mechanisms are only poorly worked out and remain one of the important new aspects of GEP endocrine cell physiology to be elucidated.

16

The Gastrointestinal Endocrine System

1. Introduction

Gut endocrinology occupies a unique place in the history of endocrinology by virtue of the discovery of *secretin* by Bayliss and Starling in 1902. They asked for a better term than ''internal secretion'' and a classicist coined ''hormone,'' used for the first time publicly in 1905. Their experiments with secretin led to the use of the word ''hormone,'' but more important, they led to formulation of the concept of bloodborne chemical messengers without intervention of the nervous system. A second gastrointestinal hormone, *gastrin,* was discovered in 1905, and a third, *cholecystokinin* (CCK), in 1928. These appeared to cover the endocrine control of the gut until a couple of decades ago. Difficulties lay in the diffuse nature of potential secreting cells. Neither collection nor identification nor extirpation/replacement studies were easy. Clinical pathologies have often stimulated experimental research, but few gut diseases of endocrine origin have been identified. Endocrine tumors of the gut and pancreas are rare; even one of the most common gastrointestinal tract disorders, duodenal ulcer, cannot be ascribed unequivocally to hypersecretion of gastrin. However, this is the case in the Zollinger–Ellison syndrome in which a gastrinoma giving hypersecretion of gastrin can be causative of duodenal ulceration. Several gut pathologies can be correlated with alteration of regulatory peptides in the enteric neurons rather than the endocrine cells. Loss of peptidergic innervation, a consequence of ganglion cell loss, is seen in the congenital Hirschsprung's disease and in Chagas' disease following infection with *Trypanosoma cruzi.* Similarly, grass sickness in horses, a wasting disease of unknown etiology occurring in the spring, is also characterized by a loss of peptidergic innervation in the bowel. In contrast, Crohn's disease with inflammation of the bowel is characterized by dramatic hyperplasia of vasoactive inhibitory polypeptide nerves. Clear advances in understanding were made possible by improved methods of protein fractionation and purification which established the necessary basis for the recent explosion in the field attributable to radioimmunoassay and immunocytochemical methodology.

2. Hormones of the Mammalian Gastrointestinal Tract

The major part of our knowledge comes from some eutherian mammals, for example, man, sheep, cattle, and various laboratory animals. Because any description of gastrointestinal (GI) hormones in nonmammalian vertebrates must have as its reference point the more precise analyses of mammals that have been gathered over many years, we start with this information.

The GI tract reaches from the *stomach* to the *anus*. In man, the *small intestine* is about 20 ft (6 m) in length, reaching from the *pyloric sphincter* of the stomach to the large intestine at the *ileocecal* junction. The small intestine has three defined areas, the *duodenum, jejunum,* and *ileum*. The *large intestine* is about 5 ft in length (1.5 m) and consists of the *cecum, colon,* and *rectum*.

The GI endocrine system is diffuse, and the endocrine cells are distributed differentially in the mucosal epithelium along the length of the digestive tract. Table I summarizes the location, mode of transmission, and general actions of the major peptides found in the mammalian GEP system and associated enteric neurons. The stomach and the colon are relatively restricted in the spectrum of endocrine cell types they exhibit, containing mostly gastrin cells (stomach) and enteroglucagon cells (colon). The small intestine, on the other hand, is quite rich in its variety of different cells that synthesize and release secretin, gastrin, cholecystokinin, motilin, gastric inhibitory polypeptide, neurotensin, and enteroglucagon. Somatostatin is found in endocrine cells scattered throughout the GI tract. These cells lie in close association with the other endocrine cells presumably affecting the latter cells via paracrine regulatory actions. Those somatostatin cells not in close proximity to other GEP endocrine cells presumably act locally to regulate the secretory or contractile activity of other enteric cells. Enteric neurons containing vasoactive intestinal polypeptide (VIP), substance P, enkephalin, bombesin, and various other peptides are located throughout the GEP system and act to modulate the secretory, vascular, and motor activity of the various regions of the gut.

2.1. General Actions of Gut Hormones

The gut hormones act to regulate the processes by which the complex constituents of ingested food are reduced to readily absorbable molecules for subsequent delivery by the blood to the peripheral tissues. This regulation involves the control of gastric acid secretion, gallbladder contraction, pancreatic bicarbonate, and enzyme secretion as well as GI tract motility. The correct function of the digestive

Table I
Distribution and Function of Gastroenteropancreatic Hormones and Neuropeptides[a]

Peptides	Tissue of origin	Possible action(s)
Hormones		
Gastrin	Antrum, upper small intestine	Stimulates gastric acid secretion and gastric mucosa growth
Pancreatic polypeptide	Pancreas	Inhibits pancreatic enzyme secretion and gallbladder contraction
Secretin	Duodenum, jejunum	Stimulates pancreatic HCO_3^- secretion
Cholecystokinin	Small intestine	Stimulates pancreatic enzyme secretion and gallbladder contraction
Gastrin-releasing peptide (bombesin)	Nonantral stomach, small intestine	Stimulates gastrin release
Motilin	Small intestine	Stimulates upper-GI-tract muscle contraction
Gastric inhibitory polypeptide (GIP)[b]	Small intestine	Inhibits gastric acid secretion, stimulates insulin release
Neurotensin	Ileum	Inhibits gastric motor activity
Enteroglucagon	Ileum, colon	Stimulates growth of intestinal mucosa
Local hormones		
Somatostatin	All areas, esp. upper GI tract and pancreas	Inhibits secretion and contraction
Neurotransmitters		
Vasoactive intestinal polypeptide (VIP)	All areas	Stimulates vasodilation, smooth muscle relaxation, and pancreatic exocrine and endocrine secretion
Substance P	All areas	Stimulates smooth muscle contraction
Enkephalin	All areas	Stimulates smooth muscle contraction, inhibits pancreatic and gastric acid secretion
Peptide histidine isoleucine	All areas	Similar to VIP

[a]Modified from Bloom, S. R., 1980, Gastro-intestinal hormones, in: *Polypeptide Hormones* (R. F. Beers, Jr., and E. G. Bassett, eds.), Raven Press, New York, pp. 353–355.
[b]GIP is also known as "glucose-dependent insulinotropic peptide."

tract is dependent on the appropriate secretory activity of the various gut endocrine cells acting in concert with other humoral and neuronal factors.

2.2. Synthesis and Secretion of Gut Hormones in Mammals

Biosynthesis and secretion of gastrointestinal hormones begins with the formation of the preprohormone on the ribosome/messenger RNA complex. Cotranslational processing occurs as the nascent peptide is translocated into the cisterna of the rough endoplasmic reticulum. The resulting prohormone is then transported to the Golgi where it is processed proteolytically and packaged in membrane-bound secretory granules. Release of the hormones is stimulated by neural, circulating, luminal, and/or local factors and presumably involves exocytotic mechanisms mediated by calcium-dependent activation of the microtubular–microfilamentous system. Gut hormones may be released into the circulation, discharged into the gut lumen, or used locally.

2.2.1. Gastrin

The principal function of gastrin is regulation of acid secretion by the parietal cells of the stomach. Acid release is associated with the ingestion of food and is dependent on the combined stimulatory effects of acetylcholine from cholinergic neurons, histamine from local sources, and gastrin from antral and duodenal cells. Acid secretion can be divided into three phases:

1. The *cephalic phase* is mediated by vagal cholinergic neurons and involves acetylcholine stimulation of acid and gastrin secretion. The trigger is sensory information including the sight, taste, smell, or anticipation of food.
2. The *gastric phase* is dependent on the presence of food in the stomach and has two components: secretion stimulated by (a) distention of the stomach and (b) constituents of food. Gastric distention by itself, however, will not produce maximal acid release, but rather components of the food, especially amino acids and partially digested protein, are necessary.
3. The *intestinal phase* begins when food enters the duodenum and continues for several hours. Both stimulation and inhibition occur. The presence of acidic food, for example, may cause the release of gastrin-releasing peptide (GRP), the mammalian counterpart to bombesin. Inhibition of acid release can be mediated by several mechanisms, including (a) the acid-induced local release of somatostatin from gastric tissue or from extragastric sources (e.g., the pancreas) and (b) the inhibitory effects of secretin, CCK, and gastric inhibitory polypeptide (GIP) acting on gastrin secretion and acid release directly. Gastrin release is also inhibited by a decline in antral pH, hence demonstrating a feedback relationship between it and acid secretion.

The release of gastrin during food ingestion is reflected in the circulating levels. A biphasic pattern of release includes an initial spike lasting about 10 min

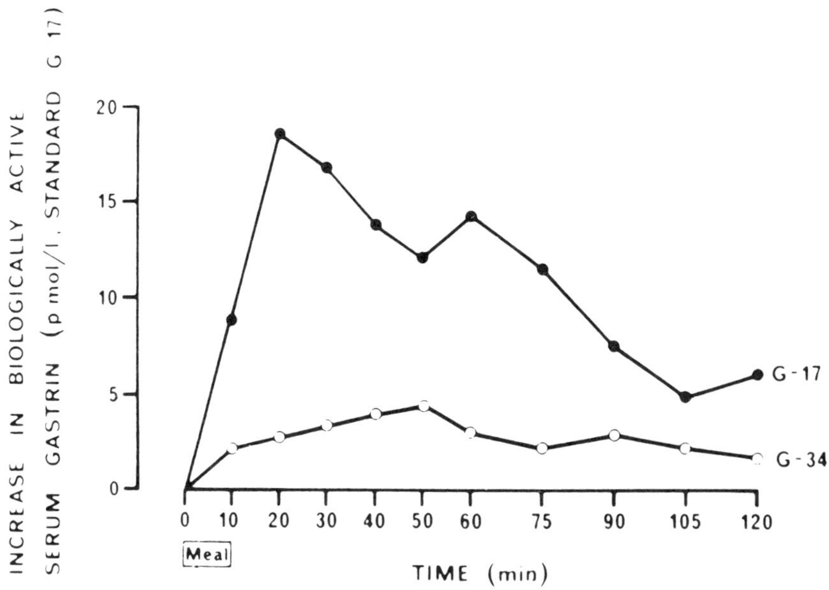

Figure 1. Increase in serum gastrin-17 (G-17) and gastrin-34 (G-34) in response to the ingestion of a meal. (From Dockray, G. J., and Taylor, I. L., 1976, Heptadecapeptide gastrin: Measurement in blood by specific radioimmunoassay. *Gastroenterology* **71**:971.)

which is twofold greater than the basal levels (Figure 1). This is followed by a prolonged second phase reaching approximately the same concentration in the plasma, but lasting over 2 hr. In the basal state gastrin-34 comprises about 75% of the circulating gastrin and is derived from antral and duodenal sources that are 90% and 60% gastrin-17, respectively. Postprandially, the level of circulating gastrin-17 rises to about 70%. During the declining phase of gastrin release, the lowering of the circulating gastrin levels may be due in part to a shift from secretion of gastrin into the circulation to secretion into the GI tract lumen (Figure 2).

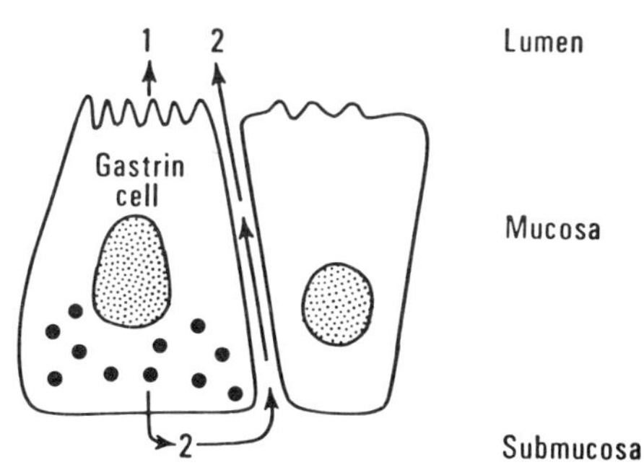

Figure 2. Intraluminal secretion of gastrin is postulated to occur by a direct route (1) or by an indirect route (2) involving the extracellular space. [From Miller, L. J., and Go, V. L. W., 1980, Intraluminal gastrointestinal hormones and their radioimmunoassay, in: *Gastrointestinal Hormones* (G. B. J. Glass, ed.), Raven Press, New York, pp. 863–874.]

2.2.2. Cholecystokinin

CCK release is dependent on a variety of different stimuli which may be presented to the duodenum. Ingestion of food causes plasma levels of CCK in the human to increase twofold within 15 min and last for 45–60 min before declining to basal levels (approximately 4 ng/ml). Proteins, peptides, and amino acids are effective CCK secretagogues, especially the amino acids leucine, phenylalanine, tryptophan, valine, and methionine, depending on the species tested. Triglycerides are also effective CCK releasers, but this is dependent on prior digestion. Hence, free fatty acids (those 10 carbons or more in length) are particularly potent secretagogues. Hydrogen ion is only a weak stimulus for CCK release with the molar response ratio of secretin to CCK being 24:1.

2.2.3. Secretin

Secretin is released from cells in the mucosa of the duodenum and jejunum in response to acidified food entering from the stomach. The magnitude of release is proportional to the amount of acid delivered to the duodenum (Figure 3). Intraduodenal infusion of acid into the canine intestine showed a threshold of pH 4.5 for the stimulation of secretin release, which is identical to the threshold for eliciting pancreatic water and bicarbonate secretion. In addition, oleate and bile salts also stimulate secretin release. Plasma levels of secretin rise from basal levels of around 5 pg/ml to about 20–30 pg/ml within 10–20 min after ingestion of a meal and remain elevated for several hours (Figure 3). Somatostatin and metenkephalin both inhibit secretin release, but no other gut hormones appear to affect its secretion.

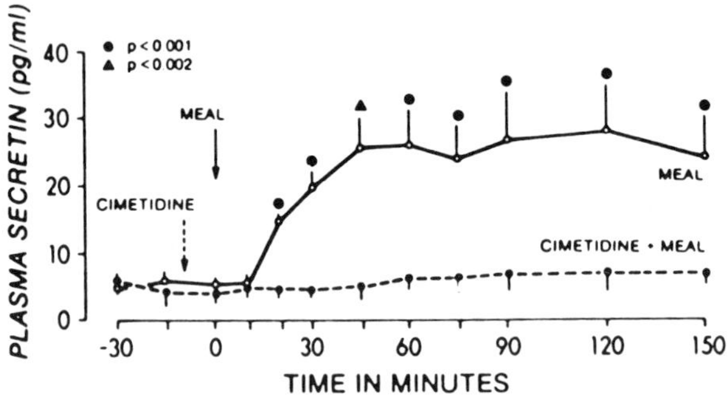

Figure 3. Plasma levels of secretin in dogs measured in response to a meal or in response to a meal plus the intravenous administration of cimetidine, a histamine blocker. The suppression of gastric acid secretion by cimetidine abolishes the postprandial rise in secretin indicating the dependence of secretin release on the acidified food entering the duodenum. (From Kim, M. S., Lee, K. Y., and Chey, W. Y., 1979, Plasma secretin concentrations in fasting and postprandial states in dog. *Am. J. Physiol.* **236:** E539–E544.)

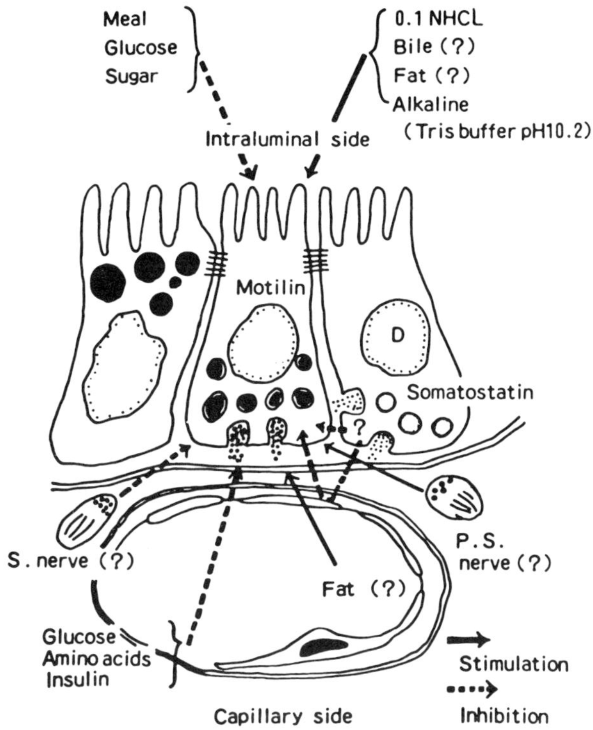

Figure 4. Diagrammatic representation of the factors regulating motilin release. [From Itoh, Z., 1980, *Recent Advances in GI Hormone Research,* Vol. 3 (Z. Itoh, ed.), Chungai Igakusha, Tokyo, pp. 152–161.]

2.2.4. Motilin

The two molecular forms of motilin are secreted in parallel from cells in the gastric and duodenal mucosa. Release can be triggered by bombesin, by fat, and by stomach distention, but it is inhibited by intraduodenal or intravenous glucose (Figure 4). In humans, intraduodenal instillation of acid increases circulating motilin whereas instillation of alkali has the opposite effect. Stimulation of motilin release does not appear to be under cholinergic control since atropine is ineffective in blocking its secretion. On the contrary, basal levels of motilin secretion do appear to be atropine sensitive. Plasma levels of motilin are elevated significantly 15 min after ingestion of a meal and peak 30–45 min postprandially, having risen from about 50 to 70 pmol/liter (Figure 5). Spikes of motilin release also occur interprandially at approximately 100-min intervals and are responsible for the interdigestive migrating contractions (see Figure 13). Somatostatin and pancreatic polypeptide are potent inhibitors of motilin release: somatostatin arising locally or from the circulation, and pancreatic polypeptide deriving from the pancreas and acting as a feedback signal generated from peripheral parts of the digestive system.

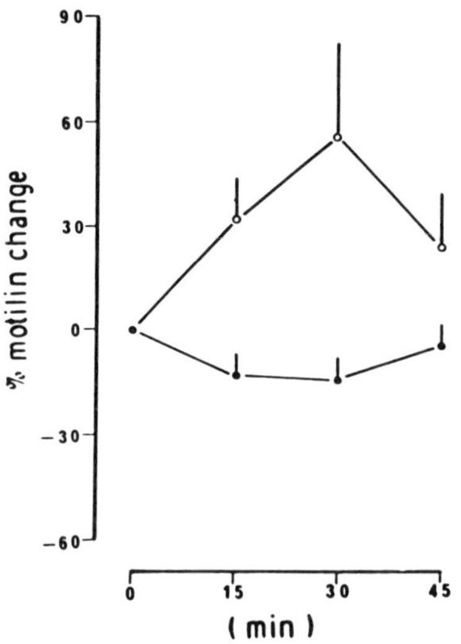

Figure 5. Percentage change in plasma motilin in response to the ingestion of breakfast in normal human subjects (open circles) and in corresponding controls where no breakfast was ingested (closed circles). [From Christofides, N. D., and Bloom, S. R., 1981, Motilin, in: *Gut Hormones* (S. R. Bloom and J. M. Polak, eds.), Churchill Livingstone, Edinburgh, pp. 273–279.]

2.2.5. Gastric Inhibitory Polypeptide

GIP is synthesized and released from cells in the duodenum and jejunum. Basal circulating levels of GIP in humans range from nondetectable to approximately 500 pg/ml, which may increase fivefold or more after a solid or liquid meal (Figure 6). Oral, but not intravenous, glucose is a potent GIP secretagogue, but triglyceride is the more so. Serum GIP rises within 5 min after oral ingestion of glucose and is sustained for 120–180 min, with a peak of about 800 pg/ml around 45–60 min. Ingestion of the same amount of fat, on the other hand, produces a rise in serum GIP in 15 minutes, which remains elevated for 3–4 hr with peak levels between 800 and 1300 pg/ml. Amino acids (orally, but not intravenously) also are GIP secretagogues. A mixture of amino acids, lacking those stimulatory for CCK release, will stimulate GIP secretion, indicating that the release of GIP is independent of CCK secretion. GIP release is also stimulated by the intraduodenal instillation of acid, but the degree varies in mammalian species. Gastrin and CCK potentiate glucose-induced GIP release but do not have any stimulatory actions of their own. Bombesin also stimulates GIP secretion whereas insulin (whose release is stimulated by GIP) suppresses fat-induced GIP release but apparently has no effect on glucose-induced GIP release. Somatostatin inhibits GIP secretion, and glucagon does likewise, but only at supraphysiological doses.

2.2.6. Enteroglucagon

Enteroglucagon is secreted from intestinal open-type GEP cells in response to a meal. The magnitude of enteroglucagon release is related to the size and composition of the meal, the size determining the length of intestine penetrated by the food

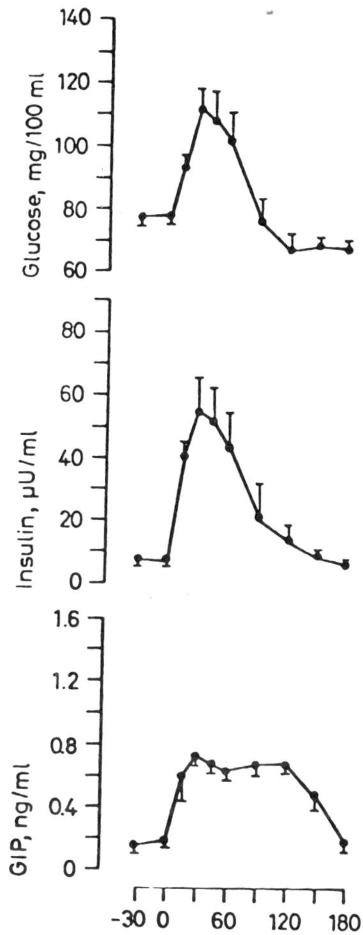

Figure 6. Serum glucose, insulin, and GIP levels in response to oral glucose in normal human males. Ingestion of glucose is one of the most potent triggers of GIP release, and the increased circulating GIP in turn augments the glucose effect on pancreatic B-cell release of insulin. [From Spitz, I. M., Leroith, D., Trestian, S., Ebert, R., and Creutzfeldt, W., 1980, Effect of acid on GIP secretion in man, in: *The Entero-Insular Axis* (W. Creutzfeldt, ed.), *Frontiers of Hormone Research,* Vol. 7, S. Karger, Basel, pp. 173–180.]

and, hence, the number of cells stimulated. Intraluminal glucose and intraluminal emulsified triglyceride appear to be the most effective enteroglucagon secretagogues. Food stimulation of release, however, does not appear to involve distention-mediated neural reflexes since stimulation of vagal or splanchnic nerves has no effect on enteroglucagon release. Circulating levels of enteroglucagon are elevated two- to threefold during pregnancy, and in the neonate, a rise in circulating enteroglucagon is observed with the transition from placental to oral feeding.

2.2.7. Neurotensin

Although neurotensin is present in central and peripheral nerves as well as in enteric nerves and endocrine cells, postprandial elevation of plasma neurotensin appears to be derived principally from the gut. Meals high in fat or infusion of lipid into the small intestine produces a significant increase in circulating neurotensin, whereas instillation of other nutrients such as glucose or amino acids has little or no effect on neurotensin release. Two peaks of neurotensin are observed after ingestion of lipid, one at 30 min and one at 90 min; by 180 min plasma levels of neurotensin have returned to baseline.

2.2.8. Substance P

Substance P is present in endocrine cells and neurons of the gut as well as being widespread throughout the central and other parts of the peripheral nervous system. Ingestion of a meal induces an increase in circulating substance P which is derived from the gut.

2.2.9. Bombesin

Circulating bombesin (GRP in birds and mammals) increases in response to ingestion of a meal. Presence of acidified food in the antrum and duodenum presumably stimulates release from the endocrine cells, but other stimuli may cause release of bombesin from enteric neurons as well.

2.2.10. Vasoactive Intestinal Polypeptide

In the mammalian gastrointestinal tract vasoactive intestinal polypeptide (VIP) is localized exclusively to the enteric neurons; hence its release is primarily neurogenic. VIP-containing neurons innervate specific anatomical sites (blood vessels, smooth muscle, epithelium) where VIP may act as a local hormone, neurotransmitter, or neuromodulator. Vagal stimulation, gastric distention, mechanical stimulation of the small intestine, or application of intraduodenal ethanol or acid all stimulate VIP release, which can be reflected in a rise in VIP in the venous effluent from these tissues (Figure 7).

2.2.11. Somatostatin

Ingestion of a meal or intragastric or intraduodenal instillation of various nutrients or acid promotes varying magnitudes of somatostatin release from the stomach and pancreas although its postprandial rise in plasma appears to be derived primarily from the pancreas. Somatostatin release from the pancreas is stimulated via several mechanisms: (1) by the same factors that stimulate insulin release during the postabsorptive state (e.g., glucose, amino acids) and (2) by signals generated from intraluminal presence of food and food constitutents. Intraluminal application of specific nutrients stimulates the release of glucose or amino acids. These signals also effect release of somatostatin from other parts of the GI tract. Experimentally, intragastric instillation of glucose, fat, or casein hydrolysate promotes somatostatin

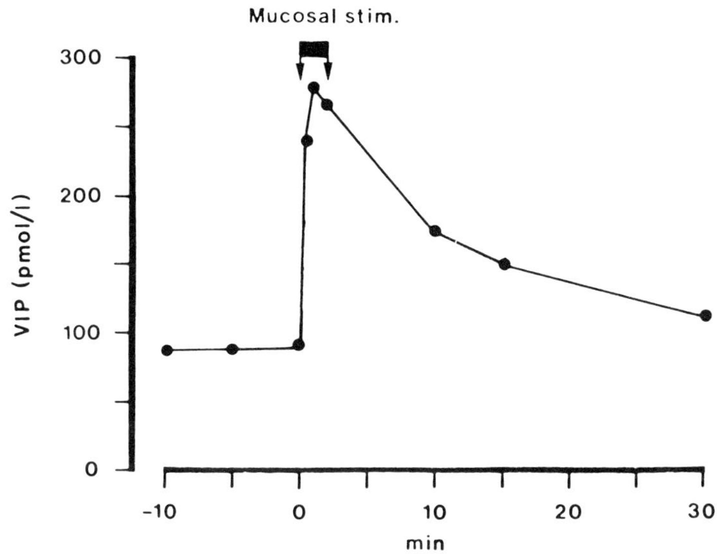

Figure 7. VIP release induced by mechanical stimulation of the small intestine mucosa in the cat. The release of VIP is from enteric neurons containing this neuropeptide and is abolished by agents that block nerve conduction. [Redrawn from Fahrenkrug, J., 1981, Physiological role of VIP in digestion, in: *Gut Hormones* (S. R. Bloom and J. M. Polak, eds.), Churchill Livingstone, Edinburgh, pp. 385–391.]

release from the fundus, antrum, and pancreas. In addition, glucose, fat, or acid put into the duodenum promotes somatostatin release though the degree is less for the stomach, whereas the converse is true for fat. Administration of acid in the stomach stimulates pancreatic and antral, but not fundic, release of somatostatin whereas intraduodenal acid promotes a sharp and significant rise in secretion from all three regions. Nutrient and acid stimulation of somatostatin release from the pancreas may be mediated by nerves or may be mediated via several gut hormones since CCK, gastrin, secretin, and GIP all stimulate somatostatin release from the pancreatic islets. Somatostatin secretion from the lower gut can also be stimulated by the instillation of nutrients although the digesta normally does not reach that far. The majority of the somatostatin secreted by gastrointestinal tissue, and not utilized locally, is released into the circulation although a small fraction also may be secreted into the GI tract lumen.

3. Possible Gut Hormones in Invertebrates and Protochordates

When we consider that the process of digestion varies only in details among the vertebrate classes, presumably, then, the basic mechanisms and patterns were developed early in vertebrate evolution. This may mean inheritance from protochordate ancestors, with regulatory mechanisms well established at the emergence of the vertebrates. Furthermore, homologs to gastroenteropancreatic (GEP) hormones have been identified (most immunocytochemically, some chemically) in a variety of subvertebrate groups. The conclusion is that counterparts of GEP endocrine cells had long been established in evolution. This is emphasized by the finding of peptides and neuropeptides characteristic of mammals in unicellular and simple multicellular organisms, both prokaryotes and eukaryotes (e.g., *Escherichia coli, Tetrahymena pyriformis, Neurospora crassa,* or *Aspergillus fumigatus*). In addition to insulin, substances similar to somatostatin, glucagon, and CCK occur. Apart from GEP hormones it is of interest to note that ACTH-, endorphin-, arginine vasotocin-, and calcitoninlike molecules have also been identified. Their possible functions are conjectural. They may transmit messages from cell to cell in a manner reminiscent of cyclic AMP or some biogenic amines which are secreted extracellularly by some unicellular organisms. Moreover, the presence of some of these peptides in prokaryotes suggests that this system of intercellular communication was in place before the divergence of plants and animals.

The presence of GEP hormone counterparts in extracts of unicellular organisms suggests that these peptides can probably be found in all invertebrate groups. Indeed, considerable evidence points to the presence of representatives of the major GEP hormone families and groups in most invertebrate species examined to date. With respect to pancreatic endocrine cell precursors, vertebrate B-cell homologs have been identified in several mollusks (*Mytilus edulis, Anodonta cygnea, Unio pictorum*) in the form of open-type secretory cells containing immunostainable insulin and located in the intestinal mucosa. Alimentary tract hormones have been identified both immunocytochemically and immunologically in a variety of species. Gastrin/CCK-like immunostainable material has been localized in ectodermal nerve cell bodies in the mouth region of the hydra, *Hydra attenuata,* and in cells within the mesoglia of tentacles and oral disc and in coarse fibers in all parts of the sea anemone, *Tealia felina*. The gastrin/CCK-like staining in these coelenterates is more CCK-like than gastrinlike whereas the material extracted from the blood and gastrointestinal tissue of the mollusks *Aplysia californica* and *Otala lactea* bears considerable structural and immunological similarities to mammalian gastrin. Among other gastrointestinal hormones, enkephalin-, secretin- and substance P-like material has been localized in cells of the alimentary tract in the giant African snail, *Achatina fulica*.

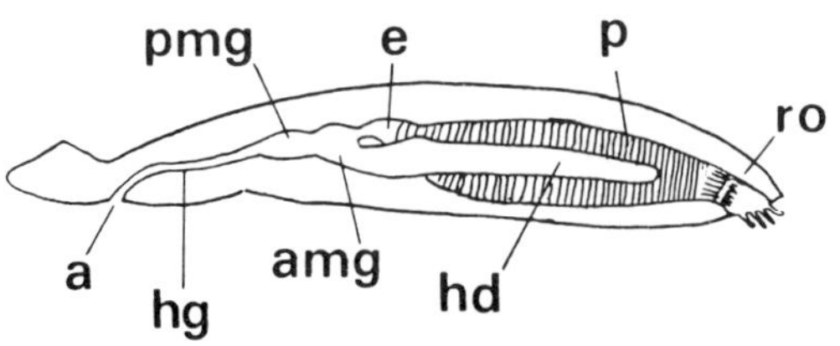

Figure 8. Diagrammatic representation of the digestive tract of the lancelet, *Branchiostoma lanceolatum*. a, Anus; hg, hindgut; pmg, posterior midgut; amg, anterior midgut; e, esophagus; hd, hepatic diverticulum; p, pharynx; ro, rotary organ.

Table II
Frequency of GEP Cells in *Branchiostoma lanceolatum*[a,b]

	Number of animals investigated	Number of animals showing a positive reaction	Esophagus	Midgut ant.	Midgut post.	Hindgut
Insulin	11	2	165	360	280	—
Somatostatin	9	7	255	744	617	7
Pancreatic polypeptide	8	5	22	480	517	107
Glucagon	7	7	144	37	324	18
Pentagastrin	7	5	134	43	314	15
Secretin	7	7	329	588	292	140
VIP	7	4	—	10	13	4
Neurotensin	11	9	3	10	8	4

[a](From Reinecke, M., 1981, Immunohistochemical localization of polypeptide hormones in endocrine cells of the digestive tract of *Branchiostoma lanceolatum*. *Cell Tissue Res.* **219:**445–456.)
[b]Average number of immunoreactive cells in different parts of the digestive tract (number/mm^2 of mucosal surface).

In the protochordates, insulin- and secretin-containing cells have been identified in the ascidian, *Styela clava,* and insulin-, gastrin-, substance P-, and somatostatin-containing cells are found in the tunicate, *Ciona intestinalis.* The digestive tract of the lancelet, *Branchiostoma lanceolatum,* contains GEP cells with a variable distribution and frequency (Figure 8, Table II) which stain for gastrin, secretin, VIP, neurotensin, and the "islet hormones," insulin, glucagon, somatostatin, and pancreatic polypeptide. These open-type cells span the thickness of the gut epithelium (Figure 9) and are found from the esophagus to the hindgut where the cells become absent close to the anus. Insofar as these extant protochordates represent vertebrate precursors, they possess most of the GEP hormone families. In animals below the cyclostomes there is, of course, no dichotomy in nomenclature between "gut endocrine" and "islet endocrine" cells as they are all associated with the gut. Separation is seen to begin in the cyclostomes.

Figure 9. Open-type GEP cells in the intestinal mucosa of the lancelet, *Branchiostoma lanceolatum.* These cells were immunostained for gastrin by the indirect immunofluorescence technique. The majority of the staining appears to be in cells at the basal portion of the epithelium, but a thin process from each cells generally extends to contact the lumen. (Micrograph courtesy of S. van Noorden.)

4. Gut Hormones in Nonmammalian Vertebrates

In submammalian vertebrates hormones of the alimentary tract have been studied less intensively than those of the endocrine pancreas. Nevertheless, bioactive, immunoreactive, or immunostainable material resembling all of the mammalian gastrointestinal hormones has been demonstrated in some or all of the various vertebrate groups below mammals. The absence of certain hormones reported for some of the lower vertebrates may well indicate that the hormone has yet to make an evolutionary appearance. However, such observations should be taken guardedly and not necessarily assumptive of its absence since, for example, the use of antisera to mammalian hormones may not permit detection of the lower vertebrate counterpart. It remains for thorough correlative studies to be undertaken in which the hormone for the species is characterized biochemically and immunochemically before concrete statements can be made concerning the presence of a hormone in any one species.

4.1. Agnathans

The cyclostomes are represented by the hagfish (Myxinidae) and the lampreys (Petromyzonidae) and constitute two evolutionary lines considered to be the oldest and possibly the most original of the extant vertebrates. Whether or not they represent highly specialized, degenerate, or neotenous forms, the agnathans provide the most primitive vertebrate species available for study having separated from the mainstream of vertebrate evolution about 500,000,000 years ago. The gastrointestinal tract of the cyclostomes is quite simplified compared with that of other vertebrates in lacking a stomach and an exocrine pancreas. Enzyme-secreting zymogen cells are found scattered throughout the intestinal mucosa, especially in the anterior region. The liver is well developed, and a gall bladder is present although it is lost secondarily in adult lampreys. Extracts from the intestines of *Myxine glutinosa, Lampetra fluviatilis,* and *Petromyzon marinus* possess CCK-like activity when tested in mammalian gallbladder or exocrine pancreas bioassay. Extracts from the intestine of the Pacific hagfish, *Eptatretus stouti,* cause contraction of the guinea pig gallbladder, but porcine CCK cannot contract the hagfish gallbladder although it does stimulate lipase release from the hagfish intestine. No gastrinlike activity has been demonstrated in the cyclostomes, presumably reflecting the absence of the stomach and hence the need to stimulate acid secretion in these animals. Immunostaining, however, has demonstrated cells reactive to gastrin and cerulein* antisera in the hagfish and lamprey intestine. Immunochemical characterization of CCK-like peptides from the gut of *Lampetra tridentata* and *L. fluviatilis* has shown the presence of two components in intestinal extracts, one major and one minor, both cross-reacting with antiserum specific to the N terminus

*Cerulein is a decapeptide isolated originally from the skin of the frog, *Hyla caerulea,* and possesses molecular features and biological properties of mammalian gastrin and CCK. See Figure 5, Chapter 15.

of gastrin or cerulein. Nevertheless, these peptides manifest CCK-like rather than gastrinlike bioactivity, which appears to be the phylogenically older action of the gastrin/CCK molecule.

Glucagonlike material resembling mammalian pancreatic and gut glucagon is localized to the same cell, which also contains gastrinlike and PP-like immunostainable material. It has been proposed that there may be a "protoprohormone" present in the lower vertebrates and invertebrates representing a single parental or ancestral molecule. "Prosecgastrin" has been suggested as a molecule that gave rise to the secretin/glucagon/VIP/GIP and the gastrin/CCK/cerulein families of hormones. It is possible that the PP family of hormones may also be represented by an ancestral molecule in these cells. With respect to other GEP hormones, secretinlike activity and somatostatin and substance P have been demonstrated in the cyclostomes, and GIP has been found in hagfish, but not lamprey, intestine by radioimmunoassay. Neurotensinlike immunoreactivity also has been localized to open-type cells scattered sparsely in *Myxine* intestine mainly in the upper regions and has been confirmed by radioimmunoassay.

4.2. Chondrichthyes—Cartilaginous Fish

Of the vertebrate groups, Chondrichthyes appeared the earliest in the evolutionary record. It is composed of two groups, the Holocephali (ratfish, rabbit fish) and the Selachii (sharks, skates, rays). Cells staining for glucagonlike and PP-like immunoreactivity are found in the ratfish (*Chimaera monstrosa*) intestine, as are somatostatin cells. Glucagon- and PP-containing cells are two distinct populations and are also separate from cells containing gastrin/CCK immunostainable material. Glucagon/glicentinlike stainable cells showing open-type morphology also are found in the gut mucosa of sharks, as are somatostatin cells. VIP-containing cells are especially numerous in the colorectal region of the shark and ray alimentary tract. Neurotensin-containing cells are sparse in the dogfish, *Squalus acanthius,* being limited to the mucosal epithelium of the spiral valves. Substance P has been identified immunocytochemically and by radioimmunoassay in elasmobranch intestinal mucosa. Open-type cells immunostainable for GIP have been found in the antrum and duodenum of the dogfish and confirmed by radioimmunoassay. CCK-like activity has been found in extracts of ratfish intestine or shark intestine and stomach which, in *Rhinobatus productus,* shows gastrinlike activity. Extracts of dogfish and skate intestine possess secretinlike activity.

4.3. Osteichthyes—Bony Fish

Information on gastrointestinal hormones in Osteichthyes is derived from the Teleostei, the majority of the bony fish, with little yet obtained from the sturgeons, bowfins, or lungfish. The teleosts possess most of the alimentary tract hormones found in mammals (see, for example, Table III). GIP is found in open-type cells of the pyloric ceca and duodenum of the daddy sculpin, *Cottus scorpio.* Secretinlike

Table III
Distribution of Immunohistochemically Localized Peptides in the Gastrointestinal Tract of 11 Teleost Species[a,b]

	Pantodon buchholzi	*Hemigrammus ocellifer*	*Idus idus*	*Brachydanio rerio*	*Gyrinocheilus aymonieri*	*Corydoras schultzei*	*Poecilia reticulata*	*Platypoecilus variatus*	*Pelmatochromis pulcher*	*Haplochromis* sp.	*Helostoma temmincki*
Stomach											
Somatostatin	−	−	−	−	⊕	⊕	−	−	⊕	−	⊕
Gastrin (N)	−	−	−	−	−	−	−	−	−	−	−
Glucagon	−	−	−	−	−	−	−	−	−	−	−
PP	−	−	−	−	−	−	−	−	−	−	−
Substance P	−	−	−	−	−	−	−	−	−	−	−
met-Enkephalin	−	⊕	⊕	−	−	−	⊕	⊕	?⊕+	⊕	−
VIP	+	−	−	−	+	−	−	−	−	+	+
Neurotensin	−	−	−	−	−	−	−	−	−	+	−
Bombesin	−	−	−	−	−	−	−	−	−	−	−
Midgut											
Somatostatin	−	−	−	−	⊕[c]	−	−	−	−	−	−
Gastrin (N)	⊕	−	−	−	−	⊕	⊕	⊕	−	⊕	−
Glucagon	⊕	−	−	−	−	⊕	−	−	−	⊕	−
PP	−	−	⊕	⊕	−	−	−	⊕	⊕	?⊕	−
Substance P	−	−	−	−	−	−	−	−	⊕	−	⊕[d]
met-Enkephalin	−	⊕	⊕	−	−	+	⊕	⊕	?⊕+	⊕	+
VIP	+	+	+	+	+	+	+	−	−	+	+
Neurotensin	−	−	−	−	−	−	+	+	−	+	+
Bombesin	−	−	−	−	−	−	−	−	−	+	−

[a](From Langer, M., Van Noorden, S., Polak, J. M., and Pearse, A. G. E., 1979, Peptide hormone-like immunoreactivity in the gastrointestinal tract and endocrine pancreas of eleven teleost species. *Cell Tissue Res.* **199:**493–508.

[b]⊕, Immunoreactive cells; +, immunoreactive nerves; −, negative. No staining was observed with antisera to insulin, GIP, secretin, motilin, CCK, or β-endorphin.

[c]Restricted to the pyloric appendage.

[d]Also in pyloric appendages.

activity has been detected in the pike, *Esox lucius,* but as its potency is greater on the turkey than the rat pancreas, it may be more VIP-like in structure. Gastrinlike and CCK-like activity has been found in extracts of the stomach and intestine, respectively, of the coho salmon, *Oncorhynchus kisutch,* and CCK-like, but not gastrinlike, activity is present in the gut extracts of the cod, *Gadus morhua,* eel, *Anguilla anguilla,* and pike, *Esox lucius.* Gastrinlike activity alone is found in gut extracts of *Lepomis machrochirus.*

Such findings present difficulties in interpretation. Immunostaining for gastrin/CCK in teleosts is demonstrated with antisera specific for the N-terminal and C-terminal portions of gastrin (the C-terminal region being common to CCK and cerulein), whereas antisera specific for the N-terminal portion of CCK will not stain in the teleost GI tract. From these results it is unclear whether gastrin and CCK are separate hormones in teleosts. Indeed, it has been proposed that the separation between them occurred at the level of the Reptilia, thence to include all the amniotes.

4.4. Amphibia—Apoda (Wormlike), Urodela (Salamanders, Newts), Anura (Frogs, Toads)

Of the three extant groups of amphibians, the urodeles and the anurans are best known with respect to GEP tract hormones. The distribution and relative frequency of GEP peptide-containing cells in the toad, *Bufo regularis,* are given in Table IV. Cells containing VIP, substance P, endorphin, and enkephalin also were shown in 10 amphibian species (*Salamandra salamandra, Ambystoma mexicanum, Bombina bombina, Alytes obstetricans, Rana temporaria, Rana esculenta, Bufo bufo, Atelopus oxyrhynchus, Hyla arborea, Hyla crepitans*), although secretin and GIP cells were not detected. In general, immunostaining for (1) gastrin/CCK-like material is found in the antrum and small intestine, (2) bombesin in the stomach, while (3) glucagon and somatostatin are distributed throughout the entire GI tract. Secretin is not found consistently by immunocytochemistry, although secretinlike activity is present in the frog stomach.

Gastrin/CCK activity in frog stomach may well be due to ceruleinlike peptides with their common C-terminal sequences. This is thought to be possible as cerulein is more potent than gastrin in stimulating acid release from the frog stomach. The picture is somewhat confusing given the three gastrin/CCK/cerulein immunoreactive peptides isolated from *Rana esculenta* and *Rana temporaria* antrum: one the size of cerulein, one the size of gastrin-34 or CCK-33, and one larger. Furthermore, gastrinlike molecules corresponding to gastrin-34 and gastrin-17 have been obtained from the antrum and duodenum of *Rana catesbeiana, Rana pipiens,* and *Necturus maculosus.* The latter genus also exhibits a gastrin-13-sized molecule and a large form similar to "big" gastrin. In *Necturus* the gastrinlike immunoreactivity lies more in the duodenum than in the antrum.

4.5. Reptilia—Chelonia (Turtles), Squamata (Lizards, Snakes), Crocodilia (Alligators, Crocodiles)

In this somewhat neglected group, we find all the gut hormones of the mammalian GEP system, although the frequency and distribution of the various cell types differ widely among species. Somatostatin cells are the most widely distributed throughout the various regions of the GI tract [found in one study of 13 species: *Emys orbicularis, Chrysemys (Pseudemys) scripta, Chelydra serpentina, Hemidactylus holbrookii, Agama agama, Python reticularis, Thamnophis sirtalis, Natrix natrix, Nerodia rhombifera, Agkistrodon piscivora, Crotalus atrox, Al-*

Table IV
Distribution of Gastroenteropancreatic Hormonal Polypeptide Cells in the Gastrointestinal Tract of the Toad, *Bufo regularis*[a,b]

	Somatostatin	Glucagon	PP	Secretin	GIP	Gastrin[d]	Neurotensin	Bombesin	PYY
Esophagus[c]	−	−	−	−	−	−	−	−	−
Stomach									
Cardiac	−	+	−	−	−	−	−	−	−
Pylorus	++++	++	+	+	+	−	−	++	−
Antrum	++	−	+++	−	+	++++	−	++++	−
Duodenum	++++	+	++	+	+	++	−	−	+
Ileum	+	+	+++	+	+	++	+	−	+
Rectum	−	−	−	−	−	−	−	−	−

[a]From El-Salhy, M., Grimelius, L., Wilander, E., Abu-Sinna, G., and Lundqvist, G., 1982, Histological and immunohistochemical studies of the endocrine cells of the gastrointestinal mucosa of the toad (*Bufo regularis*). *Histochemistry* **71**:53–65, and El-Sahly, M., Grimelius, L., Lundberg, J. M., Tatemoto, K., and Terenius, L., Immunocytochemical evidence for the occurrence of PYY, a newly isolated gut polypeptide, in endocrine cells in the gut of amphibians and reptiles. *Biomed. Res.* **3**:303–306.

[b]++++, Abundantly occurring cells; +++, numerous cells; ++, moderate number of cells; +, few cells; −, absent. No staining for insulin, VIP, CCK, substance P, motilin, enkephalin, endorphin, or ACTH was detected.

[c]Endocrine cells were stained in the esophagus, but not the rectum, by conventional histological techniques for demonstration of endocrine cells.

[d]Staining specific for C-terminal pentapeptide of gastrin, the portion in common with CCK and cerulein.

ligator mississippiensis, Caiman crocodylus crocodylus]. There also is a wide distribution of cells containing glucagon in the chelonian GI tract.

In the reptile stomach, in addition to somatostatin, gastrin/CCK cells and bombesin cells are the most common, although substance P is colocalized with bombesin in *Emys, Chelydra,* and *Alligator*. Other gut peptides occur infrequently among species so far examined. Thus, in the duodenum secretin was observed in *Chelydra, Alligator,* and *Caiman;* motilin in *Emys* and *Caiman;* and GIP in *Python*. PP has been found only in the colon of *Hemidactylus*. For the most part, separate CCK cells have not been identified, though this may reflect only a species variability, not an absolute absence. For example, most gut peptides, with the exception of VIP, substance P, and endorphin, are present in the alimentary tract of the grass lizard, *Mabuya quinquetaeniata,* including CCK distinct from gastrin (Table V). In this species, also, the recently identified hormone peptide YY (PYY) is found in a population of open-type cells in the mucosal epithelium of the duodenum and ileum. PYY cells are distinct from cells containing PP or neurotensin, though sharing some common sequences in their molecules.

It is very likely that gastrin and CCK can occur in the reptilian alimetary tract as distinct molecules, and it is at this evolutionary juncture wherein this can be demonstrated. Distinct populations of gastrin and CCK cells are seen in *Testudo graeca* and *Mabuya quinquetaeniata*. The crocodile, *Crocodylus niloticus,* offers particular interest as three gastrin/CCK-like peptides have been isolated that immunochemically do not resemble N-terminal regions of gastrin-17 or cerulein. The biological activity of only one of these peptides has been determined, and it exhibited gastrin-, but not CCK-, like activity.

4.6. Aves—Birds

The avian gastrointestinal tract is structured somewhat differently from the basic mammalian pattern in that functions performed by the mammalian stomach are carried out by the crop, proventriculus, and gizzard in the birds. The crop acts as a storage site for ingested food, the proventriculus is the site of acid and pepsinogen secretion, and the muscular gizzard mechanically disrupts the food and passes it into the duodenum. Bile and exocrine pancreatic secretions enter the small intestine at the duodenoileal junction, and at the junction of the ileum and colon paired ceca extend as blind pockets. Hormonal peptides have been identified in various regions of the avian GI tract, based primarily on fowl species. Most of the GEP hormones have been identified immunocytochemically though a few have been isolated and their structure and biological activity characterized. Cells immunoreactive with antisera specific for gastrin are, in the chicken, restricted to the the antrum equivalent, the gizzard–duodenum junction. Extracts of this in the turkey contain a molecule similar to gastrin-17, but not cross-reacting with antisera specific for the N-terminal portion. It may be derived from a precursor similar to CCK-33. Turkey jejunum extracts contain a molecule similar to CCK-8.

Secretin has been isolated and characterized from birds and, as in mammals,

Table V
Distribution and Frequency of Gastroenteropancreatic Hormonal Polypeptides in the Gastrointestinal Tract of the Grass Lizard, *Mabuya quinquetaeniata*[a,b]

	Somatostatin	Glucagon	PP	Secretin	GIP	GTP[c]	GMP[d]	CCK[e]	Neurotensin	Bombesin	Motilin	PYY
Esophagus	++	−	−	−	−	−	−	−	−	−	−	−
Stomach												
Upper part	+++	−	−	−	−	−	−	−	−	−	−	−
Lower part	+++	+++	++++	+++	−	++++	++	−	+++	++++	−	−
Duodenum	++	−	+	−	+	++	−	++	−	−	+	+
Ileum	+	−	+	−	+	++	−	+	+	−	+	+
Rectum	−	−	−	−	−	−	−	+	−	−	−	−

[a]From El-Sahly, M., and Grimelius, L., 1981, The endocrine cells of the gastrointestinal mucosa of a squamate reptile, the grass lizard (*Mabuya quinquetaeniata*). A histological and immunohistochemical study. *Biomed. Res.* **2**:639–658, and El-Sahly, M., Grimelius, L., Lundberg, J. M., Tatemoto, K., and Terenius, L., 1982, Immunocytochemical evidence for the occurrence of PYY, a newly isolated gut polypeptide, in endocrine cells in the gut of amphibians and reptiles. *Biomed. Res.* **3**:303–306.

[b]++++, Abundantly occurring cells; +++, numerous cells; ++, moderate number of cells; +, few cells; −, absent.

[c]Gastrin C-terminus immunoreactive cells.

[d]Gastrin midportion immunoreactive cells.

[e]CCK (9–20) immunoreactive cells.

promotes exocrine pancreatic water and bicarbonate secretion. Neurotensin-containing cells are found in abundance in quail (*Corturnix*) intestine concentrated especially in the pylorus and ileum, but absent in the crop, proventriculus, and gizzard. Somatostatin cells appear to be found throughout the avian gut and are particularly abundant in the duodenal–gizzard junction, an area rich in endocrine cells. Bombesin cells are found in the proventriculus and gizzard but nowhere else in the young chick GEP system. PP cells are found in the avian duodenum, jejunum, and ileum as a population of cells distinct from those containing PYY which are found in small numbers in the duodenum and jejunum. Endorphin-containing cells have also been found in the small intestine, ceca, and colon of the chicken. Gastrin-immunoreactive cells are found in the gizzard–duodenum junction; however, their presence may not be very significant in the totality of avian digestion, particularly since PP has similar biological activity and is capable of stimulating proventricular acid and protein secretion.

5. Hormonal Control of Gastrointestinal Function

Digestion is regulated by neural, paracrine, and hormonal activities that promote the appropriate and efficient function of the gastrointestinal tract. Gut hormones play a prominent role in the regulation of secretory, contractile, and motor activities of the stomach, small intestine, gallbladder, pancreas, and large intestine. The following sections demonstrate a good knowledge for mammals, but one very incomplete for nonmammalian animals. Immunocytochemistry and/or immunochemistry has been used to detect mammalian gut hormone counterparts throughout the vertebrate groups, but for most, chemical structures have yet to be determined or become available for investigations. Consequently, mammalian gut hormones have been used so that conclusions about nonmammalian vertebrates must be viewed with caution.

5.1. Hormones and Gastric Secretion

The stomach mucosa is characterized by numerous invaginations of the epithelium called gastric pits. These pits connect the gastric lumen with the various types of gastric glands (Figure 10) characteristic for the cardiac, oxyntic, and pyloric regions of the stomach. The cardiac gland area is a small region at the gastroesophageal junction and secretes mucosubstances and small amounts of electrolytes. The oxyntic gland area occupies the fundus and body of the stomach and comprises 75–80% of the total gastric mucosa area. The oxyntic glands contain several secretory cell types including the acid-secreting parietal (oxyntic) cells (Figure 11) and the pepsinogen-secreting peptic cells (Figure 11). Parietal cells are relatively large pyramidal-shaped cells characterized by large numbers of mitochondria and by intracellular canaliculi that communicate with the glandular lumen. The peptic cells make up about 25% of the oxyntic mucosa and are a typical zymogen-

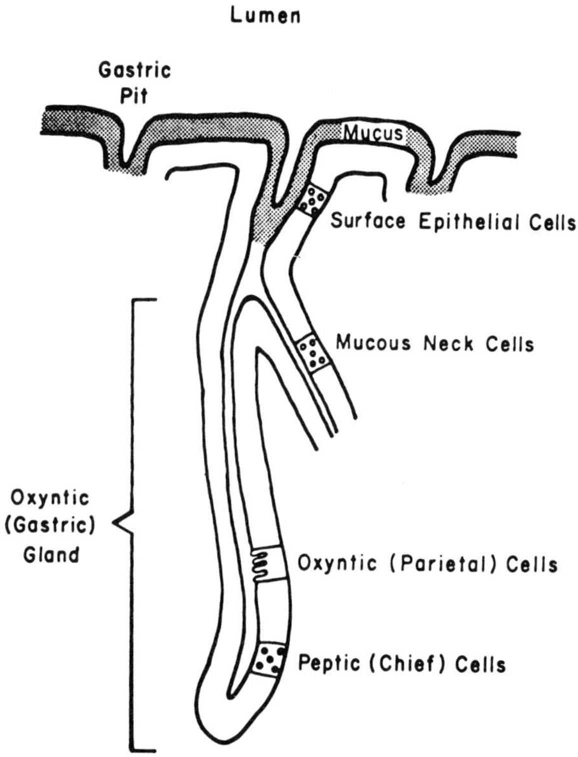

Figure 10. Diagrammatic representation of the gastric secretory unit illustrating the location of the acid-secreting oxyntic (parietal) cells and the pepsinogen-secreting peptic (chief) cells. (From Sernka, T. J., and Jacobson, E., 1979, *Gastrointestinal Physiology—The Essentials,* Williams & Wilkins, Baltimore.)

type secretory cell. The pyloric glandular area comprises about 20% of the gastric mucosa and secretes an alkaline mucus, some electrolytes, but little or no acid. The pyloric antral glands are rich in endocrine cells containing gastrin and gastrin-releasing peptide (bombesin) cells.

5.1.1. Acid Secretion

Three compounds (secretagogues) stimulate acid secretion by the parietal cells: gastrin (see above), histamine, and acetylcholine act in concert to regulate secretion and serve as potentiating or permissive agents for each other. Thus, cimetidine, a histamine receptor blocker, inhibits gastric acid response not only to histamine, but also to gastrin and acetylcholine. Likewise, anticholinergic drugs may block not only cholinergic stimulation of acid secretion but also that induced by gastrin and histamine. Of the three secretagogues, only histamine appears to act via the adenylate cyclase/cyclic AMP system although some evidence suggests that acetycholine-

or gastrin-induced acid secretion may involve the guanylate cyclase/cyclic GMP pathway.

Hydrochloric acid is released by the parietal cells of the oxyntic mucosa in response to gastrin, histamine, and acetylcholine. Acid is secreted actively into the canaliculi where the concentration of acid may rise to 170 mM or more, which is about 3×10^6 times greater than the concentration in the blood or tissues. Stimulation of acid secretion is associated with elongation of the parietal cell apical microvilli and an increase in the canalicular area at the expense of the tubulovesicular area within the cell. Upon stimulation there is a latency of about 15 min before the onset of secretion during which time the morphological changes commence. Those changes are essential to the secretory process and can be blocked by inhibitors of acid secretion. The energy required to produce and transport H^+ is derived from the aerobic metabolism of glucose and fatty acids although the ultimate source of the H^+ is still unclear. Physiological stimuli trigger a change in the redox state of the parietal cell and an increase in oxygen consumption prior to the onset of acid secretion, suggesting an early effect of the secretagogues in the mobilization of substrates.

Gastric secretion is divided into interdigestive and digestive periods, and the digestive period can be subdivided into the cephalic, gastric, and intestinal phases, which correspond to the site at which the stimuli act (see above). The presence of food in the stomach constitutes the most potent stimulus of gastric secretion through its mechanical and chemical properties. Of the chemical constituents of food, only

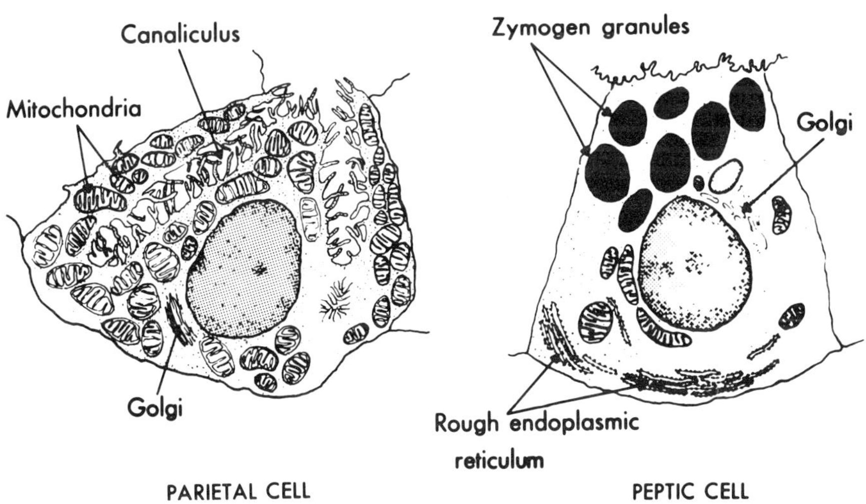

Figure 11. Comparison of the organization of the gastric parietal cells and gastric peptic cell. The acid-secreting parietal cell is characterized by the prominent canalicular system and abundant mitochondria whereas the pepsinogen-secreting peptic cell is a typical protein-secreting cell with abundant rough endoplasmic reticulum and characteristic zymogen granules. (Redrawn from Bloom, W., and Fawcett, D. W., 1975, *A Textbook of Histology,* W. B. Saunders, Philadelphia.)

amino acids and peptides stimulate acid secretion, the mechanism being through the stimulation of gastrin release. Food entering the duodenum stimulates the secretion of GRP (bombesin) and enkephalin stimulates acid secretion independently of gastrin.

Inhibition of gastric acid secretion involves locally generated hormonal signals and signals generated from the duodenum. Antral and fundic feedback inhibition of acid secretion occurs when the antral pH declines to pH 2–3, presumably by the suppression of gastrin secretion. Somatostatin, released by the presence of acid, is capable of inhibiting acid and gastrin secretion. Instillation of acid into the duodenum suppresses acid secretion and is mediated, in part, by secretin, although other duodenal hormones may be involved. Urogastrone, a 53-amino-acid-residue peptide similar to mouse epidermal growth factor and found in Brunner's glands, inhibits acid secretion. Motilin inhibits histamine- and food-induced secretion independently of antral pH and serum gastrin levels. Intraduodenal fat generates vagal and hormonal signals which suppress gastric acid secretion; GIP and VIP are both capable of inhibiting gastrin- and histamine-induced acid secretion.

5.1.2. Pepsinogen Secretion

Pepsin, a proteolytic enzyme, is found in basal gastric secretions, even in the absence of hydrochloric acid. Produced by the peptic cells in the gastric glands, pepsin is stored as pepsinogen and is secreted in a biphasic pattern as an initial high peak with a low plateau following. The inactive pepsinogen is autocatalytically converted to active pepsin in the presence of acid. Feeding, sham feeding, or local cholinergic stimulation promotes pepsinogen secretion. Cholinergic stimulation of release is both direct via acetylcholine and indirect by stimulating gastrin release.

5.1.3. Gastric Secretion in Nonmammalian Vertebrates

Gastric acid and pepsin secretion are common to almost all vertebrates. The stomach was an adaptation to macrophagus feeding used primarily as a food container; the mechanical and acid proteolytic digestive functions were acquired secondarily. The stomach may be divided into functionally distinct regions, but in some animals even specialized modifications have been developed. For example, the gizzard is a modification to provide enhanced mechanical digestion when there is a lack of masticatory function by the jaws (e.g., some fish, crocodiles, most birds).

In the lower vertebrates and birds only one granulated cell type is found in the mucosal glands. This cell secretes both acid and pepsin, although in a number of amphibians acid and pepsin are produced by separate glands. Gastric mucosal glands may be simple or branched tubular with the fish, amphibians, lizards, and snakes possessing more simply formed glands than turtles, crocodiles, and birds. Stomach contents in most vertebrates are acidified to pH 1–2, the acidity being digestive in its own right but also essential for optimal pepsin activation and activity. High acidity and low pepsin secretion can be effective in some animals because the low pH provides optimal conditions for peptic activity, but in animals where the gastric pH may be higher (less favorable for peptic activity), pepsin secretion may be increased to compensate. In general, the pH optimum for most vertebrate pepsins is around 2.

Gastrin is the dominant hormone that stimulates gastric secretion in mammals, and it may well be so in nonmammals but definitive investigations have yet to be made. Thus, in mammals sulfation of the tyrosine in gastrin is not of prime importance in stimulating acid secretion, whereas sulfation of the tyrosine in CCK is essential for full expression of CCK-like activity (generally assessed by gallbladder contractile activity). In lower vertebrates, however, the molecular requirements are often different. As pointed out previously, separate gastrin and CCK molecules (not necessarily CCK or gastrin activity) did not appear until the reptiles.

5.1.3a. Fish (Cyclostomes, Chondrichthyeans, and Teleosts)

Gastrin/CCK-containing cells have been identified in these groups, but the role of gastrin/CCK in the regulation of gastric secretion is not known. Elasmobranchs apparently have humoral control of acid secretion given the high acidity found for stomach contents in contrast to the nearly neutral pH observed in empty stomachs. In *Raja* and *Scylliorhinus* acid secretion appears to be regulated by hormonal control alone since acid secretion is unaffected by pilocarpine, atropine, acetylcholine, histamine, or various manipulations of the autonomic nervous system. Histamine and acetylcholine, however, can stimulate acid secretion in *Squalus* and *Torpedo*. Based on assays in a mammalian system CCK-like activity has been extracted from the stomach and intestine of *Squalus acanthius* and the intestine of *Chimaera monstrosa*. Whether there is any gastrin activity in these molecules awaits testing in the homologous animals, although it might not be expected in the stomachless *Chimaera*.

Gastric secretion in teleosts appears to be under nervous and hormonal control. Fishes have a cephalic and chemical phase of gastric secretion, the first phase being mediated by the vagus, the second possibly by hormones. Histamine and cholinergic drugs are potent acid secretagogues in fish and can be blocked completely by histamine receptor antagonists and partially by atropine. It has been suggested that histamine mediates the final common pathway in stimulation of gastric acid secretion in fish. Although there is physiological evidence for hormonal control of acid secretion in fish, the role of gastrin is unclear. Gastrinlike activity is demonstrable in extracts of coho salmon stomach as assayed by contraction of guinea pig antral muscle; however, neither pentagastrin nor cod gastric mucosal extracts stimulate acid secretion in the Atlantic cod (*Gadus morhua*), nor is there any effect of bombesin. VIP suppresses basal acid secretion in *Gadus*, an effect that is independent of its hypotensive action on mucosal blood flow.

5.1.3b. Amphibians

The amphibians, especially the anurans, have been popular models for studies of gastric secretion, and, consequently, more is known about them than about most other lower vertebrates. In *Rana* the foregut is divided into the esophagus, the forestomach, the corpus, and the pyloric portion. The pepsin-secreting cells are found only in the esophagus and the forestomach, and the acid-secreting cells are found mainly in the corpus. This anatomical separation of the acid- and pepsin-secreting activities is not uncommon in many amphibians, and *Proteus, Necturus, Bufo, Pelobates,* and *Triton* all show pepsin secretion restricted mainly to the esophageal glands. Distention of the stomach in the frog will stimulate both acid and pepsin secretion that is not inhibited by vagotomy or sympathetic chain section. It is, therefore, probably a local response mediated by intragastric cholinergic nerves. Acetylcholine and histamine are powerful secretagogues, and gastrin is capable of stimulating acid and pepsin secretion in frogs and acid secretion in *Necturus*. Bullfrog (*Rana catesbeiana*) gastric mucosa exhibits different responses to gastrin depending on the dosage administered. With low doses (0.1–1 μg/animal) gastrin stimulates acid release only, but at higher doses (2–20 μg/animal) it is inhibited, with increased pepsin release.

The identification of many biologically active peptides from amphibian skin, some of which have been localized to GI tract nerves or endocrine cells, has presented possibilities for other regulators of gastric secretion in amphibians (as well as other vertebrate groups). Cerulein has potent CCK-like activity in mammalian assay systems by virtue of the sulfated tyrosine at position 7 from the carboxy terminus (see Fig. 5, Chapter 15). A ceruleinlike molecule has been isolated from antral mucosa of *Rana esculenta,* where, on a molar basis, cerulein is 500–1000 times more potent than histamine, 1000–10,000 times more potent than human gastrin-17 or pentagastrin, and about 25 times more potent than CCK. Sulfated cerulein is 1000 times more potent than the unsulfated form, and gastrin-17II (sulfated) is 10 times more potent than gastrin-17I in the stimulation of acid secretion. The specificity of action of these hormones, especially given the requirement for sulfation, suggests that the isolated amphibian gastric mucosa resembles mammalian gallbladder and pancreas with respect to receptors for the gastrin/CCK family of peptides. Moreover, the high degree of sensitivity of the anuran stomach to cerulein indicates that the amphibian gastrin molecule may resemble cerulein more than mammalian gastrin.

5.1.3c. Reptiles

Acid secretion is in response to the presence of food in the stomach since an empty stomach in most reptiles is at neutral pH whereas a stomach with food present shows a pH value of 2 or below. Both histamine and parasympathetic stimulation are effective releasers of acid and pepsin release. However, vagal stimulation favors gastric juice of low volume but high pepsin content. Histamine and pilocarpine together produce a high degree of degranulation and a viscous juice with a pH of 1 in *Testudo*. The role of gastrin and other GEP hormonal peptides in the regulation of gastric secretion is virtually unknown. Outside of the identification of gastrin/CCK cells in reptile gastric mucosa and the immunochemical demonstration of the existence of separate gastrin and CCK molecules in *Testudo*, little is understood concerning the role of these hormones. Biologically active gastrin CCK peptides have been extracted and partially characterized from the stomach of *Crocodylus niloticus*, which appear to have principally gastrinlike activity based on rat bioassay systems, but nothing is known concerning the activity of these peptides in the crocodile itself. Contrary to the amphibians, *Testudo* shows little sensitivity to cerulein and ceruleinlike peptides with respect to stimulation of short-circuit current in the gastric mucosa (another assay for gastrinlike activity).

5.1.3d. Birds

The crop is a nonabsorptive structure acting as a storage site whereas the proventriculus secretes acid and pepsin producing a juice with a pH approaching 0.2–0.5. The muscular gizzard serves to grind the food and reflux it back to the proventriculus or propel it into the intestine. Histamine stimulates the secretion of acid, but not pepsin, in the chicken and is produced by the esophagus, crop, proventriculus, and gizzard. Histamine in combination with acetylcholine produces gastric secretions that are rich in pepsin, perhaps under parasympathetic control. A cephalic phase of gastric secretion has been demonstrated by sham-feeding experiments.

Gastrin-17I and pentagastrin stimulate acid secretion in birds though at doses 10–20 times higher than required for mammals. CCK (porcine) achieves this at only 2–3 times higher than that required for pancreatic enzyme secretion in mammals, suggesting that CCK-like peptides are more effective acid secretagogues in birds. Indeed, cerulein has approximately 4 times the potency of gastrin-17I in chickens. Avian gastrin, therefore, may not be that important in the regulation of proventricular secretion, especially given the restricted distribution of gastrin in the avian stomach (a narrow band at the gizzard/duodenal junction). We turn, then, to avian pancreatic polypeptide as the powerful secretagogue for proventricular acid and pepsinogen secretion. APP possesses all the attributes of a mammalian gastrin and may be the principal hormone eliciting proventricular secretion. Bombesin also stimulates proventricular secretions, acting through the release of a gastrinlike substance rather than stimulating pancreatic polypeptide (PP) release. Proventricular secretions are inhibited by secretin, serotonin, enterogastrone, and CCK.

5.2. Gastrointestinal Hormones and Pancreatic Secretion

The exocrine pancreas produces hydrolytic enzymes for digestion and a fluid rich in bicarbonate for neutralization of the acidic food entering the duodenum. Enzyme release from the acinar cells is stimulated by CCK-related peptides and by acetylcholine from cholinergic branches of the vagus whereas ductal cell electrolyte secretion is stimulated by secretin and VIP.

The pancreatic acinar cell (Figure 12) is characterized by extensive, basally located RER and by the numerous large zymogen granules located at the apical end of the cell. The well-developed Golgi produces numerous condensing vacuoles, which are thought to represent maturing zymogen granules. Biosynthesis and secretion of pancreatic enzymes proceeds via the ribosomal incorporation of amino acids as directed by messenger RNA. The leader sequence transports the nascent peptide into the cisterna of the RER and is degraded cotranslationally to produce the proenzymes (zymogens), which are transported to the Golgi for further processing (e.g., glycosylation) and for packaging via condensing vacuoles into mature zymogen granules. The process is, overall, rapid, in that within 40 min or so new enzymes appear in mature zymogen granules.

Ninety percent of the protein synthesized in the pancreas is for export, and rate of synthesis can be altered by the diet (long term) or by hormonal or cholinergic stimulation (short term). Zymogen granules are discharged by exocytosis, though the role of the microtubular/microfilamentous system is unclear. Zymogen granule contents are released into the ductal lumen having dissolved when the granule membrane fuses with the plasma membrane. The proteins secreted by the acinar cell include proteolytic enzymes (trypsinogen, chymotrypsinogen, proelastase, procarboxypeptidase A, procarboxypeptidase B), amylolytic enzymes (α-amylase), lipolytic enzymes (lipase, esterase, phospholipase, cholesterol esterase), and nucleolytic enzymes (DNase, RNase). A number of these enzymes (trypsinogen and proelastase, for example) are stored and secreted as inactive precurators (as pep-

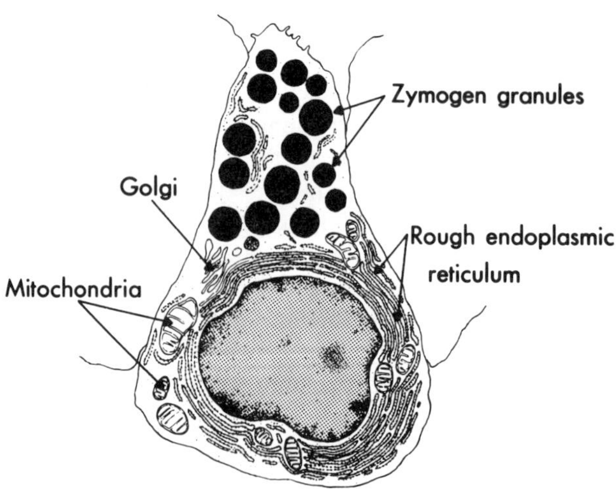

Figure 12. Diagram of a pancreatic acinar cell. Note the similarity to the gastric peptic cell. The extensive lamellar arrangement of the rough endoplasmic reticulum is particularly characteristic of this cell type.

Table VI
Composition of Pancreatic Juice[a]

Component	Concentration (mmole/kg H_2O; $\bar{X} \pm$ SEM)
Cl^- Total CO_2 (CO_2+HCO_3^-)	154 ± 10[b]
Na^+	154 ± 7
K^+	4.8 ± 0.9
$PO_4^{\equiv}$	0.8 ± 0.4
H_2O	98%

[a]Canine pancreatic juice stimulated by secretin.
[b]Chloride varies inversely with the bicarbonate concentration while the bicarbonate varies directly with the flow over a range of 25–150 meq/liter. The product of the chloride and bicarbonate concentrations, however, is constant.

sinogen in the gastric parietal cell). Activation in the gut lumen is by peptide bond cleavage in the amino-terminal region of the precursor.

Electrolytes and water are important constitutents of pancreatic fluids. The duct cells are low in RER and Golgi but relatively rich in mitochondria. Pancreatic juice is a clear alkaline fluid (Table VI) which is isosmotic with the extracellular fluid and contains the various digestive enzymes and bicarbonate for buffering the acidic digesta arriving from the stomach. The buffering is especially important given the neutral to slightly basic pH optima of most pancreatic enzymes. Electrolyte secretion follows two basic patterns in mammals. In carnivores (e.g., dog, cat) there is little or no tonic electrolyte secretion, but exogenous secretin produces a high concentration of bicarbonate in the juice upon stimulation. In contrast, in herbivores such as the sheep, rabbit, or calf, and in the omnivorous rat, there are marked levels of spontaneous secretin release but low responses to exogenous secretin and low maximal levels of bicarbonate in the juice.

The major part of pancreatic exocrine secretion depends on the release of secretin and CCK from the upper small intestine. Cholinergic input appears to modify positively the pancreatic response to secretin. Secretin and VIP are mediated intracellularly at their target tissues by the adenylate cyclase/cyclic AMP system. Both hormones increase cyclic AMP levels, theophylline potentiates submaximal doses of secretin, and dibutyryl cyclic AMP stimulates electrolyte secretion directly. Although cyclic AMP is typically associated with mobilization of intracellular calcium, the role of calcium in secretin-induced electrolyte secretion remains equivocal. CCK and acetylcholine stimulation of acinar cell enzyme secretion does, however, appear to be calcium mediated, but independently of the adenylate cyclase system. Although CCK and acetylcholine do not elevate intracellular cyclic AMP levels, the stimulatory effects of CCK or acetylcholine can be mimicked by calcium ionophores such as A23187.

Although secretin and CCK stimulate primarily electrolyte and enzyme secretion, respectively, there are some overlapping functions, and both hormones appear to act synergistically. Moreover, other GI hormones may play a role in regulating pancreatic exocrine secretion. VIP has 1/10th to 1/4th the potency of secretin in stimulating water and electrolyte secretion (oddly, equipotent in the cat). Exogenous bombesin has CCK-like effects, but it is probably acting indirectly through stimulation of CCK release. Chymodenin, a large peptide (74–76 amino acids, 8500–9500 daltons), possesses an unusual activity in that it stimulates specifically chymotrypsinogen release from the pancreatic acinar tissue.

Several hormones inhibit pancreatic enzyme and electrolyte secretion and serve as feedback regulators of secretin- and CCK-stimulated pancreatic function. Somatostatin, glucagon, neuropeptide (NPY), and met-enkephalin inhibit secretin- and CCK-induced pancreatic secretion, and somatostatin the basal secretion. The release of these inhibitors is usually generated by the presence of food in the gut and by the postprandial rise of circulating metabolites following digestion. Elevated plasma glucose, for example, causes somatostatin release. Pancreatic polypeptide from the islets stimulates (at low concentrations) or inhibits (at high concentrations) pancreatic exocrine secretion. Plasma levels of PP rise rapidly upon food ingestion, the release being mediated by vagal pathways. (See Chapter 17.) A second phase of PP release is triggered by food in the intestine and is mediated by neural and blood-borne signals. Conceivably, the first phase of PP release could enhance exocrine pancreatic secretion whereas the second phase of PP release could serve to inhibit secretin- and CCK-stimulated secretion as digestion nears completion.

5.2.1. Cyclostomes, Chondrichthyeans, and Teleosts

The exocrine pancreas makes its first appearance as a separate gland in the holocephalians. Prior to this, zymogen cells are found scattered in the mucosal epithelium of the intestine and secrete their enzymes directly into the gut lumen. The presence of secretinlike and CCK-like substances in lampreys and hagfish, and even in amphioxus and various invertebrates, indicates, nevertheless, that hormonal peptides as potential regulators of pancreatic exocrine secretion occurred before, in the evolutionarily sense, the appearance of a distinct exocrine pancreas. Their actions, however, are not known. CCK-like material extracted from cyclostome intestine stimulates mammalian gallbladder contraction and pancreatic protein secretion, and porcine CCK stimulates lipase release from the hagfish intestine. Possible secretinlike and CCK-like activity is found in a methanol-soluble fraction from extracts of river lamprey (*Lampetra fluviatilis*) or marine lamprey (*Petromyzon marinus*) when tested for stimulation of pancreatic juice secretion in the rat. Hence, the roles of secretin and CCK in hagfish and lamprey digestive processes remain to be established.

Gastrin/CCK-like peptides extracted from elasmobranchs and teleosts exhibit primarily CCK-like activity in mammalian bioassays, although gastrinlike activity has been demonstrated for materials extracted from some bony fish (*Lepomis ma-*

chrochirus, Oncorhynchus kisutch) and a chondrichthyean (*Rhinobatus productus*). The main difficulty with these and most observations to date is the lack of assessment of these biologically active preparations in the homologous species. Moreover, virtually no studies exist demonstrating the effects of the mammalian counterparts on the pancreatic enzyme secretion in many lower vertebrates. As with CCK, the effects of secretin in cartilaginous and bony fish is, at best, poorly known. Secretinlike activity has been demonstrated in representatives of the elasmobranchs and holocephalians as well as in several species of teleosts (e.g., *Esox lucius, Anguilla anguilla, Gadus morhua*) based primarily on mammalian bioassay systems. The secretinlike substance isolated from pike (*Esox lucius*) intestinal extracts stimulates the pancreatic secretory rate strongly in birds but weakly in mammals. This spectrum of activity suggests that pike secretin resembles VIP more closely than secretin with respect to its action on the avian and mammalian pancreas.

5.2.2. Amphibians and Reptiles

Gastrin/CCK peptides and secretinlike substances have been demonstrated by various means to occur in the intestines of these animals, but virtually nothing is known concerning their physiological functions. It does appear that gastrin and CCK become separate molecules in the reptiles, but there have been no studies demonstrating the effects of even the mammalian gut hormones on pancreatic electrolyte and enzyme secretion in these two groups.

5.2.3. Birds

Knowledge is wider, though not great, for this vertebrate group than for other nonmammals. Porcine CCK, CCK-8, and cerulein stimulate enzyme secretion in turkeys and chickens in doses comparable to those used in mammals. This indicates that avian and mammalian regulation of pancreatic enzyme secretion have similarities. Stimulation of electrolyte and water secretion in birds, however, may depend more on VIP than secretin. Administration of chicken secretin to chickens or turkeys produces only a weak stimulation of pancreatic juice flow, and moreover, administration of porcine CCK with the chicken secretin does not enhance juice flow over that produced by secretin alone. In mammals, postprandial levels of secretin are low, but the effects of secretin are augmented by the synergistic actions of CCK. Chicken VIP, which has been localized to intestinal endocrine cells in addition to nerves, is twice as potent as porcine VIP and 50–100 times more potent than chicken or porcine secretin in eliciting pancreatic juice flow in the turkey. The weak action of secretin combined with the absence of CCK enhancement of secretin action and the potent stimulatory effect of VIP argues against a principal role for secretin and favors one for VIP in the regulation of electrolyte and water secretion by the avian pancreas.

Peptide histidine isoleucine (PHI), another member of the secretin/glucagon family, exhibits a spectrum of activity similar to VIP with respect to the stimulation of pancreatic juice flow in birds and mammals. PHI is a potent stimulant of pancreatic secretin in the turkey but with only half the potency of chicken VIP. In the rat, PHI, like chicken and porcine VIP, is only a weak stimulant of pancreatic secretion, whereas porcine secretin is a strong stimulant in rats but a weak stimulant in birds. These observations suggest that although PHI may contribute toward the regulation of pancreatic juice flow, this may not be its primary function. Moreover, its localization to nerves, but not to endocrine cells, would indicate that its release may be regulated by stimuli different from those delivered to the VIP cells in the intestinal mucosa.

5.3. Hormones and GI Tract Motility and Contractility

5.3.1. GI Tract Motility

The digestive process is dependent on mechanical forces generated within the alimentary tract for the transport and mixing of ingested food. Movement of the contents relies on a coordinated action of smooth muscle cells, which are regulated by intrinsic myogenic factors, neural activity, and circulating and/or local hormones. The myenteric plexus (between the circular and longitudinal muscles) and the submucosal plexus are part of the intrinsic nervous system with afferents sensitive to chemical and mechanical stimuli. Efferent fibers synapse with the muscle cells and may inhibit (adrenergic fibers) or stimulate (cholinergic fibers) muscle contraction. Various neuropeptides such as VIP, bombesin, neurotensin, enkephalin, CCK-8, substance P, peptide histidine isoleucine (PHI), and NPY have been localized to the intrinsic nervous system and may serve as local hormones, neuromodulators, or neurotransmitters in the regulation of enteric contractility.

Three different types of movement are connected with contractile activity of the gut: segmenting, peristalsis, and tonus. Segmentation is a circular contraction that occurs over a 1- to 2-cm length, does not move along the bowel, but is accompanied by other contractions at short intervals. Segmenting contractions occur over a short distance at the head of a meal and are maintained regularly until the meal has passed. Peristalsis is a propagated wave of contraction that passes caudally along the gut, and tonus is the exerting of a steady compressive force on the lumen particularly at junctional areas where control points are located.

Contraction of enteric muscle is observed under two different physiological states: (1) spontaneously during the interdigestive periods when there is no food in the GI tract, and (2) in response to food in the GI tract which interrupts or superimposes contractile activity over the spontaneous activity. Spontaneous contractions are detectable in the enteric muscle during interprandial periods, and at approximately 100-min intervals there is a burst of contractile activity which originates in the stomach and runs down the small intestine. These intergestive migrating contractions or complexes (IMCs) correlate with intervals of motilin secretion which may be responsible for IMC induction (Figure 13). IMCs can be suppressed by

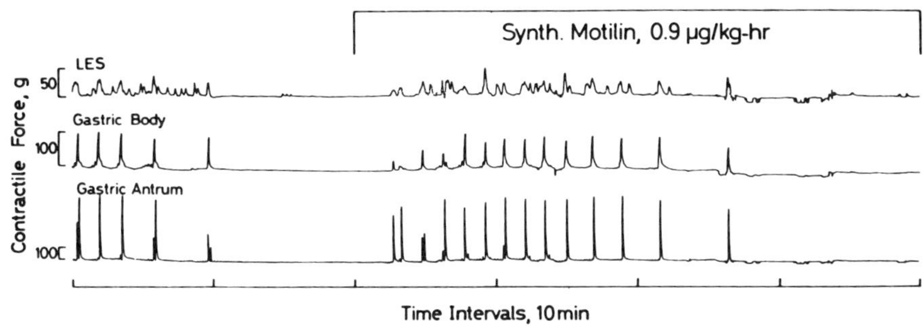

Figure 13. Inherent contractility of the lower esophageal sphincter (LES) and portions of the stomach and their response to the infusion of synthetic motilin in a dog. The contractions of the LES probably represent the most proximal component of the IMC. (From Itoh, Z., Aizawa, I., Honda, R., Hiwatashi, K., and Couch, E. F., 1978, Control of lower-esophageal-sphincter contractile activity by motilin in conscious dogs. *Am. J. Dig. Dis.* **23:**341–345.)

secretin at the level of the stomach, but not the duodenum, and physiologically they are interrupted by the ingestion of a meal.

A great many of the gut hormones have effects on basal or stimulated gastric and intestinal contractility. However, the physiological role of these observations is still uncertain as many of the responses appear to require excessive nonphysiological doses of hormone. Gastrin appears to have a physiological role in regulating gastric contractility and functions to delay gastric emptying by selectively causing contraction of the antrum thereby blocking movement of the food out of the stomach. Motilin also appears to be a physiological regulator of stomach muscle contractility stimulating gastric contraction by acting directly on the muscle cells. VIP, on the other hand, appears to be an antagonist to motilin and induces relaxation of gastric smooth muscle. In the small intestine motilin appears to be the most specific and likely regulator of intestinal contractility. It stimulates increased segmental contraction characteristic of contractility associated with digestion and is responsible for the IMCs. Gastrin, CCK, somatostatin, and substance P all stimulate small intestine contractility whereas secretin and PYY inhibit it. Certain hormones also affect large intestine contractility. CCK augments basal segmental colonic activity, and substance P stimulates contraction. Glucagon and PYY produce periods of inhibition, but motilin has no effects on colon contractile activity.

5.3.2. Gallbladder Contraction

There are other contractile forces that are important during digestion, for example, those involved in regulating the various sphincters along the length of the alimentary tract, most of which appear to be controlled by neural reflexes. In addition, there is regulation of gallbladder contraction, which is responsible for the delivery of bile into the duodenum. Bile is secreted by bile canaliculi, and by the ducts and ductules of the hepatic system, and collected in the gallbladder. This reservoir of bile is stimulated to contract primarily by the action of CCK. By virtue of their structural similarities, gastrin and gastrin-related peptides possess CCK-like activity but are generally 1000-fold less effective. CCK release is triggered by the presence of digested fat and protein in the duodenum, and increased circulating CCK elicits gallbladder smooth muscle contraction. Tension increases in the gallbladder wall followed by intermittent opening of the cholecystocystic duct, a phasic contraction of the common bile duct, and the active opening and closing of the cholechoduodenal sphincter. In humans this results in a 44% reduction in gallbladder size. CCK stimulates gallbladder contraction whereas VIP and PHI appear to cause relaxation of gallbladder muscle. Motilin, which stimulates smooth muscle contraction in other parts of the gut, has no effect on contraction of gallbladder muscle.

5.3.3. Motility and Contractile Activity in Nonmammalian Vertebrates

5.3.3a. GI Tract Motility

Regulation of motility in all vertebrates involves neural and hormonal mechanisms. The neural pathways include extrinsic and intrinsic components which increase in complexity phylogenically. The extrinsic system is composed of sympathetic and parasympathetic outflows, and in the lower vertebrates the vagal system does not appear to innervate structures beyond the stomach or proximal part of the small intestine. The cranial (vagal) outflow in lower vertebrates, moreover, appears to be primarily inhibitory as inhibitory fibers alone are found in the vagal nerves of cyclostomes, elasmobranchs, teleosts, and possibly urodele amphibians. In anuran amphibians and above, however, these inhibitory fibers are accompanied by excitatory fibers. The sympathetic (spinal) outflow in lower vertebrates appears to be primitively a mixture of excitatory and inhibitory nerves with a higher proportion of excitatory nerves than is seen in mammals. An intrinsic nervous system is present in the lower vertebrate gut and consists of submucosal and myenteric plexuses. The operation of this system is implied by the fact that peristalsis (which is observed in all vertebrates) persists when isolated from the central nervous system. The origin of peristaltic movements is not clear; distention alone cannot induce waves of peristaltic contractions, but probably a combination of distention with chemical or tactile stimulation of the mucosal epithelium, is necessary.

A great variety of neuropeptides have been demonstrated in lower vertebrate enteric nerves (see, for example, Table III), especially in the intrinsic system, but data are scarce on the role these neurotransmitters/neuromodulators and other GEP hormones play in regulating gastrointestinal motility. In the cod, substance P causes intestinal contraction, VIP and neurotensin have no effect, while other neuropeptides such as somatostatin, bombesin, and enkephalin have variable effects depending on the concentration of peptide. The gastrointestinal tract of the mudpuppy, *Necturus maculosus,* is responsive throughout its length to the actions of numerous regulatory peptides. VIP inhibits muscle tonus and/or abolishes rhythmical activity of *Necturus* enteric muscle while met-enkephalin and substance P increase tonus. Neurotensin inhibits spontaneous contractility of circular muscle but stimulates the longitudinal muscle.

5.3.3b. Gallbladder Contractility

Of the contractile activities in the GI tract of the lower vertebrates, hormonal stimulation of gallbladder contraction has been the most thoroughly studied. In mammals, CCK is the principal hormone causing gallbladder contraction, and in the submammalian vertebrates this also appears to be true. CCK-like activity appears to be the primitive, or original, action of the gastrin/CCK peptides in vertebrates as CCK-like, but not necessarily gastrinlike, activity is found in all lower vertebrates. It has been suggested that gastrinlike factors may have evolved from CCK-like substances at the level of divergence of the chondrichthyeans and osteichthyeans, although it has also been postulated that gastrin and CCK did not become separate molecules until the reptiles. CCK-like activity has been extracted from the hagfish intestine as assayed on mammalian gallbladder contraction, but, curiously enough, it has no apparent capacity for stimulating hagfish gallbladder contraction. Stimulation of CCK release is implicated in gallbladder contraction by experiments showing that lipid-rich meals stimulate gallbladder contraction and emptying in the bullhead, *Ictalurus nebulosus,* and in the carp, *Cyprinus carpio.* The molecular requirements for stimulation of coho salmon gallbladder contraction have been determined *in vitro* using mammalian peptides. In this species, as in mammals, the presence of a sulfated tyrosine is important for full expression of CCK activity. In contrast to mammals, however, the position of the sulfate does not appear to be as important since porcine CCK, CCK-8, and gastrin II are equipotent in stimulating contraction. Desulfated CCK-8 and gastrin I are also equipotent but at levels 400 times or more greater than the sulfated counterparts. The sensitivity of salmon gallbladder to CCK is nearly identical to that of the guinea pig gallbladder, which suggests that the requirement for total hormone presented to the tissue receptors for CCK has not changed significantly during evolution although the specificity of the receptor (vis-à-vis the position of the sulfated tyrosine) has changed. The ability of the CCK receptor to discriminate between sulfation positions in the gastrin/CCK peptides is a later development. Little, however, is known concerning the effects and molecular requirements for gastrin/CCK peptides and the stimulation of gallbladder contraction in amphibians and reptiles. In birds, CCK actions are similar to those seen in mammals.

5.4. The Enteroinsular Axis

Neural and hormonal signals arising from the gut stimulate pancreatic hormone release, often prior to appreciable changes in circulating levels of metabolites. Morover, hormone released in response to an ingested meal is often greater than would be elicited by an equivalent nutrient load administered intravenously. This system of gut-derived signals which evokes anticipatory and/or augmented release of pancreatic islet hormones is known as the enteroinsular axis. The presence of food in the stomach and/or duodenum is the common stimulus for activating the enteroinsular axis components and eliciting the release of all islet hormones. Although the first phase of PP release appears to be stimulated primarily by vagally mediated signals rather than changes in circulating hormones or metabolites, the second phase may be dependent on humoral signals, either metabolites or hormones (see Chapter 17). Secretin, VIP, gastrin, cerulein, CCK, and GIP all stimulate PP release from the pancreas. The presence of food in the gut lumen also stimulates pancreatic somatostatin release via neural and hormonal mediation. High-carbohydrate meals suppress, whereas high-protein meals enhance, glucagon release in magnitudes greater than seen with the equivalent amount of nutrient given intravenously. The changes seen in circulating pancreatic glucagon levels are minor and suggest a small role for the enteroinsular axis in the regulation of glucagon release from the pancreas.

Given its status as the key integrative hormone regulating metabolic processes in mammals, insulin occupies the central position in the enteroinsular axis. Insulin release after normal food ingestion is adjusted to limit the increase of plasma glucose despite a considerable intake of carbohydrate. In a human, if an equivalent amount were given intravenously, the plasma glucose would increase threefold but induce a smaller insulin release as a result of bypassing the insulinotropic signals generated by the GI tract. One would say, therefore, that regulation of insulin release by metabolites alone is suboptimal and requires the gut-derived components of the enteroinsular axis for the pancreatic B cell to respond with maximum efficiency to a nutrient load. A variety of gut hormones are known to stimulate insulin release, including secretin, gastrin, CCK, neurotensin, VIP, enteroglucagon, and GIP. Of these, GIP is now considered to be the principal and most important component of the enteroinsular axis with respect to the regulation of insulin release. GIP secretion is dependent on the ambient level of glucose and is not released until glucose is being absorbed from the small intestine. In humans, glucose in the circulation must rise 25 mg/dl above the fasting level before GIP release will occur. Because of its role in insulin secretion and its dependency on ambient levels of glucose, GIP has been renamed "glucose-dependent insulinotropic peptide" to reflect more accurately its primary physiological role. It should be noted, however, that because GIP secretion does not begin until critical plasma glucose levels have been attained postprandially, the hormone is probably not responsible for the early anticipatory rise in plasma insulin; presumably this is mediated neurally or by other hormones.

The anatomical basis for an enteroinsular axis did not exist prior to the hagfish. Only when islet tissue became separate from its origin in the gut epithelium could it be assumed that there was a need for GIP to function as an insulinotropic hormone. (One should not rule out the possibility that GIP could have been acting at a local level in the mucosal epithelium.) Radioimmunoassay and immunocytochemical investigations have demonstrated GIP in the intestine of a cyclostome (*Myxine glutinosa*), a cartilaginous fish (*Squalus acanthius*), and a bony fish (*Cottus scorpio*), but not in a protochordate (*Ciona intestinalis*). The presence of GIP in hagfish with its islet organ composed of only insulin plus a few somatostatin cells has suggested that the principal target of GIP would be the insulin cells. GIP is also found in high concentrations by radioimmunoassay in cartilaginous and bony fish stomach and has been demonstrated immunocytochemically in amphibian (*Bufo regularis*) stomach and duodenum. The finding, however, of GIP localized only to the duodenum in reptiles (*Mabuya quinquetaeniata, Python reticularis*) suggests that GIP action on gastric secretion may have been an early feature of this hormone's activities, but one that has been lost or diminished during evolution. In humans, for example, GIP has only modest effects on inhibiting acid secretion even at supraphysiological doses, and hence, an enterogastrone role for GIP is not strongly supported.

6. Conclusions

Our knowledge of hormonal control systems in submammalian vertebrate digestive processes remains very slight compared to other aspects of comparative endocrinology. From what we do know, however, the capacity for hormonal regulation of gastrointestinal function was established early in vertebrate evolution, if not before. Common regulatory mechanisms are present throughout all vertebrate groups, and counterparts to the many gut hormones isolated and characterized in mammals are found in most other vertebrates. There are variations in anatomical distribution and frequency of various hormone-containing cells and nerves as well as variation in the dependency of certain digestive functions on one or another hormone, but this is to be expected given the substantial time in which the vertebrates have been extant.

From the original three gut hormones first identified (i.e., secretin, gastrin, and CCK), the list of regulatory peptides found in the GI tract has expanded into a broad spectrum of hormones affecting all aspects of gut growth, secretion, motility, and circulation. Studies in nonmammals have made important contributions to the efforts in elucidating the various hormones involved; for example, PP was first isolated and characterized in birds, and GRP was first identified as an amphibian skin peptide, bombesin. Moreover, investigations on invertebrate neurohormones are also contributing to the process of uncovering vertebrate regulatory peptides. The tetrapeptide phenylalanyl-methionyl-arginyl-phenylalaninamide (FMRF-amide), the molluscan cardioexcitatory hormone, has recently been identified in vertebrate tissues as well.

Increased awareness of the great variety of peptide hormones has led to identification of families of hormones and has provided insight and impetus to the study of GEP hormone evolution. The gastrin/CCK/cerulein, glucagon/secretin, and PP families are well-defined groups of hormones with structural similarities but discrete anatomical locations and physiological functions. Members of the gastrin/CCK/cerulein family are perhaps the best characterized with respect to the structural determinants for specific hormonal action on target tissues. CCK-like activity appears to be the original or primitive activity for the gastrin/CCK peptides. An early feature of responsiveness by the target tissue to these peptides was the ability to discriminate between sulfated and unsulfated molecules. Later target tissues developed the capacity to discriminate between sulfation at positions 6 or 7 from the carboxyl terminus. Although gastrinlike activity has been reported for gastrin/CCK peptides in fish, separation of gastrin and CCK into separate molecules is postulated to have occurred after the amphibians. And, in fact, the CCK-like nature of these peptides does seem to persist into the amphibians, even with respect to gastrin activity, where anuran gastric mucosa is much more responsive to CCK-like than to gastrinlike molecules.

17

The Endocrine Pancreas

1. Introduction

The vertebrate endocrine pancreas consists of numerous aggregations of cells, the *islets of Langerhans,* composed of several cell types and exhibiting varying degrees of organizational complexity. In mammals, the islets are small (70–300 μ in diameter), ovoid aggregations of cells dispersed throughout the pancreatic exocrine tissue. Larger islets are located along major trunks and branches, and the intermediate and smaller islets are found further out along the vascular and ductal trees. These small endocrine organs are responsible for the elaboration and secretion of a number of hormones. Among them, insulin and glucagon are critical to the regulation of carbohydrate, fat, and protein metabolism, and ultimately to the maintenance of an appropriate blood glucose level essential for central nervous system function (see Chapter 18). Inappropriate secretion of any of the islet hormones may produce some metabolic derangement; the most familiar in the human is *diabetes mellitus,* which is associated with B-cell dysfunction.

2. Organization of the Pancreatic Islets

2.1. Islet Cells and Hormones

Over the past decade or two our concept of the mammalian islet has been transformed from one of a simple aggregation of two or three cell types known to secrete insulin and glucagon to one of a highly complex microorgan composed of four cell types: A (alpha), B (beta), D (delta), and PP (also known as F) cells. In addition to insulin and glucagon produced by the B and A cells, respectively, somatostatin and pancreatic polypeptide have been identified and accepted formally as endocrine secretory products of the D and PP cells, respectively. There is also a bewildering array of other peptide hormones remaining to be integrated into the spectrum of pancreatic islet function: cholecystokinin (CCK), gastrin, gastric inhibitory polypeptide (GIP), thyrotropin releasing hormone (TRH), luteinizing-hormone-releasing hormone (LH-RH), β-endorphin, adrenocorticotropin (ACTH), and somatocrinin (growth-hormone-releasing hormone, GH-RH) have been localized by various means to the pancreatic islets.

The relative number of constituent cell types within the islet varies with age and with the species considered. In general, however, the B cells represent the greatest proportion of the islet cells with the A, D, and PP cells occurring in declining frequencies. In the human, B cells comprise 70–80% of the islet cells except in the fetal and early postnatal pancreas, when the values are closer to 50%. A cells comprise 10–12% in the human (slightly higher in the fetal through juvenile years), and PP cells generally comprise less than 1% until after the fifth decade, when they may increase severalfold. D cells are quite well represented (35–40%) in the fetal and early postnatal years, but decline thereafter to approximately 5% or less. Various estimates for adult laboratory animals are similar in magnitude, with the B cells in rats and mice representing from 60–90% or more, the A cells 10–12%, D cells 3–5%, and PP cells 3–5% of the islet mass.

When mammalian islet populations are systematically analyzed by regions, A and PP cells are found to be distributed nonrandomly within the pancreas. Islets from certain portions of the head region are incredibly rich in PP cells, where, for example, the posterior lobe of the human pancreas exhibits islets that are 80% PP cells and the other cell types are reduced markedly (10–20% B cells, 0.5–1% A cells, 1.5% D cells). Similarly in the rat and mouse, the head region of the pancreas contains islets that are A cell-poor and PP cell-rich, double or triple the normal volume density when compared to islets from other regions. These intrapancreatic differences in islet cell constituents are attributable to the dual origin of the pancreas, i.e., the ventral and dorsal pancreatic buds which fuse to give rise to the pancreas. Presumably the PP cell-rich head region (the posterior lobe in the human) is derived from the ventral anlage, and the remainder of the pancreas, the isthmus, body, and tail, is derived from the dorsal anlage.

The precise organization of cell types within the mammalian islet has contributed to the concept and possible importance of intraislet, or paracrine, regulation (see Section 2.3). In the typical rodent islet (Figure 1), for example, a central homocellular core of B cells is surrounded by a heterocellular layer, or mantle, of A, D, and PP cells. Moreover, this mantle layer may appear capped or polarized with many of the non-B cells oriented near vascular or ductal structures. The human pancreas exhibits a less clear association among islet cell types, and the distribution of cell types within the islet has been considered as random. Nevertheless, there is an obligatory association between A and D cells with both types tending to border the large vessels and surround the more centrally located B cells, so that the human

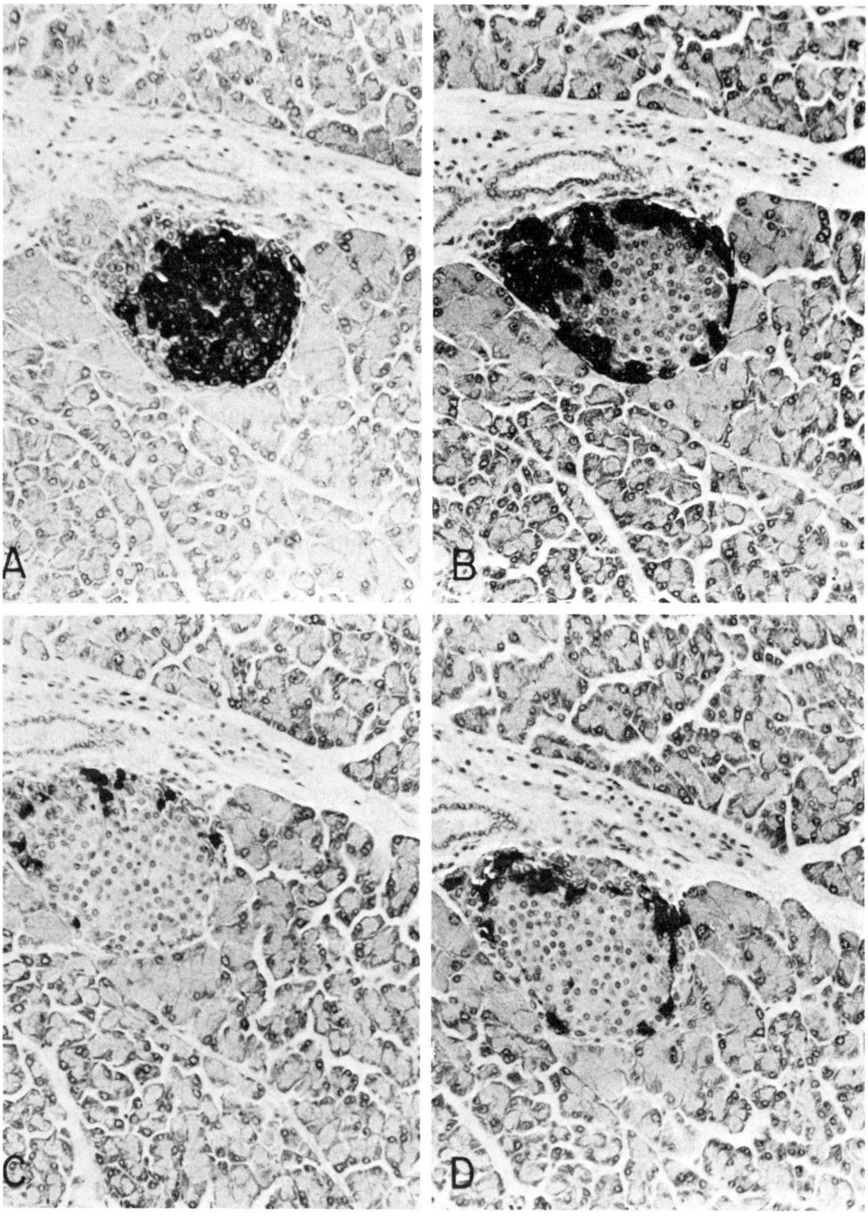

Figure 1. Adjacent sections of mouse pancreas immunostained for (A) insulin, (B) glucagon, (C) pancreatic polypeptide, and (D) somatostatin. Note that the B-cell core is surrounded by a mantle of A, D, and PP cells, an arrangement typical for many vertebrate islets. (D. A. Gapp, unpublished results.)

non-B cells into the homocellular B-cell core prior to islet degeneration. B-cell numbers decrease, and there is a concomitant increase in relative and actual A, D, and PP cell mass. In contrast, the *db* gene in the C57BL/6 strain induces an obesity (diabetes-resistant) syndrome characterized by a mild and transient hyperglycemia, hyperinsulinemia, and no overt perturbation of islet cell topography. A and D cells decline relative to the expanding B-cell mass whereas PP cells, on the other hand, maintain a relatively constant ratio compared to the B cells. The Zucker fatty (*fa*) rat, a normoglycemic hyperlipidemic model of obesity, is similar to the C57BL/6 diabetes mouse in that there is an expanding mass of B cells and little or no perturbation of islet cell topography. In contrast, however, non-B cells exhibit little evidence of compensatory growth vis-à-vis the proliferating B cells. Changes in topographical relationships and in islet cell proportions in these models of diabetes and/or obesity appear to be sequelae of the primary lesion induced by the *db* or *fa* genes rather than causative factors with syndrome development. Nevertheless, the severity of the disease state is reflected in and correlated with the degree of disruption of islet organization.

3. Comparative Aspects of Islet Morphology

As the preceding discussion indicates, the mammal exhibits complex anatomical relationships with respect to vasculature and the organization of constituent cell types. The apparent preciseness of this organization is presumed to have important implications for regulation of islet function. Yet, as we shall see, the islet, or islet organ as some prefer, was evolved from a less complex structure. Islet cells are derived from the gut endothelium and over the course of vertebrate evolution have migrated from this site to the exocrine pancreas where the various components of the islet of Langerhans assumed topographical relationships among themselves, and with respect to the pancreatic vasculature. In the next sections, the organization and constituents of islets from the various vertebrate classes are presented, and current thoughts concerning islet evolution will be discussed.

3.1. "Islet" Cells before the Vertebrates

As is the case with the gut endocrine cells, pancreatic islet cells did not arrive *de novo* with the vertebrates. Islet cell homologs have been identified in a wide variety of invertebrate and protochordate species, and it appears that all of the islet hormones and many of the other gastroenteropancreatic (GEP) hormones have phylogenetic precursors in the invertebrates. Certainly the isolation of insulin-, glucagon-, somatostatin-, and CCK-like peptides from *E. coli* and *Tetrahymena* argues in favor of a very ancient lineage for these molecules (cf. Chapter 16). Nevertheless, when searching for islet cell counterparts and their respective hormones in organisms without a separate pancreas, one looks to the gut as the principal source. Insulinlike material has been identified in the esophagus, stomach, and intestine of the whelk, *Buccinum undatum,* and in the intestine and cecum of the squid, *Eledone cirrosa.* In addition, the pyloric ceca of the starfish, *Pisaster ochraceous* and *Asterias rubens,* the hepatopancreas of the snail, *Helix pomatia,* and the stomach and intestine of the sea squirt, *Ciona intestinalis,* contain insulinlike activity based on the rat epididymal fat pad or mouse hemidiaphragm bioassays, activity that can be neutralized by antiserum to bovine inulin. Certain cells in the intestinal mucosa of several mollusks (*Mytilus edulis, Anodonta cygnea, Unio pictorum*) are open-type secretory cells containing immunostainable, insulinlike material. Moreover, a substance has been extracted from some of these mollusks that cross-reacts in an insulin radioimmunoassay and influences their carbohydrate metabolism.

Neurosecretory structures also appear to be a source of insulinlike activity. Extracts of the corpus cardiacum/corpus allatum complex from the tobacco hornworm, *Manduca sexta,* contain a substance that is immunologically similar to insulin and capable of lowering hemolymph trehalose concentration. Immunostaining for insulin in the blowfly, *Calliphora vomitaria,* shows immunoreactivity in the median neurosecretory cells and has suggested that in insects these cells represent a site of insulin production with the corpus cardiacum/corpus allatum complex serving as a storage site.

Neural structures are sources for the other "islet hormones" as well. Extracts of the *Manduca* neurosecretory system contain a material that cross-reacts with antibodies to mammalian glucagon and lowers fat body glycogen content. Somatostatin has also been localized to invertebrate neural cells (e.g., the cerebral ganglion cells in the tunicate, *Ciona intestinalis,* and in the insect, *Locusta migratorius*), and pancreatic polypeptide has been localized almost exclusively to the nervous system of invertebrates (e.g., in the earthworm, *Lumbricus terrestris,* and in the giant African snail, *Achatina fulica*) rather than in the digestive tract.

Identification of all "islet hormones" within a given invertebrate species has not often been achieved except in the protochordates. Although this may reflect the inability of antisera to mammalian hormones to recognize the appropriate hormones in invertebrate tissues, it is also possible that this reflects some evolutionary changes. For example, PP has not been found in the digestive tract of animals below the protochordates.* This relatively late evolutionary appearance of PP cells in the digestive tract could suggest that PP cells had a different physiological function previously, especially given their localization to nervous tissue. The absence of PP in the gut and islet organ of the cyclostomes coupled with the late ontogenic appearance of PP cells in the developing mammalian embryo has suggested to some that the PP cell lacked physiological function until the appearance of the exocrine pancreas. This might be possible with respect to participation in the control of digestive processes, but it seems unlikely that a hormone present in the nervous system of annelids or mollusks would be without some function, and one would expect a neuroendocrine or neurocrine function for this peptide.

It appears that all the "islet hormones" were in place in the protochordates. In particular, the lancelet, *Branchiostoma lanceolatum,* exhibits open-type GEP endo-

*PP has been localized to open-type GEP cells in the midgut of the cricket and cockroach.

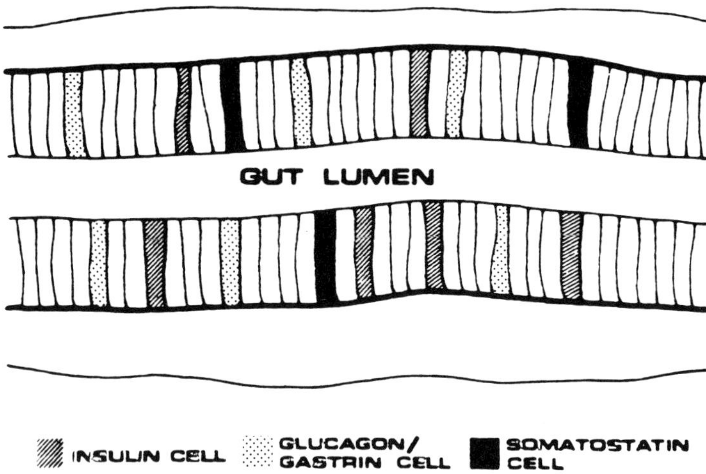

Figure 2. Diagrammatic representation of the gut of *Branchiostoma* (and also the larval lamprey) in which there is no islet organ present. "Islet" cell types represented here are the glucagon/gastrin cell (which also contains pancreatic polypeptide) and the insulin cell, both of which are open-type GEP cells with microvilli on the luminal surface (apical end of the cell) and basally located secretory granules. (From Van Noorden, S., and Falkmer, S., 1980, Gut-islet endocrinology—some evolutionary aspects. *Invest. Cell Pathol.* **3**:21–35.)

crine cells throughout most of the digestive tract with clearly identifiable populations of cells immunostaining for insulin, glucagon, somatostatin, and pancreatic polypeptide as well as several other GEP hormones (Figure 2, cf. Table II, Chapter 16). Here insulin and somatostatin immunoreactive cells are closely associated topographically, being most frequently found in the distal esophagus and the entire midgut but virtually absent in the hindgut. Pancreatic polypeptide cells are most frequent in the midgut, and glucagon immunoreactive cells are more frequent in the esophagus and posterior midgut and have been shown to contain gastrin as well (Figure 2, Chapter 16). It appears, therefore, that the lancelet may well reflect the prototypic complement of GEP cells in the type of digestive tract from which the vertebrate endocrine pancreas was to arise. Beginning with the cyclostomes, we find some "islet cells" have migrated from the intestinal mucosa to form the islet organ. This migratory process occurs before the appearance of the exocrine pancreas and will be completed in the cartilaginous fishes.

3.2. Islet Cells in the Vertebrates

3.2.1. Agnatha

The Atlantic hagfish (*Myxine glutinosa*) possesses an islet organ located at the base of the bile duct where it empties into the intestine (Figure 3). The islet organ of the river lamprey (*Lampetra fluviatilis*) is split up into three portions partly embed-

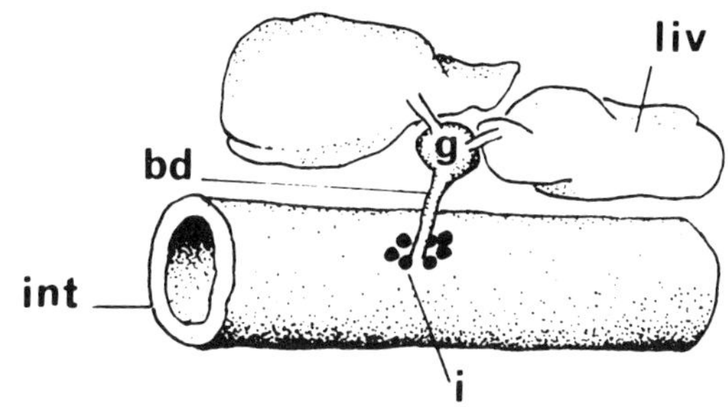

Figure 3. Anatomical location of the islet lobules in the hagfish. int, Intestine; bd, bile duct; g, gallbladder; liv, liver; i, islet lobules. [Redrawn from Epple, A., 1969, *Fish Physiology*, Vol. 2 (W. S. Hoar and D. J. Randall, eds.), Academic Press, New York, pp. 275–319.]

ded in the intestinal wall and extending into the liver. The cyclostome islet organ is composed of a series of avascular lobules or follicles in which follicular form seems to be more prevalent in the deep-sea-dwelling *M. glutinosa* than in other hagfish or the lampreys. The islet parenchyma of the cyclostomes lacks any association with an exocrine pancreas since no exocrine pancreatic epithelium or ducts have been

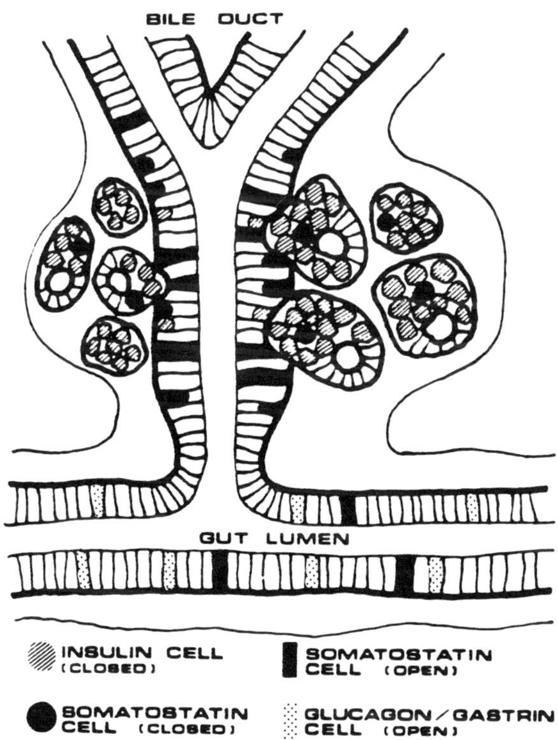

Figure 4. Representation of the gastroenteropancreatic system of the hagfish, *Myxine glutinosa*. Endocrine cells arise from the bile duct mucosa to form the islet lobules, which contain insulin and somatostatin cells in a ratio of 99:1, respectively. Glucagon/gastrin cells are restricted to the gut mucosa, and PP cells are not present. (From Van Noorden, S., and Falkmer, S., 1980, Gut-islet endocrinology—some evolutionary aspects. *Invest. Cell Pathol.* **3**:21–35.)

identified in these animals. The cellular composition of the islet organ parenchyma appears to be restricted to B (insulin) and D (somatostatin) cells, and in the hagfish, the ratio of B to D cells is 99 to 1. Cells staining for insulin and somatostatin are found also in the bile duct mucosa of *Myxine* whereas glucagon/gastrin/PP cells appear to be restricted to the intestinal epithelium (Figure 4). Cells staining for insulin are not found in the intestine of either hagfish or lampreys.

Islet lobule histogenesis in the hagfish involves proliferation of the endocrine cells from the bile duct epithelium in much the same way as islets develop from the pancreatic ductal epithelium in higher vertebrates. The endocrine cells originate from granulated cells at the basal part of the bile duct mucosa to proliferate into buds or clusters of cells. These are surrounded and separated into mature islet lobules. In the lamprey, presumptive islet cells bud from the base of the gut mucosa at the junction of the esophagus and the intestine. Although no insulin cells are found in the intestine of the adult lamprey, they are present during islet organ histogenesis in the larval lamprey and are very reminiscent of those seen in the gut of *Branchiostoma* (Figure 2).

3.2.2. Chondrichthyes–Cartilaginous Fish

The exocrine pancreas makes its first appearance in Chondrichthyes as a bilobed (dorsal and ventral) structure in the elasmobranchs, and as a single-lobed

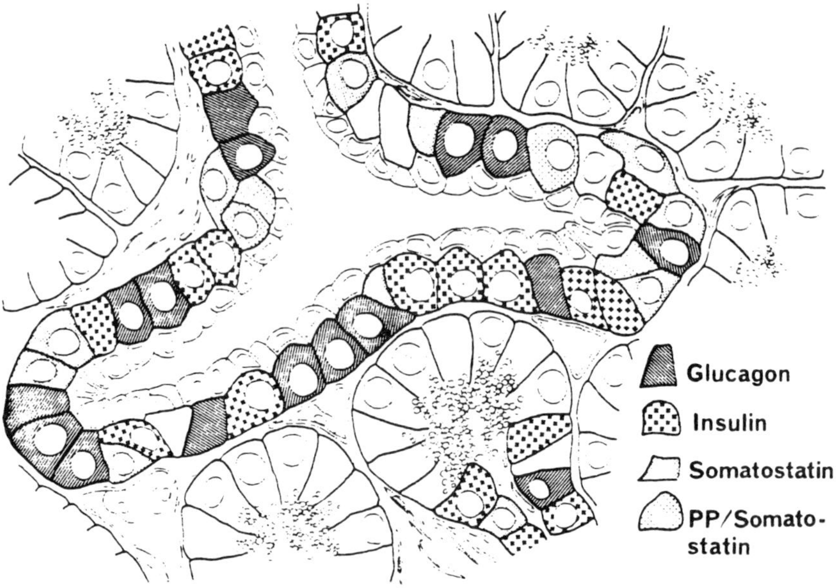

Figure 5. Pancreatic endocrine cells in the ray, *Dasyatis akajei*. Endocrine cells are arranged in a layer over the pancreatic ductal epithelium. Insulin, glucagon, and somatostatin cells comprise about 30% each of the endocrine cells, whereas the somatostatin/pancreatic polypeptide cells comprise the remaining 10%. (From Sekine, Y., and Yui, R., 1981, Immunohistochemical study of the pancreatic endocrine cells of the ray, *Dasyatis akajei*. *Arch. Histol. Jpn.* **44:**95–101.)

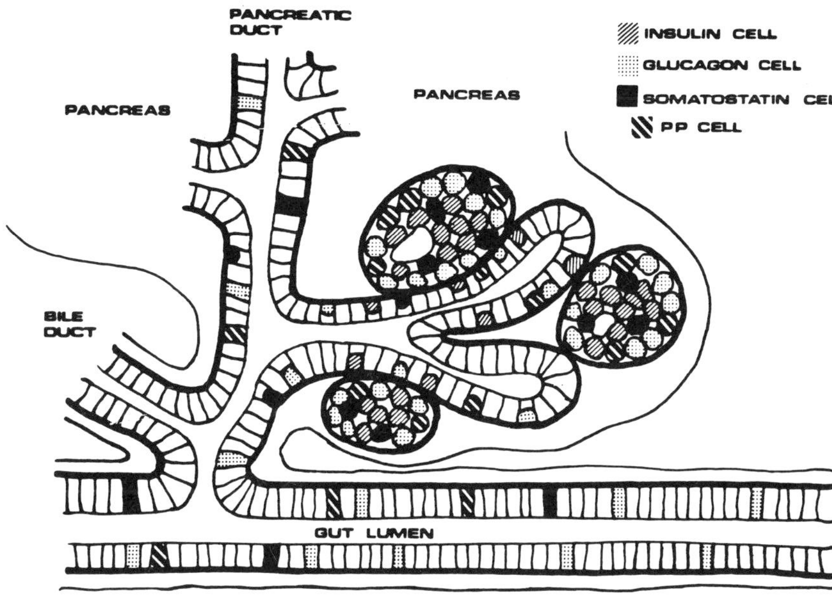

Figure 6. The gastroenteropancreatic system in the elasmobranchs representing the most advanced arrangement of islet cells in this group. Islets are composed of insulin, glucagon, somatostatin, and pancreatic polypeptide cells, and except for the insulin cells, all cell types are also present in the gut mucosal epithelium. (From Van Noorden, S., and Falkmer, S., 1980, Gut-islet endocrinology—some evolutionary aspects. *Invest. Cell Pathol.* **3**:21–35.)

compact gland attached to the liver and spleen in the ratfish. The islet cells are dispersed throughout the exocrine tissue in varying degrees of organization. There are three general categories of relationships between the islet and exocrine tissues: (1) the islet cells form a second epithelial layer around the duct cells as in *Mustelis canis* or *Dasyatis akajei* (Figure 5); (2) an intermediate condition where they form solid cords of cells budding from but contiguous with the pancreatic ducts (e.g., *Raja erinacea*); (3) an ''advanced'' arrangement appearing as islets separate from the ductal epithelium (e.g., *Squalus acanthius,* Figure 6). All four islet cell types are present in the elasmobranchs; the PP cells are now to be seen in both the gut epithelium and the islet tissue. The holocephalians represent an intermediate, though not necessarily transitional, form between cyclostomes and elasmobranchs. In the cyclostomes, A and D cells alone comprise the islet parenchyma whereas in the holocephalians the islets contain insulin, somatostatin, and glucagon cells (Figure 7), together with PP cells in intestinal mucosa and pancreatic duct epithelium. In the ratfish the islet is a three-hormone organ as the PP cells are located in the pancreatic ducts. A four-hormone islet is seen in the elasmobranchs where PP cells can be found within the islet organ. Thus the cartilaginous fish demonstrate acquisition of all four islet cells. Any changes thereafter are a variation on a four-cell presentation.

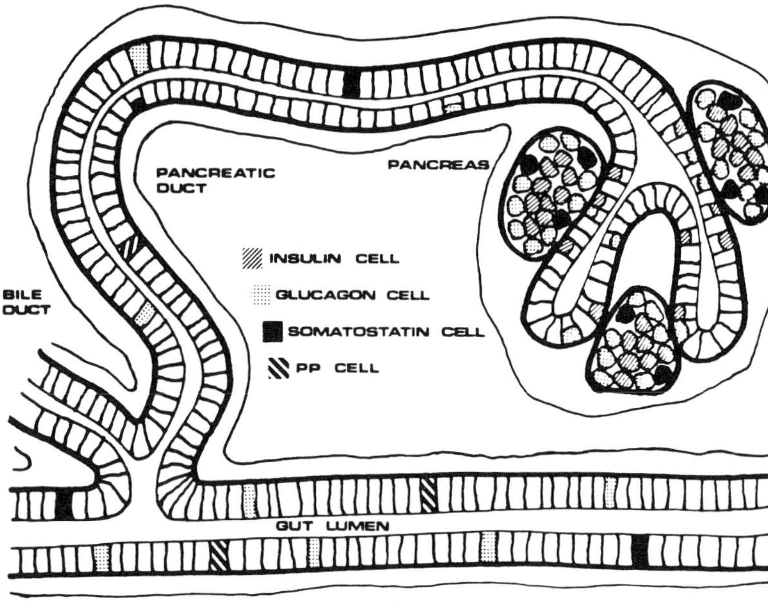

Figure 7. The gastroenteropancreatic system in the holocephalians, *Chimaera monstrosa* and *Hydrolagus collei*. A separate exocrine pancreas is present and contains islets comprised of insulin, glucagon, and somatostatin cells. The first appearance of pancreatic polypeptide cells in the vertebrates occurs in the gut mucosa and in the pancreatic ductal epithelium. (From Van Noorden, S., and Falkmer, S., 1980, Gut-islet endocrinology—some evolutionary aspects. *Invest. Cell Pathol.* **3**:21–35.)

3.2.3. Osteichthyes—Bony Fish

Of the two major subclasses of bony fish, the Sarcopterygii (lungfishes and lobe-fin fishes) are represented by few species but have been little studied in spite of their phylogenetic importance. The lungfish pancreas is a compact intraintestinal organ embedded in the dorsal wall of the foregut. The few islets present are separated from the exocrine tissue by a thick connective tissue capsule, but little is known of their constituents or organization. The compact endocrine pancreas of the crossopterygian, *Latimeria chalumnae*, exhibits islet parenchyma containing A, B, and D cells with organizational features more typical of the cartilaginous fish.

Among the Actinopterygii, the archaic survivors (sturgeons and paddlefish, bichirs and reedfish, gars and bowfins) have exocrine pancreas scattered widely throughout the body cavity, often accompanying the portal veins and bile ducts into the liver. Islets tend to be small and disseminated throughout the exocrine parenchyma without aggregation of tissue comparable to the Brockman bodies in the teleosts, except for some large islets in the intrahepatic regions of the pancreas. The composition and organization of cell types within the islet is not well known.

The major group of bony fish is the Teleostei in which the endocrine pancreas is characterized by the presence of numerous islets scattered throughout the exocrine parenchyma. One striking feature in some species is the presence of one or more principal islets, or Brockman bodies, in addition to the small islets (Figure 8). In some species, for example *Xiphophorus helleri,* a single islet will contain all the pancreatic endocrine tissue. These large aggregations of islet tissue may be found within the exocrine parenchyma or disseminated widely in the mesentery though still associated with the exocrine pancreas by strands of acinar tissue. They often lie in close association with the gallbladder or bile ducts. Cellular composition of the two size classes of islets does not appear to differ significantly. Because the Brockman bodies have been used so frequently as sources of relatively pure endocrine tissue, there exists an erroneous impression that this arrangement is typical for bony fish. On the contrary, the majority of teleosts possess a complement of small islets that are scattered throughout the exocrine parenchyma.

Immunocytochemical staining has clarified the rather confused picture of islet composition and organization which arose from past light and electron microscopic studies of teleost islets. Typically, B cells are found in the central portion of the teleost islet, often interspersed with D cells, which then form an intervening layer between the B cells and the peripherally located A and PP cells. A and PP cells often penetrate in some islets to the center of the islet, following vascular connective tissue septa. Figure 9 shows this to be the case in the islet of *Ictalurus nebulosus* with respect to the A cells whereas the PP cells remain fairly restricted to the periphery. The principal islets of the daddy sculpin, *Cottus scorpius,* and of *Xiphophorus helleri* appear to exhibit this arrangement of cell types with the PP cells being found at the outermost periphery and the A cells located just beneath them. In the anglerfish, *Lophius americanus,* however, large areas of D cells in the

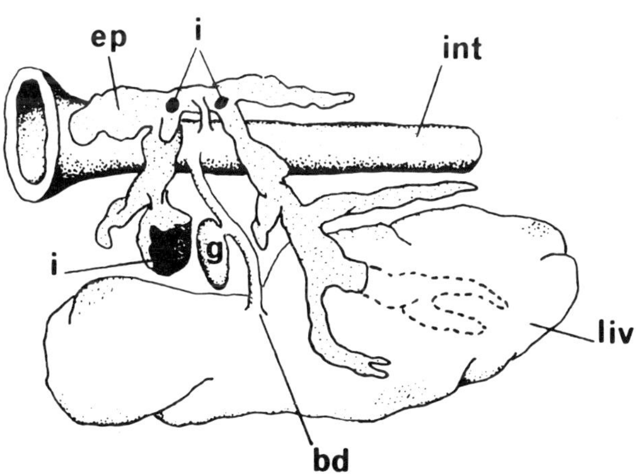

Figure 8. Anatomical relationships between the islet tissue and the exocrine pancreas in a teleost. ep, Exocrine pancreas; i, islet tissue; int, intestine; g, gallbladder; liv, liver; bd, bile duct. [Redrawn from Epple, A., 1969, *Fish Physiology,* Vol. 2 (W. S. Hoar and D. J. Randall, eds.), Academic Press, New York, pp. 275–319.]

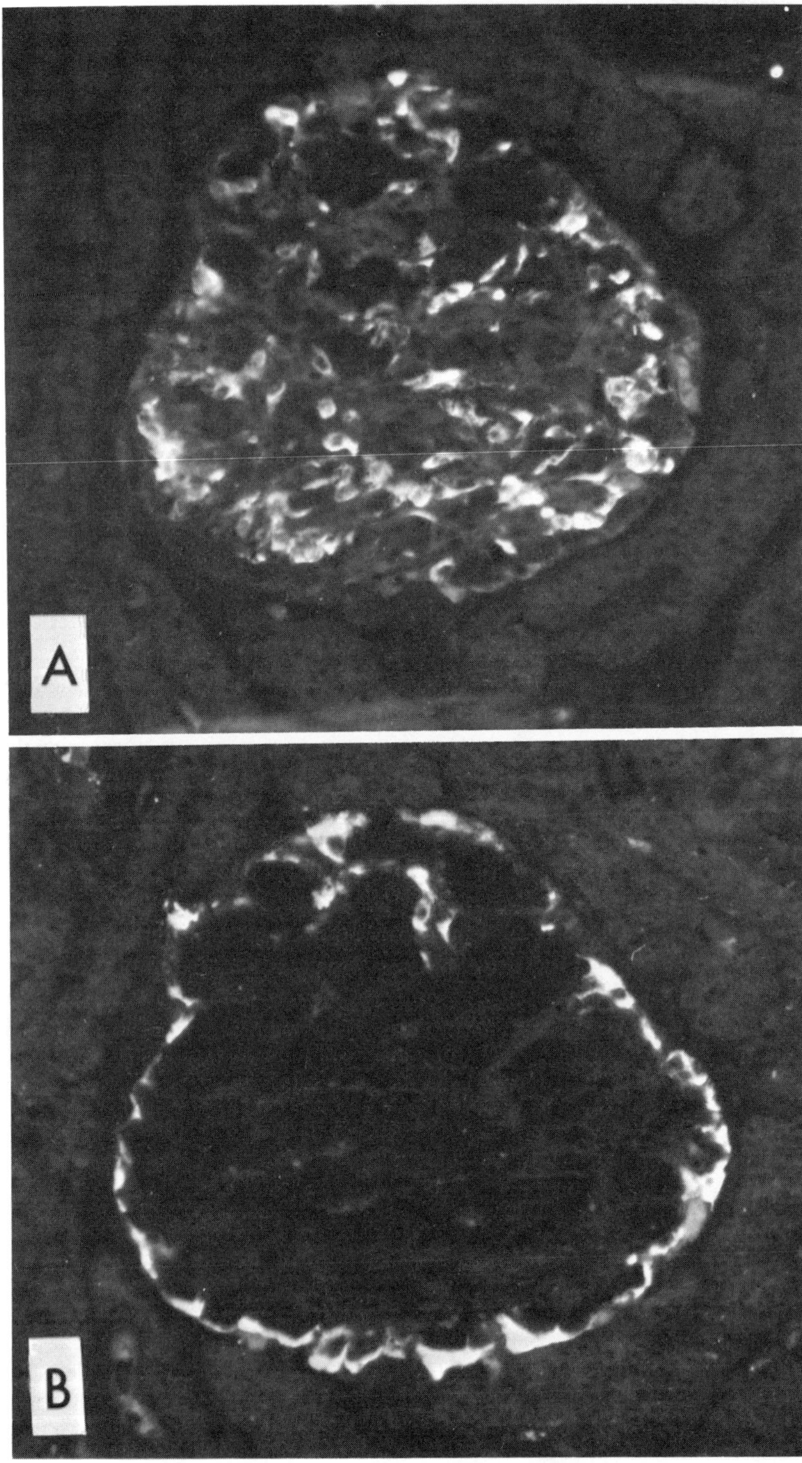
A
B

principal islet are surrounded by a rim of A cells and secondarily by B cells. The principal islet of the catfish, *Ictalurus punctatus,* on the other hand, has less regular arrangement of the cell types although the ratio is similar (9:6:4, insulin: somatostatin:glucagon) for both genera. In *Cottus* the islet tissue is concentrated into two principal islets, one situated next to the spleen and the other in the wall of the digestive tract close to the pylorus. All four islet cell types are found in the juxtapyloric islet and its surrounding smaller islets, but the juxtasplenic and its surrounding islets are devoid of PP cells. This nonrandom distribution of PP cells is comparable to the arrangement seen in the mammalian pancreas where a portion of the head region contains islets rich in PP cells, the remainder less so. This distribution dichotomy may well reflect the dual embryonic origin of the pancreas.

3.2.4. Amphibia

In all three groups (apodans, urodeles, anurans) the pancreas is a compact mammalian type with the islet tissue making up only a small portion of the total pancreatic mass. Most amphibians have at least three distinct pancreatic lobes (hepatic, duodenal, splenic), sometimes more, with the splenic lobe having the highest concentration of islet tissue and generally the largest islets. Amphibian islets exhibit a wide range of organization from lower vertebrate characteristics seen in a primitive urodele such as the mudpuppy, *Necturus maculosus,* to those of higher vertebrate type seen in some ranid and hylid frogs.

The pancreas of the toad, *Bufo arenarum,* is divided into five lobes or processes of which the gastric and free processes contain the greatest frequency and largest of the islets. The free process has islets of 25% non-B cells whereas the gastric, duodenal, and jejunal processes have islets with greater than 90% B cells. Those in the hepatic process, however, are composed exclusively of B cells. *Rana pipiens* also has five distinct pancreatic lobes or regions (Figure 10A). The gastric and splenic regions contain islets that tend to have a higher percentage of non-B cells, and the jejunal, duodenal, and hepatic regions are characterized by smaller islets with a higher proportion of B cells. The pancreas of *Bufo melanosticus,* on the other hand, is composed of anterior, middle, and posterior regions with the majority of the islets in the anterior (splenic) and middle regions.

Despite numerous studies the organization and hormone content of amphibian islet cells have not been established precisely. The one difficulty is that in most anuran and urodele species examined glucagon and pancreatic polypeptide are found contained within the same cell. Whether this observation, based on immunocytochemical staining, represents a true bihormonal cell or a problem in antibody specificity remains to be resolved. In general, however, all four islet hormones have

←

Figure 9. Adjacent sections of an islet from the catfish, *Ictalurus nebulosus,* showing immunofluorescent staining for glucagon (A) and for pancreatic polypeptide (B). The A cells are observed throughout the islet with a slight tendency toward the periphery whereas the PP cells are restricted to the extreme periphery of the islet. (Micrographs courtesy of S. Van Noorden.)

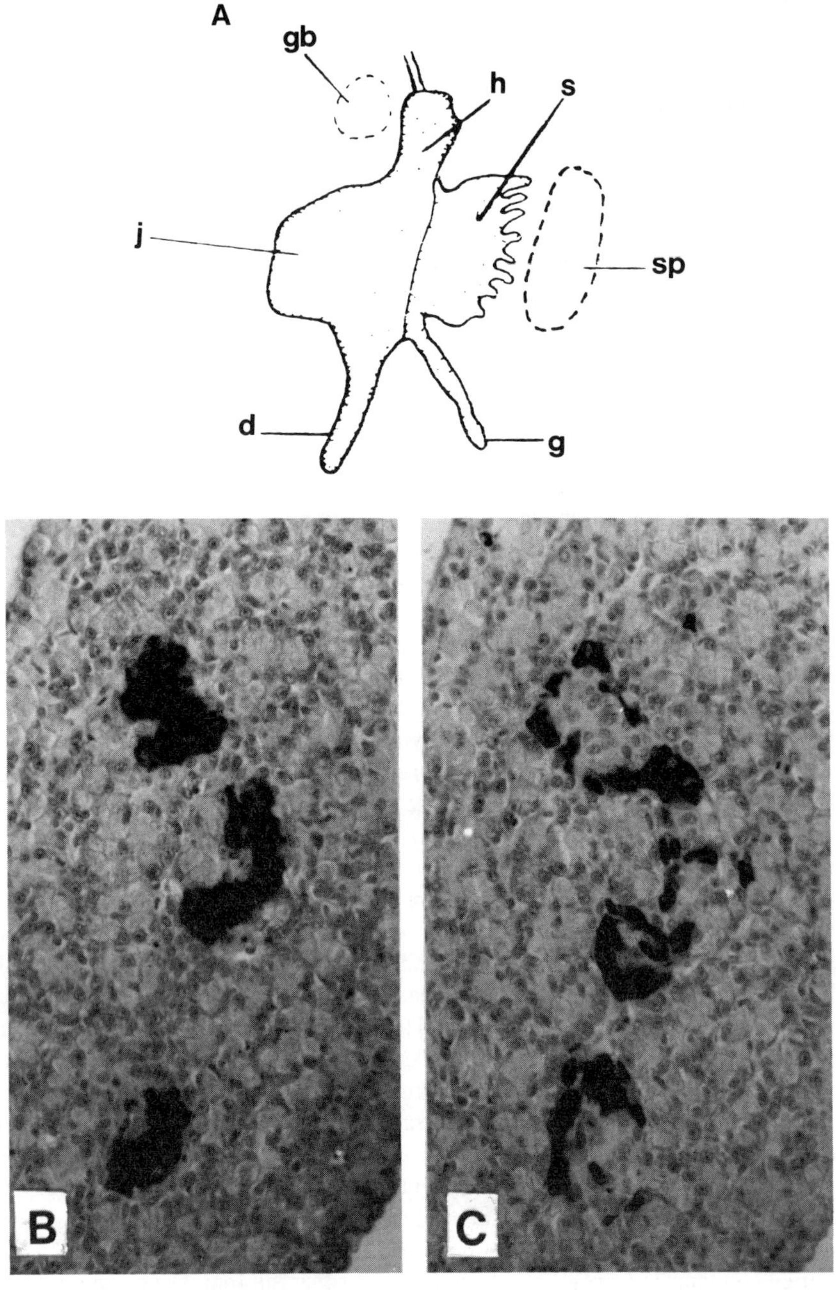

Figure 10. (A) Diagrammatic representation of the five lobes of the pancreas from the frog, *Rana pipiens*. The splenic and the gastric lobes are normally folded under the rest of the pancreatic mass. h, Hepatic lobe; j, jejunal lobe; d, duodenal lobe; s, splenic lobe; g, gastric lobe; sp, spleen; gb, gallbladder. (From unpublished results kindly provided by S. Bonner-Weir.) (B and C) Adjacent sections of pancreas from *Rana pipiens* immunostained for insulin (B) and glucagon (C) illustrating the cordlike core of the B cells with the peripherally arranged A cells. (D. A. Gapp, unpublished results.)

been localized to amphibian pancreas. Amphibian islets are usually small, and scattered single cells or small clumps of cells often may be observed in the exocrine parenchyma (e.g., *Salamandra salamandra, Necturus maculosus, Bombina bombina, Alytes obstetricans*). In amphibians exhibiting medium-sized (*Rana esculenta, Rana pipiens, Hyla arborea, Bufo bufo, Bufo arenarum*) or large islets (*Hyla crepitans*) central ribbons or cords of B cells are generally surrounded by a mantle of glucagon/PP cells and D cells. In *Rana pipiens* the islets are formed by cords of columnar B cells lining capillaries (Figure 10B). Several layers of A/PP cells surround the B cells, and D cells are found scattered throughout or around clusters of A/PP cells. About 2% of *Rana pipiens* pancreas is islet tissue, and of this 43% is comprised of B cells, 43% is comprised of A/PP cells, and 14% of D cells. Although the tendency in amphibian islets appears to involve centrally located B-cell mass, *Bufo melanosticus* islets are composed of peripheral B cells arranged in anastomosing cords which intermingle with a centrally located A-cell mass. D cells in this species are found toward the islet periphery near capillaries.

3.2.5. Reptilia

Within the compact reptilian pancreas there is a tendency in the more advanced species for the endocrine tissue to be concentrated into larger, but less numerous, islets located in the splenic portion. In some species the islets are embedded within the spleen without exocrine tissue and become analogous to the teleost Brockman

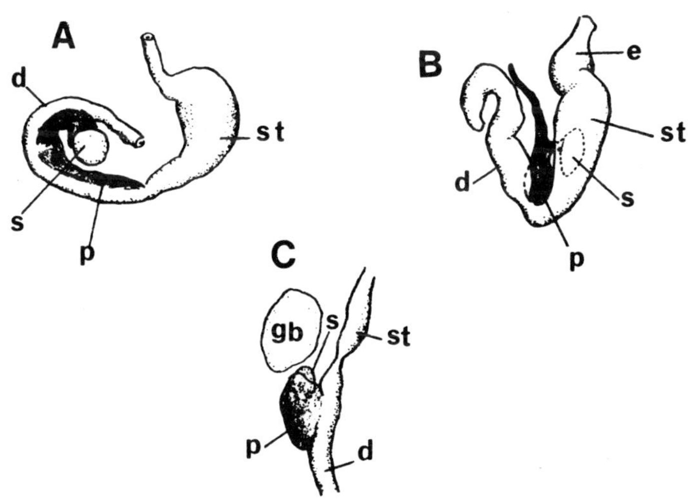

Figure 11. Diagrammatic representation of the pancreas and associated viscera of the turtle, *Chrysemys scripta* (A), the lizard, *Hemidactylus flaviviridus* (B), and the snake, *Thamnophis elegans* (C). p, Pancreas; d, duodenum; s, spleen; st, stomach; e, esophagus; gb, gallbladder. [Redrawn from Miller, M. R., and Lagios, M. D., 1970, *Biology of the Reptilia,* Vol. 3 (C. Gans and T. Parsons, eds.), Academic Press, New York, pp. 319–346.]

body. With respect to islet cell hormone content and the organization of cells within the islets, application of immunocytochemical methods has produced important new, though tentative, information.

The hard, compact pancreas of the painted turtle, *Chrysemys picta,* is triangular in cross-section, running along the duodenum from the pylorus and then leaving the duodenum to splay out over the spleen as in *Chrysemys scripta* (Figure 11A). As with the other chelonians that have been examined immunocytochemically [*Emys orbicularis, Pseudemys scripta, Chelydra serpentina*], *Chrysemys picta* islet tissue is very poorly organized and without distinct separation from the exocrine tissue. All four islet hormones are found in the pancreas of *C. picta,* with the A and B cells often found in clusters or aggregates as well as scattered individually throughout the exocrine parenchyma (Figure 12A,B). In the splenic portion, however, islets composed of a B-cell core with a loose A-cell mantle are present (Figure 12C,D). Somatostatin and PP cells are distributed throughout the pancreas but not found consistently within the islets. PP cells are absent from the splenic portion which contains the islets. Some regionalization of cell types has been observed in *Chelydra serpentina* where PP cells appear most common in the head region but are virtually lacking in the splenic portion of the pancreas. Although rigorous morphometric analyses have yet to be completed for the chelonian pancreas it appears that B cells and D cells are more prevalent in *Pseudemys scripta* whereas A and B cells appear more frequently in the pancreas of *Chrysemys picta.* Because of the loose and disseminated nature of the chelonian islet tissue, no precise topographical relationships among islet cell types have been delineated.

The pancreas of squamate reptiles (Figure 11B,C) generally exhibits islets with higher vertebrate characteristics. Islet tissue is compact and concentrated within the splenic portion of the pancreas. In lizards, the pancreas is composed of two or possibly three processes (one along the bile duct, one to the small intestine, one terminating in a lobe at the spleen). In *Hemidactylus brookii* and *Agama agama,* however, endocrine tissue appears evenly distributed throughout the pancreas. Immunocytochemical staining reveals ribbonlike islets in which insulin- and somatostatin-containing cells appear to predominate whereas glucagon-containing cells are few in number and distinct from the PP cells. In most other lizards islets are located primarily in the splenic portion of the pancreas as seen in the large splenic islets of the anole, *Anolis carolinensis,* the grass lizard, *Mabuya quinquetaeniata,* or the desert lizard, *Uromastyx aegyptia.* Moreover, A cells tend to occur in much higher frequency in lizard islets in comparison to other vertebrate groups, mammals in particular. *Varanus niloticus* exhibits large splenic islets containing predominantly A cells and very few B cells. In other regions of the *Varanus* pancreas intermediate-size islets are also predominantly A cells with small numbers of B and non-B cells although some small islets are found containing only B cells. Half of the splenic lobe islet cells in *Anolis* are A cells, and in the wormlike amphisbaenids A, B, and D cells comprise 45%, 46%, and 10% of the islet cell population, respectively.

Differences in islet cell populations within and between species have been well documented in the grass and desert lizards. *Mabuya* and *Uromastyx* both exhibit

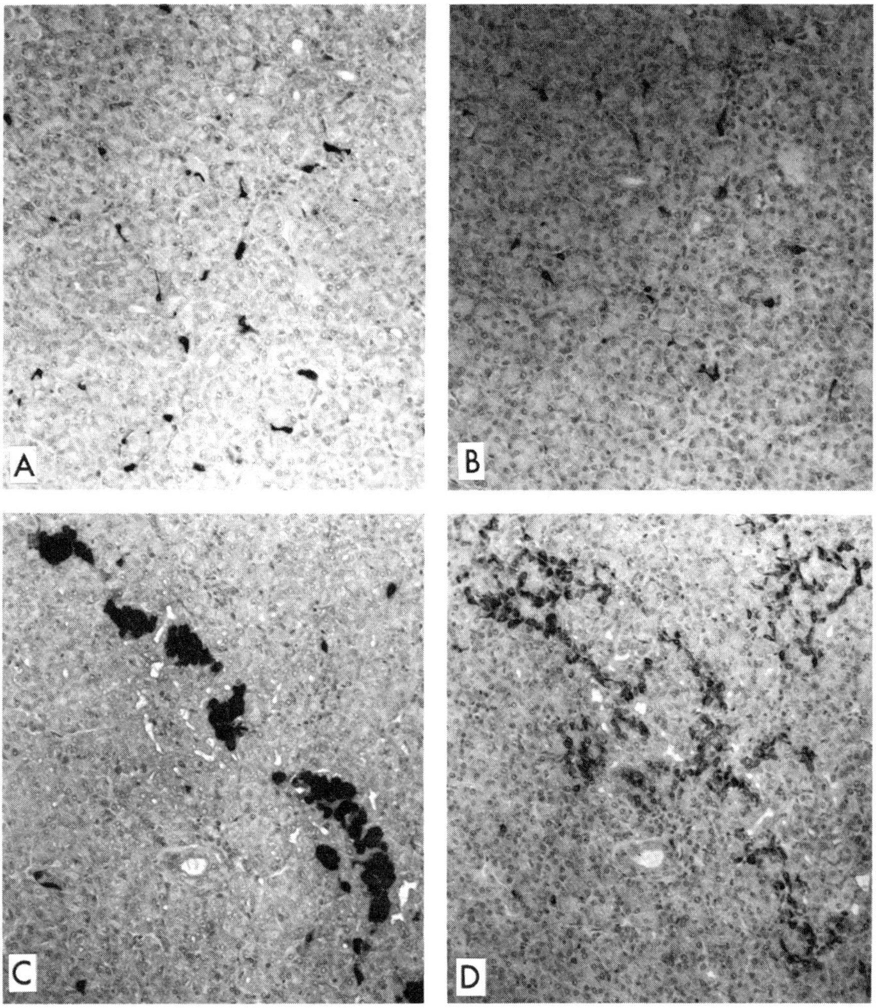

Figure 12. Adjacent sections from the duodenal (A, B) and splenic (C, D) portions from the pancreas of the painted turtle, *Chrysemys picta,* and immunostained for insulin (A, C) and glucagon (B, D). Note the lack of islets in the duodenal portion whereas in the splenic lobe distinct, although not very compact, islets are present exhibiting a B-cell core with an A-cell mantle. (D. A. Gapp, unpublished results.)

large splenic (dorsal lobe) islets which contain approximately 50% and 40% B cells, respectively. In *Mabuya* the B cells are scattered but concentrated toward the islet center; A cells are located more peripherally together with D cells. PP cells are rare and found mostly in the exocrine parenchyma. In contrast, the A cells in the splenic islets of *Uromastyx* form a large core (50% of the islet cells) whereas the B cells (40%) and D cells (6%) are localized to the periphery, the PP cells (4%) being found in the islet periphery and exocrine parenchyma. These two species show endocrine tissue arranged differently in the ventral lobe than in the splenic lobe. The ventral lobe of *Mabuya* is devoid of islets but contains scattered A, B, D, and abundant PP cells. In contrast, *Uromastyx* ventral and splenic lobes contain islets similar in content and frequency of cell type. The PP cells, however, occur in a much higher frequency and are found in the exocrine parenchyma.

The pancreas of snakes (Figure 11C) does not have obvious lobulation, but may be divided into three regions: anterior (splenic, tail), body, and head portions. Within the pancreas islet tissue is localized to the splenic pole of *Thamnophis sirtalis, Natrix natrix, Nerodia rhombifera, Agkistrodon piscivora,* and *Crotalus atrox.* An exception is seen in *Python reticularis,* where large islets occur in all regions of the pancreas. Conventional light microscopy indicates that islets are generally composed of equal numbers of A and B cells though the garter snake, *Thamnophis sirtalis,* has predominantly B cells. Islets in Ophidia are mostly large and, as with the lizards, exhibit a high proportion of A cells which may form the central core of the islet (e.g., in *Python reticularis* or *Natrix natrix*). In one species, *Crotalus atrox,* islets are found embedded in the spleen. Demonstration of PP cells in snakes has been inconsistent, one study showing that, with the exception of the Boidae, antisera to avian, bovine, or human PP could not demonstrate PP cells in the snake pancreas. Another study, however, has localized PP cells in the pancreas of the garter snake, *Thamnophis sirtalis.* This confusion may be attributable to differences in antisera specificity but, nevertheless, remains to be clarified.

The compact pancreas of the alligator, *Alligator mississippiensis,* exhibits islet tissue distributed evenly throughout the head, body, and tail (splenic) regions. Although discrete encapsulated islets are not present, cells are found in groups of clusters often appearing to be budding from the pancreatic ductal epithelium, especially B and D cells, which are often observed in close association. PP cells are present but are not concentrated in any distinct region of the pancreas. In another crocodilian, *Caiman crocodylus crocodylus,* islet tissue is organized into large islets some possibly located in the spleen.

3.2.6. Aves

The avian pancreas is best known in the fowl where it is a compact structure composed of several distinct lobes; the dorsal, ventral, and splenic are the most prominent, and in some species a small, "third" lobe is present and contiguous with the splenic lobe. Islet tissue is found in all regions of the avian pancreas and has been most studied in the domestic chicken and Japanese quail, *Coturnix coturnix.* Avian islets are of two basic types: A islets, also known as α or dark islets, are composed of A and D cells with an occasional cluster of B cells; B islets, also known as β or light islets, are composed of B and D cells, and sometimes A cells too. The A islets are restricted almost exclusively to the splenic and third lobes whereas the B islets may be found throughout all regions of the pancreas. In *Coturnix,* D cells are scattered throughout the irregular and multilobular A islets whereas in the smaller B islets, D cells are arranged around a B-cell core. Immunocytochemical assessment of the B islets of the third and splenic lobes of the chicken indicates a composition of 80% B cells, 10% A cells, and 10% D cells in contrast to a composition of 75% B cells and 25% D cells in the small islets of the dorsal and ventral lobes. PP cells are generally extrainsular in the avian pancreas or found infrequently at the islet periphery. In chickens the PP cells appear to be found in the dorsal and ventral lobes but not in the splenic or third lobes.

4. Evolutionary Considerations of Morphology

Pancreatic endocrine cells appear to have been recruited from the digestive tract mucosa to create an organ that regulates metabolic functions. During the process of islet evolution open-type paraneurons, which may have been "taste cells of the gut," were converted into closed-type cells which monitor systemic blood constituents rather than luminal contents of the intestine. Migration of "islet" cells appears to have preceded the formation of the exocrine pancreas, but once the exocrine tissue formed separate from the gut in the Chondrichthyeans, islet/exocrine association was established permanently. From this location, the islet tissue can provide the liver with a high concentration of islet hormones, monitor the posthepatic blood, and maintain an intimate association with the exocrine tissue.

The evolutionary order of arrival of the endocrine cells in the islet organ is presumed to be first insulin and somatostatin cells, then glucagon, and finally the PP cells. PP cells do not appear in vertebrates until an exocrine pancreas is present and are the last cell type to appear ontogenically in all species examined. Is this another example of "ontogeny recapitulates phylogeny"? The sequences of islet cell appearance in some animals (frog: B, A/PP, D; chick: A, B, D, PP; rat: A, B, D, PP) are not, however, totally consistent with the presumed phylogenetic sequence of islet cell appearance. Shifts in ontogenic appearance may reflect changes in functional importance of a cell type. For example, the appearance of the A cell first in the chick or rat may correlate with a developmental role hypothesized for glucagon.

If the islets of extant vertebrates reflect correctly their ancestral structure and function, then we can assume that the evolution of the islet organ was continuous during the formation of several vertebrate groups but was completed in the elasmobranchs. Although structures homologous to vertebrate islets are unknown in the invertebrates, nevertheless "islet" cell homologs are present, and the localization of A, B, D, and PP cells in the digestive tract of *Branchiostoma* provides a possible example of the immediate prevertebrate condition. With the cyclostomes, we find insulin and somatostatin cells have migrated from the gut mucosa to form islet lobules at the base of the bile duct (hagfishes) or in the duodenal wall (lampreys); glucagon/gastrin/PP cells remain in the gut mucosal epithelium. In the cartilaginous holocephalians we find development of an exocrine pancreas which contains islets composed of A, B, and D cells, and distinct PP cells are localized in the gut and the pancreatic duct epithelium. In the elasmobranchs, the endocrine tissue is comprised of all four islet cells whether organized as duct-associated endocrine parenchyma or as actual islet structures. Thus, the formation of the typical vertebrate islet with its complement of A, B, D, and PP cells is completed in the elasmobranchs with the evolutionary pathway having been traced from the gut to the pancreas by way of the pancreatic ductal epithelium (cf. Figure 4, Chapter 15).

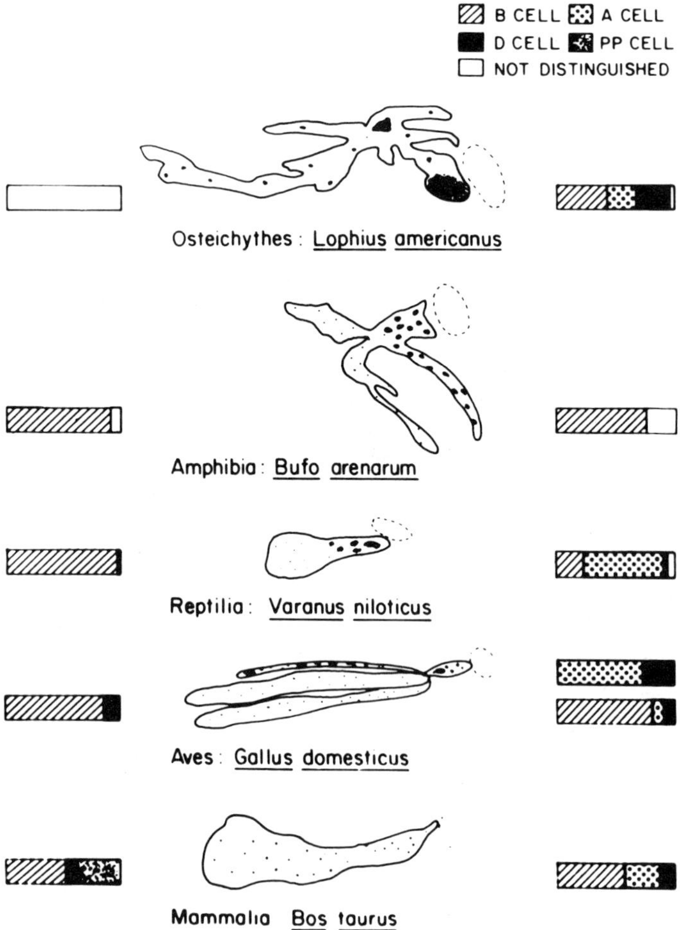

Figure 13. Diagrammatic representation of the endocrine pancreas in a number of different vertebrate groups. Schematic drawings indicate the distribution of islet tissue within the pancreas, and the bar graphs to the left and right represent the cell composition of the islets in the duodenal and splenic portions of the pancreas, respectively. The dotted oval to the right of the pancreas represents the spleen. (From Bonner-Weir, S., and Weir, G. C., 1979, The organization of the endocrine pancreas: A hypothetical unifying view of the phylogenetic differences. *Gen. Comp. Endocrinol.* **38:**28–37.)

There are several trends in islet tissue organization. Ancient vertebrat "primitive" members of higher vertebrate groups, exhibit a relatively uni dissemination of islet tissue throughout all regions of the pancreas, as ca observed in cartilaginous fish, some urodeles (e.g., *Necturus maculosus*), s snakes (e.g., *Python reticularis*), turtles, and crocodilians (e.g., *Alligator n sissippiensis*). In the more modern vertebrates, however, islet tissue is concentra into a specific region of the pancreas, as seen in the splenic islets of snakes, lizard and birds (Figure 13). Does this regional concentration of endocrine tissue hav important functional implications? It has been proposed that evolutionary dis semination of islets or islet tissue throughout the exocrine pancreas would have been advantageous for efficient regulation of exocrine tissue function by pancreatic islet cell hormones. Differentiated exocrine tissue surrounding islets ("periinsular halos"), the insuloacinar portal system, and experimental evidence demonstrating islet hormone modification of pancreatic exocrine cell secretory function support the concept of important advantages of disseminated islets. What, then, would be the benefits of a compact regionalized mass of exocrine tissue as seen in the teleost Brockman bodies or the *Anolis* splenic islets? How does the mammalian pancreas with less obvious (some say none, for example, in the human) concentration of islet tissue fit into an evolutionary scheme where the trend has been toward regional concentration of the islet mass?

Further examination of trends toward regionalization of islet tissue is best made with attention to the distribution of individual cell types. Here the dual embryonic origin of the pancreas leads to a dichotomous distribution of PP cells and, therefore, to two populations of islets. The presence of PP cells in the juxtapyloric, but not the juxtasplenic, principal islet of *Cottus scorpio* indicates that the division of the once homogeneous distribution of cell types seen in the Chondrichthyeans may have occurred in the teleosts. From this group onward, dual populations of islets (or nonuniform distribution of PP cells) are present in representatives of all vertebrate classes. In fact, in species where regional concentration of islet tissue does not occur as a pronounced feature, concentration of PP cells in the head region of the pancreas may be observed (e.g., as in *Chelydra serpentina*) and may serve a purpose similar to the more obvious concentration of the entire endocrine tissue as seen in squamates and birds. Compared to mammalian islets these large splenic islets tend to be quite rich in A cells, less so in B cells, and virtually devoid of PP cells. The smaller islets, especially those in the nonsplenic portion of the pancreas, tend to have a full complement of islet cell types, or if not, PP cells are abundant in the surrounding exocrine tissue. It is quite possible that large A cell-rich splenic islets, or the preferential aggregation of A cell-rich islet tissue into the splenic lobe, implies a different function with respect to the delivery and fate of their hormones.

The splenic islets may function as large compact, or relatively compact, endocrine glands whose purpose is to provide the portal circulation with sufficient islet hormones (principally insulin and glucagon) to satisfy hepatic and peripheral tissue needs. In contrast, the smaller islets, those with a greater frequency of PP cells (and a tendency to be A cell-poor), may well be more concerned with delivery of their hormones to the exocrine tissue for regulation of acinar cell function.

Islets from the splenic portion of the rat pancreas tend to be larger and outnumber those in the duodenal portion. Moreover, the vasculature of these larger islets appears to drain the islets directly without forming an insuloacinar portal system. In contrast, the smaller-sized islets, which are more frequent in the duodenal than in the splenic portion, exhibit vasculature that is more likely to form an insuloacinar portal system. This duality of the islets with two patterns of associated microvasculature correlating with the bimodal distribution of PP cells is further emphasized by identification of independent blood supplies for the two regions of the pancreas. The main body of the pancreas in *Chrysemys picta,* for example, is supplied by the pancreaticoduodenal artery whereas the spleen and the portion of the pancreas adjacent to it are independently supplied by the splenic artery. The juxtapyloric and juxtasplenic principal islets of *Cottus scorpio* each have their own blood supply, and in the rat the PP cell-rich islets in the head region receive blood from the pancreaticoduodenal artery whereas the remainder of the pancreas containing the PP cell-poor islets is perfused by the gastroduodenal and splenic arteries. The development of this vascular pattern in the vertebrates was established early, and possibly quite rapidly, given that the islet organ is an avascular structure in the cyclostomes and holocephalians.

Topographical relationships among islet cells have been a subject of considerable interest because of the implications for paracrine regulatory mechanisms. Typically, the mammalian islet is viewed as being composed of a homocellular B-cell core surrounded by a mantle of A, D, and PP cells (Figure 1). Variations on this theme occur in the human islet, which appears subdivided into multiple small units, or in the horse, where there is a heterocellular core surrounded by a homocellular B-cell mass. Besides a tendency for the formation of a homocellular B-cell mass (there are exceptions, of course), another typical finding in the vertebrate islet organ is the relatively close association of somatostatin-containing D cells with the B cells. Even in the protochordate *Branchiostoma lanceolatum,* such an arrangement has been observed and suggests that a paracrine relationship between B and D cells may have been established prior to the formation of the vertebrate islet organ. Such a strong relationship might also explain why the B and D cells appear simultaneously in the islet lobules of the cyclostomes when the first migratory steps from the gut were made.

The most common organization of the islet cells in the vertebrate islet appears to be patterned on a homocellular core of B cells with its surrounding mantle of non-B cells. Is this the standard vertebrate islet? If so, why then this particular arrangement? Presumably this particular organization would permit maximum effects of non-B-cell hormones on the B cells especially if, as has been proposed, the direction of vascular perfusion of the islet is from the non-B cells to the B cells. When there is a peripheral homocellular B-cell mass as in the horse, perfusion is via the heterocellular core to the B-cell mantle. What, then, about islets of birds and squamates in which the A cells comprise a substantial portion of the islet and glucagon may be relatively more important than insulin for metabolic regulation? Does this have any implication for islet organization and paracrine regulation? Although these questions remain largely unanswered, it does appear that in the variation on the theme of A- and B-cell organization, the D cells always remain in close proximity to these two cell types, possibly maximizing the number of contacts with them. Such an arrangement is consistent with a central role for somatostatin in paracrine regulation of islet cell function. What remains to be clarified, however, is a situation in an animal such as *Chrysemys picta* where the D cells do not appear intimately organized into the rather loose islet structure of the chelonian pancreas. Do the D cells act at a distance to regulate A-, B-, or PP-cell functions where there is no close cell-to-cell proximity that is observed in so many other islet organs?

It is quite apparent that we are just beginning to unravel some of the complexities of the vertebrate islet organ. Elucidation of islet cell topography and associated microvasculature in many species will provide, ultimately, a more complete picture of the functional organization of this microorgan. Application of immunocytochemical techniques is essential in interpreting previous histological and electron microscopic observations and in extending our knowledge of islet cell hormone content. Advances in the field of peptide research are providing numerous new peptides, many of which may well fit into some of the blank spaces where additional cell types for the islet have been described, though no hormones ascribed to them. In fact, peptides other than insulin, glucagon, pancreatic polypeptide, and somatostatin have been localized to the pancreas of some nonmammalian vertebrates (e.g., gastrin, GIP, and endorphin in *Uromastyx aegyptia*) and must be

Table II
Occurrence of Multiple Localization of Peptide Hormones in Islet Cells or Islet Cell Homologues Based on Immunocytochemical Staining

Species	Tissue	Hormones
Branchiostoma lanceolatum	Gut	Glucagon, gastrin
Myxine glutinosa	Gut	Glucagon, gastrin, pancreatic polypeptide
Larval lamprey	Gut	Glucagon, gastrin, pancreatic polypeptide
Dasyatis akajei	Pancreas	Somatostatin, pancreatic polypeptide
Rana pipiens[a]	Pancreas	Glucagon, pancreatic polypeptide
Uromastyx aegyptia	Pancreas	β-Endorphin, pancreatic polypeptide
Mabuya quinquetaeniata	Pancreas	β-Endorphin, pancreatic polypeptide
Chrysemys picta	Pancreas	Glucagon, peptide YY
Various mammals	Pancreas	Somatostatin, β-endorphin
		Somatostatin, gastrin (transient)
		Glucagon, α-endorphin
		Glucagon, cholecystokinin
		Glucagon, gastric inhibitory polypeptide

[a]Glucagon/pancreatic polypeptide cells appear to be found in most amphibian species examined.

integrated into the anatomical and functional organization of the islet. And finally, localization of multiple peptides to the same cell type (Table II) presents a problem of validating the true bi- or multihormonal nature of a cell type, and sorting out problems of antibody specificity, the latter being particularly critical when applying immunocytochemical methods to submammalian tissue using antisera raised against mammalian hormones.

5. Control of Synthesis and Release of Pancreatic Hormones

Hormonal secretions of the pancreatic islets are intimately involved in the regulation of metabolic processes in most vertebrates. The central role of insulin in the integration of mammalian metabolic activities and in the diabetic condition has resulted in a wealth of information concerning the pancreatic islets and especially on the synthesis and release of insulin from the pancreatic B cell. Glucose is the primary stimulus for release of insulin in man and most mammals whereas amino acids, long-chain fatty acids, and glucagon are also insulinotropic, but only in the presence of glucose. Amino acids are more effective in stimulating insulin release from the fetal B cell, which is insensitive to glucose until the perinatal period, often just postnatally, when the B cell becomes glucose sensitive.

5.1. Insulin

In mammals the circulating level of glucose is around 100 mg/dl (5.5 mM) with a range of about 75–125 mg/dl. Stimulation of insulin secretion occurs around 150–200 mg/dl (8.25–11 mM) with maximum B-cell response reached around 300 mg/dl (16.5 mM). The normal physiological response of the B cell to stimulatory levels of glucose is a biphasic release of insulin (Figure 14) in which there is a rapidly rising first phase reaching a peak several minutes after stimulation followed by an abrupt drop and then a slowly rising second phase which is sustained as long as the glucose challenge is present. The first phase is thought to represent the rapid release of insulin from the β granules which are already associated with the microtubular/microfilamentous system and represents the readily releasable pool in the B cell. The second phase is produced by a combination of the mobilization and release of additional preformed β granules and of the release of newly biosynthesized insulin.

Although glucose stimulates both the biosynthesis and release of insulin, the mechanisms controlling the two activities are different. Release of insulin involves interaction of glucose in a stereospecific manner with a putative membrane glucoreceptor that triggers electrical events at the plasma membrane, leading to the formation of action potentials and alteration of membrane permeability permitting the influx of Ca^{2+} into the B cell. Increased intracellular Ca^{2+} is required for activation of the microtubular/microfilamentous system which is responsible for translocating β granules to the cell surface. Calcium appears to be the most critical

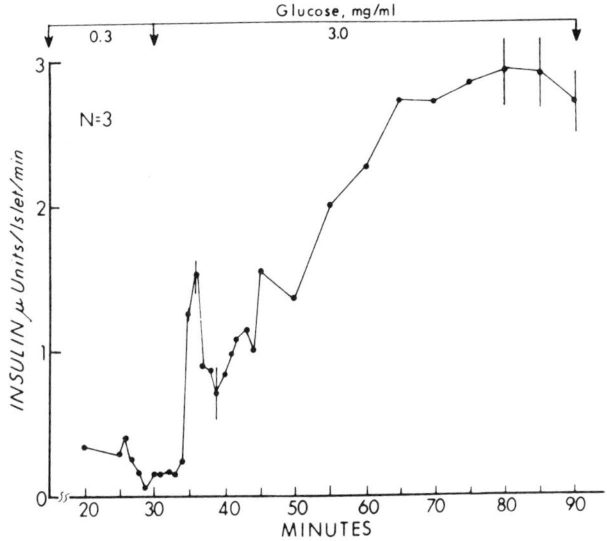

Figure 14. Biphasic pattern of insulin release from isolated, perifused rat islets stimulated with glucose at a concentration of 3 mg/ml. (From Lacy, P. E., Walker, M. M., and Fink, C. J., 1972, Participation of the microtubular system in the biphasic release of insulin. *Diabetes* **21**:987–998.)

factor for the induction of insulin release since calcium ionophores (drugs that cause Ca^{2+} translocation into the cell) are capable of stimulating insulin release in the absence of glucose. Microtubules are thought to provide the tracks along which the β granules move while the actin-containing microfilaments provide the motive force. Translocation of the β granules to the plasma membrane results in fusion of the granule membrane with the plasma membrane and dissolution of the granule contents into the extracellular space.

Under some conditions stimulating insulin release, membrane-bound adenylate cyclase is activated, which results in the generation of increased levels of cyclic AMP. Although not acting as a second messenger per se, cyclic AMP is thought to promote polymerization of tubulin monomers into microtubules and to mobilize intracellular Ca^{2+} stores helping to augment the influx of Ca^{2+} from the extracellular space. Glucose metabolism is not required for stimulation of insulin release since this can be done by certain nonmetabolizable analogs.

Conversely, insulin synthesis is a calcium-independent but sodium-dependent process which is stimulated by glucose metabolism but not by nonmetabolizable analogs of glucose. Insulin is biosynthesized as a large precursor, preproinsulin, which is proteolytically processed at several stages to form insulin and which is stored in the membrane-bound cytoplasmic β granules. Glucose stimulates the incorporation of amino acids into the growing preproinsulin molecule on the ribosomes of the RER both by increasing the translation of existing mRNA and by enhancing the rate of mRNA formation. The first part of the molecule to be synthesized is the "pre" or leader sequence which is responsible for threading the nascent peptide through the membrane and into the cisterna of the RER. The leader sequence is degraded cotranslationally, and the single-chain precursor, proinsulin, is released from the ribosome into the RER. Proinsulin is transported to the Golgi where the C peptide is removed proteolytically by trypsinlike and carboxypeptidase B-like enzymes acting at basic amino acid residue sites (Figure 15). Insulin is condensed around atoms of zinc to form a crystalline structure that is packaged in membrane and transported to the cytoplasm. The C peptide remains in the β granule and is cosecreted with insulin. Based on pulse-chase studies in isolated rat islets incubated at 37°, the $t_{1/2}$ for the appearance of labeled insulin in cytoplasmic granules is about 60 min.

5.2. Glucagon

The essential function of glucagon is to maintain an adequate level of blood glucose at all times, particularly for the central nervous system (see Chapter 18). For example, glucagon mediates 75% of the basal state for hepatic glucose produc-

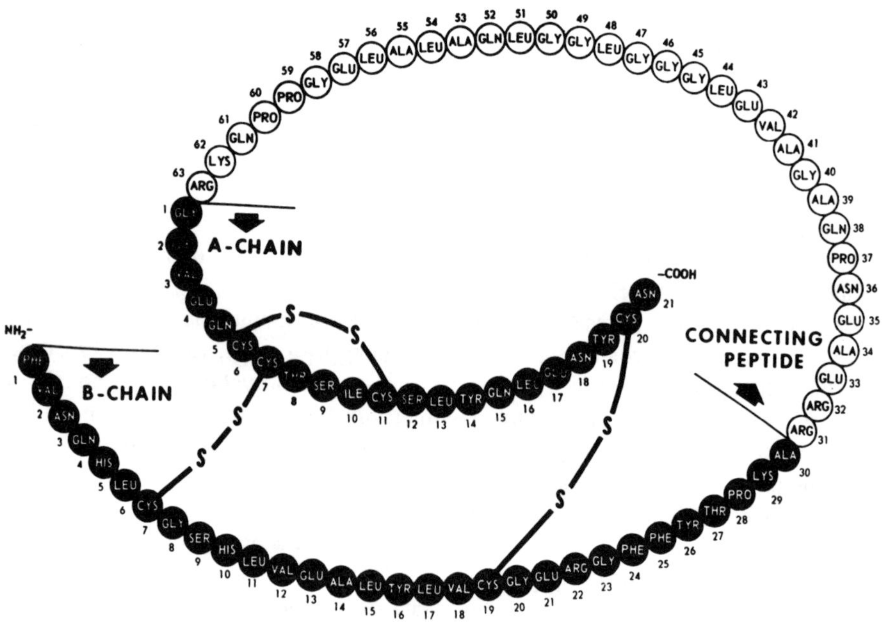

Figure 15. Amino acid sequence of the bovine proinsulin molecule. The connecting peptide (open circles) is released from proinsulin by the action of a trypsinlike endopeptidase cleaving the chain at two basic residue sites. Further action of a carboxypeptidase B-like exopeptidase removes arg-31, arg-32, lys-62, and arg-63 from the C peptide as well. (From Chance, R. E., 1969, Proinsulin and the biosynthesis of insulin. *Rec. Prog. Horm. Res.* **25**:274.)

tion in humans. Regulation of glucagon biosynthesis and release appears to involve a number of factors: metabolic, endocrine, paracrine, and neural. The generally accepted stimulus for glucagon secretion is considered to be hypoglycemia. This is based on experimental evidence and on the concept of antagonistic actions for the two major islet hormones, insulin being hypoglycemic whereas glucagon is hyperglycemic.

Glucagon secretion is sensitive to the metabolic rate within the A cell, as reflected by the intracellular levels of ATP. At low metabolic rates there is a shortage of ATP, which enhances glucagon secretion and provides a basis for the stimulatory effect of hypoglycemia. Moreover, insulin suppression of glucagon release correlates with increased intracellular ATP, presumably attributable to the insulin-mediated rise in intracellular glucose metabolism. Whether or not the A cell possesses its own glucose sensor or is dependent on the B cell for glycemia signals is unknown. Nevertheless, A cells that are not in close proximity to B cells (e.g., A cells in the gastric fundus, in glucagonomas, or in the islets of juvenile diabetics) appear unresponsive to changes in glucose concentration, suggesting that A cells may be dependent on the B cells for glycemia-mediated signals. Mixtures of amino acids, or some amino acids alone (e.g., arginine), are potent glucagon secretagogues as are free fatty acids and epinephrine. Agents that stimulate glucagon release also activate A-cell adenylate cyclase and transmembrane movements of calcium, which indicates a role for cyclic AMP and calcium in the stimulus–secretion coupling of glucagon release.

The molecule of bovine pancreatic preproglucagon (Figure 16) contains (1) a 20-amino-acid-residue leader sequence, (2) a sequence corresponding to glicentin, and (3) two additional glucagonlike peptides (GLP-I, GLP-II). Within the glicentin portion of the molecule the sequences are arranged as follows: an N-terminal 63-amino-acid peptide, the 29-residue glucagon peptide, and a C-terminal octapeptide. Proteolytic processing of this large precursor ultimately produces glucagon, which is condensed and stored in membrane-bound granules. The glicentin $_{1-63}$ fragment of the proglucagon molecule is contained within the granule and is presumably secreted along with the glucagon. The fate of GLP-I and GLP-II is unknown. As with insulin, glucagon release is mediated by microtubular/microfilamentous system activity which tranlocates the α granules to the cell membrane where granule and plasma membranes fuse and rupture to permit dissolution of granule contents into the extracellular space.

5.3. Somatostatin

The central role of somatostatin in the local control of pancreatic islet secretory function makes the understanding of the regulation of D-cell secretion particularly important. In general, stimuli that promote insulin release are also effective in promoting somatostatin release. Circulating glucose above 100 mg/dl stimulates somatostatin release in a dose-dependent manner. Amino acids (e.g., leucine or arginine) as well as CCK, GIP, secretin, gastrin, glucagon, and prostaglandin E_2

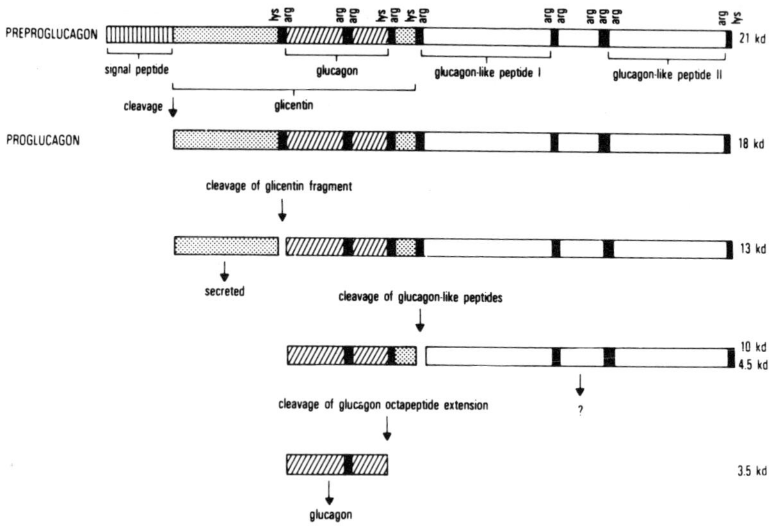

Figure 16. Proposed scheme for proteolytic processing of bovine pancreatic preproglucagon. The basic dipeptide sequences represent potential sites for cleavage. The preproglucagon molecule is larger than that found in the intestine, which is equivalent to glicentin or a C-terminally extended portion of glicentin. (From Lopez, L. C., Frazier, M. L., Su, C.-J., Kumar, A., and Saunders, G. F., 1983, Mammalian pancreatic preproglucagon contains three glucagon-related peptides. *Proc. Natl. Acad. Sci. USA* **80:**5485–5489.)

also stimulate the release of pancreatic somatostatin. Epinephrine and β-endorphin, on the other hand, inhibit somatostatin release. Fluctuations in circulating metabolites and the presence of intraluminal nutrients in the stomach or intestine stimulate pancreatic somatostatin release, the latter probably via neural mechanisms. In contrast, gastrointestinal somatostatin release appears to be regulated primarily by luminal contents alone (see Chapter 16). In the pancreatic D cell, stimulation of somatostatin release apparently involves calcium- and cyclic AMP-mediated mechanisms. Biosynthesis of somatostatin involves formation of preprosomatostatin, which is processed to form prosomatostatin (cf. Chapter 15). Processing of prosomatostatin into somatostatin-28 (predominantly in the gut) or somatostatin-14 presumably involves trypsinlike cleavage as there are basic residues at each of these sites.

5.4. Pancreatic Polypeptide

Basal levels of PP in the human are age- and sex-dependent, increasing with age and higher in males than in females. Plasma levels in diabetics are often elevated compared to nondiabetics, the degree of elevation exacerbated by ketoacidosis. Ingestion of nutrients, especially a protein-rich meal, stimulates PP release initiating a biphasic response: a rapidly rising first peak which lasts about 30 min followed by a prolonged second phase of secretion lasting several hours or more. The rapidity of PP increase after a meal suggests a gastrointestinal-mediated signal rather than a direct stimulation of the PP cell by products of digestion. Infusion of individual amino acids or triglyceride has no effect on PP release although a mixture of essential amino acids can have a modest effect, as can some gut hormones. Infusion of glucose has a suppressive effect on PP release, and hyperglycemia reduces PP-cell response to a protein meal or to intravenous administration of secretin or cerulein. Somatostatin is a potent inhibitor of PP release.

Sham feeding and gastric distention by balloon or nonnutritive material promote rapid increases of PP release; however, gastric distention is only about 10–20% as potent as the equivalent amount of food in the stomach. These effects are mediated by vagal pathways as atropine or vagotomy abolishes the responses. The first phase of PP release in response to a meal seems to be the most sensitive to vagotomy, but it appears that other cholinergic, but nonvagal, pathways exist so that vagotomized patients do show some response to nutrient ingestion. Insulin-induced hypoglycemia, the most potent stimulus for PP release, is also mediated by the vagus, as is tissue glucopenia induced by 2-deoxyglucose. Whereas the first phase of PP release appears to be due to food in the stomach, the prolonged second phase of PP release is probably caused by the generation of nutrients (e.g., essential amino acids, free fatty acids) in the intestine, which stimulate PP release via vagal and extravagal pathways and possibly by gut hormones.

Prepropancreatic polypeptide contains an 8000 to 10,000-molecular-weight precursor which is presumed to be pro-PP. Proteolytic processing of pro-PP produces PP plus a smaller peptide which is probably secreted, but whose function is unknown (Figure 17).

5.5. Neural Regulation of Islet Hormone Release

Neural mediation of hormone release from the PP cells is unique among islet cells since in most mammals neural input tends to superimpose over or modulate the direct regulation of islet hormone release by metabolites. (In submammalian vertebrates, for example in the teleosts, the regulation of islet secretion by neural mecha-

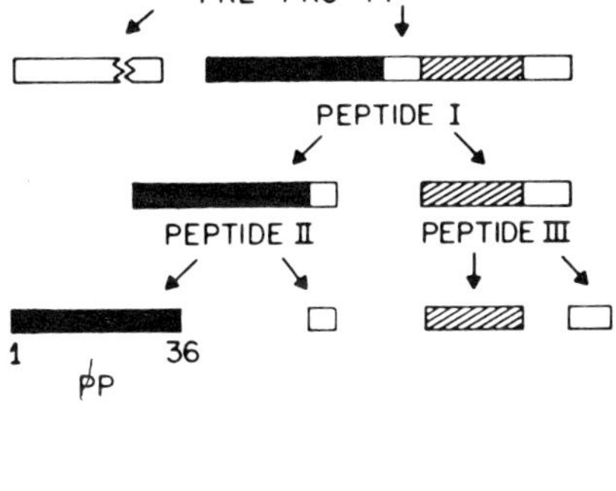

Figure 17. Hypothetical scheme for the biosynthesis of pancreatic polypeptide. Peptides I, II, and III represent biosynthetically labeled peptides identified in canine pseudoislets. The peptide represented by the cross-hatched bar and arising from peptide III is a putative peptide cosynthesized and possibly secreted with PP. [From Schwartz, T. W., and Tager, H. S., 1981, *Gut Hormones* (S. R. Bloom and J. M. Polak, eds.), Churchill Livingstone, Edinburgh, pp. 202–205.]

nisms may be considerably more important than it is in the mammals.) The pancreas is well innervated by branches of the vagus and splanchnic nerves with cholinergic, adrenergic, and peptidergic nerves terminating near or on the islets. Cholinergic stimulation promotes insulin, glucagon, somatostatin, and PP release. β-Adrenergic stimulation enhances insulin, glucagon, and somatostatin release whereas α-adrenergic stimulation inhibits insulin and somatostatin release. Certain neuropeptides also participate in regulating islet cell secretory activity. For example, VIP and CCK-4 (the C-terminal portion of gastrin and CCK, also known as tetrin) are both abundant in axon terminals impinging on or near the islets and stimulate the release of all four islet hormones. Substance P stimulates the release of insulin, somatostatin, and glucagon from the canine pancreas whereas bombesin and its mammalian counterpart, gastrin-releasing peptide, elevate plasma insulin and PP levels. TRH inhibits glucagon and PP release caused by insulin-induced hypoglycemia and can also block insulin suppression of glucagon release as well as enhance arginine-stimulated glucagon release.

6. Comparative Aspects of Pancreatic Endocrine Secretion

The mechanisms of synthesis and release of pancreatic islet hormones in nonmammalian vertebrates are similar to those in mammals. However, patterns of secretion may differ, as well as the dependence on physiological and neural stimuli for release and synthesis. Moreover, individual islet hormones may exhibit different roles in the regulation of metabolic processes in nonmammalian compared with mammalian vertebrates.

6.1. Cyclostomes

The hagfish (*Myxine glutinosa*) appears to lack a positive modulation system for control of insulin synthesis, although release is stimulated by glucose. Hagfish insulin is synthesized and secreted at a very low rate. Since the deepsea habitat of *Myxine glutinosa* makes for an ambient temperature of around 5°C (seldom above 10°C), various metabolic and cellular processes might reflect this. Pulse-chase experiments with hagfish islet tissue carried out at 11°C demonstrated a 20-hr lag phase before newly synthesized insulin passed through the B cell. After 60 hr, only 5% of the total insulin synthesized had been released. Conversion of proinsulin to insulin is temperature dependent up to 20°C; the $t_{1/2}$ for conversion at 11°C is 10 hr whereas the $t_{1/2}$ at 18°C is 8 hr. Because of the extreme sensitivity to degradation of hagfish proinsulin by trypsin, conversion of proinsulin to insulin in this species may involve different enzymes or enzymes with characteristics different from those known in the higher vertebrates. Insulin biosynthesis in *Myxine* is not stimulated by varying the concentration of glucose over the range of 0–27.8 mM. Moreover, arginine, leucine, 3-isobutyl-1-methylxanthine (IBMX), theophylline, and dibutyryl cyclic AMP have no effect on insulin biosynthesis by hagfish islet tissue.

However, release of insulin from *Myxine* islet organs is stimulated by glucose. At 16°C enhanced insulin release is observed over a range of 1–5 mM glucose, with half-maximal stimulation occurring at 3 mM. The release mechanism is sensitive to calcium and presumably also involves cyclic AMP since IBMX increases the efficacy of response to glucose. There is an inverse relationship of release and temperature where increasing the temperature from 6°C to 20°C produces a greater than twofold decline in secretion. Amino acids do not stimulate hagfish insulin release, possibly attributable to the absence of glucagon-producing cells, and, in fact, arginine inhibits insulin secretion. Epinephrine depresses insulin release but carbachol has no effect, which may reflect the lack of innervation of the hagfish islet organ. The kinetics of insulin release in *Myxine* are quite slow; at 16°C a

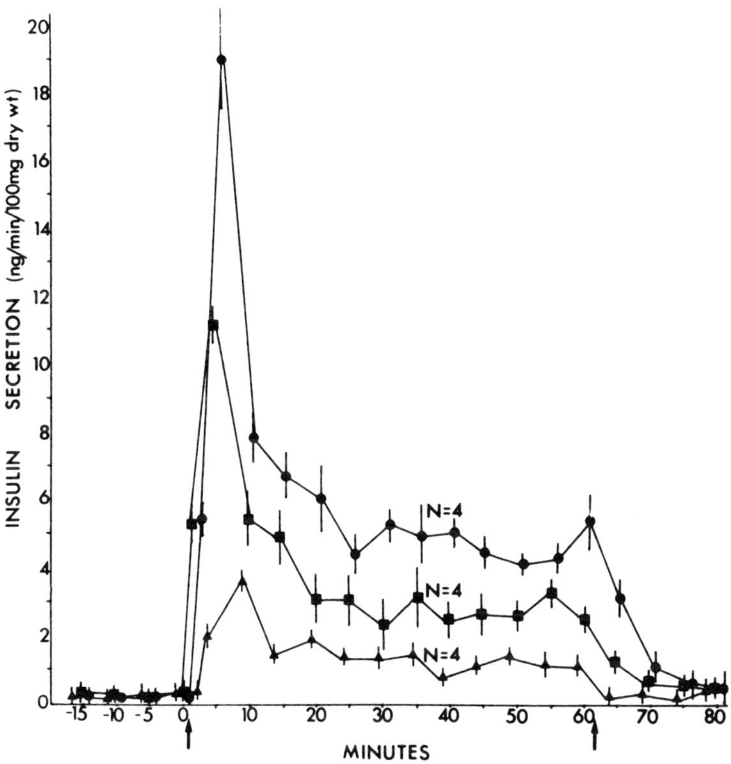

Figure 18. Insulin release from the perfused pancreas of the eel, *Anguilla anguilla*, in response to stimulation with 30 mM glucose (▲) and with 10 mM (■) or 20 mM (●) arginine. Basal release (−15 to 0 min) and the poststimulatory period (60–90 min) were measured with a glucose concentration at 2.7 mM in the perfusion medium. Note the relatively poor secretory response induced by the very high glucose concentration compared to the rapid, pronounced response induced by the two concentrations of arginine. (From Ince, B. W., 1979, Insulin secretion from the *in situ* perfused pancreas of the European silver eel, *Anguilla anguilla* L. *Gen. Comp. Endocrinol.* **37**:533–540.)

demonstrable rise in insulin secretion is not detectable until 17 min after stimulation. In other cyclostomes glucose has an equivocal effect on insulin release, appearing not to be insulinotropic in the Pacific hagfish, *Eptatretus stouti,* while stimulating insulin release in the lamprey, *Lampetra fluviatilis,* under some conditions. In contrast to the hagfish, arginine and lysine stimulate insulin release in *Lampetra.* Glucagon is also insulinotropic, which might be surprising given the lack of A cells in the lamprey islet organ. Glucagon-producing cells are present in the intestine, however, and could represent part of the enteroinsular axis in these animals.

6.2. Teleosts

Biosynthesis and secretion of islet hormones in bony fish is probably better known than for most other lower vertebrates, attributable primarily to the advantages provided by the large concentration of relatively pure endocrine tissue in the Brockman bodies of some teleosts. Radioimmunoassays specific for several piscine insulins are also an advantage not available for the studies of islet tissue in the reptiles and amphibians. The dynamics of islet hormone secretion has been determined for several piscine species, including those without principal islets. Insulin release from fish islets is stimulated by glucose and amino acids (e.g., arginine, lysine) whereas epinephrine and norepinephrine inhibit release. Both glucose and arginine stimulate a biphasic pattern of insulin release (Figures 18 and 19) although the second phase is less prominent than it is in mammals (compare to Figure 14). The second phase is temperature dependent such that perfusion of the *Ictalurus* splenic Brockman body at 26°C produced a sixfold increase in second-phase release over the release when perfused at 18°C.

The extent to which circulating glucose is the principal physiological regulator of insulin release from the fish islet is not yet completely clear. Circulating glucose levels in fish are often quite low, even negligible, but may exhibit a range of 10 mg/dl (0.6 mM) to 40 mg/dl (2.2 mM) depending on the season and/or ambient temperature. Glucose concentrations up to 300 mg/dl have been used to stimulate insulin release from fish islet tissue, but this is well out of the normal physiological range; moreover, high levels of glucose appear to inhibit insulin biosynthesis. For example, glucose at 40, 100, and 300 mg/dl stimulates insulin release, but not biosynthesis in toadfish islets, and glucose inhibits leucine incorporation into anglerfish proinsulin over the range of 12.5–300 mg/dl. Why high glucose should inhibit and not stimulate insulin biosynthesis is unclear. It may be related to the slow rate of glucose utilization by fish islets, which, at 18°C, metabolize glucose at a rate 20% that of mammalian islets at 37°C. As in mammals, glucose utilization is related to the rate of insulin biosynthesis. The threshold for glucose-induced biosynthesis in mammals is 2.75 mM, and the dose–response curve plateaus between 16.5 and 28 mM. Fish islets have a threshold at 1 mM for the induction of biosynthesis but a plateau at 2 mM even though glucose utilization increases linearly over the range of 1–10 mM. Preincubation of *Ictalurus* islet tissue for 24 hr in

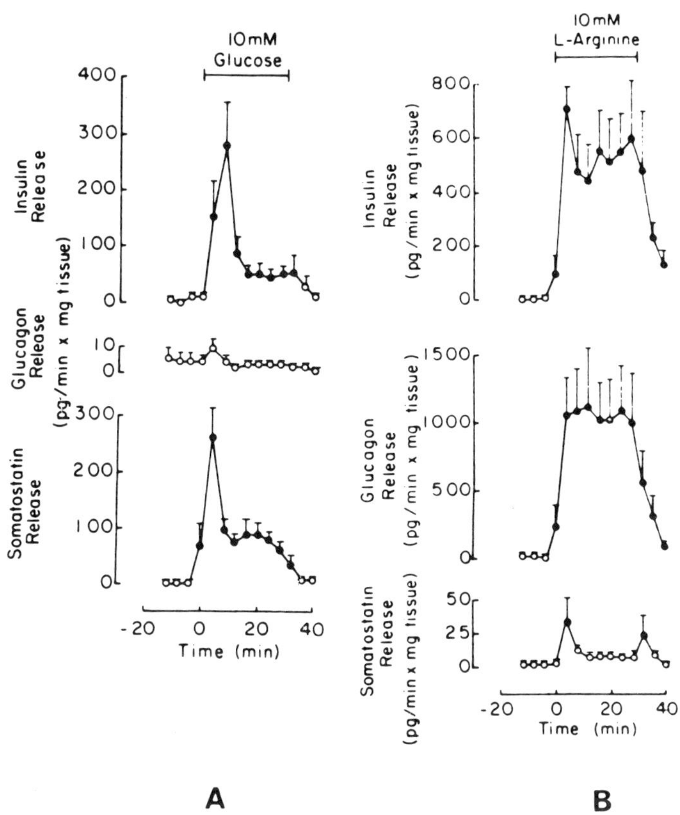

Figure 19. The effects of glucose (A) and arginine (B) on hormone release from the perfused splenic Brockman body of the channel catfish, *Ictalucus punctatus*. The glucose concentration was 2.75 mM during basal measurements and 2 mM during stimulation with arginine. Stimulation with glucose resulted in the release of 0.2%, < 0.1%, and 1% of the insulin, glucagon, and somatostatin stores, respectively, whereas stimulation with arginine produced corresponding values of 1.2%, 10%, and 0.1%, respectively. Closed circles indicate a significant difference ($p < 0.01$) from the first three baseline values. (From Ronner, P., and Scarpa, A., 1982, Isolated perfused Brockman body as a model for studying pancreatic endocrine secretion. *Am. J. Physiol.* **243:**E353–359.)

medium containing glucose will vitiate the inhibitory effect of high glucose on leucine incorporation into proinsulin. It appears, therefore, that elevated glucose does stimulate insulin biosynthesis in piscine islets if the stimulatory levels of glucose fall within the physiological range characteristic for fish and if sufficient time is permitted for adequate glucose metabolism to generate the stimulus for biosynthesis.

Incorporation of amino acid precursors into fish proinsulin is relatively rapid, but the conversion of proinsulin to insulin is somewhat slower. Incubation of catfish islet tissue *in vitro* for 30 min in the presence of labeled leucine shows incorporation

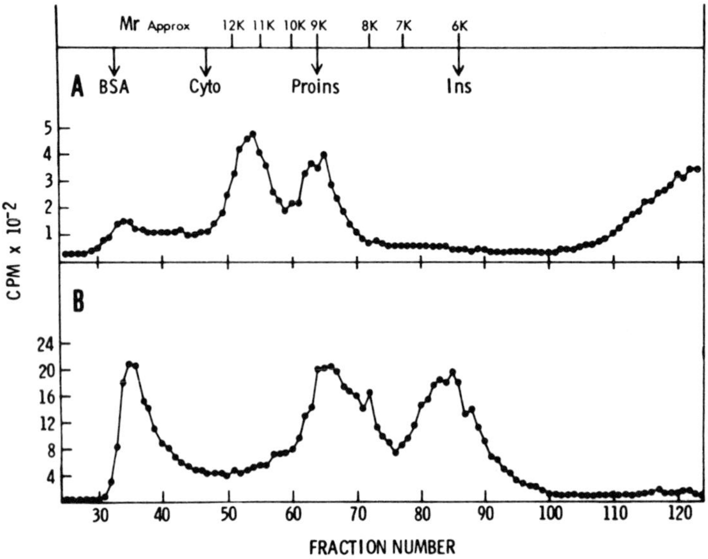

Figure 20. Separation of radiolabeled proteins from channel catfish (*Ictalurus punctatus*) islet slices incubated for 30 min (A) or 24 hr (B) in the presence of [^{3}H]leucine. Acid ethanol extractable proteins were separated on a Bio-Gel P-100 column calibrated with bovine serum albumin (BSA), cytochrome c (Cyto), bovine proinsulin (Proins), and bovine insulin (Ins). (From Albert, S. G., 1982, Insulin biosynthesis in isolated channel catfish (*Ictalurus punctatus*) islets: The effect of glucose. *Gen. Comp. Endocrinol.* **48:**167–173.)

of the radioactivity into peaks corresponding to a proinsulin precursor and to proinsulin, but not into fractions that should contain insulin (Figure 20A). After 24 hr, however, a prominent insulin peak does occur (Figure 20B). At 20°C, for example, the conversion of anglerfish proinsulin to insulin is nearly complete in about 6 hr; this is slower than in mammalian islets at 37°C, but much more rapid than conversion in *Myxine* islet organs incubated at a corresponding temperature.

Three preproglucagon molecules have been identified in the principal islets of the anglerfish, one 16,000 and two 14,000 daltons. The conversion of anglerfish proglucagon to glucagon appears to occur more slowly than the conversion of anglerfish proinsulin under the same conditions. Glucagon release from the perfused splenic islet of *Ictalurus* is biphasic; arginine is the most potent of the glucagon secretagogues. Although glucose stimulates a mild response from the catfish A cell (Figure 19), glucose over the range of 12.5–300 mg/dl inhibits glucagon secretion by anglerfish islets.

Somatostatin release from the *Ictalurus* splenic islet is stimulated by glucose and arginine, although the effect of arginine is relatively minor (Figure 19). The amount of somatostatin released by 10 mM glucose from the catfish islet is 3–5 times higher than the amount released from a rat pancreas perfused with 17 mM glucose; perhaps this is due to the high proportion of D cells (up to 35%) in the

Brockman body of the channel catfish. Although two distinct somatostatin molecules have been identified in anglerfish islets, it is not known whether there is a preferential secretion of one form or the other, nor what the difference in biological activities of the two somatostatins might be.

6.3. Amphibians

Investigations of amphibian pancreatic islet secretory function are very limited; however, several aspects of insulin release have been determined for *Amphiuma means* pancreatic tissue. Organ cultures of pancreas exhibited a dose-dependent increase in the insulin released over a range of 0–5 mg/ml glucose. Increase of culture medium calcium concentration or inclusion of a calcium ionophore also enhanced insulin release, as did glucagon at 1×10^{-5} and 1×10^{-7}M. Theophylline, caffeine, tolbutamide, and isoprenaline all stimulated insulin release whereas epinephrine inhibited it. Inclusion of alloxan in the culture medium had no effect on insulin release but did lower the tissue content of insulin. It appears, therefore, that factors influencing the amphibian pancreatic B cell as represented by *Amphiuma* are similar to those affecting most vertebrate B cells.

6.4. Reptiles

Little is known about islet hormone secretion in the reptiles. Glucose-induced insulin release has been observed in the turtle, *Chrysemys picta* (Figure 21), but the

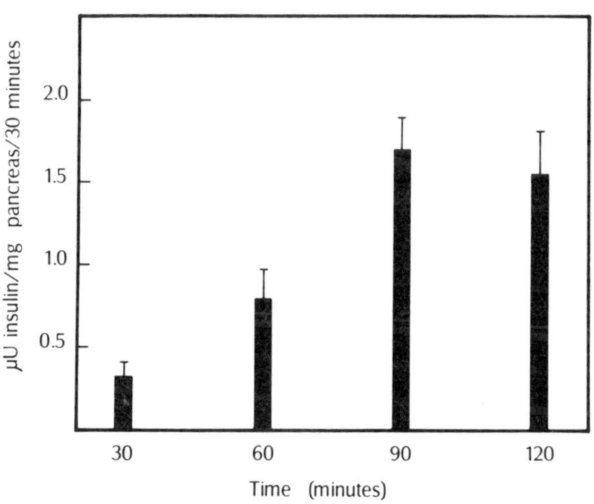

Figure 21. Insulin release by the pancreas from the painted turtle, *Chrysemys picta*. Pieces of pancreas were incubated at 30°C in medium containing glucose at 80 mg/dl for all time periods except the 30- to 60-min period when the glucose concentration was 300 mg/dl. Note that the amount of insulin released remained elevated even after return of the tissue to low glucose medium. (From D. A. Gapp, unpublished results.)

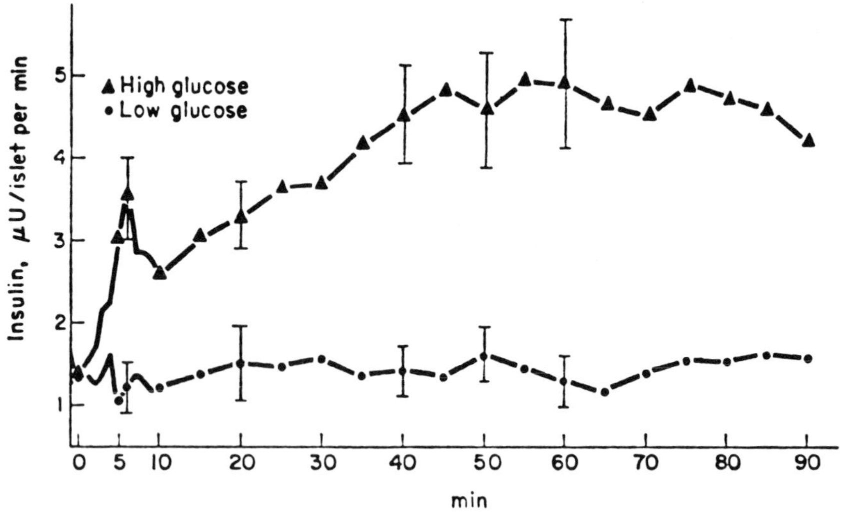

Figure 22. Biphasic pattern of insulin release from *Anolis carolinensis* splenic islets perifused with medium containing low or high (500 mg/dl) concentrations of glucose. (From Rhoten, W. B., 1973, Perifusion of saurian pancreatic islets and biphasic insulin release following glucose stimulation. *Comp. Biochem. Physiol.* **45A:**1001–1007.)

most complete, and virtually only, extensive information available on reptile pancreatic endocrine secretion comes from studies on the lizard, *Anolis carolinensis*. Perifusion of *Anolis* splenic islets with stimulatory concentrations of glucose or certain amino acids produces a biphasic pattern of insulin release (Figures 22 and 23). The level of glucose, however, required to stimulate insulin release is much higher than is required for other vertebrates, with the exception of birds. Glucose has no effect between 30 and 200 mg/dl, and release is stimulated only above 200 mg/dl. The threshold for stimulation of insulin release in mammals is between 70 and 100 mg/dl with a half-maximal value for stimulation between 125 and 170 mg/dl. Yet, in *Anolis* the threshold lies between 200 and 300 mg/dl with half-maximal response at 350 mg/dl. In comparison to mammals and a number of other vertebrate species, these high values probably reflect the high fasting blood glucose (220 mg/dl) exhibited by *Anolis*. Arginine and leucine are both effective insulin secretagogues in *Anolis* producing a biphasic pattern of release at a concentration comparable to other vertebrate species (Figure 23). Glucagon and theophylline also stimulate insulin release and cytochalasin B will enhance insulin secretion in the anole. Thus, except for the relatively high set point for glucose induction of insulin release, regulation of B-cell secretory function in *Anolis* appears to be similar to that of other vertebrates.

Glucagon secretion in *Anolis* is relatively insensitive to variations in glucose levels. Lowering the concentration of glucose from 200 to 50 mg/dl in incubation medium is required before there will be a demonstrable increase in glucagon secretion. Even at a concentration less than one-half the circulating level, glucose (100 mg/dl) plus 10 mM arginine is ineffective at stimulating glucagon release. Arginine appears to be an effective glucagon secretagogue only when in combination with glucose at 50 mg/dl, when there is a slight enhancement of glucagon release compared to 50 mg/dl glucose alone (Figure 24). Glucose up to 600 mg/dl does not exert a suppressive effect on glucagon secretion. The relatively slight effect of glucose and amino acids on the regulation of glucagon release in the anole is quite different from mammals, especially where amino acids appear to exert such a pronounced effect. This, coupled with the relatively high percentage of A cells characteristic of the *Anolis* islet, may point to a different role for glucagon in the more advanced reptiles, one possibly concerned with lipid metabolism. Nevertheless, one must not extrapolate carelessly the observations in *Anolis* to other reptiles with respect to regulation of pancreatic islet secretory function. The lizards may not reflect a general reptilian condition.

6.5. Birds

Regulation of avian islet hormone secretion is best known for several species of fowl. Perfusion of the isolated chicken pancreas with square-wave rises of glucose up to 250 mg/dl elicits no significant stimulation of insulin release whereas perfusion with 500 mg/dl glucose induces a monophasic release (the second phase

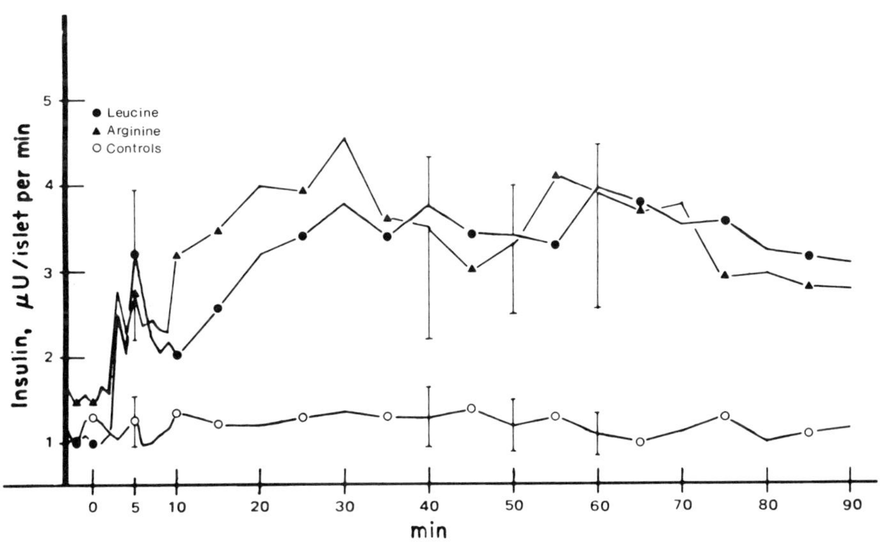

Figure 23. Biphasic pattern of insulin release from perifused splenic islets of *Anolis carolinensis* induced by 10 mM leucine or arginine in the presence of nonstimulatory levels of glucose. (From Rhoten, W. B., 1973, Insulin release from pancreatic islets of a lizard following leucine or arginine. *J. Herpetol.* **7**:207–210.)

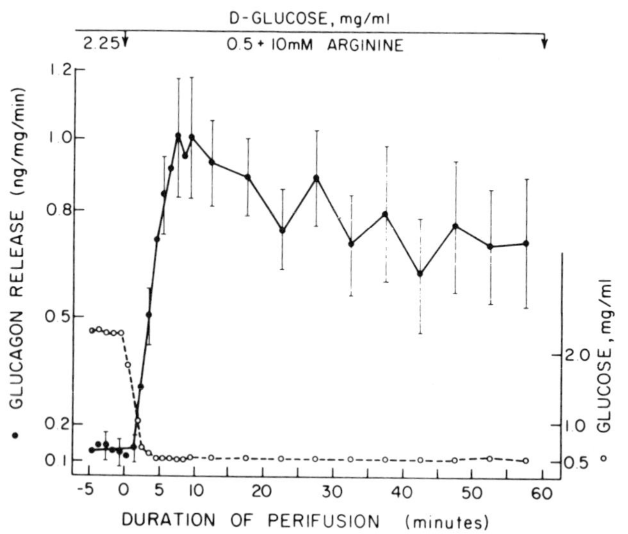

Figure 24. Glucagon released from splenic islets of *Anolis carolinensis* perifused with 0.5 mg/dl glucose and 10 mM arginine after an initial perifusion with 2.25 mg/dl glucose. (From Rhoten, W. B., 1978, Effects of glucose on glucagon secretion by the anolian splenic pancreas. *Proc. Soc. Exp. Biol. Med.* **157:**180–183.)

being absent), and only at 700 mg/dl does the typical biphasic secretory pattern become apparent. This high threshold for glucose-induced insulin release in chickens is 2–3 times that seen in mammals and may well reflect an adjustment of the relatively high circulating levels of glucose (200–300 mg/dl) characteristic for birds. Glucagon enhances B-cell response to glucose such that perfusion of the chicken pancreas with 500 mg/dl glucose in combination with glucagon (5 μg/ml) produces a typical biphasic pattern of release comparable to that seen with 700 mg/dl and with a total release which is over sixfold that induced by 500 mg/dl alone. Since glucagon circulates at rather high levels in birds (8–10 times that in mammals), it may be important for facilitating the otherwise relatively glucose-insensitive B cell. Tolbutamide stimulates insulin release in birds, but, contrary to mammals, free fatty acids appear to have no stimulatory effect on avian B-cell secretion. Arginine is a potent insulin secretagogue whereas somatostatin has a potent inhibitory effect on insulin release.

Arginine is also a potent glucagon secretagogue in birds, but unlike mammals, arginine-induced glucagon secretion in ducks is not suppressed by insulin. Free fatty acids such as oleate are inhibitors of glucagon secretion and may be of considerable physiological importance in the regulation of its release as it has pronounced lipolytic action. High levels of glucose have a suppressive effect on glucagon secretion in birds. As in mammals, the suppressive effect appears to be correlated with the state of A-cell metabolism, where low metabolic rate favors glucagon release. Tissue glucopenia resulting from 2-deoxyglucose administration in ducks

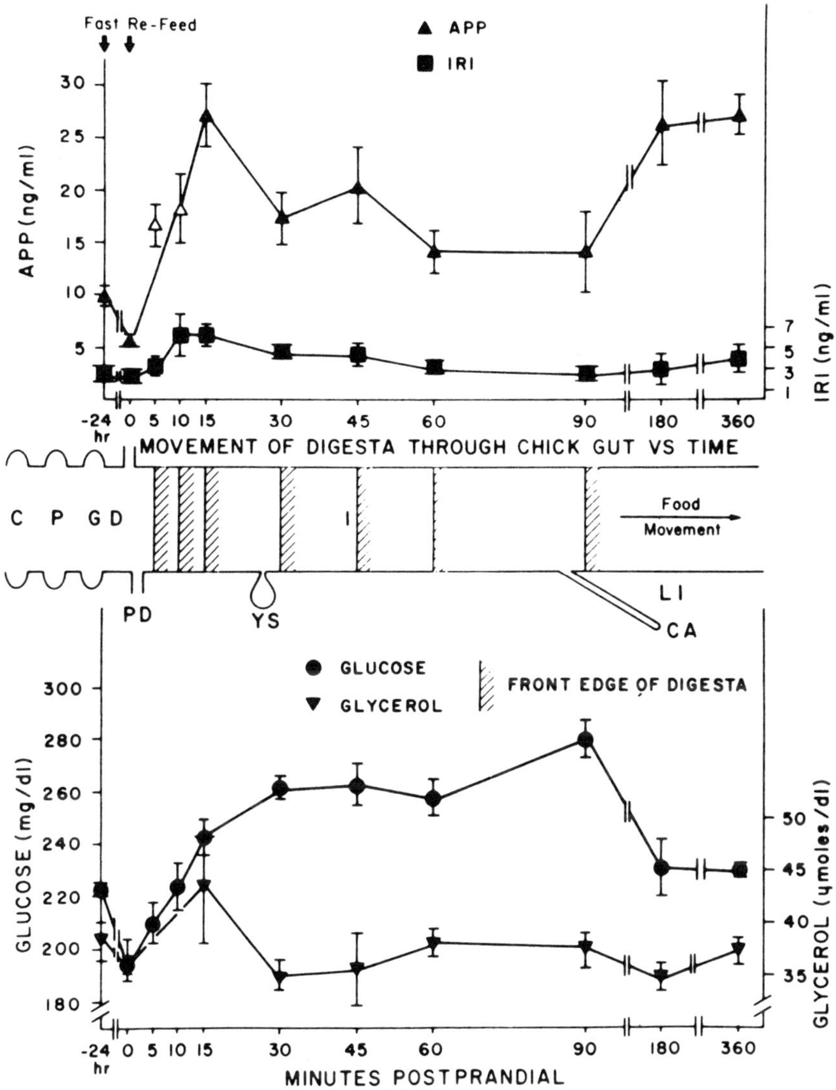

Figure 25. Changes in plasma pancreatic polpeptide (APP), insulin (IRI), glucose, and glycerol in response to starvation–refeeding in chickens. C, crop; P, proventriculus; G, gizzard; D, duodenum; I, ileum; PD, pancreatic ducts; YS, yolk sac; CA, cecum; LI, large intestine. [From Hazelwood, R. L., 1980, *Avian Endocrinology* (A. Epple and M. H. Stetson, eds.), Academic Press, New York, pp. 231–250.]

produces high circulating glucagon which can be suppressed by insulin. Perfusion of insulin in the isolated chicken pancreas also causes suppression of glucagon release although it is effective at low, but not high, glucose levels. Arginine, however, can elicit glucagon release at high glucose levels. The effects of insulin are mediated by stimulation of A-cell metabolic rate, which inhibits glucagon release induced by the low glucose but does not inhibit glucagon release stimulated by arginine. Somatostatin is a potent inhibitor of glucagon release in chickens, but, paradoxically, not in ducks, where it is stimulatory.

Somatostatin release is stimulated by glucose and arginine in birds, and when stimulated with arginine (10 mM) in the presence of nonstimulatory levels of glucose (250 mg/dl), the response is greater than with 750 mg/dl alone. Insulin and glucagon both stimulate somatostatin release in birds. The chicken pancreas appears to be strikingly sensitive to manipulation of somatostatin release compared with the mammalian pancreas.

PP release in birds appears to be regulated by mechanisms similar to those observed in mammals. PP release is elicited by both neural and humoral signals and is secreted in a biphasic manner in response to ingestion of a meal (Figure 25). In the chicken, the first phase of release is rapid, peaking within 15 min of placing food in the crop, and probably mediated entirely by neural (vagal) pathways. The second phase of release lasts several hours and does not begin until food has reached the duodenum or the upper quarter of the ileum. Proteins, and to a lesser extent certain fatty acids, are the most important nutrient stimuli and are more effective than simple distention of the gastric tract alone. Mediation includes vagal pathways plus humoral signals. For example, pentagastrin (the C-terminal pentapeptide of members of the gastrin/CCK family of hormones) is a potent secretagogue for PP and suggests hormonal signals from the gut may regulate its release. Moreover, pancreatic glucagon and enteroglucagon inhibit PP secretion and therefore may be involved in its regulation.

7. Summary and Conclusions

Our knowledge of islet secretory mechanisms in the submammalian vertebrates is very incomplete. Compared with the information about mammalian systems, much less is known for other vertebrates. However, the basic mechanisms are similar, though cyclostomes show dubious and slow control effects. For example, glucose will induce insulin secretion in most species, although there may well be other metabolites that are relatively more important to the regulation of insulin release and biosynthesis in some species; no systematic evaluation of this possibility has been reported. Glucose may be less important, for example, in the regulation of B-cell secretory activity in the advanced reptiles and birds given the relatively high circulating levels of glucose in these animals coupled with a high threshold for glucose-induced insulin release. Moreover, the relative insensitivity to insulinopenia and resistance to true insulin deficiency might suggest that insulin and carbohydrate metabolism do not occupy a central role in lizard or avian metabolism.

Regulation of glucagon secretion also appears to involve controls common to most vertebrates. Declining glucose levels and/or rising arginine levels seem to be effective in stimulating A-cell secretion in most vertebrates. Again, the lizards and birds show some variation from the mammalian patterns. In *Anolis* glucagon release is stimulated only when glucose is reduced to unphysiological levels and arginine has only a modest additive effect. In birds, the A cell appears to respond to hypoglycemia or to arginine, but it is also sensitive to the inhibitory effects of free fatty acids, of some importance given the potent lipolytic actions of glucagon in birds. Moreover, removal of A cell-rich islets by extirpation of the splenic or third lobes in the chicken produces glucagon deficiency, severe hypoglycemia, and death. These observations point to a greater role for glucagon in regulating metabolism in reptiles and birds.

Paracrine regulation of islet secretory processes is a concept that is being developed. Considerably more data are available for mammalian systems, but our knowledge of paracrine regulation below the mammals, especially among the subavian vertebrates, is exceedingly sparse and fragmentary. In chickens, glucagon stimulates insulin and somatostatin release, while suppressing pancreatic polypeptide release, and except for the duck, somastatin appears to occupy a central role by inhibiting release of all islet hormones. Insulin suppresses glucagon release while stimulating somatostatin release. Figure 26 compares the possible paracrine relationships delineated for the birds and mammals where the most obvious difference is the effect of insulin on somatostatin release—a weak inhibitor in mammals, a strong releaser in birds. Presumably, the interplay of these hormones in each of their respective species permits maximal response and delivery of the hormones needed to handle and distribute an incoming nutrient load. In subavian species, the evi-

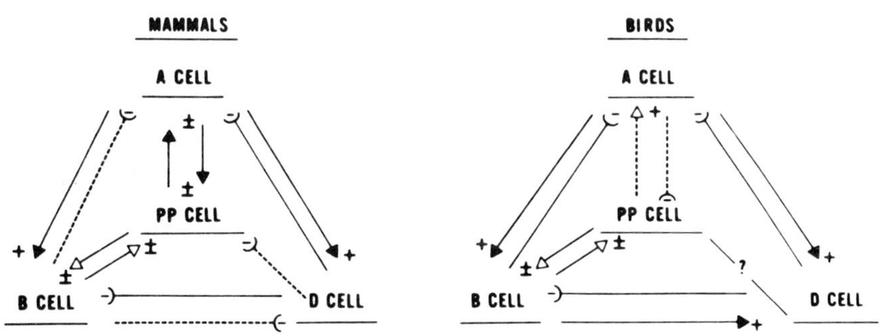

Figure 26. Comparison of possible mammalian and avian endocrine pancreatic paracrine relationships. Solid arrows, strong stimulus; open cups, demonstrable inhibition; broken lines, weak effect; ±, no consistent effect. [From Hazelwood, R. L., 1980, *Avian Endocrinology* (A. Epple and M. H. Stetson, eds.), Academic Press, New York, pp. 231–250.]

dence is very fragmentary. Glucagon stimulates insulin release by *Anolis* B cells but not by those of the eel, *Anguilla anguilla*. Somatostatin appears to inhibit insulin release from the islet organ of the Pacific hagfish, *Eptatretus stouti,* but beyond this we know very little. Moreover, paracrine relationships by definition involve local actions of hormones. How, then, do we describe paracrine relationships in species where islet cells may not be found in compact, well-organized islets and are scattered throughout the exocrine tissue? If the vascular system carries the hormones to effect intrapancreatic regulation, are we still talking about paracrine controls? It is apparent that there is much to be done in elucidating the basic mechanisms regulating islet secretory function among the submammalian vertebrates.

18

Hormonal Control of Metabolism

Vertebrates derive energy from the chemical compounds in food with minor aid from the sun or solar-heated surfaces (Figure 1). The compounds may be used for immediate metabolism, stored, or excreted. This chapter gives the basic vertebrate metabolic pathways and the hormonal influences.

1. Metabolic Pathways

Metabolism is the total of chemical reactions at the cellular level (Gr. metaboli μεταβολη = change), in a state of dynamic equilibrium, conditioned by need. The favorable conditions are donated by the constancy of the internal medium (milieu intérieur, Claude Bernard) by homeostatic mechanisms (Cannon). Molecules may be broken down—catabolism—or built up—anabolism. The central pathways of both processes comprise intermediary metabolism, for example, glucose oxidation or formation; fatty acid, phospholipid, and triglyceride biosynthesis or degradation; nucleotide and amino acid metabolism; nitrogen elimination; and cholesterol biosynthesis (Figure 2).

1.1. Carbohydrate Metabolism

Carbohydrate metabolism has two main parts, the *glycolytic pathway* (glycolysis, Embden–Myerhof pathway), proceeding from glucose to pyruvate, and the *citric acid cycle* (Krebs cycle, tricarboxylic acid cycle), involving the cyclic degradation of citrate to oxaloacetate. Both provide energy as ATP for cellular use.

Glucose homeostasis is critically important. The largest reservoir is in the circulation, readily available for use by tissues. Blood glucose concentration in

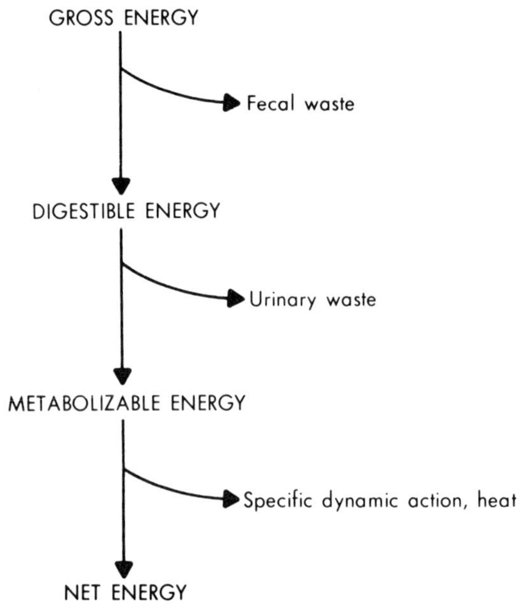

Figure 1. The flow of energy through an organism.

mammals is around 75–125 mg/dl and lower (e.g., some turtles and snakes at 50–60 mg/dl) or higher (e.g., lizards and fowl at 200 mg/dl) in other vertebrates.

The central nervous system (CNS) in mammals is particularly sensitive to changes in values of circulating glucose, and severe declination may lead to convulsions, coma, and death. Normally, all brain oxygen consumption is derived from catabolism of glucose. In extreme cases such as prolonged starvation, where glucose is limited, the brain may adapt to the metabolism of β-hydroxybutyrate (50%), acetoacetate (8%), α-amino nitrogen (12%) with glucose at about 30%. Correction of hypoglycemia depends on the rapid mobilization of glucose from liver glycogen, *glycogenolysis*, or generation of glucose from nonglucose precursors, *gluconeogenesis*. The hormones glucagon and epinephrine stimulate glycogenolysis whereas gluconeogenesis, the longer-term response, is controlled in part by glucagon and adrenal glucocorticoids.

1.2. Glucose Metabolism

Glucose is broken down to obtain energy by two series of reactions.

1. The *glycolytic pathway* (glycolysis) occurs in the cytoplasm and begins with the phosphorylation of glucose as it enters the cell. The subsequent six-carbon metabolite, fructose-1,6-diphosphate, is converted to three-carbon triose phosphates which are ultimately transformed into pyruvate. During glycolysis energy can be derived from (a) a dehydrogenation reaction—glyceraldehyde phosphate → 1,3-diphosphoglycerate—which produces hydrogen ions to pass down the electron transport system to produce ATP by oxidative phosphorylation (Figure 3), or (b)

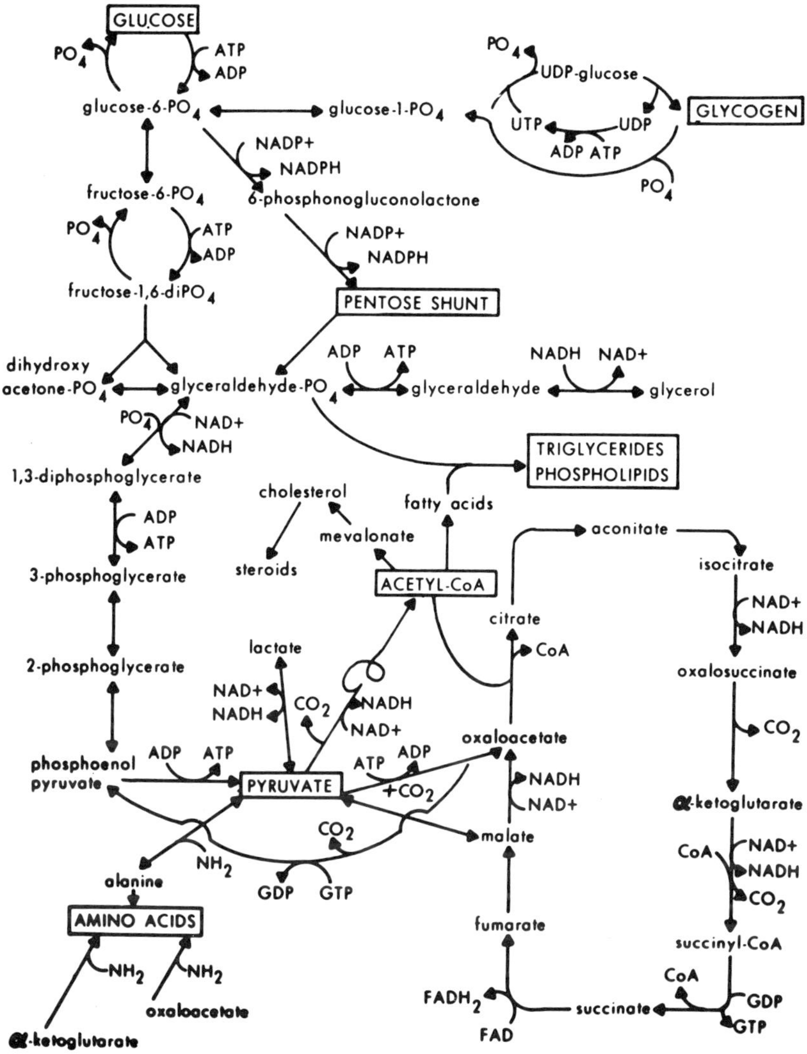

Figure 2. The major metabolic pathways of intermediary metabolism central to the metabolism of glucose. This outline illustrates the flow of carbon from glucose through the glycolytic pathway and Krebs cycle. Major branch points into other important pathways are indicated including those leading to glycogen formation, fatty acid and triglyceride biosynthesis, the amino acid pool, and the pentose shunt. Notice how the flow of carbon from the Krebs cycle can be reversed from the Krebs cycle back up the glycolytic pathway by decarboxylating oxaloacetate to form phosphoenolpyruvate.

from the reactions 1, 3-diphosphoglycerate → 3-phosphoglycerate and phosphoenolpyruvate → pyruvate where there is sufficient energy made available to form ATP directly. The three-carbon pyruvate is then converted to *acetyl coenzyme A* (CoA) (with the loss of a carbon dioxide). The two-carbon acetate from the acetyl CoA is then condensed with oxaloacetate to form citrate.

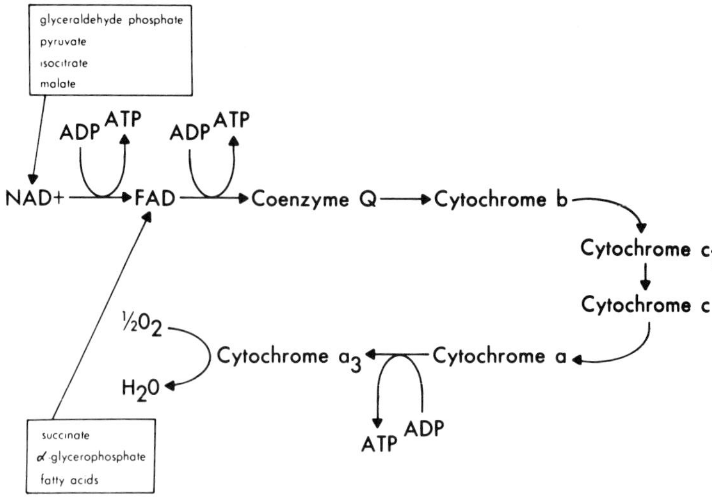

Figure 3. The electron transport pathway. This system is responsible for taking hydrogen ions generated during dehydrogenation reactions and passing them along this coupled cytochrome system to oxygen, the final acceptor. In this series of oxidative transfer steps sufficient energy is released at three sites along the pathway to enable the terminal phosphate ester–anhydride linkage to be formed in a molecule of ATP. This oxidative phosphorylation pathway is, therefore, important not only for ridding the cells of excess hydrogen ions but also for the generation of stored energy in the form of ATP.

2. This is the arbitrary beginning of the *citric acid cycle*. Enzymes for this cycle are located in the mitochondria. Here more energy as ATP is given when the citrate is converted in a stepwise manner into oxaloacetate to begin the cycle again.

The citric acid cycle represents a richer source of ATP for cellular use than the glycolytic pathway. Glycolysis can function anaerobically whereas the citric acid cycle depends on the presence of oxygen to act as the final acceptor for the electrons being passed down the electron transport pathway (Figure 3).

In summary, the net products of the degradation pathway for glucose are carbon dioxide—from the decarboxylation of pyruvate, isocitrate, α-ketoglutarate—and water—by hydrogen transferred to the electron pathway added to molecular oxygen—*plus* ATP molecules derived directly or from oxidative phosphorylation reactions which are part of the electron transport pathway. The energy made available from the oxidation of one molecule of glucose to carbon dioxide and water is a net formation of 38 molecules of ATP.

Branch points in pathways for glucose metabolism. These are numerous, to permit entry and exit gates to or from reaction sequences which are other important aspects of cell metabolism. Among them are:

1. From glucose-6-phosphate there is a branch point into glycogen biosynthesis. This allows the storage of excess glucose as glycogen, its polymer. The return pathway permits retrieval of glucose from glycogen, its stored form.

2. From glucose-6-phosphate, the carbon may also enter the hexose monophosphate pathway given in Figure 4, that is, the pentose shunt, pentose phosphate pathway. This is important for the production of riboses in nucleotide biosynthesis and for generating reducing equivalents as NADPH. These are used in various hydroxylation reactions as in the biosynthesis of cholesterol, steroids, and fatty acids.
3. Dihydroxyacetone phosphate serves as a site for entry into phospholipid and triglyceride metabolism.
4. Glycerol derived from lipolysis—the breakdown of triglycerides—can enter the pathways.

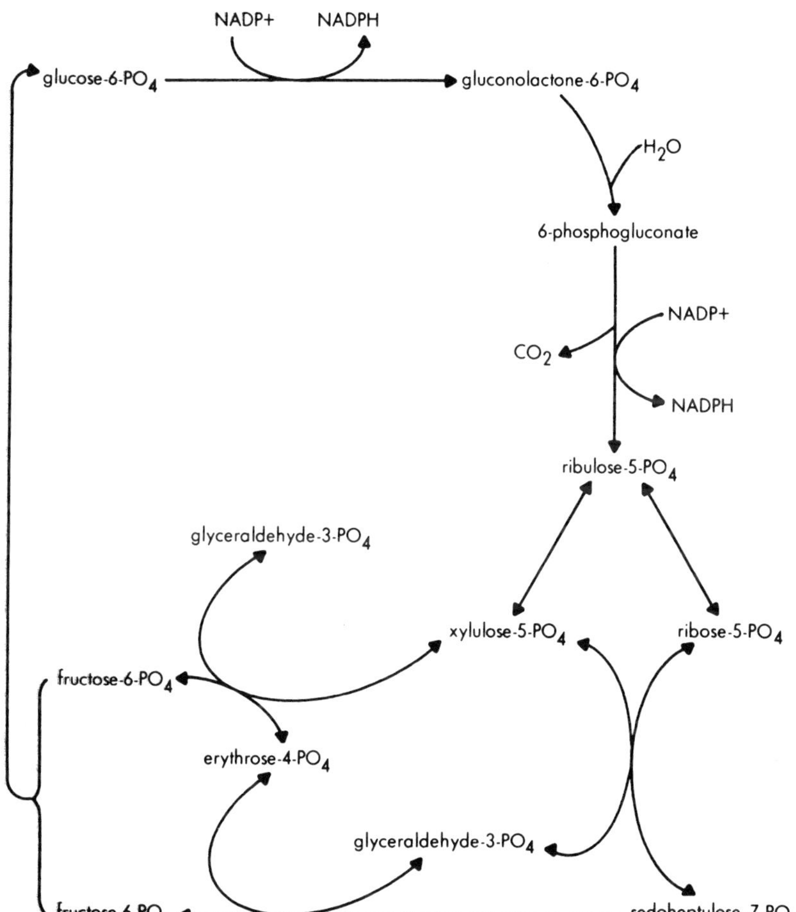

Figure 4. The pentose phosphate pathway. This pathway is an important alternate way of oxidizing glucose and is particularly important in tissues that synthesize steroids or fatty acids (e.g., adrenal cortex, gonads, mammary glands, adipose tissue). It has been estimated that from 20 to 30% of glucose oxidation in the liver occurs by this pathway whereas in muscle the amount is negligible.

5. Pyruvate provides entry into the citric acid cycle via acetyl CoA. In addition, pyruvate can be converted directly into oxaloacetate, aminated to form the amino acid alanine, or converted into lactate if the level of oxygen in the tissue is limiting, e.g., during exercise.
6. Acetyl CoA can also enter the pathways leading to fatty acid and to cholesterol biosynthesis, and the acetyl CoA derived from fatty acid catabolism can reenter here.
7. The citric acid cycle is also involved in amino acid metabolism (see below).

1.3. Fat Metabolism

Lipid metabolism is concerned primarily with the biosynthesis and degradation of triglycerides, phospholipids, waxes, sphingolipids, and sterols. Most, except triglycerides, have structural or functional roles in cells. Triglycerides are composed of free fatty acids esterified to glycerol. They store energy efficiently as fat globules, insoluble in water. Their relative energy content exceeds that of glucose and other simple sugars. For example, the net energy produced from the complete breakdown of one molecule of palmitic acid (16 carbons) is 129 ATPs, far greater than the 38 molecules of ATP derived from the oxidation of one molecule of glucose.

Free fatty acids (FFA) for triglyceride synthesis are produced endogenously by liver cells or come from dietary lipids. Their biosynthesis involves the sequential addition of two-carbon acetate units to give 16, 18, or 20 carbon fatty acids. They are passed from the liver into the circulation to be taken up by fat cells, the adipocytes. Here, three fatty acids are esterified to one glycerol molecule to form a triglyceride. Lipolysis, the catabolism of triglycerides, releases FFA which are degraded in the mitochondria by β-oxidation. These acids are oxidized for energy by muscle tissue. In the liver they give CO_2 or ketone bodies or are stored as triglycerides. With continued starvation lipolysis will proceed at twice the rate as that in the postabsorptive state, and when prolonged, hepatic ketones become the major source of energy for brain metabolism.

1.4. Protein Metabolism

Proteins may be structural (e.g., collagen, elastin) or functional (e.g., enzymes, contractile proteins), but there are no storage proteins per se. There are proteins in cells, however, especially in muscle, which give energy in times of metabolic need. There is a flux between cell proteins and intracellular and circulating amino acids (circulating at 3.0–5.5 mg/dl blood) which are readily available for cell use.

Nitrogen balance is positive when nitrogen intake exceeds excretion (growth, mammalian pregnancy). There is net movement of amino acids into cells, converted to protein. Negative nitrogen balance indicates nitrogen excretion exceeding intake (e.g., wasting diseases in man—cancer, diabetes mellitus). Here, cellular proteins are degraded into amino acids which are further catabolized or excreted.

1.5. Amino Acids and Energy Production

When carbohydrate and fat stores are depleted, amino acids serve as energy sources. Then cellular proteins are degraded to increase the available pool of amino acids which reach the liver where glucose is generated by removal of α-amino groups via transamination from amino acids such as aspartate, glutamate, and alanine to produce intermediates which can enter pathways leading to glucose production. One of the reactions is given in Figure 5 where, by transamination, aspartate converts to oxaloacetate, thence to phosphoenolpyruvate. Similarly, glutamate forms α-ketoglutarate, which can then enter the citric acid cycle.

The *gluconeogenic pathway* (Figure 6) essentially reverses the flow of carbon back up the glycolytic pathway, which occurs primarily in the liver and to a lesser extent in the kidney. Alanine from muscle is important for gluconeogenesis. The pathway is particularly significant for the CNS, which requires solely glucose for energy metabolism.

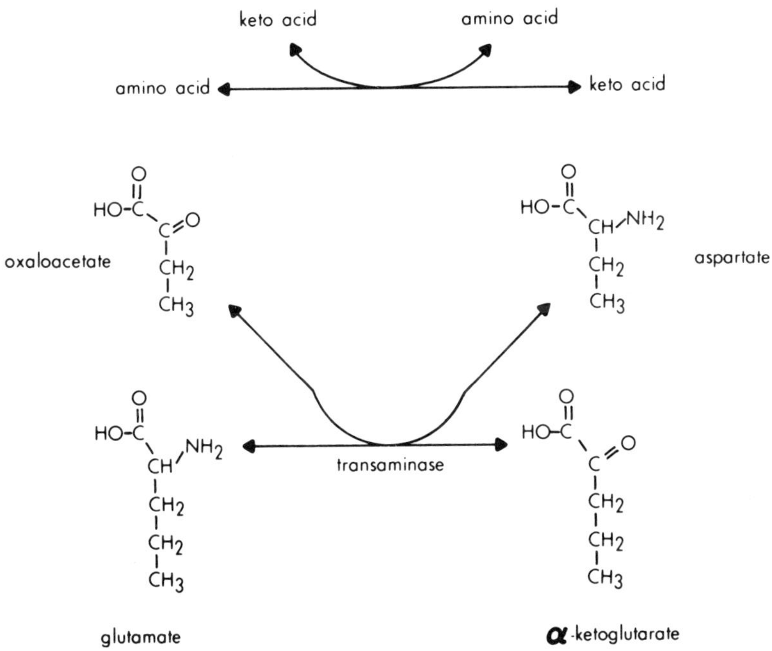

Figure 5. Transamination reactions are important to the production of intermediates from amino acids which can enter the glycolytic and citric acid cycle pathways. These reactions provide the means by which amino acid and carbohydrate metabolism are linked, especially important during gluconeogenesis when amino acids provide the major substrate for gluconeogenic pathway (Figure 6) in the form of alanine. In the specific reaction illustrated, transamination of glutamate to form α-ketoglutarate generates a molecule that is an intermediate in the citric acid cycle while the aspartate can enter the urea cycle as a mechanism to eliminate ammonia.

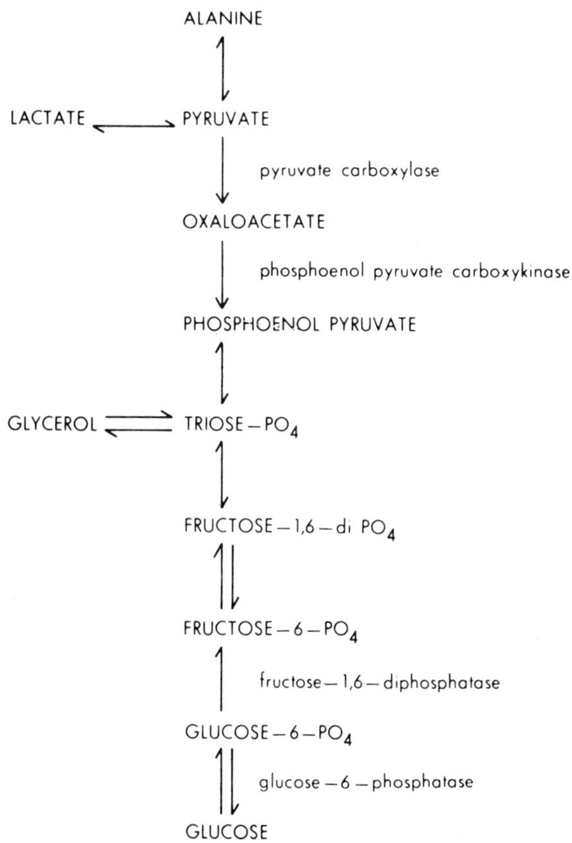

Figure 6. The gluconeogenic pathway. The gluconeogenic pathway represents essentially a reversal of the glycolytic pathway. Alanine and lactate from the muscle represent the quantitatively important substrates, glycerol and pyruvate less so. The four key gluconeogenic enzymes indicated are rate-limiting for this pathway and are inhibited by insulin and stimulated by glucocorticoids.

2. Regulation of Metabolic Pathways

Four direct mechanisms can regulate metabolic pathways: (1) substrate concentration; (2) enzyme level; (3) enzyme activity; and (4) end-product level. The first three mechanisms are employed extensively in the hormonal regulation of metabolism.

2.1. Carbohydrate Metabolism

Insulin is a key integrative hormone and promotes the uptake and utilization of glucose by most peripheral (though not nervous) tissues. Its net effect is to promote cellular metabolic pathways in both the long and short term. Insulin is *hypoglycemic* (lowers blood sugar) and is *glycolytic* and *glycogenic* (promotes breakdown of glucose—*glycolysis*—and the formation of glycogen—*glycogenesis*). Three principal target tissues are *muscle, fat* (adipose tissue), and *liver* (hepatic tissue). In these three tissues, insulin promotes the use of glucose; it stimulates glucose transport into fat and muscle. It stimulates glycogen deposition in liver and muscle, and in adipose tissue insulin stimulates lipogenesis and triglyceride formation while inhibiting hormone-induced lipolysis.

There are three *key glycolytic enzymes:* hexokinase, phosphofructokinase, and pyruvate kinase. Figure 7 shows the changes—when insulin is absent—on key glycolytic enzymes in rat liver. Peripheral cells starve as glucose transport *into* the cell is diminished, and the glycolytic pathway does not work effectively because of the lack of appropriate enzymes. Without insulin to repress the activity of enzymes for liver gluconeogenesis, carbon flow is redirected up the glycolytic pathway to give glucose. In brief, metabolism in the liver is shifted from glycolytic to gluconeogenic predominance.

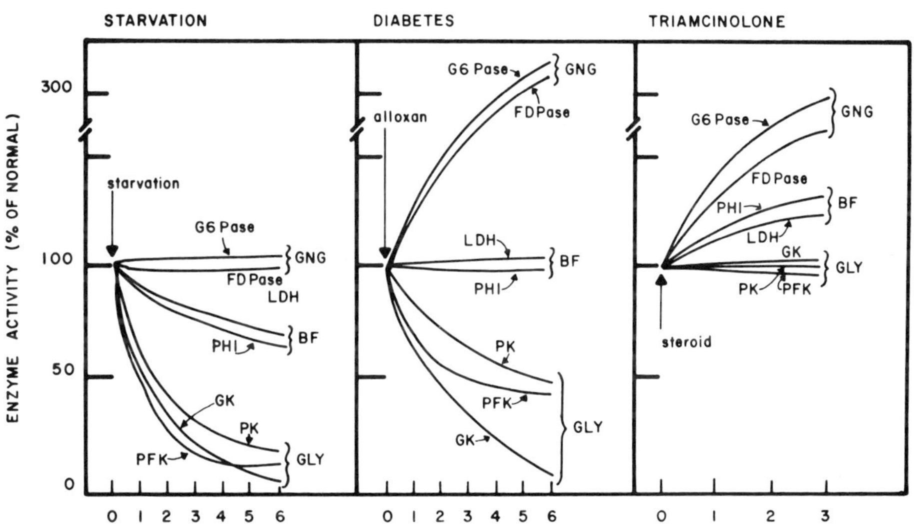

Figure 7. The role of insulin and glucocorticoids in the integration of key gluconeogenic (GNG), key glycolytic (GLY), and bifunctional (BF) enzymes in rat liver. Shown is the coordinated pattern of response of these enzyme groups which are associated with hepatic carbohydrate metabolism during starvation, diabetes, or glucocorticoid (triamcinolone) treatment. Note that in starvation the key glycolytic enzymes are low with the decrease in insulin but that the key gluconeogenic enzymes remain at normal levels. In the diabetic state, however, with the absolute lack of insulin not only are the key glycolytic enzymes down, but in the absence of insulin's inhibitory effect on the key gluconeogenic enzymes, the level of this group of enzymes is exceedingly elevated. The time scale represents days of condition or treatment. (Redrawn from Weber, G., 1972, Integrative action of insulin at the molecular level. *Israel J. Med. Sci.* **8**:325–343.)

Epinephrine and glucagon stimulate glycogenolysis and gluconeogenesis rapidly in the liver by a mechanism mediated via cyclic AMP.

Epinephrine stimulates glycogenolysis in muscle cells, but free glucose is not donated to the body. This is because the liberated glucose-1-phosphate is used by muscle for energy and because muscle lacks glucose-6-phosphatase (see scheme in Figure 7). Epinephrine stimulates ACTH, hence glucocorticoid, secretion to promote gluconeogenesis.

Glucocorticoids (adrenocortical hormones see Kime, Chapter 1) stimulate gluconeogenesis in the liver with increased synthesis of four key gluconeogenic enzymes: glucose-6-phosphatase, fructose-1,6-diphosphatase, phosphenolpyruvate carboxykinase, and pyruvate carboxylase (Figure 7).

In peripheral tissues, glucocorticoids inhibit the uptake of glucose and amino acids by fat and muscle and stimulate lipolysis and proteolysis. The free fatty and amino acids go to the liver; the former may be oxidized to CO_2 or ketone bodies or stored as triglyceride; the amino acids give glucose by the gluconeogenic pathway. Glycogen stores are increased in the hepatocytes.

2.1.1. Growth Hormone and Somatomedins

Growth hormone (GH) has its metabolic effects directly or indirectly through the somatomedins, produced primarily in the liver under the influence of GH. Somatomedins A, B, and C form a group of growth-promoting peptides of which two have insulinlike properties and are known as insulinlike growth factor I (IGF I, somatomedin C) and insulinlike growth factor II (IGF II, somatomedin A). In the adult animal, actions of GH and attendant somatomedins are not clear cut and are probably minimal in overall metabolism. Experimentally, GH has the pharmacological effect of giving a transient fall in levels of plasma glucose, amino acids, and free fatty acids. At lower doses, GH appears to decrease glucose uptake by peripheral tissues, an acute as well as a long-term effect that occurs with or without the presence of insulin. In addition, GH has an insulinotropic action independent of glucose concentration. This may be a direct effect of GH on islet β cells to give an increased sensitivity to insulin secretagogues.

2.1.2. Thyroid Hormones

These control all important factors in the regulation of basic cell metabolic rates in homeotherms. They stimulate (1) the Krebs cycle, leading to enhanced oxidative phosphorylation and hence ATP; (2) the levels of mitochondrial glycerol-3-phosphate dehydrogenase, which ultimately results in greater oxidation of cytosolic NADH and thereby increased glycolysis; (3) Na^+,K^+-ATPase activity in the plasma membrane, which accounts for oxygen consumption; and (4) an overall effect to give hyperglycemia, possibly by permissive action to allow full expression of the glycogenolytic and lipolytic functions of glucagon and catecholamines.

2.2. Hormonal Control of Fat Metabolism

Two phases are discerned: (1) lipogenesis, the biosynthesis of free fatty acids, and their coupling to glycerol to form triglyceride; and (2) lipolysis, the breakdown of triglyceride into FFA and glycerol and their subsequent catabolism. Hormonally, insulin, glucagon, and epinephrine are involved.

Insulin is a lipogenic hormone and stimulates the initial rate-limiting steps in fatty acid biosynthesis to give increased FFA via acetyl CoA carboxylase and fatty acid synthetase (see Figure 7). Thus insulin stimulates the glycolytic pathway to lead to the production of acetyl CoA substrate used in fatty acid biosynthesis.

Insulin is a "fat-sparing" hormone, inhibiting lipolysis. However, as insulin levels decline in postabsorptive states and in starvation, there is declination in the inhibition of lipolysis, and the levels of FFA rise. Glucagon and epinephrine stimulate triglyceride lipase activation, the rate-limiting enzyme in lipolysis. (Mono- and

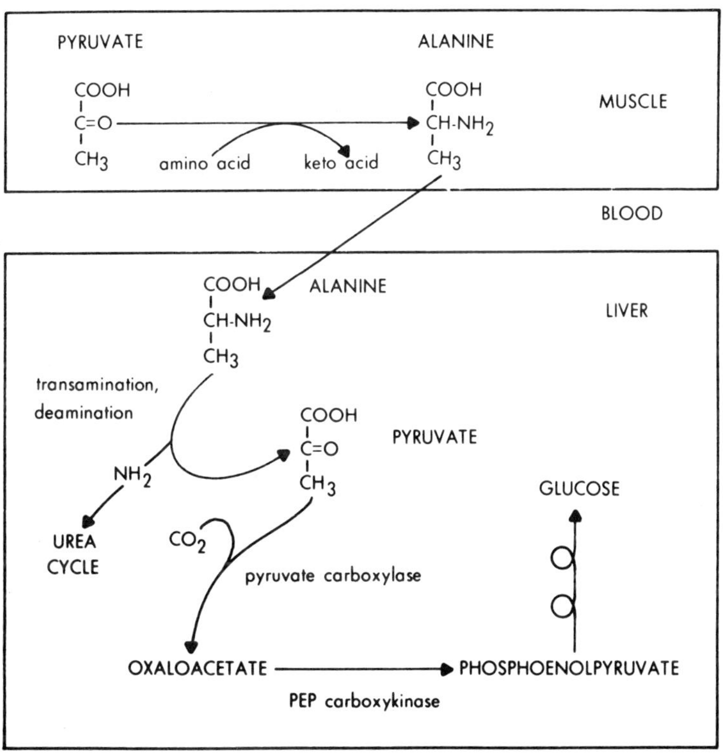

Figure 8. The pathway of alanine metabolism during gluconeogenesis. Alanine is produced in the muscle from pyruvate during gluconeogenesis. The amino groups for the amination of pyruvate are generally derived from the amino acids leucine, isoleucine, and valine. Alanine is selectively taken up by the hepatocytes, converted to pyruvate, and enters the gluconeogenic pathway as oxaloacetate.

diglyceride lipases are in excess in adipocytes and do not appear under hormonal control.) The lipolytic effects demonstrated for glucagon may be pharmacological thus leaving epinephrine as the principal lipolytic hormone. Glucocorticoids and thyroid hormone are permissive for the lipolytic action of catecholamines.

2.3. Hormonal Control of Protein Metabolism

Insulin has been termed a "protein-sparing" hormone, stimulating a positive nitrogen balance. Elevated insulin levels associated with hyperglycemia favor use of amino acids for protein synthesis. That is, carbohydrate is available so there is less use of amino acids for glucose and they can then form protein. This system is reversed in hypoglycemia, and more amino acids are used for gluconeogenesis and less are available for protein synthesis. Growth hormone, insofar as generalizations may be made, helps form protein. It accelerates transport of amino acids into cells and activates those events which depend on RNA synthesis to give, in the long term, protein synthesis. Glucocorticoids and glucagon affect protein catabolism. They mobilize protein from muscle, lymphoid tissue, and connective tissue with the resulting amino acids from protein catabolism being transported to the liver.

Glucocorticoids decrease peripheral tissue uptake of amino acids and promote hepatic trapping of amino acids. They accelerate cellular degradation of proteins and catabolism of amino acids. For example, glucocorticoids stimulate the synthesis of tyrosine aminotransferase, an enzyme important in the pathway that degrades phenylalanine and tyrosine. Glucagon also enhances the phenylalanine degradation pathway by stimulating the level of cofactor required by phenylalanine hydroxylase (Figures 7 and 8).

3. Hormonal Control of Intermediary Metabolism in Vertebrates

Control of intermediary metabolism, involving homeostasis of tissue and circulating carbohydrate, lipid, protein, and their precursor molecules, depends on several hormones mentioned previously, including insulin, glucagon, glucocorticoids, thyroid hormones, and catecholamines. The various classes of vertebrates have evolved their own spectrum of hormone dependencies, although our knowledge of these processes in nonmammalian vertebrates is quite incomplete; only recently have a number of the hormones associated with metabolic regulation been identified in these groups. Moreover, much of what we know concerning the effects of peptide hormones on metabolism has depended on the use of mammalian hormones in nonmammalian systems and therefore must be viewed with caution until homologous preparations are tested.

3.1. Insulin

For mammals insulin plays a central role in the integration of metabolic processes, but such a role has yet to be established in cyclostomes, and in birds and lizards (reptiles) insulin has no central role in blood glucose regulation. It is likely that a primary role of insulin was regulation of amino acid levels, a role taken over by growth factors perhaps, and control of glycemia was a later adoption. The importance of blood glucose level to CNS activity may have been overemphasized because of its significance to mammals: in elasmobranchs, amphibians, and some reptiles, for example, it is not critical for CNS metabolism (see Figure 9).

3.2. Glucagon

This hormone functions to mobilize carbohydrate, lipid, and protein stores to maintain adequate circulating levels of glucose. Among the vertebrates, regulation of carbohydrate metabolism by glucagon probably first appeared in the elasmobranchs; cyclostomes have no pancreatic A cells, nor are they sensitive to exogenous glucagon. However, glucagonlike peptides are present in insect neurosecretory organs and can deplete fat body lipid stores. Glucagon functions by stimulating hepatic glucose production via glycogen breakdown and gluconeogenesis; however, not all vertebrates respond in this way, and in birds particularly, glucagon is a potent lipolytic agent.

3.3. Glucocorticoids

A definite role of adrenal steroids in intermediary metabolism cannot be established in fishes, amphibia, and some reptiles. In these vertebrates adrenal steroids, both glucocorticoids (cortisol and analogs) and mineralocorticoids (aldosterone), when present, are concerned with water and ion balance.

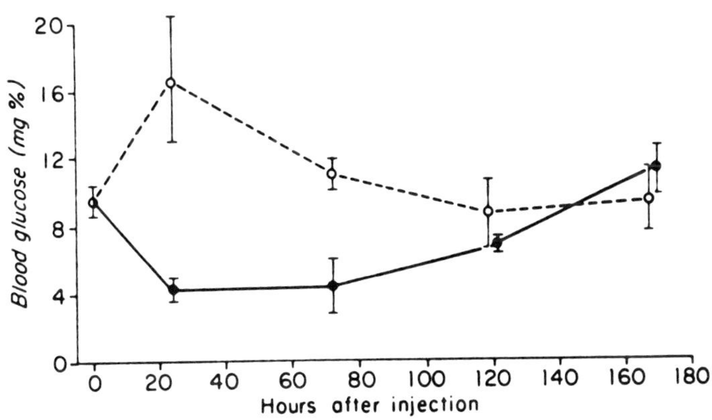

Figure 9. Effect of 0.5 IU/kg bovine insulin (closed circles) on the blood glucose levels in the Pacific hagfish, *Eptatretus stouti*. Contols = open circles. (From Inui, Y., and Gorbman, A., Sensitivity of Pacific hagfish, *Eptatretus stouti*, to mammalian insulin. *Gen. Comp. Endocrinol.* **33:**423–427.)

3.4. Catecholamines

Catecholamine production by nonneuronal cells developed in the vertebrates. Chromaffin cells, which store catecholamines, eventually became associated with and dependent on adrenal cortical cells for secretions that increased the efficiency of methylation and, hence, the production of epinephrine (E). In mammals E has assumed the role of principal lipolytic hormone, whereas in many lower vertebrates it reduces plasma FFA. In carbohydrate metabolism the hyperglycemic action of catecholamines appears ancient, since it occurs in cyclostomes, but this is effected through gluconeogenesis, not hepatic glycogenolysis as in some vertebrates. An important action of catecholamines is their ability to mobilize muscle glycogen stores, which is probably related to their role as stress hormones and associated with the need for rapid responses.

3.5. Thyroid Hormones

The role of thyroid hormones in stimulating metabolic processes in homeothermic vertebrates is established, but for poikilotherms it is now realized that only at higher temperatures is tissue respiration rate increased by thyroid hormones. Terrestrial poikilotherms can raise their body temperature above 25° by basking in the sun, for example, but the body temperature of purely aquatic species will rarely exceed 25°, so that thyroid hormones play little part in their tissue respiration.

4. Cyclostomes

In Chapter 17, it was pointed out that the islet organ of cyclostomes contained only B cells secreting insulin and D cells (somatostatin) with no cells producing glucagon. This anatomical organization and composition is reflected in their hormone physiology, particularly in the lack of tissue responsiveness to exogenous glucagon, which cannot induce hyperglycemia in cyclostomes.

4.1. Insulin

The dependence of cyclostomes on insulin for normal blood glucose levels varies, particularly between lampreys and hagfishes, but it may also be related to life-style. Both hagfishes and lampreys will dispose of a glucose load slowly, over 1–2 days, and then slowly regain normoglycemia following treatment with mammalian insulin. The hagfish, *Myxine glutinosa,* is particularly insensitive to insulin; only very high doses of mammalian insulin will induce hypoglycemia, and isletectomy is ineffective. However, homologous insulin appears to act on target tissues as in other vertebrates, stimulating incorporation of labeled glucose into muscle protein, glycogen, and neutral lipid, but having little effect on the liver, which apparently plays an insignificant role in maintenance of blood glucose. During starvation in the hagfish glycogen is depleted from liver and muscle but protein remains intact. This preferential use of carbohydrate is characteristic of some teleosts (e.g., Northern pike, Japanese eel) but different from the mammalian response.

Normal blood glucose levels in lampreys are more dependent on insulin because isletectomy or treatment with antiserum to insulin produces hypoglycemia, together with reduced heart and liver glycogen content. Although exogenous insulin causes hypoglycemia, its precise role in maintaining normoglycemia is unclear. It lowers plasma FFA and promotes muscle growth but apparently does not affect liver or muscle glycogen in lampreys. Lampreys spend most of their life-span as freshwater ammocoete larvae, with a brief period after metamorphosis as an ectoparasite in the sea, before finally returning to freshwater to spawn. During metamorphosis and migration between freshwater and seawater the lamprey does not feed but relies on breakdown of glycogen and lipid to maintain blood glucose levels. During the winter and spring of the final prespawning migration back into freshwater, carbohydrate content of muscle and body wall remains unchanged, but heart and liver glycogen levels decline and muscle fat stores are depleted. Insulin is thought to act to maintain muscle glycogen during the winter starvation, but then both insulin and heart and liver glycogen levels fall at the end of the period just prior to spawning.

4.2. Adrenal Hormones

The interrenal of cyclostomes (homolog of adrenal cortex) secretes low levels of glucocorticoids in response to ACTH stimulation. Of these, cortisol has been shown to cause hyperglycemia and increase liver glycogen stores in lampreys. E and norepinephrine (NE) are secreted by chromaffin cells (adrenal medullary cell homologs) in lampreys, and E causes a marked, but transient, hyperglycemia with a concomitant decline in glycogen synthetase. Hagfish show no such response, perhaps related to their sluggish way of life.

4.3. Thyroid Hormones

Unlike homeotherms, hagfish (*E. stouti*) and lamprey (*Entosphenus tridentatus*) respond to T_3 and T_4 by reducing blood glucose, whereas removal of thyroid hormone by antibody or propylthiouracil treatment causes hyperglycemia.

5. Elasmobranchs

All chondrichthyeans, i.e., both selachians and holocephalians, produce the full spectrum of pancreatic hormones (see Chapter 17). Although there are pronounced differences in the hormonal regulation of metabolism in the two groups, nevertheless, they are all sensitive to exogenous insulin (Figure 10). The response is intermediate between that of cyclostomes and fishes, but chondrichthyeans show considerable tolerance of low blood glucose levels; for example, the dogfish,

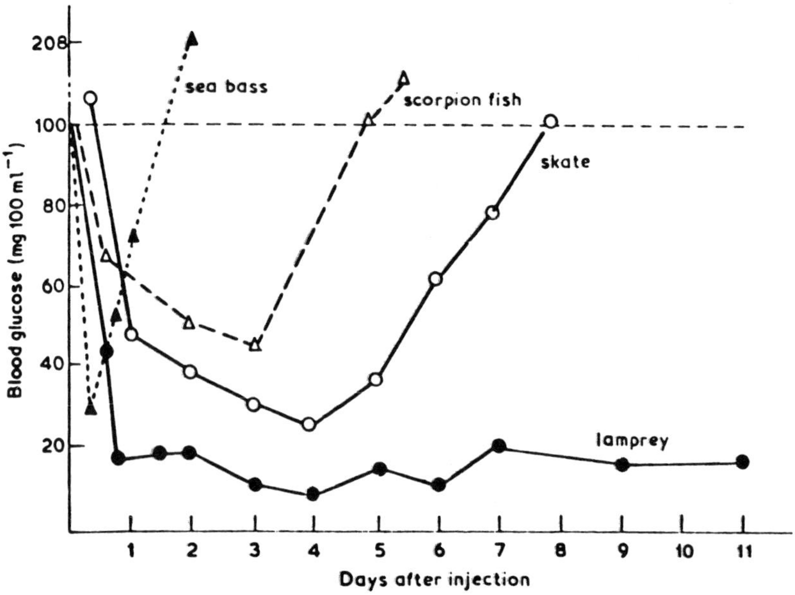

Figure 10. The effect of insulin (30–60 IU/kg) on the blood glucose concentrations in the lamprey and several species of fish. All species exhibit a profound hypoglycemia which requires several days to restore, but the lamprey remains severely hypoglycemic even 11 days after treatment. This may indicate a lack of counterregulatory hormone function in these animals. (From Hardisty, M. W., 1979, *Biology of the Cyclostomes,* Chapman and Hall, London, p. 250.)

Squalus acanthias, survives a 94% fall with levels less than 30% of normal persisting for 7–10 days or more.

5.1. Insulin and Glucagon

Liver glycogen stores are normally low and quite variable in chondrichthyeans. Glucose loading increases the store in elasmobranchs but not in the ratfish, *Hydrolagus collei.* It is not clear whether insulin stimulates glycogen storage, although it promotes glucose uptake into elasmobranch muscle. Glucagon increases blood glucose levels, but whether this is due to glycogenolysis of the small hepatic store is not clear; however, there is no effect in the ratfish, probably reflecting the absence of hepatic glycogen.

5.2. Starvation, Gluconeogenesis, and Glucocorticoids

Studies on *S. acanthus* and *R. erinacea* show that these elasmobranchs are able to maintain blood glucose levels even when fasted over a period of several weeks. Gluconeogenesis appears to be similar to the process in higher vertebrates. Alanine

Table I
Plasma Glucose and *in Vitro* Alanine Production in the Starved Spiny Dogfish, *Squalus acanthias*[a]

	Plasma glucose[b] (μmoles/ml)			Alanine production[c] (μmoles/g per 90 min)		
Days of starvation	0–1	9–14	19–25	1	12–17	20–22
	3.38 ±0.44 (11)	4.07 ±0.47 (9)	3.59 ±0.94 (8)	2.29 ±0.31 (6)	3.35 ±0.60 (6)	3.47 ±0.24[d] (7)

[a]From Leech, A. R., Goldstein, L., Cha, C.-J., and Goldstein, J. M., 1979, Alanine biosynthesis during starvation in skeletal muscle of the spiny dogfish, *Squalus acanthias*. *J. Exp. Zool.* **207**:73–80.
[b]Values represent mean ± S.E. with the number of fish per group contained in parentheses.
[c]Values represent mean ± S.E. of alanine production by pelvic fin muscles incubated at 15°C in dogfish Ringer's solution supplemented with 3 mM glucose and 3 mM isoleucine.
[d]$p < 0.05$ vs. one-day starvation.

is the only amino acid released in significant amounts by muscle and which can be synthesized *in vitro* by pelvic fin from starved skates (Table I). However, in contrast to mammals, glutamine is not released from muscle during protein catabolism.

Hormonal control of gluconeogenesis in elasmobranchs has not been worked out. Repeat doses of ACTH induce rises in blood glucose in elasmobranchs, as will glucocorticoids, but neither removal of the pituitary nor removal of the interrenal has a clear effect on liver glycogen or blood glucose concentration.

5.3. Catecholamines

Catecholamines increase circulating levels of glucose in chondrichthyeans, with NE, the principal native hormone, more effective than E. However, the mechanism is unclear, but may involve mobilization of either muscle or hepatic glycogen stores.

6. Teleosts

The successful radiation of this large and diverse group of vertebrates is reflected in the variety and complexity of indices of metabolic status. Basal blood glucose levels may range from <20 to >100 mg/dl, liver glycogen from 0.05 to >9.0 mg/100 g, and plasma FFA from <100 to >1200 meq/liter. These variations may be related to life-style (migratory or nonmigratory), age, season, reproductive state, as well as food quantity and quality. Table II shows the effect of age (re-

Table II
Plasma Glucose, Protein, Free Fatty Acids (FFA), Cortisol, and Insulin in River Breams (*Abramis brama*) according to Age[a]

	Glucose (mg/ml)	Protein (mg/ml)	FFA (mg/ml)	Cortisol (ng/ml)	Insulin (ng/ml)
Adults	1.01 ± 0.12	48.9 ± 4.1	0.73 ± 0.24	247.8 ± 53.5	0.27 ± 0.03
Juveniles	0.85 ± 0.04	41.1 ± 2.1	0.38 ± 0.08	124.1 ± 27.3	0.46 ± 0.03
			$p < 0.05$	$p < 0.01$	$p < 0.05$

[a]Values represent mean ± SEM for 10 adults and nine juveniles. (From Murat, J. C., Plisetskaya, E. M., and Woo, N. Y. S., 1981, Endocrine control of nutrition in cyclostomes and fish. *Comp. Biochem. Physiol.* **68A:**149–158)

productive versus nonreproductive) on several metabolic parameters and hormone levels in the bream, *Abramis brama,* and Figure 11 illustrates an impressive seasonal variation in glycemia and hepatic glycogen in *Spicara chryselis.* Among the numerically small migratory species the polycyclic migrators (e.g., Atlantic salmon, sturgeon) may differ from those that migrate once and then die (e.g., Pacific salmon).

The major energy store in fish is lipid, which may be deposited in the liver, as in cod, or more extensively in muscle, around the fibers, as in herring and salmon. Lipid accumulates as an end product of carbohydrate metabolism, especially prominent in sluggish anaerobic fish, and stores may increase in relation to reproductive activity, specifically just prior to gonadal maturation.

Those fish which are predominantly carnivorous derive considerable energy from dietary protein, normally in terrestrial homeotherms considered to be an inefficient energy source. Dietary carbohydrate is poorly utilized; for example, in the plaice maintained at 15° only 12–23% of glucose is oxidized in 18 hr whereas up to 10% is incorporated into glycogen; by contrast, the mouse oxidizes about 80% glucose at 37° with virtually none remaining as glucose or related small molecules. Glycogen stores in muscle and liver are generally small. The major route of glucose metabolism in fish tissues is by the glycolytic pathway. The hexose monophosphate pathway (HMP) appears to be of secondary importance but is used by certain tissues under some conditions; for example, in the brook trout, *Salvelinus fontinalis,* glycolysis and HMP are utilized during aerobic metabolism but only glycolysis during anoxia. In the electric eel HMP contributes significantly to glucose oxidation in muscle and electric tissue, but the brain catabolizes glucose by glycolysis exclusively.

6.1. Insulin

Exogenous insulin lowers blood glucose levels in all teleosts so far studied, but to varying degrees. Although insulin-induced hypoglycemia is less prolonged than in cyclostomes and chondrichthyeans, it may, nevertheless, last for several days.

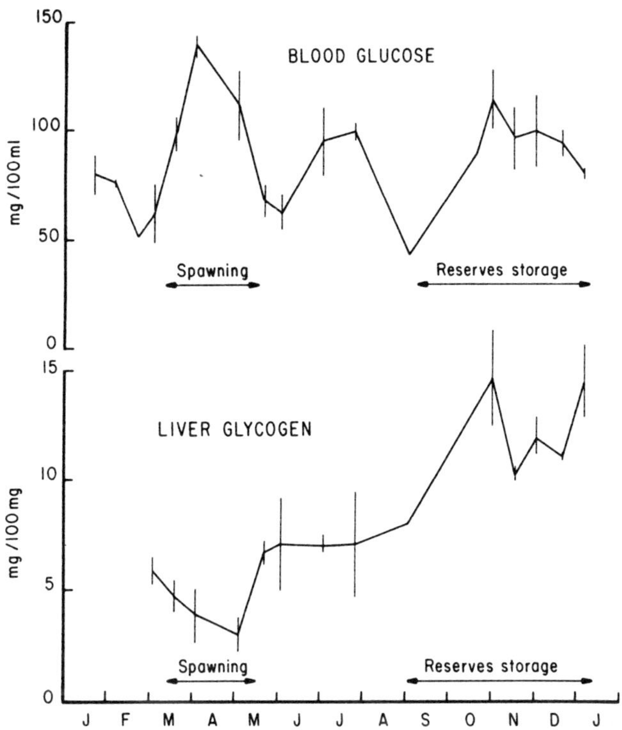

Figure 11. Yearly variations in blood glucose and liver glycogen in males of the fish *Spicara chryselis*. Changes in these parameters correlate with several important stages in the yearly cycle: in the prespawning and spawning in the spring hyperglycemia and hepatic glycogen depletion are associated with the reproductive effort, and in the fall hyperglycemia and hepatic glycogen deposition are associated with storage of energy for the winter months. (Redrawn from Fernandez, J., and Planas, J., 1980, Annual variations of some carbohydrate and lipid parameters in the fish *Spicara chryselis* during captivity. *Comp. Biochem. Physiol.* **67A:**383–389.)

However, endogenous insulin may not play a significant role in regulation of glycemia in all fish because pancreatectomy in the eel, *Anguilla rostrata,* for example, produces no specific effect on blood glucose.

The role of insulin in metabolism of glycogen in liver and muscle is unclear. *In vitro* experiments have shown stimulation of labeled glucose into liver glycogen, for example, in the toadfish (*Opsanus tau*) and pike (*Esox lucius*), but *in vivo,* exogenous insulin depletes liver glycogen in the eel (*A. anguilla*) and catfish (*Pimelodus maculatus*), depletion coinciding with maximal hypoglycemia. It is possible that the depletion of glycogen is secondary and caused by counterregulatory action (e.g., by catecholamines) prompted by hypoglycemia. In the trout, exogenous insulin has no effect on liver glycogen but does deplete lipid stores (Table III).

Actions of insulin in lipid metabolism also need clarification. The most con-

Table III
Effect of Prolonged Insulin Administration on the Level of Lipid, Protein, and Glycogen of Muscle and Liver of the Rainbow Trout, *Salmo gairdneri*[a]

	Saline-injected controls	Low insulin dose (0.05–0.5 IU/kg)	High insulin dose (0.5–5.0 IU/kg)
Skeletal muscle protein (mg/g wet wt)	180.9 ± 18.0	189.4 ± 15.7	249.6 ± 16.7*
Liver protein (mg/g wet wt)	265.5 ± 12.1	249.0 ± 12.4	282.0 ± 18.2
Skeletal muscle lipid (mg/g wet wt)	19.5 ± 2.2	21.9 ± 2.7	26.8 ± 3.2*
Liver lipid (mg/g wet wt)	51.0 ± 3.8	46.0 ± 3.2	41.0 ± 3.0*
Skeletal muscle glycogen (mg/g wet wt)	10.9 ± 1.2	9.7 ± 0.95	11.6 ± 1.1
Liver glycogen (mg/g wet wt)	12.6 ± 1.3	10.2 ± 0.90	11.2 ± 0.9

[a]Values represent mean ± SD for six fish per group treated with saline or increasing doses of insulin at 6-day intervals for 56 days.

*$p < 0.05$ compared to the saline-injected controls. (From Ablett, R. F., Sinnhuber, R. O., Holmes, R. N., and Selivonchick, D. P., 1981, The effect of prolonged administration of bovine insulin in rainbow trout (*Salmo gairdneri R.*) *Gen. Comp. Endocrinol.* **43:**211–217.)

sistent effect of exogenous insulin is to reduce the concentration of plasma FFA. This occurs in both the eel (*A. anguilla*) and goldfish (*Carassius auratus*), which also shows reduced serum phospholipid, but effects on plasma triglyceride, phospholipid total lipid, and cholesterol in the eel are equivocal. Plasma cholesterol is, however, sensitive to insulin from a teleost, the cod. Table III further illustrates some of the paradoxical effects of insulin in teleosts in that glucose uptake into muscle lipid is enhanced in the pike but such stores are depleted in the trout. In other species insulin has failed to influence fatty acid synthesis (toadfish), and to alter the activity of two key lipogenic enzymes, acetyl CoA and fatty acid synthetase (*P. maculatus, Arius felis*).

Insulin has been little investigated as a regulator of protein metabolism in teleosts. However, its effects appear to be similar to those in other vertebrates and include enhanced amino acid uptake into tissue, which correlates with a fall in plasma nitrogen. Amino acid uptake is mainly into muscle protein (see Table III) with little uptake into liver; it also antagonizes gluconeogenesis from protein catabolism. These aspects of insulin action in teleosts are worthy of more investigation because it may be a significant growth factor in these vertebrates.

6.2. Glucagon, Glucocorticoids, and Gluconeogenesis

The adaptation of fish metabolic processes to their largely carnivorous diet is further emphasized by the importance of gluconeogenesis which follows protein and amino acid catabolism. Gluconeogenesis occurs mainly in the liver and kidney and is profoundly affected by composition of the diet. In trout, for example, a high-protein/low-carbohydrate diet or starvation produces highest rates of gluconeogenesis, whereas a low-protein/high-carbohydrate diet depresses gluconeogenesis and elevates blood glucose. The ability of many fish species to withstand prolonged periods of starvation is dependent on gluconeogenesis from muscle protein, which maintains blood glucose levels. The most extensive studies have been made in *A. japonica;* in this species increased hepatic alanine and aspartate aminotransferase activities and elevated plasma amino nitrogen are associated with prolonged starvation.

Exogenous glucagon stimulates hyperglycemia in most species of fish (Figure 12), although it appears to be associated with increased hepatic glucose production rather than depletion of liver glycogen. The regulation of plasma glucose by glucagon is further dependent on gluconeogenesis, rather than mobilization of liver glycogen stores. *In vitro* glucagon can stimulate glycogen breakdown in isolated hepatocytes (*Carassius auratus*); however, glycogenolysis can only account for a proportion of the generated glucose; the rest is derived by gluconeogenesis. *In vivo*

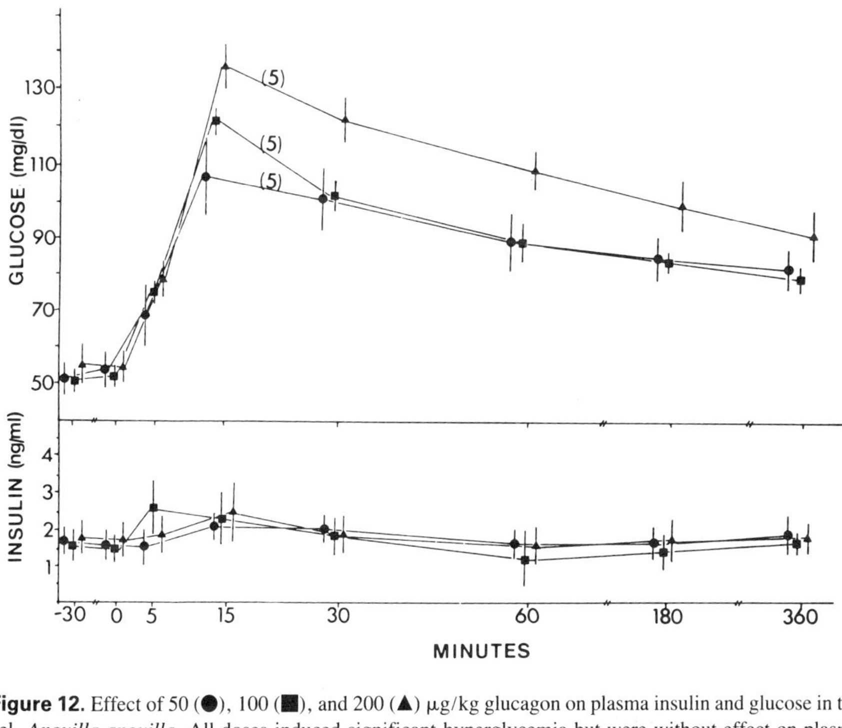

Figure 12. Effect of 50 (●), 100 (■), and 200 (▲) μg/kg glucagon on plasma insulin and glucose in the eel, *Anguilla anguilla*. All doses induced significant hyperglycemia but were without effect on plasma insulin. Values represent mean ± SEM for the number of animals indicated in parentheses. (From Ince, B. W., and Thorpe, A., 1977, Plasma insulin and glucose responses to glucagon and catecholamines in the European silver eel (*Augilla anguilla* L.). *Gen. Comp. Endocrinol.* **33**:453–459.)

Table IV
Effect of Cortisol on Metabolites in the Eel, *Anguilla rostrata*[a]

Blood	Glucose (mg/dl)	Glycerol (μmoles/ml)	Lactate (μmoles/ml)	Pyruvate (μmoles/ml)	FFA (μmoles/ml)
Control	62	0.075	0.060	0.026	1.36
	± 8	± 0.041	± 0.017	± 0.001	± 0.11
Cortisol	126	0.017	0.109	0.033	1.90
	± 6*	± 0.002	± 0.011*	± 0.004	± 0.06*
Liver	**Glycogen (mg/100 g)**	**Glycerol (μmoles/g)**	**Lactate (μmoles/g)**	**Pyruvate (μmoles/g)**	**Lipid (g/100 g)**
Control	6.86	0.038	0.101	0.126	10.34
	± 0.62	± 0.009	± 0.027	± 0.10	± 0.84
Cortisol	6.79	0.036	0.125	0.164	14.47
	± 0.41	± 0.013	± 0.019	± 0.010*	± 0.71*

[a]Values represent mean ± SE for five saline-injected (control) and six cortisol-injected (5 mg/kg per day for 10 days) animals. *$p < 0.05$ compared to controls. (From Butler, D. G., 1973, structure and function of the adrenal gland in fishes. *Am. Zool.* **13:**839–879.)

studies in the carp (*Cyprinus carpio*) show that inhibition of phosphoenolpyruvate carboxykinase with 3-mercaptopicolinic acid blunts glucagon-induced hyperglycemia, again indicating the importance of the gluconeogenic pathway. As in mammals, glucagon does not mobilize muscle glycogen.

The principal glucocorticoid of teleosts and lungfishes is cortisol, and its action on intermediary metabolism is generally similar to that in other vertebrates, in promoting gluconeogenesis in the liver and, in some cases, causing hyperglycemia (Table IV). Conversely, removal of glucocorticoids by hypophysectomy causes hypoglycemia in some species only, e.g., *Fundulus heteroclitus,* but not in *Cottus scorpio*. However, such fish would also lack thyroid and adenohypophyseal trophic hormones, which may also influence metabolism. Peripherally glucocorticoids deplete noncarbohydrate reserves of lipid and protein. Spawning sockeye salmon, for example, lose up to 96% of body fat and 53% of body protein as plasma corticosteroids rise sixfold. Starving eels and trout lose body fat and goldfish lose parietal muscle, an effect that can be mimicked by 4 days of cortisol treatment. Cortisol increases liver lipid in starved eels (Table IV), possibly derived from peripheral lipolysis. In the liver glucocorticoids are also typically associated with catabolism of amino acids, leading to gluconeogenesis and related increases in glutamate–pyruvate transaminase activity and increased nitrogen excretion as seen in goldfish and trout. Similar effects in the starved eel are evidenced by stimulated serum glutamate–oxalate transaminase activity (Table IV).

6.3. Catecholamines

Epinephrine is the predominant catecholamine of fish and, as in other vertebrates, is released in response to stress, to mobilize energy stores for immediate use, by glycogenolysis, gluconeogenesis, and lipolysis. Hyperglycemia may result from breakdown of muscle glycogen via increased activity of glycogen phosphorylase, and in the rainbow trout at least, it is not associated with hepatic glycogenolysis but may depend partially on gluconeogenesis. This is not true, however, for other fish where there is clear evidence of E acting *in vitro* through β-adrenergic receptors in isolated goldfish hepatocytes and lungfish liver to enhance glycogenolysis. Another route whereby E can cause hyperglycemia is by depressing release of insulin as in the eel.

The role of catecholamines in regulation of lipid metabolism in fish is not clear. In carp, bream, and pikeperch NE depresses plasma FFA but raises levels in carp, scorpion fish, and sea bass.

6.4. Thyroid Hormones

Although the role of thyroid hormones in homeothermic vertebrates in stimulating the rate of tissue metabolism is clear, these hormones have no defined role in intermediary metabolism in poikilotherms.

Thyroid hormones have been observed to increase or decrease blood glucose concentrations, similarly affect hepatic glycogen, and increase muscle but not brain glycogen. Apparent actions of thyroid hormones on lipid are equally incoherent. Thyroid activity is inversely related to hepatic and visceral lipid reserves, whereas plasma FFA are increased and cholesterol decreased in hyperthyroidism. However, in trout, T_4 causes a decline in plasma lipid but leaves muscle, liver, and visceral lipid unchanged, whereas juvenile coho salmon respond to prolonged T_4 treatment with increased muscle lipid accumulation.

7. Amphibia

In amphibians metabolism and its control reflect the phyletic position of the group, as well as the life history and life-style of the various members. The ability of some amphibians, both urodele (*Necturus maculosus*) and anuran (*Rana catesbeiana* and *pipiens*), to tolerate low blood glucose concentrations is similar to that of some cyclostomes, elasmobranchs, and teleosts, suggesting that their CNS too are less dependent on circulating glucose. In general, the metabolic rate in amphibians is temperature sensitive, being higher in the warmer months. Energy for metabolism is stored as glycogen in the liver and lipid (triglyceride) in fat bodies. Blood glucose, amino acids, proteins, and triglycerides fluctuate seasonally as do hormones regulating metabolic processes (Figure 13). There are differences in metabolism between adult and larval stages; for example, blood glucose in adult salamanders (*Ambystoma tigrinum*) is about 28 mg/dl compared with 18 mg/dl in larvae.

7.1. Insulin

Insulin lowers blood glucose in all amphibians examined, presumably by stimulating uptake into liver and skeletal and heart muscles and by increasing muscle

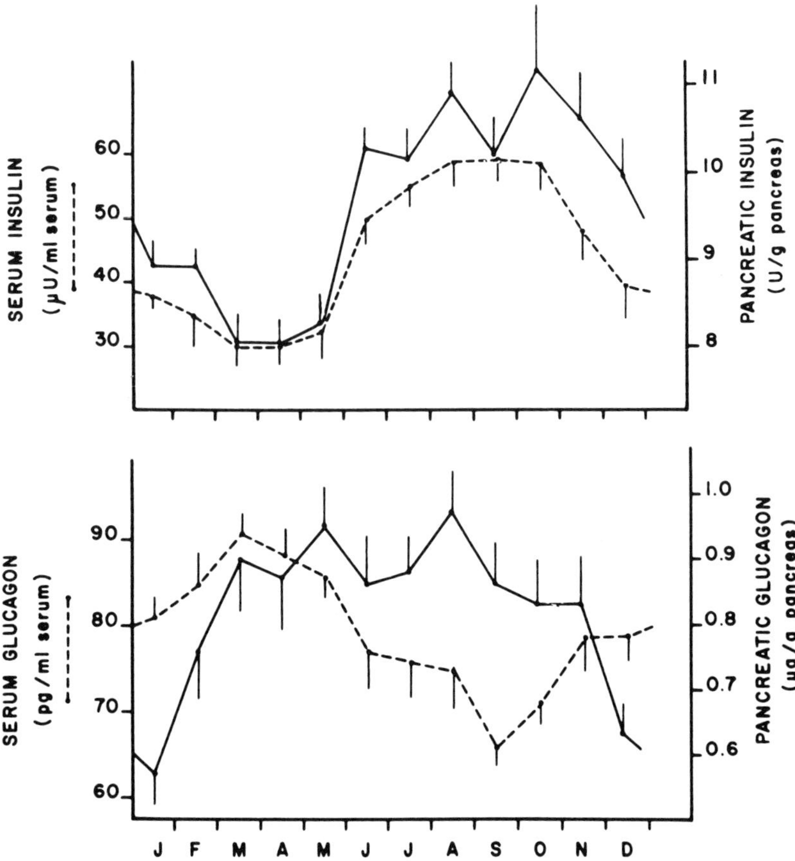

Figure 13. Seasonal changes in serum and pancreatic concentrations of insulin and glucagon in the frog, *Rana esculenta.* (Redrawn from Schlaghecke, R., and Blum, V., 1981, Seasonal variation in insulin and glucagon concentrations of *Rana esculenta* (L.). *Gen. Comp. Endocrinol.* **43:**470–483.)

glycogen and hepatic glycogen synthetase activity, as demonstrated *in vitro.* There is a dose-dependent fall in serum glucose following injections of exogenous hormone (usually mammalian) ultimately leading to convulsions, coma, and death, at very high doses. The persistence of hyperglycemia is also related to the dose level of insulin. There is no clear difference in sensitivity between larval and adult stages.

Little is known of the effects of insulin on protein and lipid metabolism, but there is a dose-dependent decline in plasma alanine and stimulated incorporation of leucine into liver protein of bullfrogs. In cold-adapted summer frogs (*Rana temporaria*) insulin lowers plasma FFA, by stimulating their reesterification, and without affecting adipose tissue lipid release. That this occurs only in cold-adapted frogs suggests seasonal variations in insulin sensitivity and activity, related to deposition of fat stores in hibernating frogs. Seasonal variations in serum and pancreatic insulin concentrations in *R. esculenta* have been observed (Figure 13), coinciding with the period after breeding and prior to hibernation. Higher levels at this stage correlate with the need to increase food stores for hibernation. Normally insulin levels are low during hibernation and glucagon elevated, so favoring mobilization of energy stores (Figure 13).

7.2. Glucagon, Glucocorticoids, and Gluconeogenesis

Amphibians are adapted to withstand long periods of caloric restrictions during hibernation or metamorphosis when tadpoles cease feeding at metamorphic climax. During long fasts glycogen and fat stores are depleted but blood glucose is maintained near nonfasted levels by gluconeogenesis using tissue protein.

7.3. Glucagon

Glucagon (mammalian) stimulates hyperglycemia in all amphibians studied. Endogenous levels of glucagon are high in the winter during hibernation (Figure 13) when energy stores need to be mobilized, but exogenous hormone has more pronounced effect in summer and fall. Glucose is mobilized by glucagon from liver glycogen stores, but in summer frogs (*R. pipiens*) the tissue appears to be less sensitive to the hormone. Mammalian glucagon has no effect on plasma FFA in *R. temporaria,* but in *Bufo marinus* the hormone decreases circulating cholesterol and stimulates ketogenesis by the liver.

7.4. Glucocorticoids

Both direct and indirect evidence indicates that corticosteroids affect blood glucose levels in amphibians. Moreover, it is clear that both corticosterone and aldosterone act as glucocorticoids. For example, in larval *Xenopus laevis* at the prometamorphic climax and juvenile stages these steroids increase hepatic and other tissue glycogen levels, although aldosterone is less effective than corticosterone. Plasma levels of these hormones vary diurnally and seasonally, as does tissue responsiveness to them.

The effect of glucocorticoids on gluconeogenic enzyme activity in amphibian species is inconsistent, as is generally true for the lower vertebrates, suggesting that the primary role of adrenal steroids is control of water and electrolyte balance, with involvement in gluconeogenesis evolving later. Corticoids affect various aspects of gluconeogenesis by stimulating alanine aminotransferase activity, lactate production, and alanine incorporation into carbohydrate pools, particularly in juvenile-stage frogs. In the axolotl plasma proteins are elevated, incorporation of labeled leucine into muscle protein is decreased, and protein in the liver is elevated following steroid or ACTH treatment. Correspondingly, urea cycle enzyme activity increases, indicating a need to eliminate excess nitrogen as the by-product of gluconeogenesis.

7.5. Catecholamines

Of the two interrenal catecholamines epinephrine is the more potent stimulator of liver and muscle glycogenolysis, leading to hyperglycemia. Epinephrine acts via the β-adrenergic receptor leading to increased cyclic AMP and glycogen phosphorylase activity. In contrast to the action of glucagon E increases circulating lactate, presumably as a result of muscle glycogenolysis. Such an effect is more pronounced in the fall, at least in *R. pipiens,* suggesting a role for E in preparation for overwintering.

7.6. Thyroid Hormones

As in fishes, the role of thyroid hormones in control of metabolism in amphibians is far from established. For example, in birds and mammals T_3 and T_4 increase oxygen consumption in tissues or whole animals; this is true for some amphibians (*R. pipens in vivo* and *Bufo woodhousii in vitro*), but others (*B. spinolosus, B. arenarum,* and *B. paracnemis*) show no response. It may be that these hormones only act over limited temperature ranges in amphibians. At 25° liver tissue from *R. tigrinum* shows higher oxygen consumption and reduced glycogen storage in response to thyroxine, but at 15° shows no response. Enhancement of the hexose monophosphate pathway may contribute to the increased metabolism, as evidenced by increased liver and fat body glucose-6-phosphate dehydrogenase and 6-phosphogluconate dehydrogenase activities in T_3- and T_4-treated *R. esculenta.*

8. Reptiles

Reptiles are the first completely terrestrial vertebrates and today are represented by four groups; Chelonia (turtles and tortoises), Squamata (lizards and snakes), Crocodilia (crocodiles and alligators), and Rhynchocephalia (sphenodon, the tuatara).

In general reptiles exhibit "metabolism in slow motion," andante rather than allegro, although the processes are similar to those in other vertebrates. Figure 14 shows the handling of glucose load by various reptiles: snakes and lizards have glucose tolerance curves similar to those of frogs and toads but dispose of the load more slowly; the whole process is even more prolonged in alligators, with maximum levels higher than in squamates and amphibia (Figure 15). Resting glucose levels vary between the groups; snakes, turtles, and alligators fall within the range of 50–100 mg/dl with variations dependent on season, stage in the life-cycle, and environmental conditions. Lizards, however, have higher levels from about 100 to >200 mg/dl. As for some fish and amphibia, glucose may not be the obligate metabolite for the CNS in all reptiles. Turtles, snakes, and alligators are sensitive to insulin and will die following insulin-induced hypoglycemia, but lizards are insensitive; for example, the skink (*Eumeces obsoletus*), the anole, and the fence lizard

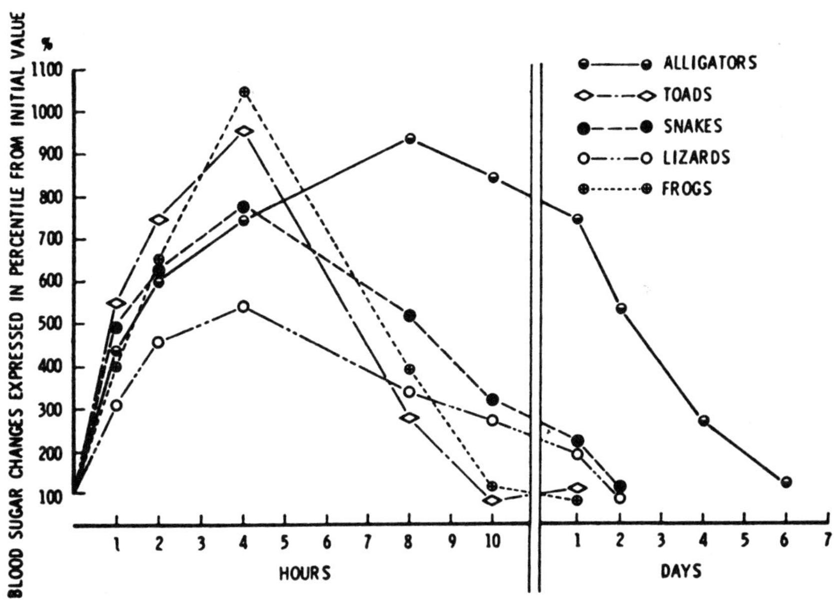

Figure 14. Variations in response to a glucose tolerance test (300 mg/100 g body weight) by various reptiles and amphibians. Note the more rapid return to normoglycemia by the amphibians as opposed to the reptiles, especially the alligator. (From Penhos, J. C., and Ramey, E., 1973, Studies on the endocrine pancreas of amphibians and reptiles. *Am. Zool.* **13**:667–698.)

(*Sceloporus occidentalis*) show only minimal response to large doses of insulin (1000 IU/kg). The slow metabolic rate and lack of dependence of CNS on glucose levels may be offset by high glycolytic capacity in both CNS and peripherally. The alligator uses glycolysis to provide immediate energy for mechanical work, prior to and independent of catecholamine-mediated reserve mobilization; moreover, muscle glycogen can be converted to lactate and ATP in a very short time. Such devices also enable diving reptiles, such as turtles and alligators, to withstand long periods of relative or absolute anoxia.

8.1. Insulin

Turtles, snakes, and alligators are sensitive to insulin and will die from severe insulin-induced hypoglycemia. The insensitivity of lizards to insulin is similar to that of birds. The effects of insulin on lipid protein metabolism have hardly been studied, but in the alligator (*A. mississippiensis*) and caiman (*Caiman latirostris*) insulin decreases both plasma and tissue amino acids, suggesting stimulated protein synthesis.

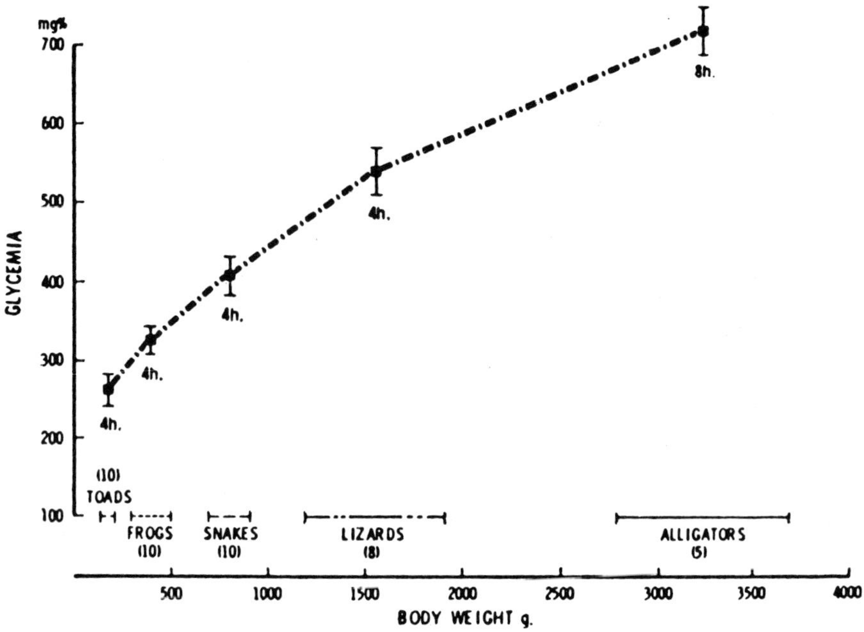

Figure 15. Maximal blood glucose levels attained in reptiles and amphibians during a glucose tolerance test (300 mg/kg body weight). (From Penhos, J. C., and Ramey, E., 1973, Studies on the endocrine pancreas of amphibians and reptiles. *Am. Zool.* **13:**667–698.)

8.2. Glucagon, Glucocorticoids, and Gluconeogenesis

Despite the clear ability of reptiles to withstand periods of hibernation or experimental fasting, little is known about the provision of energy, as glycogen or lipid stores or its mobilization during these periods. Seasonal variations of liver glycogen have been noted in representatives of two reptilian groups, lizards (*Egernia cunninghami, Dipsosaurus dorsalis, Varanus griseus,* and *A. carolinensis*), a snake (*Vipera aspis*), and a turtle (*Chrysemys picta*).

8.3. Glucagon

Glucagon appears to be hyperglycemic in reptiles, having a more prolonged effect than in amphibia (Figure 16), notably in snakes, which need more than 2 weeks to recover from a single dose of 100 μg/kg. The hyperglycemic effect is believed to be due to hepatic glycogenolysis as chronic treatment of the turtle, *Chrysemys d'orbignyi,* results in hyperglycemia and hepatic glycogen depletion, the effect being more pronounced in winter than in summer.

8.4. Glucocorticoids

The principal adrenal steroid secreted by reptilian adrenals is corticosterone, with secretory rates, and hence plasma levels, under the influence of sex, stage in the reproductive cycle, and environmental changes. In a desert iguana plasma corticosterone levels vary diurnally, reaching a peak slightly after midday, possibly related to temperature and activity cycles.

Direct and indirect evidence indicates that corticosteroids promote both hyperglycemia and glycogen deposition, but the mechanisms involved have been little studied. In the turtle, *C. picta,* adrenalectomy results in a slow decrease in liver glycogen, whereas inhibition of adrenal steroidogenesis in the lizard, *Sceloporus cyanogenys,* results in depressed blood glucose and liver glycogen levels. Exogenous glucocorticoids increase blood glucose in *A. mississippiensis* and the snake

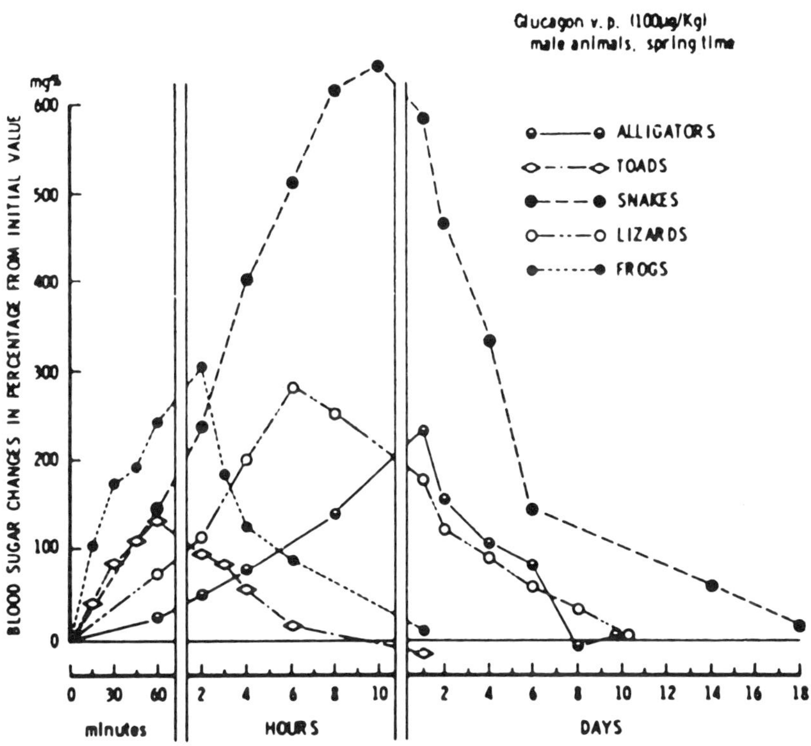

Figure 16. Effect of glucagon (100 µg/kg) on the blood glucose levels in reptiles and amphibians. Note the more rapid return to normoglycemia by the amphibians compared to the reptiles, especially the snakes where the blood glucose does not return to basal levels until more than 2 weeks after the single dose of hormone. (From Penhos, J. C., and Ramey, E., 1973, Studies on the endocrine pancreas of amphibians and reptiles. *Am. Zool.* **13:**667–698.)

Xenodon merremii and increase glycogen deposits in a number of species including *C. picta* and *A. carolinensis;* moreover, in this lizard aldosterone is also glycogenic (cf. Section 7, ''Amphibia'').

Evidence for glucocorticoid influence on gluconeogenesis and amino acid metabolism is scarce. In the alligator plasma amino acids increase in parallel with glucose levels following cortisol injections, indicating stimulation of protein catabolism, whereas cortisol stimulates tyrosine aminotransferase activity in liver of *C. picta,* but it has no effect on alanine and aspartate transaminase in the lizard, *Agama sanguinolenta.*

8.5. Catecholamines

Epinephrine is the major catecholamine of reptilian adrenals, representing 60–80% of total catecholamines; it produces hyperglycemia in a variety of turtle, lizard, and snake species, but the response of the alligator has been studied most thoroughly. Both blood glucose and lactate rise following injection of E, but lactate returns to baseline within 10 hr and glucose only after 60 hr. However, stress or strenuous exercise in alligators also induces a rise in plasma lactate, which may or may not be accompanied by an increase in glucose concentration. This short-term effect cannot be inhibited by adrenergic blocking agents, suggesting that the alligator uses the glycolytic breakdown of muscle glucose, free or derived from endogenous glycogen stores, to provide immediate energy independent of catcholamine stimulation. However, the slow mobilization of glucose by glycogenolysis appears to be due to E action at α receptors in muscle and α and β receptors in liver. The specific role of catecholamines seems to be to generate glucose for the rapid resynthesis of muscle glycogen, either by promoting hepatic glycogenolysis or by stimulating pancreatic glucagon release.

8.6. Thyroid Hormones

As in amphibians, the effect of thyroid hormones on oxidative metabolic processes in reptiles is temperature dependent. For example, T_4 stimulates oxygen consumption in *A. carolinensis* and *Lacerta muralis* at 30° compared with untreated controls, but not at temperatures around 20°. There is no clear picture of the effects of thyroid hormones on other metabolic parameters, but the euthyroid condition may help maintain normoglycemia by increasing liver and muscle glycogen and may be involved in lipid metabolism through control of liver triglyceride and FFA levels and plasma triglyceride.

9. Birds

Because of the economic importance of the domestic fowl, more is known about the metabolism of birds than for any other group of nonmammalian vertebrates. Birds are homeotherms like mammals, and, as would be expected, many aspects of avian metabolism are comparable to mammalian metabolic processes. Nevertheless, birds do show some remarkable differences from mammals and other vertebrate groups.

The major glycolytic, glycogenic, and glycogenolytic pathways for carbohydrate metabolism, the lipogenic and lipolytic pathways for lipid metabolism, and catabolic pathways for protein and amino acid metabolism are present in birds and very similar or identical to those in mammals. Energy derived from carbohydrates proceeds along the glycolytic and citric acid cycle pathways in the adult bird. During embryogenesis, however, the developing embryo lives on lipid-rich yolk, and glycolytic enzymes do not reach full activity until late in embryogenesis. The hexose monophosphate pathway is very active at this time, especially in cardiac muscle and intestinal and neural tissue, and glucose catabolism *via* this pathway exceeds that passing through the glycolytic pathway. This pathway, however, is virtually absent in adults where the NADPH necessary for fatty acid biosynthesis, steroidogenesis, or other reactions requiring reduced equivalents is probably produced by the malic enzyme. Citric acid cycle enzymes are also low in embryogenesis but rise in preparation for the conversion, at hatching, from the lipid-rich yolk to the carbohydrate-rich, grain-based diet of the adult domestic fowl. Gluconeogenesis is also important prior to hatching by providing glucose for morphogenesis. Gluconeogenic enzyme levels, however, decline significantly after hatching with the switch to a high-carbohydrate diet.

Avian lipid metabolism has some unique aspects. Adipose tissue has a very minor role in lipogenesis in the bird. The low levels of lipogenic enzymes such as acetyl–CoA carboxylase in the avian adipocyte prevent this tissue from accounting for more than 4% of the total lipid synthesis. The adipose tissue derives its lipid for storage from that exported by the liver. The liver serves as the main site for lipogenesis, probably responsible for 90–95% of the *de novo* fatty acid biosynthesis in the chick. Hepatic tolerance to dietary fat is high; over 90% of nonprotein caloric intake may be fat without inhibiting growth. The liver exports lipid primarily as low-density or β-lipoprotein.

9.1. Insulin

With the exception of some lizards, birds regulate their glycemia at a level higher than any other vertebrate. Plasma glucose rises continuously throughout chick embryogenesis to 160–180 mg/dl at hatching and continues to rise until 2 months of age, when adult levels of 200–240 mg/dl are reached. The role of insulin in regulating avian blood glucose, however, appears to be relatively secondary compared to that of glucagon. Exogenous insulin produces only a mild hypoglycemia in birds. The lethal or convulsive dose of insulin for a chicken is 5000 IU/kg compared to that of a rabbit, which is only 5 IU/kg. Pancreatectomy in birds results in a marked *hypoglycemia,* convulsion, and death unless glucose or glucagon is administered. Apparently, then, birds can tolerate the absence of insulin, but not the absence of glucagon. (Recall that the avian pancreas is very A-cell rich and that glucose is a poor stimulus for B cell insulin release.) Only in the goose does pancreatectomy produce mammallike alterations in carbohydrate metabolism.

Insulin stimulates hepatic and muscle glycogenesis, but only mild hypoglycemia. This suggests an insensitivity in peripheral tissues due perhaps to a reduced number of insulin binding sites and effector pathways in target tissues. The resistance of avian tissues to exogenous insulin (avian or mammalian) is very similar to that seen for squamate reptiles. Given the relatively poor lipogenic and glycogenic responses of adipose and hepatic tissues, respectively, to insulin, it is assumed that the mild insulin-induced hypoglycemia is attributable to insulin-induced glucose utilization by other tissues, possibly muscle.

Insulin stimulates amino acid uptake by avian tissues, which is independent of its stimulatory effect on glucose transport and protein synthesis. In the diabetic goose, insulin returns elevated plasma amino acid levels to normal.

Insulin has a variable effect on avian lipid metabolism. Table V shows that insulin stimulates glucose uptake weakly and oxidation and conversion into lipid in chicken adipocytes but has no effect on glycogenesis (the rates of glucose utilization are insufficient to contribute significantly to the hypoglycemic effect of insulin). However, in the pigeon, insulin has no effect on glycolysis or lipogenesis in adipose tissue *in vitro*. Contrary to the reaction in mammals, insulin has no antilipolytic

Table V
Effect of Insulin on Glucose Uptake and Oxidation and on Glycogen and Triglyceride Formation in Isolated Chicken Adipocytes[a]

Insulin (nM)	Glucose uptake	Glycogen formation	CO_2 formation	Triglyceride formation
0	1.03 ± 0.46	220.4 ± 19.7	254.5 ± 33.9	227.7 ± 8
0.175	—	—	—	275.9 ± 8.96****
1.75	—	—	294.3 ± 68.1	411.0 ± 95.3**
17.5	—	—	554.5 ± 236.8*	443.1 ± 69.0***
175	2.01 ± 0.69*	223.3 ± 6.7	606.3 ± 233.7**	387.4 68.7***
1750	2.04 ± 0.78*	234.6 ± 7.1	—	—

[a]Isolated chicken adipocytes were incubated in medium supplemented with 1 mM glucose and containing [^{14}C]-glucose for determination of CO_2, glycogen, and triglyceride formation. The data (mean ± SD) are expressed as μmole of glucose per gram dry weight of cells incubated for 180 min with six determinations for the glucose uptake and four determinations for all others at the concentrations of insulin indicated. *$p < 0.05$, **$p < 0.02$, ***$p < 0.001$. (From Gomez-Capilla, J. A., and Langslow, D. R., 1977, Insulin action on glucose utilisation by chicken adipocytes. *Int. J. Biochem.* **8**:417–420.)

effects in birds; in fact, injection of insulin sufficient to induce marked hypoglycemia will cause an increase in plasma FFA due to insulin stimulation of glucagon release.

9.2. Glucagon, Glucocorticoids, and Gluconeogenesis

Newly hatched chicks withstand fasting well, blood glucose falling only 25% during a 6-day fast. Residual yolk in combination with liver and muscle glycogen serve as energy stores. Adult chickens starved 8–15 days show a transient decline in blood glucose and liver glycogen followed by a progressive rise in these parameters, blood glucose exceeding the prefast levels. The increased blood glucose is attributable to gluconeogenesis fueled by amino acids from protein catabolism. Adult carnivorous birds such as the vulture or owl have hepatic gluconeogenic activity much higher than that of the chicken, indicating the importance of this pathway for glucose production when high-protein diets are consumed. Metabolism geared to gluconeogenesis also permits these birds to tolerate prolonged fasting, typical of carnivores, without changes in blood glucose. Critical key gluconeogenic enzymes such as phosphoenolpyruvate carboxykinase are increased three- to fivefold during starvation in the vulture, and even in the fed state gluconeogenic enzymes are much greater than those in the liver from the fasted chicken.

Gluconeogenesis in hepatocytes from starved chickens preferentially employs lactate and glycerol over pyruvate, alanine, and aspartate whereas in the rat alanine and glycerol are better substrates. Alanine is not as good a substrate for gluconeogenesis in the chicken as it is in mammals since the level of alanine aminotransferase in the chicken liver is low, only about 1% of that found in rat liver. Consequently, alanine may be first metabolized to lactate before being exported to the liver.

9.3. Glucagon

Glucagon is a hyperglycemic and lipolytic hormone which is perhaps the most important avian hormone for the regulation of glycemia and FFA. In isolated chicken hepatocytes glucagon stimulates glucose release by stimulating glycogenolysis and gluconeogenesis. Both effects appear to be mediated by cyclic AMP although the concentration of hormone necessary to provoke maximal response is much lower than that needed to stimulate maximal cyclic AMP production. (Avian pancreatic polypeptide is also a potent glycogenolytic hormone, but does not alter blood glucose levels.)

Glucagon facilitation of gluconeogenesis can occur in several ways: stimulation of the gluconeogenic pathway and inhibition of the glycolytic pathway. Glucagon stimulates a key gluconeogenic enzyme, phosphoenolpyruvate carboxykinase, and decreases a key glycolytic enzyme, phosphofructokinase. Glucagon suppression of phosphofructokinase activity in isolated chick hepatocytes is more pronounced in cells from starved instead of fed animals and rapidly reversible by insulin. This effect on phosphofructokinase activity in the chicken hepatocyte is independent of fructose-2,6-disphosphate levels, which otherwise appear to be integral in regulating this enzyme in mammals.

Lipid metabolism may be qualitatively more important in birds than carbohydrate metabolism, and glucagon has a very important role in avian lipid metabolism. In a number of species the plasma glucagon levels correlate more closely with plasma FFA levels than they do with blood glucose. Glucagon is a very potent adipokinetic hormone, stimulating adipose tissue triglyceride mobilization and lipolysis to cause an increase in plasma triglycerides and FFA. Glucagon-stimulated lipolysis is mediated through cyclic AMP activation of hormone-sensitive lipase and is blocked by prostaglandin E_1. In addition, glucagon inhibits adipocyte glycogenesis and lipogenesis. In the hepatocyte glucagon suppresses the incorporation of acetate into triglyceride although hepatic lipid deposition increases due to the increased availability of lipid released from the adipose stores.

9.4. Glucocorticoids

The major steroids synthesized by the adrenals of domestic species of birds are corticosterone, aldosterone, and 18-hydroxycorticosterone. Of these, only corticosterone and aldosterone are found in avian plasma. Hypophysectomy results in hypoglycemia and decreased liver glycogen which can be ameliorated with ACTH. Chronic corticosterone or hydrocortisone, but not cortisone, produces marked hyperglycemia and hepatic glycogen deposition in chickens and other domestic fowl. The increased glycogen content is attributable to stimulation of glycogenesis and not to impaired glucose utilization. Decreased body weight and protein content together with increased nitrogen excretion in these animals indicate that the glucocorticoid stimulates protein and amino acid catabolism to provide a substrate for gluconeogenesis. Glucocorticoids are lipogenic in birds causing elevated plasma lipids (triglyceride, FFA, cholesterol) and increased deposition of lipid in subcutaneous, visceral, and hepatic fat stores.

9.5. Catecholamines

Avian adrenals exhibit an extensive intermixing of cortical and medullary tissue, and the ratio of E to NE cells is highly variable. For example, the cormorant (*Phalacrocorax niger*) adrenal has 100% NE cells whereas the English sparrow (*Passer domesticus*) adrenal has 5% NE and 95% E medullary cells. Between these two extremes there is a complete range of ratios: the chicken, *Gallus domesticus*—80% NE/20% E; the pigeon, *Columbia livia*—57% NE/43% E; the golden oriole, *Oriolus oriolus*—21% NE/79% E. It has been postulated that the more "primitive" evolutionarily older forms have less efficient methylation systems and hence produce less epinephrine.

Both NE and E are glycogenolytic and lipolytic hormones leading to hyperglycemia and hyperlipidemia, respectively. Most of their effects on carbohydrate metabolism, however, originate from β-receptor stimulation by E producing hepatic and muscle glycogenolysis with subsequent hyperglycemia and lactic acidemia. In the pigeon, nevertheless, NE and E are relatively equipotent in their hyperglycemic effects although NE appears to be released more consistently in response to acute stress than does E. The rather high doses required to produce a modest elevation of FFA *in vivo* or *in vitro* preclude E as an important lipolytic hormone in birds.

E and glucagon have similar effects on isolated chicken hepatocytes; both stimulate glycogenolysis and gluconeogenesis while suppressing lipogenesis. The hormone effects are mediated by an increase in cyclic AMP, although the level of either hormone required for maximal effect produces only minimal increases in cyclic AMP levels. E is only one-third as effective as glucagon in raising intracellular cyclic AMP but can produce the same maximal level of effect. Neither the intracellular levels of cyclic AMP nor the hormone effects are altered by physiological doses of insulin or pancreatic polypeptide. Whereas 0.2–2.8 nM glucagon is required for half-maximal stimulation of glycogenolysis, 94–140 nM E is required for the same effect. Comparable figures for stimulation of gluconeogenesis are 0.48 nM for glucagon and 21 nM for E, and for suppression of lipogenesis they are 1 nM for glucagon and 70 nM for E.

9.6. Thyroid Hormone

Thyroid hormones in birds appear to have similar roles in metabolic regulation as they do in mammals. Both birds and mammals are homeotherms, and in the mammal thermogenesis is controlled by thyroid hormones through regulation of the plasma membrane Na^+/K^+-ATPase, which produces heat from the hydrolysis of ATP. This system has not been investigated yet in avian metabolism although avian body temperature appears to be under thyroid control, given that thyroidectomy reduces the ability of adult birds to produce heat. Thyroid status is directly correlated with metabolic rate in birds. Thyroidectomy produces decreased tissue oxygen consumption, associated with decreased levels of ATP, but reversible by T_3 or T_4 administration. Seasonal variations in metabolic rate, thyroid activity, and thyroid hormone effects are demonstrable. Table VI, for example, shows that, depending on the season, thyroidectomy lowers muscle respiration in birds.

With the exception of thyroid hormone's role in regulating the metabolic rate of tissues, direct effects of thyroid hormone on carbohydrate, lipid, and protein metabolism are difficult to demonstrate. More likely, many of the observed effects of T_3 or T_4 on various metabolic parameters reflect a permissive role for the hormone on various processes such as glycogenolysis or lipolysis. Nevertheless, in birds thyroid hormone produces hyperglycemia and hepatic glycogen mobilization while promoting skeletal and cardiac muscle glycogen deposition. Thyroidectomy, on the other hand, causes a mild hypoglycemia but has no effect on muscle glycogen in doves, pigeons, and ducks. In the bunting, *Embryriza bruniceps,* thyroidectomy produces hyperglycemia.

Extensive investigations in chickens and other domestic fowl have provided

Table VI
Effect of Thyroidectomy on Oxygen Consumption by Skeletal Muscle of the Common Myna, *Acridotheres tristis*[a]

Group	Oxygen consumption
Sham-operated (December)	1.0792 ± 0.037
December-thyroidectomized (1 month before study)	0.4836 ± 0.041
Sham-operated (February)	0.6364 ± 0.136
February-thyroidectomized (1 month before study)	0.5502 ± 0.060

[a]Values represent mean ± SEM oxygen consumption (μl 0_2/mg wet weight per hour) of skeletal muscle from four animals per group. All means are significantly different from the December sham-operated group. (From Thapliyal, J. P., Lal, P., Gupta B. B. P., and Sharan, S., 1983, Role of thyroid in tissue respiration of birds. *Gen. Comp. Endocrinol.* **49**: 333–343.)

some insight as to the role of thyroid hormone in regulating various aspects of metabolism. Propylthiouracil treatment (2–4 days) induces hypothyroidism in chicks which is associated with extensive liver glycogen deposition and decreased hepatic levels of glucose-6-phosphatase and cyclic AMP. These changes are associated with an increased plasma insulin:glucagon ratio, which would favor decreased hepatic glycogen phosphorylase and increased glycogen synthetase, hence glycogen deposition. Lipid metabolism is also altered by PTU-induced hypothyroidism. Plasma cholesterol, triglyceride, and phospholipid are increased as is hepatic triglyceride content, but hepatic cholesterol and phospholipid concentrations are decreased. As with the changes in carbohydrate metabolism, these alterations probably reflect an increased circulating insulin:glucagon ratio.

10. Appendix: Classification of Chordata*

Phylum—Chordata
- Subphylum—Hemichordata (Balanoglossus)
- Subphylum—Cephalochordata (Amphioxus)
- Subphylum—Urochordata (Sea squirts)
- Subphylum—Vertebrata
  - Superclass—Agnatha
    - Class—Cyclostomata
      - a. Petromyzonidae (hag-fishes)
      - b. Myxinidae (lampreys)
  - Superclass—Gnathostomata
    - Class—Chondrichthyes (Elasmobranchii)
      - a. Selachii (skates and rays)
      - b. Holocephali (chimaeras, rat, and rabbit fishes)

*Extant forms.

Class—Actinopterygii (ray-finned fishes)
- a. Chondrostei
  - i. Palaeoniscoidei (*Polypterus* and *Calamoichthys*)
  - ii. Acipensideroidei (Sturgeons and paddlefishes)
- b. Holostei (*Amia* and *Lepisosteus*)
- c. Teleostei (bony fishes)

Class—Crossopterygii (lobe-finned fishes)
- a. Osteolepids
- b. Coelacanthini (*Latimeria*)
- c. Dipnoi (lungfishes)

Class—Amphibia
- a. Anura (frogs and toads)
- b. Urodela (newts and salamanders)
- c. Apoda (limbless amphibia)

Class—Reptilia
- a. Rhynchocephalia (sphenodon)
- b. Chelonia (turtles and tortoises)
- c. Crocodilia (crocodiles and alligators)
- d. Squamata (lizards and snakes)

Class—Aves (birds)

Class—Mammalia
- a. Prototheria (echidna and platypus)
- b. Metatheria (marsupials)
- c. Eutheria (placental mammals)

ACKNOWLEDGMENTS. Preparation of a substantial portion of these chapters was completed during leave from Hamilton College and made possible by a Margaret Bundy Scott Fellowship. This leave was spent at the London University, Royal Postgraduate Medical School, Hammersmith Hospital, and I am indebted to Julia Polak and her colleagues of the Histochemistry Unit for providing a stimulating environment that was uniquely favorable for writing about peptide hormones associated with the pancreas and gastrointestinal tract. Ian Varndell and Susan van Noorden contributed generously to the illustrative material for this volume, and Abdesselm Ali-Rachedi provided many stimulating discussions.

I am very grateful for the efforts of David Tewksbury and Ann Darling of the Hamilton College Audio-Visual Services who provided numerous and excellent figures and photographs, and I thank Nancy Ernst for her efforts in manuscript preparation. I am also grateful to S. Bonner-Weir of the Joslin Clinic for providing unpublished results from her studies on the frog pancreas.

Special thanks must be given to Pearl Gapp who helped prepare and proof the manuscript and who carried out the unpublished immunostaining and secretory studies included herein.

Finally, I am in great debt to Professor Chester-Jones who has provided me opportunity and encouragement for this undertaking. This has been a task I could not have considered possible except for his expression of confidence and encouragement. I am also indebted to Dr. Pat Ingleton who, together with Prof. Chester-

Jones, has straightened out my convoluted and tortured prose and helped to organize these chapters into what I hope will be a useful contribution to the field of comparative endocrinology.

11. Selected Readings

Blundell, T. L., and Humbel, R. E., 1980, Hormone families: Pancreatic hormones and homologous growth factors. *Nature* **287**:781–787.

Bonner-Weir, S., and Weir, G. C., 1979, The organization of the endocrine pancreas: A hypothetical unifying view of the phylogenetic differences. *Gen. Comp. Endocrinol.* **38:**28–37.

Chester-Jones, I., 1976, Evolutionary aspects of the adrenal cortex and its homologues. *J. Endocrinol.* 71:1P–31P.

Coupland, R. E., 1979, Catecholamines, in: *Hormones and Evolution*, Vol. 1 (E. J. W. Barrington, ed.), Academic Press, New York, pp. 309–340.

Dockray, G. J., 1980, Gastrointestinal hormones: Nature and heterogeneity, in: *Polypeptide Hormones* (R. F. Beers and E. G. Bassett, eds.), Raven Press, New York, pp. 357–370.

Epple, A., Brinn, J. E., and Young, J. B., 1980, Evolution of pancreatic islet functions, in: *Evolution of Vertebrate Endocrine Systems* (P. K. T. Pang and A. Epple, eds.), Texas Tech Press, Lubbock, Texas, pp. 269–321.

Etkin, W., and Kim, Y. S., Role of the thyroid in vertebrate evolution, in: *Evolution of Vertebrate Endocrine Systems* (P. K. T. Pang and A. Epple, eds.), Texas Tech Press, Lubbock, Texas, pp. 233–246.

Falkmer, S., Carraway, R. E., El-Salhy, M., Emdin, S. O., Grimelius, L., Rehfeld, J. F., Reinecke, M., and Schwartz, T. F. W., 1981, Phylogeny of the GEP neuroendocrine system: A review, in: *Cellular Basis of Chemical Messengers in the Digestive System* (M. I. Grossman, M. A. B. Brazier and J. Lechago, eds.), UCLA Forum Med. Sci. 223, Academic Press, New York, pp. 21–42.

Falkmer, S., and Östberg, Y., 1977, Comparative morphology of the pancreatic islets, in: *The Diabetic Pancreas* (B. W. Volk and K. F. Wellmann, eds.), Plenum Press, New York, pp. 15–59.

Falkmer, S., and Van Noorden, S., 1983, Ontogeny and phylogeny of the glucagon cell, in: *Glucagon* (P. J. Lefebvre, ed.), Hanbuch der experimentellen Pharmakologie, Vol. 66, Part 1 (G. V. R. Born, A. Farah, H. Herken and A. D. Welch, eds.), Springer-Verlag, Heidelberg, pp. 81–119.

Fujita, T., Kobayashi, S., Yui, R., and Iwanaga, T., 1980, Evolution of neurons and paraneurons, in: *Hormones, Adaptation and Evolution* (S. Ishii, T. Hirano and M. Wada, eds.), Japan Scientific Society Press, Tokyo/Springer-Verglag, Berlin, pp. 35–43.

Hazelwood, R. L., 1980, The avian gastro-entero-pancreatic system: Structure and function, in: *Avian Endocrinology* (A. Epple and M. H. Stetson, eds.), Academic Press, New York, pp. 231–250.

Perhos, J., and Ramey, E., 1973, Studies on the endocrine pancreas of amphibians and reptiles. *Am. Zool.* **13:**667–698.

Roth, J., Le Roith, D., Shiloach, J., Rosenzweig, D. L., Lesniak, M. A., and Havrankova, 1982, The evolutionary origins of hormones, neurotransmitters, and extracellcular chemical messengers. Implications for mammalian biology. *N. Engl. J. Med.* **306:**523–527.

Sitbon, G., Strosser, M.-T., Gross, R., Laurent, F., Foltzer, C., Karmann, H., Cohen, L., Jean-Marie, P., and Mialhe, M., 1980, Endocrine factors in intermediary metabolism, with special reference to pancreatic hormones, in: *Avian Endocrinology* (A. Epple and M. H. Stetson, eds.), Academic Press, New York, pp. 251–270.

Van Noorden, S., Polak, J. M., and Pearse, A. G. E., 1980, Evolution of the gastrointestinal endocrine system, in: *Evolution of Vertebrate Endocrine Systems* (P. K. T. Pang and A. Epple, eds.), Texas Tech Press, Lubbock, Texas, pp. 247–267.

Index

Adenohypophysis [pars distalis (PD), pars intermedia (PI)]
control, 377–378
hormones, 289, 457–465
adrenocorticotropin (ACTH), 381–382, 383, 396, 461–465
from proopiocortin, 360–362
gonadotropins (GtH), 356–357, 384–387;
amphibians, 209–210, 211–212, 216–218, 280
birds, 126–130, 142–145
fish, 234–235
reptiles, 200–202, 280
follicle-stimulating (FSH) and luteinizing hormones (LH), 129, 135–136, 157, 174, 355–356, 397–398
see also Ovary, Testis, Pituitary gland
growth hormone [(GH) also called somatotropin (STH)], 289, 355–356, 380–381, 394–396
with somatomedin(s), 395, 632
β-lipotropin (LPH), 289, 363–364
see also Enkephalins
melanocorticotropins (α- and β-MSH), 289, 360–364, 396, 397
from proopiocortin, 360–364
melanocyte-stimulating hormone (MSH), control of, 382–384
prolactin [(PRL) also called lactotropic hormone (LTH)], 391–394, 457–461, 482, 487, 491
amphibians, 211, 220
birds, incubation, 174–175
control, 378–379
growth and development, 391–392
and growth hormone, 359–361
osmoregulation, 391–392, 393
reproduction, 393, 394
thyrotropin (TSH), 289, 357–359, 387–390, 397
thyrotropin-inhibitory factor (TIF), 389–390

Adrenal and interrenal glands
corticosteroids, 11–15, 38–50, 55, 95, 465–475
androgens, 38, 472
glucocorticoids (major, cortisol/corticosterone), 38, 466, 467, 470, 471, 647, 651–652, 656
mineralocorticoids (major, aldosterone, 11 deoxycorticosterone), 38, 466, 467, 470, 471
steroidogenesis (biosynthesis), 38–39, 472
Agnatha, 40–41
Amniota, 46–49
Chondrichthyes, 41–42
Osteichthyes (inc. Teleostei), 42–46
see also Steroids
structure, 118–120
ACTH, 471–472
classes
Amphibia, 111–113, 465, 470
Aves, 106–108, 465, 468–469
Mammalia, 465
Eutheria, 96–105, 467–468
Metatheria, 105
Monotremata, 105–106
Pisces, 113–118, 465, 470
Reptilia, 108–111, 465, 469
after hypophysectomy, 471, 473
Albumin, 146–148, 269, 270
APUD (amine content and/or amine precursor uptake and decarboxylation)
theory, 526–528

Behavior, 129–132, 136
Amphibia, 220–221
Aves
copulation, 168–170
courtship, 164–168
incubation, 171–175
laying, 170–171
mating, 159–161
migration, 182–185
molt, 178–182

Behavior (*cont.*)
  Aves (*cont.*)
    nesting, 162–164
    parentalism, 176–178
    territoriality, 161–162
  Pisces, 247–249
  Reptilia, 187
  see also Breeding
Brain
  nomenclature, 286
    diencephalon, 286–287
    telencephalon, 286–287
    third ventricle, 286–287
    see also Hypothalamus; Neurohemal organ
Breeding
  amphibians, 207–208
    hormones, 218–221
    seasonal, 218–219
  birds, 153–159
    annual cycles
      zones, temperate, 155–156
      zones, tropical, 156
    domestication, 156–159
    pineal gland, 158–159
    seasonality, 153–155
  fish, 228–230, 244–252, 253–254
    pineal gland, 249
  reptiles, 187–189
    cycles
      female, 193–195
      male, 195–198
      regulation, 198–202
    egg laying, parturition, 203–205
    ovarian activity, 190–192
    pineal gland, 199
Brockman bodies, teleosts
  insulin, proinsulin, 612–615

Calcitonin, 482, 486–487
Calcium, 482–488
  hormones influencing, 483–488
  role, general, 482–483
Catecholamines
  in intermediary metabolism, 636, 639
  amphibians, 648
  birds, 656–657
  reptiles, 652
  teleosts, 644–645
Caudal neurosecretory system [urophysis (Dahlgren cells)], 479–480
Classification, Chordata, 658–659
Corpuscles of Stannius, 478–479, 482, 487–488
Corticotropic intermediate lobe peptide (CLIP), 360
Corticomelanolipotropins, 289
  see Adenohypophysis

Diabetes mellitus, 583–584, 604–608
1,25-Dihydroxycholecalciferol, 482–486

Endorphins (β-endorphin family), 363–364, 535–536
Enkephalins (β-lipotropin family), 363–364, 523, 535–536, 537, 543

Gallbladder, contractility, 573–574, 574
Gastroenteropancreatic system (GEP)
  cell types, 518–523
  embryology, 526–528
  morphology, 515–518
  ontogeny, 524–528
  polypeptide family, 521–523, 528, 529, 533, 608–610, 619–620
    bombesin, 523, 533–535, 537, 543, 550
    cerulein, 529, 530
    cholecystokinin (CCK), 523, 529–530, 537, 543, 546
    enteroglucagon (EG), 531, 543, 548–549
    enteroinsular axis, 575–577
    gastric inhibitory polypeptide (GIP), 530–531, 543, 548
    gastrin family, (AG, IG), 528–530, 541, 543–546
    motilin (M), 521, 536–537, 543, 547–548
    neuropeptide tyrosine (NPY), 523
    neurotensin (N), 523, 539, 543, 549–550
    pancreatic polypeptide (PP), 521
    peptide histidine isoleucine (PHI), 523, 543, 571–572
    secretin (S), 521, 541, 546–547
    somatostatin (SRIF), 523, 536, 540, 543, 550–551, 607–608, 614–615, 620, 636
    vasoactive intestinal polypeptide (VIP), 523, 529–531, 537, 543, 550–551
    see also Glucagon, Insulin
Gills
  "chloride cells," 425–426
  in osmoregulation, 422–430
Glucagon
  classes
    amphibians, 647
    birds, 618, 620, 655–656
    elasmobranchs, 638
    reptiles, 616–617, 650
  in intermediary metabolism, 635
  nomenclature family, 528–531
  secretion, 606–607

Gonads
  development, teleosts, 223–230
  general structure, 58–67
  steroidogenic enzymes, 67–69
  see also Ovary, Testis, Pituitary gland, Hypothalamus
Gut
  contractility, motility, 572–575
  hormones
    Agnatha, 554–555
    Amphibia, 557
    Aves, 559–561
    Chondrichthyes, 555
    evolution, 551–553
    Osteichthyes, 555–557
    Reptilia, 557–559
    secretory modes, 561–567

Herring bodies, 302
Histotroph, 257–260
Homeostasis
  definition, 481, 493–495
  extracellular fluid, constancy, 413–417
  see also Osmoregulation
Hypophysiotropic hormones
  corticotropin-releasing hormone (CRH), 370–371
  growth-hormone-releasing hormone (GRH), 369
  luteinizing-hormone-releasing hormone (LH-RH), 370
  MSH-controlling factors, 371
    melanocyte-inhibiting activity (MIF, or melanostatin), 371
    melanocyte-stimulating activity (MRF, or melanoliberin), 371
  neurotransmitters (nonpeptidergic), 372–374
    acetylcholine, 372–373
    γ-aminobutyric acid (GABA), 372–373
    dopamine, 372–373
    norepinephrine [noradrenaline (NA)], 372–373
  prolactin-controlling factors, 371–372
    prolactin-release-inhibiting factor (PIF), 371
    prolactin-releasing factor (PRF), 372
  somatotropin release-inhibiting factor [somatostatin (SRIF)], 368–369, 371–372, 380–381, 387, 388
  thyrotropin-releasing hormone (TRH), 367–368, 387–390, 523, 537
Hypothalamic nuclei, 301–302
  nucleus lateralis tuberis (NLT), 315
  nucleus recessus lateralis (NRL), 318
  nucleus recessus posterioris (NRP), 318
Hypothalamic nuclei (*cont.*)
  paraventricular organ (PVO), 315, 336
  preoptic nucleus (NPO), 309, 310
Hypothalamo-pituitary-gonadal axis, 126–128
  development, 129
  gonadotropin-releasing factor (GnRH), 126–128, 174
    amphibians, 207, 208–212, 219
    birds, 126–127
    fish, 235–238, 255
    reptiles, 199–201
Hypothalamus (HT)
  binding receptors, 374
  in classes
    Agnatha, 306–312
    Amphibia, 335–338
    Aves, 347–349
    Chondrostei, 318–319
    Crossopterygii, 331–333
    Elasmobranchii, 314–316
    Holostei, 320
    Mammalia, 351–355
    Reptilia, 342–344
  extra-hypothalamic control, 375–376
  hormones, 374–390
  see also Hypophysiotropic hormones
  neurohormones
    classification, 304–305
    release, 302–304
  neurosecretory cells (NS), 301–302
  secretagogues, 374–375

Inhibin, 129, 134, 373
Insulin, 528, 531–533, 604–606
  classes
    amphibians, 615, 645–647
    birds, 617–620, 653–658
    cyclostomes, 610–612, 636–637
    elasmobranchs, 637–638
    reptiles, 615–616, 617, 618, 648–650
    teleosts, 612–614, 640–642
  in intermediary metabolism, 634–635
Islets of Langerhans
  Agnatha, 586–588
  Amphibia, 593–595
  Aves, 598–599
  cell types, 579–582
  Chondrichthyes, 588–590
  evolution, 584–586, 599–604
  neural regulation, 609–610
  Osteichthyes, 590–593
  Reptilia, 595–598
  secretions, 582–583, 604–605
  vasculature, 582

Metabolism
amino acids, 629–630
definition, 623
fat, 628, 633–634
glucose, 623–628
growth hormone (GH), 632
intermediary metabolism
amphibians, 645–648
birds, 652–658
cyclostomes, 636–637
fish, 637–639, 640–645
reptiles, 648–652
pathways (*inc.* Krebs, Embden-Myerhof), 623–630
protein, 628, 634
regulation, 630–632
somatomedins, 632
thyroid hormones, 632, 636, 637
Migration
birds, 182–185
see also Behavior

Neurohemal organs
median eminence (ME), 302–304
organum vasculosum lateralis tuberalis (OVLT), 302
structure, 302–304
Neurohypophysis
hormones, 364–367, 398–399, 448–457
arginine vasopressin (AVP), 504–505
arginine vasotocin (AVT), 504–505
isotocin, 505
mesotocin, 505
pathways, evolution, 503–506
secretions, control, 390–391
Neuropeptides, Brain–CNS actions (nomenclature), 373–374

Osmoregulation
classes
amphibians, 430–433
birds, 392, 435–441
cyclostomes, 420–421
elasmobranchs, 423–425
fish, general, 391–392, 421–423
mammals, 392–393, 441–443
reptiles, 433–435
teleosts, 425–430
mechanisms, general, 417–419
see also Calcium, Homeostasis, Prolactin, Renal function, Sodium, Water
Ovary
steroids from
Agnatha, 31–32

Ovary (*cont.*)
steroids from (*cont.*)
Amphibia, 34–36, 218
Aves, 36–37, 141–143
Chondrichthyes, 32–33
Reptilia, 36, 190–191
Teleostei, 33–34, 231
structure, 58–67, 82–93
amphibians, 215–216, 218–219
birds, 136–138, 139
fish, 224–229
reptiles, 190–195
see also Gonads, Steroids
Oviduct, 138–141, 202–207
Oviparity, 189, 192, 257–258, 268–269; *see also* Viviparity
Oviposition, 151–153, 204
Ovulation, 143–145, 189, 218, 227

Pancreas
embryology, 524–528
secretions
endocrine A and B cells, 579–583
see also Glucagon, Insulin
exocrine, 567–572
Parietal eye (reptiles), 199
Parturition (reptiles), 202–204
Pineal gland, 158–159, 199, 249
Pituitary gland
adenohypophyseal hormones, 289, 457–465
cells, structure, 289–297
amniotes, 447
classes
Amphibia, 333–335
birds, 345–347
Chondrostei, 316–318
Crossopterygii, 329–333
Elasmobranchii, 312–314
Holostei, 319–320
mammals, 350–351
reptiles, 338–342
Teleostei, 320–324
embryology, 286–289
feedback mechanisms, 376–377
neurohypophysis, 390–391, 503–506
nomenclature, 286–289
pars intermedia, 297–299
see also Adenohypophysis, Neurohypophysis
Prostaglandins, 151–153, 247, 249

Rathke's pouch, 286–288, 316, 340
Rectal gland
elasmobranchs, 425

Renal function
  kidney
    evolution, 498–501, 502
    role, 498–500
  in osmoregulation
    amphibians, 431–433
    birds, 435–441
    fish, 426–430
    mammals (Eutheria), 441–443
    reptiles, 433–435
  see also Osmoregulation
Renin-angiotensin system, 475–478, 490–491

Salt glands, 392, 435, 436, 437, 493
Sex
  behavior, 129–132, 136
  determination, 128–129
Shell
  Aves, 148–151
    calcium metabolism, 150–151
    formation, 149
    structure, 148–149
Skin, amphibian osmoregulation, 430–433
Sodium, 491–493
Somatostatin (SRIF), in hypothalamus, 368–369, 371–372, 380–381, 387, 388
Spermatogenesis, 58
  Amniotes, 75–76
  Amphibia, 212–215, 217–218
  Anamniota, 71–75
  Aves, 133–134
  Reptilia, 195–198
  Teleostei, 249–250
  see also Testis (Leydig and Sertoli cells)
Stellate-follicular cell system, 299–300
Steroidogenesis (biosynthesis), 13–17, 20, 21, 24, 28, 29, 31–37, 39, 40–49, 472
  enzymes, 15–16, 17
  pathways
    corticosteroids, 39
    estrogens, 31
    progesterone, 32
    testosterone, 20
Steroids
  catabolism, 50–52
  conjugation, 16–17
  interrelationships, 14
  major classes
    androgens, 11–15, 19–30, 55
    corticosteroids, 11–15, 38–50, 55
    estrogens, 11–15, 30–38, 55
    progestogens, 11–15, 34
  methodology, 17–19
  nomenclature, 4, 6–12
Steroids (*cont.*)
  plasma concentrations
    amphibians, 55, 216–217
    birds, 55, 130, 131, 144, 154, 166
    cyclostomes, 55
    elasmobranchs, 55
    fish, 55
    mammals, 54–56
    reptiles, 55, 190–191
  structure, 4–11
  systematic names, 6–9
  transport and action, reptiles, 201–202
  trivial names, 6–7, 9–11

Techniques, for structure
  electron microscopy, 402–404
  fluorescence, 407–408
  immunocytochemistry, 404–407
  stains, 399–402
Testis, 70–82
  Leydig cells, 58, 65, 70, 76–79, 132–136, 198, 214–215, 398
  Sertoli cells, 58, 70, 79–82, 132–136, 198, 214–215, 232, 398
  steroids, 19–29
    Agnatha, 19–22
    Amphibia, 26–27
    Aves, 28–29
    Chondrichthyes, 22–23
    Reptilia, 27
    Teleostei, 23–25
    see also Steroids, Steroidogenesis
  structure, 58–67, 70–71
    Amniota, 75–76
    Amphibia, 212–215
    Anamniota, 71–75
    Aves, 132–136
    Chondrichthyes, 252–253
    Reptilia, 189–191, 195–198
    Teleostei, 229–230, 232
Thyroid hormones
  amphibians, 650
  birds, 657–658
  reptiles, 652
  teleosts, 645
  thyroxine (T_4), 389–390
  triiodothyronine (T_3), 389–390
Trimethylamine oxide (TMAO)
  elasmobranchs, osmoregulation, 423–425

Urea
  elasmobranchs, osmoregulation, 423–425
Uric acid
  birds, 438–439
  reptiles, 433

Vitamin D_3, 484–486
Vitellogenin (yolk), vitellogenesis
  components, 270–271
  hormones required, 239–243, 277–280
  lipovitellin, 226–227, 270
  phosvitin, 270
  plasma levels, 276, 277
  precursor, phospholipoglycoprotein, 270
  synthesis, 272–275
  uptake, oocytes, 280–281
Viviparity
  classes
    Amphibia, 259, 260
    Elasmobranchii, 259, 260, 262, 263, 266
    Eutheria, 259, 263
    Metatheria, 259
    Reptilia, 192, 204–206, 260, 264–266
    Teleostei, 259, 260
Viviparity (*cont.*)
  evolution, 261, 266–268
  types
    aplacental species, histotroph, 257–259
    placental species
      chorioallantoic placenta, 257–259, 264
      choriovitelline placenta (yolk sac), 257–259, 264

Water, in osmoregulation, 488–491
  amphibians, 490–491
    Rana cancrivora, 490
  fish, 489–490
  sites, 488–489
  water drive, 491

Yolk, *see* Vitellogenin